FLUID DYNAMICS:
THEORY, COMPUTATION, AND NUMERICAL SIMULATION

Fluid Dynamics:
Theory, Computation, and Numerical Simulation
Second Edition

C. Pozrikidis

Springer

C. Pozrikidis
University of Massachusetts
Amherst, MA
USA

ISBN: 978-0-387-95869-9 e-ISBN: 978-0-387-95871-2
DOI: 10.1007/978-0-387-95871-2

Library of Congress Control Number: 2008943356

Printed on acid-free paper

9 8 7 6 5 4 3 2 1

springer.com

Preface

Ready access to computers has defined a new era in teaching and learning. The opportunity to extend the subject matter of traditional science and engineering curricula into the realm of scientific computing has become not only desirable, but also necessary. Thanks to portability and low overhead and operating cost, experimentation by numerical simulation has become a viable substitute, and occasionally the only alternative, to physical experimentation.

The new framework has necessitated the writing of texts and monographs from a modern perspective that incorporates numerical and computer programming aspects as an integral part of the discourse. Under this modern directive, methods, concepts, and ideas are presented in a unified fashion that motivates and underlines the urgency of the new elements, but neither compromises nor oversimplifies the rigor of the classical approach.

Interfacing fundamental concepts and practical methods of scientific computing can be implemented on different levels. In one approach, theory and implementation are kept complementary and presented in a sequential fashion. In another approach, the coupling involves deriving computational methods and simulation algorithms, and translating equations into computer code instructions immediately following problem formulations. Seamlessly interjecting methods of scientific computing in the traditional discourse offers a powerful venue for developing analytical skills and obtaining physical insight.

The goal of this book is to offer an introductory course in traditional and modern fluid mechanics, covering topics in a way that unifies theory, computation, computer programming, and numerical simulation. The approach is truly introductory in that only a few prerequisites are required. The intended audience includes undergraduate and entry-level graduate students, as well as a broader class of scientists, engineers, fluid dynamics and computational science enthusiasts with a general interest in computing. This book should be especially appealing to those who are making a first excursion into the world of numerical computation and computational fluid dynamics (CFD) beyond the black-box approach. This book should be an ideal text for an introductory course in fluid mechanics and CFD.

The presentation of the material is distinguished by two features. First, solution procedures and algorithms are developed immediately after problem formulations are presented, and illustrative MATLAB codes are discussed in the text. Second, numerical methods are introduced on a need-to-know basis and in order of increasing difficulty: function interpolation, function differentiation, function integration, solution of algebraic equations, finite-difference methods, etc. Computer problems at the end of each section require performing computation and simulation to study the effect of various parameters determining a flow.

In concert with the intended usage of this book as a stand-alone introductory text and as a tutorial on numerical fluid dynamics and scientific computing, only a few references are provided in the discussion. Instead, a selected compilation of introductory, advanced, and specialized texts on fluid dynamics, calculus, numerical methods, and computational fluid dynamics are listed in appendix B. The reader who wishes to focus on a particular topic is directed to these resources for further details.

A major feature of this book is the accompanying fluid dynamics software library FDLIB discussed in appendix A. The FORTRAN 77 and MATLAB programs of FDLIB explicitly illustrate how computational algorithms translate into computer instructions. The codes of FDLIB range from introductory to advanced, and the topics span a broad range of applications discussed in this text: from laminar channel flows, to vortex flows, to flow past airfoils. The MATLAB codes of FDLIB combine numerical computation, graphics display, data visualization and animation.

To run the FORTRAN 77 codes of FDLIB, a FORTRAN 77 or FORTRAN 90 compiler is required. Free compilers are available thanks to the gnu foundation. The input data is either entered from the keyboard or read from data files. The output is recorded in output files in tabular form so that it can be read and displayed using independent graphics, visualization, and animation applications on any computer platform, including Matlab.

The second edition incorporates significant improvements in substance and style. First, additional examples, solved problems, and new material have been introduced for a more comprehensive treatment of the various topics. Examples include surfactant transport and a brief introduction to compressible flow. Second, sample MATLAB programs integrating numerical computation and graphics visualization are listed and discussed in the text. A Matlab primer explaining basic programming procedures is presented in appendix C. Third, the revised text refers to the latest version of FDLIB. These improvements should render the book an accessible introductory computational fluid dynamics (CFD) resource.

The book Internet address is: `http://dehesa.freeshell.org/FD2`

I acknowledge with appreciation insightful comments by Keiko Nomura, Siggi Thoroddsen, and Mark Blyth on the manuscript of the second edition.

C. Pozrikidis

Spring, 2009

Contents

Introduction to Kinematics 1

We begin the study of fluid mechanics by pointing out the differences between fluids and solids, and by describing a fluid flow in terms of the motion of elementary fluid parcels. As the volume of a parcel becomes infinitesimal, the parcel reduces to a point particle and the average velocity of the parcel reduces to the local fluid velocity computed before the molecular nature of the fluid becomes apparent. The study of the motion and deformation of material lines and surfaces consisting of collections of point particles reveals the nature and illustrates the diversity of motion in fluid mechanics.

1.1 Fluids and solids

Casual observation of the world around us reveals objects that are classified as solids and fluids; the second category includes gases and liquids. What are the distinguishing features of these two groups? The answer may be given on a wide variety of levels: from the molecular level of the physicist, to the macroscopic level of the engineer and oceanographer, to the cosmic level of the astronomer.

From the perspective of mainstream fluid mechanics underlying this book, the single most important difference between fluids and solids is that a fluid must assume the shape of the container in which it is placed, whereas a solid is able to stand alone sustaining its own shape. As a consequence, a body of fluid is not able to resist a shearing force exerted on its surface parallel to the boundaries, and must keep deforming forever when subjected to it. In contrast, a solid is able to deform and assume a new stationary shape. Certain materials, including polymeric melts and solutions, exhibit properties that are intermediate between those of fluids and solids.

C. Pozrikidis, *Fluid Dynamics: Theory, Computation, and Numerical Simulation*,
DOI: 10.1007/978-0-387-95871-2_1,

The differences between fluids and solids can be attributed to the intensity of the forces holding the molecules together to form a coherent piece of material. The inability of a fluid to assume its own shape is due to the weakness of the potential energy associated with intermolecular forces relative to the kinetic energy associated with the vibrations of the individual molecules. The molecules are too busy vibrating to hang onto one another and thus form a long-lived crystal.

Fluids can be transformed into solids, and *vice versa*, by manipulating the relative magnitude of the potential energy due to intermolecular forces and the kinetic energy due to thermal motion. In practice, this can be done by heating or by changing the pressure of the ambient environment.

Problem

1.1.1 *Nature of a liquid/solid suspension*

Fluids containing suspended solid particles abound in nature, physiology, and technology. Examples include blood consisting of a dense suspension of red, white, and other blood cells, slurry used in the petroleum industry for the hydrodynamic transport of particulates, dough and toothpaste. Discuss whether a suspension should be classified as a fluid or solid with reference to the volume fraction of the suspended solid phase.

1.2 Fluid parcels and flow kinematics

The motion of a non-deformable solid body, called a rigid body, can be described in terms of the velocity of translation vector and the angular velocity of rotation vector, where rotation occurs around a specified center. A rigid body moves as a whole in the direction of the velocity vector, while rotating as a whole around the angular velocity vector that is pinned at the designated center of rotation. In contrast, the motion of a deformable body, such as an elastic solid or a fluid, cannot generally be described in terms of two vectors alone. A more advanced framework that allows for an extended range of motions is required.

Decomposition of a fluid into parcels

To establish the necessary generalized framework, we subdivide a body of fluid into parcels. For simplicity, we assume that all molecules comprising the parcels are identical, which means that the fluid is homogeneous. Each molecule in a certain parcel moves with its own highly fluctuating velocity. However, if the parcel exhibits a net motion, the velocities of the individual molecules are coordinated to reflect or, more accurately, give rise to the net motion. A molecule of a gas frequently collides with other molecules after having traveled a distance comparable to the mean free path. The macroscopic motion of a small fluid parcel can be described in terms of its velocity of translation, which can be quantified in terms of the average velocity of the individual molecules, as will be discussed in section 1.3. If the parcel is sufficiently small, rotation is neglected to a first approximation.

Relative parcel motion

A key observation is that the motion of a fluid can be described in terms of the relative motion of the individual fluid parcels. If all parcels move with the same velocity, the relative parcel velocity is zero and the fluid translates like a rigid body. It is possible that the velocity of the parcels is coordinated so that the fluid rotates like a rigid body as a whole around a designated center of rotation.

Consider a fluid-filled flexible rubber tube that is closed at both ends, and assume that the tube is stretched to elongate the fluid. The fluid has undergone neither translation nor rotation, but a new type of motion called deformation. Combinations of translation, rotation, and deformation whose relative strength varies with position in the fluid gives rise to a wide variety of fluid motions.

Kinematics as a field of fluid dynamics

Establishing in quantitative terms the relationship between the relative motion of fluid parcels and the structure of a flow is the main objective of kinematics. The term derives from the Greek work $\kappa\iota\nu\eta\sigma\iota\varsigma$ which means "motion." The complementary discipline of dynamics ($\delta\upsilon\nu\alpha\mu\iota\kappa\eta$) addresses the forces exerted on a fluid by an ambient surface or body force field, such as the gravitational field, as well as the forces developing inside a fluid as the result of the motion.

Problem

1.2.1 *A rolling sphere*

A sphere is rolling over a plane following a rectilinear path. How is the velocity at the center of the sphere related to the angular velocity of rotation about the center of the sphere?

1.3 Coordinates, velocity, and acceleration

To describe the motion of a molecule, we work under the auspices of classical mechanics. We begin by introducing three mutually orthogonal axes forming a Cartesian coordinate system, (x, y, z), as illustrated in figure 1.3.1. Each point in space has an associated position vector that starts at the common origin of the Cartesian axes and ends at the point. The point is identified by the values of x, y, and z, defined as the positive or negative projections of the position vector

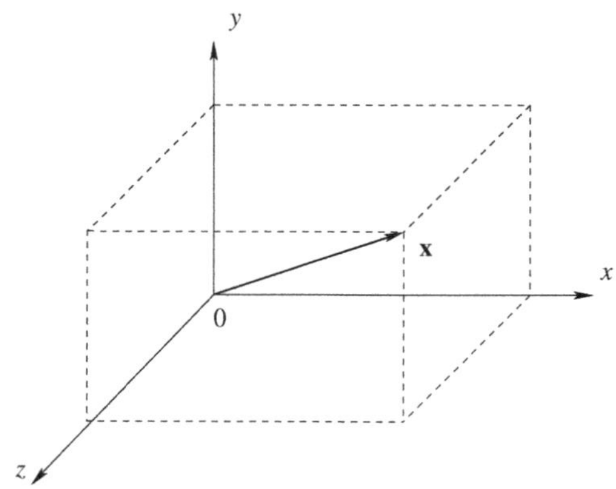

Figure 1.3.1 Three mutually orthogonal axes define a Cartesian coordinate system, (x, y, z). The arrow indicates the position vector corresponding to a point, $\mathbf{x}$.

onto the corresponding axes. In vector notation, the Cartesian coordinates are expressed by the ordered triplet

$$\mathbf{x} = (x, y, z), \tag{1.3.1}$$

where x, y, and z take values in the range $(-\infty, \infty)$. Accordingly, the Cartesian coordinates of a point have a dual interpretation: they represent a geometrical entity associated with the position vector, and they form an ordered triplet of real numbers.

Unit vectors

The three position vectors,

$$\mathbf{e}_x = (1, 0, 0), \qquad \mathbf{e}_y = (0, 1, 0), \qquad \mathbf{e}_z = (0, 0, 1), \tag{1.3.2}$$

point in the positive directions of the x, y, or z axis. The end points represented by these vectors lie on the x, y, or z axis at distances equal to one unit of length away from the origin. We say that the three vectors $\mathbf{e}_x$, $\mathbf{e}_y$, and $\mathbf{e}_z$ are mutually orthogonal Cartesian unit vectors.

Combining these definitions, we express the position vector in the form

$$\mathbf{x} = x\,\mathbf{e}_x + y\,\mathbf{e}_y + z\,\mathbf{e}_z. \tag{1.3.3}$$

In physical terms, this equation states that, to get to the point $\mathbf{x}$ departing from the origin, we may move along each one of the unit vectors $\mathbf{e}_x$, $\mathbf{e}_y$, and $\mathbf{e}_z$, by respective distances equal to x, y, and z units of length. The order of motion along the three directions is immaterial.

Velocity

Because a molecule moves with a highly fluctuating velocity, its position changes rapidly in time. Formally, we say that the coordinates of the molecule are functions of time, t, denoted by

$$x = X(t), \qquad y = Y(t), \qquad z = Z(t). \tag{1.3.4}$$

To economize our notation, we introduce the vector function

$$\mathbf{X}(t) = [\, X(t), \quad Y(t), \quad Z(t)\,], \tag{1.3.5}$$

and consolidate expressions (1.3.4) into the form

$$\mathbf{x} = \mathbf{X}(t). \tag{1.3.6}$$

By definition, the velocity of a molecule is equal to the rate of change of its position, displacement over time elapsed. If the x coordinate of the molecule has changed by an infinitesimal displacement, $\mathrm{d}X$, during an infinitesimal period of time, $\mathrm{d}t$, then, by definition $v_x = \mathrm{d}X/\mathrm{d}t$. Writing the counterparts of this equation for the y and z coordinates, we obtain

$$v_x = \frac{\mathrm{d}X}{\mathrm{d}t}, \qquad v_y = \frac{\mathrm{d}Y}{\mathrm{d}t}, \qquad v_z = \frac{\mathrm{d}Z}{\mathrm{d}t}, \tag{1.3.7}$$

which can be collected into an ordered triplet,

$$[v_x, v_y, v_z] = [\frac{\mathrm{d}X}{\mathrm{d}t}, \quad \frac{\mathrm{d}Y}{\mathrm{d}t}, \quad \frac{\mathrm{d}Z}{\mathrm{d}t}]. \tag{1.3.8}$$

In vector notation,

$$\mathbf{v} = \frac{\mathrm{d}\mathbf{X}}{\mathrm{d}t}. \tag{1.3.9}$$

We have demonstrated that the velocity of a molecule is a vector described by its three Cartesian components, v_x, v_y, and v_z. These are the positive or negative distances subtended between the projection of the last and first point of the velocity vector onto the x, y, or z axis. The distances are then multiplied by a scaling factor to acquire dimensions of velocity, length divided by time. A negative value for v_x indicates that the x coordinate of the last point of the velocity vector is lower than the x value of the first point, and the motion occurs toward the negative direction of the x axis. Similar interpretations apply to the y and z components.

In terms of the unit vectors defined in equations (1.3.2), the velocity vector is given by

$$\boxed{\mathbf{v} = v_x\,\mathbf{e}_x + v_y\,\mathbf{e}_y + v_z\,\mathbf{e}_z.} \tag{1.3.10}$$

It is evident from these definitions that the velocity vector is a free Cartesian vector, which means that it can be translated in space to any desired location. In contrast, the first point of the position vector, $\mathbf{x}$, is always pinned to the origin.

Acceleration

The acceleration vector, $\mathbf{a}$, is defined as the rate of change of the velocity vector in time,

$$\mathbf{a} = \frac{\mathrm{d}\mathbf{v}}{\mathrm{d}t} = \frac{\mathrm{d}^2\mathbf{X}}{\mathrm{d}t^2}. \tag{1.3.11}$$

By definition then,

$$\mathbf{a} = a_x\,\mathbf{e}_x + a_y\,\mathbf{e}_y + a_z\,\mathbf{e}_z, \tag{1.3.12}$$

where the Cartesian components of the acceleration are given by

$$a_x = \frac{\mathrm{d}^2X}{\mathrm{d}t^2}, \qquad a_y = \frac{\mathrm{d}^2Y}{\mathrm{d}t^2}, \qquad a_z = \frac{\mathrm{d}^2Z}{\mathrm{d}t^2}. \tag{1.3.13}$$

If the Cartesian coordinates of a molecule are constant or change linearly in time, the acceleration is zero.

1.3.1 Cylindrical polar coordinates

A point in space can be identified by the values of the ordered triplet (x, σ, φ), as illustrated in figure 1.3.2, where:

- x is the projection of the position vector onto the straight (rectilinear) x axis passing through a designated origin, taking values in the range $(-\infty, +\infty)$.
- σ is the distance of a point of interest from the x axis, taking values in the range $[0, \infty)$.
- φ is the meridional angle measured around the x axis, taking values in the range $[0, 2\pi)$. The value $\varphi = 0$ corresponds to the first and second quadrants of the xy plane.

Using elementary trigonometry, we derive relations between the Cartesian and associated polar cylindrical coordinates,

$$y = \sigma\cos\varphi, \qquad z = \sigma\sin\varphi, \tag{1.3.14}$$

and the inverse relations between the polar cylindrical and Cartesian coordinates,

$$\sigma = \sqrt{y^2 + z^2}, \qquad \varphi = \arccos\frac{y}{\sigma}. \tag{1.3.15}$$

In computing the inverse cosine function, care must be taken so that φ is a continuous function of y and σ.

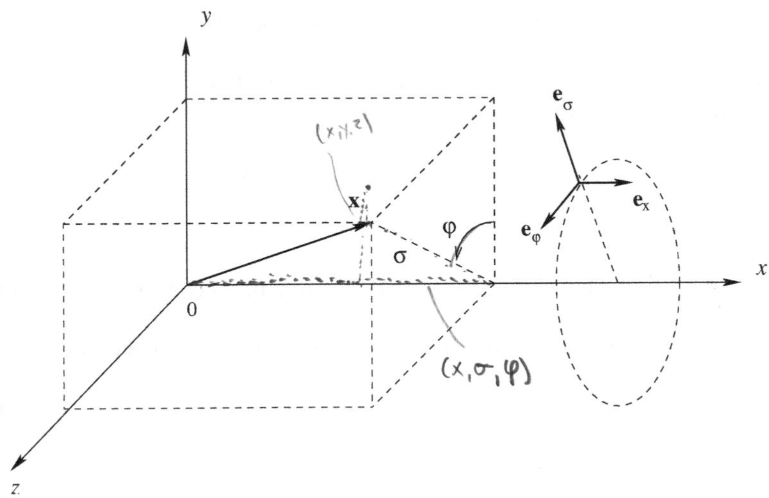

Figure 1.3.2 A system of cylindrical polar coordinates, (x, σ, φ), defined with reference to the Cartesian coordinates, (x, y, z).

Unit vectors

We consider a point in space and define three vectors of unit length, denoted by $\mathbf{e}_x$, $\mathbf{e}_\sigma$, and $\mathbf{e}_\varphi$, pointing, respectively, in the direction of the x axis, normal to the x axis, and in the meridional direction of varying angle φ, as depicted in figure 1.3.2. Note that the orientation of the unit vectors $\mathbf{e}_\sigma$ and $\mathbf{e}_\varphi$ changes with position in space, whereas the orientation of $\mathbf{e}_x$ is fixed and independent of position in space.

In terms of the first two local unit vectors, $\mathbf{e}_x$ and $\mathbf{e}_\sigma$, the position vector is given by

$$\mathbf{x} = x\,\mathbf{e}_x + \sigma\,\mathbf{e}_\sigma. \tag{1.3.16}$$

The dependence of the position vector on the meridional angle, φ, is mediated through the unit vector $\mathbf{e}_\sigma$ on the right-hand side. The absence of $\mathbf{e}_\varphi$ from the right-hand side of (1.3.16) is justified by observing that the distance from the origin, expressed by the position vector $\mathbf{x}$, is perpendicular to the third unit vector, $\mathbf{e}_\varphi$.

Correspondingly, the velocity vector can be expressed in the form

$$\mathbf{v} = v_x\,\mathbf{e}_x + v_\sigma\,\mathbf{e}_\sigma + v_\varphi\,\mathbf{e}_\varphi. \tag{1.3.17}$$

The coefficients v_x, v_σ, and v_φ are the cylindrical polar components of the velocity.

Relation to Cartesian vector components

Using elementary trigonometry, we derive relations between the Cartesian and cylindrical polar unit vectors,

$$\mathbf{e}_\sigma = \cos\varphi\,\mathbf{e}_y + \sin\varphi\,\mathbf{e}_z, \qquad \mathbf{e}_\varphi = -\sin\varphi\,\mathbf{e}_y + \cos\varphi\,\mathbf{e}_z, \tag{1.3.18}$$

and the inverse relations,

$$\mathbf{e}_y = \cos\varphi\,\mathbf{e}_\sigma - \sin\varphi\,\mathbf{e}_\varphi, \qquad \mathbf{e}_z = \sin\varphi\,\mathbf{e}_\sigma + \cos\varphi\,\mathbf{e}_\varphi. \tag{1.3.19}$$

The corresponding relations for the components of the velocity are

$$v_\sigma = \cos\varphi\,v_y + \sin\varphi\,v_z, \qquad v_\varphi = -\sin\varphi\,v_y + \cos\varphi\,v_z, \tag{1.3.20}$$

and

$$v_y = \cos\varphi\,v_\sigma - \sin\varphi\,v_\varphi, \qquad v_z = \sin\varphi\,v_\sigma + \cos\varphi\,v_\varphi. \tag{1.3.21}$$

The Cartesian and cylindrical polar components of other vectors transform in similar ways.

Rates of change

The counterparts of expressions (1.3.4) for the cylindrical polar coordinates are

$$x = X(t), \qquad \sigma = \Sigma(t), \qquad \varphi = \Phi(t). \tag{1.3.22}$$

The rate of change of the unit vectors following the motion of a molecule is given by the relations

$$\frac{\mathrm{d}\mathbf{e}_x}{\mathrm{d}t} = \mathbf{0}, \qquad \frac{\mathrm{d}\mathbf{e}_\sigma}{\mathrm{d}t} = \frac{\mathrm{d}\Phi}{\mathrm{d}t}\,\mathbf{e}_\varphi, \qquad \frac{\mathrm{d}\mathbf{e}_\varphi}{\mathrm{d}t} = -\frac{\mathrm{d}\Phi}{\mathrm{d}t}\,\mathbf{e}_\sigma. \tag{1.3.23}$$

Consistent with our earlier observation, the first unit vector, $\mathbf{e}_x$, is fixed, while the second and third unit vectors, $\mathbf{e}_\sigma$ and $\mathbf{e}_\varphi$, change with position in space.

Velocity components

Substituting expressions (1.3.22) in the right-hand side of (1.3.16), taking the time derivative and using expressions (1.3.23), we find

$$\frac{\mathrm{d}\mathbf{X}}{\mathrm{d}t} = \frac{\mathrm{d}}{\mathrm{d}t}(X\,\mathbf{e}_x + \Sigma\,\mathbf{e}_\sigma) = \frac{\mathrm{d}X}{\mathrm{d}t}\,\mathbf{e}_x + X\,\frac{\mathrm{d}\mathbf{e}_x}{\mathrm{d}t} + \frac{\mathrm{d}\Sigma}{\mathrm{d}t}\,\mathbf{e}_\sigma + \sigma\,\frac{\mathrm{d}\mathbf{e}_\sigma}{\mathrm{d}t}, \tag{1.3.24}$$

and then

$$\frac{\mathrm{d}\mathbf{X}}{\mathrm{d}t} = \frac{\mathrm{d}X}{\mathrm{d}t}\,\mathbf{e}_x + \frac{\mathrm{d}\Sigma}{\mathrm{d}t}\,\mathbf{e}_\sigma + \sigma\,\frac{\mathrm{d}\Phi}{\mathrm{d}t}\,\mathbf{e}_\varphi, \tag{1.3.25}$$

(for polar coordinates

with the understanding that $\sigma = \Sigma(t)$ on the right-hand side. Comparing this expression with the decomposition (1.3.17), we obtain the polar cylindrical components of the velocity,

$$v_x = \frac{dX}{dt}, \qquad v_\sigma = \frac{d\Sigma}{dt}, \qquad v_\varphi = \sigma \frac{d\Phi}{dt}. \tag{1.3.26}$$

Since Φ is a dimensionless function, all three right-hand sides have units of length divided by time.

Acceleration

Differentiating expression (1.3.25) with respect to time and expanding out the derivatives, we find

$$\begin{aligned}\frac{d^2\mathbf{X}}{dt^2} &= \frac{d}{dt}\Big(\frac{dX}{dt}\,\mathbf{e}_x + \frac{d\Sigma}{dt}\,\mathbf{e}_\sigma + \Sigma\,\frac{d\Phi}{dt}\,\mathbf{e}_\varphi\Big) \\ &= \frac{d^2X}{dt^2}\,\mathbf{e}_x + \frac{d^2\Sigma}{dt^2}\,\mathbf{e}_\sigma + \frac{d\Sigma}{dt}\frac{d\mathbf{e}_\sigma}{dt} + \frac{d\Sigma}{dt}\frac{d\Phi}{dt}\,\mathbf{e}_\varphi + \sigma\,\frac{d^2\Phi}{dt^2}\,\mathbf{e}_\varphi + \sigma\,\frac{d\Phi}{dt}\frac{d\mathbf{e}_\varphi}{dt},\end{aligned} \tag{1.3.27}$$

with the understanding that $\sigma = \Sigma(t)$ on the right-hand side.

Now we substitute expressions (1.3.23) and find

$$\begin{aligned}\frac{d^2\mathbf{X}}{dt^2} = {}& \frac{d^2X}{dt^2}\,\mathbf{e}_x + \frac{d^2\Sigma}{dt^2}\,\mathbf{e}_\sigma + \frac{d\Sigma}{dt}\frac{d\Phi}{dt}\,\mathbf{e}_\varphi \\ &+ \frac{d\Sigma}{dt}\frac{d\Phi}{dt}\,\mathbf{e}_\varphi + \sigma\,\frac{d^2\Phi}{dt^2}\,\mathbf{e}_\varphi - \sigma\,\frac{d\Phi}{dt}\frac{d\Phi}{dt}\,\mathbf{e}_\sigma.\end{aligned} \tag{1.3.28}$$

Finally, we consolidate the terms on the right-hand side and derive the cylindrical polar components of the acceleration vector,

$$\begin{aligned} a_x &= \frac{d^2X}{dt^2}, \\ a_\sigma &= \frac{d^2\Sigma}{dt^2} - \sigma\left(\frac{d\Phi}{dt}\right)^2, \\ a_\varphi &= \sigma\,\frac{d^2\Phi}{dt^2} + 2\,\frac{d\Sigma}{dt}\,\frac{d\Phi}{dt} = \frac{1}{\sigma}\frac{d}{dt}\Big(\Sigma^2\,\frac{d\Phi}{dt}\Big). \end{aligned} \tag{1.3.29}$$

Note that a change in the meridional angle, Φ, is accompanied by radial acceleration, a_σ.

1.3.2 Spherical polar coordinates

A point in space can be identified by the values of an ordered triplet (r, θ, φ), as illustrated in figure 1.3.3, where:

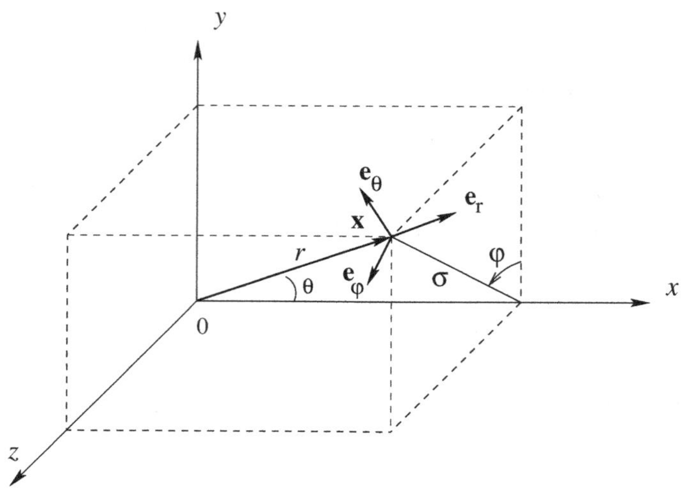

Figure 1.3.3 A system of spherical polar coordinates, (r, θ, φ), defined with reference to the Cartesian coordinates, (x, y, z).

- r is the distance from the designated origin taking values in the range $[0, \infty)$.
- θ is the azimuthal angle subtended between the x axis, the origin, and the chosen point, taking values in the range $[0, \pi]$.
- φ is the meridional angle measured around the x axis, taking values in the range $[0, 2\pi)$. The value $\varphi = 0$ corresponding to the first and second quadrants of the xy plane.

Using elementary trigonometry, we derive relations between the Cartesian, cylindrical, and spherical polar coordinates,

$$x = r\,\cos\theta, \qquad \sigma = r\,\sin\theta, \tag{1.3.30}$$

and

$$y = \sigma\,\cos\varphi = r\sin\theta\,\cos\varphi, \qquad z = \sigma\,\sin\varphi = r\sin\theta\,\sin\varphi. \tag{1.3.31}$$

The inverse relations are

$$r = \sqrt{x^2 + y^2 + z^2} = \sqrt{x^2 + \sigma^2}, \qquad \theta = \arccos\frac{x}{r}, \qquad \varphi = \arccos\frac{y}{\sigma}. \tag{1.3.32}$$

In computing the inverse cosine functions, care must be taken so that θ and φ are continuous functions of x, y, r, and σ.

Unit vectors

We consider a point in space and define three vectors of unit length, denoted by $\mathbf{e}_r$, $\mathbf{e}_\theta$, and $\mathbf{e}_\varphi$, pointing, respectively, in the radial, azimuthal, and meridional direction, as illustrated in figure 1.3.3. Note that the orientation of all three unit vectors changes with position in space; in contrast the orientation of the Cartesian unit vectors, $\mathbf{e}_x$, $\mathbf{e}_y$, and $\mathbf{e}_z$, is fixed.

In terms of the local unit vectors $\mathbf{e}_r$, $\mathbf{e}_\theta$, and $\mathbf{e}_\varphi$, the position vector is

$$\mathbf{x} = r\,\mathbf{e}_r. \tag{1.3.33}$$

The dependence on θ and φ is mediated through the unit vector $\mathbf{e}_r$ on the right-hand side. The absence of $\mathbf{e}_\theta$ and $\mathbf{e}_\varphi$ from the right-hand side of (1.3.33) is explained by observing that the distance from the origin, expressed by the position vector $\mathbf{x}$, is perpendicular to the unit vectors $\mathbf{e}_\theta$ and $\mathbf{e}_\varphi$.

Correspondingly, the velocity vector is given by

$$\mathbf{v} = v_r\,\mathbf{e}_r + v_\theta\,\mathbf{e}_\theta + v_\varphi\,\mathbf{e}_\varphi, \tag{1.3.34}$$

where the coefficients v_r, v_θ, and v_φ are the spherical polar components of the velocity.

Relation to Cartesian vector components

Using elementary trigonometry, we derive relations between the spherical polar, cylindrical polar, and Cartesian unit vectors,

$$\begin{aligned}
\mathbf{e}_r &= \cos\theta\,\mathbf{e}_x + \sin\theta\cos\varphi\,\mathbf{e}_y + \sin\theta\sin\varphi\,\mathbf{e}_z = \cos\theta\,\mathbf{e}_x + \sin\theta\,\mathbf{e}_\sigma,\\
\mathbf{e}_\theta &= -\sin\theta\,\mathbf{e}_x + \cos\theta\cos\varphi\,\mathbf{e}_y + \cos\theta\sin\varphi\,\mathbf{e}_z = -\sin\theta\,\mathbf{e}_x + \cos\theta\,\mathbf{e}_\sigma,\\
\mathbf{e}_\varphi &= -\sin\varphi\,\mathbf{e}_y + \cos\varphi\,\mathbf{e}_z.
\end{aligned} \tag{1.3.35}$$

The corresponding relations for the velocity components are

$$\begin{aligned}
v_r &= \cos\theta\,v_x + \sin\theta\cos\varphi\,v_y + \sin\theta\sin\varphi\,v_z = \cos\theta\,v_x + \sin\theta\,v_\sigma,\\
v_\theta &= -\sin\theta\,v_x + \cos\theta\cos\varphi\,v_y + \cos\theta\sin\varphi\,v_z = -\sin\theta\,v_x + \cos\theta\,v_\sigma,\\
v_\varphi &= -\sin\varphi\,v_y + \cos\varphi\,v_z.
\end{aligned} \tag{1.3.36}$$

The Cartesian and polar components of other vectors transform in similar ways.

Rates of change

The counterparts of expressions (1.3.4) for the spherical polar coordinates are

$$r = R(t), \qquad \theta = \Theta(t), \qquad \varphi = \Phi(t). \tag{1.3.37}$$

The rate of change of the unit vectors following the motion of a molecule is given by the relations

$$\left\{\begin{aligned} \frac{\mathrm{d}\mathbf{e}_r}{\mathrm{d}t} &= \frac{\mathrm{d}\Phi}{\mathrm{d}t}\sin\theta\,\mathbf{e}_\varphi + \frac{\mathrm{d}\Theta}{\mathrm{d}t}\,\mathbf{e}_\theta, \\ \frac{\mathrm{d}\mathbf{e}_\theta}{\mathrm{d}t} &= \frac{\mathrm{d}\Phi}{\mathrm{d}t}\cos\theta\,\mathbf{e}_\varphi - \frac{\mathrm{d}\Theta}{\mathrm{d}t}\,\mathbf{e}_r, \\ \frac{\mathrm{d}\mathbf{e}_\varphi}{\mathrm{d}t} &= -\frac{\mathrm{d}\Phi}{\mathrm{d}t}\cos\theta\,\mathbf{e}_\theta - \frac{\mathrm{d}\Phi}{\mathrm{d}t}\sin\theta\,\mathbf{e}_r. \end{aligned}\right\} \tag{1.3.38}$$

All three unit vectors change with position in space.

Velocity components

Substituting the first expression in (1.3.37) in the right-hand side of (1.3.47) and using (1.3.38), we find

$$\frac{\mathrm{d}\mathbf{X}}{\mathrm{d}t} = \frac{\mathrm{d}R}{\mathrm{d}t}\,\mathbf{e}_r + r\,\frac{\mathrm{d}\mathbf{e}_r}{\mathrm{d}t} = \frac{\mathrm{d}R}{\mathrm{d}t}\,\mathbf{e}_r + r\,\frac{\mathrm{d}\Phi}{\mathrm{d}t}\sin\theta\,\mathbf{e}_\varphi + r\,\frac{\mathrm{d}\Theta}{\mathrm{d}t}\,\mathbf{e}_\theta. \tag{1.3.39}$$

Comparing this expression with the decomposition (1.3.34), we derive expressions for the spherical polar components of the velocity,

$$v_r = \frac{\mathrm{d}R}{\mathrm{d}t}, \qquad v_\theta = r\,\frac{\mathrm{d}\Theta}{\mathrm{d}t}, \qquad v_\varphi = r\sin\theta\,\frac{\mathrm{d}\Phi}{\mathrm{d}t}. \tag{1.3.40}$$

Since the functions Θ and Φ are dimensionless, all three right-hand sides have units of length divided by time.

Acceleration

Differentiating expression (1.3.39) with respect to time and expanding out the derivatives, we find

$$\begin{aligned} \frac{\mathrm{d}^2\mathbf{X}}{\mathrm{d}t^2} &= \Big(\frac{\mathrm{d}^2R}{\mathrm{d}t^2}\,\mathbf{e}_r + \frac{\mathrm{d}R}{\mathrm{d}t}\frac{\mathrm{d}\mathbf{e}_r}{\mathrm{d}t}\Big) + \Big(\frac{\mathrm{d}R}{\mathrm{d}t}\frac{\mathrm{d}\Phi}{\mathrm{d}t}\sin\theta\,\mathbf{e}_\varphi + r\,\frac{\mathrm{d}^2\Phi}{\mathrm{d}t^2}\sin\theta\,\mathbf{e}_\varphi \\ &+ r\,\frac{\mathrm{d}\Phi}{\mathrm{d}t}\cos\theta\,\frac{\mathrm{d}\Theta}{\mathrm{d}t}\,\mathbf{e}_\varphi + r\,\frac{\mathrm{d}\Phi}{\mathrm{d}t}\sin\theta\frac{\mathrm{d}\mathbf{e}_\varphi}{\mathrm{d}t}\Big) + \Big(\frac{\mathrm{d}R}{\mathrm{d}t}\frac{\mathrm{d}\Theta}{\mathrm{d}t}\,\mathbf{e}_\theta + r\,\frac{\mathrm{d}^2\Theta}{\mathrm{d}t^2}\,\mathbf{e}_\theta + r\,\frac{\mathrm{d}\Theta}{\mathrm{d}t}\,\frac{\mathrm{d}\mathbf{e}_\theta}{\mathrm{d}t}\Big), \end{aligned} \tag{1.3.41}$$

where the parentheses enclose terms originating from each term on the right-hand side of (1.3.39).

Now we substitute expressions (1.3.38) and obtain

$$\frac{\mathrm{d}^2\mathbf{X}}{\mathrm{d}t^2} = \mathbf{A} + \mathbf{B} + \mathbf{C}, \tag{1.3.42}$$

where

$$\mathbf{A} = \frac{d^2R}{dt^2}\,\mathbf{e}_r + \frac{dR}{dt}\frac{d\Phi}{dt}\,\sin\theta\,\mathbf{e}_\varphi + \frac{dR}{dt}\frac{d\Theta}{dt}\,\mathbf{e}_\theta,$$

$$\mathbf{B} = \frac{dR}{dt}\frac{d\Phi}{dt}\,\sin\theta\,\mathbf{e}_\varphi + r\,\frac{d^2\Phi}{dt^2}\,\sin\theta\,\mathbf{e}_\varphi + r\,\frac{d\Phi}{dt}\frac{d\Theta}{dt}\,\cos\theta\,\mathbf{e}_\varphi$$
$$-r\left(\frac{d\Phi}{dt}\right)^2\sin\theta\cos\theta\,\mathbf{e}_\theta - r\left(\frac{d\Phi}{dt}\right)^2\sin^2\theta\,\mathbf{e}_r, \qquad (1.3.43)$$

$$\mathbf{C} = \frac{dR}{dt}\frac{d\Theta}{dt}\,\mathbf{e}_\theta + r\,\frac{d^2\Theta}{dt^2}\,\mathbf{e}_\theta + r\,\frac{d\Phi}{dt}\frac{d\Theta}{dt}\,\cos\theta\,\mathbf{e}_\varphi - r\left(\frac{d\Theta}{dt}\right)^2\mathbf{e}_r.$$

Consolidating the various terms, we derive the cylindrical polar components of the acceleration vector,

$$\left\{\begin{aligned} a_r &= \frac{d^2R}{dt^2} - r\left(\frac{d\Phi}{dt}\right)^2\sin^2\theta - r\left(\frac{d\Theta}{dt}\right)^2, \\ a_\theta &= r\,\frac{d^2\Theta}{dt^2} + 2\,\frac{dR}{dt}\,\frac{d\Theta}{dt} - r\left(\frac{d\Phi}{dt}\right)^2\sin\theta\,\cos\theta, \\ a_\varphi &= r\,\frac{d^2\Phi}{dt^2}\,\sin\theta + 2\,\frac{dR}{dt}\,\frac{d\Phi}{dt}\,\sin\theta + 2\,r\,\frac{d\Theta}{dt}\,\frac{d\Phi}{dt}\,\cos\theta. \end{aligned}\right\} \qquad (1.3.44)$$

Note that a change in the meridional angle, Φ, or azimuthal angle, Θ, is accompanied by radial acceleration.

1.3.3 Plane polar coordinates

A point in the xy plane can be identified by the values of the doublet (r, θ), where r is the distance from the origin, and θ is the angle subtended between the x axis, the origin, and the chosen point, measured in the counterclockwise direction, as illustrated in figure 1.3.4. The radial distance, r, takes values in the range $[0, \infty)$, and the polar angle, θ, takes values in the range $[0, 2\pi)$.

Using elementary trigonometry, we derive the following relations between Cartesian and plane polar coordinates,

$$x = r\,\cos\theta, \qquad y = r\,\sin\theta, \qquad (1.3.45)$$

and the inverse relations

$$r = \sqrt{x^2 + y^2}, \qquad \theta = \arccos\frac{y}{r}. \qquad (1.3.46)$$

In computing the inverse cosine function, care must be taken so that θ is a continuous function of y and r.

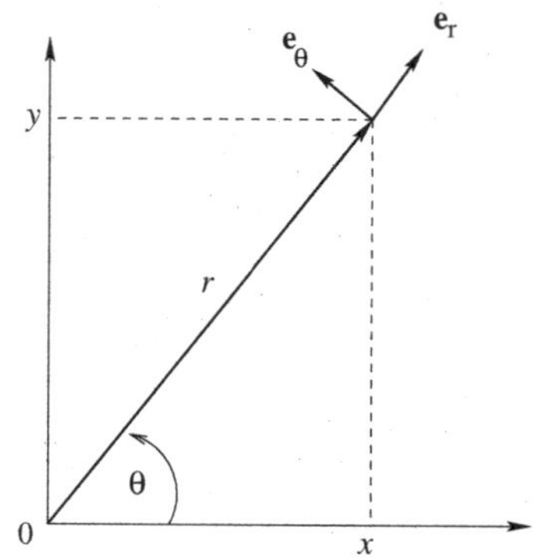

Figure 1.3.4 A system of plane polar coordinates, (r, θ), in the xy plane defined with respect to the Cartesian coordinates, (x, y).

Unit vectors

We consider a point in the xy plane and define two vectors of unit length, denoted by $\mathbf{e}_r$ and $\mathbf{e}_\theta$, pointing in the radial or polar direction, as depicted in figure 1.3.4. Note that the orientation of these unit vectors changes with position in the xy plane, whereas the orientation of the Cartesian unit vectors $\mathbf{e}_x$ and $\mathbf{e}_y$ is fixed.

In terms of the local unit vectors $\mathbf{e}_r$ and $\mathbf{e}_\theta$, the position vector is given by

$$\mathbf{x} = r\,\mathbf{e}_r, \tag{1.3.47}$$

and the velocity vector is given by

$$\mathbf{v} = v_r\,\mathbf{e}_r + v_\theta\,\mathbf{e}_\theta. \tag{1.3.48}$$

The coefficients v_r and v_θ are the plane polar components of the velocity.

Relation to Cartesian vector components

Using elementary trigonometry, we derive the following relations between the Cartesian and plane polar unit vectors,

$$\mathbf{e}_r = \cos\theta\,\mathbf{e}_x + \sin\theta\,\mathbf{e}_y, \qquad \mathbf{e}_\theta = -\sin\theta\,\mathbf{e}_x + \cos\theta\,\mathbf{e}_y, \tag{1.3.49}$$

and the inverse relations

$$\mathbf{e}_x = \cos\theta\,\mathbf{e}_r - \sin\theta\,\mathbf{e}_\theta, \qquad \mathbf{e}_y = \sin\theta\,\mathbf{e}_r + \cos\theta\,\mathbf{e}_\theta. \tag{1.3.50}$$

The corresponding relations for the velocity components are

$$v_r = \cos\theta\,v_x + \sin\theta\,v_y, \qquad v_\theta = -\sin\theta\,v_x + \cos\theta\,v_y, \tag{1.3.51}$$

and

$$v_x = \cos\theta\, v_r - \sin\theta\, v_\theta, \qquad v_y = \sin\theta\, v_r + \cos\theta\, v_\theta. \tag{1.3.52}$$

The Cartesian components of other vectors transform in similar ways.

Rates of change

The counterparts of expressions (1.3.4) for the plane polar coordinates are

$$r = R(t), \qquad \theta = \Theta(t). \tag{1.3.53}$$

The rate of change of the unit vectors following the motion of a molecule is given by the relations

$$\frac{\mathrm{d}\mathbf{e}_r}{\mathrm{d}t} = \frac{\mathrm{d}\Theta}{\mathrm{d}t}\,\mathbf{e}_\theta, \qquad \frac{\mathrm{d}\mathbf{e}_\theta}{\mathrm{d}t} = -\frac{\mathrm{d}\Theta}{\mathrm{d}t}\,\mathbf{e}_r. \tag{1.3.54}$$

Velocity components

To derive the velocity components, we substitute the first equation in (1.3.53) in the right-hand side of (1.3.47), and take the time derivative of the resulting equation. Next, we identify the left-hand side with the velocity, expand out the derivatives of the products on the right-hand side, and use the first relation in (1.3.54) to eliminate the time derivative of the radial unit vector. Comparing the result with expressions (1.3.48), we obtain the plane polar components of the velocity,

$$v_r = \frac{\mathrm{d}R}{\mathrm{d}t}, \qquad v_\theta = r\,\frac{\mathrm{d}\Theta}{\mathrm{d}t}. \tag{1.3.55}$$

Note that the right-hand sides have units of length divided by time.

Acceleration

Working as in the case of the cylindrical polar coordinates, we find that the plane polar components of the acceleration are given by

$$a_r = \frac{\mathrm{d}^2R}{\mathrm{d}t^2} - r\left(\frac{\mathrm{d}\Theta}{\mathrm{d}t}\right)^2,$$

$$a_\theta = r\,\frac{\mathrm{d}^2\Theta}{\mathrm{d}t^2} + 2\,\frac{\mathrm{d}R}{\mathrm{d}t}\,\frac{\mathrm{d}\Theta}{\mathrm{d}t} = \frac{1}{r}\frac{\mathrm{d}}{\mathrm{d}t}\Big(R^2\,\frac{\mathrm{d}\Theta}{\mathrm{d}t}\Big). \tag{1.3.56}$$

If a molecule moves along a circular path of radius r centered at the origin, $\mathrm{d}R/\mathrm{d}t = 0$, the acceleration components are

$$a_r = r\left(\frac{\mathrm{d}\Theta}{\mathrm{d}t}\right)^2 = \frac{v_\theta^2}{r}, \qquad a_\theta = r\,\frac{\mathrm{d}^2\Theta}{\mathrm{d}t^2}. \tag{1.3.57}$$

Problems

1.3.1 *Spherical polar coordinates*

Derive the inverse transformation rules shown in equations (1.3.36); that is, express the Cartesian components in terms of the spherical polar components of the velocity.

1.3.2 *Acceleration*

Derive the plane polar components of the acceleration given in (1.3.56).

1.4 Fluid velocity

Having prepared the ground for describing the motion of molecules in a fluid in quantitative terms, and turn to considering the motion of fluid parcels consisting of a large number of molecules. For simplicity, we consider a homogeneous fluid parcel consisting of identical molecules and label the N constituent molecules by the integer index i, where $i = 1, 2, \ldots, N$. Let $v_x^{(i)}$, $v_y^{(i)}$, and $v_z^{(i)}$ be the Cartesian components of the velocity of the ith molecule at a particular time instant. The corresponding components of the mean velocity are defined as

$$< v_x >= \frac{1}{N}\sum_{i=1}^{N} v_x^{(i)}, \qquad < v_y >= \frac{1}{N}\sum_{i=1}^{N} v_y^{(i)},$$
$$< v_z >= \frac{1}{N}\sum_{i=1}^{N} v_z^{(i)}, \tag{1.4.1}$$

where the pointed brackets on the left-hand sides denote averages over all molecules. Equations (1.4.1) can be combined into the vector form

$$< \mathbf{v} >= \frac{1}{N}\sum_{i=1}^{N} \mathbf{v}^{(i)}. \tag{1.4.2}$$

At a particular time t, a fluid parcel of interest is centered at the point $\mathbf{x}$. As the size of the parcel becomes smaller, the parcel tends to occupy an infinitesimal volume in space containing the point $\mathbf{x}$. In this limit, the components of the parcel velocity defined in equations (1.4.1) reduce to the corresponding components of the fluid velocity, denoted by u_x, u_y, and u_z, forming an ordered triplet,

$$\mathbf{u} = (u_x, u_y, u_z). \tag{1.4.3}$$

Continuum approximation

A conceptual difficulty undermines the physical relevance of the fluid velocity: in the limit as the size of a fluid parcel tends to zero, the number of molecules residing inside the parcel also tends to zero, and the pointed-bracket averages defined in equations (1.4.1) become ill-defined. To see this, we consider a spherical particle of radius ϵ. As ϵ tends to zero, a graph of the average molecular velocity, $< v_x >$, plotted against ϵ, will show strong fluctuations that are manifestations of random molecular motions.

To circumvent this difficulty, we adopt the continuum approximation, prescribing that, as the size of a fluid parcel tends to zero, the limit of the average molecular velocity is computed before the discrete nature of the fluid becomes apparent.

Since different choices for the designated parcel center at difference times produce different fluid velocities, the components of the velocity vector, $\mathbf{u}$, are functions of the components of the position vector, $\mathbf{x} = (x, y, z)$, and time, t. To signify this dependence, we append to u_x, u_y, and u_z a set of parentheses enclosing the four independent variables, writing

$$u_x(x, y, z, t), \qquad u_y(x, y, z, t), \qquad u_z(x, y, z, t). \tag{1.4.4}$$

In compact notation, we write

$$u_x(\mathbf{x}, t), \qquad u_y(\mathbf{x}, t), \qquad u_z(\mathbf{x}, t), \tag{1.4.5}$$

and in full vector notation, we write

$$\mathbf{u}(\mathbf{x}, t). \tag{1.4.6}$$

If the fluid translates as a rigid body in a certain direction, possibly with a time-dependent velocity, we omit $\mathbf{x}$ in the list of arguments, and write $\mathbf{u}(t)$.

For example, the Cartesian components of a velocity field are given by

$$\begin{aligned} u_x(x, y, z, t) &= a\,(y^2 + z^2) + (b + c\,t)\,x^3 y\,z + c\,e^{dxt}, \\ u_y(x, y, z, t) &= a\,(z^2 + x^2) + (b + c\,t)\,x\,y^3\,z + c\,e^{dyt}, \\ u_z(x, y, z, t) &= a\,(x^2 + y^2) + (b + c\,t)\,x\,y\,z^3 + c\,e^{dzt}, \end{aligned} \tag{1.4.7}$$

where a, b, c, and d are four constants. Velocity has dimensions of length per time L/T, and the position vector has dimensions of length, L. In order for both sides of equations (1.4.7) to have the same units, the constant a must has dimensions of inverse length-time, $1/(LT)$.

Steady flow

If a flow is steady, the components of the fluid velocity do not depend on time, and we omit t from the list of arguments in (1.4.4)–(1.4.6), writing $\mathbf{u}(\mathbf{x})$.

Two-dimensional flow

If the z component of the fluid velocity is zero, while the x and y components depend on x and y but not on z, we obtain a two-dimensional flow in the xy plane. The velocity vector lies in the xy plane at every point in this two-dimensional flow.

Swirling and axisymmetric flow

Consider the system of cylindrical polar coordinates depicted in figure 1.4.2. The cylindrical polar components of the velocity, u_σ and u_φ, are related to the Cartesian components by the counterparts of equations (1.3.20),

$$u_\sigma = \cos\varphi\, u_y + \sin\varphi\, u_z, \qquad u_\varphi = -\sin\varphi\, u_y + \cos\varphi\, u_z. \tag{1.4.8}$$

In a swirling flow, the axial and radial velocity components are zero, $u_x = 0$ and $u_\sigma = 0$, while the meridional component, u_φ, is nonzero but independent of the meridional angle, φ. The velocity vector points in the direction of the meridional angle at every point in a swirling flow.

If the meridional velocity component, u_φ, is zero at every point, while the axial and radial components, u_x and u_σ, are nonzero but independent of φ, we obtain an axially symmetric flow, also called an axisymmetric flow. The velocity vector lies in a meridional plane passing through the x axis at every point in an axisymmetric flow.

Superposing a swirling flow and an axisymmetric flow, we obtain a three-dimensional flow described as axisymmetric flow with swirling motion. All three velocity components, u_x, u_σ, and u_φ, are generally nonzero but independent of the meridional angle φ in this flow.

1.4.1 Velocity vector field, streamlines and stagnation points

Consider a flow at a certain time instant, and draw velocity vectors at a large number of points distributed in the domain of flow. The collection of these vectors defines a vector field called the velocity field. Starting at a certain point in the flow, we may draw a line that is tangential to the velocity vector at each point, as illustrated in figure 1.4.1. This generally curved three-dimensional line is called an instantaneous streamline, and a collection of streamlines composes an instantaneous streamline pattern.

Two or more streamlines can meet at a stagnation point, as illustrated in figure 1.4.1. Since the velocity is unique at each point in a flow, all velocity components must necessarily vanish at a stagnation point. A streamline must be a closed line, extend to infinity, cross a moving boundary, or terminate at a stagnation point.

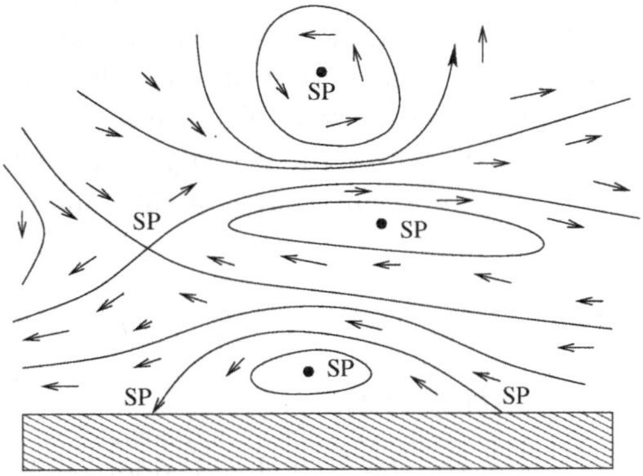

Figure 1.4.1 Illustration of a velocity vector field and associated streamline pattern in a two-dimensional flow, involving stagnation points (SP). Stagnation points may occur in the interior or at the boundaries of a flow.

Problems

1.4.1 *Dimensions of coefficients.*

Deduce the dimensions of the coefficients b, c, and d on the right-hand sides of equations (1.4.7).

1.4.2 *Streamline patterns*

Sketch streamline patterns for (a) a two-dimensional flow, (b) a swirling flow, (c) an axisymmetric flow, and (d) an axisymmetric flow with swirling motion.

1.5 Point particles and their trajectories

As the size of a fluid parcel tends to zero, the parcel reduces to an abstract entity called a point particle. In the context of continuum mechanics, a point particle is large enough to contain a large number of molecules whose average velocity is well-defined, but small enough so that its volume is infinitesimal. This means that the ratio of the volume of a point particle to the volume of the whole fluid is zero. Two consequences of this idealization are:

- A finite fluid parcel is comprised of an infinite number of point particles.
- The product of the infinite number of point particles and the infinitesimal volume of each point particle is finite and nonzero and equal to the parcel volume.

By definition, the rate of change of the position of a point particle is equal to the velocity of the fluid evaluated at the instantaneous position of the point particle. If, during an infinitesimal period of time, $\mathrm{d}t$, the x coordinate of a point particle located at the position $\mathbf{x} = \mathbf{X}$ has changed by the infinitesimal distance $\mathrm{d}X$, then $u_x = \mathrm{d}X/\mathrm{d}t$, where the velocity u_x is evaluated at $\mathbf{x} = \mathbf{X}$ at the current time, t. Writing the counterparts of this equation for the y and z components, we obtain

$$\frac{\mathrm{d}X}{\mathrm{d}t} = u_x(X(t), Y(t), Z(t), t), \qquad \frac{\mathrm{d}Y}{\mathrm{d}t} = u_y(X(t), Y(t), Z(t), t),$$
$$\frac{\mathrm{d}Z}{\mathrm{d}t} = u_z(X(t), Y(t), Z(t), t), \tag{1.5.1}$$

The first set of parentheses on the right-hand sides enclose the four scalar arguments of the velocity.

1.5.1 Path lines

Since a point particle moves with the local fluid velocity, its coordinates generally change in time according to equations (1.5.1), even if the flow is steady. Point particles in a fluid remain stationary only if they lie precisely at a stagnation point in a steady flow, or if the velocity vanishes and the fluid is macroscopically quiescent. The trajectory of a point particle is called a path line.

By definition, a streamline is tangential to the instantaneous velocity vector field at every point. The path described by a point particle moving with the frozen instantaneous velocity field is an instantaneous streamline.

1.5.2 Ordinary differential equations (ODEs)

Equations (1.5.1) comprise a system of three first-order ordinary differential equations (ODEs). If the flow is steady, the system is autonomous, meaning that there is no explicit time dependence on the right-hand side. If the flow is unsteady, the system is non-autonomous, exhibiting explicit time dependence on the right-hand side. The right-hand side of a non-autonomous system depends on time implicitly through the arguments of the dependent variables, $X(t)$, $Y(t)$, and $Z(t)$, as well as explicitly through the unsteadiness of the flow.

For example, the Cartesian velocity components of a certain steady unidirectional flow are given by

$$u_x = a\,y^2 + b\,y + c, \qquad u_y = 0, \qquad u_z = 0, \tag{1.5.2}$$

where a, b, and c are three constants with appropriate dimensions. In this case, the fluid moves along the x axis with velocity that depends on the y coordinate alone. The trajectories of point particles are straight lines described by the autonomous system of ODEs

$$\frac{\mathrm{d}X}{\mathrm{d}t} = a\,Y^2 + b\,Y + c, \qquad \frac{\mathrm{d}Y}{\mathrm{d}t} = 0, \qquad \frac{\mathrm{d}Z}{\mathrm{d}t} = 0. \tag{1.5.3}$$

The solution of these equations is readily readily found to be

$$X(t) = X_0 + (a\,Y^2 + b\,Y + c)\,t, \qquad Y(t) = Y_0, \qquad Z(t) = Z_0, \qquad (1.5.4)$$

where X_0, Y_0, and Z_0 are the coordinates of a point particle at the initial instant, $t = 0$.

However, in general, the solution of system (1.5.1) cannot be found by analytical methods, and the use of numerical methods is imperative.

1.5.3 Explicit Euler method

A numerical method for generating the trajectory of a point particle emerges by considering the change in the position of the point particle over a small time interval Δt, and replacing the differential equations (1.5.1) with the algebraic equations

$$\begin{aligned} \frac{X(t+\Delta t) - X(t)}{\Delta t} &= u_x(X(t), Y(t), Z(t), t), \\ \frac{Y(t+\Delta t) - Y(t)}{\Delta t} &= u_y(X(t), Y(t), Z(t), t), \\ \frac{Z(t+\Delta t) - Z(t)}{\Delta t} &= u_z(X(t), Y(t), Z(t), t). \end{aligned} \qquad (1.5.5)$$

To obtain these equations, we have replaced the time derivatives on the left-hand sides of equations (1.5.1) with forward finite differences. Since, by definition, the first derivative $\mathrm{d}X/\mathrm{d}t$ is equal to the limit of the ratio $[X(t+\Delta t) - X(t)]/\Delta t$ as Δt tends to zero, we expect that, as long as Δt is sufficiently small, the error introduced by replacing the derivative with a forward difference will also be reasonably small.

In fact, a Taylor series expansion shows that the magnitude of the error associated with the approximate forms (1.5.5) is comparable to the magnitude of Δt. Thus, if Δt is equal to 0.1 in some units, the error associated with the difference approximation will be on the order of 0.1 multiplied by a constant whose absolute value ranges roughly between 0.5 and 5 in corresponding units.

In vector notation, the discrete form of the differential system (1.5.1) expressed by the algebraic system (1.5.5) takes the form

$$\frac{\mathbf{X}(t+\Delta t) - \mathbf{X}(t)}{\Delta t} = \mathbf{u}(\mathbf{X}(t), t) + O(\Delta t), \qquad (1.5.6)$$

where the term $O(\Delta t)$ on the right-hand side signifies the order of the error due to the difference approximation.

Solving the first equation in (1.5.5) for $X(t+\Delta t)$, the second equation for $Y(t+\Delta t)$, and the third equation for $Z(t+\Delta t)$, we obtain

$$\begin{aligned} X(t+\Delta t) &= X(t) + \Delta t\, u_x(X(t), Y(t), Z(t), t), \\ Y(t+\Delta t) &= Y(t) + \Delta t\, u_y(X(t), Y(t), Z(t), t), \\ Z(t+\Delta t) &= Z(t) + \Delta t\, u_z(X(t), Y(t), Z(t), t). \end{aligned} \tag{1.5.7}$$

In physical terms, equations (1.5.7) state that the position of a point particle at time $t + \Delta t$ is equal to the position at the previous time, t, plus a small displacement equal to the distance traveled over the small time interval Δt. The velocity of travel has been assumed to be constant and equal to the local fluid velocity at the beginning of the time step corresponding to time t.

Algorithm

Equations (1.5.7) provide us with a scheme for computing the trajectory of a point particle according to the following algorithm implementing the explicit Euler method:

1. Specify the initial time; for example, set $t = 0$.
2. Select the size of the time step, Δt.
3. Specify the initial coordinates, X_0, Y_0, and Z_0.
4. Evaluate the velocities

$$\begin{aligned} &u_x(X(t), Y(t), Z(t), t), \quad u_y(X(t), Y(t), Z(t), t), \\ &u_z(X(t), Y(t), Z(t), t), \end{aligned} \tag{1.5.8}$$

   on the right-hand side of equations (1.5.7).
5. Evaluate the right-hand sides of (1.5.7) to obtain the new coordinates of the point particle $X(t+\Delta t)$, $Y(t+\Delta t)$, and $Z(t+\Delta t)$.
6. Reset the time to $t + \Delta t$.
7. Stop if desired, or return to execute steps 4–6.

The method is explicit in that the new position of a point particle is computed in terms of the old position using information available exclusively at the old position.

We have mentioned that approximating the derivative $\mathrm{d}\mathbf{X}/\mathrm{d}t$ with a forward difference introduces an error that is comparable to the magnitude of Δt, as shown in equations (1.5.6). Accordingly, the error in the position of the point particle after it has traveled for a time interval Δt is on the order of Δt^2. Based on the value of the exponent of Δt, we say that the explicit Euler method carries a stepwise error of second order with respect to the time step.

If N_{steps} steps are executed from time $t = 0$ to time $t = t_{final}$, the stepwise error will accumulate to an amount that is comparable to the product $N_{steps} \times \Delta t^2$. Since, by definition, $N_{steps} \times \Delta t = t_{final}$, the cumulative error will be on the order of $t_{final} \times \Delta t$. This expression shows that the cumulative error is of first order with respect to the time step. Unless Δt is sufficiently small, this level of error is hardly acceptable.

1.5.4 Modified Euler method

To reduce the magnitude of the error, we modify the explicit Euler method according to the following steps:

1. Set the initial time; for example, set $t = 0$.
2. Select the size of the time step, Δt.
3. Specify the initial coordinates, $X(0)$, $Y(0)$, and $Z(0)$.
4. Evaluate the velocities

$$\begin{aligned} &u_x(X(t), Y(t), Z(t), t), \quad u_y(X(t), Y(t), Z(t), t), \\ &u_z(X(t), Y(t), Z(t), t) \end{aligned} \tag{1.5.9}$$

   on the right-hand sides of (1.5.7), and save them for future use.
5. Evaluate the right-hand sides of (1.5.7) to obtain the predicted coordinates at time $t + \Delta t$, denoted by X^{pred}, Y^{pred}, and Z^{pred}.
6. Evaluate the velocities at the predicted position at time $t + \Delta t$,

$$\begin{aligned} &u_x^{pred}(X^{pred}, Y^{pred}, Z^{pred}, t + \Delta t), \\ &u_y^{pred}(X^{pred}, Y^{pred}, Z^{pred}, t + \Delta t), \\ &u_z^{pred}(X^{pred}, Y^{pred}, Z^{pred}, t + \Delta t). \end{aligned} \tag{1.5.10}$$

7. Compute the average of the initial and predicted velocities,

$$\begin{aligned} u_x^a &= \frac{1}{2}\, [u_x(X(t), Y(t), Z(t), t) + u_x^{pred}], \\ u_y^a &= \frac{1}{2}\, [u_y(X(t), Y(t), Z(t), t) + u_y^{pred}], \\ u_z^a &= \frac{1}{2}\, [u_z(X(t), Y(t), Z(t), t) + u_z^{pred}]. \end{aligned} \tag{1.5.11}$$

8. Compute the coordinates of the point particle at time $t+\Delta t$, by returning to the position at time t and traveling with the mean velocity computed at step 7, using the formulas

$$\begin{aligned} &X(t + \Delta t) = X(t) + \Delta t\, u_x^a, \qquad Y(t + \Delta t) = Y(t) + \Delta t\, u_y^a, \\ &Z(t + \Delta t) = Z(t) + \Delta t\, u_z^a. \end{aligned} \tag{1.5.12}$$

9. Reset the time to $t + \Delta t$.
10. Stop if desired, or return to execute steps 4–9.

The modified Euler method is a special implementation of the inclusive second-order Runge-Kutta method for solving systems of ordinary differential equations. An error analysis shows that each time step introduces a numerical error in the position of the point particle that is comparable to the cubic power of time step, Δt^3. The cumulative error is on the order of $t_{final} \times \Delta t^2$, which is much smaller than that incurred by the explicit Euler method.

To ensure a smooth particle trajectory, we may specify an approximate travel distance in each step, Δs, and adjust the time step accordingly using $\Delta t = \Delta s/u$, where u is the magnitude of the local velocity.

The following MATLAB code *path_lines* residing in directory *04_various* of FDLIB computes and plots path lines originating from specified initial points. The integration stops when a path line escapes the plotting window or returns to the initial position to form a closed loop:

```
Nsteps = 400; % number of steps
Ds = 0.01; % spatial step

xmin = -1.0; xmax = 1.0; % plotting window
ymin = -1.0; ymax = 1.0;

Xstart = [0.1, 0.2, 0.3]; % starting points of path lines
Ystart = [0.1, 0.2, 0.3];
hold on

%---
for n=1:size(Xstart') % run over starting points
%---

X0 = Xstart(n);
Y0 = Ystart(n);
Xsave = X0;
Ysave = Y0;

clear Xplot Yplot
Xplot(1) = X0;
Yplot(1) = Y0;

X = X0; Y = Y0; t = 0;

%--
for i=1:Nsteps
%--
```

```
  [ux, uy] = path_lines_vel(X,Y,t);

  um = sqrt(ux*ux+uy*uy);
  Dt = Ds/um;
  X1 = X+ux*Dt;
  Y1 = Y+uy*Dt;
  t1 = t+Dt;

  [ux1, uy1] = path_lines_vel(X1,Y1,t1);

  X = X + 0.5*Dt*(ux+ux1);
  Y = Y + 0.5*Dt*(uy+uy1);
  t = t+Dt;
  Xplot(i+1) = X;
  Yplot(i+1) = Y;

%--- stopping check for window limits and closed paths

  closed = sqrt((X-Xsave)^2+(Y-Ysave)^2);

  if(closed < 0.9*Ds | X<xmin |X>xmax | Y<ymin | Y>ymax)
    break
  end

%--
end
%--

plot(Xplot,Yplot,'.')

end   % of path lines

%---
% finish
%---

xlabel('x','fontsize',15)
ylabel('y','fontsize',15)
box
```

The velocity components are evaluated in the following companion MATLAB function:

```
function [ux,uy] = path_lines_vel(x,y,t)

%----
% evaluate the velocity
%---
```

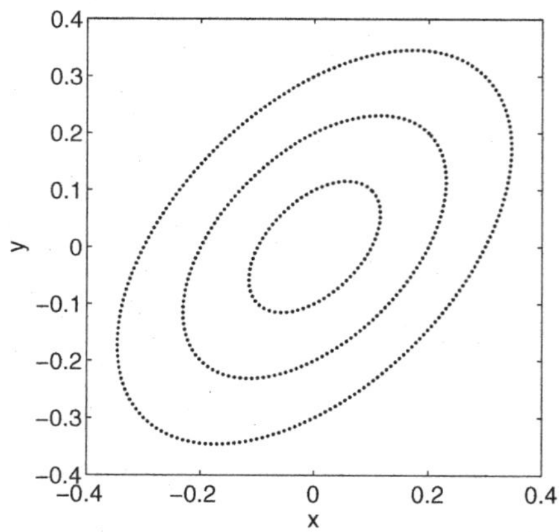

Figure 1.5.2 Path lines computed using the MATLAB code *path_lines*.

```
    Omega = 1.0;
    G = 0.5;
    alpha = 0.0;
    ux = G*x-Omega*y+0.5*alpha*x*exp(-t);
    uy = Omega*x-G*y+0.5*alpha*y*exp(t);

%----
% done
%---

return
```

Running the code *path_lines* produces the plot shown in figure 1.5.2.

1.5.5 Description in polar coordinates

In the cylindrical polar coordinates depicted in figure 1.3.2, the position of a point particle is described by the functions $x = X(t)$, $\sigma = \Sigma(t)$, and $\varphi = \Phi(t)$. Using the transformation rules given in section 1.3, we derive the differential equations

$$\frac{\mathrm{d}X}{\mathrm{d}t} = u_x(X(t), \Sigma(t), \Phi(t), t), \qquad \frac{\mathrm{d}\Sigma}{\mathrm{d}t} = u_\sigma(X(t), \Sigma(t), \Phi(t), t),$$

$$\frac{\mathrm{d}\Phi}{\mathrm{d}t} = \frac{u_\varphi(X(t), \Sigma(t), \Phi(t), t)}{\Sigma(t)}. \qquad (1.5.13)$$

In the spherical polar coordinates depicted in figure 1.3.3, the position of a point particle is described by the functions $r = R(t)$, $\theta = \Theta(t)$, and $\varphi = \Phi(t)$.

Using the transformation rules given in section 1.3, we derive the differential equations

$$\frac{\mathrm{d}R}{\mathrm{d}t} = u_r(X(t), \Theta(t), \Phi(t), t), \qquad \frac{\mathrm{d}\Theta}{\mathrm{d}t} = \frac{u_\theta(X(t), \Theta(t), \Phi(t), t)}{R(t)},$$

$$\frac{\mathrm{d}\Phi}{\mathrm{d}t} = \frac{u_\varphi(X(t), \Theta(t), \Phi(t), t)}{R(t)\,\sin\Theta(t)}. \tag{1.5.14}$$

In the plane polar coordinates depicted in figure 1.3.4, the position of a point particle is described by the functions $r = R(t)$ and $\theta = \Theta(t)$. Using the transformation rules given in section 1.3, we derive the differential equations

$$\frac{\mathrm{d}R}{\mathrm{d}t} = u_r(R(t), \Theta(t), t), \qquad \frac{\mathrm{d}\Theta}{\mathrm{d}t} = \frac{u_\theta(R(t), \Theta(t), t)}{R(t)}. \tag{1.5.15}$$

These differential equations can be integrated in time using the methods discussed previously in this section for Cartesian coordinates, including the Euler method and the modified Euler method. To deduce the position vector at any time, we use the transformation rules providing us with the Cartesian in terms of the chosen polar coordinates.

1.5.6 Streaklines

A streakline emerges by connecting the instantaneous positions of point particles that have been released into the flow from a stationary or moving source at previous times. Alternatively, the point particles may have been residing in the flow at all times, but they have been colored or tagged as they passed through the tip of a stationary or moving probe. If the flow is steady and the probe is stationary, a streakline is also a streamline.

To compute a streakline, we solve the differential equations describing the motion of the point particles after they have entered the flow or passed through the coloring probe using the methods described in this section for particle paths.

Problem

1.5.1 *Streamlines by analytical integration*

Consider a steady two-dimensional flow with velocity components

$$u_x = ax + by, \qquad u_y = bx - ay. \tag{1.5.16}$$

Deduce the dimensions of the constants a and b. Derive analytical expressions for the position of a point particle similar to those shown in (1.5.4).

Computer Problem

1.5.2 *Streamlines by numerical integration*

Program *strml* in directory *04_various* of FDLIB generates streamlines emanating from a specified collection of points, computed by the modified Euler method.

(*a*) Run the program for three velocity fields of your choice implemented in the code. Generate and discuss the structure of the streamlines patterns.

(*b*) Add to the code a new flow of your choice. Generate and discuss the corresponding streamline pattern.

1.6 Material surfaces and elementary motions

An infinite collection of point particles distributed over a surface that resides inside a fluid or at the boundaries of a flow defines a material surface. A cylindrical material surface in a two-dimensional flow can be identified by its trace in the xy plane, and a material surface of revolution in an axisymmetric flow can be identified by its trace in a meridional plane corresponding to a certain meridional angle, φ. Any patch on the surface of a cup of coffee is a material surface with a distinct identity. Under most conditions, if a material patch lies at the boundary of a fluid at a certain time, it will remain at the boundary of the fluid at all times. This means that the point particles comprising the patch are not able to penetrate the fluid.

Material parcels

A closed material surface is the boundary of a material parcel consisting of a fixed mass of fluid with a permanent identity. Under most conditions, if a material surface is located at the boundary of a material parcel at a certain time, it will remain at the boundary of the parcel at all times. To analyze the evolution of a material parcel and visualize its motion, we compute the trajectories of the point particles that lie on its boundary using analytical and numerical methods.

1.6.1 Fluid parcel rotation

Consider a two-dimensional flow in the xy plane with velocity components

$$u_x = -\Omega\, y, \qquad u_y = \Omega\, x, \tag{1.6.1}$$

where Ω is a constant with units of inverse time. In vector-matrix notation, equations (1.6.1) are collected into the form

$$[u_x \quad u_y] = [x \quad y] \cdot \begin{bmatrix} 0 & \Omega \\ -\Omega & 0 \end{bmatrix}. \tag{1.6.2}$$

According to our discussion in section 1.5, the trajectory of a point particle with Cartesian coordinates $(X(t), Y(t))$, is governed by the differential equations

$$\frac{\mathrm{d}X}{\mathrm{d}t} = -\Omega\, Y, \qquad \frac{\mathrm{d}Y}{\mathrm{d}t} = \Omega\, X, \tag{1.6.3}$$

subject to a specified initial condition $X_0 \equiv X(t=0)$ and $Y_0 \equiv Y(t=0)$. The solution is readily found to be

$$X(t) = \cos(\Omega t)\, X_0 - \sin(\Omega t)\, Y_0, \quad Y(t) = \sin(\Omega t)\, X_0 + \cos(\Omega t)\, Y_0. \tag{1.6.4}$$

In vector-matrix notation,

$$\begin{bmatrix} X(t) \\ Y(t) \end{bmatrix} = \begin{bmatrix} \cos(\Omega t) & -\sin(\Omega t) \\ \sin(\Omega t) & \cos(\Omega t) \end{bmatrix} \cdot \begin{bmatrix} X_0 \\ Y_0 \end{bmatrix}. \tag{1.6.5}$$

To deduce the nature of the motion, we refer to plane polar coordinates and find that the distance of a point particle from the origin,

$$R(t) \equiv (X(t)^2 + Y(t)^2)^{1/2}, \tag{1.6.6}$$

remains constant in time and equal to the initial distance, $R_0 \equiv R(t=0)$. The polar angle, θ, defined by the equation $\tan\theta = Y(t)/X(t)$, increases linearly in time at the rate $\mathrm{d}\theta/\mathrm{d}t = \Omega$,

$$\theta = \theta_0 + \Omega\, t, \tag{1.6.7}$$

where θ_0 is the polar angle at $t = 0$. To show this, we write

$$\tan\theta = \frac{Y(t)}{X(t)} = \frac{\sin(\Omega t)\, X_0 + \cos(\Omega t)\, Y_0}{\cos(\Omega t)\, X_0 - \sin(\Omega t)\, Y_0}$$

$$= \frac{\sin(\Omega t)\, \cos\theta_0 + \cos(\Omega t)\, \sin\theta_0}{\cos(\Omega t)\, \cos\theta_0 - \sin(\Omega t)\, \sin\theta_0} = \frac{\sin(\Omega t + \theta_0)}{\cos(\Omega t + \theta_0)} = \tan(\Omega t + \theta_0). \tag{1.6.8}$$

These results suggest that a circular material line centered at the origin rotates around the origin as a rigid body with angular velocity Ω while retaining its circular shape. Accordingly, the velocity field associated with (1.6.1) expresses rigid-body rotation around the origin in the xy plane.

1.6.2 *Fluid parcel deformation*

Now we consider a different type of two-dimensional flow in the xy plane with velocity components

$$u_x = G\, x, \qquad u_y = -G\, y, \tag{1.6.9}$$

where G is a constant with dimensions of inverse time. In vector-matrix notation,

$$[u_x \quad u_y] = [x \quad y] \cdot \begin{bmatrix} G & 0 \\ 0 & -G \end{bmatrix}. \tag{1.6.10}$$

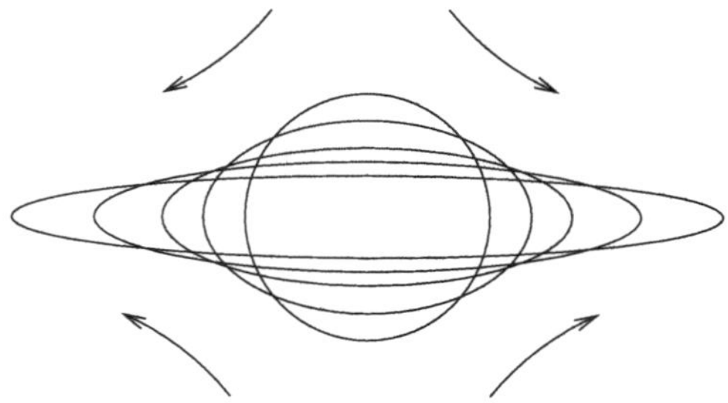

Figure 1.6.1 Deformation of a circular material line under the influence of a two-dimensional elongational flow.

In this case, the trajectory of a point particle is governed by the differential equations

$$\frac{dX}{dt} = G\,X, \qquad \frac{dY}{dt} = -G\,Y, \tag{1.6.11}$$

subject to a specified initial condition. Note that equations (1.6.11) are decoupled, as the first equation contains only X and the second equation contains only Y. The solution is readily found to be

$$X(t) = e^{Gt}\,X_0, \qquad Y(t) = e^{-Gt}\,Y_0. \tag{1.6.12}$$

In vector-matrix notation,

$$\begin{bmatrix} X(t) \\ Y(t) \end{bmatrix} = \begin{bmatrix} e^{Gt} & 0 \\ 0 & e^{-Gt} \end{bmatrix} \cdot \begin{bmatrix} X_0 \\ Y_0 \end{bmatrix}. \tag{1.6.13}$$

The evolution of a circular material line of radius a centered at the origin is illustrated in figure 1.6.1 for a positive value of G. As soon as the motion begins, the circular contour deforms into an ellipse with major axis $a\,e^{Gt}$ oriented in the x direction, and minor axis $a\,e^{-Gt}$ oriented in the y direction. Accordingly, the velocity field described by (1.6.10) describes pure deformation occurring at an exponential rate; the constant G is the rate of deformation.

More detailed consideration reveals that the area enclosed by the deforming circle remains constant in time and equal to πa^2. Thus, the deformation conserves the area of the parcel enclosed by the deforming circle.

1.6.3 Fluid parcel expansion

As a third case study, we consider a two-dimensional flow in the xy plane with velocity components

$$u_x = \frac{1}{2}\,\alpha\,x, \qquad u_y = \frac{1}{2}\,\alpha\,y, \tag{1.6.14}$$

where α is a constant with dimensions of inverse time. In vector-matrix notation,

$$[u_x \quad u_y] = [x \quad y] \cdot \begin{bmatrix} \frac{1}{2}\alpha & 0 \\ 0 & \frac{1}{2}\alpha \end{bmatrix}. \tag{1.6.15}$$

The trajectory of a point particle is governed by the decoupled differential equations

$$\frac{\mathrm{d}X}{\mathrm{d}t} = \frac{1}{2}\alpha X, \qquad \frac{\mathrm{d}Y}{\mathrm{d}t} = \frac{1}{2}\alpha Y, \tag{1.6.16}$$

subject to a specified initial condition. The solution is found by elementary methods,

$$X(t) = \exp(\frac{1}{2}\alpha t)\, X_0, \qquad Y(t) = \exp(\frac{1}{2}\alpha t)\, Y_0. \tag{1.6.17}$$

In vector-matrix notation,

$$\begin{bmatrix} X(t) \\ Y(t) \end{bmatrix} = \begin{bmatrix} \exp(\frac{1}{2}\alpha t) & 0 \\ 0 & \exp(\frac{1}{2}\alpha t) \end{bmatrix} \cdot \begin{bmatrix} X_0 \\ Y_0 \end{bmatrix}. \tag{1.6.18}$$

Based on these expressions, we deduce that a circular material line centered at the origin expands at an exponential rate while retaining its circular shape. Accordingly, the velocity field associated with equations (1.6.14) expresses isotropic expansion. If $a(t)$ is the radius of the circular material line at time t, and $a(t = 0)$ is the radius at the origin of time, then

$$a(t) = a(t = 0)\, \exp(\frac{1}{2}\alpha t). \tag{1.6.19}$$

Raising both sides to the second power and multiplying the result by π, we find that the ratio of the enclosed areas is

$$\frac{\pi a^2(t)}{\pi a^2(t = 0)} = \exp(\alpha t). \tag{1.6.20}$$

Accordingly, the constant α is the rate of areal expansion.

1.6.4 Superposition of rotation, deformation, and expansion

For future convenience, we relabel the Cartesian coordinates from (x, y) to (x', y'). Superposing the three types of motion discussed in the preceding sections, we obtain a compound velocity field with components

$$[u'_x \quad u_{y'}] = [x' \quad y'] \cdot \left(\begin{bmatrix} 0 & \Omega \\ -\Omega & 0 \end{bmatrix} + \begin{bmatrix} G & 0 \\ 0 & -G \end{bmatrix} + \begin{bmatrix} \frac{1}{2}\alpha & 0 \\ 0 & \frac{1}{2}\alpha \end{bmatrix} \right). \tag{1.6.21}$$

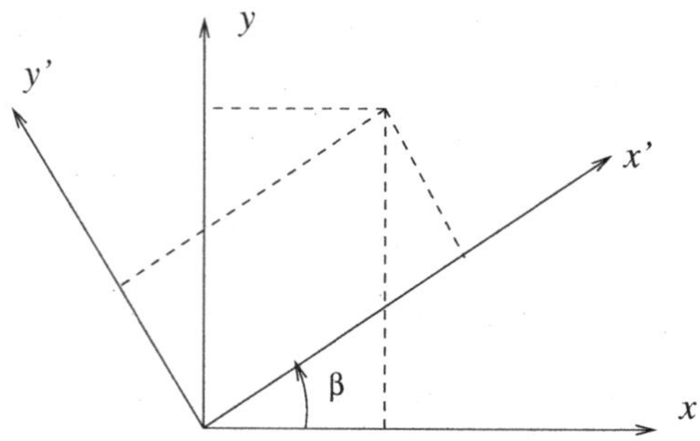

Figure 1.6.2 A system of Cartesian axes, (x', y'), arises by rotating the (x, y) axes by the angle β.

The three matrices on the right-hand side of (1.6.21) express fluid parcel rotation, pure deformation, and isotropic expansion. Summing corresponding elements, we obtain the composite vector form

$$\mathbf{u}' = \mathbf{x}' \cdot \mathbf{A}, \tag{1.6.22}$$

where $\mathbf{u}' = (u_{x'}, u_{y'})$, $\mathbf{x}' = (x', y')$, and the matrix $\mathbf{A}$ is defined as

$$\mathbf{A} = \begin{bmatrix} G + \frac{1}{2}\alpha & \Omega \\ -\Omega & -G + \frac{1}{2}\alpha \end{bmatrix}. \tag{1.6.23}$$

Because the velocity field (1.6.22) depends linearly on the position vector, the associated flow is called linear. Varying the relative magnitudes of the three parameters, Ω, G, and α, allows us to alter the character of the flow by forming hybrid forms of the three fundamental constituents.

1.6.5 Rotated coordinates

Now we observe that, although fluid parcel rotation, deformation, and expansion have been deduced with reference to the $x'y'$ system of Cartesian coordinates, expressing the position and velocity vectors in a different system of coordinates should not affect the physical nature of the motion. Motivated by this observation, we set out to generalize the velocity field described by equation (1.6.22) in a way that further clarifies its physical interpretation.

Consider a two-dimensional Cartesian system $x'y'$ that has been rotated with respect to the xy system by the angle β, as shown in figure 1.6.2. Note that the angle β is positive when the system $x'y'$ arises from the counterclockwise rotation of xy, and negative otherwise. A point in the $x'y'$ or xy plane can be identified by its primed coordinates, (x', y'), or unprimed coordinates, (x, y). Using elementary trigonometry, we find that the two sets of coordinates are related by

$$x = x' \cos\beta - y' \sin\beta, \qquad y = x' \sin\beta + y' \cos\beta. \tag{1.6.24}$$

In vector-matrix notation,

$$[x \quad y] = [x' \quad y'] \cdot \begin{bmatrix} \cos\beta & \sin\beta \\ -\sin\beta & \cos\beta \end{bmatrix}. \tag{1.6.25}$$

Rotation matrix

To simplify the analysis, we introduce the rotation matrix

$$\mathbf{R} \equiv \begin{bmatrix} \cos\beta & \sin\beta \\ -\sin\beta & \cos\beta \end{bmatrix}, \tag{1.6.26}$$

and express (1.6.25) in the form

$$\mathbf{x} = \mathbf{x}' \cdot \mathbf{R}, \tag{1.6.27}$$

where $\mathbf{x}' = (x', y')$ and $\mathbf{x} = (x, y)$. The rotation matrix $\mathbf{R}$ has two important properties. First, its determinant is equal to one. To explain the second property, we introduce the transpose of $\mathbf{R}$, which arises by interchanging the off-diagonal elements R_{12} and R_{21} to form the new matrix

$$\mathbf{R}^T = \begin{bmatrix} \cos\beta & -\sin\beta \\ \sin\beta & \cos\beta \end{bmatrix}. \tag{1.6.28}$$

Using the rules of matrix multiplication, we find

$$\mathbf{R} \cdot \mathbf{R}^T = \mathbf{I}, \qquad \mathbf{R}^T \cdot \mathbf{R} = \mathbf{I}, \tag{1.6.29}$$

where $\mathbf{I}$ is the unit or identity matrix defined as

$$\mathbf{I} = \begin{bmatrix} 1 & 0 \\ 0 & 1 \end{bmatrix}, \tag{1.6.30}$$

and the superscript T denotes the transpose.

The inverse of an arbitrary square matrix, $\mathbf{A}$, is another matrix, $\mathbf{A}^{-1}$, with the properties

$$\mathbf{A} \cdot \mathbf{A}^{-1} = \mathbf{I}, \qquad \mathbf{A}^{-1} \cdot \mathbf{A} = \mathbf{I}. \tag{1.6.31}$$

If the inverse matrix, $\mathbf{A}^{-1}$, is equal to the matrix transpose, $\mathbf{A}^T$, the matrix $\mathbf{A}$ is orthogonal. In light of this definition, equations (1.6.29) ensure that the rotation matrix, $\mathbf{R}$, is orthogonal.

Velocities

Working in a similar fashion, we find that the components of the velocity vector in the xy and $x'y'$ coordinates are related by the counterpart of equation (1.6.27),

$$\mathbf{u} = \mathbf{u}' \cdot \mathbf{R}, \tag{1.6.32}$$

where $\mathbf{u}' = (u_{x'}, u_{y'})$ and $\mathbf{u} = (u_x, u_y)$.

Having made the necessary preparations, we multiply both sides of equation (1.6.22) by the rotation matrix, $\mathbf{R}$, and exploit the first orthogonality property in (1.6.29) to obtain the equivalent form

$$\mathbf{u}' \cdot \mathbf{R} = \mathbf{x}' \cdot \mathbf{A} \cdot \mathbf{R} = \mathbf{x}' \cdot \mathbf{I} \cdot \mathbf{A} \cdot \mathbf{R} = \mathbf{x}' \cdot \mathbf{R} \cdot \mathbf{R}^T \mathbf{A} \cdot \mathbf{R}. \tag{1.6.33}$$

Using equations (1.6.27) and (1.6.32), we obtain

$$\mathbf{u} = \mathbf{x} \cdot \mathbf{B}, \tag{1.6.34}$$

where

$$\mathbf{B} \equiv \mathbf{R}^T \cdot \mathbf{A} \cdot \mathbf{R} \tag{1.6.35}$$

is a new matrix. Substituting (1.6.23), (1.6.26), and (1.6.28) in the right-hand side of (1.6.35), and using the trigonometric identities

$$\cos(2\beta) = \cos^2\beta - \sin^2\beta, \qquad \sin(2\beta) = 2\,\sin\beta\,\cos\beta, \tag{1.6.36}$$

we derive the explicit form

$$\mathbf{B} = \begin{bmatrix} G\,\cos(2\beta) + \frac{1}{2}\,\alpha & -G\,\sin(2\beta) + \Omega \\ -G\,\sin(2\beta) - \Omega & -G\,\cos(2\beta) + \frac{1}{2}\,\alpha \end{bmatrix}. \tag{1.6.37}$$

Note that, when $\beta = 0$ or π, the matrix $\mathbf{B}$ reduces to the matrix $\mathbf{A}$ given in equation (1.6.23). The four elements of the matrix $\mathbf{B}$ are defined in terms of the three flow parameters Ω, G, and α, and rotation angle, β.

1.6.6 Flow decomposition

In practice, we are interested in the inverse problem. Given the four elements of the matrix $\mathbf{B}$, obtained by laboratory measurements or numerical computation, we want to evaluate the four parameters Ω, G, and α, and β, and thereby deduce, respectively, the rate of rotation, the rate of deformation, the rate of expansion, and the direction of deformation.

To be more specific, we consider a linear flow with velocity components

$$[u_x \quad u_y] = [x \quad y] \cdot \begin{bmatrix} a & b \\ c & d \end{bmatrix}, \tag{1.6.38}$$

where a, b, c, and d are four constants with dimensions of inverse time. Setting the components of the matrix on the right-hand side of (1.6.38) equal to the corresponding components of the matrix $\mathbf{B}$ on the right-hand side of (1.6.37), we obtain a nonlinear system of four trigonometric equations for the four unknowns Ω, G, α, and β. The solution can be found most readily according to the following steps.

First, we decompose the matrix shown on the right-hand side of (1.6.38) into three constituents,

$$\mathbf{B} = \begin{bmatrix} a & b \\ c & d \end{bmatrix} = \frac{1}{2}\begin{bmatrix} 0 & b-c \\ c-b & 0 \end{bmatrix} + \frac{1}{2}\begin{bmatrix} a-d & b+c \\ c+b & d-a \end{bmatrix} + \frac{1}{2}\begin{bmatrix} a+d & 0 \\ 0 & a+d \end{bmatrix}. \tag{1.6.39}$$

The first matrix on the right-hand side of (1.6.39) is antisymmetric or skew-symmetric, meaning that the 12 component is equal to the negative of the 21 component. The second matrix is symmetric with zero trace; the trace of an arbitrary matrix is defined as the sum of the diagonal elements. The third matrix is diagonal and isotropic, meaning that the two diagonal elements are identical. The decomposition into these three components is unique.

With reference to the first matrix, we deduce the rate of rotation,

$$\Omega = \frac{b-c}{2}. \tag{1.6.40}$$

With reference to the third matrix, we deduce the rate of expansion,

$$\alpha = a + d. \tag{1.6.41}$$

To compute the remaining two unknowns, G and β, we consider the second matrix on the right-hand side of (1.6.39), defined as

$$\mathbf{E} = \frac{1}{2}\begin{bmatrix} a-d & b+c \\ c+b & d-a \end{bmatrix}. \tag{1.6.42}$$

An eigenvalue of $\mathbf{E}$, denoted by λ, and the corresponding eigenvector

$$\mathbf{f} = \begin{bmatrix} f_x \\ f_y \end{bmatrix}, \tag{1.6.43}$$

satisfy the equation

$$\mathbf{E} \cdot \mathbf{f} = \lambda\, \mathbf{f}, \tag{1.6.44}$$

or

$$(\mathbf{E} - \lambda\, \mathbf{I}) \cdot \mathbf{f} = \mathbf{0}, \tag{1.6.45}$$

where $\mathbf{I}$ is the 2×2 unit matrix. The eigenvalues of $\mathbf{E}$ are found by setting the determinant of the following shifted matrix to zero,

$$\mathbf{E} - \lambda \mathbf{I} = \frac{1}{2} \begin{bmatrix} a-d-2\lambda & b+c \\ c+b & d-a-2\lambda \end{bmatrix}. \tag{1.6.46}$$

Formulating the determinant, we obtain the quadratic equation

$$[\frac{1}{2}(a-d) - \lambda]\,[\frac{1}{2}(d-a) - \lambda] - \frac{1}{4}\,(b+c)^2 = 0, \tag{1.6.47}$$

whose roots are found to be

$$\lambda = \pm\frac{1}{2}\,\sqrt{(a-d)^2 + (b+c)^2}. \tag{1.6.48}$$

The corresponding eigenvectors are found by solving the homogeneous system (1.6.45). Having computed the eigenvalues and eigenvectors of the matrix $\mathbf{E}$, we recover G and β by setting

$$G = \lambda, \tag{1.6.49}$$

with the plus or minus sign selected on the right-hand side of (1.6.48), and then identify β with the angle subtended between the corresponding eigenvector $\mathbf{f}$ and the x axis; that is, we compute the angle β from the equation $\tan\beta = f_y/f_x$.

As an example, we consider a two-dimensional linear velocity field with velocity components

$$u_x(x,y,t) = w(t)\,(2\,x - y), \qquad u_y(x,y,t) = w(t)\,(-3\,x + 3\,y), \tag{1.6.50}$$

where $w(t)$ is an arbitrary function of time. The four time-dependent parameters a, b, c and d introduced in (1.6.38) are

$$a = 2\,w(t), \qquad b = -3\,w(t), \qquad c = -w(t), \qquad d = 3\,w(t). \tag{1.6.51}$$

Thus,

$$\mathbf{B} = w(t) \begin{bmatrix} 2 & -3 \\ -1 & 3 \end{bmatrix}. \tag{1.6.52}$$

Carrying out the decomposition shown in equation (1.6.39), we find

$$\mathbf{B} = w(t)\,\frac{1}{2} \begin{bmatrix} 0 & -2 \\ 2 & 0 \end{bmatrix} + w(t)\,\frac{1}{2} \begin{bmatrix} -1 & -4 \\ -4 & 1 \end{bmatrix} + w(t)\,\frac{1}{2} \begin{bmatrix} 5 & 0 \\ 0 & 5 \end{bmatrix}. \tag{1.6.53}$$

Using equations (1.6.40) and (1.6.41), we find that the rate of rotation is $\Omega = -w(t)$, and the rate of expansion is $\alpha = 5\,w(t)$. The symmetric matrix $\mathbf{E}$ defined in equation (1.6.42) is given by the second term on the right-hand side of (1.6.53),

$$\mathbf{E} = w(t)\,\frac{1}{2} \begin{bmatrix} -1 & -4 \\ -4 & 1 \end{bmatrix}. \tag{1.6.54}$$

The eigenvalues of $\mathbf{E}$ are found by setting the determinant of the following matrix to zero,

$$\mathbf{E} - \lambda \mathbf{I} = \begin{bmatrix} -\frac{1}{2}\, w(t) - \lambda & -2\, w(t) \\ -2\, w(t) & \frac{1}{2}\, w(t) - \lambda \end{bmatrix}. \tag{1.6.55}$$

The roots of the resulting quadratic equation are found to be

$$\lambda = \pm \frac{\sqrt{17}}{2}\, w(t). \tag{1.6.56}$$

Either one of these values can be identified with the rate of extension, G, as indicated by equation (1.6.49).

Substituting expressions (1.6.56) in (1.6.45), we find

$$\begin{bmatrix} -\frac{1}{2}\,(1 \pm \sqrt{17}) & -2 \\ -2 & -\frac{1}{2}(-1 \pm \sqrt{17}) \end{bmatrix} \cdot \begin{bmatrix} f_x \\ f_y \end{bmatrix} = \begin{bmatrix} 0 \\ 0 \end{bmatrix}. \tag{1.6.57}$$

In fact, the two scalar equations comprising system (1.6.57) are identical. Using the first equation, we find

$$\frac{f_y}{f_x} = -\frac{1 \pm \sqrt{17}}{4}. \tag{1.6.58}$$

Executing the instructions given in the paragraph following equation (1.6.48), we finally set $\tan\beta = f_y/f_x$, and find

$$\beta = -\arctan\frac{1 \pm \sqrt{17}}{4}. \tag{1.6.59}$$

In summary, we have managed to extract the rate of rotation, rate of expansion, two rates of deformation, and the corresponding eigenvectors in a linear two-dimensional flow.

Problems

1.6.1 *Material lines*

A collection of point particles distributed along a line in a flow defines a material line. Explain why, if the flow is steady, a material line that lies on a streamline at a certain time will remain on the streamline at all times.

1.6.2 *Rotation of coordinates*

Derive two equations that relate the old coordinates, (x', y'), to the new coordinates, (x, y), and then express them in vector form similar to that shown in equation (1.6.25).

1.6.3 *Flow decomposition*

Carry out the decomposition of a two-dimensional flow with velocity components $u_x(x, y, t) = w(t)\,(2\,x + 3\,y)$ and $u_y(x, y, t) = w(t)\,(-x - 2\,y)$, where $w(t)$ is a given function of time.

1.7 Interpolation

In practice, the components of the fluid velocity are rarely given in explicit form, as in equations (1.4.7) and (1.5.2). Instead, they are either measured in the laboratory with velocity probes, or computed by numerical methods at data points located in the domain of flow. Typically, but not always, the data points are located at the nodes of a grid defined by the intersections of straight or curved lines in two dimensions, or by the intersection of planar or curved surfaces in three dimensions. The velocity at an arbitrary point is then obtained by a numerical procedure called function interpolation.

A Cartesian grid is defined by the intersection of straight lines that are normal to the x and y axes in two dimensions, and by the intersection of planes that are normal to x, y, and z axis in three dimensions. A one-dimensional, a two-dimensional, and a three-dimensional Cartesian grid with evenly spaced grid lines are illustrated in figure 1.7.1.

1.7.1 Interpolation in one dimension

To prepare the ground for computing the components of the velocity at an arbitrary point in a flow from specified grid values, we discuss methods for interpolating a function, $f(x)$, of one independent variable, x.

Let us assume that the values of a function, $f(x)$, are available at $N + 1$ nodes of a one-dimensional grid located at x_i, where $i = 1, \ldots, N + 1$, and $x_1 < x_2 < \cdots < x_{N+1}$, as shown in figure 1.7.1(a). Effectively, we are provided with a three-column table of $N + 1$ entries listing i, x_i, and $f(x_i)$; for simplicity, we denote $f(x_i)$ by f_i. Our goal is to compute the value of the function $f(x)$ at a point x that does not necessarily coincide with a node.

A set of $N + 1$ nodes define N intervals, where the ith interval starts at the ith node and ends at the $i + 1$ node. Suppose that the point x lies inside the kth interval subtended between the nodes x_k and x_{k+1}. A simple way of finding the value of k is by computing the products $p_i = (x - x_i)(x - x_{i+1})$ for all values of i. The appropriate value of k is the unique value of i for which p_i is negative.

There are better and faster methods of finding the label of the host interval k. For example, in the method of logarithmic search, we first examine whether the point x lies on the left or on the right of the mid-point of the interpolation

(a)

(b)

(c)

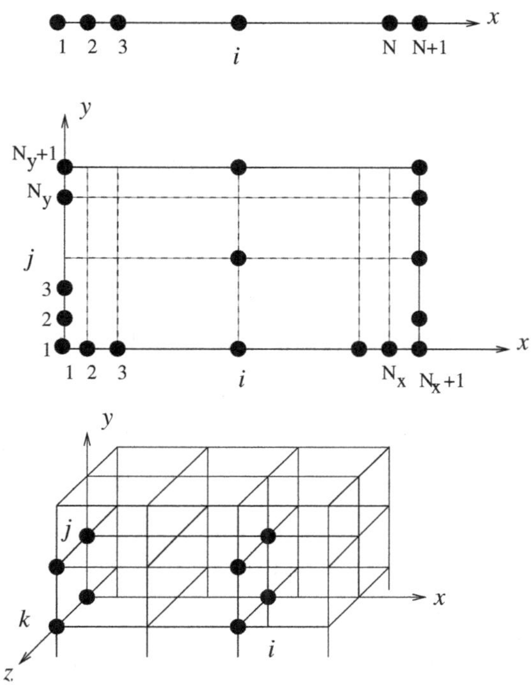

Figure 1.7.1 Illustration of (a) a one-dimensional, (b) a two-dimensional, and (c) a three-dimensional Cartesian grid with evenly spaced grid lines.

domain (x_1, x_{N+1}). Having found the host half-interval, we repeat the process until the host sub-interval has been reduced to the kth interval.

Linear interpolation

To compute the value $f(x)$, we may approximate the graph of the function $f(x)$ in the interval (x_k, x_{k+1}) with a straight line, and require that the straight line interpolates through the two data points corresponding to the doublets (x_k, f_k) and (x_{k+1}, f_{k+1}), as illustrated in figure 1.7.2. In mathematical terms, we approximate the function $f(x)$ over the interval (x_k, x_{k+1}) with a linear function expressed by a first-degree polynomial,

$$P_1^{(k)}(x) = a^{(k)}(x - x_k) + b^{(k)}, \tag{1.7.1}$$

where the coefficient $a^{(k)}$ is the slope and the constant $b^{(k)}$ is the intercept. To facilitate forthcoming algebraic manipulations, the polynomial has been expressed in terms of the shifted monomial $x - x_k$ rather than the unshifted monomial x.

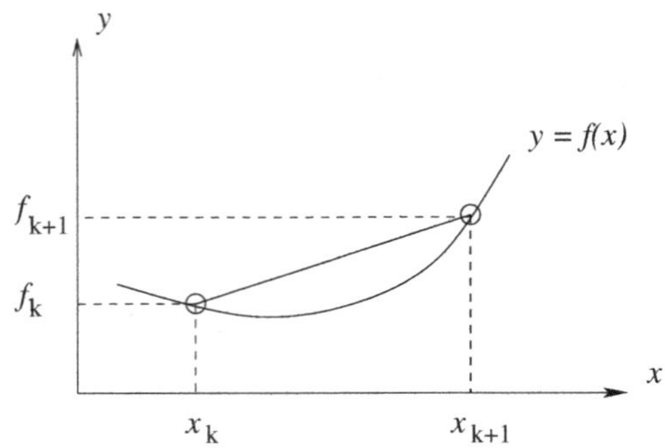

Figure 1.7.2 Local approximation of a function with a linear interpolating function represented by the straight line.

To compute the constants $a^{(k)}$ and $b^{(k)}$, we require the interpolation conditions

$$P_1^{(k)}(x_k) = b^{(k)} = f_k,$$
$$P_1^{(k)}(x_{k+1}) = a^{(k)}(x_{k+1} - x_k) + b^{(k)} = f_{k+1}, \qquad (1.7.2)$$

ensuring that the graph of the polynomial passes through the data labeled k and $k+1$. Solving the system of the two linear equations (1.7.2) for the two monomial coefficients, we find

$$a^{(k)} = \frac{f_{k+1} - f_k}{x_{k+1} - x_k}, \qquad b^{(k)} = f_k. \qquad (1.7.3)$$

To compute the linear polynomial $P_1^{(k)}(x)$, we first calculate the coefficients $a^{(k)}$ and $b^{(k)}$ using equations (1.7.3), and then evaluate the right-hand side of (1.7.1) for a desired value of x that lies between x_k and x_{k+1}. The result will be a reasonable approximation to the unknown value $f(x)$.

Quadratic interpolation

Interpolation based on the straight-line approximation overlooks the curvature of the graph of the function $f(x)$. For better accuracy, we approximate the function $f(x)$ with a parabola defined in the interval (x_k, x_{k+1}), as depicted in figure 1.7.3. In mathematical terms, we approximate the function $f(x)$ with a quadratic function expressed by the second-degree polynomial

$$P_2^{(k)}(x) = a^{(k)}\,(x - x_k)^2 + b^{(k)}(x - x_k) + c^{(k)}. \qquad (1.7.4)$$

To simplify the forthcoming algebraic manipulations, we have again expressed the polynomial in terms of the shifted monomial $x - x_k$ rather than the unshifted monomial x.

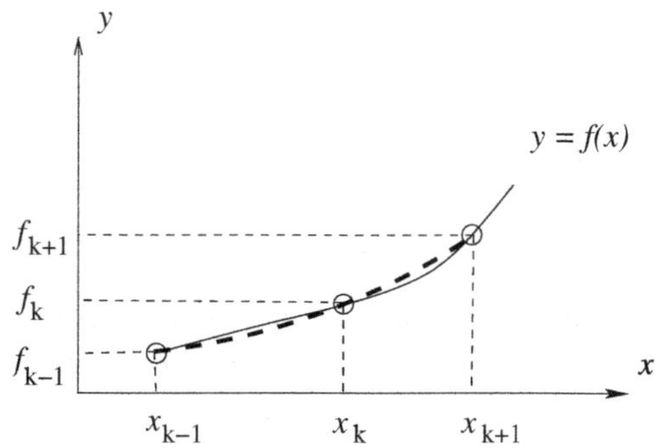

Figure 1.7.3 Local approximation of a function with a parabolic interpolating function represented by the thick dashed line.

To compute the three constants, $a^{(k)}$, $b^{(k)}$, and $c^{(k)}$, we require three equations. First, we demand that the parabola interpolates through the two data points (x_k, f_k) and (x_{k+1}, f_{k+1}), and obtain the interpolation conditions

$$\begin{aligned} P_2^{(k)}(x_k) &= c^{(k)} = f_k, \\ P_2^{(k)}(x_{k+1}) &= a^{(k)}(x_{k+1} - x_k)^2 + b^{(k)}(x_{k+1} - x_k) + c^{(k)} = f_{k+1}. \end{aligned} \tag{1.7.5}$$

One more datum point is required, and we may choose either the backward point, (x_{k-1}, f_{k-1}), or the forward point, (x_{k+2}, f_{k+2}). The backward choice provides us with the condition

$$P_2^{(k)}(x_{k-1}) = a^{(k)}(x_{k-1} - x_k)^2 + b^{(k)}(x_{k-1} - x_k) + c^{(k)} = f_{k-1}. \tag{1.7.6}$$

With the choice expressed by equations (1.7.5) and (1.7.6), the coefficients of the binomial are found to be

$$a^{(k)} = \frac{\frac{f_{k+1}-f_k}{h_k} - \frac{f_k - f_{k-1}}{h_{k-1}}}{h_k + h_{k-1}}, \qquad b^{(k)} = \frac{h_{k-1}\frac{f_{k+1}-f_k}{h_k} + h_k\frac{f_k - f_{k-1}}{h_{k-1}}}{h_k + h_{k-1}},$$
$$c^{(k)} = f_k, \tag{1.7.7}$$

where $h_{k-1} = x_k - x_{k-1}$ and $h_k = x_{k+1} - x_k$ are the lengths of the backward and forward intervals.

When the data points are spaced evenly, $h_{k-1} = h_k = h$, we obtain the simplified expressions

$$a^{(k)} = \frac{f_{k+1} - 2f_k + f_{k-1}}{2h^2}, \qquad b^{(k)} = \frac{f_{k+1} - f_{k-1}}{2h}, \qquad c^{(k)} = f_k. \tag{1.7.8}$$

To compute $P_2^{(k)}(x)$, we first calculate the coefficients $a^{(k)}$, $b^{(k)}$ and $c^{(k)}$ using the preceding equations, and then evaluate the right-hand side of (1.7.4). The result will be an approximation to $f(x)$ that is improved over that computed by linear interpolation.

1.7.2 Interpolation in two dimensions

Next, we consider a function of two independent variables x and y. For the present purposes, a function of two variables is an engine that receives a pair of numbers, x and y, and produces a new number, $f(x, y)$.

Assume that we are given the values of a function, $f(x, y)$, at the nodes of a two-dimensional grid defined by the intersections of x-level lines x_i, where $i = 1, \ldots, N_x + 1$, and y-level lines y_j, where $j = 1, \ldots, N_y + 1$, as shown in figure 1.7.1(*b*). A grid node is identified by the values of two indices, i and j, forming an ordered integer doublet, (i, j). The value of the function $f(x, y)$ at the (i, j) node is equal to $f(x_i, y_j)$. Our goal is to compute the value of f at a point (x, y) that is not necessarily a node.

The following MATLAB function *grid_2d* residing in directory *rec_2d* inside directory *02_grids* of FDLIB generates evenly spaced grid lines and grid points in a rectangular domain in the xy plane confined between $a_x \le x \le b_x$ and $a_y \le y \le b_y$, with N_x intervals in the x direction and N_y intervals in the y direction:

```
function [glx,gly,gx,gy] = grid_2d(ax,bx,ay,by,Nx,Ny)

%-----
% grid spacing
%-----

Dx = (bx-ax)/Nx; Dy = (by-ay)/Ny;

%---
% generate the grid lines
%---

for i=1:Nx+1
  glx(i) = ax+(i-1.0)*Dx;
end

for j=1:Ny+1
  gly(j) = ay+(j-1.0)*Dy;
end

%---
% generate the grid points
%---
```

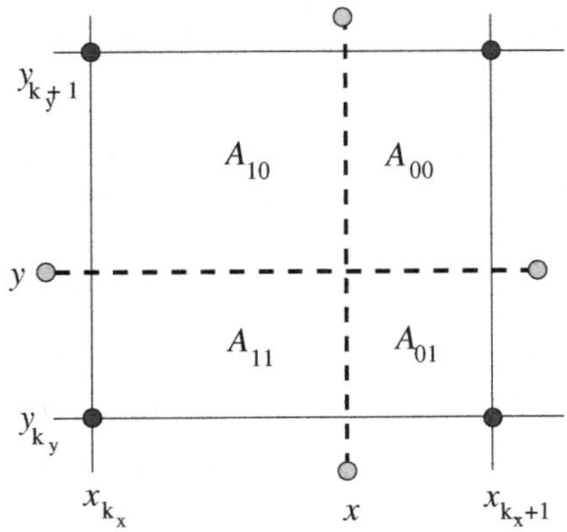

Figure 1.7.4 Bilinear interpolation of a scalar function through a rectangular grid.

```
for i=1:Nx+1
  for j=1:Ny+1
    gx(i,j) = glx(i);
    gy(i,j) = gly(j);
  end
end

%-----
% done
%-----

return
```

Suppose that a value of x lies inside the k_xth x-interval confined between the x_{k_x} and x_{k_x+1} x-level lines, and a value of y lies inside the k_yth y-interval confined between the y_{k_y} and y_{k_y+1} y-level lines, as shown in figure 1.7.4. The values of k_x and k_y can be found by the methods discussed in section 1.7.1 for one-dimensional interpolation.

Bilinear interpolation

An approximation to $f(x, y)$ can be obtained by replacing the unknown function, $f(x, y)$, with a function, $P_{BL}^{k_x,k_y}(x, y)$, defined over the rectangular domain that is confined between the x-level lines $x = x_{k_x}$, $x = x_{k_x+1}$, and y-level lines $y = y_{k_y}$, and $y = y_{k_y+1}$. The bilinear function $P_{BL}^{k_x,k_y}(x, y)$ is distinguished by the following properties:

- For a fixed value of x, call it x_0, the function $P_{BL}^{k_x,k_y}(x_0, y)$ varies linearly with respect to y.

• For a fixed value of y, call it y_0, the function $P_{BL}^{k_x,k_y}(x, y_0)$ varies linearly with respect to x.

• The following four interpolation conditions ensuring that the bilinear function reproduces the neighboring grid values are met:

$$\begin{aligned} P_{BL}^{k_x,k_y}(x_{k_x}, y_{k_y}) &= f(x_{k_x}, y_{k_y}), \\ P_{BL}^{k_x,k_y}(x_{k_x}, y_{k_y+1}) &= f(x_{k_x}, y_{k_y+1}), \\ P_{BL}^{k_x,k_y}(x_{k_x+1}, y_{k_y}) &= f(x_{k_x+1}, y_{k_y}), \\ P_{BL}^{k_x,k_y}(x_{k_x+1}, y_{k_y+1}) &= f(x_{k_x+1}, y_{k_y+1}). \end{aligned} \tag{1.7.9}$$

The first and second properties require that the bilinear function has the functional form

$$P_{BL}^{k_x,k_y}(x, y) = (a_x^{k_x,k_y}\, x + b_x^{k_x,k_y})\,(a_y^{k_x,k_y}\, y + b_y^{k_x,k_y}). \tag{1.7.10}$$

To evaluate the four constants, $a_x^{k_x,k_y}$, $b_x^{k_x,k_y}$, $a_y^{k_x,k_y}$, and $b_y^{k_x,k_y}$, we use the four interpolation conditions (1.7.9), finding

$$\begin{aligned} P_{BL}^{k_x,k_y}(x, y) = {} & w_{00}^{k_x,k_y}(x, y)\, f(x_{k_x}, y_{k_y}) + w_{10}^{k_x,k_y}(x, y)\, f(x_{k_x+1}, y_{k_y}) \\ & + w_{01}^{k_x,k_y}(x, y)\, f(x_{k_x}, y_{k_y+1}) + w_{11}^{k_x,k_y}(x, y)\, f(x_{k_x+1}, y_{k_y+1}). \end{aligned} \tag{1.7.11}$$

We have introduced the position-dependent interpolation weights

$$\begin{aligned} w_{00}^{k_x,k_y}(x, y) &= \frac{A_{00}}{A}, \qquad & w_{01}^{k_x,k_y}(x, y) &= \frac{A_{10}}{A}, \\ w_{10}^{k_x,k_y}(x, y) &= \frac{A_{01}}{A}, \qquad & w_{11}^{k_x,k_y}(x, y) &= \frac{A_{11}}{A}. \end{aligned} \tag{1.7.12}$$

The numerators, A_{00}, A_{10}, A_{01}, and A_{11}, are the areas of the four sub-rectangles depicted in figure 1.7.4, given by

$$\begin{aligned} A_{00} &= (x_{k_x+1} - x)(y_{k_y+1} - y), \qquad & A_{10} &= (x - x_{k_x})(y_{k_y+1} - y), \\ A_{01} &= (x_{k_x+1} - x)(y - y_{k_y}), \qquad & A_{11} &= (x - x_{k_x})(y - y_{k_y}), \end{aligned} \tag{1.7.13}$$

and the common denominator, A, is the area of the interpolation rectangle, given by

$$A = (x_{k_x+1} - x_{k_x})(y_{k_y+1} - y_{k_y}). \tag{1.7.14}$$

It is reassuring to observe that the sum of the four interpolation weights given in (1.7.12) is equal to unity independent of the values of x and y,

$$w_{00}^{k_x,k_y}(x, y) + w_{10}^{k_x,k_y}(x, y) + w_{01}^{k_x,k_y}(x, y) + w_{11}^{k_x,k_y}(x, y) = 1. \tag{1.7.15}$$

This property guarantees that, if the four participating grid values are equal,

$$f(x_{k_x}, y_{k_y}) = f(x_{k_x}, y_{k_y+1}) = f(x_{k_x+1}, y_{k_y}) = f(x_{k_x+1}, y_{k_y+1}) = a, \qquad (1.7.16)$$

then bilinear interpolation based on (1.7.11) produces

$$P_{BL}^{k_x,k_y}(x,y) = a\,(w_{00}^{k_x,k_y} + w_{10}^{k_x,k_y} + w_{01}^{k_x,k_y} + w_{11}^{k_x,k_y}) = a, \qquad (1.7.17)$$

as required.

1.7.3 Interpolation of the velocity in a two-dimensional flow

Returning to fluid mechanics, we consider a two-dimensional flow in the xy plane and specify the values of the x and y velocity components, u_x and u_y, at the nodes of a two-dimensional Cartesian grid. To obtain the corresponding values at an arbitrary point (x, y), we employ bilinear interpolation, finding

$$\begin{aligned} u_x(x,y) &= w_{00}^{k_x,k_y}(x,y)\, u_x(x_{k_x}, y_{k_y}) + w_{10}^{k_x,k_y}(x,y)\, u_x(x_{k_x+1}, y_{k_y}) \\ &+ w_{01}^{k_x,k_y}(x,y)\, u_x(x_{k_x}, y_{k_y+1}) + w_{11}^{k_x,k_y}(x,y)\, u_x(x_{k_x+1}, y_{k_y+1}), \end{aligned}$$

$$\begin{aligned} u_y(x,y) &= w_{00}^{k_x,k_y}(x,y)\, u_y(x_{k_x}, y_{k_y}) + w_{10}^{k_x,k_y}(x,y)\, u_y(x_{k_x+1}, y_{k_y}) \\ &+ w_{01}^{k_x,k_y}(x,y)\, u_y(x_{k_x}, y_{k_y+1}) + w_{11}^{k_x,k_y}(x,y)\, u_y(x_{k_x+1}, y_{k_y+1}). \end{aligned} \qquad (1.7.18)$$

The following MATLAB function *rec_2d_int* residing in directory *rec_2d* inside directory *02_grids* of FDLIB performs the interpolation in a rectangular domain confined by $a_x \le x \le b_x$ and $a_y \le y \le b_y$:

```
function [ux,uy] = rec_2d_int ...
...
(ax,bx ...          % x end points
,ay,by ...          % y end points
,Nx,Ny ...          % grid size
,glx,gly ...          % grid lines
,gux,guy ...          % grid velocity
,x,y ...         % interpolation point
)

%-----------------------------
% Bilinear interpolation of the velocity
% at the point (x, y)
%-----------------------------

%-----------------------------
% locate the x and y intervals
%-----------------------------
```

```
for kx=1:Nx
  prod = (x-glx(kx))*(x-glx(kx+1));
  if(prod<0) break; end
end

for ky=1:Ny
  prod = (y-gly(ky))*(y-gly(ky+1));
  if(prod<0) break; end
end

%----------------------
% interpolation weights
%----------------------

A00 = (glx(kx+1)-x)*(gly(ky+1)-y);
A01 = (glx(kx+1)-x)*(y-gly(ky));
A10 = (x-glx(kx))*(gly(ky+1)-y);
A11 = (x-glx(kx))*(y-gly(ky));
A = (glx(kx+1)-glx(kx))*(gly(ky+1)-gly(ky));

w00 = A00/A; w01 = A01/A; w10 = A10/A; w11 = A11/A;

%--------------
% interpolation
%--------------

ux = w00 * gux(kx ,ky) + w01 * gux(kx ,ky+1) ...
   + w10 * gux(kx+1,ky) + w11 * gux(kx+1,ky+1);
uy = w00 * guy(kx ,ky) + w01 * guy(kx ,ky+1) ...
   + w10 * guy(kx+1,ky) + w11 * guy(kx+1,ky+1);

%---
% done
%---

return
```

The following MATLAB code *rec 2d* residing in directory *02 grids* of FDLIB generates a grid and calls this function to produce a velocity vector field:

```
%==================
% rec_2d:  velocity vector field by interpolation
%==================

ax = -1.3; bx = 1.3; ay = -1.0; by = 1.0;

Nx = 16; Ny = 8;    % grid size
```

```
%-----
% generate the grid
%-----

[ glx,gly,gx,gy ] = grid_2d (ax,bx,ay,by,Nx,Ny);

%---
% specify the grid velocities
%---

for i=1:Nx+1
  for j=1:Ny+1
    gux(i,j) = gx(i,j);
    guy(i,j) =-gy(i,j);
  end
end

%---
% velocity vector field
% in the middle of each cell
%---

for i=1:Nx
  x = 0.5*(glx(i)+glx(i+1));
  for j=1:Ny
    y = 0.5*(gly(j)+gly(j+1));
    plot(x,y,'g.')

    [ux,uy] = rec_2d_int ...
    ...
      (ax,bx ...  % x end points
      ,ay,by ...  % y end points
      ,Nx,Ny ...  % grid size
      ,glx,gly ...  % grid lines
      ,gux,guy ...  % grid velocity
      ,x,y ...  % interpolation point
      );

    vector = draw_arrow_2d(x,y,ux/Nx,uy/Nx);
    plot(vector(:,1),vector(:,2));
  end
end

%----
% finish plotting
%----

plot([ax, bx, bx, ax, ax],[ay, ay, by, by, ay])
axis([ax-0.1, bx+0.1, ay-0.1, by+0.1])
```

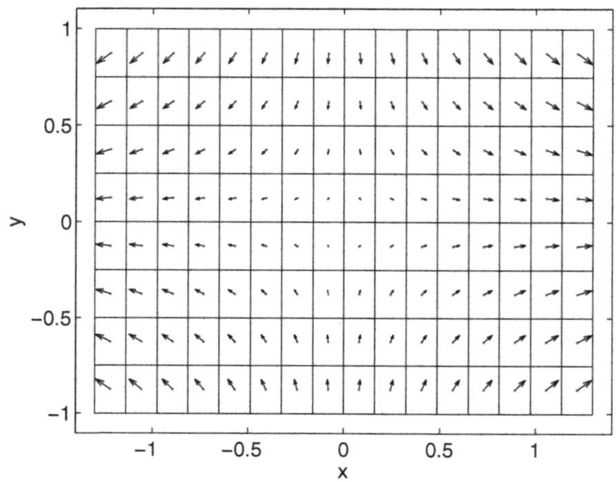

Figure 1.7.5 Bilinear interpolation of a velocity vector field through a Cartesian grid in a two-dimensional flow.

```
axis equal
xlabel('x','fontsize',15)
ylabel('y','fontsize',15)

%----
% done
%----
```

The graphics produced by the code for the specified velocity field is shown in figure 1.7.5.

The following MATLAB function *draw_arrow_2d* is invoked to generate nice looking arrows:

```
function vector = draw_arrow_2d(x1,y1,dx,dy)

%-------------------------------------
% Generate coordinates for plotting
% a five-point arrow starting at the point
% (x1,x2) and ending at the point (x1+dx, y1+dy)
%
% SYMBOLS:
% --------
%
% dx, dy:  arrow vector
%
```

```
% vector(i,1):  x-coordinate of the ith point
% vector(i,2):  y-coordinate of the ith point
% where i=1,2,3,4,5
%
% angle:  angle of arrow tip in radians
% tip:  length of arrow tip sides
% as a fraction of the arrow length
%----------------------------------------

angle = 0.3; tip = 0.50;

%--------
% prepare
%--------

x2 = x1+dx; y2 = y1+dy;
cs = cos(angle); sn = sin(angle);
dxi = -dx; dyi = -dy;

%------------
% five points
%------------

vector(1,1) = x1; vector(1,2) = y1;
vector(2,1) = x2; vector(2,2) = y2;
vector(3,1) = x2+( dxi*cs+dyi*sn)*tip;
vector(3,2) = y2+(-dxi*sn+dyi*cs)*tip;
vector(4,1) = x2; vector(4,2) = y2;
vector(5,1) = x2+(dxi*cs-dyi*sn)*tip;
vector(5,2) = y2+(dxi*sn+dyi*cs)*tip;

%-----
% done
%-----

return
```

MATLAB offers the graphics function *quiver*that also generates velocity vector fields.

1.7.4 Streamlines by interpolation

Our ability to interpolate the velocity components at any point in a flow from specified grid values allows us to generate particle paths and streamlines without having available explicit expressions for the velocity in analytical form. In computational fluid dynamics (CFD), grid values are computed by solving the equations governing the fluid motion using a variety of numerical methods. Examples are discussed in chapters 3 and 8.

The following MATLAB code *rec_2d_strml* residing in directory *rec_2d* inside directory *02_grids* of FDLIB generates and draws streamlines:

```
%=========
% streamlines in a rectangular cavity
%
% confined in ax<x<bx and ay<y<by
%=========

ax = -1.2; bx = 1.2;
ay = -1.0; by = 1.0;

Nx = 16; Ny = 20;    % grid size

Nmax = 400; % maximum steps along a streamline
Ds = 0.020; % travel distance in each step

%-----
% generate the grid
%-----

[ glx,gly,gx,gy ] = grid_2d(ax,bx,ay,by,Nx,Ny);

%---
% specify the grid velocities
%---

for i=1:Nx+1
  for j=1:Ny+1
    gux(i,j) =-gy(i,j);
    guy(i,j) = gx(i,j);
  end
end

%---------------------------
% initial streamline points
%---------------------------

x0 = [0.05 0.1 0.2 0.3 0.4 0.5 0.6 0.7 0.8 0.9 1.0 1.2];
Nstr=size(x0');

for i=1:Nstr
  y0(i) = 0.001;
end

%----------------------
% loop over streamlines
%---------------------
```

```
for l=1:Nstr

  xn = x0(l);
  yn = y0(l);
  xstr(1) = xn;   % new point
  ystr(1) = yn;   % new point

%----
% integrate by the modified Euler method
%----

  for i=2:Nmax   % step in time

  [ux,uy] = rec_2d_int ...
  ...
  (ax,bx ...
  ,ay,by ...
  ,Nx,Ny ...
  ,glx,gly ...
  ,gux,guy ...
  ,xn,yn ...
  );

  Umag = sqrt(ux*ux+uy*uy);
  Dt = Ds/Umag;
  x1 = xn + Dt*ux;
  y1 = yn + Dt*uy;

  [ux1,uy1] = rec_2d_int ...
  ...
  (ax,bx ...
  ,ay,by ...
  ,Nx,Ny ...
  ,glx,gly ...
  ,gux,guy ...
  ,x1,y1 ...
  );

  Dth = 0.5*Dt;
  xn = xn + Dth*(ux+ux1);
  yn = yn + Dth*(uy+uy1);
  xstr(i) = xn;
  ystr(i) = yn;

%----
% check for a closed streamline after 5 steps
%---
```

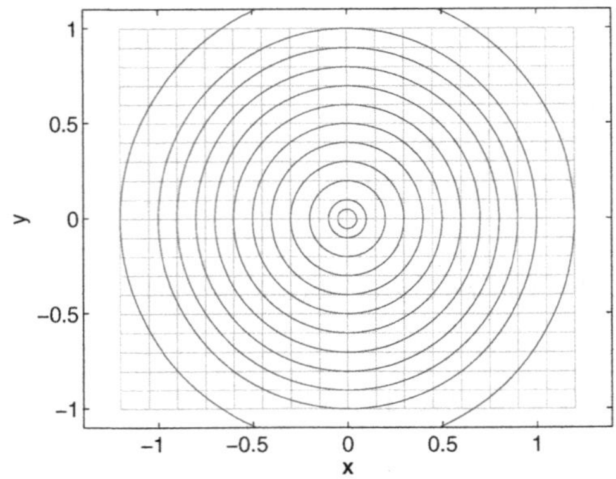

Figure 1.7.6 Streamlines of a rotary flow produced by interpolation from an underlying Cartesian grid.

```
  if(i>5)
    Dist = sqrt((xn-xstr(1))^2+(yn-ystr(1))^2);
    if(Dist<Ds)
      xstr(i+1) = xstr(1);
      ystr(i+1) = ystr(1);
      break;
    end
  end

  end   % over time

  plot(xstr,ystr,'r-')
  clear xstr ystr

end   % of run over streamlines

%---
% finish
%---

axis([ax-0.1, bx+0.1, ay-0.1, by+0.1])
axis equal
xlabel('x','fontsize',15)
ylabel('y','fontsize',15)
```

The graphics produced by the code for the specified velocity field is shown in figure 1.7.6.

Problems

1.7.1 *Quadratic interpolation*

Solve the linear system of three equations (1.7.5) and (1.7.6) to derive formulas (1.7.7). *Hint*: Compute first the coefficient $c^{(k)}$ using the first of equations (1.7.5).

1.7.2 *Forward-point parabolic interpolation*

Consider the parabolic interpolation of a function of one variable, $f(x)$, as discussed in the text. Forward interpolation uses the interpolation condition

$$P_2^{(k)}(x_{k+2}) = a^{(k)}(x_{k+2} - x_k)^2 + b^{(k)}(x_{k+2} - x_k) + c^{(k)} = f_{k+2}, \quad (1.7.19)$$

in place of (1.7.6). Derive expressions for the coefficients $a^{(k)}$, $b^{(k)}$, and $c^{(k)}$ in terms of the grid values f_k, f_{k+1}, and f_{k+2}, and the interval sizes h_k and h_{k+1}. Then derive simplified expressions when h_k and h_{k+1} are both equal to h.

1.7.3 *Trilinear interpolation*

Consider a function of three variables, $f(x, y, z)$. Extend the method of bilinear interpolation of a function of two variables discussed in the text, to the method of trilinear interpolation that generates the value of f at an arbitrary point (x, y, z) using the values of f at the nodes of a three-dimensional Cartesian grid. The interpolation formula should be the counterpart of (1.7.11) with properly defined interpolation weights.

Computer Problems

1.7.4 *Bilinear interpolation*

Run the code *rec_2d* for a velocity field of your choice. Confirm that the interpolated values are identical to the specified grid values, and prepare a plot of the velocity vector field similar to that displayed in figure 1.7.5. Discuss the structure of the flow.

1.7.5 *Streamlines by interpolation*

Run the code *rec_2d_strml* for a velocity field of your choice. Generate, plot, and discuss the streamline pattern.

More on Kinematics

2

We continue the study of kinematics by considering in more detail the behavior of fluid parcels, by deriving expressions for the volumetric and mass flow rate across surfaces drawn in a fluid, and by developing numerical methods for evaluating kinematic variables of interest in terms of derivatives and integrals of the velocity. Mass conservation and physical conditions imposed at boundaries introduce mathematical constraints that motivate the description of a flow in terms of ancillary functions that expedite the mathematical analysis and considerably simplify the numerical computation.

2.1 Fundamental modes of fluid parcel motion

In chapter 1, we pointed out that the nature of the motion of a small fluid parcel is determined by the relative motion of point particles residing inside the parcel and over the parcel boundary. If variations in the point particle velocity are negligible compared to the average point particle velocity, the parcel exhibits rigid-body translation. Significant variations in the point particle velocity generate further general types of motion including local rotation, deformation, and isotropic expansion.

To study the relative motion of point particles in the vicinity of a certain point, $\mathbf{x}_0 = (x_0, y_0, z_0)$, we consider differences in the corresponding velocity components evaluated at a point $\mathbf{x} = (x, y, z)$ that lies close to $\mathbf{x}_0$, and at the chosen point, $\mathbf{x}_0$. If the differences are small compared to the distance between

C. Pozrikidis, *Fluid Dynamics: Theory, Computation, and Numerical Simulation*,
DOI: 10.1007/978-0-387-95871-2_2,

the points $\mathbf{x}$ and $\mathbf{x}_0$, both measured in proper units, the relative motion is negligible. If the differences are substantial, the relative motion is significant and needs to be properly analyzed.

2.1.1 Function linearization

To prepare the ground for our analysis, we consider a scalar function of three independent variables that receives a triplet of numbers, (x, y, z), and produces a number, $f(x, y, z)$. If the function f is locally well behaved, and if the point $\mathbf{x}$ lies near the point $\mathbf{x}_0$, then we expect that the value $f(x, y, z)$ will be close to the value $f(x_0, y_0, z_0)$. Stated differently, in the limit as $\mathbf{x}$ tends to $\mathbf{x}_0$, meaning that all three scalar differences $x - x_0$, $y - y_0$, and $z - z_0$ tend to zero, the difference in the function values $f(x, y, z) - f(x_0, y_0, z_0)$ will vanish.

The variable point, $\mathbf{x}$, may approach the fixed point, $\mathbf{x}_0$, from many different directions. Selecting the direction that is parallel to the x axis, we set $\mathbf{x} = (x, y_0, z_0)$, and consider the limit of the difference $f(x, y_0, z_0) - f(x_0, y_0, z_0)$ as $x - x_0$ tends to zero. Because the function f has been assumed well behaved, the ratio of the differences,

$$\frac{f(x, y_0, z_0) - f(x_0, y_0, z_0)}{x - x_0}, \qquad (2.1.1)$$

tends to a finite number defined as the first partial derivative of the function f with respect to the variable x evaluated at the point $\mathbf{x}_0$, denoted by $(\partial f/\partial x)(\mathbf{x}_0)$. Elementary calculus ensures that the partial derivative can be computed using the usual rules of differentiation of a function of one variable with respect to x, regarding all other independent variables as constant. For example, if $f = xyz$, then $\partial f/\partial x = yz$, and $(\partial f/\partial x)(\mathbf{x}_0) = y_0 z_0$,

Setting the fraction shown in (2.1.1) equal to $(\partial f/\partial x)(\mathbf{x}_0)$ and solving the resulting equation for $f(x, y_0, z_0)$, we find

$$f(x, y_0, z_0) \simeq f(x_0, y_0, z_0) + (x - x_0) \left(\frac{\partial f}{\partial x}\right)_{\mathbf{x}_0}. \qquad (2.1.2)$$

It is important to bear in mind that this equation is exact only in the limit as $\Delta x \equiv x - x_0$ tends to zero. For small but non-infinitesimal values of Δx, the difference between the left- and right-hand sides is on the order of Δx^2, which is small compared to Δx. For example, if Δx is equal to 0.01 is some units, then Δx^2 is equal to 0.0001 in corresponding units.

The point $\mathbf{x}$ may also approach the point $\mathbf{x}_0$ along the y or z axis, yielding the counterparts of equation (2.1.2),

$$f(x_0, y, z_0) \simeq f(x_0, y_0, z_0) + (y - y_0) \left(\frac{\partial f}{\partial y}\right)_{\mathbf{x}_0}, \qquad (2.1.3)$$

and

$$f(x_0, y_0, z) \simeq f(x_0, y_0, z_0) + (z - z_0) \left(\frac{\partial f}{\partial z}\right)_{\mathbf{x}_0}. \tag{2.1.4}$$

Combining the arguments that led us to equations (2.1.2)–(2.1.4), we let the point $\mathbf{x}$ approach the point $\mathbf{x}_0$ from an arbitrary direction and derive the approximation

$$\begin{aligned} f(x, y, z) &\simeq f(x_0, y_0, z_0) \\ &+(x - x_0) \left(\frac{\partial f}{\partial x}\right)_{\mathbf{x}_0} + (y - y_0) \left(\frac{\partial f}{\partial y}\right)_{\mathbf{x}_0} + (z - z_0) \left(\frac{\partial f}{\partial z}\right)_{\mathbf{x}_0}. \end{aligned} \tag{2.1.5}$$

We pause to emphasize that relation (2.1.5) is exact only in the limit as all three spatial differences, $\Delta x = x - x_0$, $\Delta y = y - y_0$, and $\Delta z = z - z_0$ tend to zero. For small but non-infinitesimal values of any of these differences, the left-hand side of (2.1.5) differs from the right-hand side by an amount that is generally on the order of the maximum of Δx^2, Δy^2, or Δz^2.

Taylor series

Equation (2.1.5) can be made exact for any value of Δx, Δy, or Δz, by adding to the right-hand side a term called the remainder. As all three differences Δx, Δy, and Δz tend to zero, the remainder vanishes faster than these differences. Elementary calculus shows that, if Δx, Δy, and Δz are sufficiently small, the remainder can be expressed as an infinite series involving products of powers of Δx, Δy, and Δz, called the Taylor series of the function f about the point $\mathbf{x}_0$.

The process of deriving (2.1.5) is called linearization of the function f about the point $\mathbf{x}_0$. The linearized form (2.1.5) states that, in the immediate vicinity of a point $\mathbf{x}_0$, any regular function resembles a linear function of the shifted monomials Δx, Δy, and Δz. If all three first partial derivatives happen to vanish at the point $\mathbf{x}_0$, then f behaves like a quadratic function; however, this is a rare exception.

Gradient of a scalar function

To economize our notation, we introduce the gradient of a function, f, denoted by ∇f, defined as the vector of the three partial derivatives,

$$\nabla f \equiv \mathbf{e}_x \frac{\partial f}{\partial x} + \mathbf{e}_y \frac{\partial f}{\partial y} + \mathbf{e}_z \frac{\partial f}{\partial z}, \tag{2.1.6}$$

where $\mathbf{e}_x$, $\mathbf{e}_y$, and $\mathbf{e}_z$ are the unit vectors along the x, y, and z axes. The symbol ∇ is a vector operator called the del or gradient operator, defined as

$$\nabla = \mathbf{e}_x \frac{\partial}{\partial x} + \mathbf{e}_y \frac{\partial}{\partial y} + \mathbf{e}_z \frac{\partial}{\partial z}. \tag{2.1.7}$$

Unlike a regular vector, ∇ may not stand alone, but must operate on a scalar function of position from the left to acquire a meaningful interpretation.

Inner vector product

As a second preliminary, we define the inner product of a pair of three-dimensional vectors, $\mathbf{f} = (f_x, f_y, f_z)$ and $\mathbf{g} = (g_x, g_y, g_z)$, as the scalar

$$\mathbf{f} \cdot \mathbf{g} = f_x\, g_x + f_y\, g_y + f_z\, g_z. \tag{2.1.8}$$

In index notation,

$$\mathbf{f} \cdot \mathbf{g} \equiv f_i\, g_i, \tag{2.1.9}$$

where summation of the repeated index i is implied over x, y, and z, according to Einstein's repeated-index summation convention: if an index appears twice in a product, then summation of that index is implied over its range.

Interpretation of the inner vector product

It can be shown using the rule of cosines that the inner product defined in (2.1.8) is equal to the product of (a) the length of the first vector, $\mathbf{f}$, (b) the length of the second vector, $\mathbf{g}$, and (c) the cosine of the angle subtended between the two vectors. If the angle is equal to $\pi/2$, which means that the two vectors are orthogonal, the cosine of the angle is zero and the inner product vanishes. If the angle is zero, which means the two vectors are parallel, the inner product is equal to the product of the vector lengths. If both $\mathbf{f}$ and $\mathbf{g}$ are unit vectors, meaning that their lengths are equal to one unit of length, then the inner product is equal to the cosine of the angle subtended between them.

Linearized expansion in compact form

Using the preceding definitions, we state equation (2.1.5) in a compact vector form

$$f(\mathbf{x}) \simeq f(\mathbf{x}_0) + (\mathbf{x} - \mathbf{x}_0) \cdot (\nabla f)_{\mathbf{x}_0}, \tag{2.1.10}$$

where the subscript $\mathbf{x}_0$ signifies that the gradient, ∇f, is evaluated at the point $\mathbf{x}_0$. The second term on the right-hand side of (2.1.10) is the inner product of the distance vector, $\mathbf{x} - \mathbf{x}_0$, and the gradient vector ∇f evaluated at $\mathbf{x}_0$.

2.1.2 Velocity gradient tensor

To derive the linearized form of the velocity field in the vicinity of a point, $\mathbf{x}_0$, we identify the function f with the x, y, or z velocity component, u_x, u_y, or u_z, and obtain

$$
\begin{aligned}
u_x(\mathbf{x}) &\simeq u_x(\mathbf{x}_0) + (x - x_0)\left(\frac{\partial u_x}{\partial x}\right)_{\mathbf{x}_0} + (y - y_0)\left(\frac{\partial u_x}{\partial y}\right)_{\mathbf{x}_0} + (z - z_0)\left(\frac{\partial u_x}{\partial z}\right)_{\mathbf{x}_0}, \\
u_y(\mathbf{x}) &\simeq u_x(\mathbf{x}_0) + (x - x_0)\left(\frac{\partial u_y}{\partial x}\right)_{\mathbf{x}_0} + (y - y_0)\left(\frac{\partial u_y}{\partial y}\right)_{\mathbf{x}_0} + (z - z_0)\left(\frac{\partial u_y}{\partial z}\right)_{\mathbf{x}_0}, \\
u_z(\mathbf{x}) &\simeq u_z(\mathbf{x}_0) + (x - x_0)\left(\frac{\partial u_z}{\partial x}\right)_{\mathbf{x}_0} + (y - y_0)\left(\frac{\partial u_z}{\partial y}\right)_{\mathbf{x}_0} + (z - z_0)\left(\frac{\partial u_z}{\partial z}\right)_{\mathbf{x}_0}.
\end{aligned}
\tag{2.1.11}
$$

Collecting these equations into a unified vector form, we obtain

$$
\mathbf{u}(\mathbf{x}) \simeq \mathbf{u}(\mathbf{x}_0) + (\mathbf{x} - \mathbf{x}_0) \cdot \mathbf{A}(\mathbf{x}_0), \tag{2.1.12}
$$

where $\mathbf{A}$ is a 3×3 matrix called the velocity gradient tensor,

$$
\mathbf{A} = \begin{bmatrix} \frac{\partial u_x}{\partial x} & \frac{\partial u_y}{\partial x} & \frac{\partial u_z}{\partial x} \\ \frac{\partial u_x}{\partial y} & \frac{\partial u_y}{\partial y} & \frac{\partial u_z}{\partial y} \\ \frac{\partial u_x}{\partial z} & \frac{\partial u_y}{\partial z} & \frac{\partial u_z}{\partial z} \end{bmatrix}. \tag{2.1.13}
$$

The notation $\mathbf{A}(\mathbf{x}_0)$ in (2.1.12) emphasizes that the nine components of the velocity gradient tensor are evaluated at the chosen point $\mathbf{x}_0$ around which linearization has taken place.

As an example, the velocity gradient corresponding to the velocity field expressed by equations (1.4.7) is

$$
\mathbf{A} = \begin{bmatrix} (b + ct)\, 3\, x^2\, y\, z + c\, d\, t\, e^{d\, x\, t} & 2\, a\, x + (b + ct)\, y^3\, z & 2\, a\, x + (b + ct)\, y\, z^3 \\ 2\, ay + (b + ct) x^3 z & (b + ct) 3 y^2 x z + cdt\, e^{d\, y\, t} & 2ay + (b + ct) x\, z^3 \\ 2\, az + (b + ct) x^3 y & 2\, az + (b + ct) x y^3 & (b + ct)\, 3 z^2 x y + cdt\, e^{d\, z\, t} \end{bmatrix}. \tag{2.1.14}
$$

Placing the point $\mathbf{x}_0$ at the location $(x_0, y_0, z_0) = (1, 0, 0)$, we find

$$
\mathbf{A}(\mathbf{x}_0) = \begin{bmatrix} cd\, t\, e^{dt} & 2\, a & 2\, a \\ 0 & cd\, t & 0 \\ 0 & 0 & cd\, t \end{bmatrix}. \tag{2.1.15}
$$

Thus, in the vicinity of the point $(1, 0, 0)$, the flow expressed by equations (1.4.7) can be approximated with a linear flow described by

$$\begin{aligned} u_x(x,y,z) &\simeq u_x(1,0,0) + c\,d\,t\,e^{d\,t}\,(x-1), \\ u_y(x,y,z) &\simeq u_y(1,0,0) + 2\,a\,(x-1) + cd\,ty, \\ u_z(x,y,z) &\simeq u_z(1,0,0) + 2\,a\,(x-1) + cd\,tz. \end{aligned} \tag{2.1.16}$$

The right-hand sides of equations (2.1.16) are linear functions of the spatial coordinates x, y, and z, but not necessarily time, t.

What is a tensor?

The velocity gradient tensor is a matrix containing the three first partial derivatives of the three components of the velocity with respect to x, y, and z, a total of nine scalar elements. Why have we called this matrix a tensor?

A tensor is a matrix whose elements are physical entities evaluated with reference to a chosen system of Cartesian or polar coordinates. If the coordinate system is changed, for example, by translation or rotation, the elements of the matrix will also change to reflect the new directions. This change is analogous to that undergone by the components of the position or velocity vector, occurring when a new system of coordinates is introduced, as discussed in section 1.5.

If the elements of the matrix corresponding to the new system are related to the elements corresponding to the old system by certain rules discussed in texts of matrix calculus and continuum mechanics mechanics,[1] then the matrix is called a tensor. Establishing whether or not a matrix is a tensor is important in deriving physical laws relating matrices with different physical interpretations.

2.1.3 Relative motion of point particles

According to equation (2.1.12), the motion of a point particle near a point, $\mathbf{x}_0$, is governed by the equation

$$\frac{\mathrm{d}\mathbf{X}}{\mathrm{d}t} = \mathbf{u}(\mathbf{X}) \simeq \mathbf{u}(\mathbf{x}_0) + (\mathbf{X} - \mathbf{x}_0) \cdot \mathbf{A}(\mathbf{x}_0), \tag{2.1.17}$$

where $\mathbf{X}$ is the position of the point particle, and $\mathbf{u}(\mathbf{X})$ is the point-particle velocity, which is equal to the local and instantaneous fluid velocity.

The first term on the right-hand side of (2.1.17) states that a point particle located at the point $\mathbf{X}$ translates with the velocity of the point particle located at $\mathbf{x}_0$. The second term expresses the relative motion with respect to the point particle located at $\mathbf{x}_0$. Different velocity gradient tensors, $\mathbf{A}(\mathbf{x}_0)$, represent different types of relative motion. Our goal is to delineate the nature of this motion in terms of the components of $\mathbf{A}(\mathbf{x}_0)$.

[1] POZRIKIDIS, C., 1997, *Introduction to Theoretical and Computational Fluid Dynamics*, Oxford University Press.

2.1.4 *Fundamental motions in two-dimensional flow*

We begin by considering a two-dimensional flow in the xy plane and introduce the 2×2 velocity gradient tensor

$$\mathbf{A} = \begin{bmatrix} \frac{\partial u_x}{\partial x} & \frac{\partial u_y}{\partial x} \\ \\ \frac{\partial u_x}{\partial y} & \frac{\partial u_y}{\partial y} \end{bmatrix}. \tag{2.1.18}$$

In section 1.6, we studied the velocity field associated with the linear flow expressed by equation (1.6.38). Comparing equations (2.1.12) and (2.1.18) with equation (1.6.38), we set

$$a = \frac{\partial u_x}{\partial x}, \qquad b = \frac{\partial u_y}{\partial x}, \qquad c = \frac{\partial u_x}{\partial y}, \qquad d = \frac{\partial u_y}{\partial y}, \tag{2.1.19}$$

where all partial derivatives are evaluated at $\mathbf{x}_0$. To study the properties of the linearized flow, we carry out the decomposition shown in equation (1.6.39), setting

$$\mathbf{A} = \mathbf{\Xi} + \mathbf{E} + \frac{\alpha}{2} \begin{bmatrix} 1 & 0 \\ 0 & 1 \end{bmatrix}, \tag{2.1.20}$$

where

$$\mathbf{\Xi} \equiv \frac{1}{2} \begin{bmatrix} 0 & \frac{\partial u_y}{\partial x} - \frac{\partial u_x}{\partial y} \\ \frac{\partial u_x}{\partial y} - \frac{\partial u_y}{\partial x} & 0 \end{bmatrix} \tag{2.1.21}$$

is a skew-symmetric matrix with zero trace called the vorticity tensor,

$$\mathbf{E} \equiv \frac{1}{2} \begin{bmatrix} \frac{\partial u_x}{\partial x} - \frac{\partial u_y}{\partial y} & \frac{\partial u_y}{\partial x} + \frac{\partial u_x}{\partial y} \\ \\ \frac{\partial u_x}{\partial y} + \frac{\partial u_y}{\partial x} & \frac{\partial u_y}{\partial y} - \frac{\partial u_x}{\partial x} \end{bmatrix} \tag{2.1.22}$$

is a symmetric matrix with zero trace called the rate of deformation tensor, and the scalar

$$\alpha = \frac{\partial u_x}{\partial x} + \frac{\partial u_y}{\partial y} \tag{2.1.23}$$

is the areal rate of expansion.

Expansion

The results of section 1.6 suggest that a fluid parcel centered at the point $\mathbf{x}_0$ expands isotropically with an areal rate of expansion that is equal to the right-hand side of (2.1.23) evaluated at $\mathbf{x}_0$, as illustrated in figure 2.1.1.

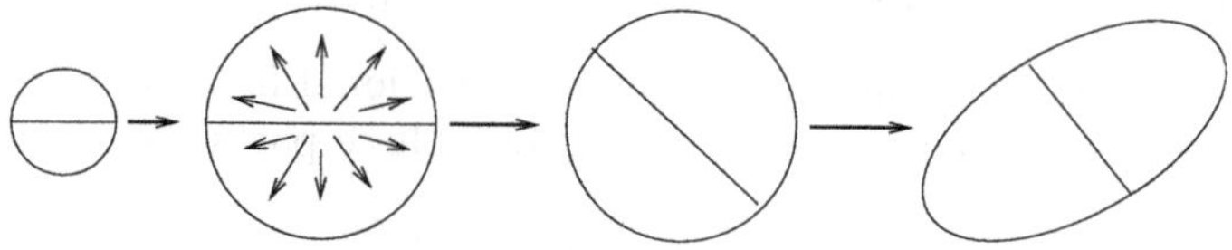

Figure 2.1.1 Expansion, rotation, and deformation of a small discoidal fluid parcel occurring during an infinitesimal period of time in a two-dimensional flow.

Rotation

Referring to equation (1.6.40), we find that a fluid parcel centered at the point $\mathbf{x}_0$ rotates in the xy plane around the point $\mathbf{x}_0$ with angular velocity

$$\Omega = \frac{1}{2}\left(\frac{\partial u_y}{\partial x} - \frac{\partial u_x}{\partial y}\right), \tag{2.1.24}$$

where the right-hand side is evaluated at $\mathbf{x}_0$, as shown in figure 2.1.1. When Ω is positive, the parcel rotates in the counterclockwise direction; when Ω is negative, the parcel rotates in the clockwise direction.

Deformation

The discussion in section 1.6 suggests that the flow associated with the rate of deformation tensor, $\mathbf{E}$, expresses pure deformation, as illustrated in figure 2.1.1. To compute the rate of deformation, G, we consider the eigenvalues of $\mathbf{E}$. Denoting $E_{xx} \equiv E_{11}$, introducing a similar notation for the other components, and taking into account that $E_{xx} + E_{yy} = 0$ and $E_{xy} = E_{yx}$, we find the eigenvalues

$$G = \pm(E_{xx}^2 + E_{xy}^2)^{1/2}. \tag{2.1.25}$$

The corresponding eigenvectors define the principal directions of the rate of strain. It can be shown that, because $\mathbf{E}$ is symmetric, the two eigenvectors are mutually orthogonal.

An eigenvalue of the rate of strain tensor expresses the rate of deformation of a circular fluid parcel centered at the point $\mathbf{x}_0$, in the direction of the associated eigenvector. A theorem of matrix calculus ensures that the sum of the eigenvalues is equal to the sum of the diagonal elements of $\mathbf{E}$, which is equal to zero. Because of this property, the deformation conserves the area of the parcel during the motion.

2.1.5 Fundamental motions in three-dimensional flow

To extend the preceding analysis to three-dimensional flow, we decompose the three-dimensional velocity gradient tensor into three parts, as

$$\mathbf{A} = \mathbf{\Xi} + \mathbf{E} + \frac{\alpha}{3} \begin{bmatrix} 1 & 0 & 0 \\ 0 & 1 & 0 \\ 0 & 0 & 1 \end{bmatrix}, \tag{2.1.26}$$

where

$$\mathbf{\Xi} \equiv \frac{1}{2} \begin{bmatrix} 0 & \frac{\partial u_y}{\partial x} - \frac{\partial u_x}{\partial y} & \frac{\partial u_z}{\partial x} - \frac{\partial u_x}{\partial z} \\ \frac{\partial u_x}{\partial y} - \frac{\partial u_y}{\partial x} & 0 & \frac{\partial u_z}{\partial y} - \frac{\partial u_y}{\partial z} \\ \frac{\partial u_x}{\partial z} - \frac{\partial u_z}{\partial x} & \frac{\partial u_y}{\partial z} - \frac{\partial u_z}{\partial y} & 0 \end{bmatrix} \tag{2.1.27}$$

is the skew-symmetric vorticity tensor,

$$\mathbf{E} \equiv \begin{bmatrix} \frac{\partial u_x}{\partial x} - \frac{1}{3}\alpha & \frac{1}{2}\left(\frac{\partial u_y}{\partial x} + \frac{\partial u_x}{\partial y}\right) & \frac{1}{2}\left(\frac{\partial u_z}{\partial x} + \frac{\partial u_x}{\partial z}\right) \\ \frac{1}{2}\left(\frac{\partial u_x}{\partial y} + \frac{\partial u_y}{\partial x}\right) & \frac{\partial u_y}{\partial y} - \frac{1}{3}\alpha & \frac{1}{2}\left(\frac{\partial u_z}{\partial y} + \frac{\partial u_y}{\partial z}\right) \\ \frac{1}{2}\left(\frac{\partial u_x}{\partial z} + \frac{\partial u_z}{\partial x}\right) & \frac{1}{2}\left(\frac{\partial u_y}{\partial z} + \frac{\partial u_z}{\partial y}\right) & \frac{\partial u_z}{\partial z} - \frac{1}{3}\alpha \end{bmatrix} \tag{2.1.28}$$

is the symmetric and traceless rate of deformation tensor, and the scalar coefficient

$$\alpha \equiv \frac{\partial u_x}{\partial x} + \frac{\partial u_y}{\partial y} + \frac{\partial u_z}{\partial z} \tag{2.1.29}$$

is the volumetric rate of expansion. In compact notation,

$$\mathbf{A} = \mathbf{\Xi} + \mathbf{E} + \frac{\alpha}{3}\mathbf{I}, \tag{2.1.30}$$

where $\mathbf{I}$ is the 3×3 identity matrix.

The three terms on the right-hand side of (2.1.30) express, respectively, isotropic expansion, rotation, and pure deformation, as illustrated in figure 2.1.2. Because of the fundamental significance of these motions, these terms deserve and will be given individual attention in sections 2.2–2.4.

2.1.6 Gradient in polar coordinates

We have defined the velocity gradient tensor as the gradient of the velocity. Expressions for the gradient operator in polar coordinates can be obtained using geometrical transformation rules combined with the chain rule of differentiation.

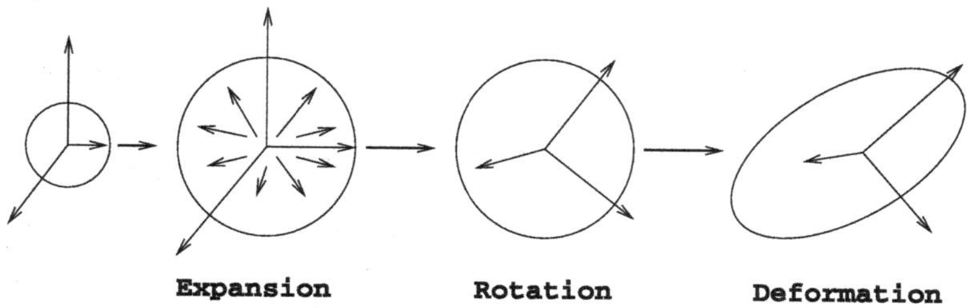

Figure 2.1.2 Expansion, rotation, and deformation of a small spherical fluid parcel occurring during an infinitesimal period of time in a three-dimensional flow.

Cylindrical polar coordinates

In the cylindrical polar coordinates depicted in figure 1.3.2, the gradient of a scalar function, f, is determined by its cylindrical polar components F_x, F_σ, and F_φ, as

$$\nabla f = F_x \, \mathbf{e}_x + F_\sigma \, \mathbf{e}_\sigma + F_\varphi \, \mathbf{e}_\varphi. \tag{2.1.31}$$

Using the transformation rules shown in equations (1.3.20), we find

$$F_\sigma = \cos\varphi \, \frac{\partial f}{\partial y} + \sin\varphi \, \frac{\partial f}{\partial z}, \qquad F_\varphi = -\sin\varphi \, \frac{\partial f}{\partial y} + \cos\varphi \, \frac{\partial f}{\partial z}. \tag{2.1.32}$$

To express the derivatives with respect to y and z in terms of derivatives with respect to cylindrical polar coordinates, we use the chain rule of differentiation along with the coordinate transformation rules (1.3.14) and (1.3.15), and find

$$\begin{aligned} \left(\frac{\partial f}{\partial y}\right)_{x,z} &= \left(\frac{\partial f}{\partial x}\right)_{\sigma,\varphi} \left(\frac{\partial x}{\partial y}\right)_{x,z} + \left(\frac{\partial f}{\partial \sigma}\right)_{x,\varphi} \left(\frac{\partial \sigma}{\partial y}\right)_{x,z} + \left(\frac{\partial f}{\partial \varphi}\right)_{x,\sigma} \left(\frac{\partial \varphi}{\partial y}\right)_{x,z} \\ &= \cos\varphi \left(\frac{\partial f}{\partial \sigma}\right)_{x,\varphi} - \frac{\sin\varphi}{\sigma} \left(\frac{\partial f}{\partial \varphi}\right)_{x,\sigma}, \end{aligned} \tag{2.1.33}$$

and

$$\begin{aligned} \left(\frac{\partial f}{\partial z}\right)_{x,y} &= \left(\frac{\partial f}{\partial x}\right)_{\sigma,\varphi} \left(\frac{\partial x}{\partial z}\right)_{x,y} + \left(\frac{\partial f}{\partial \sigma}\right)_{x,\varphi} \left(\frac{\partial \sigma}{\partial z}\right)_{x,y} + \left(\frac{\partial f}{\partial \varphi}\right)_{x,\sigma} \left(\frac{\partial \varphi}{\partial z}\right)_{x,y} \\ &= \sin\varphi \left(\frac{\partial f}{\partial \sigma}\right)_{x,\varphi} + \frac{\cos\varphi}{\sigma} \left(\frac{\partial f}{\partial \varphi}\right)_{x,\sigma}. \end{aligned} \tag{2.1.34}$$

Substituting relations (2.1.33) and (2.1.34) in the right-hand sides of relations (2.1.32), we find

$$F_x = \frac{\partial f}{\partial x}, \qquad F_\sigma = \frac{\partial f}{\partial \sigma}, \qquad F_\varphi = \frac{1}{\sigma} \frac{\partial f}{\partial \varphi}. \tag{2.1.35}$$

Equations (2.1.35) illustrate that the polar components of the gradient are equal to the partial derivatives with respect to the corresponding variables multiplied by an appropriate scaling factor.

Spherical polar coordinates

In the spherical polar coordinates depicted in figure 1.3.3, the gradient of a scalar function, f, is defined by its spherical polar components F_r, F_θ, and F_φ, as

$$\nabla f = F_r \, \mathbf{e}_r + F_\theta \, \mathbf{e}_\theta + F_\varphi \, \mathbf{e}_\varphi. \tag{2.1.36}$$

Working as in the case of cylindrical coordinates, we find

$$F_r = \frac{\partial f}{\partial r}, \qquad F_\theta = \frac{1}{r} \frac{\partial f}{\partial \theta}, \qquad F_\varphi = \frac{1}{r \sin\theta} \frac{\partial f}{\partial \varphi}. \tag{2.1.37}$$

Note that the expression for F_φ is consistent with that given in the third relation of (2.1.35), subject to the substitution $\sigma = r \sin\theta$.

Plane polar coordinates

In the plane polar coordinates depicted in figure 1.3.4, the gradient of a scalar function, f, is defined by its plane polar components F_r and F_θ, as

$$\nabla f = F_r \, \mathbf{e}_r + F_\theta \, \mathbf{e}_\theta. \tag{2.1.38}$$

Working as in the case of cylindrical coordinates, we find

$$F_r = \frac{\partial f}{\partial r}, \qquad F_\theta = \frac{1}{r} \frac{\partial f}{\partial \theta}. \tag{2.1.39}$$

Problems

2.1.1 *Inner vector product*

Prove the interpretation of the inner vector product described after equation (2.1.9). *Hint:* Use the law of cosines.

2.1.2 *Decomposition of a linearized flow*

(*a*) Linearize the velocity described by equations (1.5.2) around the origin of the y axis, and then decompose the velocity gradient tensor of the linearized flow into the three modes shown on the right-hand side of (2.1.26).

(*b*) Decompose the velocity gradient tensor of the linearized flow expressed by equations (2.1.16) into the three modes shown on the right-hand side of (2.1.26).

2.2 Fluid parcel expansion

The velocity field associated with the third term on the right-hand side of (2.1.26) is described by

$$\mathbf{u}^{exp}(\mathbf{x}) = \alpha(\mathbf{x}_0)\,\frac{1}{3}\,(\mathbf{x}-\mathbf{x}_0)\cdot\begin{bmatrix} 1 & 0 & 0 \\ 0 & 1 & 0 \\ 0 & 0 & 1 \end{bmatrix}. \tag{2.2.1}$$

Under the influence of this field, a spherical fluid parcel centered at the point $\mathbf{x}_0$ expands when the coefficient $\alpha(\mathbf{x}_0)$ is positive, and contracts when the coefficient $\alpha(\mathbf{x}_0)$ is negative, always retaining the spherical shape.

To see this more clearly, we consider the motion of a point particle that lies at the surface of the spherical parcel. Using (2.2.1), we find that the radius of the parcel, $a(t)$, is given by

$$\frac{a(t)}{a(t=0)} = \exp(\frac{1}{3}\alpha t). \tag{2.2.2}$$

Raising both sides to the third power and multiplying the result by the factor $4\pi/3$, we find that the ratio of the instantaneous to the initial parcel volume is

$$\frac{\frac{4\pi}{3}\,a^3(t)}{\frac{4\pi}{3}\,a^3(t=0)} = \exp(\alpha t). \tag{2.2.3}$$

This result explains why the constant α is called the rate of volumetric expansion.

Divergence of the velocity field

The rate of expansion defined in equation (2.1.29) can be expressed in a compact form that simplifies the notation. Taking the inner product of the del operator defined in (2.1.7) and the velocity, we find

$$\nabla\cdot\mathbf{u} \equiv \frac{\partial u_x}{\partial x} + \frac{\partial u_y}{\partial y} + \frac{\partial u_z}{\partial z}. \tag{2.2.4}$$

In index notation,

$$\nabla\cdot\mathbf{u} \equiv \frac{\partial u_i}{\partial x_i}, \tag{2.2.5}$$

where summation of the repeated index i is implied over x, y, and z. In the case of two-dimensional flow in the xy plane, the derivative of u_z with respect to z does not appear. Accordingly, we write

$$\alpha = \nabla\cdot\mathbf{u}. \tag{2.2.6}$$

The right-hand side of (2.2.6) is defined as the divergence of the velocity.

Solenoidal velocity fields

We have found that the rate of volumetric expansion at a point in a three-dimensional flow and the areal rate of expansion at a point in a two-dimensional flow are equal to the divergence of the velocity evaluated at that point. If the divergence of the velocity vanishes at every point in the flow, with the physical consequence that no fluid parcel undergoes expansion but only translation, rotation, and deformation, then the velocity field is called solenoidal.

Problem

2.2.1 *Rate of expansion*

Derive the rate of expansion of the flow described by equations (1.4.7), and then evaluate it at the point $\mathbf{x}_0 = (1, 0, 1)$.

2.3 Fluid parcel rotation and vorticity

The velocity field associated with the first term on the right-hand side of (2.1.26) is given by

$$\mathbf{u}^{rot}(x, y, z) = (\mathbf{x} - \mathbf{x}_0) \cdot \mathbf{\Xi}(\mathbf{x}_0), \tag{2.3.1}$$

where $\mathbf{\Xi}$ is the vorticity tensor defined in equation (2.1.27).

A planar fluid parcel in a two-dimensional flow in the xy plane rotates only around the z axis. In contrast, a three-dimensional fluid parcel in a three-dimensional flow may rotate around any arbitrary axis passing through the designated center of rotation, $\mathbf{x}_0$, and pointing in any arbitrary direction.

The orientation, magnitude, and direction of rotation define an angular velocity vector, $\mathbf{\Omega}$, whose components can be deduced from the three upper triangular or three lower triangular entries of the vorticity tensor shown on the left-hand side of (2.1.27), and are given by

$$\Omega_x = \frac{1}{2}\left(\frac{\partial u_z}{\partial y} - \frac{\partial u_y}{\partial z}\right), \qquad \Omega_y = \frac{1}{2}\left(\frac{\partial u_z}{\partial x} - \frac{\partial u_x}{\partial z}\right),$$

$$\Omega_z = \frac{1}{2}\left(\frac{\partial u_y}{\partial x} - \frac{\partial u_x}{\partial y}\right), \tag{2.3.2}$$

where the right-hand sides are evaluated at the designated parcel center, $\mathbf{x}_0$. As we look down into the vector $\mathbf{\Omega}$ from the tip of its arrow, the fluid rotates in the clockwise direction.

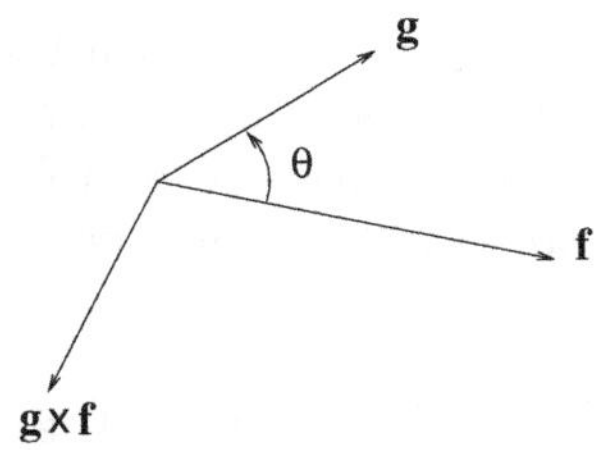

Figure 2.3.1 The outer product of two vectors, **f** and **g**, is a new vector perpendicular to their plane.

Equation (2.3.1) can be recast into a compact form in terms of the angular velocity vector, as

$$\mathbf{u}^{rot}(x,y,z) = (\mathbf{x} - \mathbf{x}_0) \cdot \begin{bmatrix} 0 & \Omega_z & \Omega_y \\ -\Omega_z & 0 & \Omega_x \\ -\Omega_y & -\Omega_x & 0 \end{bmatrix}, \tag{2.3.3}$$

where $\boldsymbol{\Omega}$ derives from the velocity by means of (2.3.2).

The three components of the angular velocity vector arise by combining selected partial derivatives of the components of the velocity in a particular fashion. Stated differently, the angular velocity vector field arises from the velocity field by operating with a differential operator, just as the rate of expansion arises from the velocity field by operating with the divergence operator $(\nabla\cdot)$, as discussed in section 2.2.

Outer vector product

To identify the differential operator that generates the point particle angular velocity field from the velocity field according to equations (2.3.2), we introduce the outer vector product. Consider a vector, $\mathbf{f} = (f_x, f_y, f_z)$, and another vector, $\mathbf{g} = (g_x, g_y, g_z)$. The outer product of the first vector with the second vector, stated in this particular order, is a new vector denoted by $\mathbf{f} \times \mathbf{g}$, defined as

$$\mathbf{f} \times \mathbf{g} = (f_y\, g_z - f_z\, g_y)\, \mathbf{e}_x + (f_z\, g_x - f_x\, g_z)\, \mathbf{e}_y + (f_x\, g_y - f_y\, g_x)\, \mathbf{e}_z, \tag{2.3.4}$$

where $\mathbf{e}_x$, $\mathbf{e}_y$, and $\mathbf{e}_z$ are unit vectors along the x, y, and z axis.

Interpretation of the outer vector product

It can be shown that the vector $\mathbf{f} \times \mathbf{g}$ is normal to the plane defined by $\mathbf{f}$ and $\mathbf{g}$, as illustrated in figure 2.3.1. The magnitude of $\mathbf{f} \times \mathbf{g}$ is equal to the product of (a) the length of the vector $\mathbf{f}$, (b) the length of the vector $\mathbf{g}$, and (c) the absolute value of the sine of the angle, θ, subtended between the two vectors. The orientation of $\mathbf{f} \times \mathbf{g}$ is chosen such that, as we look down at the plane defined by $\mathbf{f}$ and $\mathbf{g}$ toward the negative direction of $\mathbf{f} \times \mathbf{g}$, the angle θ measured in the counterclockwise direction from $\mathbf{f}$ is less than π. If θ is equal to 0 or π, the two vectors are parallel or anti-parallel, the sine of the angle is zero, and the outer product vanishes.

The directions defined by the three vectors $\mathbf{f}$, $\mathbf{g}$, and $\mathbf{f} \times \mathbf{g}$, arranged in this particular order, form a right-handed system of coordinates. This is another way of saying that $\mathbf{f} \times \mathbf{g}$ arises from $\mathbf{f}$ and $\mathbf{g}$ according to the right-hand rule.

Now invoking the definition of the cross product, we recast equation (2.3.3) into the form

$$\mathbf{u}^{rot}(x, y, z) = \boldsymbol{\Omega} \times (\mathbf{x} - \mathbf{x}_0), \tag{2.3.5}$$

which describes rigid-body rotation around the point $\mathbf{x}_0$ with angular velocity $\boldsymbol{\Omega}$, in agreement with the previously stated physical interpretation.

2.3.1 Curl and vorticity

Taking the outer product of the del operator and the velocity field, we obtain the curl of the velocity defined as the vorticity,

$$\boldsymbol{\omega} \equiv \nabla \times \mathbf{u} = (\frac{\partial u_z}{\partial y} - \frac{\partial u_y}{\partial z})\, \mathbf{e}_x + (\frac{\partial u_x}{\partial z} - \frac{\partial u_z}{\partial x})\, \mathbf{e}_y + (\frac{\partial u_y}{\partial x} - \frac{\partial u_x}{\partial y})\, \mathbf{e}_z. \tag{2.3.6}$$

Comparing equation (2.3.6) with equations (2.3.2), we find

$$\boldsymbol{\Omega} = \frac{1}{2}\,\boldsymbol{\omega}, \tag{2.3.7}$$

which shows that the angular velocity vector is equal to half the vorticity vector, or half the curl of the velocity.

Irrotational flow

If the curl of a velocity field vanishes at every point in a flow, with the consequence that no spherical fluid parcel undergoes rotation, then the velocity field is called irrotational. The properties and computation of irrotational flow will be discussed in chapter 3, and then again in chapter 12 in the context of aerodynamics.

The alternating tensor

The long expression on the right-hand side of equation (2.3.4) defining the outer vector product is cumbersome. To simplify the notation, we introduce the three-index alternating tensor ϵ_{ijk}, defined as follows:

1. If $i = j$, or $j = k$, or $k = i$, then $\epsilon_{ijk} = 0$. For example, $\epsilon_{xxy} = \epsilon_{zyz} = \epsilon_{zyy} = 0$.
2. If i, j, and k are all different, then $\epsilon_{ijk} = \pm 1$. The plus sign applies when the triplet ijk is a cyclic permutation of xyz, and the minus sign applies otherwise. For example, $\epsilon_{xyz} = \epsilon_{zxy} = \epsilon_{yzx} = 1$, but $\epsilon_{xzy} = -1$.

Two important properties of the alternating tensor stemming from its definition are

$$\epsilon_{ijk}\, \epsilon_{ljk} = 2\, \delta_{il}, \tag{2.3.8}$$

where summation of the repeated indices j and k is implied on the left-hand side, and

$$\epsilon_{ijk}\, \epsilon_{lmk} = \delta_{il}\, \delta_{jm} - \delta_{im}\, \delta_{jl}, \tag{2.3.9}$$

where summation of the repeated index k is implied on the left-hand side. Kronecker's delta δ_{ij} represents the identity matrix: $\delta_{ij} = 1$ if $i = j$, or 0 if $i \neq j$. Additional properties of the alternating tensor are listed in problem 2.3.2.

In terms of the alternating tensor, the ith component of the outer product $\mathbf{f} \times \mathbf{g}$ defined in equation (2.3.4) is given by

$$(\mathbf{f} \times \mathbf{g})_i = \epsilon_{ijk}\, f_j\, g_k, \tag{2.3.10}$$

where summation of the two repeated indices j and k is implied on the right-hand side.

Using the definition (2.3.6), we find that the ith component of the vorticity is given by

$$\omega_i = \epsilon_{ijk}\, \frac{\partial u_k}{\partial x_j}. \tag{2.3.11}$$

Straightforward manipulation of (2.3.11) provides us with an expression for the vorticity vector in terms of the vorticity tensor,

$$\begin{aligned} \omega_i &= \frac{1}{2}\, (\epsilon_{ijk} \frac{\partial u_k}{\partial x_j} + \epsilon_{ijk} \frac{\partial u_k}{\partial x_j}) = \frac{1}{2}\, (\epsilon_{ijk} \frac{\partial u_k}{\partial x_j} - \epsilon_{ikj} \frac{\partial u_k}{\partial x_j}) \\ &= \frac{1}{2}\, (\epsilon_{ijk} \frac{\partial u_k}{\partial x_j} - \epsilon_{ijk} \frac{\partial u_j}{\partial x_k}) = \epsilon_{ijk} \frac{1}{2} (\frac{\partial u_k}{\partial x_j} - \frac{\partial u_j}{\partial x_k}) = \epsilon_{ijk}\, \Xi_{jk}. \end{aligned} \tag{2.3.12}$$

The converse relationship is

$$\Xi_{ij} = \frac{1}{2}\, \epsilon_{ijk}\, \omega_k \tag{2.3.13}$$

(problem 2.3.3).

2.3.2 Two-dimensional flow

Consider a two-dimensional flow in the xy plane. Inspecting the right-hand side of (2.3.6) shows that the x and y components of the vorticity vanish. The vorticity vector is then parallel to the z axis and thus perpendicular to the plane of the flow,

$$\boldsymbol{\omega} = \omega_z \, \mathbf{e}_z, \tag{2.3.14}$$

where $\mathbf{e}_z$ is the unit vector along the z axis. The scalar ω_z is the strength of the vorticity, defined as

$$\omega_z = \frac{\partial u_y}{\partial x} - \frac{\partial u_x}{\partial y}. \tag{2.3.15}$$

In the case of simple shear flow with constant shear rate k, $u_x = ky$, $u_y = 0$ and $\omega_z = -k$.

Using the transformation rules discussed in section 1.1, we find that the strength of the vorticity in the plane polar coordinates depicted in figure 1.1.4 is given by

$$\omega_z = \frac{1}{r}\Big(\frac{\partial(r u_\theta)}{\partial r} - \frac{\partial u_r}{\partial \theta}\Big). \tag{2.3.16}$$

In the case of rigid-body rotation with angular velocity Ω, $u_\theta = \Omega \, r \; u_r = 0$, and $\omega_z = \frac{1}{2}\,\Omega$.

2.3.3 Axisymmetric flow

Next, we consider axisymmetric flow without swirling motion and refer to the polar cylindrical coordinates (x, σ, φ) depicted in figure 1.1.2 and to the spherical polar coordinates (r, θ, φ) depicted in figure 1.1.3.

A fluid patch that lies in a meridional plane of constant angle φ is able to rotate only around an axis that is perpendicular to this plane and thus points in the direction of increasing or decreasing meridional angle, φ. This observation suggests that the vorticity vector takes the form

$$\boldsymbol{\omega} = \omega_\varphi \, \mathbf{e}_\varphi, \tag{2.3.17}$$

where $\mathbf{e}_\varphi$ is the unit vector pointing in the meridional direction, and ω_φ is the strength of the vorticity given by

$$\omega_\varphi = \frac{\partial u_\sigma}{\partial x} - \frac{\partial u_x}{\partial \sigma} = \frac{1}{r}\Big(\frac{\partial(r u_\theta)}{\partial r} - \frac{\partial u_r}{\partial \theta}\Big). \tag{2.3.18}$$

Note that the expression in spherical polar coordinates given on the right-hand side of (2.3.18) is identical to that in plane polar coordinates given in (2.3.16).

Problems

2.3.1 *Properties of the outer vector product*

(*a*) Show that $\mathbf{f} \times \mathbf{g} = -\mathbf{g} \times \mathbf{f}$, where the outer vector product is defined in equation (2.3.4).

(*b*) The outer vector product of two vectors, $\mathbf{f}$ and $\mathbf{g}$, can be identified with the determinant of a matrix,

$$\mathbf{f} \times \mathbf{g} = \det(\begin{bmatrix} \mathbf{e}_x & \mathbf{e}_y & \mathbf{e}_z \\ f_x & f_y & f_z \\ g_x & g_y & g_z \end{bmatrix}). \tag{2.3.19}$$

Show that this rule is consistent with the definition of the curl in (2.3.6).

2.3.2 *Properties of Kronecker's delta and alternating tensor*

Prove the properties

$$\delta_{ii} = 3, \qquad \epsilon_{ljk}\,\delta_{jk} = 0, \qquad a_j\,\delta_{jk} = a_k, \qquad A_{lj}\,\delta_{jk} = A_{lk}, \tag{2.3.20}$$

where δ_{ij} is Kronecker's delta representing the 3×3 identity matrix, $\mathbf{a}$ is an arbitrary vector, $\mathbf{A}$ is an arbitrary matrix, and summation is implied over a repeated index.

2.3.3 *Relation between the vorticity tensor and vector*

Prove relation (2.3.13). *Hint:* Express the vorticity in terms of the velocity as shown in (2.3.11), and then use property (2.3.9).

2.3.4 *The vorticity field is solenoidal*

Show that $\nabla \cdot \boldsymbol{\omega} = 0$; that is, the vorticity field is solenoidal.

2.4 Fluid parcel deformation

The velocity field associated with the second term on the right-hand side of (2.1.26) is

$$\mathbf{u}^{def}(x, y, z) = (\mathbf{x} - \mathbf{x}_0) \cdot \mathbf{E}(\mathbf{x}_0), \tag{2.4.1}$$

where $\mathbf{E}$ is the symmetric and traceless rate of deformation tensor defined in equation (2.1.22).

To develop insights into the nature of the motion described by (2.4.1), we consider a special case where $\mathbf{E}(\mathbf{x}_0)$ is diagonal, given by

$$\mathbf{E}(\mathbf{x}_0) = \begin{bmatrix} \frac{\partial u_x}{\partial x} - \frac{1}{3}\alpha & 0 & 0 \\ 0 & \frac{\partial u_y}{\partial y} - \frac{1}{3}\alpha & 0 \\ 0 & 0 & \frac{\partial u_z}{\partial z} - \frac{1}{3}\alpha \end{bmatrix}, \tag{2.4.2}$$

with the understanding that the right-hand side is evaluated at the point $\mathbf{x}_0$. The matrix (2.4.2) has three real eigenvalues equal to the diagonal elements. The corresponding eigenvectors point along the x, y, or z axes.

Cursory inspection reveals that, under the action of the flow described by (2.4.1) and (2.4.2), a spherical fluid parcel centered at the point $\mathbf{x}_0$ deforms into an ellipsoidal shape while preserving its volume, as illustrated in figure 2.1.2. The three eigenvalues of the rate of deformation tensor express the rate of deformation in three principal directions corresponding to the eigenvectors. If an eigenvalue is negative, the parcel is compressed in the corresponding direction to obtain an oblate shape.

More generally, the rate of deformation tensor has three real eigenvalues, λ_1, λ_2, and λ_3, that are found by setting the determinant of the following matrix to zero,

$$\mathbf{E} - \lambda \mathbf{I} = \begin{bmatrix} E_{xx} - \lambda & E_{xy} & E_{xz} \\ E_{yx} & E_{yy} - \lambda & E_{yz} \\ E_{zx} & E_{zy} & E_{zz} - \lambda \end{bmatrix}, \tag{2.4.3}$$

and then computing the roots of the emerging cubic equation for λ. It can be shown that, because $\mathbf{E}$ is symmetric, all three eigenvalues are real and each eigenvalue has a distinct corresponding eigenvector. Moreover, the three eigenvectors are mutually orthogonal, pointing in the principal directions of the rate of strain.

Under the action of the flow stated in (2.4.1), a spherical fluid parcel centered at the point $\mathbf{x}_0$ deforms to obtain an ellipsoidal shape whose axes are generally inclined with respect to the x, y, and z directions. The three axes of the ellipsoid are parallel to the eigenvectors of $\mathbf{E}$, and the respective rates of deformation are equal to the corresponding eigenvalues. A theorem of matrix calculus states that the sum of the eigenvalues is equal to the sum of the diagonal elements of $\mathbf{E}$, which is zero. Because of this property, the deformation conserves the parcel volume.

Computation of the rate of strain

Setting the determinant of the matrix (2.4.3) to zero, we obtain the cubic equation

$$\lambda^3 + a\,\lambda^2 + b\,\lambda + c = 0, \tag{2.4.4}$$

where

$$\begin{aligned} a &= -\mathrm{trace}(\mathbf{E}) = -(E_{xx} + E_{yy} + E_{zz}), \\ b &= (E_{yy}E_{zz} - E_{yz}E_{zy}) + (E_{xx}E_{zz} - E_{xz}E_{zx}) + (E_{xx}E_{yy} - E_{xy}E_{yx}), \\ c &= -\det(\mathbf{E}), \end{aligned} \tag{2.4.5}$$

and det stands for the determinant. Using Cardano's formulas, we find that the three roots of (2.4.4) are given by

$$\lambda_1 = -\frac{a}{3} + d\,\cos\frac{\chi}{3}, \qquad \lambda_2 = -\frac{a}{3} - d\,\cos\frac{\chi-\pi}{3},$$
$$\lambda_3 = -\frac{a}{3} - d\,\cos\frac{\chi+\pi}{3}, \tag{2.4.6}$$

where

$$d = 2\,\sqrt{|p|/3}, \qquad \chi = \arccos\Big(-\frac{1}{2\,(|p|/3)^{3/2}}\Big),$$
$$p = b - \frac{1}{3}\,a^2, \qquad q = c + \frac{2}{27}\,a^3 - \frac{1}{3}\,ab. \tag{2.4.7}$$

In the present case, because $a = 0$, we obtain the simplified expressions

$$\lambda_1 = d\,\cos\frac{\chi}{3}, \qquad \lambda_2 = -d\,\cos\frac{\chi-\pi}{3}, \qquad \lambda_3 = -d\,\cos\frac{\chi+\pi}{3}, \tag{2.4.8}$$

where

$$d = 2\sqrt{\frac{|b|}{3}}, \qquad \chi = \arccos\Big(-\frac{c}{2\,(\frac{|b|}{3})^{3/2}}\Big). \tag{2.4.9}$$

Once the eigenvalues have been found, the eigenvectors are computed by solving a homogeneous system of three equations for three unknowns. For example, the eigenvector $\mathbf{e}^{(1)} = (e_x^{(1)}, e_y^{(1)}, e_z^{(1)})$ corresponding to the eigenvalue λ_1 is found by solving the homogeneous system

$$(\mathbf{E} - \lambda_1 \mathbf{I}) \cdot \mathbf{e}^{(1)} = \mathbf{0}, \tag{2.4.10}$$

which can be restated as

$$\begin{aligned}
(E_{xx} - \lambda_1)\,e_x^{(1)} + E_{xy}\,e_y^{(1)} &= -E_{xz}\,e_z^{(1)}, \\
E_{yx}\,e_x^{(1)} + (E_{yy} - \lambda_1)\,e_y^{(1)} &= -E_{yz}\,e_z^{(1)}, \\
E_{zx}\,e_x^{(1)} + E_{zy}\,e_y^{(1)} &= -(E_{zz} - \lambda_1)\,e_z^{(1)}.
\end{aligned} \tag{2.4.11}$$

To solve system (2.4.11), we assign an arbitrary value to $e_z^{(1)}$, evaluate the first two right-hand sides, and solve the first two equations for $e_x^{(1)}$ and $e_y^{(1)}$ using, for example, Cramer's rule. The solution is guaranteed to also satisfy the third equation. A solution will not exist if the eigenvector is perpendicular to the z axis, in which case $e_z^{(1)}$ is zero. If this occurs, we simply transfer the term involving $e_x^{(1)}$ or $e_y^{(1)}$ to the right-hand side instead, and solve for the other two components.

Problems

2.4.1 *Properties of eigenvalues*

(*a*) Confirm that the sum of the three eigenvalues given in (2.4.6) is equal to the trace of **E**.

(*b*) Confirm that the product of the three eigenvalues given in (2.4.6) is equal to the determinant of **E**.

(*c*) Confirm that, when **E** is diagonal, formulas (2.4.6) identify the eigenvalues with the diagonal elements.

Computer Problem

2.4.2 *Eigenvalues and eigenvectors*

Subdirectory *05_eigen* inside directory *01_num_meth* of FDLIB contains program *eigen33* that computes the eigenvalues of a 3×3 matrix. Use the program to compute the eigenvalues and eigenvectors of the rate of deformation tensor corresponding to the linearized flow (2.1.16) for $a = 1\ \text{s}^{-1}$ and $cdt = 2\ \text{s}^{-1}$.

2.5 Numerical differentiation

We have mentioned that, in practice, the components of the velocity field are hardly ever given in analytical form by way of mathematical expressions. Instead, their values are either measured in the laboratory with probes, or computed by numerical methods at data points or grid nodes located in the domain of flow. The partial derivatives of the velocity are then recovered by a numerical procedure called numerical differentiation.

2.5.1 Numerical differentiation in one dimension

As a prelude to computing the partial derivatives of the components of the velocity from specified grid values, we consider computing the first derivative of a function, $f(x)$, of one independent variable, x, defined on a grid. Assume that we are given the values of $f(x)$ at $N+1$ nodes of a one-dimensional grid with nodes located at x_i, $i = 1, \ldots, N+1$, where $x_1 < x_2 < \cdots < x_{N+1}$, as shown in figure 1.7.1(*a*). Our goal is to compute the derivative, $\mathrm{d}f/\mathrm{d}x$, at a point, x, that lies in the kth interval subtended between the nodes x_k and x_{k+1}.

First-order differentiation

In the simplest approach, we replace the graph of the function $f(x)$ in the interval (x_k, x_{k+1}) with a straight line, as shown in figure 1.7.2, and approximate

df/dx with the slope. Using equations (1.7.1) and (1.7.3), we derive the finite-difference approximation

$$f'(x) \simeq \frac{f_{k+1} - f_k}{x_{k+1} - x_k}, \tag{2.5.1}$$

where a prime denotes a derivative with respect to x. Now identifying the evaluation point x with the grid point x_k, we obtain the forward-difference approximation

$$f'(x_k) \simeq \frac{f_{k+1} - f_k}{x_{k+1} - x_k}, \tag{2.5.2}$$

The error associated with this approximation is proportional to the interval size, h.

Using instead the straight-line approximation for the $k-1$ interval, we obtain the backward-difference approximation

$$f'(x_k) \simeq \frac{f_k - f_{k-1}}{x_k - x_{k-1}}. \tag{2.5.3}$$

Formulas (2.5.2) and (2.5.3) carry a comparable amount of error due to the straight-line approximation.

To evaluate the derivative at the first point, $f'(x_1)$, we use a forward difference; to evaluate the derivative at the last point, $f'(x_{N+1})$, we use a backward difference; to evaluate $f'(x_i)$ at an interior grid point, where $i = 2, \ldots, N$, we use either a forward or a backward difference, whichever is deemed more convenient or appropriate.

Second-order differentiation

Numerical differentiation based on linear interpolation neglects the curvature of the graph of the function $f(x)$. To improve the accuracy of the interpolation, we approximate $f(x)$ with a parabolic function defined in the interval (x_k, x_{k+1}), as depicted in figure 1.7.3, and then approximate the slope of the function, f', with the slope of the parabola. Differentiating (1.7.4), we derive the second-order finite-difference approximation

$$f'(x) \simeq 2\,a^{(k)}\,(x - x_k) + b^{(k)}, \tag{2.5.4}$$

where the coefficients $a^{(k)}$ and $b^{(k)}$ are given in (1.7.7).

Identifying the evaluation point, x, with the grid point, x_k, we obtain the centered-difference approximation

$$f'(x_k) \simeq b^{(k)}. \tag{2.5.5}$$

When the grid points are spaced evenly, $x_k - x_{k-1} = x_{k+1} - x_k = h$, where h is the grid spacing, we obtain the simple form

$$f'(x_k) \simeq \frac{f_{k+1} - f_{k-1}}{2h}. \tag{2.5.6}$$

The error associated with this approximation is proportional to the square of the interval size, h^2.

The parabolic approximation allows us to also obtain an estimate for the second derivative, $\mathrm{d}^2 f/\mathrm{d}x^2$. Differentiating (1.7.4) twice with respect to x, we derive the finite-difference approximation

$$f''(x) \simeq 2a^{(k)}, \tag{2.5.7}$$

where the coefficient $a^{(k)}$ is given in (1.7.7). When the grid points are spaced evenly along the x axis with separation h, we obtain the simpler formula

$$f''(x) \simeq \frac{f_{k+1} - 2f_k + f_{k-1}}{h^2}. \tag{2.5.8}$$

Identifying the evaluation point, x, with the grid point, x_k, we obtain the centered-difference approximation

$$f''(x_k) \simeq \frac{f_{k+1} - 2f_k + f_{k-1}}{h^2}. \tag{2.5.9}$$

The error associated with this approximation is proportional to the square of the interval size, h^2.

2.5.2 Numerical differentiation in two dimensions

Consider the computation of the first partial derivatives of a function, $f(x, y)$, of two independent variables, x and y, $\partial f/\partial x$ and $\partial f/\partial y$, from given values of the function at the nodes of a two-dimensional grid defined by the intersection of the x-level lines x_i, $i = 1, \ldots, N_x+1$, and y-level lines y_j, $j = 1, \ldots, N_y+1$, as illustrated in figure 1.7.2(*b*). The value of x lies in the k_xth x-interval confined between the x_{k_x} and x_{k_x+1} x-level lines, and the values of y lies in the k_yth y-interval confined between the y_{k_y} and y_{k_y+1} y-level lines.

First-order differentiation

Using the method of bilinear interpolation discussed in section 1.7, we approximate the first partial derivatives of the function $f(x, y)$ with the partial derivatives of the bilinear function defined in equation (1.7.11). Considering the derivative with respect to x, we obtain the forward-difference approximation

$$\left(\frac{\partial f}{\partial x}\right)_{(x,y)} \simeq \left(\frac{\partial P_{BL}^{k_x,k_y}}{\partial x}\right)_{(x,y)}$$

$$= \left(\frac{\partial w_{00}^{k_x,k_y}}{\partial x}\right)_{(x,y)} f(x_{k_x}, y_{k_y}) + \left(\frac{\partial w_{10}^{k_x,k_y}}{\partial x}\right)_{(x,y)} f(x_{k_x+1}, y_{k_y})$$

$$+\left(\frac{\partial w_{01}^{k_x,k_y}}{\partial x}\right)_{(x,y)} f(x_{k_x}, y_{k_y+1}) + \left(\frac{\partial w_{11}^{k_x,k_y}}{\partial x}\right)_{(x,y)} f(x_{k_x+1}, y_{k_y+1}). \tag{2.5.10}$$

Using expressions (1.7.12) and (1.7.13), we find

$$\left(\frac{\partial f}{\partial x}\right)_{(x,y)} = -\frac{y_{k_y+1} - y}{A} f(x_{k_x}, y_{k_y}) + \frac{y_{k_y+1} - y}{A} f(x_{k_x+1}, y_{k_y})$$

$$-\frac{y - y_{k_y}}{A} f(x_{k_x}, y_{k_y+1}) + \frac{y - y_{k_y}}{A} f(x_{k_x+1}, y_{k_y+1}). \tag{2.5.11}$$

Recalling that $A = (x_{k_x+1} - x_{k_x})(y_{k_y+1} - y_{k_y})$, as given in (1.7.14), we derive the first-order forward-difference approximation for the southwestern grid point,

$$\left(\frac{\partial f}{\partial x}\right)_{(x_{k_x}, y_{k_y})} \simeq \frac{f(x_{k_x+1}, y_{k_y}) - f(x_{k_x}, y_{k_y})}{x_{k_x+1} - x_{k_x}}. \tag{2.5.12}$$

A similar approximation for the y derivative yields

$$\left(\frac{\partial f}{\partial y}\right)_{(x_{k_x}, y_{k_y})} \simeq \frac{f(x_{k_x}, y_{k_y+1}) - f(x_{k_x}, y_{k_y})}{y_{k_y+1} - y_{k_y}}. \tag{2.5.13}$$

Both formulas express forward-difference approximations with respect to x or y.

Second-order differentiation

Second-order centered-difference formulas for evaluating the first partial derivative of a function at a grid point can be derived based on the one-dimensional formula (2.5.5). Using the expression for the coefficient $b^{(k)}$ given in equations (1.7.7), we obtain

$$\left(\frac{\partial f}{\partial x}\right)_{(x_{k_x}, y_{k_y})} \simeq \frac{(x_{k_x} - x_{k_x-1}) \frac{f_{k_x+1,k_y} - f_{k_x,k_y}}{x_{k_x+1} - x_{k_x}} + (x_{k_x+1} - x_{k_x}) \frac{f_{k_x,k_y} - f_{k_x-1,k_y}}{x_k - x_{k_x-1}}}{x_{k_x+1} - x_{k_x-1}}. \tag{2.5.14}$$

The corresponding expression for the derivative with respect to y is

$$\left(\frac{\partial f}{\partial y}\right)_{(x_{k_x}, y_{k_y})} \simeq \frac{(y_{k_y} - y_{k_y-1}) \frac{f_{k_x,k_y+1} - f_{k_x,k_y}}{y_{k_y+1} - y_{k_y}} + (y_{k_y+1} - y_{k_y}) \frac{f_{k_x,k_y} - f_{k_x,k_y-1}}{y_{k_y} - yk-1}}{y_{k_y+1} - y_{k_y-1}}. \tag{2.5.15}$$

When the grid lines are spaced evenly, $x_{k_x} - x_{k_x-1} = x_{k_x+1} - x_{k_x} = h_x$, and $y_{k_y} - y_{k_y-1} = y_{k_y+1} - y_{k_y} = h_y$, we obtain the simpler forms

$$\left(\frac{\partial f}{\partial x}\right)_{(x_{k_x}, y_{k_y})} = \frac{f_{k_x+1,k_y} - f_{k_x-1,k_y}}{2h_x} \tag{2.5.16}$$

and

$$\left(\frac{\partial f}{\partial y}\right)_{(x_{k_x}, y_{k_y})} = \frac{f_{k_x,k_y+1} - f_{k_x,k_y-1}}{2\, h_y}, \tag{2.5.17}$$

expressing centered-difference approximations in x or y.

2.5.3 Velocity gradient and related functions

The formulas derived previously in this section can be used to obtain approximations to the elements of the velocity gradient tensor, rate of deformation tensor, vorticity vector, and rate of expansion, from specified values of the velocity at grid points. To illustrate the methodology, we consider a two-dimensional flow and employ a uniform grid with constant x and y grid spacings equal to h_x and h_y. Using the second-order centered-difference approximations (2.5.16) and (2.5.2), we find that the rate of expansion can be approximated with the finite-difference formula

$$\left(\nabla \cdot \mathbf{u}\right)_{(x_{k_x}, y_{k_y})} = \frac{(u_x)_{k_x+1,k_y} - (u_x)_{k_x-1,k_y}}{2h_x} + \frac{(u_y)_{k_x,k_y+1} - (u_y)_{k_x,k_y-1}}{2h_y}. \tag{2.5.18}$$

The corresponding finite-difference approximation for the z component of the vorticity is

$$\omega_z(x_{k_x}, y_{k_y}) = \frac{(u_y)_{k_x+1,k_y} - (u_y)_{k_x-1,k_y}}{2h_x} - \frac{(u_x)_{k_x,k_y+1} - (u_x)_{k_x,k_y-1}}{2h_y}. \tag{2.5.19}$$

Similar finite-difference approximations can be derived for the elements of the rate of deformation tensor, and subsequently used to obtain approximations to its eigenvalues and eigenvectors.

The following MATLAB function *rec_2d_vgt* residing in directory *rec_2d* inside directory *02_grids* of FDLIB computes the velocity gradient tensor at the nodes of a two-dimensional Cartesian grid:

```
function [Axx,Axy,Ayx,Ayy] = rec_2d_vgt ...
...
    (glx,gly,Nx,Ny,gux,guy)

%-------------------------------
% compute the velocity gradient
% tensor A_ij at the grid points
%-------------------------------

%---------------
% interior points
% compute derivatives by central differences
%---------------

for i=2:Nx
  for j=2:Ny
    Axx(i,j) = (gux(i+1,j)-gux(i-1,j))/(glx(i+1)-glx(i-1));
    Axy(i,j) = (guy(i+1,j)-guy(i-1,j))/(glx(i+1)-glx(i-1));
    Ayx(i,j) = (gux(i,j+1)-gux(i,j-1))/(gly(j+1)-gly(j-1));
    Ayy(i,j) = (guy(i,j+1)-guy(i,j-1))/(gly(j+1)-gly(j-1));
  end
end

%---------
% left wall
% compute derivatives by central or forward differences
%---------

i=1;
for j=2:Ny
  Axx(i,j) = (gux(i+1,j)-gux(i,j))/(glx(i+1)-glx(i));
  Axy(i,j) = (guy(i+1,j)-guy(i,j))/(glx(i+1)-glx(i));
  Ayx(i,j) = (gux(i,j+1)-gux(i,j-1))/(gly(j+1)-gly(j-1));
  Ayy(i,j) = (guy(i,j+1)-guy(i,j-1))/(gly(j+1)-gly(j-1));
end

%---------
% bottom wall
% compute derivatives by central or forward differences
%---------

j=1;
for i=2:Nx
  Axx(i,j) = (gux(i+1,j)-gux(i-1,j))/(glx(i+1)-glx(i-1));
  Axy(i,j) = (guy(i+1,j)-guy(i-1,j))/(glx(i+1)-glx(i-1));
  Ayx(i,j) = (gux(i,j+1)-gux(i,j))/(gly(j+1)-gly(j));
  Ayy(i,j) = (guy(i,j+1)-guy(i,j))/(gly(j+1)-gly(j));
end
```

```
%----------
% right wall
% compute derivatives by central or backward differences
%----------

i=Nx+1;
for j=2:Ny
  Axx(i,j) = (gux(i,j)-gux(i-1,j))/(glx(i)-glx(i-1));
  Axy(i,j) = (guy(i,j)-guy(i-1,j))/(glx(i)-glx(i-1));
  Ayx(i,j) = (gux(i,j+1)-gux(i,j-1))/(gly(j+1)-gly(j-1));
  Ayy(i,j) = (guy(i,j+1)-guy(i,j-1))/(gly(j+1)-gly(j-1));
end

%----------
% top wall
% compute derivatives by central or backward differences
%----------

j=Ny+1;
for i=2:Nx
  Axx(i,j) = (gux(i+1,j)-gux(i-1,j))/(glx(i+1)-glx(i-1));
  Axy(i,j) = (guy(i+1,j)-guy(i-1,j))/(glx(i+1)-glx(i-1));
  Ayx(i,j) = (gux(i,j) -gux(i,j-1))/(gly(j) -gly(j-1));
  Ayy(i,j) = (guy(i,j) -guy(i,j-1))/(gly(j) -gly(j-1));
end

%-------------------
% four corner points
%-------------------

i=1;
j=1;
Axx(i,j) = (gux(i+1,j)-gux(i,j))/(glx(i+1)-glx(i));
Axy(i,j) = (guy(i+1,j)-guy(i,j))/(glx(i+1)-glx(i));
Ayx(i,j) = (gux(i,j+1)-gux(i,j))/(gly(j+1)-gly(j));
Ayy(i,j) = (guy(i,j+1)-guy(i,j))/(gly(j+1)-gly(j));

i=Nx+1;
j=1;
Axx(i,j) = (gux(i,j)-gux(i-1,j))/(glx(i)-glx(i-1));
Axy(i,j) = (guy(i,j)-guy(i-1,j))/(glx(i)-glx(i-1));
Ayx(i,j) = (gux(i,j+1)-gux(i,j))/(gly(j+1)-gly(j));
Ayy(i,j) = (guy(i,j+1)-guy(i,j))/(gly(j+1)-gly(j));

i=Nx+1;
j=Ny+1;
Axx(i,j) = (gux(i,j)-gux(i-1,j))/(glx(i)-glx(i-1));
Axy(i,j) = (guy(i,j)-guy(i-1,j))/(glx(i)-glx(i-1));
Ayx(i,j) = (gux(i,j)-gux(i,j-1))/(gly(j)-gly(j-1));
```

```
Ayy(i,j) = (guy(i,j)-guy(i,j-1))/(gly(j)-gly(j-1));

i=1;
j=Ny+1;
Axx(i,j) = (gux(i+1,j)-gux(i,j))/(glx(i+1)-glx(i));
Axy(i,j) = (guy(i+1,j)-guy(i,j))/(glx(i+1)-glx(i));
Ayx(i,j) = (gux(i,j)-gux(i,j-1))/(gly(j)-gly(j-1));
Ayy(i,j) = (guy(i,j)-guy(i,j-1))/(gly(j)-gly(j-1));

%-----
% done
%-----

return
```

The following MATLAB code appended to code *rec_2d* discussed in section 1.10 residing in directory *rec_2d* inside directory *02_grids* of FDLIB computes various flow variables:

```
%---
% specify the grid velocities
%---

for i=1:Nx+1
  for j=1:Ny+1
    px = gx(i,j); py = gy(i,j);
    wnx = 2*pi/(bx-ax); wny = 2*pi/(by-ay);
    gux(i,j) = cos(wnx*px)*cos(wny*py);
    guy(i,j) = sin(wnx*px)*sin(wny*py);
  end
end

%---
% velocity gradient tensor
%----

[Axx,Axy,Ayx,Ayy] = rec_2d_vgt (glx,gly,Nx,Ny,gux,guy);

%---
% compute the rate of expansion
%     the rate of strain tensor
%     the strains
%     the vorticity
%---

for i=1:Nx+1
  for j=1:Ny+1
    roe(i,j) = Axx(i,j)+Ayy(i,j);    % rate of expansion
    omega(i,j) = Axy(i,j)-Ayx(i,j);    % vorticity
```

```
    Exx(i,j) = Axx(i,j)-0.5*roe(i,j);   % rate of deformation
    Exy(i,j) = 0.5*(Axy(i,j)+Ayx(i,j));   % rate of deformation
    Eyx(i,j) = Exy(i,j);   % rate of deformation
    Eyy(i,j) = Ayy(i,j)-0.5*roe(i,j);   % rate of deformation
    det = 4.0*(Exx(i,j)^2+Exy(i,j)^2);   % eigenvalues
    srd = sqrt(det);
    strain1(i,j) = 0.5*srd; strain2(i,j) = -0.5*srd;

%---
% compute the eigenvectors of the rate of strain
%---

   if(abs(Exy(i,j))<0.0001)   % E is diagonal

   if(abs(Exx(i,j)-strain1(i,j))>0.0001)
     egv1x(i,j) = 0.0; egv1y(i,j) = 1.0;
   else
     egv1x(i,j) = 1.0; egv1y(i,j) = 0.0;
   end
   if(abs(Exx(i,j)-strain2(i,j))>0.0001)
     egv2x(i,j) = 0.0; egv2y(i,j) = 1.0;
   else
     egv2x(i,j) = 1.0; egv2y(i,j) = 0.0;
   end

   else   % E is not diagonal

   egv1x(i,j) = 1.0;
   egv1y(i,j) = -(Exx(i,j)-strain1(i,j))/Exy(i,j);
   egv2x(i,j) = 1.0;
   egv2y(i,j) = -(Exx(i,j)-strain2(i,j))/Exy(i,j);

   end

%---
% normalize the eigenvectors
%---

   fc1 = 1.0/sqrt(egv1x(i,j)^2+egv1y(i,j)^2);
   egv1_x(i,j) = fc1*egv1x(i,j); egv1_y(i,j) = fc1*egv1y(i,j);
   fc2 = 1.0/sqrt(egv2x(i,j)^2+egv2y(i,j)^2);
   egv2_x(i,j) = fc2*egv2x(i,j); egv2_y(i,j) = fc2*egv2y(i,j);

   end
end

%---
% plotting
%---
```

```
figure
mesh(glx,gly,omega')
xlabel('x','fontsize',15) ylabel('y','fontsize',15)
zlabel('\omega','fontsize',15)

figure
mesh(glx,gly,roe')
xlabel('x','fontsize',15); ylabel('y','fontsize',15)
zlabel('\alpha','fontsize',15)

figure
mesh(glx,gly,strain1')
xlabel('x','fontsize',15)
ylabel('y','fontsize',15)
zlabel('s\_1','fontsize',15)

figure
mesh(glx,gly,strain2')
xlabel('x','fontsize',15); ylabel('y','fontsize',15)
zlabel('s\_2','fontsize',15)

figure
hold on
[glx,gly,gx,gy] = grid_2d (ax,bx,ay,by,Nx,Ny);
for i=1:Nx+1
  for j=1:Ny+1
    vector = draw_arrow_2d ...
      (gx(i,j),gy(i,j),egv1x(i,j)/Ny,egv1y(i,j)/Ny);
    plot(vector(:,1),vector(:,2));
  end
end
xlabel('x','fontsize',15); ylabel('y','fontsize',15)
box

figure
hold on
[glx,gly,gx,gy] = grid_2d (ax,bx,ay,by,Nx,Ny);
for i=1:Nx+1
  for j=1:Ny+1
    vector = draw_arrow_2d ...
      (gx(i,j),gy(i,j),egv2x(i,j)/Ny,egv2y(i,j)/Ny);
    plot(vector(:,1),vector(:,2));
  end
end
xlabel('x','fontsize',15); ylabel('y','fontsize',15)
```

The graphics generated by the code for the coded velocity field is shown in figure 2.5.1.

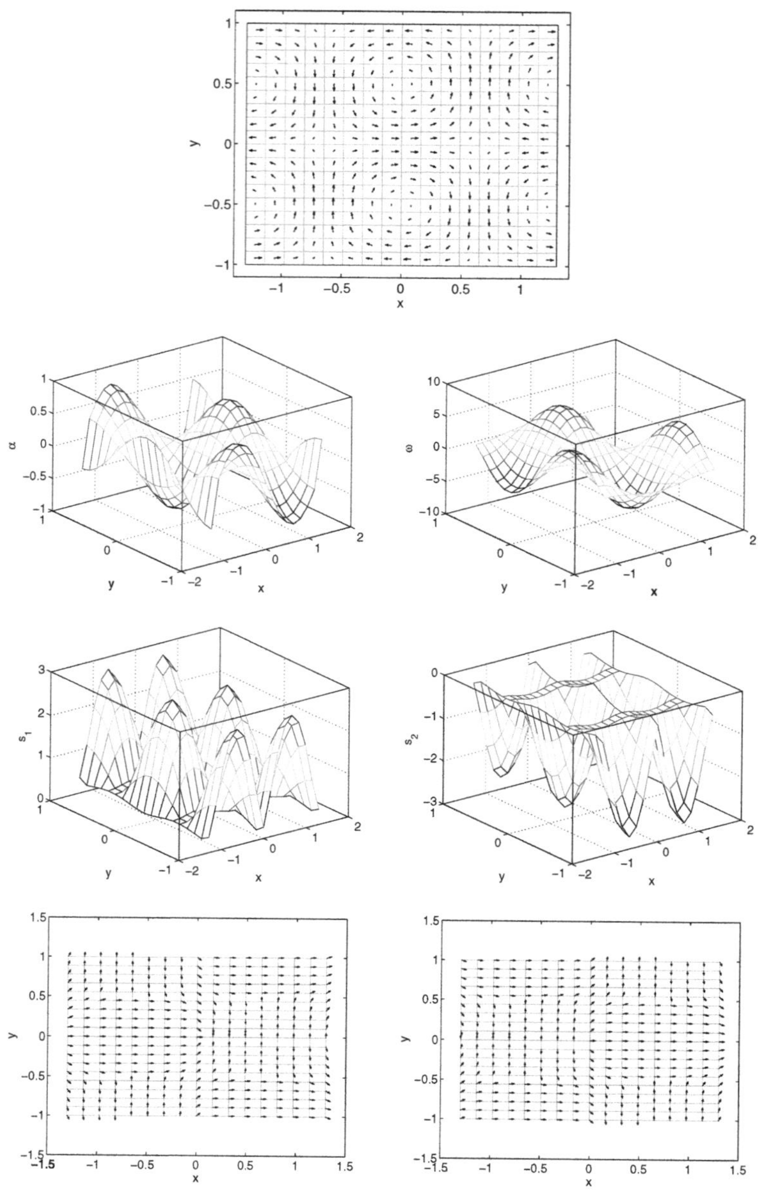

Figure 2.5.1 Velocity vector field, rate of expansion, vorticity, principal strains, and principal eigenvectors produced by numerical differentiation on a Cartesian grid.

Problem

2.5.1 *Numerical differentiation*

Use formula (2.5.9) to evaluate the second derivative of the exponential function $f(x) = e^x$ at $x = 0$ in terms of the values of $f(x)$ at $x = -h, 0, h$, for $h =$ 0.16, 0.08, 0.04, 0.02, and 0.01. Compute and plot the difference between the numerical value and the exact value against h on a linear-log scale. Assess and discuss the slope of the graph.

Computer Problem

2.5.2 *Numerical differentiation of a two-dimensional flow*

Run the code *rec_2d* for a velocity field of your choice. Prepare and discuss plots of the vorticity, eigenvalues, and eigenvectors of the rate of strain tensor.

2.6 Flow rate

Consider a two-dimensional flow in the xy plane, and draw a line that resides completely inside the fluid. At any instant, point particles cross the line generating a net, positive or negative, areal flow rate in a designated direction. Our goal is to quantify this flow rate in terms of the shape of the line and the fluid velocity.

Unit tangent and unit normal vectors

First, we consider an open line beginning at a point, A, and ending at another point, B, as shown in figure 2.6.1(a). As a preliminary, we introduce the unit tangent vector, $\mathbf{t} = (t_x, t_y)$, defined as the vector that is tangential to the line at a point, subject to the normalization condition $t_x^2 + t_y^2 = 1$. The direction of $\mathbf{t}$ is chosen such that, if we start moving along the line from point A in the direction of $\mathbf{t}$, we will finally end up at point B.

Next, we introduce the unit normal vector, denoted by $\mathbf{n} = (n_x, n_y)$, defined as the vector that is perpendicular to the line at every point, The magnitude of $\mathbf{n}$ is equal to unity, $n_x^2 + n_y^2 = 1$, and the orientation of $\mathbf{n}$ is such that the tangent vector $\mathbf{t}$ arises by rotating $\mathbf{n}$ around the z axis in the counterclockwise direction by an angle equal to $\pi/2$.

Now we consider an infinitesimal section of the line beginning at the point $\mathbf{x}$ and ending at the point $\mathbf{x} + \mathrm{d}\mathbf{x}$, where the differential displacement, $\mathrm{d}\mathbf{x} = (\mathrm{d}x, \mathrm{d}y)$ is parallel to, and points in the direction of the tangent vector, $\mathbf{t}$. The components of the unit tangent vector and unit normal vector are

$$t_x = \frac{\mathrm{d}x}{\mathrm{d}l}, \qquad t_y = \frac{\mathrm{d}y}{\mathrm{d}l}, \tag{2.6.1}$$

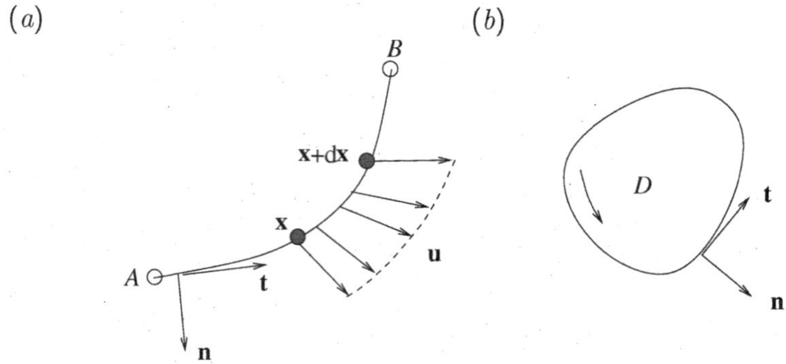

Figure 2.6.1 (*a*) An open line with end points A and B is used to define the local areal flow rate and flux in a two-dimensional flow. When the end points coincide, we obtain the closed line shown in (*b*).

and

$$n_x = \frac{dy}{dl}, \qquad n_y = -\frac{dx}{dl}, \tag{2.6.2}$$

where

$$dl = \sqrt{dx^2 + dy^2} = \sqrt{1 + \left(\frac{dy}{dx}\right)^2}\, |dx| \tag{2.6.3}$$

is the differential arc length of the infinitesimal section of the line.

Normal and tangential velocities

Next, we consider a group of adjacent point particles deployed along the infinitesimal arc length dl. During an infinitesimal period of time dt, the point particles move to a new position, thus allowing other point particles located behind or in front of them to cross the line into the other side. To compute the net area of fluid that has crossed the infinitesimal arc length dl, we resolve the velocity of the point particles into a normal and a tangential component, writing

$$\mathbf{u} = u_n\,\mathbf{n} + u_t\,\mathbf{t}. \tag{2.6.4}$$

The normal and tangential velocities, u_n and u_t, can be computed readily in terms of the inner vector product defined in equation (2.1.8).

First, we note that, because $\mathbf{t}$ and $\mathbf{n}$ are mutually orthogonal, their inner product is zero,

$$\mathbf{t} \cdot \mathbf{n} = \mathbf{n} \cdot \mathbf{t} = 0. \tag{2.6.5}$$

To confirm this, we simply substitute (2.6.1) and (2.6.2) in the right-hand side of (2.1.8). Second, we recall that, being unit vectors, the magnitudes of **n** and **t** are equal to unity,

$$\mathbf{n} \cdot \mathbf{n} = 1, \qquad \mathbf{t} \cdot \mathbf{t} = 1. \tag{2.6.6}$$

To confirm this, we substitute (2.6.1) or (2.6.2) in the right-hand side of (2.1.8), and use (2.6.3). Third, we take the inner product of the unit normal vector with both sides of (2.6.4), and use (2.6.5) and (2.6.6) to find

$$u_n = \mathbf{u} \cdot \mathbf{n} = u_x\, n_x + u_y\, n_y. \tag{2.6.7}$$

Finally, we take the inner product of the unit tangent vector with both sides of (2.6.4), and use (2.6.5) and (2.6.6) to find

$$u_t = \mathbf{u} \cdot \mathbf{t} = u_x\, t_x + u_y\, t_y. \tag{2.6.8}$$

2.6.1 Areal flow rate and flux

The local areal flow rate across the infinitesimal section, $\mathrm{d}l$, is defined as the area of fluid, $\mathrm{d}a$, that crosses this section during an infinitesimal period of time, $\mathrm{d}t$, given by

$$\frac{\mathrm{d}a}{\mathrm{d}t} = u_n\, \mathrm{d}l, \tag{2.6.9}$$

Why have we selected the normal component of the velocity on the right-hand side? The answer becomes evident by observing that, if the normal component vanishes, the particles move tangentially to the line and fluid does not cross the line.

The corresponding local areal flux, q, is defined as the ratio of the local areal flow rate, $\mathrm{d}a/\mathrm{d}t$, to the infinitesimal length of the line across which transport takes place,

$$q \equiv \frac{\mathrm{d}a}{\mathrm{d}t\, \mathrm{d}l} = u_n. \tag{2.6.10}$$

Thus, the local areal flux is equal to the normal component of the fluid velocity.

Substituting (2.6.7) in (2.6.9) and using (2.6.2), we find

$$\frac{\mathrm{d}a}{\mathrm{d}t} = u_n\, \mathrm{d}l = q\, \mathrm{d}l = (u_x\, n_x + u_y\, n_y)\, \mathrm{d}l = u_x\, \mathrm{d}y - u_y\, \mathrm{d}x, \tag{2.6.11}$$

which allows us to evaluate the local areal flow rate in terms of the components of the velocity.

2.6.2 Areal flow rate across a line

To compute the areal flow rate across the open line depicted in figure 2.6.1(a), denoted by Q, we subdivide the line into an infinite collection of infinitesimal sections with differential length dl, and add all contributions. In mathematical terms, we integrate the local areal flux along the line with respect to arc length, finding

$$Q = \int_A^B q\,\mathrm{d}l = \int_A^B \frac{\mathrm{d}a}{\mathrm{d}t\,\mathrm{d}l}\,\mathrm{d}l = \int_A^B \frac{\mathrm{d}a}{\mathrm{d}t} = \int_A^B (u_x n_x + u_y n_y)\,\mathrm{d}l$$
$$= \int_A^B u_n\,\mathrm{d}l = \int_A^B (u_x \mathrm{d}y - u_y \mathrm{d}x). \tag{2.6.12}$$

The expression on the right-hand side of (2.6.12) allows us to evaluate Q in terms of the geometry of the line and the two components of the velocity.

The integral representation is also applicable for the closed line depicted in figure 2.6.1(b). In that case, the last point B simply coincides with the first point A, yielding a closed integral,

$$Q = \oint (u_x \mathrm{d}y - u_y \mathrm{d}x). \tag{2.6.13}$$

Analytical integration

If a line has a simple shape and the components of the velocity are known functions of position with simple forms, the integrals in (2.6.12) can be computed by analytical methods. As an example, we consider a line that has the shape of a section of a circle of radius R centered at the point $\mathbf{x}_c$, with end points corresponding to the polar angles θ_A and θ_B. Points along the circular arc are described by

$$x = x_c + R\,\cos\theta, \qquad y = y_c + R\,\sin\theta. \tag{2.6.14}$$

Differentiating these equations with respect to θ, we find

$$\mathrm{d}x = -R\,\sin\theta\,\mathrm{d}\theta, \qquad \mathrm{d}y = R\,\cos\theta\,\mathrm{d}\theta. \tag{2.6.15}$$

Substituting these expressions in the last integral in (2.6.12), we obtain

$$Q = R\int_{\theta_A}^{\theta_B} (u_x \cos\theta + u_y \sin\theta)\,\mathrm{d}\theta. \tag{2.6.16}$$

Substituting stipulated expressions for the velocity components in terms of the angle θ, and obtain an integral representation in terms of θ. As an example, we assume

$$u_x = \frac{\alpha}{2\pi r}\cos\theta, \qquad u_y = \frac{\alpha}{2\pi r}\sin\theta, \tag{2.6.17}$$

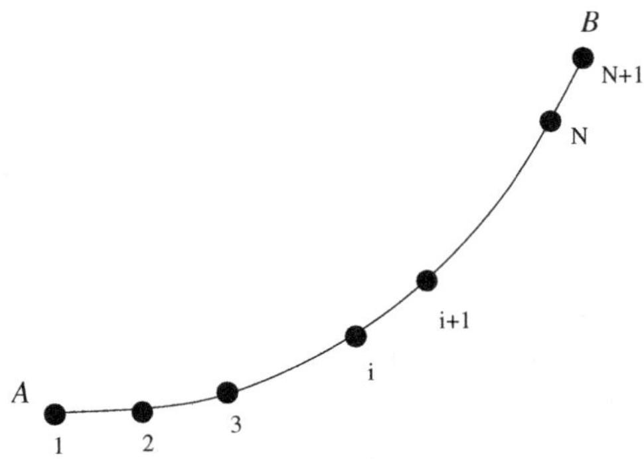

Figure 2.6.2 An array of points deployed along a section of a line in the xy plane is used to compute the areal flow rate by numerical methods.

where α is a constant, and r is the distance from the origin. The flow rate is given by

$$Q = R \frac{\alpha}{2\pi R} \int_{\theta_A}^{\theta_B} \mathrm{d}\theta = \frac{\alpha}{2\pi} (\theta_B - \theta_A). \tag{2.6.18}$$

If the circular segment forms a complete circle and the integration is performed in the counterclockwise direction from $\theta_A = \theta_0$ to $\theta_B = 2\pi - \theta_0$, where θ_0 is an arbitrary angle, then $Q = \alpha$ independent of the radius R.

2.6.3 Numerical integration

Under most conditions, we will not be able to compute the line integrals in (2.6.12) exactly by analytical methods and we must resort to numerical approximations. To perform numerical integration, we mark the location of the line with $N+1$ sequential nodes denoted by $\mathbf{x}_i$, where $i = 1, \ldots, N+1$, as depicted in figure 2.6.2. The first node coincides with the first point, A, and the last node coincides with the last point, B. If the line is closed, the first node numbered 1 coincides with the last node numbered $N+1$.

Next, we approximate the shape of the line between two successive nodes numbered i and $i+1$ with a straight segment passing through these nodes, denoted by E_i, where E stands for "element." The union of the N elements forms a polygonal line, called a polyline, beginning at the first point, A, and ending at the last point, B.

Trapezoidal rule

A key step in developing a numerical approximation is the replacement of the line integrals in (2.6.12) with the sum of integrals over the elements, and the approximation of the velocity components over each element with the average of the values at the element end points. With these simplifications, the last integral in (2.6.12) takes the form

$$Q = \sum_{i=1}^{N} \Big[\frac{u_x(\mathbf{x}_i) + u_x(\mathbf{x}_{i+1})}{2} (y_{i+1} - y_i) - \frac{u_y(\mathbf{x}_i) + u_y(\mathbf{x}_{i+1})}{2} (x_{i+1} - x_i) \Big]. \tag{2.6.19}$$

Writing out the sum and rearranging, we find

$$\begin{aligned} Q = {} & \frac{1}{2} \left[u_x(\mathbf{x}_1)(y_2 - y_1) - u_y(\mathbf{x}_1)(x_2 - x_1) \right] \\ & + \frac{1}{2} \sum_{i=2}^{N} [u_x(\mathbf{x}_i)(y_{i+1} - y_{i-1}) - u_y(\mathbf{x}_i)(x_{i+1} - x_{i-1})] \\ & + \frac{1}{2} \left[u_x(\mathbf{x}_{N+1})(y_{N+1} - y_N) - u_y(\mathbf{x}_{N+1})(x_{N+1} - x_N) \right]. \end{aligned} \tag{2.6.20}$$

If the line is closed, nodes 1 and $N + 1$ coincide, and the first and last contributions on the right-hand side of (2.6.20) combine to yield the simpler form

$$Q = \frac{1}{2} \sum_{i=1}^{N} [\, u_x(\mathbf{x}_i)(y_{i+1} - y_{i-1}) - u_y(\mathbf{x}_i)(x_{i+1} - x_{i-1}) \,], \tag{2.6.21}$$

where the wrapped point numbered 0 coincides with the penultimate point numbered N.

The computation of the right-hand sides of (2.6.20) and (2.6.21) requires knowledge of the velocity components at the nodes. In practice, these are either given explicitly or computed by interpolation from grid values, as discussed in section 1.7.

2.6.4 The Gauss divergence theorem in two dimensions

Consider a closed loop in the xy plane, C, and a vector function of position $\mathbf{h} = (h_x, h_y)$, where $h_x(x, y)$ and $h_y(x, y)$ are two scalar functions. The normal component of $\mathbf{h}$ along C is given by the inner product

$$h_n \equiv \mathbf{h} \cdot \mathbf{n} = h_x \, n_x + h_y \, n_y. \tag{2.6.22}$$

The divergence of $\mathbf{h}$ is a scalar function of position given by

$$\nabla \cdot \mathbf{h} = \frac{\partial h_x}{\partial x} + \frac{\partial h_y}{\partial y}. \tag{2.6.23}$$

The Gauss divergence theorem states that the line integral of the normal component, h_n, along C is equal to the integral of the divergence of $\mathbf{h}$ over the area D enclosed by C,

$$\oint_C \mathbf{h} \cdot \mathbf{n} \, \mathrm{d}l = \iint_D \nabla \cdot \mathbf{h} \, \mathrm{d}x \, \mathrm{d}y, \tag{2.6.24}$$

where $\mathbf{n}$ is the unit vector normal to C pointing outward.

Now consider the areal flow rate across a closed loop. Applying (2.6.24) with $\mathbf{h} = \mathbf{u}$, we find that the flow rate across this loop is equal to the areal integral of the divergence of the velocity over D,

$$Q = \oint_C \mathbf{u} \cdot \mathbf{n} \, \mathrm{d}l = \iint_D \nabla \cdot \mathbf{u} \, \mathrm{d}x \, \mathrm{d}y, \tag{2.6.25}$$

where the unit normal vector points outward, as shown in figure 2.6.1(*b*). The right-hand side of (2.6.25) allows us to compute the instantaneous flow rate across a closed loop in terms of the integral of the rate of expansion over the enclosed area.

Incompressible fluids

It is clear from expression (2.6.25) that, if the velocity field is solenoidal, that is, the divergence of the velocity vanishes at every point, $\nabla \cdot \mathbf{u} = 0$, then the areal flow rate across any closed loop is zero. In physical terms, fluid parcels deform and rotate but do not expand, and the amount of fluid entering the area enclosed by a stationary closed loop is equal to the amount of fluid exiting the loop during any period of time.

2.6.5 Flow rate in a three-dimensional flow

The preceding discussion for two-dimensional flow can be extended in a straightforward fashion to three-dimensional flow. To carry out this extension, we simply replace the line integrals along open or closed loops with surface integrals over open or closed surfaces residing inside the flow.

Working in this fashion, we find that the flow rate across an open or closed surface, H, is given by the surface integral

$$Q = \iint_H \mathbf{u} \cdot \mathbf{n} \, \mathrm{d}S. \tag{2.6.26}$$

The unit normal vector, $\mathbf{n}$, and the differential area of a surface element, $\mathrm{d}S$, are depicted in figure 2.6.3.

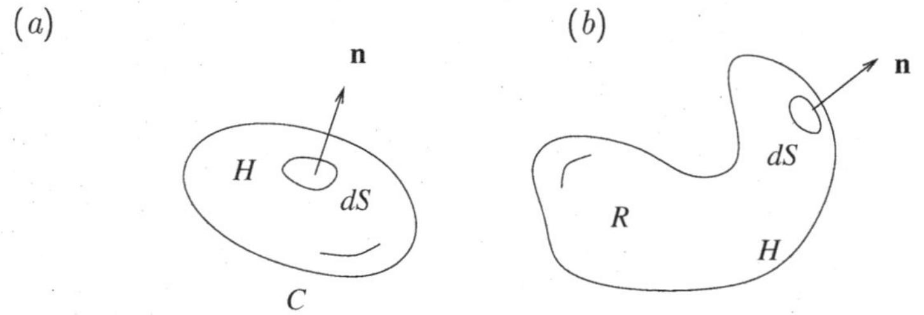

Figure 2.6.3 Illustration of (a) an open surface in a three-dimensional flow bounded by a closed line, C, and (b) a closed surface enclosing a volume, R, used to define the areal flow rate and establish the Gauss divergence theorem.

2.6.6 Gauss divergence theorem in three dimensions

Consider a closed surface, H, and a vector function of position $\mathbf{h} = (h_x, h_y, h_z)$. The normal component of $\mathbf{h}$ over H is given by the inner product

$$h_n \equiv \mathbf{h} \cdot \mathbf{n} = h_x n_x + h_y n_y + h_z n_z, \tag{2.6.27}$$

and the divergence of $\mathbf{h}$ is defined as

$$\nabla \cdot \mathbf{h} \equiv \frac{\partial h_x}{\partial x} + \frac{\partial h_y}{\partial y} + \frac{\partial h_z}{\partial z}. \tag{2.6.28}$$

The Gauss divergence theorem states that the surface integral of h_n over H is equal to the integral of the divergence of $\mathbf{h}$ over the volume R enclosed by H,

$$\iint_H \mathbf{h} \cdot \mathbf{n} \, \mathrm{d}S = \iiint_R \nabla \cdot \mathbf{h} \, \mathrm{d}V, \tag{2.6.29}$$

where $\mathbf{n}$ is the unit vector normal to H pointing outward, and $\mathrm{d}V = \mathrm{d}x\,\mathrm{d}y\,\mathrm{d}z$ is the differential volume expressed in Cartesian coordinates.

Now consider the flow rate across the closed surface H, given in (2.6.26). Applying (2.6.29) for $\mathbf{h} = \mathbf{u}$, we find

$$Q = \iiint_R \nabla \cdot \mathbf{u} \, \mathrm{d}V. \tag{2.6.30}$$

This expression shows that, if the velocity field is solenoidal, the volumetric flow rate across any closed surface enclosing fluid alone is required to vanish.

2.6.7 Axisymmetric flow

Next, we consider an axisymmetric flow and draw a line that begins at a point, A, and ends at another point, B, in a meridional plane, as illustrated in figure

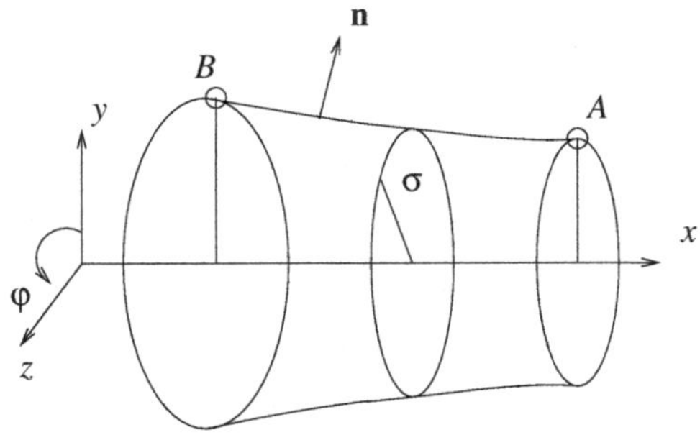

Figure 2.6.4 Illustration of an axisymmetric surface whose trace in a meridional plane is an open line beginning at a point A, and ending at another point B.

2.6.4. The flow rate across the axisymmetric surface that arises by rotating the line around the x axis is given by

$$Q = \int_A^B (u_x\, n_x + u_\sigma\, n_\sigma)\, 2\pi\sigma\, \mathrm{d}l = \int_A^B 2\pi\sigma\, (u_x\, \mathrm{d}\sigma - u_\sigma\, \mathrm{d}x), \qquad (2.6.31)$$

where $\mathrm{d}l$ is the differential arc length along the generating line. Expression (2.6.31) arises by adding the fluxes across all elementary axisymmetric surfaces confined between two parallel planes that are perpendicular to the x axis and are separated by the infinitesimal distance $\mathrm{d}x$ corresponding to the arc length $\mathrm{d}l$, taking into consideration that the surface area of an elementary surface centered at a ring of radius σ is equal to $2\pi\sigma \mathrm{d}l$.

Computer Problem

2.6.1 *Flow rate across an ellipse*

Consider a closed loop in the xy plane having the shape of a horizontal ellipse centered at the point $\mathbf{x}_c$, with major and minor semi-axes equal to a and b. The elliptical shape is described in parametric form by the equations $x = x_c + a \cos\chi$ and $y = y_c + b \sin\chi$, where χ is the native parameter of the ellipse ranging from 0 to 2π. The velocity components are given by $u_x = \alpha \cos\theta/(2\pi r)$ and $u_y = \alpha \sin\theta/(2\pi r)$, where α is a constant, and (r, θ) are plane polar coördinates in the xy plane with origin at the center of the ellipse.

Write a program that computes the flow rate across the ellipse using a numerical method based on equation (2.6.21). Perform computations for ellipses with aspect $a/b = 1$, 2, 4, and 8, in each case for $N = 8$, 16, 32, and 64. Discuss the results of your computations.

2.7 Mass conservation

In section 1.5, we defined a point particle as an idealized entity arising in the limit as the size of a small fluid parcel becomes smaller and eventually infinitesimal. In this limit, the ratio of the mass of the parcel to the volume of the parcel tends to a nonzero and non-infinite value defined as the fluid density, ρ. To indicate that the density is a function of position and time, we write $\rho(x, y, z, t)$ or $\rho(\mathbf{x}, t)$, with the understanding that the density at a particular point in a flow is equal to the density of the point particle that happens to be at that position at the current time.

2.7.1 Mass flux and mass flow rate

Consider a two-dimensional flow in the xy plane and draw a line beginning at a point A and ending at another point B, as illustrated in figure 2.6.1(a). At any instant, point particles cross this line generating a net mass flow rate in a specified direction. Our goal is to quantify this mass flow rate in terms of the shape of the line and the velocity and density distributions in the flow.

Repeating the analysis of section 2.6, we find that the mass flux across an infinitesimal section of a line is given by the following counterpart of equation (2.6.10),

$$q_m = \rho\, u_n, \tag{2.7.1}$$

where u_n is the component of the velocity normal to the line. The mass flow rate across a line beginning at a point, A, and ending at another point, B, is given by the following counterpart of equation (2.6.12),

$$Q_m = \int_A^B q_m \,\mathrm{d}l = \int_A^B \rho\,\mathbf{u}\cdot\mathbf{n}\,\mathrm{d}l = \int_A^B \rho\,(u_x \mathrm{d}y - u_y \mathrm{d}x). \tag{2.7.2}$$

The integrals in (2.7.2) can be computed by analytical or numerical methods, as discussed in section 2.6.

2.7.2 Mass flow rate across a closed line

The mass flow rate outward from a closed line can be expressed in terms of a closed line integral as

$$Q_m = \oint q_m \,\mathrm{d}l = \oint \rho\,\mathbf{u}\cdot\mathbf{n}\,\mathrm{d}l = \oint \rho\,(u_x \mathrm{d}y - u_y \mathrm{d}x), \tag{2.7.3}$$

where the unit normal vector $\mathbf{n}$ points into the exterior of the area enclosed by the closed line, as depicted in figure 2.6.1(b).

The Gauss divergence theorem expressed by equation (2.6.24) states that the line integral in (2.7.3) is equal to the integral of the divergence of the velocity multiplied by the fluid density over the area D enclosed by the line,

$$Q_m = \iint_D \nabla \cdot (\rho \mathbf{u}) \, dx \, dy, \tag{2.7.4}$$

where

$$\nabla \cdot (\rho \, \mathbf{u}) = \frac{\partial(\rho \, u_x)}{\partial x} + \frac{\partial(\rho \, u_y)}{\partial y}. \tag{2.7.5}$$

For future reference, we expand the derivatives of the products on the left-hand side of (2.7.5) using the rules of product differentiation, finding

$$\nabla \cdot (\rho \, \mathbf{u}) = \frac{\partial \rho}{\partial x} u_x + \frac{\partial u_x}{\partial x} \rho + \frac{\partial \rho}{\partial y} u_y + \frac{\partial u_y}{\partial y} \rho = \mathbf{u} \cdot \nabla \rho + \rho \, \nabla \cdot \mathbf{u}. \tag{2.7.6}$$

We have introduced the vector of the first partial derivatives of the density,

$$\nabla \rho = \left(\frac{\partial \rho}{\partial x}, \; \frac{\partial \rho}{\partial y}\right), \tag{2.7.7}$$

defined as the gradient of the density.

2.7.3 The continuity equation

The first principle of thermodynamics mandates that the rate of change of the mass residing inside an area, D, enclosed by a stationary closed line, is equal to the mass flow rate inward across the line or to the negative of the mass flow rate outward across the line. If the outward mass flow rate is positive, the rate of the change of mass enclosed by the line is negative, reflecting a reduction in time.

In terms of the mass flow rate defined in equation (2.7.3) and expressed as an areal integral in equation (2.7.4), mass conservation requires

$$\frac{d}{dt} \iint_D \rho \, dx \, dy = -Q_m = -\iint_D \nabla \cdot (\rho \, \mathbf{u}) \, dx \, dy. \tag{2.7.8}$$

Since the area D is fixed in space, we can interchange the order of time differentiation and space integration on the left-hand side of (2.7.8), and then combine the two integrals to find

$$\iint_D \left(\frac{\partial \rho}{\partial t} + \nabla \cdot (\rho \, \mathbf{u}) \right) dx \, dy = 0. \tag{2.7.9}$$

Since the shape of the area D is arbitrary, the integrand on the right-hand side of (2.7.9) must vanish, yielding a partial differential equation expressing mass conservation,

$$\frac{\partial \rho}{\partial t} + \nabla \cdot (\rho \, \mathbf{u}) = 0, \tag{2.7.10}$$

called the continuity equation. This nomenclature emphasizes that, in the absence of singularities in the form of sources and sinks, mass neither appears nor disappears in the flow, and the fluid must move in a continuous fashion.

Combining equations (2.7.6) and (2.7.10), we derive an alternative form of the continuity equation,

$$\frac{\partial \rho}{\partial t} + \mathbf{u} \cdot \nabla \rho + \rho \nabla \cdot \mathbf{u} = 0, \tag{2.7.11}$$

involving the density gradient and the rate of expansion.

2.7.4 Three-dimensional flow

Our discussion for two-dimensional flow can be generalized in a straightforward fashion to three-dimensional flow. To carry out this extension, we replace the line integrals with surface integrals over a closed or open surface. The mass flow rate across an open or closed surface R depicted in figure 2.6.3 is given by the surface integral

$$Q_m = \iint_R \rho \, \mathbf{u} \cdot \mathbf{n} \, \mathrm{d}S. \tag{2.7.12}$$

If the surface is closed and the unit normal vector points outward, as shown in figure 2.6.3(*b*), we may use the divergence theorem to recast the surface integral on the right-hand side of (2.7.12) into an integral of the rate of expansion over the volume R enclosed by the surface,

$$Q_m = \iiint_R \mathbf{u} \cdot (\rho \, \mathbf{u}) \, \mathrm{d}V. \tag{2.7.13}$$

The continuity equation expressed by (2.7.10) or (2.7.11) stands true, with the understanding that $\nabla \rho$ is the three-dimensional density gradient with components

$$\nabla \rho = (\frac{\partial \rho}{\partial x}, \; \frac{\partial \rho}{\partial y}, \; \frac{\partial \rho}{\partial z}). \tag{2.7.14}$$

2.7.5 Rigid-body translation

When a fluid translates as a rigid body, the fluid velocity $\mathbf{u}$ has the constant but possibly time-dependent value $\mathbf{U}(t)$. In this case, the continuity equation (2.7.10) simplifies to the linear convection equation

$$\frac{\partial \rho}{\partial t} + \mathbf{U} \cdot \nabla \rho = 0. \tag{2.7.15}$$

Consider a steady flow where $\mathbf{U}$ is independent of time. Using equation (2.7.15), we find that, if $\rho_0(\mathbf{x})$ is the density field at $t = 0$, then

$$\rho(\mathbf{x}, t) = \rho_0(\mathbf{x} - \mathbf{U}\, t) \tag{2.7.16}$$

will be the density field at any other time, t. Physically, the density at the point $\mathbf{x} = \mathbf{x}_0 - \mathbf{U}\, t$ at time t is equal to the density at the point $\mathbf{x}_0$ at $t = 0$. In physical terms, the density field is "convected" by the uniform flow.

To confirm (2.7.16), we define the auxiliary variables $w_x \equiv x - U_x t$, $w_y \equiv y - U_y t$, and $w_z \equiv z - U_z t$, and use the chain rule to write

$$\begin{aligned} \frac{\partial \rho_0}{\partial t} &= \frac{\partial \rho_0}{\partial w_x}\frac{\partial w_x}{\partial t} + \frac{\partial \rho_0}{\partial w_y}\frac{\partial w_y}{\partial t} + \frac{\partial \rho_0}{\partial w_z}\frac{\partial w_z}{\partial t} \\ &= \frac{\partial \rho_0}{\partial w_x}(-U_x) + \frac{\partial \rho_0}{\partial w_y}(-U_y) + \frac{\partial \rho_0}{\partial w_z}(-U_z). \end{aligned} \tag{2.7.17}$$

The proof follows by observing that $\partial\rho_0/\partial w_x = \partial\rho_0/\partial x$, $\partial\rho_0/\partial w_y = \partial\rho_0/\partial y$, and $\partial\rho_0/\partial w_z = \partial\rho_0/\partial z$.

2.7.6 Evolution equation for the density

The continuity equation can be regarded as an evolution equation for the density, determined by the velocity. To see this, we recast equation (2.7.10) into the form

$$\frac{\partial \rho}{\partial t} = -\nabla \cdot (\rho \mathbf{u}). \tag{2.7.18}$$

Evaluating the right-hand side of (2.7.18) at a certain point, x, in terms of the instantaneous velocity and density, we obtain an expression for the local and current rate of change of the density in time.

Let us consider the change in density occurring during a small time interval, Δt, following the present time t. Evaluating both sides of equation (2.7.18) at the point $\mathbf{x}$ and approximating the right-hand side with a first-order forward difference, we find

$$\frac{\rho(\mathbf{x}, t + \Delta t) - \rho(\mathbf{x}, t)}{\Delta t} = -\nabla \cdot (\rho \mathbf{u}), \tag{2.7.19}$$

where the right-hand side is evaluated at $(\mathbf{x}, t)$. Solving for $\rho(\mathbf{x}, t + \Delta t)$, we obtain

$$\rho(\mathbf{x}, t + \Delta t) = \rho(\mathbf{x}, t) - \Delta t\, \nabla \cdot (\rho \mathbf{u}), \tag{2.7.20}$$

which provides us with an explicit expression for $\rho(\mathbf{x}, t + \Delta t)$ in terms of values at the current time, t.

Finite-difference method

In practice, equation (2.7.18) is typically solved by numerical methods. Consider an idealized one-dimensional flow along the x axis representing, for example, flow along a conduit with a known axial velocity, $u(x,t)$. The one-dimensional version of the continuity equation (2.7.18) is

$$\frac{\partial \rho}{\partial t} = -\frac{\partial(\rho u)}{\partial x}. \tag{2.7.21}$$

The solution must be found inside the interval $a \leq x \leq b$, subject to an initial condition that specifies the density distribution at the designated origin of time, $\rho(x, t = 0)$, and a boundary condition that specifies the density at the left end of the domain of flow, $x = a$.

To develop the numerical method, we divide the domain into N intervals defined by $N+1$ nodes x_i, $i = 1, \ldots, N+1$, as depicted in figure 1.7.1(a). The first node coincides with the left end point, $x = a$, and the last node coincides with the right end point, $x = b$. Our goal is to generate the values of ρ at the nodes, at a sequence of time intervals separated by the time interval Δt. To simplify the notation, we denote the density at the ith node at the kth time level corresponding to time $t_k = k\,\Delta t$, by ρ_i^k. Evaluating both sides of (2.7.21) at the ith node at the kth time level, and approximating the time derivative on the left-hand side with a first-order forward difference and the spatial derivative on the right-hand side with a first-order backward difference, we derive the finite-difference approximation

$$\frac{\rho_i^{k+1} - \rho_i^k}{\Delta t} = -\frac{(\rho u)_i^k - (\rho u)_{i-1}^k}{x_i - x_{i-1}}. \tag{2.7.22}$$

Solving for ρ_i^{k+1}, we obtain

$$\rho_i^{k+1} = \rho_i^k - \frac{\Delta t}{x_i - x_{i-1}} \left[\, (\rho u)_i^k - (\rho u)_{i-1}^k \,\right]. \tag{2.7.23}$$

The numerical method involves the following steps:

1. Specify the initial values ρ_i^0 for $i = 1, \ldots, N+1$.
2. Use equation (2.7.23) to compute ρ_i^1 for $i = 2, \ldots, N+1$.
3. Use the left end boundary value to set the value of ρ_1^1.
4. Use equation (2.7.23) to compute ρ_i^2 for $i = 2, \ldots, N+1$.
5. Use the left end boundary value to set the value of ρ_1^2.
6. Stop, or continue for further steps.

Note that a boundary condition at the right end of the solution domain is not required. Numerical analysis shows that the success of this method depends on the size of the time step and sign of the convection velocity, u.

Problems

2.7.1 *Convection under constant velocity*

Consider the one-dimensional flow discussed in the text where the density field is governed by (2.7.21), with u being a constant. Sketch a profile of the density distribution along the x axis at the initial time $t = 0$ and at a subsequent time.

2.7.2 *Steady state*

Consider a steady one-dimensional flow with a specified velocity distribution, $u(x)$. Derive an expression for the density distribution at steady state based on (2.7.21). Discuss the behavior of the density when the velocity is zero at a point.

Computer Problem

2.7.3 *Finite-difference method*

Consider a steady, one-dimensional, periodic flow along the x axis with velocity distribution $u(x) = U\,(1 + \epsilon \cos(2\pi x/L))$, where U is a constant velocity, ϵ is a specified dimensionless constant, and L is the period.

Write a computer program that uses the numerical method discussed in the text to compute the evolution of the density over one spatial period L, subject to a uniform initial distribution. Run the program for $\epsilon = 0.0$, 0.2, 0.4, and 0.8, prepare graphs of the density distribution at different times, and discuss the behavior of the solution at long times.

2.8 Properties of point particles

The physical properties of a homogeneous fluid parcel consisting of a single chemical species are determined by the number of molecules, the kinetic energy, the potential energy, and the thermal energy of the molecules that comprise the parcel. Each one of these physical properties is extensive, in that the larger the parcel volume, the larger the magnitude of the physical property. As the size of a parcel tends to zero, the ratio between the value of an extensive property and the parcel volume tends to a limit that may be regarded as an intensive physical property of the point particle emerging from the parcel immediately before the molecular nature of the fluid becomes apparent.

For example, we have already seen that, as the volume of a parcel tends to zero, the ratio of the mass of the parcel to the volume of the parcel tends to a finite limit defined as the fluid density, ρ. Similarly, the ratio of the number of molecules residing within the parcel to the volume of the parcel tends to the molecular number density, and the ratio of the potential energy of the molecules to the volume of the parcel tends to the specific potential energy.

2.8.1 The material derivative

To prepare the ground for establishing evolution laws governing the motion and physical state of a fluid, we seek corresponding laws determining the rate of change of physical and kinematic properties of point particles moving with the local fluid velocity. Kinematic properties include the point particle velocity and its first time derivative defined as the point particle acceleration, the vorticity, and the rate of strain. A key concept is the material derivative, defined as the rate of change of a physical or kinematic property following a point particle. Our first objective is to derive an expression for the material derivative in terms of Eulerian derivatives; that is, partial derivatives with respect to time and spatial coordinates.

Consider the material derivative of the density of a point particle which, at a certain time $t_{,0}$, is located at the point $\mathbf{x}_0$. In three-dimensional flow, the density is a function of four independent variables, including the three Cartesian coordinates, (x, y, z), determining position in space, and time, t. We begin by linearizing the density field $\rho(x, y, z, t)$ around the hyperpoint (x_0, y_0, z_0, t_0), as discussed in section 2.1. Adding time dependence to equation (2.1.5), and identifying the generic function f with the density, we obtain the linearized form

$$\rho(x,y,z,t) \simeq \rho(x_0,y_0,z_0,t_0) + (t-t_0)\left(\frac{\partial\rho}{\partial t}\right)_{\mathbf{x}_0,t_0} \tag{2.8.1}$$

$$+(x-x_0)\left(\frac{\partial\rho}{\partial x}\right)_{\mathbf{x}_0,t_0} + (y-y_0)\left(\frac{\partial\rho}{\partial y}\right)_{\mathbf{x}_0,t_0} + (z-z_0)\left(\frac{\partial\rho}{dz}\right)_{\mathbf{x}_0,t_0}.$$

Next, we bring the first term on the right-hand side, $\rho(x_0, y_0, z_0, t_0)$, to the left-hand side and divide both sides of the resulting equation by the time lapse $t - t_0$, to find

$$\frac{\rho(x,y,z,t)-\rho(x_0,y_0,z_0,t_0)}{t-t_0} = \left(\frac{\partial\rho}{\partial t}\right)_{\mathbf{x}_0,t_0} + \frac{x-x_0}{t-t_0}\left(\frac{\partial\rho}{\partial x}\right)_{\mathbf{x}_0,t_0}$$

$$+\frac{y-y_0}{t-t_0}\left(\frac{\partial\rho}{\partial y}\right)_{\mathbf{x}_0,t_0} + \frac{z-z_0}{t-t_0}\left(\frac{\partial\rho}{\partial z}\right)_{\mathbf{x}_0,t_0}. \tag{2.8.2}$$

The second key step involves the judicious choice of the point $\mathbf{x}$. This point is selected so that, if a point particle is located at the position $\mathbf{x}_0$ at time t_0, then the same point particle is located at the position $\mathbf{x}$ at the later time t. By definition then, the left-hand side of (2.8.2) reduces to the material derivative. Since the point particle moves with the fluid velocity, the three fractions on the right-hand side of (2.8.2) reduce to the three components of the fluid velocity, u_x, u_y, and u_z. Denoting the material derivative by $\mathrm{D}/\mathrm{D}t$, we find

$$\frac{\mathrm{D}\rho}{\mathrm{D}t}(\mathbf{x}_0, t_0) = \left(\frac{\partial \rho}{\partial t} + u_x \frac{\partial \rho}{\partial x} + u_y \frac{\partial \rho}{\partial y} + u_z \frac{\partial \rho}{\partial z} \right)_{\mathbf{x}_0, t_0}. \tag{2.8.3}$$

In terms of the density gradient defined in (2.7.14), equation (2.8.3) takes the simpler form

$$\frac{\mathrm{D}\rho}{\mathrm{D}t} = \frac{\partial \rho}{\partial t} + \mathbf{u} \cdot \nabla \rho, \tag{2.8.4}$$

where both sides are evaluated at the arbitrary point, $\mathbf{x}_0$, at an arbitrary time instant, t_0.

Equation (2.8.3) allows us to compute the material derivative of the density in terms of Eulerian derivatives; that is, in terms of partial derivatives of the density with respect to time and spatial coordinates, x, y, and z. In numerical practice, the partial derivatives are computed using finite-difference approximations, as discussed in section 2.5.

2.8.2 *The continuity equation*

Comparing equations (2.8.4) and (2.7.11), we find that, in terms of the material derivative of the density, the continuity equation takes the form

$$\frac{\mathrm{D}\rho}{\mathrm{D}t} + \rho \nabla \cdot \mathbf{u} = 0, \tag{2.8.5}$$

which reveals that the rate of change of the density of a point particle is determined exclusively by the local rate of expansion. However, the inverse interpretation is physically more appropriate: the structure of the velocity field is partly determined by the rate of change of the density of all point particles.

Consider a small fluid parcel with volume δV_p density ρ, and mass $\delta m_p = \rho\, \delta V_p$. Mass conservation requires $\mathrm{D}\, \delta m_p/\mathrm{D}t = 0$. Expanding the material derivative (problem 2.8.1), we find

$$\frac{\mathrm{D}(\rho\, \delta V_p)}{\mathrm{D}t} = \delta V_p \frac{\mathrm{D}\rho}{\mathrm{D}t} + \rho \frac{\mathrm{D}\, \delta V_p}{\mathrm{D}t} = 0. \tag{2.8.6}$$

Using the continuity equation (2.8.5) and rearranging, we find

$$\frac{1}{\delta V_p} \frac{\mathrm{D}\, \delta V_p}{\mathrm{D}t} = \nabla \cdot \mathbf{u}, \tag{2.8.7}$$

which reinforces our interpretation of the divergence of the velocity as the rate of volumetric expansion.

2.8.3 Point particle acceleration

The acceleration of a point particle, $\mathbf{a}$, is the rate of change of the point particle velocity. Invoking the definition of the material derivative, we find that the x component of the acceleration is equal to the material derivative of the x component of the point particle velocity, which is equal to the local fluid velocity, $a_x = \mathrm{D}u_x/\mathrm{D}t$. Similar arguments reveal that $a_y = \mathrm{D}u_y/\mathrm{D}t$ and $a_z = \mathrm{D}u_z/\mathrm{D}t$.

Replacing ρ in equation (2.8.4) with u_x, u_y, or u_z, we find that the three Cartesian components of the point particle acceleration are given by

$$\begin{aligned} a_x &\equiv \frac{\mathrm{D}u_x}{\mathrm{D}t} = \frac{\partial u_x}{\partial t} + \mathbf{u}\cdot\nabla u_x = \frac{\partial u_x}{\partial t} + u_x\frac{\partial u_x}{\partial x} + u_y\frac{\partial u_x}{\partial y} + u_z\frac{\partial u_x}{\partial z}, \\ a_y &\equiv \frac{\mathrm{D}u_y}{\mathrm{D}t} = \frac{\partial u_y}{\partial t} + \mathbf{u}\cdot\nabla u_y = \frac{\partial u_y}{\partial t} + u_x\frac{\partial u_y}{\partial x} + u_y\frac{\partial u_y}{\partial y} + u_z\frac{\partial u_y}{\partial z}, \\ a_z &\equiv \frac{\mathrm{D}u_z}{\mathrm{D}t} = \frac{\partial u_z}{\partial t} + \mathbf{u}\cdot\nabla u_z = \frac{\partial u_z}{\partial t} + u_x\frac{\partial u_z}{\partial x} + u_y\frac{\partial u_z}{\partial y} + u_z\frac{\partial u_z}{\partial z}. \end{aligned} \tag{2.8.8}$$

The three scalar equations (2.8.8) can be conveniently collected into a compact vector form,

$$\frac{\mathrm{D}\mathbf{u}}{\mathrm{D}t} = \frac{\partial \mathbf{u}}{\partial t} + \mathbf{u}\cdot\mathbf{A} = \frac{\partial \mathbf{u}}{\partial t} + \mathbf{u}\cdot\nabla\mathbf{u}, \tag{2.8.9}$$

where $\mathbf{A} = \nabla\mathbf{u}$ is the velocity gradient tensor defined in equation (2.1.13), with components $A_{ij} = \partial u_j/\partial x_i$. In index notation, the jth component of (2.8.9) takes the form

$$\frac{\mathrm{D}u_j}{\mathrm{D}t} = \frac{\partial u_j}{\partial t} + u_i\,\frac{\partial u_j}{\partial x_i}, \tag{2.8.10}$$

where summation is implied over the repeated index i.

The linear momentum of a small fluid parcel is defined as the product of the mass of the parcel, $\delta m_p = \rho\,\delta V_p$ and the parcel velocity, $\mathbf{u}$. Requiring mass conservation, we find that the rate of change of the linear momentum can be expressed in terms of the point particle acceleration in the form

$$\frac{\mathrm{D}\,(\delta m_p \mathbf{u})}{\mathrm{D}t} = \frac{\mathrm{D}\mathbf{u}}{\mathrm{D}t}\,\rho\,\delta V_p = \frac{\mathrm{D}\mathbf{u}}{\mathrm{D}t}\,\delta m_p. \tag{2.8.11}$$

Thus, the mass of an infinitesimal parcel can be extracted from the material derivative.

Cylindrical polar coordinates

In the system of cylindrical polar coordinates defined in figure 1.3.2, the point particle acceleration is expressed in terms of its cylindrical polar components, a_x, a_σ, and a_φ, as

$$\mathbf{a} = a_x\,\mathbf{e}_x + a_\sigma\,\mathbf{e}_\sigma + a_\varphi\,\mathbf{e}_\varphi. \tag{2.8.12}$$

Using the transformation rules shown in (1.3.20), we find

$$a_\sigma = \cos\varphi\,a_y + \sin\varphi\,a_z, \qquad a_\varphi = -\sin\varphi\,a_y + \cos\varphi\,a_z. \tag{2.8.13}$$

Substituting the right-hand sides of the second and third relations in (2.8.8) in the right-hand sides of (2.8.13), and then using the chain rule of differentiation to convert derivatives with respect to x, y, and z to derivatives with respect to x, σ, and φ in the resulting equations as well as in the first equation in (2.8.8), we obtain

$$\begin{aligned}
a_x &= \frac{\partial u_x}{\partial t} + u_x\frac{\partial u_x}{\partial x} + u_\sigma\frac{\partial u_x}{\partial \sigma} + \frac{u_\varphi}{\sigma}\frac{\partial u_x}{\partial \varphi},\\
a_\sigma &= \frac{\partial u_\sigma}{\partial t} + u_x\frac{\partial u_\sigma}{\partial x} + u_\sigma\frac{\partial u_\sigma}{\partial \sigma} + \frac{u_\varphi}{\sigma}\frac{\partial u_\sigma}{\partial \varphi} - \frac{u_\varphi^2}{\sigma},\\
a_\varphi &= \frac{\partial u_\varphi}{\partial t} + u_x\frac{\partial u_\varphi}{\partial x} + u_\sigma\frac{\partial u_\varphi}{\partial \sigma} + \frac{u_\varphi}{\sigma}\frac{\partial u_\varphi}{\partial \varphi} + \frac{u_\sigma u_\varphi}{\sigma}.
\end{aligned} \tag{2.8.14}$$

Using the expression for the gradient of a function in cylindrical polar coordinates defined in equations (2.1.31) and (2.1.35), we recast expressions (2.8.14) into the more compact form

$$\begin{aligned}
a_x &= \frac{\partial u_x}{\partial t} + \mathbf{u}\cdot\nabla u_x = \frac{\mathrm{D}u_x}{\mathrm{D}t},\\
a_\sigma &= \frac{\partial u_\sigma}{\partial t} + \mathbf{u}\cdot\nabla u_\sigma - \frac{u_\varphi^2}{\sigma} = \frac{\mathrm{D}u_\sigma}{\mathrm{D}t} - \frac{u_\varphi^2}{\sigma},\\
a_\varphi &= \frac{\partial u_\varphi}{\partial t} + \mathbf{u}\cdot\nabla u_\varphi + \frac{u_\sigma u_\varphi}{\sigma} = \frac{\mathrm{D}u_\sigma}{\mathrm{D}t} + \frac{u_\varphi u_\varphi}{\sigma}.
\end{aligned} \tag{2.8.15}$$

These expressions illustrate that the polar cylindrical components of the acceleration are not simply equal to the material derivative of the corresponding polar components of the velocity.

Spherical polar coordinates

In the spherical polar coordinates depicted in figure 1.3.3, the point particle acceleration is expressed in terms of its spherical polar components, a_r, a_θ, and a_φ, as

$$\mathbf{a} = a_r\,\mathbf{e}_r + a_\theta\,\mathbf{e}_\theta + a_\varphi\,\mathbf{e}_\varphi. \tag{2.8.16}$$

Working as previously for the cylindrical polar coordinates, we find the somewhat more involved expressions

$$\begin{aligned}
a_r &= \frac{\partial u_r}{\partial t} + u_r \frac{\partial u_r}{\partial r} + \frac{u_\theta}{r}\frac{\partial u_r}{\partial \theta} + \frac{u_\varphi}{r\sin\theta}\frac{\partial u_r}{\partial \varphi} - \frac{u_\theta^2 + u_\varphi^2}{r} \\
&= \frac{\mathrm{D}u_r}{\mathrm{D}t} - \frac{u_\theta^2 + u_\varphi^2}{r}, \\
a_\theta &= \frac{\partial u_\theta}{\partial t} + u_r \frac{\partial u_\theta}{\partial r} + \frac{u_\theta}{r}\frac{\partial u_\theta}{\partial \theta} + \frac{u_\varphi}{r\sin\theta}\frac{\partial u_\theta}{\partial \varphi} + \frac{u_r u_\theta}{r} - \frac{u_\varphi^2 \cot\theta}{r} \\
&= \frac{\mathrm{D}u_\theta}{\mathrm{D}t} + \frac{u_r u_\theta}{r} - \frac{u_\varphi^2 \cot\theta}{r}, \\
a_\varphi &= \frac{\partial u_\varphi}{\partial t} + u_r \frac{\partial u_\varphi}{\partial r} + \frac{u_\theta}{r}\frac{\partial u_\varphi}{\partial \theta} + \frac{u_\varphi}{r\sin\theta}\frac{\partial u_\varphi}{\partial \varphi} + \frac{u_r u_\varphi}{r} + \frac{u_\theta u_\varphi}{r}\cot\theta \\
&= \frac{\mathrm{D}u_\theta}{\mathrm{D}t} + \frac{u_r u_\varphi}{r} + \frac{u_\theta u_\varphi}{r}\cot\theta.
\end{aligned} \tag{2.8.17}$$

Plane polar coordinates

In the system of plane polar coordinates depicted in figure 1.3.4, the point particle acceleration is expressed in terms of its plane polar components, a_r and a_θ, as

$$\mathbf{a} = a_r\,\mathbf{e}_r + a_\theta\,\mathbf{e}_\theta. \tag{2.8.18}$$

Working in the familiar way, we find

$$\begin{aligned}
a_r &= \frac{\partial u_r}{\partial t} + u_r \frac{\partial u_r}{\partial \theta} + \frac{u_\theta}{r}\frac{\partial u_r}{\partial \theta} - \frac{u_\theta^2}{r} = \frac{\mathrm{D}u_r}{\mathrm{D}t} - \frac{u_\theta^2}{r}, \\
a_\theta &= \frac{\partial u_\theta}{\partial t} + u_r \frac{\partial u_\theta}{\partial r} + \frac{u_\theta}{r}\frac{\partial u_\theta}{\partial \theta} + \frac{u_r u_\theta}{r} = \frac{\mathrm{D}u_\theta}{\mathrm{D}t} + \frac{u_r u_\theta}{r}.
\end{aligned} \tag{2.8.19}$$

Acceleration at a point with zero vorticity

If all components of the vorticity vector are zero at a point in a flow, the velocity gradient tensor is symmetric at that point. Consequently, selected partial derivatives of the velocity must be such that the three terms enclosed by the parentheses on the right-hand side of (2.3.6) are zero, yielding

$$\frac{\partial u_z}{\partial y} = \frac{\partial u_y}{\partial z}, \qquad \frac{\partial u_x}{\partial z} = \frac{\partial u_z}{\partial x}, \qquad \frac{\partial u_y}{\partial x} = \frac{\partial u_x}{\partial y}. \tag{2.8.20}$$

The sum of the last three terms on the right-hand side of the first equation in (2.8.8) may then be written as

$$u_x \frac{\partial u_x}{\partial x} + u_y \frac{\partial u_y}{\partial x} + u_z \frac{\partial u_z}{\partial x} = \frac{1}{2}\frac{\partial u_x^2}{\partial x} + \frac{1}{2}\frac{\partial u_y^2}{\partial x} + \frac{1}{2}\frac{\partial u_z^2}{\partial x} = \frac{1}{2}\frac{\partial (u_x^2 + u_y^2 + u_z^2)}{\partial x}. \qquad (2.8.21)$$

Working in a similar fashion with the y and z components and collecting the derived expressions into a vector form, we obtain

$$\mathbf{u} \cdot \nabla \mathbf{u} = \frac{1}{2} \nabla u^2 = (\frac{\partial u^2}{\partial x}, \quad \frac{\partial u^2}{\partial y}, \quad \frac{\partial u^2}{\partial z}), \qquad (2.8.22)$$

where $u^2 \equiv u_x^2 + u_y^2 + u_z^2$ is the square of the magnitude of the velocity, and ∇u^2 is its gradient. The point particle acceleration may thus be expressed in the alternative form

$$\mathbf{a} \equiv \frac{\mathrm{D}\mathbf{u}}{\mathrm{D}t} = \frac{\partial \mathbf{u}}{\partial t} + \frac{1}{2} \nabla u^2. \qquad (2.8.23)$$

The first term on the right-hand side of (2.8.23) is zero in a steady flow, and the point particle acceleration is equal to half the gradient of the square of the magnitude of the local velocity, which is a measure of the local kinetic energy of the fluid. We conclude that the acceleration is oriented in the direction of maximum change of kinetic energy indicated by the gradient.

In chapter 6, we will see that the simplified expression (2.8.23) serves as a point of departure for the theoretical analysis and numerical computation of irrotational flow.

Problems

2.8.1 *Properties of the material derivative*

Consider two scalar physical or kinematic fluid properties, such as the density or a component of the velocity, denoted, respectively, by f and g. Show that the following usual rule of product differentiation applies,

$$\frac{\mathrm{D}(fg)}{\mathrm{D}t} = g\,\frac{\mathrm{D}f}{\mathrm{D}t} + f\,\frac{\mathrm{D}g}{\mathrm{D}t}. \qquad (2.8.24)$$

2.8.2 *Point particle acceleration in rotational flow*

Show that the counterpart of equation (2.8.22) at a point where the vorticity is not necessarily zero, is

$$\mathbf{u} \cdot \nabla \mathbf{u} = \frac{1}{2} \nabla u^2 - \mathbf{u} \times \boldsymbol{\omega}. \qquad (2.8.25)$$

How does this expression simplify at a point where velocity and vorticity vectors are parallel?

2.8.3 *Point particle motion in one-dimensional flow*

Consider an idealized one-dimensional flow along the x axis with velocity $u(x,t)$ satisfying the inviscid Burgers equation

$$\frac{\partial u}{\partial t} + u\,\frac{\partial u}{\partial x} = 0. \tag{2.8.26}$$

Explain why point particles in this flow travel with a time-independent velocity that is equal to the velocity assigned to them at the initial instant; different point particles may travel with different velocities.

2.9 Incompressible fluids and stream functions

If the volume of a fluid parcel is preserved as the parcel is convected in a flow, the fluid residing inside the parcel is incompressible. In contrast, if the volume of the parcel changes in time, the fluid residing inside the parcel is compressible. Conservation of mass requires that the mass of the fluid is conserved irrespective of whether the fluid is compressible or incompressible.

Since both the mass and the volume of an incompressible fluid parcel are conserved during the motion, the density of the point particles comprising the parcel must remain constant in time. Using the definition of the material derivative, we derive the mathematical statement of incompressibility

$$\frac{\mathrm{D}\rho}{\mathrm{D}t} = 0. \tag{2.9.1}$$

It is important to bear in mind that the density of an incompressible fluid is not necessarily uniform throughout the domain of flow. Different point particles may have different densities, but the density of each individual point particle is conserved during the motion.

Using equation (2.9.1), we find that the continuity equation (2.8.5) for an incompressible fluid simplifies to

$$\nabla \cdot \mathbf{u} = 0, \tag{2.9.2}$$

stating that the velocity field is solenoidal, and the rate of expansion α defined in equation (2.2.6) is equal to zero: an incompressible fluid parcel undergoes translation, rotation, and isochoric deformation, but not expansion. "Isochoric" is composed from the Greek words $\iota\sigma o\varsigma$ which means equal, and the word $\chi\omega\rho o\varsigma$ which means volume or space.

It is important to keep in mind that the stipulation (2.9.1) is the defining property of an incompressible fluid, and the simplified form of the continuity equation (2.9.2) is a mere consequence of mass conservation.

2.9.1 Mathematical consequences of incompressibility

Equation (2.9.2) states that the x, y, and z components of the velocity of an incompressible fluid may not be prescribed arbitrarily, but must be such that the differential constraint imposed on them by the requirement that the velocity field be solenoidal is satisfied throughout the domain of flow at all times. In contrast, the three components of the velocity of a compressible fluid can be arbitrary; the density of the point particles will then be adjusted to ensure mass conservation, as dictated by the continuity equation.

A second important consequence of incompressibility is that, because the evolution of the density is governed by the kinematic constraint (2.9.1), an equation of state relating the pressure to the density and temperature is not required. The significance of this consequence will be discussed in chapters 4 and 8 in the context of hydrodynamics.

2.9.2 Stream function for two-dimensional flow

The continuity equation for two-dimensional flow in the xy plane stated in (2.9.2) takes the form

$$\frac{\partial u_x}{\partial x} + \frac{\partial u_y}{\partial y} = 0. \tag{2.9.3}$$

In computing the velocity field of an incompressible fluid using analytical or numerical methods, it is convenient to satisfy this constraint at the outset and concentrate on satisfying boundary conditions and other constraints that arise by balancing forces and torques, as will be discussed in later chapters.

To achieve this, we express the two components of the velocity in terms of a scalar function ψ, called the stream function, as

$$u_x = \frac{\partial \psi}{\partial y}, \qquad u_y = -\frac{\partial \psi}{\partial x}. \tag{2.9.4}$$

If the two velocity components derive from ψ by means of equations (2.9.4), then the satisfaction of the incompressibility constraint (2.9.3) is guaranteed. To see this, we substitute (2.9.4) in (2.9.3) and find

$$\frac{\partial^2 \psi}{\partial x\, \partial y} - \frac{\partial^2 \psi}{\partial y\, \partial x} = 0. \tag{2.9.5}$$

Since the order of partial differentiation with respect to the two independent spatial variables x and y is immaterial, the equality is satisfied.

For example, we consider a two-dimensional flow with velocity components $u_x = Gx$ and $u_y = -Gy$, where G is a constant. Substituting these expressions in (2.9.3), we confirm that the fluid is incompressible. The stream function corresponding to this flow is given by $\psi = Gxy + c$, where c is an unspecified constant.

Non-uniqueness of the stream function

The example discussed in the last paragraph illustrates that the stream function of a specified two-dimensional flow is not unique. Cursory inspection of equation (2.9.4) shows that an arbitrary constant may be added to a particular stream function to yield another perfectly acceptable stream function expressing the same flow. However, this ambiguity is neither essential nor alarming. In performing analytical or numerical computation, the arbitrary constant simply provides us with one degree of freedom that can be used to simplify numerical and algebraic manipulations.

Physical interpretation

Consider the areal flow rate Q across a line beginning at a point, A, and ending at another point, B, as illustrated in figure 2.6.1(a). Substituting expressions (2.9.4) in the right-hand side of equation (2.6.12), we find

$$Q = \int_A^B \left(\frac{\partial \psi}{\partial y}\,\mathrm{d}y + \frac{\partial \psi}{\partial x}\,\mathrm{d}x\right) = \int_A^B \mathrm{d}\psi = \psi_B - \psi_A, \tag{2.9.6}$$

where ψ_A and ψ_B are the values of the stream function at points A and B.

Equation (2.9.6) shows that the difference in the values of the stream function between two points is equal to the areal flow rate across a line that begins at the first point and ends at the second point. Because the fluid is incompressible, the flow rate is independent of the actual shape of the line, provided that the line begins and ends at the specified points.

Vorticity

The z component of the vorticity in two-dimensional flow in the xy plane was given in equation (2.3.15) in terms of derivatives of the velocity. Substituting expressions (2.9.4) in the right-hand side, we find

$$\omega_z = -\left(\frac{\partial^2 \psi}{\partial x^2} + \frac{\partial^2 \psi}{\partial y^2}\right) \equiv -\nabla^2 \psi, \tag{2.9.7}$$

where $\nabla^2 \equiv \partial^2/\partial x^2 + \partial^2/\partial y^2$ is the Laplacian operator, to be discussed further in section 3.2. Thus, the z component of the vorticity is equal to the negative of the Laplacian of the stream function.

If the stream function satisfies Laplace's equation, $\nabla^2 \psi = 0$, the velocity field is solenoidal and the flow is irrotational. A function that satisfies Laplace's equation is called harmonic.

Plane polar coordinates

Departing from equations (2.9.4) and (2.3.15) and using the rules of coordinate transformation, we derive the velocity components of a two-dimensional flow in plane polar coordinates, (r, θ),

$$u_r = \frac{1}{r}\frac{\partial \psi}{\partial \theta}, \qquad u_\theta = -\frac{\partial \psi}{\partial r}, \tag{2.9.8}$$

and the strength of the vorticity,

$$\omega_z = -\frac{1}{r}\frac{\partial}{\partial r}\left(r\frac{\partial \psi}{\partial r}\right) - \frac{1}{r^2}\frac{\partial^2 \psi}{\partial \theta^2} \equiv -\nabla^2\psi, \tag{2.9.9}$$

where

$$\nabla^2 \equiv \frac{\partial^2}{\partial x^2} + \frac{\partial^2}{\partial y^2} = \frac{1}{r}\frac{\partial}{\partial r}\left(r\frac{\partial}{\partial r}\right) + \frac{1}{r^2}\frac{\partial^2}{\partial \theta^2} \tag{2.9.10}$$

is the Laplacian operator in plane polar coordinates.

2.9.3 Stream function for axisymmetric flow

In the case of axisymmetric flow without swirling motion, we express all dependent and independent variables in (2.9.2) in cylindrical polar coordinates. Carrying out a fair amount of algebra using the chain rule, we find that the continuity equation takes the form of a constraint on the axial and radial components of the velocity,

$$\frac{\partial u_x}{\partial x} + \frac{1}{\sigma}\frac{\partial(\sigma u_\sigma)}{\partial \sigma} = 0. \tag{2.9.11}$$

To ensure the satisfaction of this equation, we express the axial and radial components of the velocity in terms of the axisymmetric stream function, ψ, also called the Stokes stream function, defined by the equations

$$u_x = \frac{1}{\sigma}\frac{\partial \psi}{\partial \sigma}, \qquad u_\sigma = -\frac{1}{\sigma}\frac{\partial \psi}{\partial x}. \tag{2.9.12}$$

Straightforward algebra shows that the velocity components given in (2.9.12) satisfy the continuity equation (2.9.11) for any choice of ψ.

As an example, we consider an axisymmetric flow with velocity components $u_x = Gx$ and $u_\sigma = -\frac{1}{2}G\sigma$, where G is a constant. Substituting these expressions in (2.9.11), we readily confirm that the left-hand side vanishes and the fluid is incompressible. The corresponding stream function is given by $\psi = \frac{1}{2}Gx\sigma^2 + c$, where c is an unspecified constant.

Physical interpretation

Working as previously for two-dimensional flow, we find that the volumetric flow rate across an axisymmetric surface whose trace in a meridional plane starts at a point, A, and ends at another point, B, as illustrated in figure 2.6.4, is $Q = \psi_B - \psi_A$ (problem 2.9.2). This result is consistent with the dimensions of the axisymmetric stream function, velocity multiplied by length squared, evident from equations (2.9.12). In contrast, the two-dimensional stream function has units of velocity multiplied by length.

Vorticity

The meridional component of the vorticity in an axisymmetric flow was given in equation (2.3.18) in terms of derivatives of the cylindrical polar components of the velocity. Using expressions (2.9.12), we find

$$\omega_\varphi = -\frac{1}{\sigma}\, E^2\psi = -\frac{1}{\sigma}\,\Big(\frac{\partial^2\psi}{\partial x^2} + \frac{\partial^2\psi}{\partial \sigma^2} - \frac{1}{\sigma}\frac{\partial\psi}{\partial\sigma}\Big), \tag{2.9.13}$$

where E^2 is a second-order linear differential operator defined as

$$E^2 \equiv \frac{\partial^2}{\partial x^2} + \frac{\partial^2}{\partial \sigma^2} - \frac{1}{\sigma}\frac{\partial}{\partial\sigma}. \tag{2.9.14}$$

If the stream function is such that the right-hand side of (2.9.13) is zero throughout the domain of flow, the flow is irrotational.

Spherical polar coordinates

Departing from equations (2.9.12) and (2.3.18) and using the rules of coordinate transformation, we derive the velocity components in spherical polar coordinates, (r, θ, φ),

$$u_r = \frac{1}{r^2\sin\theta}\,\frac{\partial\psi}{\partial\theta}, \qquad u_\theta = -\frac{1}{r\sin\theta}\,\frac{\partial\psi}{\partial r}. \tag{2.9.15}$$

The azimuthal component of the vorticity is given by

$$\omega_\varphi = -\frac{1}{r\sin\theta}\, E^2\psi, \tag{2.9.16}$$

where E^2 is the second-order differential operator defined in (2.9.14). In spherical polar coordinates,

$$E^2 \equiv \frac{\partial^2}{\partial r^2} + \frac{\sin\theta}{r^2}\frac{\partial}{\partial\theta}\Big(\frac{1}{\sin\theta}\frac{\partial}{\partial\theta}\Big) = \frac{\partial^2}{\partial r^2} + \frac{1}{r^2}\frac{\partial^2}{\partial\theta^2} - \frac{\cot\theta}{r^2}\frac{\partial}{\partial\theta}. \tag{2.9.17}$$

Problems

2.9.1 *Stream function for two-dimensional flow*

Derive the Cartesian components of the velocity and the strength of the vorticity corresponding to the stream functions (a) $\psi = \frac{1}{2}\, a\, y^2$, and (b) $\psi = \frac{1}{2}\, a\,(x^2 - y^2)$, where a is a constant. Discuss the nature of the flow

2.9.2 *Stream function of axisymmetric flow*

Substitute expressions (2.9.12) in the right-hand side of (2.6.31) and perform the integration to show that $Q = \psi_B - \psi_A$, as stated in the text.

2.10 Kinematic conditions at boundaries

In real life, a flow occurs in a domain that is bounded by stationary or moving surfaces with different constitutions and physical properties. Examples include the flow in an internal combustion engine generated by the motion of an engine piston, the flow induced by the motion of an aircraft or ground vehicle, the flow induced by the sedimentation of an aerosol particle in the atmosphere, the flow induced by a small bubble rising in a carbonated beverage, and the flow induced by the motion of an elephant running through the Savannah to escape a mouse.

In the context of flow kinematics, boundaries are classified into the following four main categories:

1. *Impermeable solid boundaries:* examples include the surface of a rigid or flexible solid body, such as a vibrating radio antenna or a swimming microorganism.

2. *Permeable solid boundaries:* examples include the surface of a porous medium, such as a rock bed or a biological tissue composed of cells with gaps in the intervening spaces.

3. *Sharp interfaces between immiscible fluids:* examples include the free surface of the ocean, and the interface between oil and vinegar in an Italian salad dressing.

4. *Diffuse interfaces between miscible fluids:* examples include the fuzzy edge of a river discharging into the ocean, and the ambiguous edge of a smoke ring rising in still air.

2.10.1 The no-penetration boundary condition

By definition, a point particle moving with the fluid velocity may not cross an impermeable solid boundary or a sharp interface between two immiscible fluids, but is required to lie on one side at all times. As a consequence, the velocity of a point particle that lies on a stationary or moving impermeable boundary or sharp interface must be consistent with, but not necessarily equal to, the velocity of the boundary or interface. To ensure compatibility, we require the no-penetration boundary condition.

Impermeable solid boundaries

Consider a flow that is bounded by an impermeable solid, but not necessarily rigid, boundary. The no-penetration condition requires that the component of the fluid velocity normal to the boundary is equal to the component of the boundary velocity normal to its instantaneous shape. The tangential component of the velocity can be arbitrary. If the boundary is stationary, the normal component of the fluid velocity must vanish.

To derive the mathematical statement of the no-penetration condition, we introduce the unit vector normal to the boundary at a point, $\mathbf{n}$, and the velocity of the boundary, $\mathbf{v}^B$, where the orientation of $\mathbf{n}$ is left unspecified. If the boundary is stationary, the boundary velocity is zero, $\mathbf{v}^B = \mathbf{0}$; if the boundary translates as a rigid body, $\mathbf{v}^B$ is constant; if the boundary rotates as a rigid body or exhibit some type of deformation, $\mathbf{v}^B$ is a function of position, as will be discussed later in this section. In all cases, the no-penetration boundary condition requires

$$\mathbf{u} \cdot \mathbf{n} = \mathbf{v}^B \cdot \mathbf{n}, \tag{2.10.1}$$

where both sides are evaluated at a point on the boundary.

Rigid-body motion

Consider an impermeable rigid boundary translating with velocity $\mathbf{U}^B$ and rotating with angular velocity $\boldsymbol{\Omega}^B$ around a specified center of rotation, $\mathbf{x}_R$. The angular velocity vector, $\boldsymbol{\Omega}^B$, passes through the center of rotation, $\mathbf{x}_R$, and its magnitude and orientation express the rate and direction of rotation. As we look down at the angular velocity vector from above, the body rotates in the counterclockwise direction.

In terms of the velocity of translation and angular velocity of rotation, the velocity at a point $\mathbf{x}$ that lies on the boundary is given by the expression

$$\mathbf{v}^B = \mathbf{U}^B + \boldsymbol{\Omega}^B \times (\mathbf{x} - \mathbf{x}_R), \tag{2.10.2}$$

where $\times$ denotes the outer vector product defined in equation (2.3.4). In component form,

$$\begin{aligned}\mathbf{v}^B = \mathbf{e}_x \,[\, U_x^B + \Omega_y^B\,(z - z_R) - \Omega_z^B\,(y - y_R)\,] \\ + \mathbf{e}_y \,[\, U_y^B + \Omega_z^B\,(x - x_R) - \Omega_x^B\,(z - z_R)\,] \\ + \mathbf{e}_z \,[\, U_z^B + \Omega_x^B\,(y - y_R) - \Omega_y^B\,(x - x_R)\,],\end{aligned} \tag{2.10.3}$$

where $\mathbf{e}_x$, $\mathbf{e}_y$, and $\mathbf{e}_z$ are unit vectors parallel to the x, y, or z axis. In the case of two-dimensional flow in the xy plane, the z component of the velocity is zero, $U_z^B = 0$, and the angular velocity vector is parallel to the z axis, $\Omega_x^B = 0$ and $\Omega_y^B = 0$, yielding the simplified form

$$\mathbf{v}^B = \mathbf{e}_x \left[U_x^B - \Omega_z^B (y - y_R) \right] + \mathbf{e}_y \left[U_y^B + \Omega_z^B (x - x_R) \right]. \qquad (2.10.4)$$

The no-penetration boundary condition arises by substituting expression (2.10.3) or (2.10.4) in the right-hand side of (2.10.1), respectively, for three-dimensional and two-dimensional flow. If the boundary is stationary, $\mathbf{v}^B = \mathbf{0}$, we obtain the simple form

$$\mathbf{u} \cdot \mathbf{n} = \mathbf{0}, \qquad (2.10.5)$$

where the direction of the unit normal vector, $\mathbf{n}$, is unspecified.

The no-penetration condition in terms of the stream function

Next, we consider an incompressible fluid in two-dimensional flow and express the velocity in terms of the stream function, ψ, defined in equations (2.9.4). The no-penetration condition (2.10.1) requires

$$\mathbf{u} \cdot \mathbf{n} = u_x \, n_x + u_y \, n_y = \frac{\partial \psi}{\partial y} \, n_x - \frac{\partial \psi}{\partial x} \, n_y = \mathbf{v}^B \cdot \mathbf{n}. \qquad (2.10.6)$$

Substituting expressions (2.6.2) for the components of the normal vector in terms of differential displacements along the boundary, we obtain

$$\frac{\partial \psi}{\partial y} \frac{\mathrm{d}y}{\mathrm{d}l} + \frac{\partial \psi}{\partial x} \frac{\mathrm{d}x}{\mathrm{d}l} = \frac{\mathrm{d}\psi}{\mathrm{d}l} = \mathbf{v}^B \cdot \mathbf{n}, \qquad (2.10.7)$$

where $\mathrm{d}l$ is the differential arc length measured along the boundary from an arbitrary origin.

If the boundary is stationary, the right-hand side of (2.10.7) is zero, $\mathrm{d}\psi/\mathrm{d}l = 0$, and the stream function is constant over the boundary. The no-penetration condition takes the form

$$\psi = \psi_0, \qquad (2.10.8)$$

where the constant ψ_0 is either assigned arbitrarily or computed as part of the solution.

Similar arguments reveal that the stream function is constant over an impermeable stationary boundary in axisymmetric flow (problem 2.10.2(*b*)).

Sharp interfaces

Next, we consider the no-penetration condition over a stationary or moving sharp interface separating two immiscible fluids. Physical arguments require that the normal component of the fluid velocity on one side of the interface must be equal to the normal component of the velocity on the other side of the interface. The tangential velocities may be different.

To derive the mathematical statement of the no-slip condition, we introduce the velocity on one side of the interface, denoted by $\mathbf{u}^{(1)}$, and the velocity on the other side of the interface, denoted by $\mathbf{u}^{(2)}$, and require

$$\mathbf{u}^{(1)} \cdot \mathbf{n} = \mathbf{u}^{(2)} \cdot \mathbf{n}. \tag{2.10.9}$$

Both sides are evaluated at a point on the interface, and the direction of the unit normal vector, $\mathbf{n}$, is unspecified.

Problems

2.10.1 *Changing the center of rotation*

The center of rotation of a rigid body can be placed at any arbitrary position. Suppose that we choose a point, $\mathbf{x}'_R$, instead of the point $\mathbf{x}_R$ discussed in the text. The counterpart of equation (2.10.2) is

$$\mathbf{v}^B = \mathbf{U}'^B + \mathbf{\Omega}'^B \times (\mathbf{x} - \mathbf{x}'_R). \tag{2.10.10}$$

Set the right-hand side of (2.10.10) equal to the right-hand side of (2.10.2) to derive expressions for $\mathbf{U}'^B$ and $\mathbf{\Omega}'^B$ in terms of $\mathbf{U}^B$ and $\mathbf{\Omega}^B$, and *vice versa.*

2.10.2 *Stream functions*

(*a*) Use the no-penetration boundary condition to derive an expression for the stream function over a translating but non-rotating two-dimensional impermeable boundary.

(*b*) Show that the no-penetration condition over a stationary boundary in axisymmetric flow takes the form expressed by (2.10.8).

3 Flow Computation based on Kinematics

Flows can be classified according to the vorticity distribution as irrotational flows if the vorticity vanishes throughout the domain of flow, vortex flows dominated by the presence of compact regions of concentrated vorticity embedded in an otherwise irrotational fluid, and rotational flows if the vorticity is significant throughout the domain of flow. In this chapter, we discuss the kinematic structure and mathematical description of the simplest class of irrotational flows. Following the mathematical analysis, we develop finite-difference methods for computing the velocity field from knowledge of the velocity distribution at the boundaries, and then derive a class of elementary irrotational flows that serve as fundamental building blocks for generating desired solutions. Complementary building blocks associated with elementary vortex flows will provide us with additional elementary units that allow us to address a broader class of irrotational flows where the fluid exhibits circulatory motion.

3.1 Flow classification based on kinematics

In chapters 1 and 2, we discussed the general kinematic features of a flow with reference to the motion of fluid parcels and infinitesimal point particles. To make further progress, we establish a taxonomy by classifying flows according to some sensible criteria. Examples of classifications include internal and external flows, inviscid and viscous flows, subsonic and supersonic flows. On the basis of kinematics alone, flows are classified into three main categories including irrotational flows, flows containing compact regions of intense vorticity

C. Pozrikidis, *Fluid Dynamics: Theory, Computation, and Numerical Simulation*,
DOI: 10.1007/978-0-387-95871-2_3,

embedded in an otherwise perfectly or nearly irrotational fluid, and rotational flows with distributed vorticity.

Irrotational flow

The first category includes flows where the vorticity vector vanishes, and the magnitude of the vorticity is zero throughout the domain of flow. According to our discussion in chapter 2, small spherical fluid parcels in a three-dimensional irrotational flow and discoidal fluid parcels in a two-dimensional irrotational flow translate and deform but do not rotate.

A perfectly irrotational flow is a mathematical idealization. In practice, a small amount of vorticity is always present, and a real-life nominal irrotational flow is a nearly but not perfectly irrotational flow. An example is high-speed flow past a slender airfoil under conditions of no-stall, as will be discussed in chapter 12 in the context of aerodynamics.

Vortex flows

The second category includes flows that contain well-defined compact regions where the magnitude of the vorticity is significant, embedded in an otherwise irrotational fluid. The vortical flow regions cannot be neglected without introducing serious discrepancies and compromising the physics of the flow under consideration. In practice, regions of intense vorticity appear in the form of narrow layers, thin filaments, wakes behind bluff bodies, tornadoes and swirls. A vortex flow familiar to the aircraft traveler is the flow associated with a high-speed jet emerging from a turbine engine.

Rotational flows

The third category includes flows where the vorticity is significant throughout the domain of flow. The distinction between vortex flows and rotational flows is somewhat vague, and some flows can be classified into both categories. However, we will see that vortex flows can be analyzed and computed using a special class of numerical methods, called vortex methods for flow simulation. The availability of these methods provides us with a useful criterion for the distinction between vortex and rotational flows.

Flows in nature and technology

Most flows in nature and technology are rotational. Examples include the flow due to a small particle settling in the atmosphere, the flow through a turbine engine, and blood flow in the heart and through large blood vessels and small capillaries. High-speed flows develop regions of concentrated vorticity and are typically classified as vortex flows. High-speed turbulent flows contain random collections of rapidly evolving vortices, called eddies or coherent structures, embedded in a low- or moderate-vorticity background fluid.

Irrotational flows are simplified models of vortex flows that emerge by neglecting the regions of concentrated vorticity, or else by shifting the actual boundaries of the flow to the edges of the vortex regions, thereby placing them outside the domain of flow.

Considerable physical insight and practical experience are necessary to accurately predict whether a flow will develop to become irrotational, rotational, or a vortex flow. Insights can be gained by studying model flows that are amenable to analytical and simple numerical methods, and by analyzing the laws governing the generation and evolution of the vorticity field from a given initial state.

Flow computation

The difficulty of computing the structure of a flow increases as we transition from irrotational flows, to vortex flows, to rotational flows. Exceptions to this general rule arise in special cases. Our discussion of analytical and computational methods for flow computation begins in this chapter by considering the most amenable class of irrotational flows. In the context of kinematics alone, the problem can be stated as follows: given the boundary geometry and the velocity distribution over the boundaries, compute the structure of a steady irrotational flow or the evolution of an unsteady irrotational flow from a specified initial state.

Problem

3.1.1 *Flow classification*

Suggest a possible way of classifying flows according to sensible criteria apart from the those discussed in the text.

3.2 Irrotational flow and the velocity potential

The vorticity of a three-dimensional flow was defined in equation (2.3.6) as the curl of the velocity. The z component of the vorticity in a two-dimensional flow in the xy plane was given in equation (2.3.15), and the meridional strength of the vorticity in an axisymmetric flow was given in equation (2.3.18). If a flow is irrotational, the structure of the velocity field must be such that the right-hand sides of these equations are zero.

3.2.1 Two-dimensional flow

Consider a two-dimensional irrotational flow in the xy plane. Setting the left-hand side of equation (2.3.15) to zero, we obtain a constraint on selected partial derivatives of the velocity,

$$\frac{\partial u_y}{\partial x} = \frac{\partial u_x}{\partial y}. \tag{3.2.1}$$

To compute a two-dimensional irrotational flow, we may attempt to obtain the two velocity components individually subject to constraints imposed by the continuity equation and boundary conditions, while ensuring that (3.2.1) is fulfilled at every point in the flow. Alternatively, we may choose to satisfy (3.2.1) at the outset and then concentrate on fulfilling the rest of the requirements. It should not be surprising that the second approach is much more expedient in both theoretical analysis and numerical computation.

The velocity potential

The key idea is to introduce a new scalar function, ϕ, called the velocity potential, such that the two velocity components of a two-dimensional flow derive from the relations

$$u_x = \frac{\partial \phi}{\partial x}, \qquad u_y = \frac{\partial \phi}{\partial y}. \tag{3.2.2}$$

In vector notation, equations (3.2.2) are collected into the compact form

$$\mathbf{u} = \nabla\phi, \tag{3.2.3}$$

where $\nabla\phi = (\partial\phi/\partial x, \partial\phi/\partial y)$ is the two-dimensional gradient of the potential. The velocity, and thus the velocity potential, ϕ, is a function of position, (x, y, z), and, in the case of unsteady flow, time, t. Inspection of (3.2.2) reveals that the velocity potential has dimensions of velocity multiplied by length, which amounts to length squared divided by time.

It is a straightforward exercise to confirm that, if the velocity components derive from ϕ in terms of equations (3.2.2), then the irrotationality constraint (3.2.1) is automatically satisfied. Substituting expressions (3.2.2) in (3.2.1), we find

$$\frac{\partial^2 \phi}{\partial x\,\partial y} = \frac{\partial^2 \phi}{\partial y\,\partial x}. \tag{3.2.4}$$

Since the order of partial differentiation with respect to two independent spatial variables, x and y, can be interchanged, relation (3.2.1) is satisfied. Accordingly, an irrotational flow is also a potential flow, and *vice versa.*

The velocity potential of a certain irrotational flow is not unique. An arbitrary constant can be added to a particular potential to produce another perfectly acceptable potential. However, this ambiguity is neither essential nor alarming. In performing analytical or numerical computation, the arbitrary constant is determined by introducing a proper constraint.

Deriving the potential

Given the velocity field of an irrotational flow, we can derive the corresponding potential by integrating the system of differential equations (3.2.2), where the left-hand sides are treated as a known. As an example, we consider unidirectional streaming flow with velocity components

$$u_x = V_x, \qquad u_y = V_y, \tag{3.2.5}$$

where V_x and V_y are two constant velocities. Integrating the first equation in (3.2.2), we find that the potential must take the form

$$\phi = x\,V_x + f(y), \tag{3.2.6}$$

where $f(y)$ is an unknown function of y. Differentiating both sides of this equation with respect to y and using the second equation in (3.2.2), we find $\mathrm{d}f/\mathrm{d}y = V_y$, which can be integrated to give $f(y) = y\,V_y + c$, where c is an arbitrary constant. Combining these expressions, we find that the velocity potential corresponding to (3.2.5) is

$$\phi = x\,V_x + y\,V_y + c. \tag{3.2.7}$$

In agreement with our previous observation, the velocity potential is defined uniquely only up to an arbitrary constant, c.

Computation of the potential based on kinematics

The automatic satisfaction of the irrotationality constraint (3.2.1) by way of the velocity potential is helpful, but we still require one equation, or a system of equations, that will allow us to compute the potential. Normally, these equations would have to be derived by considering forces and torques exerted on the surfaces and over the volume of small fluid parcels, as will be discussed in subsequent chapters with reference to the more general class of rotational flows. Fortunately, this is not necessary in the case of irrotational flows. Given the boundary distribution of the velocity, an irrotational flow can be computed on the basis of kinematics alone.

3.2.2 Incompressible fluids and the harmonic potential

If the fluid is incompressible, mass conservation requires that the velocity field is solenoidal, which means that the velocity components must satisfy the constraint expressed by the continuity equation (2.9.2). In the case of two-dimensional flow,

$$\frac{\partial u_x}{\partial x} + \frac{\partial u_y}{\partial y} = 0. \tag{3.2.8}$$

Substituting expressions (3.2.2) in (2.9.3) we obtain Laplace's equation in two dimensions,

$$\frac{\partial^2 \phi}{\partial x^2} + \frac{\partial^2 \phi}{\partial y^2} = 0. \tag{3.2.9}$$

It is convenient to define the two-dimensional Laplacian operator,

$$\nabla^2 \equiv \frac{\partial^2}{\partial x^2} + \frac{\partial^2}{\partial y^2}, \tag{3.2.10}$$

and recast (3.2.9) into the more compact form

$$\nabla^2 \phi = 0. \tag{3.2.11}$$

A function that satisfies Laplace's equation (3.2.11) is called harmonic.

It is instructive to derive Laplace's equation working in vector notation. Substituting (3.2.3) in (2.9.2), we find

$$\nabla \cdot \mathbf{u} = \nabla \cdot (\nabla \phi) \equiv \nabla^2 \phi = 0, \tag{3.2.12}$$

which identifies the Laplacian operator with the divergence of the gradient, $\nabla^2 = \nabla \cdot \nabla$.

Laplace's equation arises in a broad range of contexts under and beyond the auspices of fluid mechanics. For example, Laplace's equation governs the steady-state distribution of temperature in a conductive material, such as a fin or a cooling plate.

Quasi-steady state

Laplace's equation (3.2.11) articulates a statement of mass conservation for an incompressible fluid. Although time does not appear explicitly in this equation, the velocity field and thus the velocity potential may change in time. The absence of explicit time dependence classifies an irrotational flow of an incompressible fluid as a quasi-steady flow. This terminology implies that the instantaneous structure of the flow depends on the instantaneous boundary geometry and boundary conditions, and is independent of the motion at previous times. Thus, if all boundaries are stationary at a particular time, the fluid will also be stationary at that time, independent of the history of the fluid and boundary motion.

3.2.3 Three-dimensional flow

The foregoing analysis can be extended in a straightforward fashion to three-dimensional flow. The velocity components of a three-dimensional flow derive from the velocity potential by the equations

$$u_x = \frac{\partial \phi}{\partial x}, \qquad u_y = \frac{\partial \phi}{\partial y}, \qquad u_z = \frac{\partial \phi}{\partial z}. \tag{3.2.13}$$

The velocity components and thus the potential, ϕ, are functions of position (x, y, z) and time, t, in the case of unsteady flow.

If the fluid is incompressible, the velocity potential satisfies the counterpart of Laplace's equation (3.2.12) for three-dimensional flow,

$$\nabla \cdot bu = \nabla \cdot (\nabla \phi) \equiv \nabla^2 \phi \equiv \frac{\partial^2 \phi}{\partial x^2} + \frac{\partial^2 \phi}{\partial y^2} + \frac{\partial^2 \phi}{\partial z^2} = 0, \tag{3.2.14}$$

where

$$\nabla^2 \equiv \frac{\partial^2}{\partial x^2} + \frac{\partial^2}{\partial y^2} + \frac{\partial^2}{\partial z^2} \tag{3.2.15}$$

is the Laplacian operator in three dimensions.

3.2.4 Boundary conditions

Laplace's equation for a function, ϕ, in two or three dimensions is a second-order, elliptic partial differential equation. One implication of this classification is that, in order to compute the solution, we must specify one scalar boundary condition for ϕ, one of its first partial derivatives, or a combination thereof, along each boundary.

Impermeable boundaries

Over an impermeable boundary, we require the no-penetration condition discussed in section 2.10. If the boundary is stationary, $\mathbf{u} \cdot \mathbf{n} = 0$, where $\mathbf{n}$ is the unit vector normal to the boundary pointing either into or outward from the domain of flow. Using equations (3.2.2), we find that, in the case of two-dimensional flow,

$$n_x \frac{\partial \phi}{\partial x} + n_y \frac{\partial \phi}{\partial y} = 0. \tag{3.2.16}$$

This is truly a boundary condition for the normal component of the gradient of the potential, which is equal to the derivative with respect to distance normal to the boundary, called a Neumann boundary condition. Because the right-hand side of (3.2.16) is zero, this boundary condition is classified as homogeneous.

Permeable boundaries

Over a permeable boundary, we may specify the tangential component of the velocity and allow the normal component to arise as part of the solution. To implement this condition in the case of two-dimensional flow, we introduce the unit vector tangential to the boundary, $\mathbf{t}$, defined in equations (2.6.1). The tangential component of the velocity in the direction of $\mathbf{t}$ is given by the inner product

$$u_t \equiv \mathbf{u} \cdot \mathbf{t} = t_x \frac{\partial \phi}{\partial x} + t_y \frac{\partial \phi}{\partial y} = \frac{\mathrm{d}x}{\mathrm{d}l} \frac{\partial \phi}{\partial x} + \frac{\mathrm{d}y}{\mathrm{d}l} \frac{\partial \phi}{\partial y} = \frac{\mathrm{d}\phi}{\mathrm{d}l}, \tag{3.2.17}$$

where l is the arc length measured in the direction of $\mathbf{t}$. If the distribution of ϕ over the boundary is known, the right-hand side of (3.2.17) can be computed by differentiating the potential with respect to arc length by analytical or numerical methods. This observation suggests that, instead of specifying the tangential component of the velocity, we may specify the boundary distribution of the potential. A boundary condition for the distribution of the potential is a Dirichlet boundary condition.

A word of caution is here in order. If the flow is bounded by a number of disconnected boundaries, replacing the boundary condition for the tangential velocity with a boundary condition for the distribution of the potential is permissible only over one boundary; otherwise, inconsistencies may arise.

3.2.5 Cylindrical polar coordinates

Consider a three-dimensional irrotational flow, and introduce cylindrical polar coordinates, (x, σ, φ), as shown in figure 1.3.2. Using expressions (2.1.35), we find that the cylindrical polar components of the velocity are

$$u_x = \frac{\partial \phi}{\partial x}, \qquad u_\sigma = \frac{\partial \phi}{\partial \sigma}, \qquad u_\varphi = \frac{1}{\sigma}\frac{\partial \phi}{\partial \varphi}. \tag{3.2.18}$$

Laplace's equation for the harmonic potential reads

$$\nabla^2 \phi \equiv \frac{\partial^2 \phi}{\partial x^2} + \frac{1}{\sigma}\frac{\partial}{\partial \sigma}\Big(\sigma \frac{\partial \phi}{\partial \sigma}\Big) + \frac{1}{\sigma^2}\frac{\partial^2 \phi}{\partial \varphi^2} = 0. \tag{3.2.19}$$

If the flow is axisymmetric, the velocity potential is a function of x and σ but not φ, as required for u_φ to vanish.

3.2.6 Spherical polar coordinates

Now we consider a three-dimensional irrotational flow and introduce spherical polar coordinates, (r, θ, φ), as shown in figure 1.3.3. Using relations (2.1.37), we find that the spherical polar components of the velocity are

$$u_r = \frac{\partial \phi}{\partial r}, \qquad u_\theta = \frac{1}{r}\frac{\partial \phi}{\partial \theta}, \qquad u_\varphi = \frac{1}{r \sin\theta}\frac{\partial \phi}{\partial \varphi}. \tag{3.2.20}$$

Laplace's equation for the harmonic potential takes the form

$$\nabla^2 \phi \equiv \frac{1}{r^2}\frac{\partial}{\partial r}\Big(r^2 \frac{\partial \phi}{\partial r}\Big) + \frac{1}{r^2\,\sin\theta}\frac{\partial}{\partial \theta}\Big(\sin\theta\,\frac{\partial \phi}{\partial \theta}\Big) + \frac{1}{r^2\,\sin^2\theta}\frac{\partial^2 \phi}{\partial \varphi^2} = 0. \tag{3.2.21}$$

If the flow is axisymmetric, the velocity potential is a function of r and θ but not φ, as required for u_φ to vanish.

3.2.7 Plane polar coordinates

Finally, we consider a two-dimensional irrotational flow, and introduce plane polar coordinates, (r, θ), as shown in figure 1.3.4. Using relations (2.1.39), we find that the plane polar components of the velocity are

$$u_r = \frac{\partial \phi}{\partial r}, \qquad u_\theta = \frac{1}{r}\frac{\partial \phi}{\partial \theta}. \tag{3.2.22}$$

Laplace's equation for the harmonic potential takes the form

$$\nabla^2 \phi \equiv \frac{1}{r}\frac{\partial}{\partial r}\Big(r\frac{\partial \phi}{\partial r}\Big) + \frac{1}{r^2}\frac{\partial^2 \phi}{\partial \theta^2} = 0. \tag{3.2.23}$$

Note that this equation derives from (3.2.19) by replacing σ with r, φ with θ, and then discarding the x dependence.

Problems

3.2.1 *Deriving the velocity potential*

(a) Consider a two-dimensional flow with velocity components

$$u_x = V\,\cos(kx)\,e^{-ky}, \qquad u_y = -V\,\sin(kx)\,e^{-ky}, \tag{3.2.24}$$

where V and k are two constants. Confirm that this flow is irrotational, derive the corresponding velocity potential, investigate whether or not the potential is harmonic, and explain why. Sketch the streamline pattern and discuss the structure of the flow and the physical interpretation of the constant k.

(b) Consider a three-dimensional flow with velocity components

$$\begin{aligned} u_x &= V\,\frac{k_x}{k}\,\cos(k_x x)\,\sin(k_y y)\,e^{-kz}, \\ u_y &= V\,\frac{k_y}{k}\,\sin(k_x x)\,\cos(k_y y)\,e^{-kz}, \\ u_z &= -V\,\sin(k_x x)\,\sin(k_y y)\,e^{-kz}, \end{aligned} \tag{3.2.25}$$

where V, k_x, and k_y, are three constants, and $k = (k_x^2 + k_y^2)^{1/2}$. This is the three-dimensional counterpart of the two-dimensional flow discussed in (a). Confirm that this flow is irrotational, derive the corresponding velocity potential, investigate whether or not the potential is harmonic, and explain why. Discuss the structure of the flow and the physical interpretation of the constants k_x and k_y.

(c) Explain why it is not possible to find a velocity potential for simple shear flow along the x axis varying along the y axis whose velocity components are $u_x = ky$, $u_y = 0$, and $u_z = 0$, where k is a constant with dimensions of inverse time called the shear rate.

3.2.2 *Irrotational flow in cylindrical polar coordinates*

Verify by direct substitution that the potential

$$\phi = U x \,(1 + \frac{1}{2} \frac{a^3}{(x^2 + \sigma^2)^{3/2}}) \tag{3.2.26}$$

satisfies Laplace's equation (3.2.19), where U and a are two constants. Discuss the structure of the axisymmetric flow represented by this potential.

3.2.3 *Irrotational flow in spherical polar coordinates*

Verify by direct substitution that the potential

$$\phi = U\, r\, \cos\theta \,(1 + \frac{1}{2} \frac{a^3}{r^3}) \tag{3.2.27}$$

satisfies Laplace's equation (3.2.21), where U and a are two constants. Discuss the structure of the axisymmetric flow represented by this potential.

3.2.4 *Irrotational flow in plane polar coordinates*

Verify by direct substitution that the potential

$$\phi = U\, r\, \cos\theta \,(1 + \frac{a^2}{r^2}) + \frac{\kappa}{2\pi}\theta, \tag{3.2.28}$$

satisfies Laplace's equation (3.2.23), where κ, U, and a are three constants. Discuss the structure of the corresponding two-dimensional flow.

3.3 Finite-difference methods

In practice, Laplace's equation for the harmonic potential is solved using a variety of numerical methods. To illustrate the implementation of the finite-difference method, we consider a two-dimensional potential flow in the xy plane inside a rectangular domain confined between

$$a_x \le x \le b_x, \qquad a_y \le y \le b_y, \tag{3.3.1}$$

as illustrated in figure 3.3.1. The left, bottom, and right walls are impermeable, whereas the top wall is exposed to an external flow.

3.3.1 Boundary conditions

Before attempting to compute the solution, we must specify boundary conditions. Over the left wall, the unit vector normal to the wall pointing into the fluid is given by $\mathbf{n} = (1, 0)$. Accordingly, the no-penetration condition (3.2.16) provides us with the Neumann boundary condition

$$\frac{\partial \phi}{\partial x} = 0 \quad \text{at} \quad x = a_x. \tag{3.3.2}$$

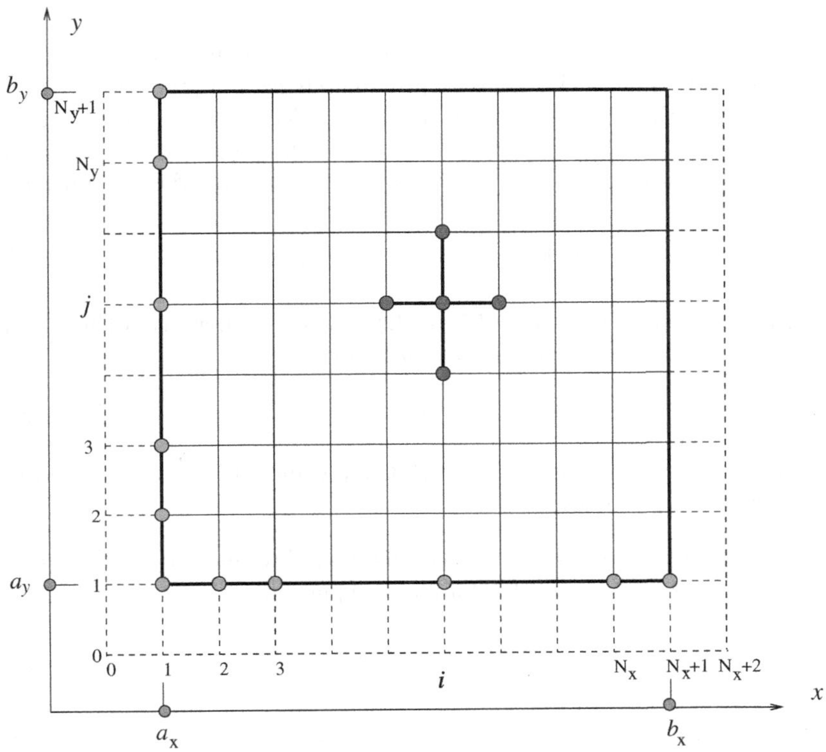

Figure 3.3.1 Illustration of a Cartesian grid used to compute the harmonic potential of an irrotational flow in a rectangular domain. The grid nodes are parametrized by two indices, i and j, where $i = 1, \ldots N_x + 1$ and $j = 1, \ldots N_y + 1$. Phantom grid lines are introduced at $i = 0$, $i = N_x + 2$, and $j = 0$, to implement the Neumann boundary conditions. The solution is found by solving Laplace's equation using a finite-difference method. The five-point stencil indicates the nodal pattern used to approximate the Laplacian at an interior node.

Over the bottom wall, the unit vector normal to the wall pointing into the flow is $\mathbf{n} = (0, 1)$. Accordingly, the no-penetration condition (3.2.16) provides us with the Neumann boundary condition

$$\frac{\partial \phi}{\partial y} = 0 \qquad \text{at} \qquad y = a_y. \tag{3.3.3}$$

Over the right wall, the unit vector normal to the wall pointing into the flow is $\mathbf{n} = (-1, 0)$. Accordingly, the no-penetration condition (3.2.16) provides us with the Neumann boundary condition

$$\frac{\partial \phi}{\partial x} = 0 \qquad \text{at} \qquad x = b_x. \tag{3.3.4}$$

Over the top wall, we stipulate that the tangential component of the velocity is constant and equal to V. This is just one choice adopted for the purpose of illustration; other choices can be made to reflect different flow conditions. Since the top wall is parallel to the x axis, the unit tangent vector is $\mathbf{t} = (1, 0)$. Accordingly, expression (3.2.17) provides us with the boundary condition

$$u_t \equiv \mathbf{u} \cdot \mathbf{t} = \frac{\partial \phi}{\partial x} = V \qquad \text{at} \qquad y = b_y. \tag{3.3.5}$$

Straightforward integration of (3.3.5) with respect to x shows that this condition is equivalent to a Dirichlet boundary condition,

$$\phi = V x + c \qquad \text{at} \qquad y = b_y. \tag{3.3.6}$$

The constant, c, can be assigned an arbitrary value without any consequences on the structure of the flow.

The problem formulation is now complete, and we may proceed to compute the solution. Our task is to solve Laplace's equation

$$\frac{\partial^2 \phi}{\partial x^2} + \frac{\partial^2 \phi}{\partial y^2} = 0, \tag{3.3.7}$$

subject to the four boundary conditions expressed by equations (3.3.2)–(3.3.4) and (3.3.6).

3.3.2 Finite-difference grid

We begin implementing the finite-difference method by dividing the x interval $[a_x, b_x]$ into N_x evenly spaced sub-intervals separated by the spacing $\Delta x = (b_x - a_x)/N_x$, and draw vertical grid lines at $x = x_i$, where

$$x_i = a_x + (i - 1)\,\Delta x, \tag{3.3.8}$$

for $i = 1, \ldots, N_x + 1$, as shown in figure 3.3.1. Similarly, we divide the y interval $[a_y, b_y]$ into N_y evenly spaced sub-intervals separated by the spacing $\Delta y = (b_y - a_y)/N_y$, and draw horizontal grid lines at $y = y_j$, where

$$y_j = a_y + (j - 1)\,\Delta y, \tag{3.3.9}$$

for $j = 1, \ldots, N_y + 1$, as shown in figure 3.3.1. The intersections of vertical and horizontal grid lines define grid points or nodes. For convenience, we denote the value of the harmonic potential ϕ at the (i, j) node as

$$\phi_{i,j} \equiv \phi(x_i, y_j). \tag{3.3.10}$$

The Dirichlet boundary condition (3.3.6) provides us with the values

$$\phi_{i,N_y+1} = V x_i. \tag{3.3.11}$$

Without loss of generality, we have made the arbitrary choice $c = 0$. Our objective is to compute the remaining unknown values, $\phi_{i,j}$, at the grid points $i = 1, \ldots, N_x + 1$ and $j = 1, \ldots, N_y$, comprising a set of $N_u = (N_x + 1)N_y$ unknowns.

3.3.3 Finite-difference discretization

To build a system of equations for the unknowns, we require the satisfaction of Laplace's equation (3.2.9) at the (i, j) node, and approximate the second partial derivatives with finite differences. Introducing the approximations implemented in formula (2.5.9), we write

$$\left(\frac{\partial^2 \phi}{\partial x^2}\right)_{i,j} \simeq \frac{\phi_{i+1,j} - 2\,\phi_{i,j} + \phi_{i-1,j}}{\Delta x^2}, \tag{3.3.12}$$

and

$$\left(\frac{\partial^2 \phi}{\partial y^2}\right)_{i,j} \simeq \frac{\phi_{i,j+1} - 2\,\phi_{i,j} + \phi_{i,j-1}}{\Delta y^2}. \tag{3.3.13}$$

These approximations transform the differential equation (3.3.7) into an algebraic equation,

$$\frac{\phi_{i+1,j} - 2\,\phi_{i,j} + \phi_{i-1,j}}{\Delta x^2} + \frac{\phi_{i,j+1} - 2\,\phi_{i,j} + \phi_{i,j-1}}{\Delta y^2} = 0. \tag{3.3.14}$$

Rearranging the left-hand side, we obtain

$$\phi_{i+1,j} - 2\,(1+\beta)\,\phi_{i,j} + \phi_{i-1,j} + \beta\,\phi_{i,j+1} + \beta\,\phi_{i,j-1} = 0, \tag{3.3.15}$$

where $\beta \equiv (\Delta x/\Delta y)^2$ is the square of the grid spacing ratio.

We can apply equation (3.3.14) or its equivalent version (3.3.15) at the interior nodes, $i = 2, \ldots, N_x$ and $j = 2, \ldots N_y$, and thus obtain a system of $N_{fde} = (N_x - 1)(N_y - 1)$ difference equations. However, we cannot apply the equation at a boundary node, as one grid point involved in the finite difference approximation will lie outside the flow domain. We must somehow generate

$$N_u - N_{fde} = (N_x + 1)N_y - (N_x - 1)(N_y - 1) = N_x + 2N_y - 1 \tag{3.3.16}$$

additional equations.

The missing equations originate from the Neumann boundary condition on the left, bottom, and right walls where the no-penetration condition is prescribed. One way to implement these boundary conditions with an error that is comparable to that of the finite-difference approximations (3.3.12) and (3.3.13), is to extend the domain of solution beyond the physical boundaries of the flow and introduce fictitious or phantom nodes located at

$$x = x_0 = a_x - \Delta x, \quad y = y_0 = a_y - \Delta y, \quad x = x_{N_x+2} = b_x + \Delta x. \tag{3.3.17}$$

Having introduced these extensions, we apply the second-order finite difference approximation (2.5.6) to recast the Neumann boundary condition into the discrete form

$$\frac{\phi_{2,j} - \phi_{0,j}}{2\,\Delta x} = 0 \tag{3.3.18}$$

for $j = 1, \ldots, N_y$, corresponding to the left wall,

$$\frac{\phi_{i,2} - \phi_{i,0}}{2\,\Delta y} = 0 \tag{3.3.19}$$

for $i = 1, \ldots, N_x + 1$, corresponding to the bottom wall, and

$$\frac{\phi_{N_x+2,j} - \phi_{N_x,j}}{2\,\Delta x} = 0 \tag{3.3.20}$$

for $j = 1, \ldots, N_y$, corresponding to the right wall.

Algebraic balance

To this end, we pause to count the number of unknowns and ensure that it matches the number of available equations. First, we note that the difference equation (3.3.14) or (3.3.15) may now be applied at the interior and boundary nodes for $i = 1, \ldots, N_x + 1$ and $j = 1, \ldots, N_y$, to yield $(N_x + 1)N_y$ equations. Adding to these equations the $N_x + 2N_y + 1$ boundary conditions (3.3.18)–(3.3.20), we obtain $(N_x + 1)(N_y + 1) + 2N_y$ additional equations. The total number of equations matches the number of unknowns, including the values of ϕ at the $(N_x + 1)N_y$ interior and boundary nodes and the values of ϕ at the $2N_y + N_x + 1$ phantom nodes.

3.3.4 Compilation of a linear system

To formalize the method, we collect the interior and boundary unknowns into a long vector $\mathbf{w}$ consisting of row-blocks, beginning from the bottom,

$$\begin{aligned} \mathbf{w} = [\; & \phi_{1,1}, \phi_{2,1}, \ldots \phi_{N_x+1,1}, \\ & \phi_{1,2}, \phi_{2,2}, \ldots \phi_{N_x+1,2}, \\ & \ldots, \\ & \phi_{1,N_y-1}, \phi_{2,N_y-1}, \ldots \phi_{N_x+1,N_y-1}, \\ & \phi_{1,N_y}, \phi_{2,N_y}, \ldots \phi_{N_x+1,N_y} \;]. \end{aligned} \tag{3.3.21}$$

Next, we apply the finite-difference equation (3.3.15) successively at boundary and interior nodes. Without loss of generality, we scan the grid points row-by-row starting from the bottom; a column-by-column compilation would also be acceptable.

Southwestern corner node

For the southwestern corner node $(1, 1)$, we obtain the finite-difference equation

$$\phi_{2,1} - 2\,(1+\beta)\,\phi_{1,1} + \phi_{0,1} + \beta\,\phi_{1,2} + \beta\,\phi_{1,0} = 0. \tag{3.3.22}$$

Boundary condition (3.3.18) for $j = 1$ requires $\phi_{2,1} = \phi_{0,1}$, and boundary condition (3.3.17) for $j = 1$ requires $\phi_{1,2} = \phi_{1,0}$. Using these equations to eliminate $\phi_{0,1}$ and $\phi_{1,0}$ in favor of $\phi_{2,1}$ and $\phi_{1,2}$ on the right-hand side of (3.3.22), we find

$$2\,\phi_{2,1} - 2\,(1+\beta)\,\phi_{1,1} + 2\,\beta\,\phi_{1,2} = 0. \tag{3.3.23}$$

For future reference, we express this equation in the form of the inner product of a vector $\mathbf{a}^{(1,1)}$ and the vector of unknowns $\mathbf{w}$ defined in (3.3.21), as

$$\mathbf{a}^{(1,1)} \cdot \mathbf{w} = 0, \tag{3.3.24}$$

where

$$\mathbf{a}^{(1,1)} = \Big[-2(1+\beta), 2, 0, \ldots, 0 \,\Big|\, 2\beta, 0, \ldots, 0 \,\Big|\, 0, \ldots, 0 \,\Big|\, \cdots \,\Big|\, 0, \ldots, 0 \,\Big] \tag{3.3.25}$$

is a sparse block vector. Each one of the N_y blocks on the right-hand side of (3.3.25) has $N_x + 1$ entries.

Southwestern bordering node

Next, we consider the boundary node $(2, 1)$ and obtain

$$\phi_{3,1} - 2\,(1+\beta)\,\phi_{2,1} + \phi_{1,1} + \beta\,\phi_{2,2} + \beta\,\phi_{2,0} = 0. \tag{3.3.26}$$

Boundary condition (3.3.19) applied for $i = 2$ requires $\phi_{2,2} = \phi_{2,0}$. Using this equation to eliminate $\phi_{2,0}$ in favor of $\phi_{2,2}$ on the right-hand side of (3.3.26), we find

$$\phi_{3,1} - 2\,(1+\beta)\,\phi_{2,1} + \phi_{1,1} + 2\,\beta\,\phi_{2,2} = 0. \tag{3.3.27}$$

For future reference, we express this equation in the form of an inner vector product,

$$\mathbf{a}^{(2,1)} \cdot \mathbf{w} = 0. \tag{3.3.28}$$

where

$$\mathbf{a}^{(2,1)} = \Big[\, 1, -2(1+\beta), 1, 0, \ldots, 0 \,\Big|\, 0, 2\beta, 0, \ldots, 0 \,\Big|\, 0, \ldots, 0 \,\Big|\, \ldots \,\Big|\, 0, \ldots, 0 \,\Big] \tag{3.3.29}$$

is a sparse block vector. Each one of the N_y blocks on the right-hand side of (3.3.29) has $N_x + 1$ entries.

Other nodes

Continuing in this fashion, we build the rest of the vectors $\mathbf{a}^{(i,j)}$ for $i = 1, \ldots N_x + 1$ and $j = 1, \ldots N_y - 1$, until we have reached the penultimate row corresponding to $j = N_y$. In simplifying the finite-difference equations for this row, we take into consideration not only the Neumann boundary conditions (3.3.18) and (3.3.20) for the side walls, but also the Dirichlet condition (3.3.11) for the top wall. For example, considering the northwestern node $(1, N_y)$, we obtain the difference equation

$$-2\,(1+\beta)\,\phi_{1,N_y} + 2\,\phi_{2,N_y} + \beta\,\phi_{1,N_y-1} = -\beta\, V\, x_1, \tag{3.3.30}$$

which can be expressed in the form of the inner product

$$\mathbf{a}^{(1,N_y)} \cdot \mathbf{w} = -\beta\, V\, x_1, \tag{3.3.31}$$

where

$$\mathbf{a}^{(1,N_y)} = \Big[0, \ldots, 0 \,\Big|\, \ldots \,\Big|\, 0, \ldots, 0 \,\Big|\, \beta, 0, \ldots, 0 \,\Big|\, -2(1+\beta), 2, 0, \ldots, 0 \,\Big] \tag{3.3.32}$$

is a sparse block vector. Each one of the N_y blocks on the right-hand side of (3.3.32) has $N_x + 1$ entries.

Assembly

Finally, we collect equations (3.3.24), (3.3.28), (3.3.31) and their counterparts for the rest of the interior and boundary nodes into a large system of equations,

$$\mathbf{A} \cdot \mathbf{w} = \mathbf{b}. \tag{3.3.33}$$

The first row of the matrix $\mathbf{A}$ is the vector $\mathbf{a}^{(1,1)}$ defined in (3.3.25); the second row is the vector $\mathbf{a}^{(2,1)}$ defined in (3.3.29); subsequent rows have similar identities. The block vector $\mathbf{b}$ on the right-hand side of (3.3.33) is given by

$$\mathbf{b} = \Big[0, \ldots, 0 \,\Big|\, \ldots \,\Big|\, 0, \ldots, 0 \,\Big|\, -\beta V x_1, -\beta V x_2, \ldots, -\beta V x_{N_x+1} \,\Big]. \tag{3.3.34}$$

The matrix $\mathbf{A}$ is composed of N_y vertical and N_y horizontal partitions in the block tridiagonal form

$$\mathbf{A} = \begin{bmatrix} \mathbf{T} & 2\,\mathbf{D} & \mathbf{0} & \cdots & \cdots & \cdots & \mathbf{0} \\ \mathbf{D} & \mathbf{T} & \mathbf{D} & \mathbf{0} & \cdots & \cdots & \mathbf{0} \\ \mathbf{0} & \mathbf{D} & \mathbf{T} & \mathbf{D} & \mathbf{0} & \cdots & \mathbf{0} \\ \cdots & \cdots & \cdots & \cdots & \cdots & \cdots & \cdots \\ \mathbf{0} & \mathbf{0} & \cdots & \mathbf{D} & \mathbf{T} & \mathbf{D} & \mathbf{0} \\ \mathbf{0} & \mathbf{0} & \cdots & \mathbf{0} & \mathbf{D} & \mathbf{T} & \mathbf{D} \\ \mathbf{0} & \mathbf{0} & \mathbf{0} & \cdots & \mathbf{0} & \mathbf{D} & \mathbf{T} \end{bmatrix}. \qquad (3.3.35)$$

The factor of two in front of the $\mathbf{D}$ block in the first row is due to the Neumann boundary condition. We have introduced the $(N_x + 1) \times (N_x + 1)$ tridiagonal matrix

$$\mathbf{T} = \begin{bmatrix} -2(1+\beta) & 2 & 0 & 0 & \cdots & 0 & 0 \\ 1 & -2(1+\beta) & 1 & 0 & \cdots & 0 & 0 \\ 0 & 1 & -2(1+\beta) & 1 & \cdots & 0 & 0 \\ \cdots & \cdots & \cdots & \cdots & \cdots & \cdots & \cdots \\ 0 & 0 & \cdots & \cdots & 1 & -2(1+\beta) & 1 \\ 0 & 0 & \cdots & \cdots & 0 & 2 & -2(1+\beta) \end{bmatrix}, \qquad (3.3.36)$$

and the $(N_x + 1) \times (N_x + 1)$ diagonal matrix

$$\mathbf{D} = \begin{bmatrix} \beta & 0 & 0 & 0 & \cdots & 0 & 0 \\ 0 & \beta & 0 & 0 & \cdots & 0 & 0 \\ \cdots & \cdots & \cdots & \cdots & \cdots & \cdots & \cdots \\ 0 & 0 & \cdots & \cdots & 0 & \beta & 0 \\ 0 & 0 & \cdots & \cdots & 0 & 0 & \beta \end{bmatrix}. \qquad (3.3.37)$$

Note that the super- and sub-diagonal elements of $\mathbf{T}$ are equal to unity, except for the elements in the first the last rows that are equal to two. The origin of these irregular elements can be traced back to the Neumann boundary condition. Cursory inspection reveals that all elements of the matrix $\mathbf{A}$ are zero, except for the elements along five diagonal lines. Because of the dominant presence of zeros, the matrix $\mathbf{A}$ is sparse.

The following MATLAB function *cvt_2d_fdm* located in directory *cvt_2d* inside directory *07_ptf* of FDLIB generates the coefficient matrix $\mathbf{A}$ for a specified grid size:

```
function A = cvt_2d_fdm (Nx,Ny,beta)

%--------------------------------
% Generate the coefficient matrix
% of a linear system.
%--------------------------------
```

```
N = Ny*(Nx+1);   % matrix size
A = zeros(N,N);
cf = -2.0*(1.0+beta);

%------------------------
% set the five diagonals
%------------------------

A(1,1) = cf;   % first row
A(1,2) = 2.0;
A(1,Nx+2) = 2.0*beta;

for i=2:Nx+1
  A(i,i) = cf;   % first block
  A(i,i+1) = 1.0;
  A(i,i-1) = 1.0;
  A(i,i+Nx+1) = 2.0*beta;
end

for i=Nx+2:N-Nx-1   % intermediate blocks
  A(i,i) = cf;
  A(i,i+1) = 1.0;
  A(i,i-1) = 1.0;
  A(i,i+Nx+1) = beta;
  A(i,i-Nx-1) = beta;
end

for i=N-Nx:N-1
  A(i,i) = cf;   % last block
  A(i,i+1) = 1.0;
  A(i,i-1) = 1.0;
  A(i,i-Nx-1) = beta;
end

A(N,N) = cf;   % last row
A(N,N-1) = 2.0;
A(N,N-Nx-1) = beta;

%------------------------------
% reset the ones to twos and
% the faulty betas to zeros
%------------------------------

for i=2:Ny   % run over horizontal partitions
  loc = (i-1)*(Nx+1)+1;
  A(loc,loc-1) = 0.0;
  A(loc,loc+1) = 2.0;
end
```

```
for i=1:Ny-1   % run over horizontal partitions
  loc = i*(Nx+1);
  A(loc,loc-1) = 2.0;
  A(loc,loc+1) = 0.0;
end

%-----
% done
%-----

return
```

For $N_x = 2$, $N_y = 3$, and $\beta = 1$, the code produces the matrix:

```
|-4   2   0  |  2   0   0  |  0   0   0 |
| 1  -4   1  |  0   2   0  |  0   0   0 |
| 0   2  -4  |  0   0   2  |  0   0   0 |
| ------------------------------------- |
| 1   0   0  | -4   2   0  |  1   0   0 |
| 0   1   0  |  1  -4   1  |  0   1   0 |
| 0   0   1  |  0   2  -4  |  0   0   1 |
| ------------------------------------- |
| 0   0   0  |  1   0   0  | -4   2   0 |
| 0   0   0  |  0   1   0  |  1  -4   1 |
| 0   0   0  |  0   0   1  |  0   2  -4 |
```

Solution of the linear system

We have formulated the problem in terms of the linear system of equations (3.3.33) for the vector $\mathbf{w}$ defined in (3.3.21). Our next task is to solve this system by numerical methods. Once this has been accomplished, the velocity components at the grid nodes arise from the partial derivatives of the potential computed by finite-difference methods.

The following MATLAB code *cvt_2d* located in directory *07_ptf* of FDLIB assembles and solves the linear system using a numerical method implemented in an internal MATLAB function stated as a vector-by-matrix division:

```
ax = 0.0; bx = 1.0;
ay = 0.0; by = 0.4;
veltop = 1.0;
Nx = 16; Ny = 32;

%--------
% prepare
%--------
```

```
Dx = (bx-ax)/Nx;    % grid spacing
Dy = (by-ay)/Ny;    % grid spacing
beta = (Dx/Dy)^2;
N = Ny*(Nx+1);    % system size

%------------------
% generate the grid
%------------------

[glx,gly,gx,gy] = grid_2d (ax,bx,ay,by,Nx,Ny);

%--------------------------------------
% specify the potential at the top row
%--------------------------------------

for i=1:Nx+1
  phitop(i) = veltop*glx(i);
end

%-------------------
% coefficient matrix
%-------------------

A = cvt_2d_fdm (Nx,Ny,beta);

%----------------
% right-hand side
%----------------

for i=1:N-Nx-1
  rhs(i) = 0.0;
end
for i=1:Nx+1
  rhs(N-Nx-1+i) = -beta*phitop(i);
end

%------------------------
% solve the linear system
%------------------------

sln = rhs/A';

%-------------------------
% assign solution to nodes
%-------------------------

Ic = 1;    % counter
```

```
for j=1:Ny
  for i=1:Nx+1
    ptl(i,j) = sln(Ic);
    Ic = Ic+1;
  end
end

for i=1:Nx+1
  ptl(i,Ny+1) = phitop(i);
end

%------------
% surface plot
%------------

surf(glx,gly,ptl')
xlabel('x','fontsize',15)
ylabel('y','fontsize',15)
zlabel('\phi','fontsize',15)

%----------------------------------------
% compute the velocity at the internal nodes
% by numerical differentiation
% using central differences
%----------------------------------------

for i=2:Nx
  for j=2:Ny
    gux(i,j) = (ptl(i+1,j) -ptl(i-1,j)) ...
      /(gx(i+1,j)-gx(i-1,j));
    guy(i,j) = (ptl(i,j+1) -ptl(i,j-1)) ...
      /(gy(i,j+1)-gy(i,j-1));
  end
end

%--------------------------------
% compute the velocity on the walls
% by numerical differentiation
%--------------------------------

%---
% left wall:  i=1
%---

for j=2:Ny
  gux(1,j) = 0.0;
  guy(1,j) = ( ptl(1,j+1)- ptl(1,j-1)) ...
    /(gy(1,j+1)-gy(1,j-1));
end
```

```
%---
% bottom wall:  j=1
%---

for i=2:Nx
  gux(i,1) = ( ptl(i+1,1)- ptl(i-1,1)) ...
    /(gx(i+1,1)-gx(i-1,1));
  guy(i,1) = 0.0;
end

%---
% right wall:  i=Nx+1
%---

for j=2:Ny
  gux(Nx+1,j) = 0.0;
  guy(Nx+1,j) = ( ptl(Nx+1,j+1) -ptl(Nx+1,j-1)) ...
    /(gy(Nx+1,j+1)-gy(Nx+1,j-1));
end

%---
% top wall:  j=Ny+1
%---

for i=2:Nx
  gux(i,Ny+1) = veltop;
  guy(i,Ny+1) = ( ptl(i,Ny+1)- ptl(i,Ny)) ...
    /(gy(i,Ny+1)-gy(i,Ny));
end

%---
% four corners
% velocity is zero or singular;
% set it to zero
%---

gux(1 ,1) = 0.0; gux(Nx+1,1) = 0.0;
gux(1 ,Ny+1) = 0.0; gux(Nx+1,Ny+1) = 0.0;
guy(1 ,1) = 0.0; guy(Nx+1,1) = 0.0;
guy(1 ,Ny+1) = 0.0; guy(Nx+1,Ny+1) = 0.0;

%---
% graphics
%---

figure
hold on
quiver(gx,gy,gux,guy)
```

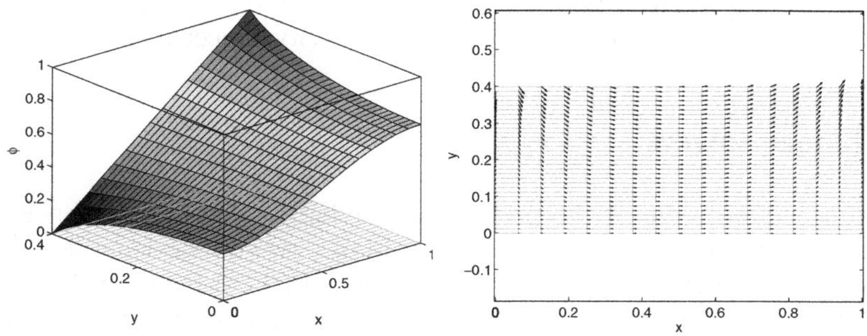

Figure 3.3.2 Surface plot of the potential (left) and velocity vector field (right) of a two-dimensional potential flow in a rectangular cavity computed by a finite-difference method.

```
axis equal
xlabel('x','fontsize',15)
ylabel('y','fontsize',15)
box
```

The graphics produced by the code is shown in figure 3.3.2. The velocity vector field is visualized by the MATLAB function *quiver*.

Problems

3.3.1 *Explicit form of a linear system*

Write the explicit form of the linear system (3.3.33) for discretization level $N_x = 3$ and $N_y = 3$.

3.3.2 *Neumann boundary conditions all around*

Derive the counterpart of the linear system (3.3.33) when the no-penetration boundary condition is applied over all four walls. Specifically, write out the components of the unknown vector $\mathbf{w}$, constant vector $\mathbf{b}$, and coefficient matrix $\mathbf{A}$, in a form that is analogous to that displayed in (3.3.35). Then confirm that the sum of the elements in each row of the matrix $\mathbf{A}$ is zero. Based on this result, explain why the matrix $\mathbf{A}$ is singular; that is, its determinant is equal to zero.

3.3.3 *Irrotational flow in a cavity*

(*a*) Run the code *cvt_2d* for a cavity with length to depth ratio equal to unity and discretization level as high as you can afford. Plot the velocity-vector field and discuss the structure of the flow.

(*b*) Repeat (*a*) for a cavity with length to depth ratio equal to 4.0. Discuss the dependence of the structure of the flow on the cavity aspect ratio.

3.4 Linear solvers

In section 3.3, we reduced the problem of solving Laplace's equation for the harmonic potential in the rectangular domain of a two-dimensional flow to the problem of solving the linear system of equations (3.3.33) for the values of the potential at the nodes of a finite-difference grid deployed over the domain of flow. The reduction was carried out by implementing finite-difference approximations. The solution of the linear system was finally found using a internal MATLAB function. Since such systems arise in broad range of applications within and beyond fluid dynamics, in this section, we discuss pertinent numerical methods in a generalized framework.

Consider a system of N linear algebraic equations for N unknown scalars, $x_1, x_2, \ldots, x_N$,

$$\begin{aligned} &A_{1,1}\, x_1 + A_{1,2}\, x_2 + \cdots + A_{1,N-1}\, x_{N-1} + A_{1,N}\, x_N = b_1, \\ &A_{2,1}\, x_1 + A_{2,2}\, x_2 + \cdots + A_{2,N-1}\, x_{N-1} + A_{2,N}\, x_N = b_2, \\ &\ldots, \\ &A_{N,1}\, x_1 + A_{N,2}\, x_2 + \cdots + A_{N,N-1} x_{N-1} + A_{N,N}\, x_N = b_N, \end{aligned} \tag{3.4.1}$$

where $A_{i,j}$, for $i = 1, 2, \ldots, N$ and $j = 1, 2, \ldots, N$ are given coefficients, and b_i are given constants. In matrix notation, the system takes the compact form

$$\mathbf{A} \cdot \mathbf{x} = \mathbf{b}, \tag{3.4.2}$$

where $\mathbf{A}$ is an $N \times N$ coefficient matrix,

$$\mathbf{A} = \begin{bmatrix} A_{1,1} & A_{1,2} & \cdots & A_{1,N-1} & A_{1,N} \\ A_{2,1} & A_{2,2} & \cdots & A_{2,N-1} & A_{2,N} \\ \cdots & \cdots & \cdots & \cdots & \cdots \\ A_{N-1,1} & A_{N-1,2} & \cdots & A_{N-1,N-1} & A_{N-1,N} \\ A_{N,1} & A_{N,2} & \cdots & A_{N,N-1} & A_{N,N} \end{bmatrix}, \tag{3.4.3}$$

and $\mathbf{b}$ is an N-dimensional vector,

$$\mathbf{b} = \begin{bmatrix} b_1 \\ b_2 \\ \vdots \\ b_{N-1} \\ b_N \end{bmatrix}. \tag{3.4.4}$$

3.4.1 Gauss elimination

A general procedure for solving system (3.4.2) relies on the method of Gauss elimination. The basic idea is to solve the first equation in (3.4.2) for the first unknown, x_1, and use the expression thus obtained to eliminate x_1 from all subsequent equations. We then retain the first equation as is, and replace all subsequent equations with their descendants that do not contain x_1. At the second stage, we solve the second equation for the second unknown, x_2, and use the expression thus obtained to eliminate x_2 from all subsequent equations. We then retain the first and second equations, and replace all subsequent equations with their descendants that do not contain x_1 or x_2. Continuing in this fashion, we arrive at the last equation, which contains only the last unknown, x_N.

Having completed the elimination, we compute the unknowns by the method of backward substitution. First, we solve the last equation for x_N, which thus becomes a known. Second, we solve the penultimate equation for x_{N-1}, which also becomes a known. Continuing in the backward direction, we scan the reduced system until we have computed all unknowns.

Pivoting

Immediately before the mth equation has been solved for the mth unknown, where $m = 1, 2, \ldots, N-1$, the linear system has the form

$$\begin{bmatrix} A_{1,1}^{(m)} & A_{1,2}^{(m)} & \cdots & \cdots & \cdots & A_{1,N}^{(m)} \\ 0 & A_{2,2}^{(m)} & \cdots & \cdots & \cdots & A_{2,N}^{(m)} \\ 0 & 0 & \cdots & \cdots & \cdots & \cdots \\ 0 & 0 & A_{m-1,m-1}^{(m)} & A_{m-1,m}^{(m)} & \cdots & A_{m-1,N}^{(m)} \\ 0 & \cdots & 0 & A_{m,m}^{(m)} & \cdots & A_{m,N}^{(m)} \\ 0 & \cdots & 0 & \cdots & \cdots & \cdots \\ 0 & \cdots & 0 & A_{N,m}^{(m)} & \cdots & A_{N,N}^{(m)} \end{bmatrix} \begin{bmatrix} x_1 \\ x_2 \\ x_3 \\ \vdots \\ x_{N-1} \\ x_N \end{bmatrix} = \begin{bmatrix} b_1^{(m)} \\ b_2^{(m)} \\ b_3^{(m)} \\ \vdots \\ b_{N-1}^{(m)} \\ b_N^{(m)} \end{bmatrix}, \tag{3.4.5}$$

where $A_{i,j}^{(m)}$ are intermediate coefficients and $b_i^{(m)}$ are intermediate right-hand sides. The first equation of (3.4.5) is identical to the first equation of (3.4.2) for any value of m. Subsequent equations are different, except at the first step corresponding to $m = 1$.

A difficulty arises when the diagonal element, $A_{m,m}^{(m)}$, is nearly or precisely zero, for then we may no longer solve the mth equation for x_m, as required. However, the failure of the method does not imply that the system does not have a unique solution. To circumvent this difficulty, we simply rearrange the equations or relabel the unknowns so as to bring the mth unknown to the mth equation, using the method of pivoting. If there is no way we can make this happen, the matrix $\mathbf{A}$ is singular and the linear system has either no solution or an infinite number of solutions.

In the method of row pivoting, potential difficulties are bypassed by switching the mth equation in the system (3.4.5) with the subsequent kth equation, where $k > m$; the value of k is chosen so that $|A_{k,m}^{(m)}|$ is the maximum value of the elements in the mth column below the diagonal, $A_{i,m}^{(m)}$, for $i \geq m$. If $A_{i,m}^{(m)} = 0$ for all $i \geq m$, the matrix $\mathbf{A}$ is singular and the system under consideration does not have a unique solution.

3.4.2 A menagerie of other methods

In practice, the size of system (3.3.33) can be on the order of $10^4 \times 10^4$ or even higher, corresponding to discretization levels N_x and N_y on the order of 10^2. For such large systems, the method of Gauss elimination requires a prohibitive computational time. The practical need for solving systems of large size has motivated the development of a host of powerful methods for general or specific applications.[1]

Iterative methods

For sparse systems of large size, iterative methods are appropriate. The main idea is to split the coefficient matrix $\mathbf{A}$ into two matrices, $\mathbf{A}'$ and $\mathbf{A}''$, writing

$$\mathbf{A} = \mathbf{A}' - \mathbf{A}'', \tag{3.4.6}$$

and then recast the system (3.4.2) into the form

$$\mathbf{A}' \cdot \mathbf{x} = \mathbf{A}'' \cdot \mathbf{x} + \mathbf{b}. \tag{3.4.7}$$

The procedure involves guessing the solution, $\mathbf{x}$, computing the right-hand side of (3.4.7), and solving for $\mathbf{x}$ on the left-hand side. The advantage of this approach is that, if the splitting (3.4.6) is done craftily, it much easier to solve (3.4.7) than (3.4.2) for $\mathbf{x}$ on the left-hand side. The computation is repeated until the value of $\mathbf{x}$ used to compute the right-hand side of (3.4.7) is virtually identical to that arising by solving the linear system (3.4.7). Examples of iterative methods are the Jacobi, the Gauss-Seidel, and the successive over-relaxation (SOR) method.

A different class of iterative methods search for the solution vector, $\mathbf{x}$, by making steps in the N-dimensional space toward craftily designed or even optimal directions. Directory *03_lin_eq* inside directory *01_num_meth* of FDLIB contains programs that implement the conjugate and biconjugate gradients methods. The multi-grid method is another powerful technique for solving systems of linear equations arising from finite-difference and related discretization.

[1] POZRIKIDIS, C., 2008, *Numerical Computation in Science and Engineering*, Second Edition, Oxford University Press.

Computer Problem

3.4.1 *Gauss elimination*

Program *gel* in directory *03_lin_eq* inside directory *01_num_meth* of FDLIB solves a system of linear equations using the method of Gauss elimination with row pivoting. Use the program to solve a system of your choice and verify the accuracy of the solution.

3.5 Two-dimensional point sources and point-source dipoles

Laplace's equation for the harmonic velocity potential–equation (3.2.11) for two-dimensional flow or equation (3.2.14) for three-dimensional flow–is linear. This means that if ϕ_1 and ϕ_2 are two harmonic potentials representing two elementary flows, any linear combination of them,

$$\phi = a\,\phi_1 + b\,\phi_2, \tag{3.5.1}$$

will also be a harmonic potential representing a hybrid flow, where a and b are two constants.

3.5.1 Function superposition and fundamental solutions

The linearity of Laplace's equation allows us to generate exact and approximate solutions by the method of superposition. The key idea is to introduce a family of harmonic potentials playing the role of basis functions, also called fundamental solutions, and then use them as building blocks to generate further solutions. For example, if ϕ_1 and ϕ_2 are two such fundamental solutions, a desired solution can be expressed by the right-hand side of (3.5.1), and the two constants a and b can be adjusted to satisfy the boundary conditions.

Various families of fundamental solutions are available for flows in infinite and semi-infinite domains, and for singly, doubly, or triply periodic flows. The most general class of fundamental solutions consists of the fundamental singularities of potential flow.

3.5.2 Two-dimensional point source

Imagine that an incompressible fluid is discharged into an infinite pool through the walls of an infinite perforated cylinder, thereby generating a radial flow in the xy plane outward from the inlet point. In plane polar coordinates centered at the discharge point, $\mathbf{x}_0 = (x_0, y_0)$, the radial and polar components of the velocity at the point $\mathbf{x} = (x, y)$ are given by

$$u_r(x, y) = \frac{m}{2\pi}\,\frac{1}{r}, \qquad u_\theta(x, y) = 0, \tag{3.5.2}$$

where

$$r = [\,(x - x_0)^2 + (y - y_0)\,]^{1/2} \tag{3.5.3}$$

is the distance of the field point, $\mathbf{x}$, from the discharge point, $\mathbf{x}_0$, and m is a constant expressing the rate of areal discharge; the units of m are velocity multiplied by length. The flow described by equations (3.5.2) is attributed to a two-dimensional point source, and the rate of areal discharge m is the strength of the point source. If m is negative, we obtain a point source with negative strength described as a point sink.

The radial velocity of the flow due to a point source decays as the inverse of the distance from the point of discharge, r, for the following reason. Since the fluid is incompressible, the flow rate Q across any circular loop or radius R centered at the point of discharge must be independent of the loop radius. To verify that the velocity field (3.5.2) satisfies this constraint, we use expression (2.6.16) and find

$$Q = R \int_0^{2\pi} u_r \,\mathrm{d}\theta = \frac{mR}{2\pi} \int_0^{2\pi} \frac{1}{r} \,\mathrm{d}\theta = \frac{mR}{2\pi} \frac{1}{R} \int_0^{2\pi} \mathrm{d}\theta = m, \tag{3.5.4}$$

as required. If we had set, for example, $u_r = m/(2\pi r^k)$, where the exponent k is not equal to unity, the restriction of constant areal flow rate associated with an incompressible fluid would not be satisfied.

Singular behavior of the point source

As the distance from the point source r tends to zero, the right-hand side of the radial velocity in (3.5.2) tends to infinity. This singular behavior is a manifestation of the idealized nature of the flow due to a point source, and explains why the point source is classified as a singularity.

In practice, the flow expressed by (3.5.2) is valid only for $r > b$, where b is the radius of the cylinder discharging the fluid. Extending the domain of flow all the way up to the center of the cylinder located at $\mathbf{x}_0$, we allow for a mathematical singularity.

Velocity potential

The velocity potential of a two-dimensional point source, ϕ^{2D-PS}, is related to the velocity components according to equations (3.2.22),

$$u_r = \frac{\partial \phi^{2D-PS}}{\partial r} = \frac{m}{2\pi} \frac{1}{r}, \qquad u_\theta = \frac{\partial \phi^{2D-PS}}{\partial \theta} = 0. \tag{3.5.5}$$

Integrating the first equation and using the second equation to evaluate the integration constant, we find

$$\phi^{2D-PS} = \frac{m}{2\pi} \ln r = \frac{m}{4\pi} \ln[(x - x_0)^2 + (y - y_0)^2]. \tag{3.5.6}$$

Straightforward differentiation confirms that ϕ^{2D-PS} satisfies Laplace's equation in two dimensions at every point, except at the singular point $\mathbf{x}_0$ where the potential and its derivatives are not defined.

Cartesian components of the velocity

To derive the Cartesian components of the velocity, we take the partial derivatives of ϕ^{2D-PS} with respect to x or y, and find

$$u_x^{2D-PS} = \frac{\partial \phi^{2D-PS}}{\partial x} = \frac{m}{2\pi} \frac{x - x_0}{(x-x_0)^2 + (y-y_0)^2},$$

$$u_y^{2D-PS} = \frac{\partial \phi^{2D-PS}}{\partial y} = \frac{m}{2\pi} \frac{y - y_0}{(x-x_0)^2 + (y-y_0)^2}. \tag{3.5.7}$$

The streamlines of the flow due to a point source are radial straight lines emanating from the singular point, $\mathbf{x}_0$.

Point source embedded in uniform flow

As an application, consider the superposition of uniform flow along the x axis with velocity V_x, and the flow due to a point source with strength m situated at the origin, $x_0 = 0$ and $y_0 = 0$. Using the potential (3.2.7) for uniform flow with the constant c set to zero and the potential (3.5.6) for the point source, we find that the potential of the composite flow is

$$\phi = x\, V_x + \frac{m}{4\pi} \ln(x^2 + y^2). \tag{3.5.8}$$

The associated Cartesian components of the velocity are

$$u_x = V_x + \frac{m}{2\pi} \frac{x}{x^2 + y^2} = V_x \left(1 + \frac{m}{2\pi a V_x} \frac{\frac{x}{a}}{(\frac{x}{a})^2 + (\frac{y}{a})^2} \right),$$

$$u_y = \frac{m}{2\pi} \frac{y}{x^2 + y^2} = V_x \frac{m}{2\pi a V_x} \frac{\frac{y}{a}}{(\frac{x}{a})^2 + (\frac{y}{a})^2}, \tag{3.5.9}$$

where a is an arbitrary length.

To study the structure of the flow, we introduce dimensionless variables,

$$\hat{x} = \frac{x}{a}, \qquad \hat{y} = \frac{y}{a}, \qquad \hat{u}_x = \frac{u_x}{V_x}, \qquad \hat{u}_y = \frac{u_y}{V_x}, \tag{3.5.10}$$

and recast equations (3.5.9) into the dimensionless form

$$\hat{u}_x = 1 + \beta \frac{\hat{x}}{\hat{x}^2 + \hat{y}^2}, \qquad \hat{u}_y = \beta \frac{\hat{y}}{\hat{x}^2 + \hat{y}^2}. \tag{3.5.11}$$

where $\beta = m/(2\pi a V_x)$ is a dimensionless parameter expressing the strength of the point source relative to the magnitude of the incident flow. Equations

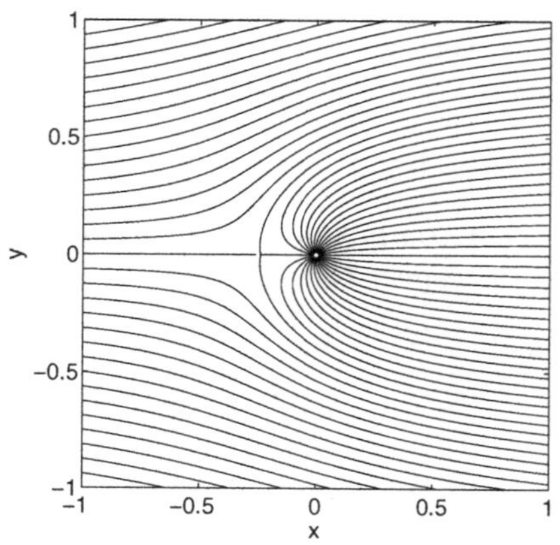

Figure 3.5.1 Streamline pattern of the flow due to the superposition of streaming (uniform) flow along the x axis and the flow due to a two-dimensional point source.

(3.5.11) reveal that the structure of the flow is determined by the value of the parameter β.

Figure 3.5.1 illustrates the streamline pattern for $\beta = 0.25$, revealing that the velocity potential (3.5.8) describes uniform flow along the x axis past a semi-infinite body whose surface can be identified with two streamlines emanating from a stagnation point on the x axis. Using the first equation in (3.5.11), we find that the x component of the velocity on the x axis is zero when $\hat{x} = -\beta$. Thus, the larger the value of β expressing the strength of the point source, the farther the stagnation point is located from the origin.

3.5.3 *Two-dimensional point-source dipole*

Next, we consider the flow due to the superposition of a point source with strength m located at the point $(x_0 + a, y_0)$, and a point sink with strength $-m$ located at the point $(x_0 - a, y_0)$, where a is a specified distance, as shown in figure 3.5.2(a). Using expression (3.5.6), we find that the harmonic potential induced by these singularities is

$$\begin{aligned}\phi &= \frac{m}{4\pi}\ln\{[x-(x_0+a)]^2+(y-y_0)^2\} - \frac{m}{4\pi}\ln\{[x-(x_0-a)]^2+(y-y_0)^2\}\\ &= \frac{m}{4\pi}\ln\frac{[x-(x_0+a)]^2+(y-y_0)^2}{[x-(x_0-a)]^2+(y-y_0)^2}.\end{aligned} \tag{3.5.12}$$

The Cartesian components of the fluid velocity are

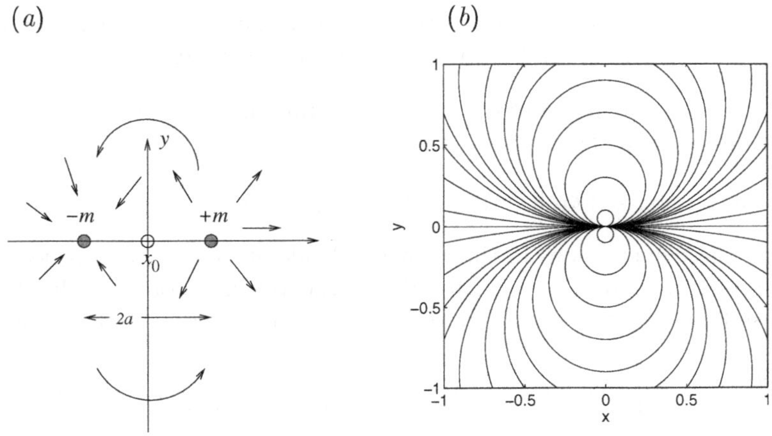

Figure 3.5.2 (*a*) A point source and a point sink merge to yield a point-source dipole. (*b*) Streamline pattern due to a two-dimensional dipole pointing along the x axis.

$$u_x = \frac{\partial \phi}{\partial x} = \frac{m}{2\pi}\Big[\frac{x-(x_0+a)}{[x-(x_0+a)]^2+(y-y_0)^2} - \frac{x-(x_0-a)}{[x-(x_0-a)]^2+(y-y_0)^2}\Big],$$

$$u_y = \frac{\partial \phi}{\partial y} = \frac{m}{2\pi}\Big[\frac{y-y_0}{[x-(x_0+a)]^2+(y-y_0)^2} - \frac{y-y_0}{[x-(x_0-a)]^2+(y-y_0)^2}\Big]. \tag{3.5.13}$$

Now we hold the position of the field point (x, y) fixed and reduce the distance between the two singularities; that is, we let a tend to zero. In this limit, the flow due to the point sink tends to cancel the flow due to the point source. However, if the strength, m, also increases like the inverse of the distance between the two singularities, $2a$, then a nontrivial flow due to a point source dipole will arise in the limit.

To derive the flow due to a point-source dipole, we recast the expression on the right-hand side of (3.5.12) into the form

$$\phi = \frac{m}{4\pi}\Big[\ln(1 - a\frac{2(x-x_0)-a}{(x-x_0)^2+(y-y_0)^2}) \tag{3.5.14}$$

$$-\ln(1 + a\frac{2(x-x_0)+a}{(x-x_0)^2+(y-y_0)^2})\Big] = \frac{m}{4\pi}[\ln(1-\epsilon_1) - \ln(1+\epsilon_2)],$$

where

$$\epsilon_1 \equiv a\,\frac{2(x-x_0)-a}{(x-x_0)^2+(y-y_0)^2}, \qquad \epsilon_2 \equiv a\,\frac{2(x-x_0)+a}{(x-x_0)^2+(y-y_0)^2} \tag{3.5.15}$$

are dimensionless numbers. As the distance, a, becomes smaller than the distance between the points $\mathbf{x}$ and $\mathbf{x}_0$, both ϵ_1 and ϵ_2 tend to zero.

A Taylor series expansion of the logarithmic function $\ln w$ about the point $w = 1$ provides us with the approximations

$$\ln(1 - \epsilon_1) = -\epsilon_1 + \cdots, \qquad \ln(1 + \epsilon_2) = \epsilon_2 + \cdots. \tag{3.5.16}$$

Substituting these expressions in the right-hand side of (3.5.14) and neglecting the terms represented by the dots, we obtain the velocity potential due to a point source dipole located at the point (x_0, y_0) and oriented along the x axis,

$$\phi^{2D-PSD-x} = -\frac{m}{4\pi}(\epsilon_1 + \epsilon_2) = -\frac{d_x}{2\pi}\,\frac{x - x_0}{(x - x_0)^2 + (y - y_0)^2}, \tag{3.5.17}$$

where

$$d_x \equiv 2ma \tag{3.5.18}$$

is the strength of the dipole.

Comparing (3.5.6), (3.5.17), and (3.5.18), we find

$$\phi^{2D-PSD-x} = 2a\,\frac{\partial \phi^{2D-PS}}{\partial x_0}, \tag{3.5.19}$$

which shows that the potential due to a point-source dipole oriented along the x axis arises by differentiating the potential due to a point source with respect to the x coordinate of the singular point, x_0. This property classifies the dipole as a derivative singularity descending from the point source.

The velocity components associated with a dipole oriented along the x axis are given by

$$\begin{aligned} u_x^{2D-PSD-x} &= \frac{\partial \phi^{2D-PSD-x}}{\partial x} = 2a\,\frac{\partial^2 \phi^{2D-PS}}{\partial x\,\partial x_0} = -2a\,\frac{\partial^2 \phi^{2D-PS}}{\partial x^2}, \\ u_y^{2D-PSD-x} &= \frac{\partial \phi^{2D-PSD-x}}{\partial y} = 2a\,\frac{\partial^2 \phi^{2D-PS}}{\partial y\,\partial x_0} = -2a\,\frac{\partial^2 \phi^{2D-PS}}{\partial x\,\partial y}, \end{aligned} \tag{3.5.20}$$

yielding

$$\begin{aligned} u_x^{2D-PSD-x} &= \frac{d_x}{2\pi}\Big[-\frac{1}{(x - x_0)^2 + (y - y_0)^2} + \frac{2\,(x - x_0)^2}{[(x - x_0)^2 + (y - y_0)^2]^2}\Big], \\ u_y^{2D-PSD-x} &= \frac{d_x}{2\pi}\,\frac{2\,(x - x_0)(y - y_0)}{[(x - x_0)^2 + (y - y_0)^2]^2}. \end{aligned} \tag{3.5.21}$$

The associated streamline pattern is shown in figure 3.5.2(b).

Dipole along the y axis

Working in a similar fashion, we derive the flow due to a point-source dipole with strength d_y oriented along the y axis. The associated harmonic potential is

$$\phi^{2D-PSD-y} = 2a\,\frac{\partial \phi^{2D-PS}}{\partial y_0} = -\frac{d_y}{2\pi}\,\frac{y-y_0}{(x-x_0)^2+(y-y_0)^2}, \tag{3.5.22}$$

where $d_y = 2ma$. The corresponding Cartesian components of the velocity are

$$\begin{aligned} u_x^{2D-PSD-y} &= \frac{\partial \phi^{2D-PSD-y}}{\partial x} = 2a\,\frac{\partial^2 \phi^{2D-PS}}{\partial x\,\partial y_0} = -2a\,\frac{\partial^2 \phi^{2D-PS}}{\partial x\,\partial y}, \\ u_y^{2D-PSD-y} &= \frac{\partial \phi^{2D-PSD-y}}{\partial y} = 2a\,\frac{\partial^2 \phi^{2D-PS}}{\partial y\,\partial y_0} = -2a\,\frac{\partial^2 \phi^{2D-PS}}{\partial y^2}, \end{aligned} \tag{3.5.23}$$

yielding

$$\begin{aligned} u_x^{2D-PSD-x} &= \frac{d_y}{2\pi}\,\frac{2\,(x-x_0)(y-y_0)}{[(x-x_0)^2+(y-y_0)^2]^2}, \\ u_x^{2D-PSD-y} &= \frac{d_y}{2\pi}\Big[-\frac{1}{(x-x_0)^2+(y-y_0)^2} + \frac{2\,(y-y_0)^2}{[(x-x_0)^2+(y-y_0)^2]^2}\Big]. \end{aligned} \tag{3.5.24}$$

The streamline pattern is found by rotating the pattern shown in figure 3.5.2(*b*) by 90° around the location of the dipole.

General dipole orientation

Combining equations (3.5.17) and (3.5.22), we find that the harmonic potential due to a potential dipole with vectorial strength $\mathbf{d} \equiv (d_x, d_y)$ located at the point $\mathbf{x}_0$ is given by

$$\phi^{2D-PSD} = \mathbf{d}\cdot\mathbf{\Phi}^{2D-PSD}, \tag{3.5.25}$$

where the vector function $\mathbf{\Phi}^{2D-PSD}$ is defined as

$$\mathbf{\Phi}^{2D-PSD} \equiv -\frac{1}{2\pi}\,\frac{1}{(x-x_0)^2+(y-y_0)^2}\begin{bmatrix} x-x_0 \\ y-y_0 \end{bmatrix}. \tag{3.5.26}$$

The velocity field can be expressed in the corresponding form

$$\mathbf{u} = -\mathbf{d}\cdot\mathbf{U}^{2D-PSD}, \tag{3.5.27}$$

where $\mathbf{U}^{2D-PSD}$ is a 2×2 matrix function of position (problem 3.5.2).

3.5.4 Flow past a circular cylinder

As an application, we consider the superposition of uniform flow along the x axis with velocity V_x, and flow due to a point-source dipole located at the origin and oriented along the x axis. Using the potential (3.2.7) with $c = 0$ for the uniform flow and the potential (3.5.17) with $x_0 = 0$ and $y_0 = 0$ for the dipole, we derive the potential of the composite flow,

$$\phi(x, y) = V_x x - \frac{d_x}{2\pi} \frac{x}{x^2 + y^2} = V_x x \left(1 - \frac{d_x}{2\pi V_x} \frac{1}{r^2}\right) \tag{3.5.28}$$

or

$$\phi(r, \theta) = V_x \left(r - \frac{d_x}{2\pi V_x} \frac{1}{r}\right) \cos\theta, \tag{3.5.29}$$

where $r = \sqrt{x^2 + y^2}$ is the distance of the field point, (x, y), from the center of the cylinder, and θ is the polar angle measured in the counterclockwise direction around the center of the cylinder, defined such that $x = r\cos\theta$.

Now using the expression for the radial component of the velocity in terms of the potential given in the first of equations (3.2.22), we find

$$u_r = \frac{\partial \phi}{\partial r} = V_x \left(1 + \frac{d_x}{2\pi V_x} \frac{1}{r^2}\right) \cos\theta. \tag{3.5.30}$$

The sum inside the parentheses on the right-hand side of (3.5.30) is zero at the radial distance $r = [-d_x/(2\pi V_x)]^{1/2}$. Conversely, if the strength of the dipole has the value

$$d_x = -2\pi V_x a^2, \tag{3.5.31}$$

the radial velocity is zero at the radial distance $r = a$.

It is evident that the potential (3.5.29) with d_x evaluated from expression (3.5.31) describes uniform flow with velocity V_x past a circular cylinder of radius a centered at the origin, with the no-penetration condition satisfied over the surface of the cylinder. Substituting (3.5.31) in (3.5.29), we derive the explicit solution

$$\phi = V_x \left(r + \frac{a^2}{r}\right) \cos\theta. \tag{3.5.32}$$

The corresponding Cartesian components of the velocity are

$$u_x = V_x \left(1 + \frac{a^2}{r^4} - 2\frac{a^2}{r^2} x^2\right), \qquad u_y = -2\, V_x \frac{a^2}{r^4} xy. \tag{3.5.33}$$

The associated streamline pattern is shown in figure 3.5.3(a).

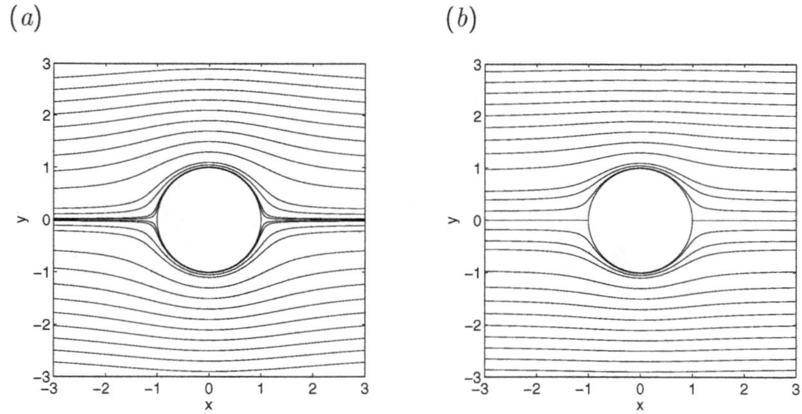

Figure 3.5.3 Streamline pattern of (*a*) uniform (streaming) flow past a circular cylinder with vanishing circulation around the cylinder, and (*b*) uniform (streaming) flow past a sphere.

3.5.5 Sources and dipoles in the presence of boundaries

When the domain of flow is bounded by an impermeable surface, the flow due to a point source or point source dipole must be enhanced with a complementary flow whose purpose is to satisfy the no-penetration boundary condition. For simple boundary geometries, the complementary flow can be identified with the flow generated by singularities located at image positions. Directory *lgf_2d* inside directory *07_ptf* of FDLIB contains a collection of subroutines that evaluate the harmonic potential and associated velocity field for several boundary geometries.

Point source above a wall

In the case of a point source placed above a plane wall located at $y = w$, the complementary flow is generated by reflecting the point source with respect to the wall. If a primary point source with strength m is located at the point (x_0, y_0), then an image point source with equal strength is located at the point $(x_0, 2w - y_0)$. The streamline pattern is depicted in figure 3.5.4(*a*).

Point source outside a circular cylinder

In the case of a point source located outside a circular cylinder of radius a centered at the point (x_c, y_c), the complementary flow is generated by two image point sources. The first image point source is located at the inverse point of the primary point source with respect to the cylinder. If a primary point source with strength m is located at (x_0, y_0), then an image point source with

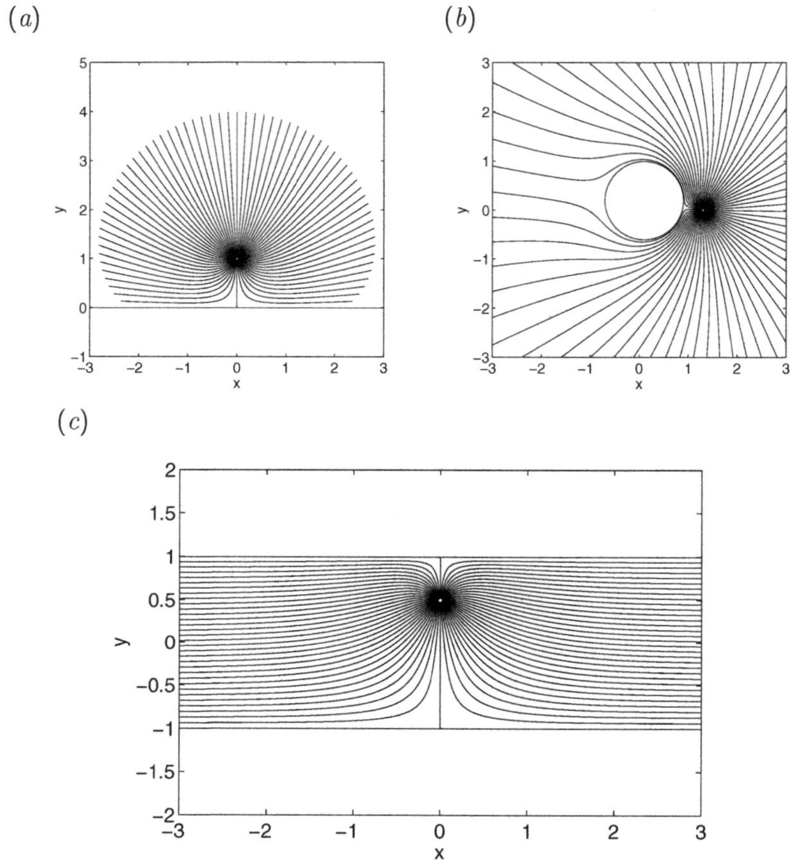

Figure 3.5.4 Streamline pattern of the flow due to a two-dimensional point source (a) above a plane wall, (b) in front of a circular cylinder, and (c) between two parallel plates.

the same strength is located at the point

$$x_0^I = x_c + (x_0 - x_c)\,\frac{a^2}{d^2}, \qquad y_0^I = y_c + (y_0 - y_c)\,\frac{a^2}{d^2}, \tag{3.5.34}$$

where $d = [\,(x_0 - x_c)^2 + (y_0 - y_c)^2\,]^{1/2}$ is the distance of the primary point source from the center of the cylinder. A second image point source with strength $-m$ is located at the center of the cylinder. Note that the sum of the strengths of the image singularities is zero to ensure that a net flow rate across the surface of the cylinder does not arise. The streamline pattern of the induced flow is shown in figure 3.5.4(b).

Problems

3.5.1 *Oblique streaming flow past a circular cylinder*

Derive an expression for the harmonic potential and Cartesian components of the velocity of oblique streaming flow with uniform velocity $u_x = V_x$, $u_y = V_y$ past a circular cylinder of radius a centered at the origin.

3.5.2 *Flow due to a point-source dipole*

Use expressions (3.5.21) and (3.5.24) to derive the explicit form of the matrix $\mathbf{U}^{2D-PSD}$ introduced in (3.5.27).

3.5.3 *Stream functions*

Derive the stream function associated with (a) a two-dimensional point source, and (b) a two-dimensional point-source dipole pointing in the x or y direction.

Computer Problem

3.5.4 *Point source in a semi-infinite rectangular strip*

Directory *strml* inside directory *04_various* of FDLIB contains a program that generates the streamline pattern of the flow induced by a point source for several boundary geometries. Three examples are illustrated in figure 3.5.4. Run the program to generate the streamline pattern of the flow due to a point source in a semi-infinite rectangular strip, and discuss the structure of the flow.

3.6 Three-dimensional point sources and point-source dipoles

The fundamental solutions derived in section 3.5 for two-dimensional potential flow can be extended in a straightforward fashion to three-dimensional flow.

3.6.1 Three-dimensional point source

The harmonic potential due to a three-dimensional point source with strength m located at the point $\mathbf{x}_0 = (x_0, y_0, z_0)$ is

$$\phi^{3D-PS} = -\frac{m}{4\pi}\,\frac{1}{r}, \tag{3.6.1}$$

where $r = [\,(x-x_0)^2 + (y-y_0)^2 + (z-z_0)^2\,]^{1/2}$ is the distance of the field point, $\mathbf{x}$, from the point source, $\mathbf{x}_0$. The corresponding Cartesian components of the velocity are

$$u_x^{3D-PS} = \frac{m}{4\pi}\,\frac{x-x_0}{r^3}, \qquad u_y^{3D-PS} = \frac{m}{4\pi}\,\frac{y-y_0}{r^3}, \qquad u_z^{3D-PS} = \frac{m}{4\pi}\,\frac{z-z_0}{r^3}. \tag{3.6.2}$$

The streamlines are radial straight lines emanating from the singular point, $\mathbf{x}_0$.

3.6.2 Three-dimensional point-source dipole

The harmonic potential due to a three-dimensional point-source dipole oriented along the x, y, or z axis is given, respectively, by

$$\phi^{3D-PSD-x} = -\frac{d_x}{4\pi}\,\frac{x-x_0}{r^3}, \qquad \phi^{3D-PSD-y} = -\frac{d_y}{4\pi}\,\frac{y-y_0}{r^3},$$

$$\phi^{3D-PSD-z} = -\frac{d_z}{4\pi}\,\frac{z-z_0}{r^3}, \qquad (3.6.3)$$

where d_x, d_y, and d_z are the directional strengths of the dipole.

The corresponding velocity components are found by straightforward differentiation with respect to x, y, and z. For a dipole oriented along the x axis,

$$u_x^{3D-PSD-x} = \frac{\partial\phi^{3D-PSD-x}}{\partial x} = \frac{d_x}{4\pi}\Big[-\frac{1}{r^3}+\frac{3\,(x-x_0)^2}{r^5}\Big],$$

$$u_y^{3D-PSD-x} = \frac{\partial\phi^{3D-PSD-x}}{\partial y} = \frac{d_x}{4\pi}\,\frac{3\,(x-x_0)(y-y_0)}{r^5}, \qquad (3.6.4)$$

$$u_z^{3D-PSD-x} = \frac{\partial\phi^{3D-PSD-x}}{\partial z} = \frac{d_x}{4\pi}\,\frac{3\,(x-x_0)(z-z_0)}{r^5}.$$

The streamline pattern in the xy plane is qualitatively similar, but not identical, to that shown in figure 3.5.2(b) for two-dimensional flow. For a dipole oriented along the y axis,

$$u_x^{3D-PSD-y} = \frac{\partial\phi^{3D-PSD-y}}{\partial x} = \frac{d_y}{4\pi}\,\frac{3\,(y-y_0)(x-x_0)}{r^5},$$

$$u_y^{3D-PSD-y} = \frac{\partial\phi^{3D-PSD-y}}{\partial y} = \frac{d_y}{4\pi}\Big[-\frac{1}{r^3}+\frac{3\,(y-y_0)^2}{r^5}\Big], \qquad (3.6.5)$$

$$u_z^{3D-PSD-y} = \frac{\partial\phi^{3D-PSD-y}}{\partial z} = \frac{d_y}{4\pi}\,\frac{3\,(y-y_0)(z-z_0)}{r^5}.$$

For a dipole oriented along the z axis,

$$u_x^{3D-PSD-z} = \frac{\partial\phi^{3D-PSD-z}}{\partial x} = \frac{d_z}{4\pi}\,\frac{3\,(z-z_0)(x-x_0)}{r^5},$$

$$u_y^{3D-PSD-z} = \frac{\partial\phi^{3D-PSD-z}}{\partial y} = \frac{d_z}{4\pi}\,\frac{3\,(z-z_0)(y-y_0)}{r^5}, \qquad (3.6.6)$$

$$u_z^{3D-PSD-z} = \frac{\partial\phi^{3D-PSD-z}}{\partial z} = \frac{d_z}{4\pi}\Big[-\frac{1}{r^3}+\frac{3\,(z-z_0)^2}{r^5}\Big].$$

Expressions (3.6.4)–(3.6.6) can be conveniently placed into a compact vector-matrix form, as discussed in problem 3.6.1.

3.6.3 Streaming flow past a sphere

As an application, we consider the superposition of uniform flow along the x axis with velocity V_x, and the flow due to a three-dimensional point-source dipole located at the origin, $x_0 = 0$, $y_0 = 0$, and $z_0 = 0$, and pointing along the x axis. Using expression (3.2.7) with $c = 0$ for the uniform flow and expression (3.6.1) for the point-source dipole, we find that the potential of the composite axisymmetric flow is

$$\phi = V_x x - \frac{d_x}{4\pi}\frac{x}{r^3} = V_x x\,(1 - \frac{d_x}{4\pi V_x}\frac{1}{r^3}), \tag{3.6.7}$$

or

$$\phi(r,\theta) = V_x\,(r - \frac{d_x}{4\pi V_x}\frac{1}{r^2}\,)\,\cos\theta, \tag{3.6.8}$$

where $r = [x^2+y^2+z^2]^{1/2}$ is the distance from the origin, and θ is the azimuthal angle defined such that $x = r\cos\theta$.

Using the first equation in (3.2.20), we find that the radial component of the velocity is

$$u_r = \frac{\partial\phi}{\partial r} = V_x\,(\,1 + \frac{d_x}{2\pi V_x}\frac{1}{r^3}\,)\,\cos\theta. \tag{3.6.9}$$

The sum inside the parentheses on the right-hand side of (3.6.9) is zero at the radial distance $r = [-d_x/(2\pi V_x)]^{1/3}$. Conversely, if the strength of the dipole has the value

$$d_x = -2\pi\,V_x\,a^3, \tag{3.6.10}$$

then the radial velocity vanishes at the radial distance $r = a$. These observations suggest that the potential (3.6.8) with d_x evaluated from expression (3.6.10) describes uniform flow along the x axis with velocity V_x past a stationary sphere of radius a centered at the origin. Substituting (3.6.10) in (3.6.8), we obtain an explicit expression for the potential,

$$\phi = V_x\,(r + \frac{1}{2}\frac{a^3}{r^2})\,\cos\theta. \tag{3.6.11}$$

The corresponding Cartesian components of the velocity are

$$u_x = V_x\left[1 + \frac{1}{2}\,a^3(\frac{1}{r^3} - \frac{3x^2}{r^5})\right], \qquad u_y = -V_x\,\frac{3}{2}\,a^3\,\frac{xy}{r^5}, \qquad u_z = -V_x\,\frac{3}{2}\,a^3\,\frac{xz}{r^5}. \tag{3.6.12}$$

The streamline pattern in a meridional plane is shown in figure 3.6.3(b).

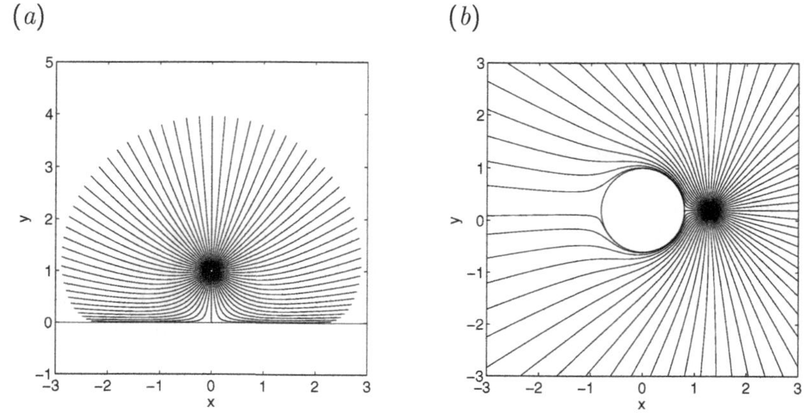

Figure 3.6.1 Streamline pattern in a meridional plane of the flow due to a three-dimensional point source (a) above a plane wall, and (b) outside a sphere.

3.6.4 Sources and dipoles in the presence of boundaries

To account for the presence of boundaries, we introduce a complementary flow whose purpose is to ensure the satisfaction of the no-penetration boundary condition, as discussed in section 3.5.5 for two-dimensional flow. For simple boundary geometries, the complementary flow can be identified with the flow produced by singularities located at image positions outside the domain of flow.

Directory *lgf_3d* inside directory *07_ptf* of FDLIB contains a collection of subroutines that evaluate the velocity field for several boundary geometries. Figure 3.6.1(a) shows the streamline pattern of the flow due to a point source located above a plane wall, and figure 3.6.1(b) shows the streamline pattern of the flow due to a point source outside a sphere. In the first case, the complementary flow is due to a reflected point source.

Problems

3.6.1 *Flow due to a three-dimensional point-source dipole*

Express the potential and velocity field of a three-dimensional point source dipole in terms of (a) the vectorial strength of the dipole, (b) a three-component vector function, $\mathbf{\Phi}^{3D-PSD}$, and (c) a 3×3 matrix function, $\mathbf{U}^{3D-PSD}$.

3.6.2 *Stream functions*

Introduce cylindrical polar coordinates with origin at the location of a three-dimensional point source or point-source dipole, and derive expressions for the axisymmetric stream function.

3.7 Point vortices and line vortices

Consider a long circular cylinder parallel to the z axis immersed in an infinite fluid and rotating around its axis with a constant angular velocity, thereby generating a two-dimensional swirling flow. In plane polar coordinates with origin at the center of the cylinder located at $\mathbf{x}_0 = (x_0, y_0)$, the radial and angular velocity components are

$$u_r(x, y) = 0, \qquad u_\theta(x, y) = \frac{\kappa}{2\pi}\frac{1}{r}, \tag{3.7.1}$$

where $r = [\,(x - x_0)^2 + (y - y_0)\,]^{1/2}$ is the distance of the point where the velocity is evaluated, $\mathbf{x} = (x, y)$, from the center of the cylinder, and κ is a constant with units of velocity multiplied by length. The magnitude and sign of κ express, respectively, the strength of the flow and the direction of rotation. If κ is positive, point particles in the flow rotate around the cylinder in the counterclockwise direction. If κ is negative, point particles in the flow rotate around the cylinder in the clockwise direction.

The magnitude of the angular component of the velocity, u_θ, decays like $1/r$. If the fluid rotated as a rigid body with angular velocity Ω around the point $\mathbf{x}_0$, the velocity u_θ would be increasing linearly with respect to radial distance, r, as $u_\theta = \Omega r$. Clearly, the velocity field expressed by (3.7.1) represents a flow that is different than rigid-body rotation.

Point vortex singularity

The flow described by equations (3.7.1) is physically meaningful only in the exterior of the cylinder. Neglecting the surface of the cylinder and extending the domain of flow all the way up to the center of the cylinder, we obtain a singular flow described as the flow due to a point vortex with strength κ, as illustrated in figure 3.7.1(a). The singularity occurs because, as the distance of an observation point from the point vortex, r, tends to zero, the magnitude of the velocity becomes infinite.

To confirm that the flow due to a point vortex is irrotational, we substitute expressions (3.7.1) in (2.3.16) and find that the z component of the vorticity vanishes at every point in the flow, except at the location of the point vortex, $\mathbf{x}_0$, where a singularity appears. These properties classify the point vortex as a singularity of two-dimensional irrotational flow.

A seemingly paradoxical behavior can be noted. Because the flow is irrotational at every point except at the location of the point vortex, small circular fluid parcels not containing the point vortex translate and deform but do not rotate around their center, and yet the fluid exhibits a net circulatory motion. However, these features are not inconsistent, and their apparent but not essential contradiction serves to underscore that global circulatory motion does not necessarily imply rotational flow.

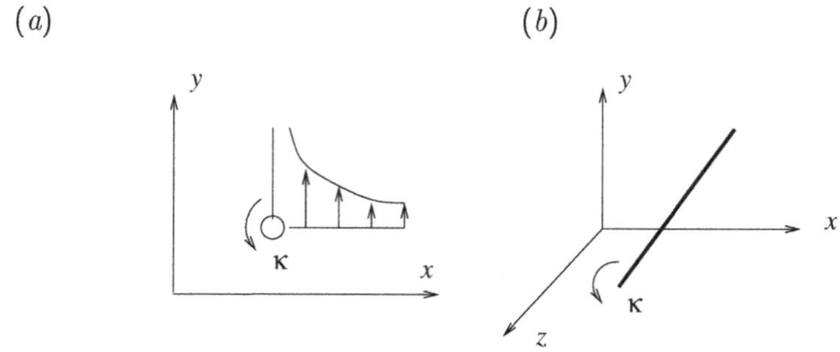

Figure 3.7.1 (a) Two-dimensional, and (b) three-dimensional perspectives of a point vortex with positive strength representing a rectilinear line vortex parallel to the z axis.

3.7.1 The potential of irrotational circulatory flow

The presence of circulation has important implications on our ability to describe a flow in terms of a velocity potential. To see this, we use equations (3.2.22) and find that the potential due to a point vortex, ϕ^{PV}, satisfies the equations

$$\frac{\partial \phi^{PV}}{\partial r} = 0, \qquad \frac{1}{r}\frac{\partial \phi^{PV}}{\partial \theta} = \frac{\kappa}{2\pi}\frac{1}{r}. \tag{3.7.2}$$

Integrating these equations, we find

$$\phi^{PV} = \frac{\kappa}{2\pi}\,\theta, \tag{3.7.3}$$

where θ is the polar angle measured around the point vortex in the counterclockwise direction. An arbitrary but irrelevant constant can be added to the right-hand side of (3.7.3). The corresponding stream function is

$$\psi^{PV} = -\frac{\kappa}{2\pi}\ln r. \tag{3.7.4}$$

According to expression (3.7.3), as we move around the point vortex on a circular path in the counterclockwise direction, the potential increases in proportion to the angle θ. But then, as we return to the point of departure, because θ has increased by 2π, the potential undergoes a jump with respect to the initial value, equal to κ. We can continue traveling around the point vortex for one more turn, only to find that each time we perform a complete rotation, the potential undergoes a jump equal to κ. This observation illustrates that the potential associated with a point vortex is multi-valued. Moreover, since our point of departure is arbitrary, the potential is multi-valued at every point in the flow.

We have discovered by example that circulatory motion is associated with a multi-valued potential and *vice versa.* In practice, a multi-valued potential is too ambiguous to handle by analytical and numerical methods. To circumvent this difficulty, we may decompose the potential into an easy multi-valued part and a harder complementary single-valued part; we specify the former, and extract the latter by analytical or numerical computation. The implementation of this method will be discussed in chapter 12 in the context of aerodynamics.

In an alternative approach, we introduce an artificial boundary residing inside the fluid called a branch cut, and work under the assumption that the potential has two different values on either side. If the flow does not exhibit net circulatory motion, the two values will be identical.

3.7.2 Flow past a circular cylinder

To illustrate the usefulness of the point vortex singularity, we consider uniform flow past a circular cylinder discussed earlier in section 3.5. Equation (3.5.32) provides us with the single-value harmonic potential in the absence of circulatory motion around the cylinder. To allow for circulatory motion, we add to the right-hand side of (3.5.32) the potential due to a point vortex situated at the center of cylinder given in equation (3.7.3), obtaining

$$\phi = V_x \left(r + \frac{a^2}{r} \right) \cos\theta + \frac{\kappa}{2\pi}\, \theta. \tag{3.7.5}$$

The corresponding Cartesian components of the velocity are

$$u_x = V_x \left(1 + \frac{a^2}{r^2} - \frac{2\,x^2 a^2}{r^4} \right) - \frac{\kappa}{2\pi}\,\frac{y}{r^2}, \qquad u_y = -V_x\,\frac{2\,xy\,a^2}{r^4} + \frac{\kappa}{2\pi}\,\frac{x}{r^2}. \tag{3.7.6}$$

Because the radial component of the velocity, $u_r = \partial\phi/\partial r$, is zero at the surface of the cylinder located at $r = a$, the no-penetration condition is satisfied.

Rearranging (3.7.5), we obtain the dimensionless form

$$\phi = V_x a \left(\cos\theta\,(\hat{r} + \frac{1}{\hat{r}}) - 2\beta\,\theta \right), \tag{3.7.7}$$

where $\hat{r} = r/a$ is the dimensionless radial distance, $\hat{r} = 1$ corresponds to the cylinder surface, and

$$\beta \equiv -\frac{\kappa}{4\pi V_x a} \tag{3.7.8}$$

is a dimensionless circulation number. When $\beta = 0$, the circulation vanishes. Expression (3.7.7) reveals that the structure of the flow is determined by the dimensionless parameter β.

The tangential component of the velocity on the surface of the cylinder is

$$u_\theta(r = a) = -2\,V_x\,\sin\theta + \frac{\kappa}{2\pi a} = -2\,V_x\,(\sin\theta + \beta). \tag{3.7.9}$$

We note that the magnitude of the velocity is zero when $\theta = \arcsin(-\beta)$, and conclude that stagnation points develop on the surface on the cylinder when $-1 \leq \beta \leq 1$.

When $\beta = 0$, two stagnation points occur in on the horizontal mid-surface of the cylinder at $\theta = 0$ and π. As β increases from zero to unity, the stagnation points move downward and finally merge at lowest point of the cylinder, $\theta = -\pi/2$. When β exceeds the value of unity, the merged stagnation points move off the surface of the cylinder and into the flow. Streamline patterns for $\beta = 0.5$, 1.0, and 1.2 illustrating this transition are shown in figure 3.7.2.

3.7.3 Circulation

Having discussed the effect of the circulation around a cylinder placed in a uniform flow, we proceed to generalize the concept of circulation to more general flow configurations. Consider a two-dimensional flow in the xy plane, and draw a simple closed loop inside the flow. If the loop encloses fluid alone and no boundaries, it is reducible. If the loop encloses fluid and one or more boundaries, it is irreducible. The distinguishing feature of a reducible loop is that it can be shrunk to a point without crossing flow boundaries. One reducible and three irreducible loops are depicted in figure 3.7.3.

Next, we select a point on a reducible or irreducible loop and introduce the unit tangent vector pointing in the counterclockwise direction, $\mathbf{t} = (t_x, t_y)$, as shown in figure 3.7.3. The inner product of the velocity and the unit tangent vector is given by the right-hand side of (2.6.8). The circulation around the loop is defined as the line integral of the tangential component of the velocity with respect to arc length around the loop, l,

$$C \equiv \oint_{\mathcal{L}} u_t\,\mathrm{d}l = \oint_{\mathcal{L}} \mathbf{u}\cdot\mathbf{t}\,\mathrm{d}l, \tag{3.7.10}$$

where $\mathcal{L}$ denotes the loop, and $\mathrm{d}l = \sqrt{\mathrm{d}x^2 + \mathrm{d}y^2}$.

Reducible loops

Stokes's circulation theorem discussed in section 11.1 states that, if no point vortices reside inside the area enclosed by a reducible loop, the circulation around the loop is equal to the strength of the vorticity integrated over the area of fluid enclosed by the loop, $\mathcal{D}$,

$$C = \iint_{\mathcal{D}} \omega_z\,\mathrm{d}x\,\mathrm{d}y. \tag{3.7.11}$$

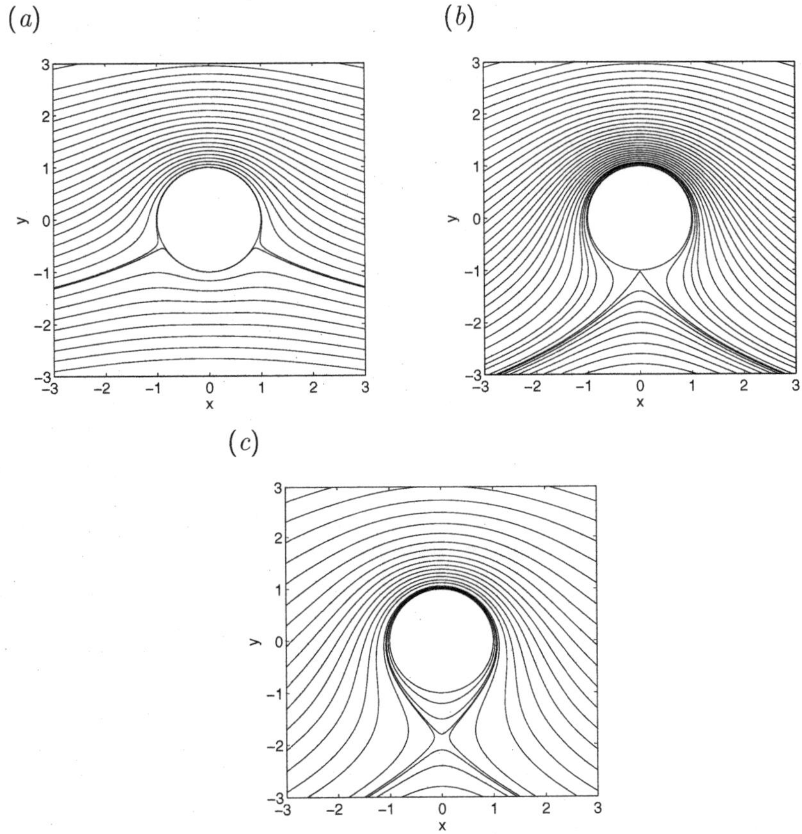

Figure 3.7.2 Streamline pattern of uniform (streaming) flow past a circular cylinder with different degrees of circulation around the cylinder determined by the dimensionless parameter β defined in equation (3.7.8); (*a*) $\beta = 0.5$, (*b*) 1.0, and (*c*) 1.2. The streamline pattern in the absence of circulation, β=0, is shown in figure 3.5.3(*a*).

In this case, the right-hand sides of (3.7.10) and (3.7.11) are equal. One consequence of Stokes's circulation theorem is that, because the vorticity in an irrotational flow vanishes at every point, the circulation around any reducible loop drawn in an irrotational flow is zero. Important implications of this property will be discussed in chapter 11 in the context of vortex dynamics.

Reducible loops enclosing point vortices

The circulation around a reducible loop that encloses a collection of N point vortices with strength $\kappa_1, \kappa_2, \ldots, \kappa_N$, is equal to the sum of the strengths of the

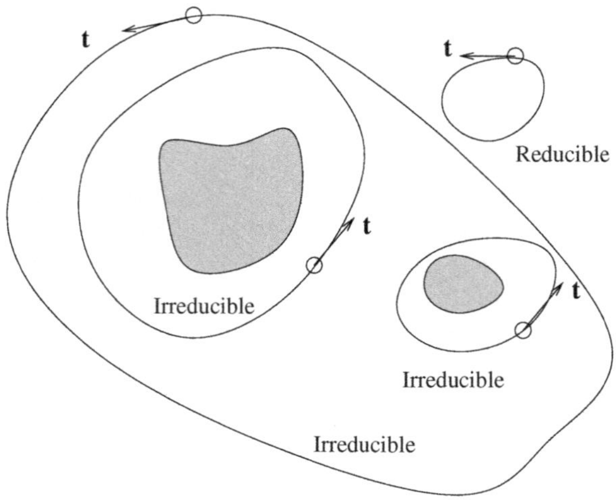

Figure 3.7.3 Illustration of reducible and irreducible loops in a two-dimensional flow. The shaded areas represent flow boundaries.

point vortices,

$$C = \sum_{i=1}^{N} \kappa_i. \tag{3.7.12}$$

If some point vortices have positive strength and other point vortices have negative strength so that the sum of the strengths zero, the circulation around the loop is zero.

As an example, we consider uniform flow past a circular cylinder described by the potential shown in (3.7.5). To confirm that the circulation around any loop that encloses the cylinder is equal to κ, we compute the circulation around a loop of radius R centered at the cylinder, and find the expected result

$$C \equiv \oint u_t \, \mathrm{d}l = \oint u_\theta \, R \, \mathrm{d}\theta = R \oint \frac{1}{r} \frac{\partial \phi}{\partial \theta} \, \mathrm{d}\theta = R \frac{1}{R} \oint \frac{\kappa}{2\pi} \, \mathrm{d}\theta = \kappa. \tag{3.7.13}$$

Irreducible loops

The circulation around a loop that encloses one boundary or more than one boundaries of a two-dimensional irrotational flow can be arbitrary. In practice, the amount of circulation is set up internally during a start up period when the flow develops from the state of rest. The circulation established around a moving body is of central interest in aerodynamics, as will be discussed in chapter 12.

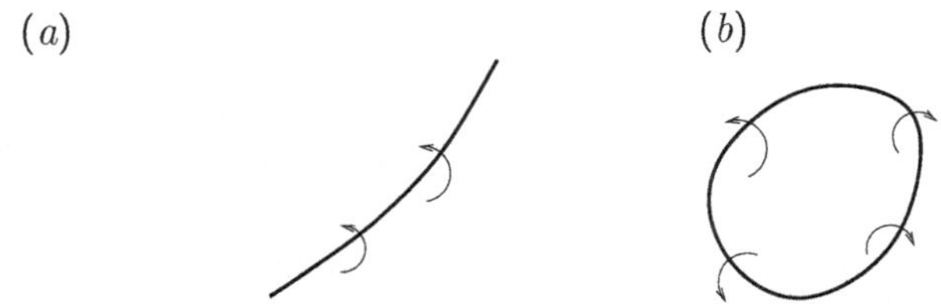

Figure 3.7.4 Illustration of (a) an infinite, and (b) a closed line vortex in a three-dimensional flow.

3.7.4 Line vortices in three-dimensional flow

Viewed from a three-dimensional perspective, a point vortex in the xy plane appears like a rectilinear line vortex parallel to the z axis, as shown in figure 3.7.1(b). Deforming this rectilinear line vortex or merging the two ends to form a loop, we obtain a curved three-dimensional line vortex in a three-dimensional flow. One example familiar to smokers is a closed line vortex with a circular or wobbly shape, called a line vortex ring.

A line vortex can be infinite, as illustrated in figure 3.7.4(a), or closed, as illustrated in figure 3.7.4(b), but may not suddenly end in the interior of the fluid. In real life, a fluid is always bounded by a rigid or deformable surface, and an infinite line vortex inevitably ends at the boundaries. The analysis and computation of the flow associated with, or induced by, a three-dimensional line vortex defines an important field of fluid dynamics with important applications in turbulent motion and aerodynamics, as discussed in chapters 11 and 12.

Problems

3.7.1 *Circulation around a loop in the xy plane*

Consider a closed loop in the xy plane performing m turns around a point vortex with strength κ. Explain why the circulation around this loop is equal to $m\kappa$.

3.7.2 *Point vortex dipole*

Just as the point-source dipole arises from a point source/sink dipole, as discussed in section 3.5, a point-vortex dipole arises from a point vortex with positive strength and a point vortex with negative strength of equal magnitude, in the limit as the distance between the two point vortices tends to zero while their strength increases by the inverse proportion. The harmonic potential associated with a point vortex dipole oriented along the x or y axis is given, respectively, by

$$\phi^{PVD-x}(x, y) = \lambda_x \frac{\partial \phi^{PV}}{\partial x_0}, \qquad \phi^{PVD-y}(x, y) = \lambda_y \frac{\partial \phi^{PV}}{\partial y_0}, \tag{3.7.14}$$

where λ_x and λ_y are the components of the vectorial strength of the point-vortex dipole in the x and y direction.

Carry out the differentiations on the right-hand sides of (3.7.14) and compare the resulting expressions with those shown in equations (3.5.17) and (3.5.22) for the two-dimensional point-source dipole. Based on this comparison, establish a relationship between the flow due to a point-vortex dipole and the flow due to a point-source dipole.

3.7.3 *Irreducible loops in three-dimensional flow*

(*a*) Consider a three-dimensional domain of flow extending to infinity and bounded internally by a toroidal boundary having the shape of a donut. Show that this flow contains irreducible loops that may not be shrunk to a point without crossing flow boundaries.

(*b*) Invent another three-dimensional domain of flow containing irreducible loops.

Forces and Stresses 4

In previous chapters, we have discussed the kinematic structure of a flow but made no reference to the external action that is necessary to establish a flow and to the physical mechanism that is necessary to sustain the motion of the fluid. To address these issues, we now turn our attention to the hydrodynamic forces developing in a fluid as a result of the motion, and introduce constitutive equations relating the stresses developing at the surface of infinitesimal fluid parcels to the parcel motion and deformation. The constitutive equations will then be incorporated into an integrated theoretical framework that allows us to compute the structure of a steady flow and the evolution of an unsteady flow from a specified initial state.

4.1 Body forces and surface forces

Two types of forces are exerted on any coherent material: a homogeneous forces acting on its volume, and a surface forces acting on its boundaries.

4.1.1 Body forces

A fluid parcel, like any material, is subject to a force mediated by an ambient gravitational, electrical, electromagnetic, or any other external field. Electrical and electromagnetic forces arise when the fluid is electrically charged or contains molecules or particles of a polarized material. Under the influence of these fields, the molecules residing inside the parcel are acted upon individually and

C. Pozrikidis, *Fluid Dynamics: Theory, Computation, and Numerical Simulation*,
DOI: 10.1007/978-0-387-95871-2_4,

independently by a force that may be constant or vary with position in the parcel. The sum of the forces exerted on the individual molecules amounts to a net body force that is proportional to the number of molecules residing in the parcel, and thus to the parcel volume.

Let $\mathrm{d}F_p$ be the gravitational force exerted on a small fluid parcel with volume $\mathrm{d}V_p$, density ρ, and mass $\mathrm{d}m_p = \rho\,\mathrm{d}V_p$. By definition,

$$\mathrm{d}\mathbf{F}_p = \mathbf{g}\,\rho\,\mathrm{d}V_p, \tag{4.1.1}$$

where $\mathbf{g}$ is the acceleration of gravity. The right-hand side of (4.1.1) has units of acceleration multiplied by mass, which amounts to force. One distinguishing feature of the body force due to gravity is that it is independent of the molecular motions. This means that a certain mass of fluid weighs the same, independent of whether the fluid is stationary or flows.

4.1.2 Surface forces

A different type of force arises at the surface of a fluid parcel and at the boundaries of a flow, such as the surface of a bubble rising through an ambient liquid or the hood of a moving vehicle. More generally, a surface force can be defined on any fictitious surface drawn inside the bulk of a fluid or over the boundary. Understanding the physical origin of the surface force requires consideration of molecular motions and necessitates a distinction between gases and liquids. A key idea is the equivalence between local hydrodynamic force and rate of exchange of momentum between adjacent fluid layers due to molecular excursions.

Gases

To understand the origin of surface forces developing in a gas, we draw a surface in the interior of the gas and consider the momentum of the molecules crossing the surface from either side. The momentum normal to the surface is responsible for a normal force. If the molecules move with different average tangential velocities on either side of the surface, the net transport of tangential momentum is responsible for a tangential surface force necessary to accelerate or decelerate the molecules. In the present context, the average velocity of a molecule can be identified with the velocity of the fluid at the location where a molecule last underwent a collision with one of its peers. The effective force field due to the tangential surface force slows down fast moving molecules as they move toward regions of slower moving fluid.

Liquids

The physical origin of surface forces developing in a liquid is somewhat different. The molecules of a liquid perform oscillatory motions around a mean position with an amplitude that is determined by their distance from closely spaced neighbors. Occasional excursions into vacant spots are responsible for momentum transport attributed to the action of a surface force.

Problem

4.1.1 *Friction*

The friction on a body sliding over a horizontal surface imparts to the body a tangential surface force. Does this frictional force depend on the contact area?

4.2 Traction and the stress tensor

Consider a small surface with area $\mathrm{d}S$ centered at a point, $\mathbf{x} = (x, y, z)$, in a stationary or moving fluid, as illustrated in figure 4.2.1. The designated outer side of the surface is indicated by the direction of the unit vector normal to the surface at the point $\mathbf{x}$, denoted by $\mathbf{n} = (n_x, n_y, n_z)$. According to our discussion in section 4.1, a body of fluid whose instantaneous boundary includes the small surface under consideration experiences a surface force, $\mathrm{d}\mathbf{F}^S$. This surface force can point in any direction; that is, it may have a component normal to the surface, and a component tangential to the surface.

Traction

The ratio between the surface force, $\mathrm{d}\mathbf{F}^S$, and the area of the surface, $\mathrm{d}S$, is the average stress exerted on the surface. As the surface area $\mathrm{d}S$ becomes infinitesimal, the average stress tends to a limit defined as the traction exerted on an infinitesimal surface, $\mathbf{f}$. Thus, by definition,

$$\mathbf{f} \equiv \frac{\mathrm{d}\mathbf{F}^S}{\mathrm{d}S}. \tag{4.2.1}$$

The three scalar components of the traction have units of force per area, which amounts to stress. Rearranging equation (4.2.1), we obtain an expression for the surface force exerted on an infinitesimal surface in terms of the traction,

$$\mathrm{d}\mathbf{F}^S = \mathbf{f}\,\mathrm{d}S. \tag{4.2.2}$$

Integrating the traction over a surface area we obtain a resultant force.

It is clear from relation (4.2.1) that the traction is defined only when (*a*) the location, and (*b*) the side of the infinitesimal surface upon which the traction is exerted are specified, respectively, in terms of the coordinates of the center-point, $\mathbf{x}$, and orientation of the unit normal vector, $\mathbf{n}$. This requirement is signified by writing

$$\mathbf{f}(\mathbf{x}, \mathbf{n}), \tag{4.2.3}$$

where the parentheses enclose the arguments of the three scalar components of the traction. If the flow is unsteady, or the position and orientation of the surface change in time, time should be added to the arguments on the right-hand side of (4.2.3).

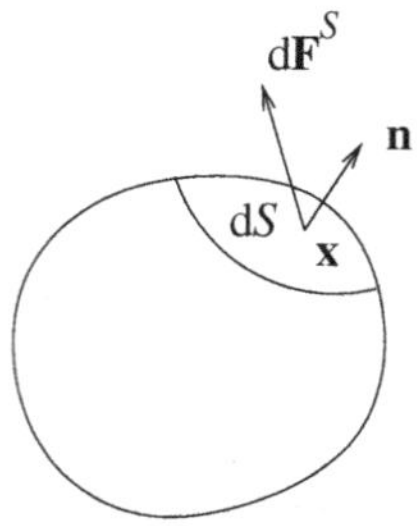

Figure 4.2.1 Illustration of a small section on the surface of a fluid parcel used to define the hydrodynamic traction exerted on the parcel.

The stress tensor

The traction exerted on a small surface that is perpendicular to the x, y or z axis, is denoted, respectively, by

$$\mathbf{f}^{(x)} = (f_x^{(x)}, f_y^{(x)}, f_z^{(x)}), \qquad \mathbf{f}^{(y)} = (f_x^{(y)}, f_y^{(y)}, f_z^{(y)}),$$
$$\mathbf{f}^{(z)} = (f_x^{(z)}, f_y^{(z)}, f_z^{(z)}), \tag{4.2.4}$$

where the normal vector points in the positive directions of these axes, as depicted in figure 4.2.2. Stacking these vectors on top of one another in a particular order, we obtain the 3×3 stress tensor

$$\boldsymbol{\sigma} \equiv \begin{bmatrix} f_x^{(x)} & f_y^{(x)} & f_z^{(x)} \\ f_x^{(y)} & f_y^{(y)} & f_z^{(y)} \\ f_x^{(z)} & f_y^{(z)} & f_z^{(z)} \end{bmatrix}. \tag{4.2.5}$$

Next, we introduce the standard two-index notation for the components of the stress tensor,

$$\sigma_{ij} \equiv f_j^{(i)}, \tag{4.2.6}$$

where $i, j = x, y, z$. The first index of σ_{ij} indicates the component of the normal vector on the infinitesimal surface upon which the traction is exerted, and the second index indicates the component of the corresponding traction. With the convention expressed by (4.2.6), the stress tensor takes the form

$$\boldsymbol{\sigma} = \begin{bmatrix} \sigma_{xx} & \sigma_{xy} & \sigma_{xz} \\ \sigma_{yx} & \sigma_{yy} & \sigma_{yz} \\ \sigma_{zx} & \sigma_{zy} & \sigma_{zz} \end{bmatrix}. \tag{4.2.7}$$

In the case of two-dimensional flow in the xy plane, the stresses are encapsulated in the 2×2 stress tensor

$$\boldsymbol{\sigma} = \begin{bmatrix} \sigma_{xx} & \sigma_{xy} \\ \sigma_{yx} & \sigma_{yy} \end{bmatrix}. \tag{4.2.8}$$

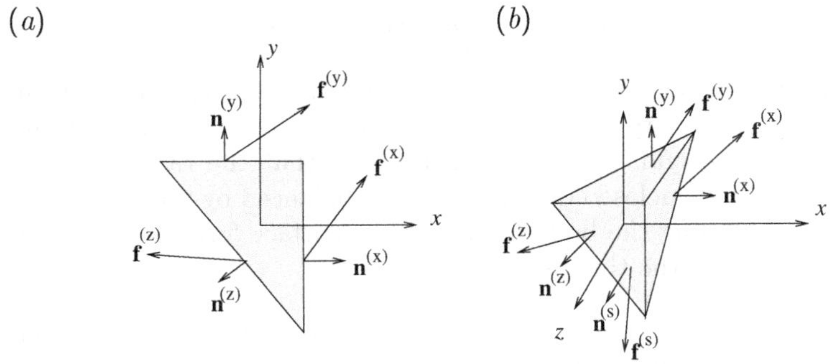

Figure 4.2.2 (a) Illustration of a triangular fluid parcel in a two-dimensional flow, and (b) a polyhedral fluid parcel in a three-dimensional fluid. These parcels are used as devices for computing the traction exerted on an arbitrary surface in terms of (a) the unit vector normal to the surface, and (b) the stress tensor.

We will see that the five omitted components involving the subscript z are either constant or zero.

Traction in terms of the stress tensor

Next, we demonstrate that the dependence of the traction on the position vector and normal vector, displayed symbolically in (4.2.3), can be decoupled in a simple fashion yielding

$$\mathbf{f}(\mathbf{x}, \mathbf{n}) = \mathbf{n}(\mathbf{x}) \cdot \boldsymbol{\sigma}(\mathbf{x}). \tag{4.2.9}$$

In index notation,

$$f_j(\mathbf{x}, \mathbf{n}) = n_i \sigma_{ij} = n_x \sigma_{xj} + n_y \sigma_{yj} + n_z \sigma_{zj}, \tag{4.2.10}$$

where summation of the repeated index i is implied in the middle expression of (4.2.10), and the index j is free to vary over x, y, and z.

An important consequence of (4.2.9) is that, if the nine components of the stress tensor are known at a point, the traction exerted on any infinitesimal surface centered at that point can be evaluated in terms of the unit normal vector, merely by carrying out a vector-matrix multiplication. To confirm that expression (4.2.9) is consistent with the foregoing definitions, we choose $\mathbf{n} = (1, 0, 0)$ and carry out the vector-matrix multiplication on the right-hand side of (4.2.10) to find $\mathbf{f} = \mathbf{f}^{(x)}$, as required. Working in a similar fashion with $\mathbf{n} = (0, 1, 0)$ and $\mathbf{n} = (0, 0, 1)$, we find $\mathbf{f} = \mathbf{f}^{(y)}$ and $\mathbf{f} = \mathbf{f}^{(z)}$, as required.

It remains to show that (4.2.9) holds true for general orientations of the unit normal vector, $\mathbf{n}$. For simplicity, we present the proof for two-dimensional flow in the xy plane with reference to the 2×2 stress tensor defined in (4.2.8).

Force balance on a small triangular parcel

We begin by considering a small area of fluid enclosed by an infinitesimal triangle with two sides perpendicular to the x and y axis, as depicted in figure 4.2.2(a). Newton's second law of motion requires that the rate of change of momentum of the fluid enclosed by the triangle is balanced by the forces exerted on the triangle, including the body force and the surface force associated with the traction exerted on the three sides.

The momentum of the parcel and the body force exerted on the parcel are both proportional to the area of the triangle, $\frac{1}{2}\,\Delta x\,\Delta y$. The surface force exerted on the vertical side is equal to $\mathbf{f}^{(x)}\Delta y$, the surface force exerted on the horizontal side is equal to $\mathbf{f}^{(y)}\Delta x$, and the surface force exerted on the slanted side is equal to $\mathbf{f}^{(s)}\Delta l$, where Δl is the length of the slanted side, $\Delta l = (\Delta x^2 + \Delta y^2)^{1/2}$.

In the limit as Δx and Δy tend to zero, the fluid momentum and body force become negligible compared to the surface force exerted on the sides, and the sum of the three surface forces vanish. Setting the x and y components of the sum to zero, we find

$$f_x^{(s)}\,\Delta l + \sigma_{xx}\,\Delta y + \sigma_{yx}\,\Delta x = 0,$$
$$f_y^{(s)}\,\Delta l + \sigma_{xy}\,\Delta y + \sigma_{yy}\,\Delta x = 0. \tag{4.2.11}$$

Using elementary trigonometry, we find that the components of the unit vector normal to the slanted side of the triangle pointing outward, $\mathbf{n}^{(s)}$, are given by

$$n_x^{(s)} = -\frac{\Delta y}{\Delta l}, \qquad n_y^{(s)} = -\frac{\Delta x}{\Delta l}. \tag{4.2.12}$$

Combining equations (4.2.11) and (4.2.12), we find

$$f_x^{(s)} = n_x^{(s)}\,\sigma_{xx} + n_y^{(s)}\,\sigma_{yx}, \qquad f_y^{(s)} = n_x^{(s)}\,\sigma_{xy} + n_y^{(s)}\,\sigma_{yy}, \tag{4.2.13}$$

which are precisely the x and y components of (4.2.9).

To carry out an analogous proof for three-dimensional flow, we consider the forces exerted on the sides and over the volume of the polyhedral fluid parcel illustrated in figure 4.2.2(b), and work in similar ways (problem 4.2.1).

4.2.1 Traction on either side of a fluid surface

Next, we consider a thin fluid layer with a designated outer side indicated by the direction of the unit normal vector, $\mathbf{n}$, and a designated inner side indicated by the direction of the opposite normal vector, $\mathbf{n}^{inner} = -\mathbf{n}$, as illustrated in figure 4.2.3. Balancing the rate of change of momentum of the fluid residing inside the thin layer with the forces exerted on the layer, and repeating the

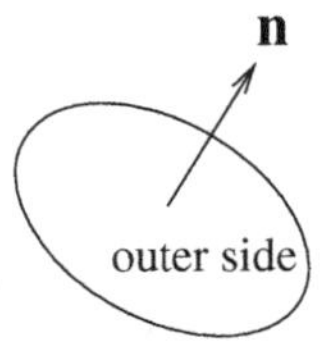

Figure 4.2.3 Illustration of a thin fluid layer with a designated inner and outer side.

preceding arguments on the insignificance of the fluid momentum and body force compared to the surface force, we derive the force balance equation

$$\mathbf{f}^{outer} + \mathbf{f}^{inner} = \mathbf{0}, \tag{4.2.14}$$

which is a statement of Newton's third law of action and reaction: the force exerted on one body by a second body is equal in magnitude and opposite in direction to that exerted by the second body on the first body.

It is reassuring to confirm that expression (4.2.9) is consistent with the physical law expressed by (4.2.14). Substituting the former in the latter, we obtain

$$\mathbf{n} \cdot \boldsymbol{\sigma} + \mathbf{n}^{inner} \cdot \boldsymbol{\sigma} = \mathbf{0}, \tag{4.2.15}$$

which is true in light of the definition $\mathbf{n}^{inner} = -\mathbf{n}$. More generally, a prerequisite for the satisfaction of (4.2.14) is the property

$$\mathbf{f}(\mathbf{x}, -\mathbf{n}) = -\mathbf{f}(\mathbf{x}, \mathbf{n}), \tag{4.2.16}$$

which is clearly satisfied by the right-hand side of (4.2.9).

4.2.2 Traction on a boundary

Now we consider a small fluid surface residing at the boundary of a flow. The outer side of the surface is indicated by the unit normal vector $\mathbf{n}^{outward}$ pointing into the boundary, as illustrated in figure 4.2.4. Newton's third law of action and reaction requires that the traction exerted on the surface should balance the traction exerted by the fluid on the boundary, $\mathbf{f}^B$,

$$\mathbf{f}^B + \mathbf{n}^{outward} \cdot \boldsymbol{\sigma} = \mathbf{0}. \tag{4.2.17}$$

In terms of the inward unit normal vector pointing into the fluid, $\mathbf{n}^{inward} = -\mathbf{n}^{outward}$, we obtain

$$\mathbf{f}^B = \mathbf{n}^{inward} \cdot \boldsymbol{\sigma}. \tag{4.2.18}$$

Expression (4.2.18) allows us to compute the traction exerted on a boundary in terms of the stress tensor evaluated at the boundary.

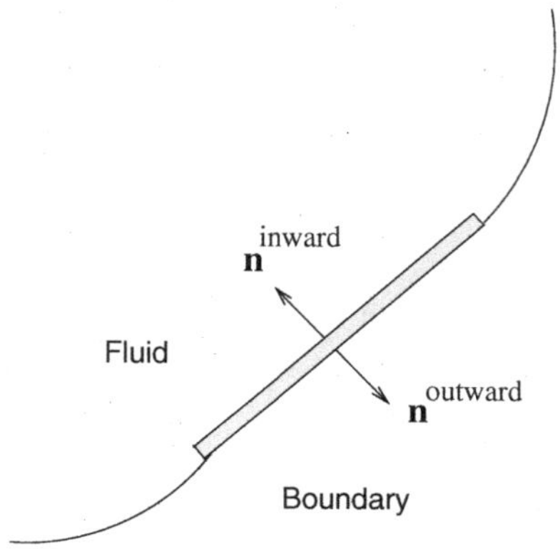

Figure 4.2.4 Illustration of a thin fluid layer adjacent to a boundary used to define the hydrodynamic force exerted on the boundary.

4.2.3 Symmetry of the stress tensor

The torque with respect to the point, $\mathbf{x}_0$, due to a force, $\mathbf{F}$, applied at a point, $\mathbf{x}$, is defined by the outer vector product

$$\mathbf{T}(\mathbf{x}_0) \equiv (\mathbf{x} - \mathbf{x}_0) \times \mathbf{F}. \tag{4.2.19}$$

A fundamental law of mechanics originating from Newton's second law of motion requires that the rate of change of angular momentum of a fluid parcel must be balanced by the torque exerted on the fluid parcel, including the torque due to the body force and the torque due to surface force.

Applying this law for a rectangular fluid parcel whose sides are parallel to the x, y, and z, axes, we find, that, in the absence of a body force that induces a torque, the tangential component of the traction in the jth direction exerted on the side that is perpendicular to the ith axis must be equal to the tangential component of the traction in the ith direction exerted on the side that is perpendicular to the jth axis, otherwise an imbalance will arise (problem 4.2.2). Thus,

$$f_j^{(i)} = f_i^{(j)}, \tag{4.2.20}$$

stating that the stress tensor is symmetric,

$$\sigma_{ij} = \sigma_{ji}. \tag{4.2.21}$$

It is important to emphasize that the stress tensor is symmetric only in the absence of an externally induced torque; that is, in the absence of an external force field causing point particles to spin around an axis. This condition will be tacitly assumed in the remainder of this book.

Problems

4.2.1 *Traction in three-dimensional flow*

Prove expression (4.2.9) for three-dimensional flow. *Hint:* Perform a force balance over the polyhedral volume depicted in figure 4.2.2(b).

4.2.2 *Symmetry of the stress tensor*

Demonstrate the symmetry of the stress tensor for two-dimensional flow in the absence of an externally induced torque.

4.3 Traction jump across a fluid interface

Equation (4.2.14) states that the traction exerted on one side of a surface drawn inside a fluid is equal in magnitude and opposite in direction to that exerted on the other side. To derive this relation, we have performed a force balance over a thin fluid layer centered at the surface, considering the force exerted along the edges infinitesimal. If the fluid residing inside this layer is homogeneous, the edge force scales with the layer thickness and is negligible compared to the surface force exerted on the two sides.

However, if a thin layer is centered at an interface between two different fluids instead of a regular surface residing inside a homogeneous fluid, differences in the magnitude of intermolecular forces on either side of the layer generate an effective edge force that does not scale with the layer thickness.

Interfacial tension

The interfacial edge force can be expressed in terms of the interfacial tension or surface tension, γ, defined as the tangential force per differential length exerted around the edge of a section of an interface. In reality, the surface tension is the integral of the normal stress exerted in a plane that is normal to the interface over a thin interfacial layer. Interfaces with a membrane-like constitution exhibit tangential as well as normal interfacial tensions and possibly bending moments.

Consider a simple interface separating two fluids. The surface tension pulls the interfacial layer in a direction that is tangential to the interface and normal to the edges. The magnitude of the surface tension depends on the local temperature and molecular constitution of the interface determined by the concentration of surface active substances residing over interface, called surfactants, as will be discussed in section 4.10. The higher the temperature or the concentration of a surfactant, the lower the surface tension. Surfactants are often

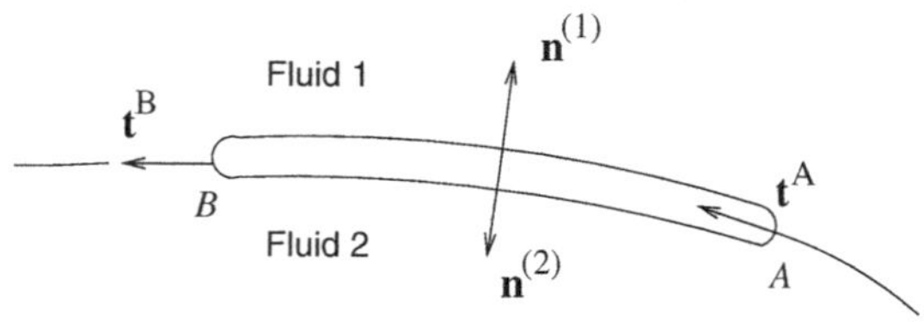

Figure 4.3.1 Illustration of forces exerted on a thin fluid layer centered on a two-dimensional interface, including the hydrodynamic force due to the fluid stresses and the force due to the surface tension.

added to liquids to lower the surface tension and achieve a desired effect. A dish or laundry detergent is a common household surfactant used to lower the strength of the forces anchoring particles to a soiled surface.

4.3.1 Force balance at a two-dimensional interface

Figure 4.3.1 illustrates a small section of a two-dimensional interface with length Δl. Surface tension pulls the layer at the two edges in directions that are tangential to the interface at the two end points, A and B. Balancing the surface force due to the stress tensor and the edge forces, we obtain

$$[\mathbf{n}^{(1)} \cdot \boldsymbol{\sigma}^{(1)}]\,\Delta l + [\mathbf{n}^{(2)} \cdot \boldsymbol{\sigma}^{(2)}]\,\Delta l - \gamma^A\,\mathbf{t}^A + \gamma^B\,\mathbf{t}^B = \mathbf{0}. \tag{4.3.1}$$

The unit normal vector, $\mathbf{n}^{(1)}$, points into the fluid labeled 1, whereas the unit normal vector, $\mathbf{n}^{(2)} = -\mathbf{n}^{(1)}$, points into the fluid labeled 2. Expressing the second normal vector in terms of the first normal vector and rearranging the resulting expression, we find

$$\mathbf{n}^{(1)} \cdot (\boldsymbol{\sigma}^{(1)} - \boldsymbol{\sigma}^{(2)}) = -\frac{\gamma^B\,\mathbf{t}^B - \gamma^A\,\mathbf{t}^A}{\Delta l}. \tag{4.3.2}$$

If the surface tension is uniform, $\gamma^B = \gamma^A$. If the interface is flat, $\mathbf{t}^B = \mathbf{t}^A$, the vectorial difference $\mathbf{t}^B - \mathbf{t}^A$ is zero, and the right-hand side of (4.3.2) vanishes. Equation (4.3.2) then requires that the traction is continuous across the interface.

As the length of the segment, Δl, tends to zero, the fraction on the right-hand side of (4.3.2) tends to the derivative of the product $\gamma\,\mathbf{t}$ with respect to arc length, l, measured in the direction of the tangent vector $\mathbf{t}$ from an arbitrary origin,

$$\mathbf{n}^{(1)} \cdot (\boldsymbol{\sigma}^{(1)} - \boldsymbol{\sigma}^{(2)}) = -\frac{\mathrm{d}(\gamma\mathbf{t})}{\mathrm{d}l}. \tag{4.3.3}$$

Expanding the derivative of the product on the right-hand side, we find

$$\mathbf{n}^{(1)} \cdot (\boldsymbol{\sigma}^{(1)} - \boldsymbol{\sigma}^{(2)}) = -\gamma\,\frac{\mathrm{d}\mathbf{t}}{\mathrm{d}l} - \frac{\mathrm{d}\gamma}{\mathrm{d}l}\,\mathbf{t}. \tag{4.3.4}$$

The second term on the right-hand side contributes a traction discontinuity tangential to the interface, known as the Marangoni traction. If the surface tension is uniform over the interface, the Marangoni traction does not appear.

To interpret the first term on the right-hand side of (4.3.4), we consider the difference between the two nearly equal tangential vectors $\mathbf{t}^B$ and $\mathbf{t}^A$. As the arc length Δl tends to zero, the difference between these vectors tends to a new vector directed normal to the interface. More precisely, in this limit, the ratio $(\mathbf{t}^B - \mathbf{t}^A)/\Delta l$ tends to the derivative

$$\frac{\mathrm{d}\mathbf{t}}{\mathrm{d}l} = -\kappa\,\mathbf{n}^{(1)}, \tag{4.3.5}$$

where κ is the positive or negative curvature of the interface. By definition, $\kappa = 1/R$, where R is the positive or negative radius of curvature of the interface. Conversely,

$$\frac{\mathrm{d}\mathbf{n}^{(1)}}{\mathrm{d}l} = \kappa\,\mathbf{t}. \tag{4.3.6}$$

Equations (4.3.5) and (4.3.6) comprise the Frenet relations.

Substituting (4.3.5) in (4.3.4) and rearranging, we find

$$\Delta\mathbf{f} \equiv \mathbf{n}^{(1)} \cdot (\boldsymbol{\sigma}^{(1)} - \boldsymbol{\sigma}^{(2)}) = \gamma\,\kappa\,\mathbf{n}^{(1)} - \frac{\mathrm{d}\gamma}{\mathrm{d}l}\,\mathbf{t}. \tag{4.3.7}$$

The first term on the right-hand side of (4.3.7) contributes a traction discontinuity normal to the interface. If either the curvature of the interface or the surface tension vanishes, the normal stress is continuous across the interface.

Local coordinates

As an application, we consider the jump in traction across the interface depicted in figure 4.3.2(a). The origin of the Cartesian axes has been set at a point on the interface, the x axis is tangential to the interface, and the y axis is normal to the interface pointing into the fluid labeled 1. At the origin of the Cartesian axes, the components of the unit normal vector, $\mathbf{n}^{(1)}$, are $n_x^{(1)} = 0$ and $n_y^{(1)} = 1$, and the jump in the interfacial traction is

$$\Delta\mathbf{f} = (\sigma_{yx}^{(1)} - \sigma_{yx}^{(2)})\,\mathbf{e}_x + (\sigma_{yy}^{(1)} - \sigma_{yy}^{(2)})\,\mathbf{e}_y, \tag{4.3.8}$$

where $\mathbf{e}_x$ and $\mathbf{e}_y$ are, respectively, the unit vectors parallel to the x and y axes.

At the origin, $\mathbf{t} = \mathbf{e}_x$ and $\mathbf{n}^{(1)} = \mathbf{e}_y$. Comparing the right-hand sides of (4.3.8) and (4.3.7), we derive an expression for the jump in the shear stress,

$$\sigma_{yx}^{(1)} - \sigma_{yx}^{(2)} = -\frac{\mathrm{d}\gamma}{\mathrm{d}l}, \tag{4.3.9}$$

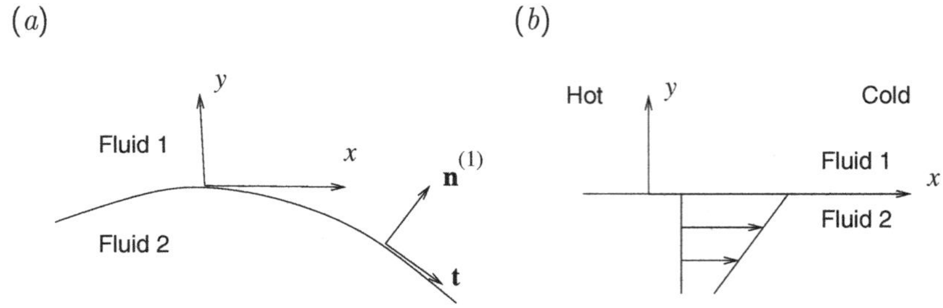

Figure 4.3.2 (*a*) A local coordinate system with the x axis tangential to a two-dimensional interface at a point is used to evaluate the jump in the traction across the interface. (*b*) A differentially heated interface drives a thermocapillary flow.

and another expression for the jump in the normal stress,

$$\sigma_{yy}^{(1)} - \sigma_{yy}^{(2)} = \gamma\,\kappa. \tag{4.3.10}$$

Next, we consider a liquid layer that is hot at the left end and cold at the right end, as illustrated in figure 4.3.2(*b*). In this case, $\mathrm{d}T/\mathrm{d}x < 0$, $\mathrm{d}\gamma/\mathrm{d}x > 0$, and equation (4.3.9) yields

$$\sigma_{yx}^{(1)} - \sigma_{yx}^{(2)} = -\frac{\mathrm{d}\gamma}{\mathrm{d}x} < 0. \tag{4.3.11}$$

Because the shear stress is insignificant in the gas above the layer, $\sigma_{yx}^{(1)} = 0$, and

$$\sigma_{yx}^{(2)} = \mu_2\,\frac{\mathrm{d}u_x}{\mathrm{d}y} > 0, \tag{4.3.12}$$

which is consistent with the velocity profile drawn in figure 4.3.2(*b*). Physically, the high surface tension at the cold end pulls the fluid against the low surface tension at the hot end. A flow induced by temperature differences causing variations in surface tension is called a thermocapillary flow.

Curvature of a circle

We have seen that the curvature of an interface determines the jump in the traction due to surface tension. To develop expertise in the computation of the curvature, we consider a circle of radius a centered at the origin, described in parametric form by the equations $x = a\cos\theta$ and $y = a\sin\theta$, where θ is the polar angle measured around the circle center, varying from 0 to 2π. The components of the unit tangent vector pointing in the counterclockwise direction are

$$t_x = \frac{\mathrm{d}x}{\mathrm{d}l}, \qquad t_y = \frac{\mathrm{d}y}{\mathrm{d}l}, \tag{4.3.13}$$

where $\mathrm{d}l = (\mathrm{d}x^2 + \mathrm{d}y^2)^{1/2}$ is the differential arc length measured in the counterclockwise direction. Using the parametric representation, we find

$$\begin{aligned} &\mathrm{d}x = -a \sin\theta \,\mathrm{d}\theta, \qquad \mathrm{d}y = a \cos\theta \,\mathrm{d}\theta, \qquad \mathrm{d}l = a\,\mathrm{d}\theta, \\ &t_x = -\sin\theta, \qquad t_y = \cos\theta. \end{aligned} \tag{4.3.14}$$

Based on these formulas, we compute

$$\frac{\mathrm{d}t_x}{\mathrm{d}l} = \frac{\mathrm{d}(-\sin\theta)}{\mathrm{d}(a\theta)} = -\frac{1}{a}\cos\theta, \qquad \frac{\mathrm{d}t_y}{\mathrm{d}l} = \frac{\mathrm{d}(\cos\theta)}{\mathrm{d}(a\theta)} = -\frac{1}{a}\sin\theta. \tag{4.3.15}$$

In unified vector form,

$$\frac{\mathrm{d}\mathbf{t}}{\mathrm{d}l} = -\frac{1}{a}\mathbf{n}^{(1)}, \tag{4.3.16}$$

where $\mathbf{n}^{(1)} = (\cos\theta, \sin\theta)$ is the unit vector normal to the circle pointing into the exterior. Comparing (4.3.16) with (4.3.5), we confirm that the curvature of the circle is equal to the inverse of its radius.

Formulas for the curvature

Taking the inner product of both sides of equation (4.3.5) with the unit normal vector $\mathbf{n}^{(1)}$, we find

$$\kappa = -\mathbf{n}^{(1)} \cdot \frac{\mathrm{d}\mathbf{t}}{\mathrm{d}l} = -\frac{\mathrm{d}x}{\mathrm{d}l}\,\mathbf{n}^{(1)} \cdot \frac{\mathrm{d}\mathbf{t}}{\mathrm{d}x}. \tag{4.3.17}$$

If an interface does not turn upon itself but has a monotonic shape, the position of the interface can be described by a single-valued function, $y = f(x)$. Using elementary geometry, we find

$$\begin{aligned} &\mathbf{n}^{(1)} = \frac{1}{(1+f'^2)^{1/2}}\,(-f'\,\mathbf{e}_x + \mathbf{e}_y), \\ &\mathbf{t} = \frac{1}{(1+f'^2)^{1/2}}\,(\mathbf{e}_x + f'\,\mathbf{e}_y), \qquad \frac{\mathrm{d}l}{\mathrm{d}x} = (1+f'^2)^{1/2}, \end{aligned} \tag{4.3.18}$$

where a prime denotes a derivative with respect to x. Substituting expressions (4.3.18) in the right-hand side of (4.3.17) and simplifying, we derive an expression for the curvature,

$$\kappa = -\frac{f''}{(1+f'^2)^{3/2}} = \frac{1}{f'}\Big(\frac{1}{(1+f'^2)^{1/2}}\Big)'. \tag{4.3.19}$$

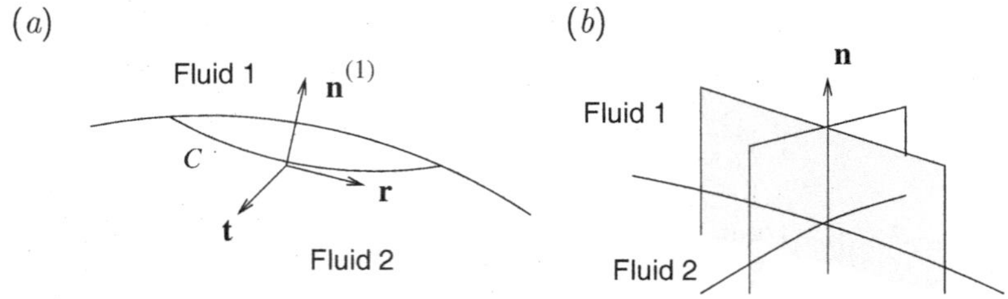

Figure 4.3.3 (*a*) Forces exerted on a thin fluid layer centered at a three-dimensional interface include the hydrodynamic force due to the fluid stresses and the force due to the surface tension. (*b*) The mean curvature of a three-dimensional surface is equal to the average of the directional curvatures in two perpendicular planes containing the normal vector.

The slope angle, θ, is defined by the equation

$$\tan\theta = f'. \tag{4.3.20}$$

We note that the fraction on the right-hand side of (4.3.19) is equal to $|\cos\theta|$ and derive the alternative expression

$$\kappa = \frac{1}{f'}\frac{\mathrm{d}|\cos\theta|}{\mathrm{d}x}. \tag{4.3.21}$$

The curvature of an interface that is described parametrically by the functions $x = X(\xi)$ and $y = Y(\xi)$ is given by

$$\kappa = \frac{X_{\xi\xi}\,Y_\xi - Y_{\xi\xi}\,X_\xi}{(X_\xi^2 + Y_\xi^2)^{3/2}}, \tag{4.3.22}$$

where a subscript denotes a derivative with respect to the parametric variable, ξ. Formula (4.3.19) arises by setting $\xi = x$.

4.3.2 Force balance at a three-dimensional interface

To derive the counterpart of the dynamic interfacial condition (4.3.7) for a three-dimensional interface, we consider a thin material patch straddling the interface, as illustrated in figure 4.3.3. Let $\mathbf{n}^{(1)}$ be the unit vector normal to the interface pointing into fluid 1, and $\mathbf{r}$ be the unit vector tangential to the layer edge. Surface tension pulls the layer in the direction of the unit vector $\mathbf{t}$ that is tangential to the interface and normal to both $\mathbf{n}^{(1)}$ and $\mathbf{r}$. Recalling the geometrical interpretation of the outer vector product discussed in section 2.3, we write $\mathbf{t} = \mathbf{r} \times \mathbf{n}$.

Next, we balance the surface force due to the fluid stress and the edge force due to the surface tension, and find

$$[\mathbf{n}^{(1)} \cdot \boldsymbol{\sigma}^{(1)}]\, \Delta S + [\mathbf{n}^{(2)} \cdot \boldsymbol{\sigma}^{(2)}]\, \Delta S + \oint_C \gamma\, \mathbf{t}\, \mathrm{d}l = \mathbf{0}, \tag{4.3.23}$$

where ΔS is the surface area of the layer, and l is the arc length around the layer edge, C. Equation (4.3.23) is the three-dimensional counterpart of equation (4.3.1). Putting $\mathbf{n}^{(2)} = -\mathbf{n}^{(1)}$ and rearranging, we find

$$\mathbf{n}^{(1)} \cdot (\boldsymbol{\sigma}^{(1)} - \boldsymbol{\sigma}^{(2)}) = -\frac{1}{\Delta S} \oint_C \gamma\, \mathbf{t}\, \mathrm{d}l. \tag{4.3.24}$$

In the limit as ΔS tends to zero and the loop C shrinks to a point, equation (4.3.24) reduces to

$$\Delta \mathbf{f} \equiv \mathbf{n}^{(1)} \cdot (\boldsymbol{\sigma}^{(1)} - \boldsymbol{\sigma}^{(2)}) = \gamma\, 2\, \kappa_m\, \mathbf{n}^{(1)} - \frac{\partial \gamma}{\partial \hat{l}}\, \hat{\mathbf{t}}, \tag{4.3.25}$$

where:

- κ_m is the mean curvature of the interface defined in terms of the surface divergence of the normal vector discussed later in this section.
- $\hat{\mathbf{t}}$ is the unit vector tangent to the interface pointing in the direction where the surface tension changes most rapidly; $\hat{l}$ is the arc length measured in the direction of the tangential vector, $\hat{\mathbf{t}}$; and $\partial\gamma/\partial\hat{l}$ is the corresponding maximum rate of change of the surface tension with respect to arc length.

The first term on the right-hand side of (4.3.25) expresses a discontinuity in the normal direction, whereas the second term expresses a discontinuity in the tangential direction identified as the Marangoni traction. If the x axis is normal to the interface and the yz plane is tangential to the interface at a point, then

$$\Delta \mathbf{f} = \gamma\, 2\kappa_m\, \mathbf{e}_x - \frac{\partial \gamma}{\partial y}\, \mathbf{e}_y - \frac{\partial \gamma}{\partial z}\, \mathbf{e}_z, \tag{4.3.26}$$

where $\mathbf{e}_y$ and $\mathbf{e}_z$ are unit vectors.

Mean curvature

If the yz plane is tangential to an interface at a point, the mean curvature at that point is given by the surface divergence

$$2\kappa_m \equiv \nabla_s \cdot \mathbf{n} = \frac{\partial n_y}{\partial y} + \frac{\partial n_z}{\partial z}. \tag{4.3.27}$$

Requiring $n_x^2 + n_y^2 + n_z^2 = 1$ yields $\partial n_x/\partial x = 0$, which shows that the normal derivative of the normal component of the normal vector is zero. This property allows us to write the more general expression

$$2\kappa_m \equiv \nabla \cdot \mathbf{n} = \frac{\partial n_x}{\partial x} + \frac{\partial n_y}{\partial y} + \frac{\partial n_z}{\partial z}, \tag{4.3.28}$$

pertinent to an arbitrary system of Cartesian coordinates whose axes are not necessarily tangential or normal to the interface.

If an interface is described by the function $F(x, y, z) = 0$, the unit vector normal to the interface is

$$\mathbf{n} = \frac{1}{|\nabla F|} \nabla F, \tag{4.3.29}$$

and the mean curvature is given by

$$2\kappa_m = \nabla \cdot \Big(\frac{1}{|\nabla F|} \nabla F \Big) = \frac{1}{|\nabla F|} \nabla^2 F - \frac{1}{|\nabla F|^3} \nabla F \cdot (\nabla\nabla F) \cdot \nabla F. \tag{4.3.30}$$

If the interface is described explicitly by the function $z = f(x, y)$, we set $F = z - f(x, y)$, and find

$$2\kappa_m = -\frac{(1 + f_y^2)\, f_{xx} - 2 f_x f_y f_{xy} + (1 + f_x^2)\, f_{yy}}{(1 + f_x^2 + f_y^2)^{3/2}}, \tag{4.3.31}$$

where the subscript x denotes a derivative with respect to x, and the subscript y denotes a derivative with respect to y. For a nearly flat interface,

$$2\kappa_m \simeq -(f_{xx} + f_{yy}). \tag{4.3.32}$$

If an interface is described in spherical polar coordinates, (r, θ, φ), as $r = f(\theta, \varphi)$, the unit normal vector is

$$\nabla F = \mathbf{e}_r - \frac{f_\theta}{r}\, \mathbf{e}_\theta - \frac{f_\varphi}{r \sin\theta}\, \mathbf{e}_\varphi. \tag{4.3.33}$$

For a nearly spherical interface, the mean curvature can be approximated by

$$2\kappa_m \simeq \frac{2}{r} - \frac{\cot\theta}{r^2} f_\theta - \frac{1}{r^2} f_{\theta\theta} - \frac{1}{r^2 \sin^2\theta} f_{\varphi\varphi}. \tag{4.3.34}$$

Directional curvatures

To compute the mean curvature of a three-dimensional interface, we consider the traces of the interface in two orthogonal planes that are normal to the interface at a point and thus contain the normal vector, as depicted in figure

4.3.3(*b*). If κ_1 and κ_2 are the curvatures of the two traces at that point computed using formula (4.3.5) with the x and y axes residing in the two planes, then the mean curvature of the interface is

$$\kappa_m = \frac{1}{2}(\kappa_1 + \kappa_2). \tag{4.3.35}$$

A theorem due to Euler states that the mean value of the directional curvatures is independent of the orientation of the two planes, provided only that the planes are mutually orthogonal.

Principal curvatures

There is a particular orientation of the normal plane corresponding to maximum directional curvature, κ_{max}, and a conjugate orthogonal orientation corresponding to minimum directional curvature, κ_{min}. These are the principal curvatures of the interface at the chosen point. Euler's theorem states that the curvature in an arbitrary direction is related to the principal curvatures by

$$\kappa = \kappa_{max} \cos^2 \alpha + \kappa_{min} \sin^2 \alpha, \tag{4.3.36}$$

where α is the angle subtended between (*a*) the tangential vector pointing in a chosen direction, and (*b*) the tangential vector pointing in the direction of maximum curvature.

4.3.3 Axisymmetric interfaces

Next, we consider axisymmetric interfaces, as shown in figure 4.3.4. The mean curvature is the average of the two principal curvatures: one is the curvature of the trace of the interface in the σx plane denoted by κ_1, and the second is the curvature of the trace of the interface in the orthogonal plane denoted by κ_2.

First description

The position of the interface in a meridional plane can be described by the function $\sigma = w(x)$. The unit normal vector is

$$\mathbf{n} = \frac{1}{\sqrt{1+w'^2}} (\mathbf{e}_\sigma - w' \, \mathbf{e}_x), \tag{4.3.37}$$

where a prime denotes a derivative with respect to x. The mean curvature is given by the divergence of the normal vector,

$$2\kappa_m = \frac{\partial n_x}{\partial x} + \frac{1}{\sigma} \frac{\partial(\sigma \, n_\sigma)}{\partial \sigma} = -\Big(\frac{w'}{\sqrt{1+w'^2}}\Big)' + \frac{1}{\sigma}\frac{\partial}{\partial \sigma}\Big(\frac{\sigma}{\sqrt{1+w'^2}}\Big). \tag{4.3.38}$$

Carrying out the differentiations, we find

$$2\kappa_m = -\frac{w''}{(1+w'^2)^{3/2}} + \frac{1}{w}\,\frac{1}{\sqrt{1+w'^2}}. \tag{4.3.39}$$

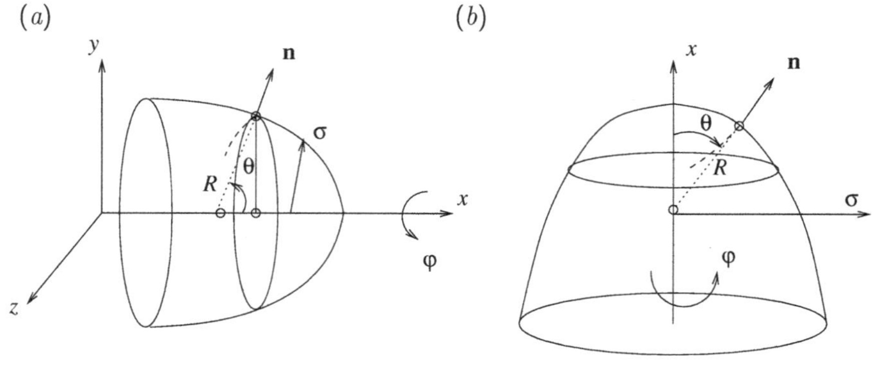

Figure 4.3.4 Illustration of an axisymmetric interface. The second principal curvature is the curvature of the dashed line representing the trace of the interface in a plane that is normal to the interface and normal to the xy plane.

The first term on the right-hand side is the principal curvature in a meridional plane. The second term is the second principal curvature,

$$\kappa_2 = \frac{1}{R_2}, \qquad R_2 = \sigma\sqrt{1+w'^2} = \frac{\sigma}{\sin\theta} = \frac{\sigma}{n_\sigma}, \tag{4.3.40}$$

where R_2 is the second principal radius of curvature, and the angle θ is defined in figure 4.3.4(a). We have found that R_2 is the signed distance between (a) the point where the curvature is evaluated, and (b) the intersection of the extension of the normal vector and the x axis. If n_σ is negative, R_2 is also negative.

Second description

Alternatively, the position of an interface can be described by a function, $x = f(\sigma)$, as shown in figure 4.3.4(b). The unit normal vector is

$$\mathbf{n} = \frac{1}{\sqrt{1+f'^2}}\,(\mathbf{e}_x - f'\,\mathbf{e}_\sigma), \tag{4.3.41}$$

where a prime denotes a derivative with respect to σ. The mean curvature is given by the divergence of the normal vector,

$$2\kappa_m = \frac{\partial n_x}{\partial x} + \frac{1}{\sigma}\frac{\partial(\sigma\, n_\sigma)}{\partial\sigma} = -\frac{1}{\sigma}\Big(\frac{\sigma f'}{\sqrt{1+f'^2}}\Big)'. \tag{4.3.42}$$

Carrying out the differentiations, we find

$$2\kappa_m = -\frac{f''}{(1+f'^2)^{3/2}} - \frac{1}{\sigma}\frac{f'}{(1+f'^2)^{1/2}}. \tag{4.3.43}$$

The first term on the right-hand side is the principal curvature in a meridional plane,

$$\kappa_1 = -\frac{f''}{(1+f'^2)^{3/2}} = \frac{1}{f'}\Big(\frac{1}{(1+f'^2)^{1/2}}\Big)'. \qquad (4.3.44)$$

The second term is the second principal curvature,

$$\kappa_2 = \frac{1}{R_2}, \qquad R_2 = -\frac{\sigma}{f'}\sqrt{1+f'^2} = \frac{\sigma}{\sin\theta} = \frac{\sigma}{n_\sigma}, \qquad (4.3.45)$$

where R_2 is the second principal radius of curvature, and the angle θ is defined in figure 4.3.4(b).

Problems

4.3.1 *Curvature of an ellipse*

Consider a horizontal ellipse centered at the origin of the xy plane, described in parametric form by the equations $x = a\cos\chi$ and $y = b\sin\chi$, where χ is the natural parameter of the ellipse varying between 0 and 2π, and a, b are the ellipse semi-axes. Derive an expression for the curvature of the ellipse in terms of a, b, and χ. Confirm that, as b tends to a, the curvature of the ellipse reduces to that of a circle.

4.3.2 *Mean curvature*

(a) Based on formula (4.3.35) and its accompanying interpretation discussed in the text, show that the mean curvature of a sphere of radius a is equal to $\kappa_m = 1/a$, whereas the mean curvature of a circular cylinder of radius a is equal to $\kappa_m = 1/(2a)$.

(b) The sphere and the circular cylinder are two shapes with constant mean curvature. Discuss one additional shape.

4.3.3 *Jump in traction in local coordinates*

Derive the counterparts of equations (4.3.9) and (4.3.10) for a three-dimensional interface.

Computer Problems

4.3.4 *Computation of the curvature*

A line in the xy plane is described by a set of $N+1$ marker points with coordinates (x_i, y_i), where $i = 1, \ldots, N+1$. An approximation to the components of the tangent vector at the ith point is provided by the central-difference formulas

$$t_x^{(i)} = \frac{x_{i+1} - x_{i-1}}{\Delta l^{(i)}}, \qquad t_y^{(i)} = \frac{y_{i+1} - y_{i-1}}{\Delta l^{(i)}}, \tag{4.3.46}$$

where

$$\Delta l^{(i)} = [(x_{i+1} - x_{i-1})^2 + (y_{i+1} - y_{i-1})^2]^{1/2}. \tag{4.3.47}$$

The derivatives of the components of the tangent vector with respect to arc length can be approximated with the corresponding formulas

$$\frac{\mathrm{d}t_x^{(i)}}{\mathrm{d}l} = \frac{t_x^{(i+1)} - t_x^{(i-1)}}{\Delta l^{(i)}}, \qquad \frac{\mathrm{d}t_y^{(i)}}{\mathrm{d}l} = \frac{t_y^{(i+1)} - t_y^{(i-1)}}{\Delta l^{(i)}}. \tag{4.3.48}$$

The components of the outward normal vector at the marker points are given by

$$n_x^{(i)} = t_y^{(i)}, \qquad n_y^{(i)} = -t_x^{(i)}. \tag{4.3.49}$$

Write a computer program that reads or generates the coordinates of a set of marker points, computes the right-hand sides of (4.3.46)–(4.3.49), and then evaluates the curvature at the marker points from the expression $\kappa = -(\mathrm{d}\mathbf{t}/\mathrm{d}l) \cdot \mathbf{n}$. Perform a series of computations with marker points distributed evenly along a circle, and compare the numerically computed with the exact curvatures.

4.3.5 *Motion induced by curvature*

Interfaces exhibit a variety of motions under the influence of surface tension. In a simplified model, point particles distributed along a two-dimensional interface move normal to the interface with velocity that is proportional to the local curvature. If $\mathbf{X}^{(i)}$ is the position of the ith marker point, then the motion of the marker point is described by the vectorial differential equation

$$\frac{\mathrm{d}\mathbf{X}^{(i)}}{\mathrm{d}t} = \kappa\, [\mathbf{n}^{(1)}]^{(i)}, \tag{4.3.50}$$

where t stands for time, $\mathbf{n}^{(1)}$ is the outward normal vector, and κ is the curvature.

Write a program that computes the motion of marker points distributed over an interface using the finite-difference approximations discussed in problem 4.3.4 and the modified Euler method for integrating in time the differential equations (4.3.50). Run the program to compute the evolution of marker points distributed along a circle or an ellipse with axes ratio equal to two. Discuss the nature of the motion in each case.

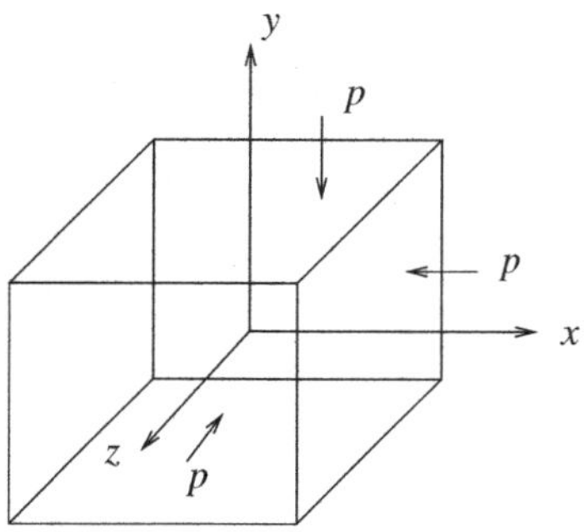

Figure 4.4.1 The traction exerted at the three sides of a cubic parcel of a stationary fluid has only a normal component defined in terms of the thermodynamic pressure.

4.4 Stresses in a fluid at rest

If a fluid does not exhibit macroscopic motion as seen by a stationary observer–that is, the observable fluid velocity is zero–the molecules are in a state of dynamic equilibrium determined by the physical conditions prevailing in their immediate environment.

Consider a cubic fluid parcel with its six faces perpendicular to the x, y, or z axis, as illustrated in figure 4.4.1. In the absence of macroscopic fluid motion, the traction exerted on the sides that are perpendicular to the x axis must be directed normal to this side; otherwise, an unresolved tangential component pointing in a physically indeterminate direction will arise. In the notation of section 4.2,

$$f_x^{(x)} \neq 0, \qquad f_y^{(x)} = 0, \qquad f_z^{(x)} = 0. \tag{4.4.1}$$

Similar arguments can be made to show that the traction exerted on the sides that are perpendicular to the y or z axis are directed normal to these sides,

$$\begin{aligned} f_x^{(y)} &= 0, \qquad f_y^{(y)} \neq 0, \qquad f_z^{(y)} = 0, \\ f_x^{(z)} &= 0, \qquad f_y^{(z)} = 0, \qquad f_z^{(z)} \neq 0. \end{aligned} \tag{4.4.2}$$

If the size of the cubic parcel is infinitesimal, the fluid residing inside the parcel is perfectly or nearly homogeneous, and the non-vanishing components of the tractions, $f_x^{(x)}$, $f_y^{(y)}$, and $f_z^{(z)}$, must be identical. By definition, the common value of these normal components is equal to the negative of the pressure, p,

$$f_x^{(x)} = f_y^{(y)} = f_z^{(z)} \equiv -p. \tag{4.4.3}$$

Thus, in hydrostatics, the stress tensor introduced in equation (4.2.7) is defined exclusively in terms of the pressure, and is given by

$$\boldsymbol{\sigma} = \begin{bmatrix} -p & 0 & 0 \\ 0 & -p & 0 \\ 0 & 0 & -p \end{bmatrix} = -p \begin{bmatrix} 1 & 0 & 0 \\ 0 & 1 & 0 \\ 0 & 0 & 1 \end{bmatrix}. \tag{4.4.4}$$

In compact notation,

$$\boldsymbol{\sigma} = -p\,\mathbf{I}, \tag{4.4.5}$$

where $\mathbf{I}$ is the unit or identity matrix shown on the right-hand side of (4.4.4).

Traction on a surface

As an application, we use expression (4.4.5) to evaluate the traction exerted on a surface residing in a stationary fluid. Substituting (4.4.5) in formula (4.2.9), we find

$$\mathbf{f}(\mathbf{x}, \mathbf{n}) = \mathbf{n} \cdot (-p\,\mathbf{I}) = -p\,\mathbf{n} \cdot \mathbf{I} = -p\,\mathbf{n}. \tag{4.4.6}$$

The last equality arises from the identity $\mathbf{n} \cdot \mathbf{I} = \mathbf{n}$. Equation (4.4.6) shows that the traction exerted on a surface in hydrostatics is directed normal to the surface; the tangential component is identically zero.

Traction on a boundary

As a second application, we use (4.4.5) to evaluate the traction exerted on a boundary confining a fluid. Using formula (4.2.18), we find

$$\mathbf{f}^B = \mathbf{n}^{inward} \cdot (-p\,\mathbf{I}) = -p\,\mathbf{n}^{inward} \cdot \mathbf{I} = -p\,\mathbf{n}^{inward}. \tag{4.4.7}$$

The last equality arises from the identity $\mathbf{n}^{inward} \cdot \mathbf{I} = \mathbf{n}^{inward}$. Thus, the traction exerted on a fluid boundary in hydrostatics is directed normal to the boundary; the tangential component is identically zero.

4.4.1 Pressure from molecular motions

The hydrostatic pressure distribution established in a fluid at rest cannot be computed working exclusively in the context of fluid mechanics. Additional information concerning the relationship between the density and the pressure for the particular fluid under consideration is required.

Gases

Molecular thermodynamics states that the pressure of a small gas parcel is determined by (a) the number of molecules residing inside the parcel expressed by the local density, ρ, (b) the kinetic energy of the molecules determined by the absolute temperature, T, and (c) the nature and intensity of the intermolecular forces due to an intermolecular potential. For an ideal gas, intermolecular forces are negligible and the pressure derives from the density and temperature in terms of the ideal gas law,

$$p = \frac{\rho R T}{M}, \tag{4.4.8}$$

where M is the gram-molecular mass defined as the mass of one mole comprised of a collection of N_A molecules; $N_A = 6.022 \times 10^{26}$ is the Avogadro number; $R = 8.314 \times 10^3$ kg m^2/(sec^2 kmole K) is the ideal gas constant; T is Kelvin's absolute temperature, equal to the Celsius centigrade temperature reduced by 273 units. The gram-molecular weight of an element is equal to the atomic weight of the element listed in the periodic table, expressed in gram.

Liquids

Because liquids are nearly incompressible, thermodynamics allows us to compute the pressure under the assumption that the density is a function of temperature alone, independent of pressure. The computation of the hydrostatic pressure distribution in gases and liquids will be discussed in more detail in chapter 5.

4.4.2 Jump in the pressure across an interface

Equations (4.3.7) and (4.3.25) provide us with expressions for the jump in the traction across a two- or three-dimensional interface. If the fluids on either side of the interface are stationary, the corresponding stress tensors are given by (4.4.5) in term of the pressures, and the jump in the traction is

$$\Delta \mathbf{f} \equiv \mathbf{n}^{(1)} \cdot (\boldsymbol{\sigma}^{(1)} - \boldsymbol{\sigma}^{(2)}) = \mathbf{n}^{(1)} \cdot [\, -p^{(1)} \mathbf{I} - (-p^{(2)} \mathbf{I})\,], \tag{4.4.9}$$

yielding

$$\Delta \mathbf{f} = (p^{(2)} - p^{(1)})\, \mathbf{n}^{(1)}, \tag{4.4.10}$$

where the superscript 1 or 2 signifies the choice of fluid. Assuming that the surface tension is uniform, we compare the right-hand side of (4.4.10) with the right-hand side of the force equilibrium equation (4.3.7) for a two-dimensional interface, and find

$$p^{(2)} - p^{(1)} = \gamma\, \kappa, \tag{4.4.11}$$

where κ is the mean curvature of the interface. Working in a similar fashion for a three-dimensional interface using (4.3.25), we find

$$p^{(2)} - p^{(1)} = \gamma\, 2\, \kappa_m, \tag{4.4.12}$$

where κ_m is the mean curvature of the interface.

We have shown that the jump in the pressure across a two-dimensional interface is equal to the product of the surface tension and the curvature of the interface, and the jump in the pressure across a three-dimensional interface is equal to the product of the surface tension and twice the mean curvature of the interface.

Laplace's law

As an application, we compute the jump in pressure across a spherical interface of radius a, representing the surface of a liquid drop or bubble. Designating the outer fluid as 1 and the inner fluid as 2, we find that the mean curvature is $\kappa_m = 1/a$. Consequently, the pressure jump across the spherical interface is given by Laplace's law,

$$p^{(2)} - p^{(1)} = \frac{2\gamma}{a}. \tag{4.4.13}$$

Problems

4.4.1 *Jump in pressure across a two-dimensional interface*

Derive an expression for the jump in pressure across a circular interface of radius a representing the trace of a cylindrical thread in the xy plane.

4.4.2 *Curvature of a soap film*

Explain why the mean curvature of a thin soap film confined attached to a wire frame must be zero.

4.5 Constitutive equations

In the absence of macroscopically observable fluid motion, the traction exerted on a specified side of a small fluid surface is given by equation (4.4.6) in terms of the pressure. If the fluid exhibits macroscopic motion, this equation is modified in two ways. First, the normal component of the traction is complemented by a new contribution that depends on the physical properties of the fluid and the nature of the local fluid motion. Second, a tangential traction component is established.

To understand how these new contributions arise, it is helpful to consider the tractions developing in two complementary types of flow: an extensional flow where the fluid stretches and elongates, and a channel flow where the fluid is sheared due to boundary motion.

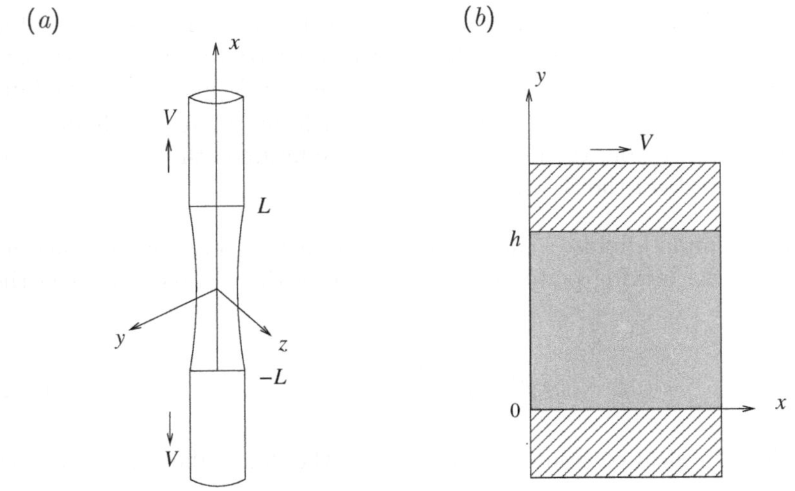

Figure 4.5.1 (*a*) Stretching of a liquid bridge subtended between two coaxial cylinders that are pulled apart along their axes. (*b*) Shear flow in a two-dimensional channel confined between two parallel plates; the motion is due to the parallel translation of the upper plate.

Stretching of a thread

In one experiment, a thread of a liquid is suspended between two rods forming an axisymmetric bridge, and the rods are pulled apart extending the thread, as illustrated in figure 4.5.1(*a*). A force is required to pull the rods apart and thus overcome the normal component of the hydrodynamic traction imparted by the fluid to the tips of the rods, $f_x^{(x)}$. The faster the rods are pulled apart, the higher the magnitude of the traction; the greater the distance between the rods, the lower the magnitude of the traction.

For most common fluids, a linear relationship exists between the traction, the velocity of the rods and the inverse of their distance, in the form

$$f_x^{(x)} = -p + 2\,\mu_{ext}\,\frac{V}{L}, \tag{4.5.1}$$

where p is the pressure discussed in section 4.4 in the context of hydrostatics, V is the velocity of the rods, L is half the distance between the rods, and μ_{ext} is a physical constant associated with the fluid called the extensional viscosity.

Shearing of a layer

In a second experiment, a fluid is placed in a channel confined between two parallel plates. The upper plate translates in the x direction parallel to itself with constant velocity V while the lower plate is held stationary, as depicted in figure 4.5.1(*b*). A force in the x direction must be exerted on the upper plate to balance the tangential component of the traction developing due to the fluid motion, $f_x^{(y)}$. The faster the velocity of the translating plate, the higher the magnitude of the traction; the greater the distance between the two plates, the lower the magnitude of the traction.

For most common fluids, a linear relationship exists between the traction, the velocity of the moving plate, and the inverse of the distance between the plates, h,

$$f_x^{(y)} = \mu_{sh} \frac{V}{h}, \tag{4.5.2}$$

where μ_{sh} is a physical constant associated with the fluid called the shear viscosity.

4.5.1 Simple fluids

We have demonstrated that stresses develop in a fluid as a result of the motion. To proceed further, we consider the tractions developing at the surface of a small fluid parcel and argue the following properties characterizing a simple fluid: (*a*) if the parcel translates or rotates as a rigid body, tractions do not develop over the parcel surface; (*b*) tractions develop only when the parcel deforms; (*c*) the distribution of traction over the parcel surface at a particular time depends only on the type and rate of deformation of the parcel at that particular time.

Our analysis of kinematics in chapter 2 revealed that a small spherical fluid parcel deforms to obtain an ellipsoidal shape whose axes are parallel to the three eigenvectors of the rate of deformation tensor defined in equation (2.1.28). The rates of deformation are equal to the corresponding eigenvalues. With these observations as a point of departure, we proceed to relate the stress tensor to the physical properties of the fluid and to the structure of the velocity field by means of a constitutive equation.

4.5.2 Incompressible Newtonian fluids

The constitutive equation for an incompressible Newtonian fluid is

$$\boldsymbol{\sigma} = -p\mathbf{I} + \mu\, 2\mathbf{E}, \tag{4.5.3}$$

where p is the pressure, the coefficient μ is the fluid viscosity, sometimes also called the dynamic viscosity, and $\mathbf{E}$ is the rate of deformation tensor defined in equation (2.1.28). Note that the Newtonian constitutive relation respects the symmetry of the stress tensor discussed at the end of section 4.2. Explicitly,

Temperature °C	*Water* cp ≡ 10^{-2}g/(cm sec)	*Air* cp
20	1.002	0.0181
40	0.653	0.0191
80	0.355	0.0209

Table 4.5.1 The viscosity of water and air at three temperatures; cp stands for centipoise, which is one hundredth of the viscosity unit poise defined as 1 g/(cm sec).

the components of the stress tensor are given by the matrix equation

$$\begin{bmatrix} \sigma_{xx} & \sigma_{xy} & \sigma_{xz} \\ \sigma_{yx} & \sigma_{yy} & \sigma_{yz} \\ \sigma_{zx} & \sigma_{zy} & \sigma_{zz} \end{bmatrix} = \begin{bmatrix} -p + 2\mu\,\frac{\partial u_x}{\partial x} & \mu\,(\frac{\partial u_y}{\partial x} + \frac{\partial u_x}{\partial y}) & \mu\,(\frac{\partial u_z}{\partial x} + \frac{\partial u_x}{\partial z}) \\ \mu\,(\frac{\partial u_x}{\partial y} + \frac{\partial u_y}{\partial x}) & -p + 2\mu\,\frac{\partial u_y}{\partial y} & \mu\,(\frac{\partial u_z}{\partial y} + \frac{\partial u_y}{\partial z}) \\ \mu\,(\frac{\partial u_x}{\partial z} + \frac{\partial u_z}{\partial x}) & \mu\,(\frac{\partial u_y}{\partial z} + \frac{\partial u_z}{\partial y}) & -p + 2\mu\,\frac{\partial u_z}{\partial z} \end{bmatrix}. \tag{4.5.4}$$

Table 4.5.1 summarizes the viscosity of water and air at three temperatures.

In the absence of flow, we recover the hydrostatic stress tensor defined in equation (4.4.5), involving the pressure alone.

Unidirectional shear flow

As an example, we consider flow in a two-dimensional channel confined between two parallel plane walls. The motion is generated by the translation of the upper wall, as depicted in figure 4.5.1(*b*). Physical intuition suggests that, at low and moderate velocities, the fluid will translate only in the x direction with a position-dependent velocity u_x varying in the y direction; to signify this dependence, we write $u_x(y)$. Using equation (2.1.22), we find that the rate of deformation tensor is

$$\mathbf{E} = \frac{1}{2} \begin{bmatrix} 0 & \mathrm{d}u_x/\mathrm{d}y \\ \mathrm{d}u_x/\mathrm{d}y & 0 \end{bmatrix}. \tag{4.5.5}$$

Substituting this expression in the right-hand side of (4.5.3), we obtain the stress tensor

$$\boldsymbol{\sigma} = \begin{bmatrix} -p & \mu\,\mathrm{d}u_x/\mathrm{d}y \\ \mu\,\mathrm{d}u_x/\mathrm{d}y & -p \end{bmatrix}. \tag{4.5.6}$$

The x component of the traction exerted on a fluid surface that is perpendicular to the y axis, identified as the shear stress, is

$$f_x^{(y)} = \sigma_{yx} = \mu \frac{\mathrm{d}u_x}{\mathrm{d}y}. \tag{4.5.7}$$

Physically, this traction is attributed to the friction experienced by adjacent fluid layers as they slide over one another with slightly different velocities.

4.5.3 Viscosity

Strictly speaking, the viscosity of a fluid is a proportionality coefficient relating the stress tensor to the rate of deformation tensor, as shown in equation (4.5.3). However, it is reassuring to know that this mathematical definition established by phenomenological observation has a firm physical foundation. Thus, the viscosity is a genuine physical property of the fluid dependent on the local physical conditions, including temperature.

As the temperature is raised, the viscosity of liquids is reduced whereas the viscosity of gases increases, as shown in table 4.5.1. This dichotomy is a reflection of the different physical mechanisms responsible for the development of stresses in these two fundamental classes of fluids. In the case of liquids, the viscosity is due to occasional molecular excursions from a mean position into neighboring empty sites. In the cases of gases, the viscosity is due to relentless molecular excursions from regions of high velocity to regions of low velocity in the course of random motion due to thermal fluctuations.

Viscosity of a gas

To demonstrate the relation between molecular and macroscopic fluid motion, we consider a gas in unidirectional shear flow and derive an expression for the viscosity in terms of molecular properties. In the simplest kinetic theory, the molecules are modeled as rigid spheres moving with the local fluid velocity defined in section 1.4, and with a randomly fluctuating component. The average magnitude of the fluctuating component is

$$\bar{v} = \sqrt{\frac{8\,k_B T}{\pi M}}, \tag{4.5.8}$$

where k_B is the Boltzmann constant, T is the absolute temperature, and M is the gram-molecular mass. In the course of the random motion, two molecules occasionally collide having traveled an average distance equal to the mean free path, λ.

Consider a macroscopically stationary gas with vanishing fluid velocity. Using principles of statistical mechanics, we find that the number of molecules crossing an infinitesimal surface area per infinitesimal time as a result of the fluctuating motion, Z, is proportional to (a) the number of molecules per infinitesimal volume defined as the number density n, and (b) the average magnitude of the fluctuating velocity, $\bar{v}$. It can be shown that

$$Z = \frac{1}{4} n \bar{v}. \tag{4.5.9}$$

A molecule crossing a surface at a particular instant has collided with another molecule above or below the surface at an average distance a. Using principles of statistical mechanics, we find

$$a = \frac{2}{3} \lambda, \tag{4.5.10}$$

where λ is the mean free path. Relations (4.5.9) and (4.5.10) have been derived taking into consideration that, since the molecules move randomly in all directions, only one component of the velocity brings them toward the crossing surface under consideration. Relations (4.5.8)–(4.5.10) also hold true when the fluid exhibits macroscopic motion, provided that the molecular velocity is computed relative to the average velocity at the position where a molecule last underwent a collision.

Figure 4.5.2 shows a schematic illustration of the instantaneous distribution of molecules in a gas undergoing unidirectional shear flow. Without loss of generality, we have assumed that the fluid velocity increases in the positive direction of the y axis. In the course of the motion, gas molecules cross a horizontal plane corresponding to a certain value of y, drawn with the heavy horizontal line, from either side. Because the x velocity of molecules crossing from above is higher than the x velocity of molecules crossing from below, x momentum is transferred toward the negative direction of the y axis. The rate of transport of x momentum across a surface that is perpendicular to the y axis amounts to a hydrodynamic traction, $f_x^{(y)}$.

The rate of momentum transport defined in the last paragraph can be quantified by setting

$$f_x^{(y)} = -M\,(Z^- u_x^- - Z^+ u_x^+), \tag{4.5.11}$$

where u_x is the fluid velocity, and the superscripts $+$ and $-$ indicate that the underlying variable is evaluated at a distance equal to a above or below the transport surface. Effectively, the collection of molecules crossing the y plane during an infinitesimal period of time are represented by model molecules with two important properties: (a) the model molecules last underwent a collision at a distance a above or below the y plane, and (b) the model molecules move with an average velocity that is equal to the local fluid velocity evaluated at the position of the last collision.

Because the flow is unidirectional, the average fluid velocity normal to a horizontal plane is zero, and the number of molecules crossing the y plane from

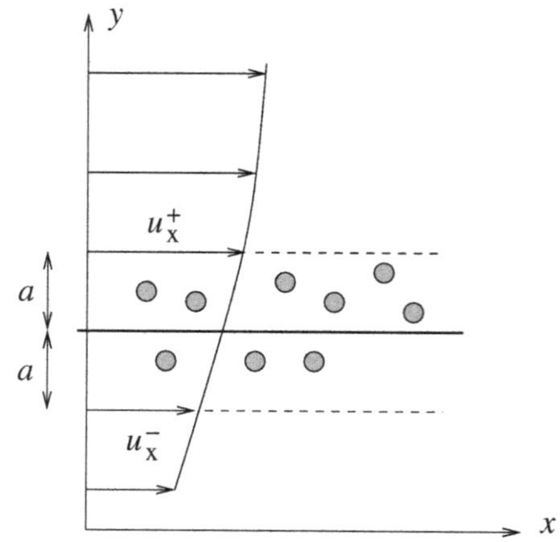

Figure 4.5.2 A molecular model of a gas in shear flow is used to derive an expression for the viscosity in terms of molecular properties, as shown in equation (4.5.14).

either side during an infinitesimal time period are equal. Combining equations (4.5.9)–(4.5.11), we find

$$f_x^{(y)} = M\,\frac{1}{4}\,n\bar{v}\,\frac{u_x^+ - u_x^-}{2a}\,\frac{4}{3}\,\lambda. \tag{4.5.12}$$

Since a is small compared to the macroscopic length scale of the shear flow, the fraction enclosed by the parentheses on the right-hand side of (4.5.12) can be approximated with a derivative, yielding

$$f_x^{(y)} = \frac{1}{3}\,nM\bar{v}\lambda\,\frac{du_x}{dy}. \tag{4.5.13}$$

Comparing equation (4.5.13) with (4.5.7), we derive an expression for the viscosity,

$$\mu = \frac{1}{3}\,nM\bar{v}\lambda. \tag{4.5.14}$$

We have obtained the Newtonian constitutive equation from molecular considerations and derived a prediction for the viscosity of a gas in terms of molecular properties.

4.5.4 Ideal fluids

If the viscosity of a fluid vanishes, the fluid is frictionless and is called ideal. The stress tensor in an ideal fluid is given by a simplified version of (4.5.3),

$$\boldsymbol{\sigma} = -p\,\mathbf{I}. \tag{4.5.15}$$

However, in practice, no fluid is ideal, and the absence of viscosity should be interpreted strictly as insignificance of hydrodynamic forces or stresses associated with the fluid viscosity. Viscous stresses are always important near solid boundaries, as discussed in chapter 10. The formal requirement for viscous stresses to be negligible will be discussed in chapter 6.

4.5.5 Significance of the pressure in incompressible fluids

The physical interpretation of the pressure in the Newtonian constitutive equation (4.5.3), or any other constitutive equation, is not clear. Strictly speaking, the pressure is defined in terms of the trace of the stress tensor,

$$p = -\frac{1}{3}\,\mathrm{trace}(\boldsymbol{\sigma}) = -\frac{1}{3}\,(\sigma_{xx} + \sigma_{yy} + \sigma_{zz}). \tag{4.5.16}$$

All we can say with confidence is that, as a fluid becomes quiescent, this dynamic pressure reduces to the hydrostatic pressure computed from consideration of random molecular motions.

4.5.6 Pressure in compressible fluids

In the case of compressible fluids, an equation relating the pressure to the density and to the temperature can be derived working in the framework of equilibrium thermodynamics.

Consider a cylindrical fluid parcel of a compressible gas with volume V. To reduce the volume of the parcel by the differential amount, $\mathrm{d}V$, we apply an external pressure, p, by way of an ideal frictionless piston. The differential work required to carry out this reversible process is

$$\delta W = -p\,\mathrm{d}V = -p\,n\,\mathrm{d}v, \tag{4.5.17}$$

where n is the number of moles contained in the parcel, and v is the gram-molecular volume of the gas. In our experiment, $\mathrm{d}V$ and $\mathrm{d}v$ are both negative, while $\mathrm{d}W$ is positive. The notation δW emphasizes that the differential work can be computed only after the frictional properties of the piston have been specified; in our discussion, the piston is frictionless. In formal thermodynamics, we say that δW is an inexact differential.

Part of the work is spent to increase the temperature of the parcel, T, by a differential amount, $\mathrm{d}T$, and therefore the internal (thermal) energy of the parcel, U, by a differential amount, $\mathrm{d}U$. The remainder of the work escapes as a (negative) process-dependent heat loss, δQ. Energy conservation for a closed system in the absence of significant kinetic or potential energy requires $\mathrm{d}U = \delta W + \delta Q$ or

$$\delta W = \mathrm{d}U - \delta Q. \tag{4.5.18}$$

To quantify the heat loss, we write for a reversible process

$$\delta Q = T\,\mathrm{d}S, \tag{4.5.19}$$

where S is the entropy, and find

$$\delta W = \mathrm{d}U - T\,\mathrm{d}S. \tag{4.5.20}$$

Dividing this equation by the number of moles of the gas, n, we obtain

$$\delta w = \mathrm{d}u - T\,\mathrm{d}s, \tag{4.5.21}$$

where

$$w = \frac{W}{n}, \qquad u = \frac{U}{n}, \qquad s = \frac{S}{n} \tag{4.5.22}$$

are the specific work, specific internal energy, and specific entropy.

Now substituting $\delta w = -p\,\mathrm{d}v$ in (4.5.21) and rearranging, we derive a process-independent differential relation,

$$\mathrm{d}u = T\,\mathrm{d}s - p\,\mathrm{d}v. \tag{4.5.23}$$

If the specific volume of a gas is changed by the same small amount according to two different processes, the changes in the specific internal energy and specific entropy will be such that equation (4.5.23) is satisfied in both cases.

Ideal gases

The specific internal energy of an arbitrary gas depends on the temperature, T, and specific volume, v. Because of the absence of intermolecular forces, the specific internal energy of an ideal gas depends on temperature alone, and the change in the specific internal energy is

$$\mathrm{d}u = c_v\,\mathrm{d}T, \tag{4.5.24}$$

where c_v is the specific heat capacity under constant volume. Combining the last two equations and rearranging, we find

$$T\,\mathrm{d}s = c_v\mathrm{d}T + p\,\mathrm{d}v, \tag{4.5.25}$$

showing that, $c_v/T = (\partial s/\partial T)_v$.

Solving the equation of state, $pv = RT$, for the pressure, we find

$$\delta w = -p\,\mathrm{d}v = -RT\,\frac{\mathrm{d}v}{v}. \tag{4.5.26}$$

Substituting this expression in (4.5.25), we find

$$-RT\,\frac{\mathrm{d}v}{v} = c_v\,\mathrm{d}T - T\,\mathrm{d}s. \tag{4.5.27}$$

Rearranging, we derive an expression for the change in the specific entropy,

$$ds = c_v \frac{dT}{T} + R \frac{dv}{v}. \tag{4.5.28}$$

Next, we assume that the heat capacity is constant and integrate (4.5.28) between two states 1 and 2 to derive an expression for the difference in entropy,

$$\Delta s \equiv s_2 - s_1 = c_v \ln \frac{T_2}{T_1} + R \ln \frac{v_2}{v_1}, \tag{4.5.29}$$

or

$$\Delta s = \ln \left[\left(\frac{T_2}{T_1} \right)^{c_v} \left(\frac{v_2}{v_1} \right)^{R} \right]. \tag{4.5.30}$$

In the case of constant pressure, $p_2 = p_1$, $v_2/v_1 = T_2/T_1$, and

$$\Delta s = c_p \ln \frac{T_2}{T_1}, \tag{4.5.31}$$

where $c_p = c_v + R$ is the specific heat capacity under constant pressure.

In the case of a constant entropy (isentropic) reversible process, $s_2 = s_1$, equation (4.5.30) yields

$$\left(\frac{T_2}{T_1} \right)^{c_v} \left(\frac{v_2}{v_1} \right)^{R} = 1, \tag{4.5.32}$$

and thus

$$T^{c_v} v^R = c, \tag{4.5.33}$$

where c is a constant. This equation provides us with a relation between the specific volume and temperature. Introducing the density, $\rho = 1/v$, we write

$$\frac{T^{c_v}}{\rho^R} = c. \tag{4.5.34}$$

Using the ideal gas law, $p = \rho R T$, to eliminate the temperature, we obtain a relation between the density and pressure in an ideal compressible gas in isentropic transition,

$$\frac{p}{\rho^k} = d, \tag{4.5.35}$$

where d is a new constant and

$$k \equiv \frac{c_v + R}{c_v} = \frac{c_p}{c_v}. \tag{4.5.36}$$

The speed of sound, c, is given by

$$c^2 = \left(\frac{\partial p}{\partial \rho}\right)_s. \tag{4.5.37}$$

Using (4.5.35), we find

$$c^2 = d\,k\,\rho^{k-1} = \frac{kp}{\rho} = kRT. \tag{4.5.38}$$

The higher the temperature, the faster the speed of sound.

Problems

4.5.1 *Flow in a channel.*

Consider steady unidirectional flow in a channel due to the translation of the upper wall, as depicted in figure 4.5.1(b).

(a) Perform a force balance over a rectangular fluid layer confined between two y levels to show that, if the pressure is uniform, the shear stress $f_x^{(y)}$ must be independent of y.

(b) Having established that $f_x^{(y)}$ is constant, solve the first-order differential equation (4.5.7) for u_x in terms of y subject to the no-slip boundary conditions $u_x(y=0)=0$ and $u_x(y=h)=V$, and evaluate $f_x^{(y)}$ in terms of μ, V, and the channel width, h.

4.5.2 *Extensional flow*

(a) Consider a two-dimensional extensional flow in the xy plane with velocity components $u_x = kx$ and $u_y = -ky$, where k is the constant rate of extension with dimensions of inverse time. The corresponding pressure field is uniform throughout the domain of flow. Confirm that the fluid is incompressible, sketch the streamline pattern, and evaluate the stress tensor.

(b) Repeat (a) for axisymmetric extensional flow with Cartesian velocity components $u_x = kx$, $u_y = -\frac{1}{2}\,ky$, and $u_z = -\frac{1}{2}\,kz$.

(c) The axisymmetric extensional flow describes the motion inside the thread illustrated in figure 4.5.1(a). Assuming that the fluid is Newtonian, compute the force necessary to pull the rods apart with velocity V in terms of the half-length of the thread, L, the fluid viscosity, μ, and the cross-sectional area of the rods, A.

4.6 Simple non-Newtonian fluids

The Newtonian constitutive law for an incompressible fluid shown in equation (4.5.3) describes the stresses developing in a fluid consisting of small molecules.

Fluids containing or consisting of macromolecules, such as polymeric solutions and melts and fluids containing suspended rigid or deformable particles, exhibit more complicated behavior described by more involved constitutive equations. Examples include pastes, bubbly liquids, and biological fluids such as blood.

To derive a constitutive equation for a non-Newtonian fluid, we consider the motion of a fluid parcel and seek to establish a relation between the instantaneous traction exerted on the parcel surface and the history of the parcel deformation. In the simplest class of materials, the traction depends only on the instantaneous rate of parcel deformation expressed by the rate of deformation tensor, **E**.

A distinguishing feature of a non-Newtonian fluid is that the relation between the stress tensor, $\boldsymbol{\sigma}$, and the rate of deformation tensor, **E**, is nonlinear. In contrast, the corresponding relation for a Newtonian fluid is linear.

4.6.1 Unidirectional shear flow

In the case of unidirectional shear flow, the Newtonian shear stress (4.5.7) can be generalized by allowing the viscosity to depend on the magnitude of the shear rate, $|\mathrm{d}u_x/\mathrm{d}y|$. If the viscosity decreases as the shear rate increases, the fluid is shear-thinning or pseudo-plastic. If the viscosity increases as the shear rate increases, the fluid is shear-thickening or dilatant.

Physically, the dependence of the viscosity on the shear rate is attributed to changes in the configuration of molecules, changes in the shape and relative position of particles suspended in the fluid, and to the spontaneous formation of an internal microstructure due to intermolecular force fields and other particle interactions.

Power-law fluids

The shear stress developing in a certain class of non-Newtonian fluids in unidirectional shear flow can be described by the Ostwald-de Waele model. In this model, the viscosity is proportional to the magnitude of the shear rate raised to a power,

$$\mu = \mu_0 \left| \frac{\mathrm{d}u_x}{\mathrm{d}y} \right|^{n-1}, \tag{4.6.1}$$

where μ_0 is a constant, and n is the power-law exponent. When $n = 1$, we obtain a Newtonian fluid with viscosity μ_0; when $n < 1$, we obtain a shear-thinning fluid; when $n > 1$, we obtain a shear-thickening fluid. Substituting (4.6.1) in (4.5.7), we derive an expression for the shear stress,

$$f_x^{(y)} = \sigma_{yx} = \mu_0 \left| \frac{\mathrm{d}u_x}{\mathrm{d}y} \right|^{n-1} \frac{\mathrm{d}u_x}{\mathrm{d}y}. \tag{4.6.2}$$

Channel flow

As an application, we consider flow in a channel due to the translation of the upper wall with velocity V, as illustrated in figure 4.5.1(b). Performing a force balance over a rectangular fluid layer, we find that, if the pressure is uniform, the shear stress is independent of y and the right-hand side of (4.6.2) is constant (problem 4.5.1). The fluid velocity at the upper wall located at $y = h$ is equal to the wall velocity, V, and the fluid velocity at the lower wall located at $y = 0$ is zero. Integrating equation $\mathrm{d}u_x/\mathrm{d}y = c$, where c is the constant shear rate, and using the aforementioned boundary conditions, we derive a linear velocity profile with shear rate $\mathrm{d}u_x/\mathrm{d}y = V/h$, independent of the value of n.

Although the velocity profile is linear for any value of n, the magnitude of the shear stress does depend on n, as shown in equation (4.6.2). This distinction emphasizes that the kinematic appearance of a flow does not necessarily reflect the magnitude of the stresses developing in the fluid. Two flows that are kinematically identical may support different stress fields.

Yield-stress fluids

A class of heterogeneous fluids called Bingham plastics flow only when the shear stress established due to the motion lies above a certain threshold. Examples include pastes and concentrated suspensions of fine particles. An idealized constitutive equation between stress and shear rate for this class of materials is

$$\frac{\mathrm{d}u_x}{\mathrm{d}y} = 0 \qquad \text{if} \qquad |\sigma_{yx}| < \tau_0, \tag{4.6.3}$$

and

$$f_x^{(y)} = \sigma_{yx} = \tau_0 + \mu\,\frac{\mathrm{d}u_x}{\mathrm{d}y} \qquad \text{if} \qquad |\sigma_{yx}| > \tau_0, \tag{4.6.4}$$

where μ is the viscosity, and τ_0 is the yield stress. The relation between the shear stress and the shear rate is represented by the solid line in figure 4.6.1.

As an application, we consider the familiar unidirectional flow in a channel confined between two parallel walls, generated by exerting on the upper wall a force, F, parallel to the x axis. If the fluid is a Bingham plastic whose rheological behavior is described by equations (4.6.3) and (4.6.4), a shear flow across the whole of the channel will be established only if the externally imposed force F per unit length of the channel, L, counteracting the shear stress, $\sigma_{yx} = F/L$, is greater than the yield stress threshold, τ_0.

Assuming that this occurs, we treat σ_{yx} as a constant, solve equation (4.6.4) for $\mathrm{d}u_x/\mathrm{d}y$, and then integrate with respect to y subject to the boundary condition $u_x(y = 0) = 0$ to obtain a linear velocity profile,

$$u_x = \frac{y}{\mu}\,\Big(\frac{F}{L} - \tau_0\Big). \tag{4.6.5}$$

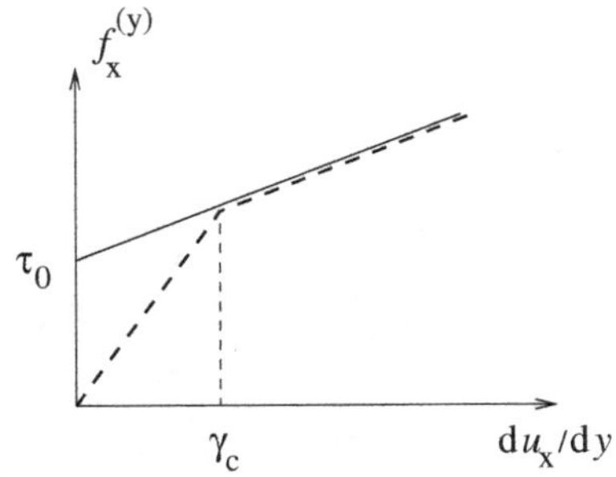

Figure 4.6.1 Rheological response of a Bingham plastic showing a yield-stress behavior in unidirectional flow.

The velocity at the upper wall is

$$u_x(y = h) = \frac{h}{\mu}\left(\frac{F}{L} - \tau_0\right) = V. \tag{4.6.6}$$

In practice, equation (4.6.6) allows us to estimate the values of the physical constants μ and τ_0 from laboratory observation.

Problem

4.6.1 *Yield-stress fluid*

The relation between the shear stress and the shear rate for a class of yield-stress fluids is described by the dashed line in figure 4.6.1, where the constant γ_c is the critical shear rate.

(*a*) State the equations describing this rheological behavior.

(*b*) Compute the shear stress established in a channel with parallel walls, where the upper wall translates with velocity V while the lower wall is held stationary.

4.7 Stresses in polar coordinates

We have discussed tractions and stresses in Cartesian coordinates. In practice, it is often convenient to work in cylindrical, spherical, or plane polar coordinates, with the benefit of reduced algebraic manipulations and easy implementation of boundary conditions. In this section, we define the components of the stress tensor in polar coordinates, and then relate them to the pressure and to the corresponding components of the rate of deformation tensor using the constitutive equation for incompressible Newtonian fluids.

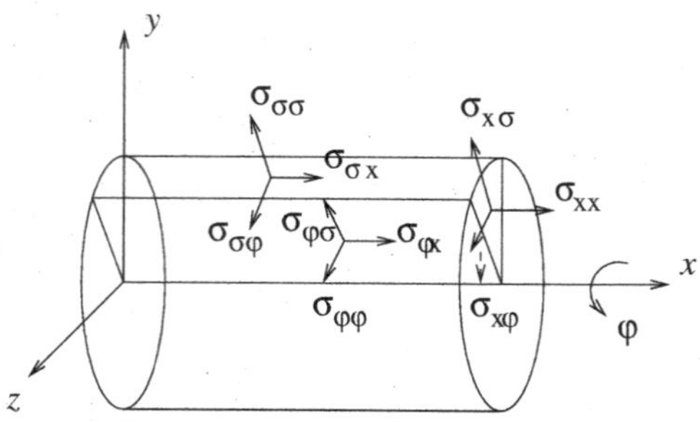

Figure 4.7.1 Components of the stress tensor in cylindrical polar coordinates.

4.7.1 Cylindrical polar coordinates

Consider the cylindrical polar coordinates, (x, σ, φ), depicted in figure 4.7.1. The traction exerted on a small surface that is perpendicular to the x axis, $\mathbf{f}^{(x)}$, acting on the side that faces the positive direction of the x axis, can be resolved into its cylindrical polar components as

$$\mathbf{f}^{(x)} = f_x^{(x)}\, \mathbf{e}_x + f_\sigma^{(x)}\, \mathbf{e}_\sigma + f_\varphi^{(x)}\, \mathbf{e}_\varphi, \tag{4.7.1}$$

where $\mathbf{e}_x$, $\mathbf{e}_\sigma$, and $\mathbf{e}_\varphi$ are unit vectors pointing, respectively, in the axial, radial, and meridional direction. Note that the orientation of $\mathbf{e}_x$ is constant, whereas the orientation of $\mathbf{e}_\sigma$ and $\mathbf{e}_\varphi$ changes with position in the flow.

The traction exerted on a small surface that is perpendicular to the distance from the x axis, $\mathbf{f}^{(\sigma)}$, and is thus parallel to the axial and meridional directions at the designated center of the surface, can be resolved into corresponding components as

$$\mathbf{f}^{(\sigma)} = f_x^{(\sigma)}\, \mathbf{e}_x + f_\sigma^{(\sigma)}\, \mathbf{e}_\sigma + f_\varphi^{(\sigma)}\, \mathbf{e}_\varphi. \tag{4.7.2}$$

The traction exerted on a small surface that is normal to the meridional direction, $\mathbf{f}^{(\varphi)}$, can be resolved as

$$\mathbf{f}^{(\varphi)} = f_x^{(\varphi)}\, \mathbf{e}_x + f_\sigma^{(\varphi)}\, \mathbf{e}_\sigma + f_\varphi^{(\varphi)}\, \mathbf{e}_\varphi. \tag{4.7.3}$$

Stacking the coefficients of the unit vectors on the right-hand sides of (4.7.1)–(4.7.3) on top of one another in a particular order, we obtain the stress tensor in cylindrical polar coordinates,

$$\boldsymbol{\sigma} \equiv \begin{bmatrix} f_x^{(x)} & f_\sigma^{(x)} & f_\varphi^{(x)} \\ f_x^{(\sigma)} & f_\sigma^{(\sigma)} & f_\varphi^{(\sigma)} \\ f_x^{(\varphi)} & f_\sigma^{(\varphi)} & f_\varphi^{(\varphi)} \end{bmatrix}. \tag{4.7.4}$$

$$\sigma_{xx} = -p + 2\mu \frac{\partial u_x}{\partial x}$$

$$\sigma_{x\sigma} = \sigma_{\sigma x} = \mu \left(\frac{\partial u_x}{\partial \sigma} + \frac{\partial u_\sigma}{\partial x} \right)$$

$$\sigma_{x\varphi} = \sigma_{\varphi x} = \mu \left(\frac{\partial u_\varphi}{\partial x} + \frac{1}{\sigma} \frac{\partial u_x}{\partial \varphi} \right)$$

$$\sigma_{\sigma\sigma} = -p + 2\mu \frac{\partial u_\sigma}{\partial \sigma}$$

$$\sigma_{\sigma\varphi} = \sigma_{\varphi\sigma} = \mu \left[\sigma \frac{\partial}{\partial \sigma} \left(\frac{u_\varphi}{\sigma} \right) + \frac{1}{\sigma} \frac{\partial u_\sigma}{\partial \varphi} \right]$$

$$\sigma_{\varphi\varphi} = -p + 2\,\mu \left(\frac{1}{\sigma} \frac{\partial u_\varphi}{\partial \varphi} + \frac{u_\sigma}{\sigma} \right)$$

Table 4.7.1 Components of the stress tensor for a Newtonian fluid in cylindrical polar coordinates. Note that the stress tensor remains symmetric in these coordinates.

To simplify the notation, we define

$$\sigma_{\alpha\beta} \equiv f_\beta^{(\alpha)}, \tag{4.7.5}$$

where Greek indices stand for x, σ, or φ. With the convention expressed by (4.7.5), the stress tensor is given by

$$\boldsymbol{\sigma} = \begin{bmatrix} \sigma_{xx} & \sigma_{x\sigma} & \sigma_{x\varphi} \\ \sigma_{\sigma x} & \sigma_{\sigma\sigma} & \sigma_{\sigma\varphi} \\ \sigma_{\varphi x} & \sigma_{\varphi\sigma} & \sigma_{\varphi\varphi} \end{bmatrix}. \tag{4.7.6}$$

Newtonian fluids

The stress components in an incompressible Newtonian fluid derive from the constitutive equation (4.5.3) as shown in table 4.7.1. To derive these relations, we write

$$\mathbf{f}^{(x)} = \sigma_{xx}\,\mathbf{e}_x + \sigma_{xy}\,\mathbf{e}_y + \sigma_{xz}\,\mathbf{e}_z,$$

$$\mathbf{f}^{(\sigma)} = \mathbf{f}^{(y)}\,\cos\varphi + \mathbf{f}^{(z)}\,\sin\varphi$$
$$= (\sigma_{yx}\,\mathbf{e}_x + \sigma_{yy}\,\mathbf{e}_y + \sigma_{yz}\,\mathbf{e}_z)\,\cos\varphi + (\sigma_{zx}\,\mathbf{e}_x + \sigma_{zy}\,\mathbf{e}_y + \sigma_{zz}\,\mathbf{e}_z)\,\sin\varphi, \tag{4.7.7}$$

$$\mathbf{f}^{(\varphi)} = -\mathbf{f}^{(y)}\,\sin\varphi + \mathbf{f}^{(z)}\,\cos\varphi$$
$$= -(\sigma_{yx}\,\mathbf{e}_x + \sigma_{yy}\,\mathbf{e}_y + \sigma_{yz}\,\mathbf{e}_z)\,\sin\varphi + (\sigma_{zx}\,\mathbf{e}_x + \sigma_{zy}\,\mathbf{e}_y + \sigma_{zz}\,\mathbf{e}_z)\,\cos\varphi,$$

and substitute

$$\mathbf{e}_y = \mathbf{e}_\sigma\,\cos\varphi - \mathbf{e}_\varphi\,\sin\varphi, \qquad \mathbf{e}_z = \mathbf{e}_\sigma\,\sin\varphi + \mathbf{e}_\varphi\,\sin\varphi. \tag{4.7.8}$$

Consolidating terms multiplying the unit cylindrical polar vectors in the first equation of (4.7.7), we obtain

$$\sigma_{x\sigma} = \sigma_{xy}\,\cos\varphi + \sigma_{xz}\,\sin\varphi$$
$$= \mu\left(\cos\varphi\,\frac{\partial u_x}{\partial y} + \sin\varphi\,\frac{\partial u_x}{\partial z}\right) + \mu\,\frac{\partial(u_y\,\cos\varphi + u_z\,\sin\varphi)}{\partial x} \tag{4.7.9}$$

and

$$\sigma_{x\varphi} = -\sigma_{xy}\,\sin\varphi + \sigma_{xz}\,\cos\varphi$$
$$= \mu\left(-\sin\varphi\,\frac{\partial u_x}{\partial y} + \cos\varphi\,\frac{\partial u_x}{\partial z}\right) + \mu\,\frac{\partial(-u_y\,\sin\varphi + u_z\,\cos\varphi)}{\partial x}, \tag{4.7.10}$$

which reproduce the second and third relations in table 4.7.1. The rest of the relations are derived working in a similar fashion.

4.7.2 Spherical polar coordinates

Consider a system of spherical polar coordinates, (r, θ, φ), depicted in figure 4.7.2. The traction exerted on a small surface that is normal to the distance from the origin, acting on the side of the surface that faces away from the origin, $\mathbf{f}^{(r)}$, can be resolved into its spherical polar components as

$$\mathbf{f}^{(r)} = f_r^{(r)}\,\mathbf{e}_r + f_\theta^{(r)}\,\mathbf{e}_\theta + f_\varphi^{(r)}\,\mathbf{e}_\varphi, \tag{4.7.11}$$

where $\mathbf{e}_x$, $\mathbf{e}_\theta$, and $\mathbf{e}_\varphi$ are unit vectors pointing in the radial, azimuthal, and meridional directions.

The traction exerted on a small surface that is normal to the azimuthal direction corresponding to the angle θ, and is thus parallel to the radial and meridional directions, $\mathbf{f}^{(\theta)}$, can be similarly resolved as

$$\mathbf{f}^{(\theta)} = f_r^{(\theta)}\,\mathbf{e}_r + f_\theta^{(\theta)}\,\mathbf{e}_\theta + f_\varphi^{(\theta)}\,\mathbf{e}_\varphi. \tag{4.7.12}$$

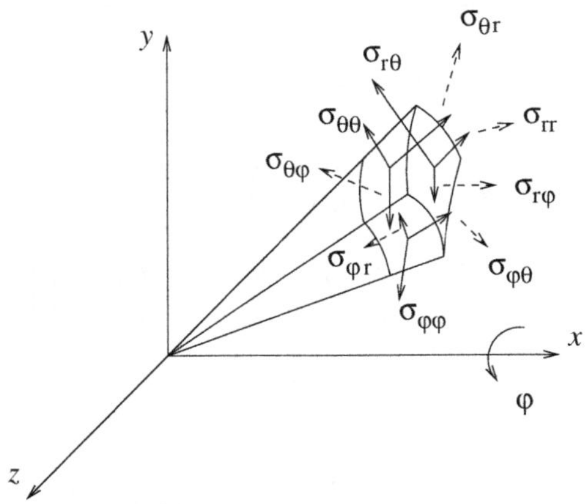

Figure 4.7.2 Components of the stress tensor in spherical polar coordinates.

The traction exerted on a small surface that is normal to the meridional direction, and is thus parallel to the radial and azimuthal directions, $\mathbf{f}^{(\varphi)}$, can be resolved as

$$\mathbf{f}^{(\varphi)} = f_r^{(\varphi)}\,\mathbf{e}_r + f_\theta^{(\varphi)}\,\mathbf{e}_\theta + f_\varphi^{(\varphi)}\,\mathbf{e}_\varphi. \tag{4.7.13}$$

Stacking the coefficients of the unit vectors on the right-hand sides of (4.7.11)–(4.7.13) on top of one another in a particular order, we obtain the stress tensor in spherical polar coordinates

$$\boldsymbol{\sigma} \equiv \begin{bmatrix} f_r^{(r)} & f_\theta^{(r)} & f_\varphi^{(r)} \\ f_r^{(\theta)} & f_\theta^{(\theta)} & f_\varphi^{(\theta)} \\ f_r^{(\varphi)} & f_\theta^{(\varphi)} & f_\varphi^{(\varphi)} \end{bmatrix}. \tag{4.7.14}$$

To simplify the notation, we introduce the standard two-index notation for the stress tensor, writing

$$\sigma_{\alpha\beta} \equiv f_\beta^{(\alpha)}, \tag{4.7.15}$$

where Greek indices stand for r, θ, or φ. With the convention expressed by (4.7.15), the stress tensor is given by

$$\boldsymbol{\sigma} = \begin{bmatrix} \sigma_{rr} & \sigma_{r\theta} & \sigma_{r\varphi} \\ \sigma_{\theta r} & \sigma_{\theta\theta} & \sigma_{\theta\varphi} \\ \sigma_{\varphi r} & \sigma_{\varphi\theta} & \sigma_{\varphi\varphi} \end{bmatrix}. \tag{4.7.16}$$

$$\sigma_{rr} = -p + 2\,\mu\,\frac{\partial u_r}{\partial r}$$

$$\sigma_{r\theta} = \sigma_{\theta r} = \mu\left[r\frac{\partial}{\partial r}\Big(\frac{u_\theta}{r}\Big) + \frac{1}{r}\frac{\partial u_r}{\partial \theta}\right]$$

$$\sigma_{r\varphi} = \sigma_{\varphi r} = \mu\left[\frac{1}{r\sin\theta}\frac{\partial u_r}{\partial \varphi} + r\frac{\partial}{\partial r}\Big(\frac{u_\varphi}{r}\Big)\right]$$

$$\sigma_{\theta\theta} = -p + 2\,\mu\Big(\frac{1}{r}\frac{\partial u_\theta}{\partial \theta} + \frac{u_r}{r}\Big)$$

$$\sigma_{\theta\varphi} = \sigma_{\varphi\theta} = \mu\left[\frac{\sin\theta}{r}\frac{\partial}{\partial \theta}\Big(\frac{u_\varphi}{\sin\theta}\Big) + \frac{1}{r\sin\theta}\frac{\partial u_\theta}{\partial \varphi}\right]$$

$$\sigma_{\varphi\varphi} = -p + \mu\,\frac{2}{r\sin\theta}\,(\,\frac{\partial u_\varphi}{\partial \varphi} + u_r\sin\theta + u_\theta\cos\theta\,)$$

Table 4.7.2 Components of the stress tensor for a Newtonian fluid in spherical polar coordinates. Note that the stress tensor remains symmetric in these coordinates.

Newtonian fluids

The stress components for an incompressible Newtonian fluid derive from the constitutive equation (4.5.3) as shown in table 4.7.2.

4.7.3 Plane polar coordinates

Consider a two-dimensional flow in the xy plane and refer to the plane polar coordinates, (r, θ), depicted in figure 4.7.3. The traction exerted on a small segment that is normal to the distance from the origin, acting on the side facing away from the origin, $\mathbf{f}^{(r)}$, can be resolved into its plane polar components as

$$\mathbf{f}^{(r)} = f_r^{(r)}\,\mathbf{e}_r + f_\theta^{(r)}\,\mathbf{e}_\theta, \tag{4.7.17}$$

where $\mathbf{e}_r$ and $\mathbf{e}_\theta$ are unit vectors pointing in the radial and polar direction.

The traction exerted on a small surface that is normal to the direction of the polar angle θ, and is thus parallel to the distance from the origin, $\mathbf{f}^{(\theta)}$, can be similarly resolved as

$$\mathbf{f}^{(\theta)} = f_r^{(\theta)}\,\mathbf{e}_r + f_\theta^{(\theta)}\,\mathbf{e}_\theta. \tag{4.7.18}$$

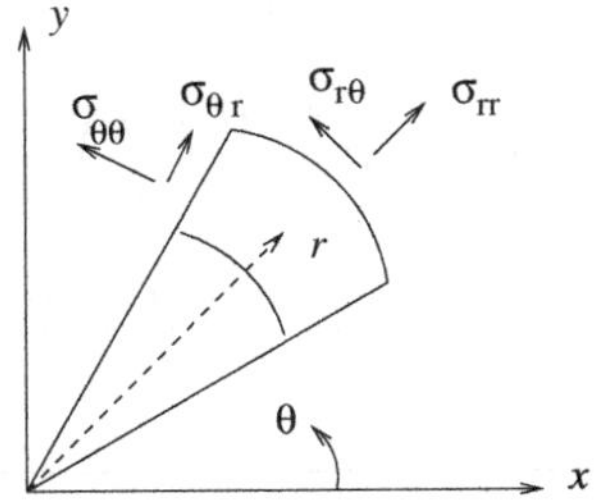

Figure 4.7.3 Components of the stress tensor in plane polar coordinates.

Stacking the coefficients of the unit vectors on the right-hand sides of (4.7.17) and (4.7.18) on top of one another in a particular order, we obtain the stress tensor in plane polar coordinates,

$$\boldsymbol{\sigma} \equiv \begin{bmatrix} f_r^{(r)} & f_\theta^{(r)} \\ f_r^{(\theta)} & f_\theta^{(\theta)} \end{bmatrix}. \tag{4.7.19}$$

Next, we introduce the familiar two-index notation,

$$\sigma_{\alpha\beta} \equiv f_\beta^{(\alpha)}, \tag{4.7.20}$$

where Greek indices stand for r or θ. With the convention expressed by (4.7.20), the stress tensor is given by

$$\boldsymbol{\sigma} = \begin{bmatrix} \sigma_{rr} & \sigma_{r\theta} \\ \sigma_{\theta r} & \sigma_{\theta\theta} \end{bmatrix}. \tag{4.7.21}$$

Newtonian fluids

The stress components for an incompressible Newtonian fluid derive from the constitutive equation (4.5.3) as shown in table 4.7.3.

Problem

4.7.1 *Plane polar coordinates*

Derive the Newtonian constitutive equations shown in table 4.7.3 working as in the text for the cylindrical polar coordinates.

$$\sigma_{rr} = -p + 2\mu \frac{\partial u_r}{\partial r}$$

$$\sigma_{r\theta} = \sigma_{\theta r} = \mu \left[r \frac{\partial}{\partial r}\left(\frac{u_\theta}{r}\right) + \frac{1}{r}\frac{\partial u_r}{\partial \theta} \right]$$

$$\sigma_{\theta\theta} = -p + 2\mu \left(\frac{1}{r}\frac{\partial u_\theta}{\partial \theta} + \frac{u_r}{r} \right)$$

Table 4.7.3 Components of the stress tensor for a Newtonian fluid in plane polar coordinates. Note that the stress tensor remains symmetric in these coordinates.

4.8 Boundary conditions for the tangential velocity

In section 2.10, we discussed the no-penetration boundary condition over impermeable boundaries and interfaces between immiscible fluids, involving the normal component of the fluid velocity. Viscous fluids obey an additional boundary condition concerning the tangential component of the velocity.

4.8.1 No-slip boundary condition

Under most conditions, the vast majority of fluids satisfy the no-slip boundary condition requiring that (*a*) the tangential component of the fluid velocity over a solid boundary is equal to the tangential component of the boundary velocity, and (*b*) the tangential component of the fluid velocity is continuous across an interface between two immiscible fluids. Combined with the no-slip condition, the no-penetration condition requires that the fluid velocity is equal to the local velocity of an impermeable solid boundary, and continuous across an interface. The no-slip boundary condition has been confirmed in the overwhelming majority of applications, and is the standard choice in mainstream fluid dynamics.

The physical origin of the no-slip boundary condition over a solid surface has not been established with certainty. One theory argues that the molecules of a fluid next to a solid surface are adsorbed on the surface for a short period of time, and are then desorbed to be ejected into the fluid. This process slows down the fluid and renders the tangential component of the fluid velocity equal to the corresponding component of the boundary velocity. Another theory argues that the true boundary condition is the condition of vanishing shear stress, and the no-slip boundary condition arises due to microscopic boundary roughness. Thus, a perfectly smooth boundary would allow the fluid to slip.

4.8.2 Slip boundary condition

Exceptions to the no-slip boundary condition arise in the flow of rarefied gases, in high-pressure flow of polymer melts, in the flow near a three-phase contact line where a solid meets two liquids or a gas and a liquid, and in the flow past interfaces consisting of dual or multiple molecular layers that may exhibit relative motion yielding a discontinuous macroscopic velocity. The no-slip boundary condition is sometimes relaxed in numerical simulations to prevent singularities stemming from excessive idealization. One example is the development of an infinite force on a sharp plate scraping fluid off a flat surface.

Consider steady unidirectional flow in a channel with parallel walls driven by the parallel translation of the upper wall along the x axis with velocity V, as illustrated in figure 4.5.1(b). To implement the slip boundary condition, we allow the fluid to slip over the lower wall such that the slip velocity, $u_x(y=0)$, is related to the wall shear stress by

$$u_x(y=0) = L\mu\beta\,\sigma_{yx}(y=0) = \frac{L}{\beta}\left(\frac{\partial u_x}{\partial y}\right)_{y=0}, \tag{4.8.1}$$

where L is a reference length identified, for example, with the channel width, and the constant β is the slip coefficient. As β tends infinity, the slip velocity tends to zero and the no-slip boundary condition prevails. The slip length is defined as $\lambda_s = L/\beta$.

In the case of a rarefied gas, the slip coefficient, β, and slip length, λ_s, can be rigorously related to the molecular mean free path, λ, by the Maxwell relation

$$\frac{\lambda}{\lambda_s} = \beta_p\, K_n = \frac{\sigma}{2-\sigma}, \tag{4.8.2}$$

where $K_n \equiv \lambda/L$ is the Knudsen number, and σ is the tangential momentum accommodation coefficient (TMAC) expressing the fraction of molecules that undergo diffusive instead of specular reflection. The limit $\sigma = 2$ yields the no-slip boundary condition, $\beta \to \infty$, whereas the limit $\sigma = 0$ yields the perfect-slip boundary condition, $\beta \to 0$.

Problem

4.8.1 *Flow in a channel with slip*

In the case of channel flow illustrated in figure 4.5.1(b), the pressure is uniform and the shear stress σ_{yx} is constant, independent of y. Assuming that the slip condition applies at the upper and lower walls, derive an expression for σ_{yx} and for the velocity profile of a Newtonian fluid in terms of V, h, μ, and l.

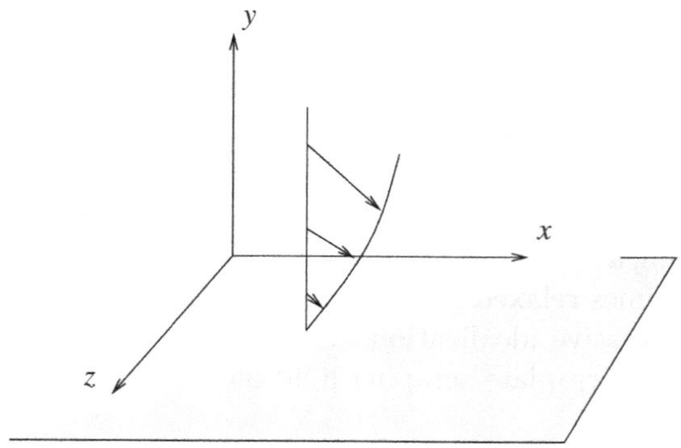

Figure 4.9.1 Flow over a plane wall with the no-slip boundary condition. The wall shear stress is proportional to the slope of the velocity profile with respect to distance normal to the wall. The normal stress is equal to the negative of the pressure.

4.9 Wall stresses in Newtonian fluids

Combining the no-slip boundary condition discussed in section 4.8 with the no-penetration boundary condition discussed in section 2.10, we derive remarkably simple expressions for the Newtonian traction exerted on a solid surface, called the wall stress, amenable to simple physical interpretation.

Consider a flow above a stationary flat solid surface located at $y = 0$, as illustrated in figure 4.9.1. The no-slip boundary condition requires that the tangential components of the velocity, and thus their derivatives with respect to x and z, vanish over the surface,

$$\frac{\partial u_x}{\partial x} = 0, \qquad \frac{\partial u_z}{\partial x} = 0, \qquad \frac{\partial u_x}{\partial z} = 0, \qquad \frac{\partial u_z}{\partial z} = 0, \tag{4.9.1}$$

where the partial derivatives are evaluated at $y = 0$. The no-penetration condition requires that the normal component of the velocity, and thus its derivatives with respect to x and z, also vanish over the surface,

$$\frac{\partial u_y}{\partial x} = 0, \qquad \frac{\partial u_y}{\partial z} = 0, \tag{4.9.2}$$

where the partial derivatives are evaluated at $y = 0$.

4.9.1 Shear stress

The two components of the Newtonian shear stress exerted on the surface are given by

$$\sigma_{yx} = \mu\,(\frac{\partial u_x}{\partial y} + \frac{\partial u_y}{\partial x}), \qquad \sigma_{yz} = \mu\,(\frac{\partial u_z}{\partial y} + \frac{\partial u_y}{\partial z}), \tag{4.9.3}$$

evaluated at $y = 0$. Using (4.9.2) to simplify (4.9.3), we find

$$\sigma_{yx} = \mu \frac{\partial u_x}{\partial y}, \qquad \sigma_{yz} = \mu \frac{\partial u_z}{\partial y}, \tag{4.9.4}$$

evaluated at $y = 0$. Equations (4.9.4) show that the wall shear stress is simply equal to the slope of the tangential velocity with respect to distance normal to the wall, multiplied by the fluid viscosity.

4.9.2 Normal stress

The Newtonian normal stress exerted on the solid surface is given by

$$\sigma_{yy} = -p + 2\,\mu \frac{\partial u_y}{\partial y}, \tag{4.9.5}$$

evaluated at $y = 0$, where p is the pressure. Since the fluid has been assumed incompressible, we may use the continuity equation (2.9.2) to write

$$\frac{\partial u_y}{\partial y} = -\frac{\partial u_x}{\partial x} - \frac{\partial u_z}{\partial z}, \tag{4.9.6}$$

evaluated at $y = 0$, and then invoke the first and fourth of equations (4.9.1) to find $\partial u_y / \partial y = 0$. Expression (4.9.5) thus simplifies to

$$\sigma_{yy} = -p, \tag{4.9.7}$$

which shows that the normal stress exerted on a solid surface is equal to the negative of the pressure.

Generalization

The results displayed in equations (4.9.4) and (4.9.7) remain valid even when the surface translates with a constant or time-dependent velocity. Moreover, these equation also apply when the surface is curved, provided that the zx plane is tangential to the surface, and the y axis is normal to the surface at the position when the shear or normal stress is evaluated.

Problem

4.9.1 *Vorticity at a no-slip surface*

Show that the component of the vorticity vector normal to an impermeable wall vanishes, and thus the vortex lines are tangential to the surface. *Hint:* Use the second and third of equations (4.9.1).

4.10 Interfacial surfactant transport

An impure interface between two immiscible fluids is sometimes occupied by a molecular layer of a surfactant affecting the local surface tension, γ. Dividing the number of surfactant molecules residing inside an infinitesimal patch centered at a given point by the patch surface area, we obtain the surface concentration of the surfactant, Γ.

The higher the surfactant concentration, the lower the surface tension. For small surfactant concentrations, a linear surface equation of state can be assumed relating γ to Γ according to Gibbs' law,

$$\gamma = \frac{\gamma_0}{1-\beta}\,(1-\beta\,\frac{\Gamma}{\Gamma_0}), \tag{4.10.1}$$

where Γ_0 and γ_0 are, respectively, a reference surfactant concentration and the corresponding surface tension. The dimensionless physical constant $\beta = \Gamma_0 RT/\gamma_c$ expresses the sensitivity of the surface tension to the surfactant concentration; R is the ideal gas constant, T is the absolute temperature, and γ_c is the surface tension of a clean interface that is devoid of surfactants. More involved surface equations of state for moderate and large surfactant concentrations near saturation are available.

The molecules of an insoluble surfactant are convected and diffuse over the interface, but do not enter the bulk of the fluid. Our objective is to derive an evolution equation for the surface surfactant concentration determining the local surface tension and thus the jump in the traction across an interface.

4.10.1 Two-dimensional interfaces

Consider a chain of material point particles distributed along the inner or outer side of a two-dimensional interface. We begin by labeling the point particles using a parameter, ξ, so that their position can be described in parametric form as $\mathbf{X}(\xi)$, as shown in figure 4.10.1. Let l be the arc length along the interface measured from an arbitrary point particle labeled ξ_0. The number of surfactant molecules residing inside a test section of the interface confined between ξ_0 and ξ is

$$n(\xi,t) = \int_{l(\xi_0,t)}^{l(\xi,t)} \Gamma(\xi',t)\,\mathrm{d}l(\xi') = \int_{\xi_0}^{\xi} \Gamma(\xi',t)\,\frac{\partial l}{\partial \xi'}\,\mathrm{d}\xi', \tag{4.10.2}$$

where ξ' is an integration variable.

Conservation of the total number of surfactant molecules inside the test section requires

$$\frac{\partial n}{\partial t} = q(\xi_0) - q(\xi), \tag{4.10.3}$$

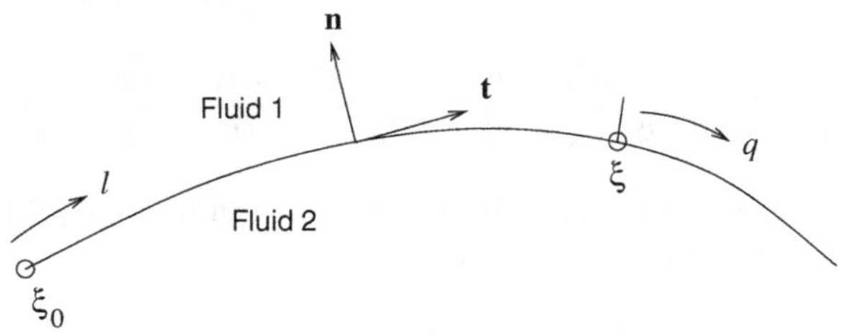

Figure 4.10.1 Point particles along a two-dimensional interface are identified by a parameter, ξ.

where q is the flux of surfactant molecules along the interface due to diffusion, and the time derivative is taken keeping ξ fixed. Substituting the expression for n from the last integral of (4.10.2) and transferring the derivative inside the integral, we find

$$\int_{\xi_0}^{\xi} \frac{\mathrm{D}}{\mathrm{D}t}\Big(\Gamma(\xi', t)\,\frac{\partial l}{\partial \xi'}\Big)\,\mathrm{d}\xi' = q(\xi_0) - q(\xi), \tag{4.10.4}$$

where $\mathrm{D}/\mathrm{D}t$ is the material derivative. Now taking the limit as ξ tends to ξ_0, we derive the differential equation

$$\frac{\mathrm{D}}{\mathrm{D}t}\Big(\Gamma\,\frac{\partial l}{\partial \xi}\Big) = -\frac{\partial q}{\partial \xi}. \tag{4.10.5}$$

Expanding the material derivative on the left-hand side, we find

$$\frac{\mathrm{D}\Gamma}{\mathrm{D}t}\,\frac{\partial l}{\partial \xi} + \Gamma\,\frac{\mathrm{D}}{\mathrm{D}t}\Big(\frac{\partial l}{\partial \xi}\Big) = -\frac{\partial q}{\partial \xi}. \tag{4.10.6}$$

Next, we use the Pythagorean theorem to write

$$\frac{\partial l}{\partial \xi} = \Big[\Big(\frac{\partial X}{\partial \xi}\Big)^2 + \Big(\frac{\partial Y}{\partial \xi}\Big)^2\Big]^{1/2}, \tag{4.10.7}$$

and compute the material derivative

$$\frac{\mathrm{D}}{\mathrm{D}t}\Big(\frac{\partial l}{\partial \xi}\Big) = \frac{1}{2}\,\frac{1}{\partial l/\partial \xi}\,\Big[2\,\frac{\partial X}{\partial \xi}\,\frac{\mathrm{D}}{\mathrm{D}t}\Big(\frac{\partial X}{\partial \xi}\Big) + 2\,\frac{\partial Y}{\partial \xi}\,\frac{\mathrm{D}}{\mathrm{D}t}\Big(\frac{\partial Y}{\partial \xi}\Big)\Big]. \tag{4.10.8}$$

Interchanging the material and the ξ derivative, and setting $\mathrm{D}X/\mathrm{D}t = u_x$ and $\mathrm{D}Y/\mathrm{D}t = u_y$, we find

$$\frac{\mathrm{D}}{\mathrm{D}t}\Big(\frac{\partial l}{\partial \xi}\Big) = \frac{1}{\partial l/\partial \xi}\,\Big(\frac{\partial X}{\partial \xi}\,\frac{\partial u_x}{\partial \xi} + \frac{\partial Y}{\partial \xi}\,\frac{\partial u_y}{\partial \xi}\Big), \tag{4.10.9}$$

where $\mathbf{u} = (u_x, u_y)$ is the fluid velocity. Rearranging, we find

$$\frac{\mathrm{D}}{\mathrm{D}t}\Big(\frac{\partial l}{\partial \xi}\Big) = \frac{1}{\partial l/\partial \xi}\,\frac{\partial \mathbf{X}}{\partial \xi}\cdot\frac{\partial \mathbf{u}}{\partial \xi} = \frac{\partial l}{\partial \xi}\,\frac{\partial \mathbf{X}}{\partial l}\cdot\frac{\partial \mathbf{u}}{\partial l} = \frac{\partial l}{\partial \xi}\,\mathbf{t}\cdot\frac{\partial \mathbf{u}}{\partial l}, \tag{4.10.10}$$

where $\mathbf{t} = \partial \mathbf{X}/\partial l$ is the unit tangent vector shown in figure 4.10.1. Substituting the final expression in (4.10.6), we find

$$\frac{\mathrm{D}\Gamma}{\mathrm{D}t} + \Gamma\,\mathbf{t}\cdot\frac{\partial \mathbf{u}}{\partial l} = -\frac{\partial q}{\partial l}, \tag{4.10.11}$$

which is the targeted evolution equation for the surfactant concentration.

Fick's law

The diffusive flux can be described by Fick's law,

$$q = -D_s\,\frac{\partial \Gamma}{\partial l}, \tag{4.10.12}$$

where D_s is the surfactant surface diffusivity. Substituting this expression in (4.10.11), we derive a convection–diffusion equation,

$$\frac{\mathrm{D}\Gamma}{\mathrm{D}t} + \Gamma\,\mathbf{t}\cdot\frac{\partial \mathbf{u}}{\partial l} = \frac{\partial}{\partial l}\Big(D_s\,\frac{\partial \Gamma}{\partial l}\Big). \tag{4.10.13}$$

In practice, the surfactant diffusivity is typically small.

Stretching and expansion

It is illuminating to decompose the velocity into a tangential and a normal component,

$$\mathbf{u} = u_t\,\mathbf{t} + u_n\,\mathbf{n}. \tag{4.10.14}$$

Noting that $\mathbf{t}\cdot\mathbf{n} = 0$, $\mathbf{t}\cdot\mathbf{t} = 1$, and $\mathbf{n}\cdot\mathbf{n} = 1$, and using the Frenet relations (4.3.5) and (4.3.6), we compute

$$\mathbf{t}\cdot\frac{\partial \mathbf{u}}{\partial l} = \frac{\partial u_t}{\partial l} + u_n\,\mathbf{t}\cdot\frac{\partial \mathbf{n}}{\partial l} = \frac{\partial u_t}{\partial l} + \kappa\,u_n, \tag{4.10.15}$$

where κ is the interfacial curvature. Substituting this expression in (4.10.13), we find

$$\frac{\mathrm{D}\Gamma}{\mathrm{D}t} + \Gamma\,(\,\frac{\partial u_t}{\partial l} + \kappa\,u_n\,) = \frac{\partial}{\partial l}\Big(D_s\,\frac{\partial \Gamma}{\partial l}\Big). \tag{4.10.16}$$

The first and second terms inside the parentheses on the left-hand side express, respectively, the change in the surfactant concentration due to interfacial stretching, and the change in the surfactant concentration due to interfacial expansion.

Stretching of a flat interface

As an application, we consider a flat interface along the x axis that is stretched uniformly under the influence of a tangential velocity field, u_x. Identifying l with x and setting $\kappa = 0$, we find that the transport equation (4.10.16) reduces to

$$\frac{\mathrm{D}\Gamma}{\mathrm{D}t} + \Gamma \frac{\partial u_x}{\partial x} = \frac{\partial}{\partial x}\Big(D_s \frac{\partial \Gamma}{\partial x}\Big). \tag{4.10.17}$$

We can resolve the material derivative into Eulerian derivatives with respect to x and t, obtaining

$$\frac{\partial \Gamma}{\partial t} + u_x \frac{\partial \Gamma}{\partial x} + \Gamma \frac{\partial u_x}{\partial x} = \frac{\partial}{\partial x}\Big(D_s \frac{\partial \Gamma}{\partial x}\Big), \tag{4.10.18}$$

or

$$\frac{\partial \Gamma}{\partial t} + \frac{\partial (u_x \Gamma)}{\partial x} = \frac{\partial}{\partial x}\Big(D_s \frac{\partial \Gamma}{\partial x}\Big). \tag{4.10.19}$$

This equation could have been derived directly by performing a surfactant molecular balance over a differential control volume along the x axis, taking into consideration the convective and diffusive flux.

Consider a uniformly stretched interface where $u_x = kx$, and k is a constant rate of extension. If the surfactant concentration is uniform at the initial instant, it will remain uniform at all times, governed by the linear equation

$$\frac{\mathrm{d}\Gamma}{\mathrm{d}t} + k\,\Gamma = 0. \tag{4.10.20}$$

The solution reveals that the surfactant concentration decreases exponentially due to dilution, $\Gamma(t) = \Gamma(t = 0)\exp(-kt)$.

Expansion of a circular interface

As a second application, we consider a cylindrical interface with a circular cross-section of radius a centered at the origin, expanding under the influence of a uniform radial velocity in the absence of circumferential motion. In plane polar coordinates, (r, θ), the transport equation (4.10.16) with constant diffusivity becomes

$$\frac{\mathrm{D}\Gamma}{\mathrm{D}t} + \Gamma \frac{u_r}{a} = \frac{D_s}{a^2} \frac{\partial^2 \Gamma}{\partial \theta^2}. \tag{4.10.21}$$

Resolving the material derivative into Eulerian derivatives with respect to r and t, we obtain

$$\frac{\partial \Gamma}{\partial t} + \Gamma \frac{u_r}{a} = \frac{D_s}{a^2} \frac{\partial^2 \Gamma}{\partial \theta^2}. \tag{4.10.22}$$

If the surfactant concentration is uniform at the initial instant, it will remain uniform at all times, governed by the linear equation

$$\frac{\mathrm{d}\Gamma}{\mathrm{d}t} + \Gamma\,\frac{u_r}{a} = 0. \tag{4.10.23}$$

In the case of expansion, $u_r > 0$, the surfactant concentration decreases exponentially due to dilution.

Interfacial markers

The material derivative expresses the rate of change of the surfactant concentration following the motion of point particles residing on either side of an interface. In numerical practice, it may be expedient to follow the motion of interfacial marker points moving with the normal component of the fluid velocity and with an arbitrary tangential velocity, v_t. If $v_t = 0$, the marker points move normal to the interface at any instant. The velocity of a marker point is then

$$\mathbf{v} = u_n\,\mathbf{n} + v_t\,\mathbf{t}. \tag{4.10.24}$$

By definition,

$$\frac{\mathrm{D}\Gamma}{\mathrm{D}t} = \frac{\mathrm{d}\Gamma}{\mathrm{d}t} + (u_t - v_t)\,\frac{\partial\Gamma}{\partial l}, \tag{4.10.25}$$

where $\mathrm{d}/\mathrm{d}t$ is the rate of change of the surfactant concentration following the marker points. Substituting this expression in (4.10.16), we find

$$\frac{\mathrm{d}\Gamma}{\mathrm{d}t} + (u_t - v_t)\,\frac{\partial\Gamma}{\partial l} + \Gamma\,(\,\frac{\partial u_t}{\partial l} + \kappa\,u_n\,) = \frac{\partial}{\partial l}\Big(D_s\,\frac{\partial\Gamma}{\partial l}\Big), \tag{4.10.26}$$

which can be restated as

$$\frac{\mathrm{d}\Gamma}{\mathrm{d}t} + \frac{\partial(u_t\Gamma)}{\partial l} - v_t\,\frac{\partial\Gamma}{\partial l} + \Gamma\,\kappa\,u_n = \frac{\partial}{\partial l}\Big(D_s\,\frac{\partial\Gamma}{\partial l}\Big). \tag{4.10.27}$$

The second term on the left-hand side is the derivative of the interfacial convective flux.

4.10.2 Axisymmetric interfaces

Next, we consider a chain of material point particles distributed along the inner or outer side of the trace of an axisymmetric interface in a meridional plane, and label the point particles using a parameter, ξ, so that their position in the meridional plane is described in parametric form by $\mathbf{X}(\xi)$. Let l be the arc length along the trace of the interface measured from an arbitrary point particle labeled ξ_0, as illustrated in figure 4.10.2.

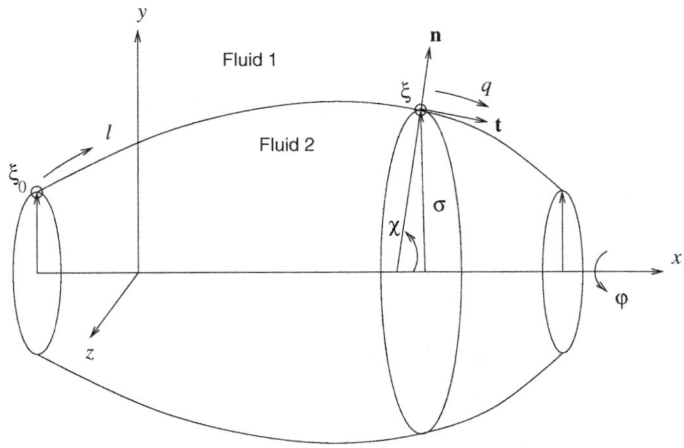

Figure 4.10.2 Point particles along the trace of an axisymmetric interface in a meridional plane are identified by a parameter ξ. The angle χ is subtended between the x axis and the straight line defined by the extension of the normal vector.

To derive an evolution equation for the surfactant concentration, we introduce cylindrical polar coordinates (x, σ, φ), and express the number of surfactant molecules inside a ring-like material section of the interface confined between ξ_0 and ξ as

$$n(\xi, t) = 2\pi \int_{l(\xi_0,t)}^{l(\xi,t)} \Gamma(\xi', t)\, \sigma(\xi')\, \mathrm{d}l(\xi') = 2\pi \int_{\xi_0}^{\xi} \Gamma(\xi', t)\, \frac{\partial l}{\partial \xi'}\, \sigma(\xi')\mathrm{d}\xi'. \tag{4.10.28}$$

Conservation of the total number of surfactant molecules inside the test section requires

$$\frac{\partial n}{\partial t} = 2\pi\sigma_0\, q(\xi_0) - 2\pi\sigma q(\xi), \tag{4.10.29}$$

where q is the flux of surfactant molecules along the interface by diffusion, and the time derivative is taken keeping ξ fixed. The counterpart of the balance equation (4.10.5) is

$$\frac{\mathrm{D}}{\mathrm{D}t}\Big(\Gamma(\xi, t)\, \sigma(\xi, t)\, \frac{\partial l}{\partial \xi}\Big) = -\frac{\partial(\sigma q)}{\partial \xi}, \tag{4.10.30}$$

and the counterpart of equation (4.10.11) is

$$\frac{\mathrm{D}\Gamma}{\mathrm{D}t} + \Gamma\,(\,\mathbf{t} \cdot \frac{\partial \mathbf{u}}{\partial l} + \frac{u_\sigma}{\sigma}\,) = -\frac{1}{\sigma}\,\frac{\partial(\sigma q)}{\partial l}. \tag{4.10.31}$$

In deriving this equation, we have set $\mathrm{D}\sigma/\mathrm{D}t = u_\sigma$. In terms of the normal and tangential velocities,

$$\frac{\mathrm{D}\Gamma}{\mathrm{D}t} + \Gamma\,(\frac{\partial u_t}{\partial l} + \kappa\, u_n + \frac{u_\sigma}{\sigma}) = -\frac{1}{\sigma}\,\frac{\partial(\sigma q)}{\partial l}, \tag{4.10.32}$$

where κ is the curvature of the interface in a meridional plane. Substituting

$$u_\sigma = u_n\,\sin\chi - u_t\,\cos\chi, \tag{4.10.33}$$

we find

$$\frac{\mathrm{D}\Gamma}{\mathrm{D}t} + \Gamma\left[\frac{\partial u_t}{\partial l} - \frac{\cos\chi}{\sigma}\,u_t + (\kappa + \frac{\sin\chi}{\sigma})\,u_n\right] = -\frac{1}{\sigma}\,\frac{\partial(\sigma q)}{\partial l}, \tag{4.10.34}$$

where the angle χ is defined in figure 4.10.2. The sum of the terms inside the innermost parentheses on the left-hand side is twice the mean curvature of the interface, $2\kappa_m$. The first two terms inside the square brackets can be consolidated, yielding the final form

$$\frac{\mathrm{D}\Gamma}{\mathrm{D}t} + \Gamma\left(\frac{1}{\sigma}\,\frac{\partial(\sigma u_t)}{\partial l} + 2\,\kappa_m\,u_n\right) = -\frac{1}{\sigma}\,\frac{\partial(\sigma q)}{\partial l}. \tag{4.10.35}$$

The first term inside the large parentheses on the left-hand side expresses the rate of change of the surface area of an axisymmetric material ring.

Marker points

An evolution equation for interfacial marker points can be derived working as in section 4.10.1 for a two-dimensional interface. The result is

$$\frac{\mathrm{d}\Gamma}{\mathrm{d}t} + (u_t - v_t)\,\frac{\partial\Gamma}{\mathrm{d}l} + \Gamma\left(\frac{1}{\sigma}\,\frac{\partial(\sigma u_t)}{\partial l} + 2\,\kappa_m\,u_n\right) = -\frac{1}{\sigma}\,\frac{\partial(\sigma q)}{\partial l}, \tag{4.10.36}$$

which can be restated as

$$\frac{\mathrm{d}\Gamma}{\mathrm{d}t} + \frac{1}{\sigma}\,\frac{\partial(\sigma u_t\Gamma)}{\partial l} - v_t\,\frac{\partial\Gamma}{\partial l} + \Gamma\,2\,\kappa_m\,u_n = -\frac{1}{\sigma}\,\frac{\partial(\sigma q)}{\partial l}. \tag{4.10.37}$$

When $v_t = 0$, the marker points move normal to the interface and the third term on the right-hand side of (4.10.37) does not appear.

4.10.3 Three-dimensional interfaces

The equations derived in this section for two-dimensional and axisymmetric interfaces can be generalized to three-dimensional interfaces.[1] The normal component of the marker-point velocity over a three-dimensional interface must be

[1] YON, S. & POZRIKIDIS, C., 1998, A finite-volume / boundary-element method for interfacial flow in the presence of surfactants, with applications to shear flow past a viscous drop, *Computers & Fluids* **27**, 879–902.

equal to the normal component of the fluid velocity, but the tangential component can be arbitrary. The general form of marker point velocity is then

$$\mathbf{v} = u_n \, \mathbf{n} + \mathbf{v}^t, \tag{4.10.38}$$

where $\mathbf{v}^t$ is an arbitrary tangential component. When $\mathbf{v}^t = \mathbf{0}$, the marker points move with the fluid velocity normal to the interface. When $\mathbf{v}^t = \mathbf{u} - u_n \, \mathbf{n}$, the marker points are point particles moving with the fluid velocity.

Adopting Fick's law for the surface diffusion, we find that the evolution of the concentration of an immiscible surfactant following the motion of interfacial marker points takes the form

$$\frac{\mathrm{d}\Gamma}{\mathrm{d}t} + \nabla_s \cdot (\mathbf{u}_s \Gamma) - \mathbf{v}^t \cdot \nabla_s \Gamma + \Gamma \, 2 \, \kappa_m \, \mathbf{u} \cdot \mathbf{n} = \nabla_s \cdot (D_s \nabla_s \Gamma). \tag{4.10.39}$$

To define the various terms, we introduce the tangential projection matrix

$$P_{ij} = \delta_{ij} - n_i n_j, \tag{4.10.40}$$

where δ_{ij} is Kronecker's delta representing the identity matrix; in matrix notation, $\mathbf{P} = \mathbf{I} - \mathbf{nn}$. Subject to this definition, $\mathbf{u}_s = \mathbf{P} \cdot \mathbf{u}$ is the component of the fluid velocity tangential to the interface, and $\nabla_s \equiv \mathbf{P} \cdot \nabla$ is the surface gradient.

In the case of two-dimensional flow depicted in figure 4.10.1 or axisymmetric flow depicted in figure 4.10.2, equation (4.10.39) reduces, respectively, to (4.10.27) or (4.10.37), by setting $\mathbf{v}^t = v_t \mathbf{t}$.

Problem

4.10.1 *Transport on a flat interface*

Simplify equation (4.10.39) for a flat interface in the xy plane.

Hydrostatics 5

The simplest state of a fluid is the state of rest. The macroscopically observable velocity vanishes, and the forces developing in the fluid are described in terms of the pressure field established in response to a body force. The subject of hydrostatics encompasses two main topics: the computation of forces exerted on immersed surfaces and submerged bodies, and the study of the shapes of interfaces separating two stationary, translating, or rotating fluids. Although the problem statement and mathematical formulation is straightforward in both cases, deriving solutions for all but the simplest configurations requires the use of numerical methods for solving algebraic, ordinary, and partial differential equations.

5.1 Equilibrium of pressure and body forces

Consider a parcel of a stationary fluid, as illustrated in figure 5.1.1(a). Newton's second law of motion requires that, in the absence of a macroscopically observable flow, the sum of the forces exerted on the parcel should balance to zero.

In chapter 4, we saw that two kinds of forces are exerted on the parcel: a body force due to the gravitational or another force field mediated by long-range molecular interactions, and a surface force associated with the hydrodynamic traction.

C. Pozrikidis, *Fluid Dynamics: Theory, Computation, and Numerical Simulation*,
DOI: 10.1007/978-0-387-95871-2_5,

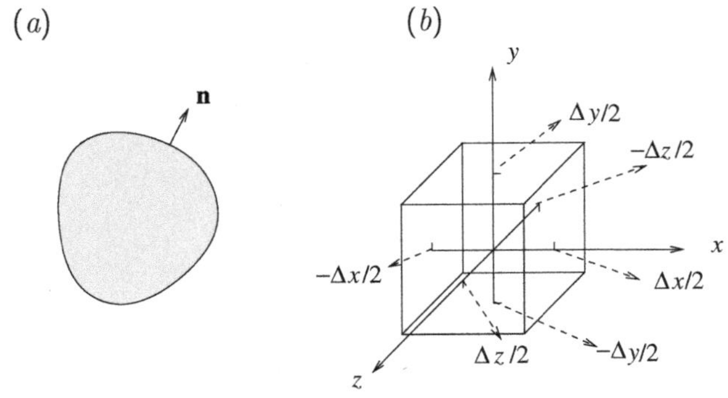

Figure 5.1.1 (*a*) Schematic illustration of a parcel of a stationary fluid, showing the outward unit normal vector. (*b*) A parcel with a rectangular parallelepiped shape serves as a control volume for deriving the differential equations of hydrostatics.

Body force

The body force due to gravity can be expressed as an integral over the volume of the parcel involving the possibly position-dependent fluid density, ρ, in the form

$$\mathbf{F}^B = \iiint_{Parcel} \rho\,\mathbf{g}\,\mathrm{d}V, \tag{5.1.1}$$

where $\mathbf{g} = (g_x, g_y, g_z)$ is the acceleration of gravity vector. On the surface of the earth, the magnitude of $\mathbf{g}$ has the approximate value $|\mathbf{g}| \equiv g = 9.80665$ m/sec^2.

Surface force

The surface force can be expressed in terms of the traction exerted on the parcel surface, $\mathbf{f}$ in the corresponding form

$$\mathbf{F}^S = \iint_{Parcel} \mathbf{f}\,\mathrm{d}S. \tag{5.1.2}$$

In the absence of fluid motion, the traction is due to the pressure alone pushing the surface of the parcel toward the interior. If $\mathbf{n}$ is the unit vector normal to the parcel surface pointing outward, as illustrated in figure 5.1.1(a), then

$$\mathbf{f} = -p\,\mathbf{n}. \tag{5.1.3}$$

The minus sign on the right-hand side accounts for the opposite orientation of the normal vector and normal traction due to the pressure. Substituting (5.1.3) in (5.1.2), we derive an integral expression for the surface force in terms of the pressure,

$$\mathbf{F}^S = - \iint_{Parcel} p\,\mathbf{n}\,\mathrm{d}S. \tag{5.1.4}$$

Force equilibrium

Setting the sum of the body force given in (5.1.1) and the surface force given in (5.1.4) to zero, we obtain the equilibrium condition

$$\iiint_{Parcel} \rho\,\mathbf{g}\,\mathrm{d}V - \iint_{Parcel} p\,\mathbf{n}\,\mathrm{d}S = \mathbf{0}. \tag{5.1.5}$$

The three scalar components of this vectorial equation are

$$\begin{aligned} \iiint_{Parcel} \rho\, g_x\,\mathrm{d}V &= \iint_{Parcel} p\, n_x\,\mathrm{d}S, \\ \iiint_{Parcel} \rho\, g_y\,\mathrm{d}V &= \iint_{Parcel} p\, n_y\,\mathrm{d}S, \\ \iiint_{Parcel} \rho\, g_z\,\mathrm{d}V &= \iint_{Parcel} p\, n_z\,\mathrm{d}S. \end{aligned} \tag{5.1.6}$$

5.1.1 *Equilibrium of an infinitesimal parcel*

Next, we consider a small fluid parcel having the shape of a rectangular parallelepiped centered at the origin, as illustrated in figure 5.1.1(*b*). The six flat sides of the parcel are perpendicular to the x, y, or z axis, the lengths of the three sides are, respectively, equal to Δx, Δy, and Δz, and the volume of the parcel is equal to $\Delta V = \Delta x\, \Delta y\, \Delta z$. Because the size of the parcel is small, density variations over the parcel volume can be neglected and the volume integrals on the left-hand side of equations (5.1.6) can be approximated with the products

$$\rho_0\, g_x\, \Delta V, \qquad \rho_0\, g_y\, \Delta V, \qquad \rho_0\, g_z\, \Delta V, \tag{5.1.7}$$

where ρ_0 is the density of the fluid at the center of the parcel, located at the origin.

Next, we consider the surface integral on the left-hand side of the first equation in (5.1.6). The x component of the normal vector vanishes on all sides, except on the two sides that are perpendicular to the x axis, located at $x = \frac{1}{2}\Delta x$, and $x = -\frac{1}{2}\Delta x$, designated as the first and second side. On the first side $n_x = 1$, and on the second side $n_x = -1$. Because the parcel size is small, variations in pressure over each side can be neglected and the pressure over a side can be approximated with the value at the side center. Subject to this approximation, the surface integrals on the right-hand side of the first equation in (5.1.6) over the first and second side are, respectively, equal to

$$p(x = \frac{1}{2}\Delta x, y = 0, z = 0)\,\Delta y\,\Delta z, \quad -p(x = -\frac{1}{2}\Delta x, y = 0, z = 0)\,\Delta y\,\Delta z, \tag{5.1.8}$$

where the parentheses enclose arguments. Adding these two contributions, we obtain

$$\Big[p(x = \frac{1}{2}\Delta x, y = 0, z = 0) - p(x = -\frac{1}{2}\Delta x, y = 0, z = 0)\Big]\,\Delta y\,\Delta z. \tag{5.1.9}$$

In the limit as Δx tends to zero, the ratio of the differences

$$\frac{p(x = \frac{1}{2}\Delta x, y = 0, z = 0) - p(x = -\frac{1}{2}\Delta x, y = 0, z = 0)}{\frac{1}{2}\Delta x - (-\frac{1}{2}\Delta x)}$$
$$= \frac{p(x = \frac{1}{2}\Delta x, y = 0, z = 0) - p(x = -\frac{1}{2}\Delta x, y = 0, z = 0)}{\Delta x} \tag{5.1.10}$$

tends to the partial derivative $\partial p/\partial x$ evaluated at the origin, and the difference (5.1.9) reduces to

$$\frac{\partial p}{\partial x}(x = 0, y = 0, z = 0)\,\Delta x\,\Delta y\,\Delta z = \frac{\partial p}{\partial x}(x = 0, y = 0, z = 0)\,\Delta V. \tag{5.1.11}$$

Substituting (5.1.11) along with the first approximate form in (5.1.7) in the x component of the force balance (5.1.6), and simplifying by eliminating ΔV on both sides, we obtain the differential equation

$$\rho\, g_x = \frac{\partial p}{\partial x}, \tag{5.1.12}$$

where the density and the partial derivative of the pressure are evaluated at the origin. However, since the location of the origin is arbitrary, equation (5.1.12) is valid at every point in the fluid.

Working in a similar fashion with the second and third hydrostatic equilibrium equations stated in (5.1.6), we obtain the corresponding differential equations

$$\rho\, g_y = \frac{\partial p}{\partial y}, \qquad \rho\, g_z = \frac{\partial p}{\partial z}. \tag{5.1.13}$$

The three scalar equations (5.1.12) and (5.1.13) can be collected into a compact vector form,

$$\rho\,\mathbf{g} = \nabla p, \tag{5.1.14}$$

where $\nabla p = (\partial p/\partial x, \partial p/\partial y, \partial p/\partial z)$ is the pressure gradient. In physical terms, the differential equation (5.1.14) expresses a balance between gravitational and pressure force in hydrostatics.

Equation (5.1.14) provides us with a basis for computing the distribution of pressure and density in a fluid, subject to additional conditions concerning the physical properties of the fluid required by thermodynamics. Specifically, given the density field, or a relation between the density and the pressure, equation (5.1.14) allows us to compute the corresponding pressure and *vice versa*. To this end, we make a distinction between compressible gases and incompressible liquids.

5.1.2 Gases in hydrostatics

The density of a gas, ρ, is related to the pressure, p, and temperature, T, by an equation of state provided by thermodynamics. For an ideal gas,

$$\rho = \frac{Mp}{RT}, \tag{5.1.15}$$

where M is the gram-molecular weight, and R is the ideal-gas constant, as discussed in section 4.4. Substituting (5.1.15) in (5.1.14) and rearranging, we obtain a vectorial equation involving the pressure and temperature,

$$\frac{M}{RT}\,\mathbf{g} = \frac{1}{p}\,\nabla p. \tag{5.1.16}$$

The x component of equation (5.1.16) reads

$$\frac{M g_x}{RT} = \frac{1}{p}\frac{\partial p}{\partial x} = \frac{\partial \ln p}{\partial x}. \tag{5.1.17}$$

When the temperature of the fluid is uniform, we integrate (5.1.17) with respect to x and find

$$\ln p = \frac{M g_x}{RT}\,x + f_x(y,z), \tag{5.1.18}$$

where $f_x(y,z)$ is an unknown function. Working in a similar fashion with the y and z components of (5.1.16) under the assumption of uniform temperature, we find

$$\ln p = \frac{M g_y}{RT}\,y + f_y(x,z), \qquad \ln p = \frac{M g_z}{RT}\,z + f_z(x,y). \tag{5.1.19}$$

Combining the last three equations we obtain the pressure distribution

$$\ln p = \frac{M}{RT}\,(g_x\,x + g_y\,y + g_z\,z) + \ln c, \tag{5.1.20}$$

where c is a positive, possibly time-dependent, constant with units of pressure determined by requiring an appropriate boundary condition.

Expressing the term in the parentheses on the right-hand side of (5.1.20) in terms of the inner product of the gravity vector and the position vector and transferring the last term to the left-hand side, we obtain the compact form

$$\ln \frac{p}{c} = \frac{M}{RT}\,\mathbf{g}\cdot\mathbf{x}, \tag{5.1.21}$$

which describes the pressure distribution in an ideal gas with uniform temperature.

Pressure distribution in the atmosphere

As an application, we consider the pressure distribution in the atmosphere regarded as an ideal gas with molecular weight $M = 28.97$ kg/kmole, at temperature 25°C corresponding to the absolute temperature $T = 298$ K. In Cartesian coordinates with origin at sea level, where the y axis points upward and the x and z axes are horizontal, the components of the acceleration of gravity vector are given by $g_x = 0$, $g_y = -g$, and $g_z = 0$, where $g = 9.80665$ m/sec^2. Equation (5.1.21) simplifies to

$$\ln \frac{p}{c} = -\frac{Mg}{RT}\,y, \tag{5.1.22}$$

where $c = P_0$ is the pressure at sea level. Solving for p, we derive the exponentially decaying field

$$p = P_0 \exp\left(-\frac{Mg}{RT}\,y\right). \tag{5.1.23}$$

Taking $P_0 = 1.0$ atm $= 1.0133 \times 10^5$ Pascal $= 1.0133 \times 10^5$kg m^{-1} sec^{-2}, we find that the pressure at the elevation of $y = 1$ km$=$ 1000 m is

$$p = 1.0 \exp\left(-\frac{28.97 \times 9.80665}{8.314 \times 10^3 \;\times 298}\,1000\right) \text{atm} = 0.892\ \text{atm}. \tag{5.1.24}$$

The corresponding density distribution is found by substituting the pressure distribution (5.1.23) in the right-hand side of the equation of state (5.1.15).

5.1.3 Liquids in hydrostatics

Because liquids at low and moderate pressures are nearly incompressible, the density is a physical property determined primarily by the temperature. Working as in the case of gases but treating the density as a constant, we find that the pressure distribution is given by the counterpart of equation (5.1.21)

$$p = \rho\,(g_x\,x + g_y\,y + g_z\,z) + c = \rho\,\mathbf{g}\cdot\mathbf{x} + c, \tag{5.1.25}$$

where c is a constant with units of pressure determined by an appropriate boundary condition.

Pressure distribution in a pool

As an application, we consider the pressure distribution in a liquid pool with a horizontal surface. In Cartesian coordinates where the y axis is perpendicular to the pool surface pointing in the vertical direction upward and the x and z axes are horizontal, the components of the acceleration of gravity vector are $g_x = 0$, $g_y = -g$, and $g_z = 0$, where g is the magnitude of the acceleration of gravity. Equation (5.1.25) then simplifies to

$$p = -\rho\, g\, y + c. \tag{5.1.26}$$

Setting the origin of the y axis at the pool surface where the liquid pressure is equal to the atmospheric pressure, P_{atm}, we find $c = P_{atm}$.

Manometer

The pressure distribution given in (5.1.25) also applies when the liquid occupies a convoluted domain, provided that the fluid is contiguous and uninterrupted. In practice, this property is exploited to compute the pressure difference across the two ends of a tube in terms of the difference in the levels of a liquid column placed inside the tube. A simple device serving this purpose is the U-tube manometer illustrated in figure 5.1.2. The pressure distribution in the liquid inside the U-tube manometer is given by equation (5.1.26). Applying this equation at the two ends of the liquid located at $y = y_1$ and y_2 and subtracting the resulting expressions, we find

$$\Delta p \equiv p(y_1) - p(y_2) = \rho\, g\, (y_2 - y_1). \tag{5.1.27}$$

If the tube is exposed to the atmosphere at the first end, $p(y_1) = P_{atm}$, then $p(y_2) = P_{atm} + \rho\, g\, h$ where $h \equiv y_1 - y_2$ is the readily measurable rise of the liquid column in the manometer.

Problems

5.1.1 *Hydrostatic pressure distribution*

(*a*) Derive the pressure distribution in an incompressible liquid given in equation (5.1.25).

(*b*) Derive the pressure distribution in an ideal gas occupying the semi-infinite region $y > 0$ when the temperature decreases exponentially with distance as $T = T_0 - \Delta T\,(1 - e^{-\alpha y})$, where T_0, ΔT, and α are three specified constants. The gravity vector points in the negative direction of the y axis.

5.1.2 *Function of an aircraft altimeter*

The temperature in the lower part of the troposphere extending 10 km above the surface of the earth decreases at a nearly linear rate as $T = T_0 - \alpha\, y$, where

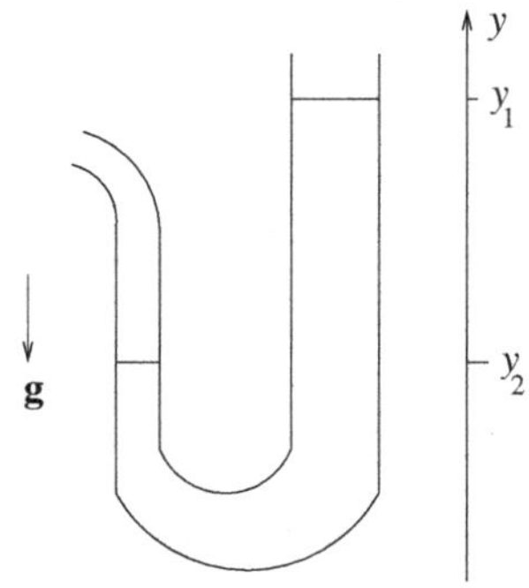

Figure 5.1.2 Illustration of a U-tube manometer. The pressure distribution in the fluid is described by the equations of hydrostatics even if the fluid has a convoluted shape, as long as it remains contiguous and uninterrupted.

T_0 is the temperature at the surface of the earth positioned at $y = 0$, and α is the lapse rate. In North America, $\alpha = 6.5$ K/km.

(a) Assuming that the atmosphere behaves like an ideal gas, derive the pressure distribution

$$p = P_0 \left(1 - \frac{\alpha}{T_0} y\right)^{\beta}, \tag{5.1.28}$$

and evaluate the dimensionless exponent $\beta \equiv Mg/(R\alpha)$, where P_0 is the pressure at sea level. Solving (5.1.28) for the elevation y, we find

$$y = \frac{T_0}{\beta} \left[1 - \left(\frac{p}{P_0}\right)^{1/\beta}\right]. \tag{5.1.29}$$

This equation is used for calibrating aircraft altimeters; that is, for translating pressure measured with a barometer into altitude.

(b) Show that, as α tends to zero, in which case the temperature distribution tends to become constant, the pressure distribution (5.1.28) reduces to that shown in (5.1.23).

5.1.3 *How many molecules inside a volume of gas?*

How many molecules are there inside one cubic centimeter (1 milliliter) of a gas under atmospheric pressure and temperature 25° C?

5.2 Force exerted on immersed surfaces

To compute the hydrostatic surface force exerted on a surface that either contains or is immersed in a stationary fluid, as illustrated in figure 5.2.1, we repeat

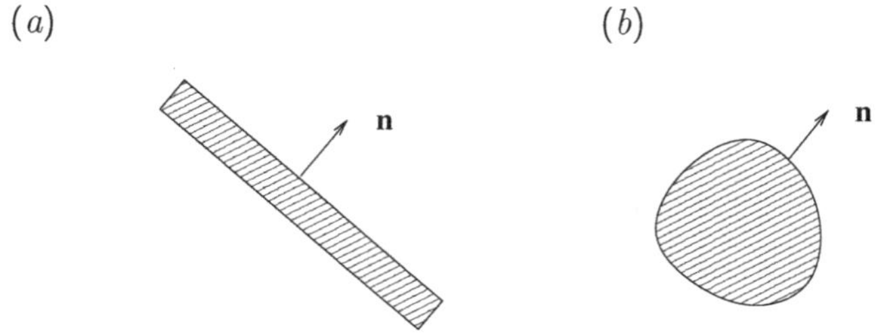

Figure 5.2.1 Illustration of a surface that (*a*) contains, or (*b*) is immersed in a stationary fluid.

the arguments the led us to equation (5.1.4) and find

$$\mathbf{F}^S = -\iint p\,\mathbf{n}\,\mathrm{d}S, \tag{5.2.1}$$

where $\mathbf{n}$ is the unit vector normal to the surface pointing into the fluid, and the integration is performed over the surface. To evaluate the right-hand side of (5.2.1), we must first compute the pressure distribution in the fluid as discussed in section 5.1, and then evaluate the integral by analytical or numerical methods.

5.2.1 A sphere floating on a flat interface

As an application, we consider the force exerted on a sphere of radius a floating on the flat surface of a liquid underneath a zero-density gas, as depicted in figure 5.2.2. In spherical polar coordinates with origin at the center of the sphere with the x axis pointing upward, the circular contact line where the surface of the liquid meets the sphere is located at $\theta = \beta$.

Symmetry requires that the horizontal component of the surface force exerted on the sphere should vanish. The vertical component of the surface force is given by

$$F_x^S = -\iint p\,n_x\,\mathrm{d}S, \tag{5.2.2}$$

where $n_x = \cos\theta$ is the x component of the unit normal vector.

The pressure distribution is given by equation (5.1.25) with gravity components $g_x = -g$, $g_y = 0$, and $g_z = 0$, yielding $p = -\rho\,g\,x + c$. To compute the constant c, we require that the pressure at the contact line is equal to the atmospheric pressure, $p(x = a\cos\beta) = P_{atm}$, and find $P_{atm} = -\rho\,g\,a\cos\beta + c$, which can be rearranged to give $c = \rho\,g\,a\cos\beta + P_{atm}$. Writing $x = a\cos\theta$, we find that the pressure distribution over the sphere is given by

$$p = -\rho\,g\,a\,(\cos\theta - \cos\beta) + P_{atm}. \tag{5.2.3}$$

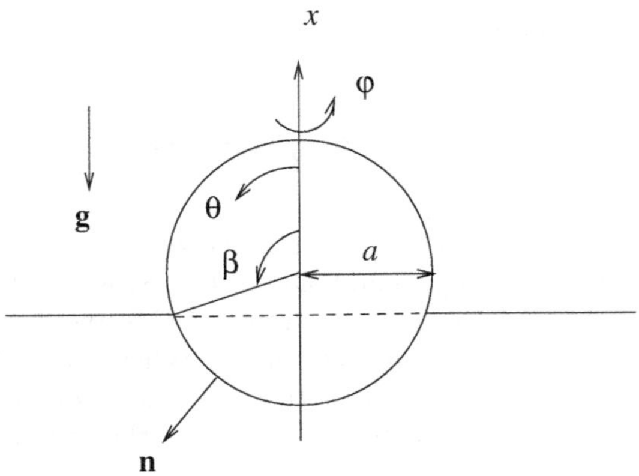

Figure 5.2.2 A sphere floating in the flat surface of a liquid with floating angle β. The dashed line represents the contact line.

Substituting the pressure distribution (5.2.3) in the integral on the right-hand side of (5.2.2), we find that the force exerted on the sphere by the liquid is given by

$$F_x^S = \iint [\rho\, g\, a\, (\cos\theta - \cos\beta) - P_{atm}] \cos\theta \,\mathrm{d}S. \tag{5.2.4}$$

The differential surface area of the sphere can be expressed in the form $\mathrm{d}S = (\sigma \mathrm{d}\varphi)(a\mathrm{d}\theta)$, where $\sigma = a \sin\theta$ is the distance of a point on the surface of the sphere from the x axis, and φ is the meridional angle. Substituting this expression in the right-hand side of (5.2.4) and integrating with respect to φ, we find

$$F_x^S = 2\pi a^2 \int_\beta^\pi [\rho\, g\, a\, (\cos\theta - \cos\beta) - P_{atm}] \cos\theta \,\sin\theta \,\mathrm{d}\theta. \tag{5.2.5}$$

Next, we set $\sin\theta \,\mathrm{d}\theta = -\mathrm{d}\cos\theta$ and carry out the integration on the right-hand side to find

$$F_x^S = \pi a^2 [\rho g a \,\frac{1}{3}\,(2 + 3\cos\beta - \cos^3\beta) + P_{atm}\,(1 - \cos^2\beta)]. \tag{5.2.6}$$

Working in a similar fashion, we find that the x component of the force due to the atmospheric pressure exerted on the non-immersed portion of the sphere subtended between $\theta = 0$ and β is given by

$$F_x^{atm} = -2\pi a^2 \int_0^\beta P_{atm} \cos\theta \,\sin\theta \,\mathrm{d}\theta = -\pi a^2 P_{atm} (1 - \cos^2\beta). \tag{5.2.7}$$

Adding the two contributions (5.2.6) and (5.2.7), we obtain the buoyancy force exerted on the sphere,

$$F_x^{BNC} \equiv F_x^S + F_x^{atm} = \rho g \, [a^3 \, \frac{\pi}{3} \, (2 + 3 \cos \beta - \cos^3 \beta)]. \tag{5.2.8}$$

It can be shown by elementary trigonometry that the term enclosed by the square brackets on the right-hand side of (5.2.8) is equal to the immersed volume of the sphere underneath the surface of the liquid, which is equal to the volume of the fluid displaced by the sphere. When the sphere is completely immersed, $\beta = 0$, the term enclosed by the parentheses on the right-hand side of (5.2.8) is equal to four, and the term enclosed by the square brackets is equal to the volume of the sphere, $4\pi a^3/3$.

Equation (5.2.8) states that the hydrostatic force exerted on a floating sphere is equal in magnitude and opposite in direction to the weight of the fluid displaced by the sphere. In section 5.3, we will see that this is a more general result applicable to arbitrarily shaped floating or immersed objects.

Computation of the floating angle

The floating angle, β, is determined by the weight of the sphere; the heavier the sphere, the smaller the angle. There is a critical weight where β becomes equal to zero and the sphere is completely submerged. To compute the floating angle corresponding to a certain weight W, we set W equal to the buoyancy force given in (5.2.8) and rearrange to obtain a cubic equation for $\cos \beta$,

$$\cos^3 \beta - 3 \cos \beta + 2 \, (2 \, s - 1) = 0, \tag{5.2.9}$$

where

$$s \equiv \frac{W}{\rho g \, \frac{4\pi a^3}{3}} \tag{5.2.10}$$

is a dimensionless constant. The fraction in the denominator of (5.2.10) is equal to the volume of the sphere. Thus, if the sphere is made of a homogeneous material with density ρ_B, then $s = \rho_B/\rho$ is the density ratio. A neutrally buoyant sphere corresponds to $s = 1$, in which case $\cos \beta = 1$ and $\beta = 0$ satisfies equation (5.2.9), as expected.

Newton's method

A variety of numerical methods are available for solving the nonlinear algebraic equation (5.2.9) for β. In practice, Newton's method strikes an optimal balance between conceptual simplicity and numerical efficiency. To formalize the method, we define the variable $q \equiv \cos \beta$, and express (5.2.9) in the generic form

$$f(q) = 0, \tag{5.2.11}$$

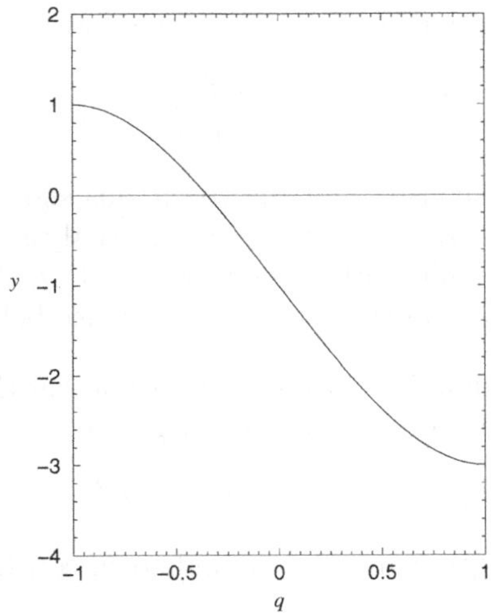

Figure 5.2.3 Graph of the function $f(q)$ defined in equation (5.2.12) whose root, Q, is required. To compute the root using Newton's method, we make an initial guess, $q^{(0)}$, and then improve it moving along the tangent to the graph toward the q axis.

where

$$f(q) \equiv q^3 - 3\,q + 2\,(2s - 1). \tag{5.2.12}$$

A graph of the function $f(q)$ for $s = 0.25$ is shown in figure 5.2.3. The requisite value of q is located at the intersection of the graph of $f(q)$ and the q axis, denoted by Q and satisfying $f(Q) = 0$.

To implement Newton's method, we make an initial guess for the desired root Q, denoted by $q^{(0)}$, and then generate a sequence of improvements working as follows. Near the point $q^{(0)}$, the function $f(q)$ can be approximated with a linear function that arises by expanding $f(q)$ in a Taylor series about $q^{(0)}$. Discarding all nonlinear terms, we obtain the approximate form

$$f(q) \simeq f(q^{(0)}) + \left(\frac{\mathrm{d}f}{\mathrm{d}q}\right)_{q=q^{(0)}} (q - q^{(0)}). \tag{5.2.13}$$

Setting $f(q) = 0$, solving for the q inside the parentheses on the right-hand side of (5.2.13) and denoting the solution by $q^{(1)}$, we obtain the improved value

$$q^{(1)} = q^{(0)} - \left(\frac{f}{f'}\right)_{q=q^{(0)}}, \tag{5.2.14}$$

where a prime denotes a derivative with respect to q. In the case of a floating sphere, $(\partial f/\partial q)(q) = 3\,q^2 - 3$. The process is then repeated, yielding a sequence of successive approximations based on the recursive formula

$$q^{(k+1)} = q^{(k)} - \left(\frac{f}{f'}\right)_{q=q^{(k)}}, \tag{5.2.15}$$

for $k = 0, 1, \ldots$. Erroneously omitting the minus sign on the right-hand side of (5.2.15) is a common source of frustration. Analysis shows that, as long as the initial guess $q^{(0)}$ is sufficiently close to the root Q, the sequence defined by (5.2.15) converges to Q, and the rate of convergence behaves as follows.

If the graph of the function $f(q)$ is not horizontal at the root, $f'(q = Q) \neq 0$, the rate of convergence is quadratic, meaning that

$$q^{(k+1)} - Q \simeq \delta\, (q^{(k)} - Q)^2, \tag{5.2.16}$$

where $\delta = f''(Q)/[2f'(Q)]$ is an *a priori* unknown constant.

Equation (5.2.16) states that the magnitude of the error in the current iteration, expressed by the left-hand side, is roughly equal to the square of the magnitude of the error in the previous iteration multiplied by a constant. Consequently, if the initial error, $q^{(0)} - Q$, is sufficiently small, the magnitude of the error, $q^{(k)} - Q$, will keep decreasing during the iterations, no matter how large the value of the coefficient δ. A prerequisite is that the initial guess is close enough to the root so that (5.2.16) applies.

If the graph of the function $f(q)$ is horizontal at the root, $(\mathrm{d}f/\mathrm{d}q)_{q=Q} = 0$ and the root is multiple, the rate of convergence is linear, meaning that

$$q^{(k+1)} - Q \simeq \frac{m-1}{m}\, (q^{(k)} - Q), \tag{5.2.17}$$

where m is the multiplicity of the root; for a double root, $m = 2$.

Equation (5.2.17) states that the magnitude of the error in the current iteration is roughly equal to that in the previous iteration multiplied by the positive coefficient $(m-1)/m$, which is less than unity for any value of $m > 1$. Consequently, the error $|q^{(k)} - Q|$ will keep decreasing during the iterations as long as the initial guess is close enough to the root for (5.2.17) to apply.

Problems

5.2.1 *Pycnometer*

A pycnometer is an antiquated device used to measure the specific gravity of a liquid, defined as the ratio of the density of the liquid to the density of water. In practice, this is done by reading the level of the free surface on a scale printed on a vertical tube attached to a spherical flask floating on the liquid, as illustrated in figure 5.2.4. Pycnometer derives from the Greek word $\pi\upsilon\kappa\nu o\tau\eta\tau\alpha$, which

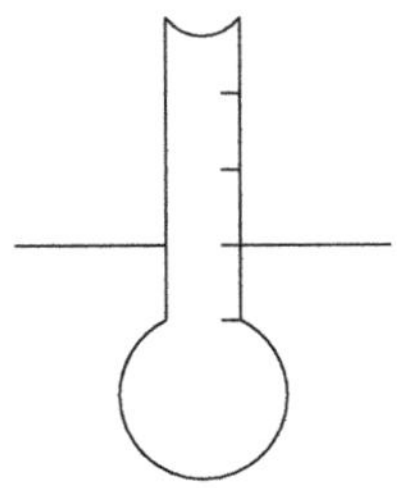

Figure 5.2.4 A pycnometer is used to measure the specific gravity of a liquid defined as the ratio between the density of the liquid to the density of water.

means density. Derive an equation that allows us to calibrate a pycnometer based on the known density of water.

5.2.2 *A sphere at the interface between two fluids*

Derive the counterpart of expression (5.2.8) for a sphere straddling the interface between a lower fluid with density ρ_2 and an upper fluid with density ρ_1.

Computer Problem

5.2.3 *Floating sphere*

(*a*) Directory *104_nl_eq* inside directory *01_num_meth* of FDLIB includes program *nonlinear* that implements Newton's method for solving one nonlinear equation. Use the program to solve equation (5.2.9) and plot the floating angle β against the dimensionless parameter s defined in equation (5.2.10). Discuss the rate of convergence of the iterations in light of equations (5.2.16) and (5.2.17).

(*b*) Directory *104_nl_eq* inside directory *01_num_meth* of FDLIB includes program *cubic* that computes the three roots of a cubic equation using Cardano's formula. Use the program to solve equation (5.2.9) and prepare a plot of the floating angle β against the dimensionless parameter s.

5.3 Archimedes' principle

Consider the force exerted on a body with arbitrary shape immersed in a stationary fluid. Using equation (5.2.1), we find that the surface force exerted on the body is given by

$$\mathbf{F}^S = -\iint_{Body} p\,\mathbf{n}\,\mathrm{d}S, \tag{5.3.1}$$

where $\mathbf{n}$ is the unit vector normal to the body pointing into the fluid. It would appear that the computation of the integral on the right-hand side of (5.3.1) requires detailed knowledge of the geometry of the body. However, if the fluid is incompressible, the integral can be evaluated in a generic fashion, yielding a remarkably simple expression for the force in terms of the body volume alone.

Substituting the pressure distribution for an incompressible fluid given in (5.1.25) in the right-hand side of (5.3.1), we find

$$\mathbf{F}^S = -\iint_{Body} [\rho\,(g_x\,x + g_y\,y + g_z\,z) + P_0]\,\mathbf{n}\,\mathrm{d}S. \tag{5.3.2}$$

A key observation in evaluating the surface integral is that the integrand is the product of the unit normal vector and a scalar function that is *linear* with respect to the components of the position vector.

To see how the evaluation of the integral can be simplified, we consider a body having the shape of a rectangular parallelepiped, as illustrated in figure 5.1.1(b). The six flat sides of the body are perpendicular to the x, y, or z axis, the lengths of the edges are equal to Δx, Δy, and Δz, and the volume of the body is equal to $V_B = \Delta x\,\Delta y\,\Delta z$. The unit normal vector is constant over each one of the six sides. For example, over the side that is perpendicular to the x axis and faces the positive direction of the x axis, $\mathbf{n} = (1, 0, 0)$.

Taking into consideration this and similar simplifications, we evaluate the integral on the right-hand side of (5.3.2) without any approximation, finding

$$\mathbf{F}^S = -\rho\,V_B\,\mathbf{g}, \tag{5.3.3}$$

which expresses Archimedes' principle, stating that the force exerted on an immersed body by the ambient fluid is equal in magnitude and opposite in direction to the weight of the fluid displaced by the body.

To compute the integral on the right-hand side of (5.3.2) over an arbitrarily shaped body, we subdivide the volume of the body into small rectangular parallelepipeds, and approximate the surface of the body with the collection of the faces of the parallelepipeds that are wetted by the fluid. Because of cancellations, the sum of the integrals over the faces of all elementary parallelepipeds is equal to the sum of the integrals of the faces that are wetted by the fluid. Summing the contributions, we find that the force exerted on the body is given by (5.3.3) independent of the shape of the body. We have found that Archimedes's principles stands true for arbitrarily shaped bodies.

The Gauss divergence theorem

Formula (5.3.3) can be derived rigorously using the Gauss divergence theorem in three dimensions stated in equation (2.6.29). Selecting $h_x = f$, $h_y = 0$, and $h_z = 0$ to form the vector function $\mathbf{h} = (f, 0, 0)$, we obtain

$$\iint_H f\, n_x \, \mathrm{d}S = \iiint_R \frac{\partial f}{\partial x} \, \mathrm{d}V, \tag{5.3.4}$$

where f is a scalar function of position. The complementary selections $\mathbf{h} = (0, f, 0)$ and $\mathbf{h} = (0, 0, f)$ yield the corresponding identities

$$\iint_H f\, n_y \, \mathrm{d}S = \iiint_R \frac{\partial f}{\partial y} \, \mathrm{d}V, \qquad \iint_H f\, n_z \, \mathrm{d}S = \iiint_R \frac{\partial f}{\partial z} \, \mathrm{d}V. \tag{5.3.5}$$

Relations (5.3.4) and (5.3.5) can be collected into the vector identity

$$\iint_H f\, \mathbf{n} \, \mathrm{d}S = \iiint_R \nabla f \, \mathrm{d}V, \tag{5.3.6}$$

where $\nabla f = (\partial f/\partial x, \partial f/\partial y, \partial f/\partial z)$ is the gradient of f.

Now comparing (5.3.6) with (5.3.2), we set

$$f \equiv \rho\,(g_x\, x + g_y\, y + g_z\, z) + P_0, \tag{5.3.7}$$

compute

$$\nabla f = \rho\,(g_x, g_y, g_z) = \rho\, \mathbf{g}, \tag{5.3.8}$$

and find that the surface force is given by

$$\begin{aligned} \mathbf{F}^S &= - \iint_{Body} \Big[\rho\,(g_x\, x + g_y\, y + g_z\, z) + P_0 \Big] \, \mathbf{n} \, \mathrm{d}S \\ &= -\rho\, \mathbf{g} \iiint_{Body} \mathrm{d}V = -\rho\, \mathbf{g}\, V_B, \end{aligned} \tag{5.3.9}$$

which reproduces and confirms equation (5.3.3).

5.3.1 Net force on a submerged body

The mass of a body with volume V_B made of a homogeneous material with density ρ_B is $m_B = \rho_B\, V_B$. The weight of the body is

$$\mathbf{W} = \rho_B\, V_B\, \mathbf{g}, \tag{5.3.10}$$

where $\mathbf{g}$ is the acceleration of gravity. Adding the weight to the buoyancy force given in (5.3.3), we find that the net force exerted on an immersed body is

$$\mathbf{F} = \mathbf{F}^S + \mathbf{W} = (\rho_B - \rho)\, V_B\, \mathbf{g}. \tag{5.3.11}$$

The density of a neutrally buoyant body is equal to the density of the ambient fluid, and the right-hand side of (5.3.11) vanishes, yielding a zero net force.

5.3.2 Moments

The moment of the surface force about a chosen point, $\mathbf{x}_0$, is found by integrating the moments of the traction expressed in terms of the pressure,

$$\mathbf{M}^S = \iint_{Body} (\mathbf{x} - \mathbf{x}_0) \times (-p\,\mathbf{n})\,\mathrm{d}S. \tag{5.3.12}$$

where $\times$ denotes the outer vector product. Substituting the hydrostatic pressure distribution, we find

$$\mathbf{M}^S = -\rho \iint_{Body} (\mathbf{g}\cdot\mathbf{x} + P_0)(\mathbf{x} - \mathbf{x}_0) \times \mathbf{n}\,\mathrm{d}S. \tag{5.3.13}$$

To evaluate this integral, we resort to the Gauss divergence theorem in three dimensions stated in equation (2.6.29). Setting $\mathbf{h} = \mathbf{a} \times \mathbf{G}$, where $\mathbf{a}$ is a constant vector and $\mathbf{G}$ is a differentiable function, and then discarding the arbitrary constant $\mathbf{a}$, we obtain the new identity

$$\iint_H \nabla \times \mathbf{G}\,\mathrm{d}V = \iint_R \mathbf{n} \times \mathbf{G}\,\mathrm{d}S. \tag{5.3.14}$$

Now setting $\mathbf{G} = (\mathbf{g}\cdot\mathbf{x} + P_0)(\mathbf{x} - \mathbf{x}_0)$, we find

$$\mathbf{M}^S = \rho \iiint_{Body} \nabla \times [(\mathbf{g}\cdot\mathbf{x} + P_0)(\mathbf{x} - \mathbf{x}_0)]\,\mathrm{d}V. \tag{5.3.15}$$

Using a vector identity, we write

$$\mathbf{M}^S = \rho \iiint_{Body} [\,\nabla(\mathbf{g}\cdot\mathbf{x}) \times (\mathbf{x} - \mathbf{x}_0) + (\mathbf{g}\cdot\mathbf{x} + P_0)\,\nabla \times \mathbf{x}\,]\,\mathrm{d}V, \tag{5.3.16}$$

and note that $\nabla(\mathbf{g}\cdot\mathbf{x}) = \mathbf{g}$ and $\nabla \times \mathbf{x} = \mathbf{0}$ to find

$$\mathbf{M}^S = \rho\,\mathbf{g} \times \iiint_{Body} (\mathbf{x} - \mathbf{x}_0)\,\mathrm{d}V. \tag{5.3.17}$$

If $\mathbf{x}_0$ is the center of mass of the homogeneous fluid displaced by the body,

$$\mathbf{x}_c = \frac{1}{V_B} \iiint_{Body} \mathbf{x}\,\mathrm{d}V, \tag{5.3.18}$$

the surface moment is zero, where V_B is the volume of the body.

The moment of the body force due to gravity is

$$\mathbf{M}^B = -\mathbf{g} \times \iiint_{Body} \rho_B\,(\mathbf{x} - \mathbf{x}_0)\,\mathrm{d}V, \tag{5.3.19}$$

and the total moment is

$$\mathbf{M} = \mathbf{M}^S + \mathbf{M}^B = \mathbf{g} \iiint_{Body} (\rho - \rho_B)\,(\mathbf{x} - \mathbf{x}_0)\,\mathrm{d}V, \tag{5.3.20}$$

The buoyancy force passes through the center of mass of the fluid displaced by the body, whereas the weight of the body passes through the center of mass of the body. If the former lies above the latter, the body is in a state of stable equilibrium and will remain stationary. In the opposite case, the body will spontaneously rotate to reach a stable configuration.

Problem

5.3.1 *Applications of the Gauss divergence theorem*

(*a*) Apply (5.3.6) for a constant function f and discuss your results.

(*b*) Show that the center of gravity of a homogeneous body can be computed in terms of a surface integral as

$$\mathbf{x}_c = \frac{1}{2V_B} \iint_{Body} (x^2 + y^2 + z^2)\,\mathbf{n}\,\mathrm{d}S. \tag{5.3.21}$$

5.4 Interfacial shapes

Consider two superposed stationary incompressible fluids separated by an infinite horizontal interface located at $y = y_I$, as illustrated in figure 5.4.1. The acceleration of gravity points toward the negative direction of the y axis. The upper fluid is designated as fluid 1, and the lower fluid is designated as fluid 2.

Using the general expression for the pressure distribution in an incompressible liquid given in equation (5.1.25), and setting $g_y = -g$, we find that the pressure distributions in the two fluids are given by

$$p^{(1)}(y) = -\rho_1\, g\, y + c_1, \qquad p^{(2)}(y) = -\rho_2\, g\, y + c_2. \tag{5.4.1}$$

The constants c_1 and c_2 are related by the condition for the jump in the traction across an interface with constant surface tension stated in equation (4.4.11).

Since in this case the curvature of the interface vanishes, $\kappa = 0$, condition (4.4.11) requires that the pressure is continuous across the interface,

$$p^{(1)}(y = y_I) = p^{(2)}(y = y_I). \tag{5.4.2}$$

Substituting the pressure distributions (5.4.1) in (5.4.2), we find

$$-\rho_1\, g\, y_I + c_1 = -\rho_2\, g\, y_I + c_2, \tag{5.4.3}$$

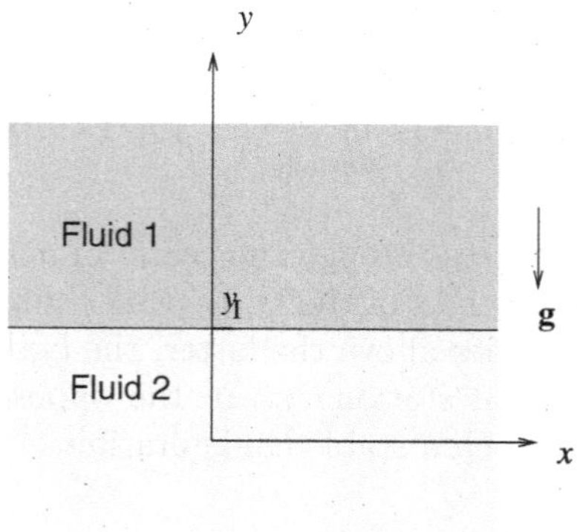

Figure 5.4.1 Illustration of an infinite horizontal interface separating two stationary fluids.

which can be rearranged to give

$$c_2 = c_1 + (\rho_2 - \rho_1)\, g\, y_I. \tag{5.4.4}$$

One of the two constants c_1 or c_2 is determined by requiring an appropriate boundary condition far from the interface, and the second constant follows from (5.4.4). For example, if the pressure on the upper side of the interface is equal to the atmospheric pressure P_{atm}, then $c_1 = P_{atm} + \rho_1\, g\, y_I$ and $c_2 = P_{atm} + \rho_2\, g\, y_I$.

5.4.1 Curved interfaces

In practice, the flat interface depicted in figure 5.4.1 terminates at a side wall, as illustrated in figure 5.4.2. Further examples of terminated interfaces are depicted in figure 5.5.1 illustrating a semi-infinite interface ending at an inclined plate, in figure 5.6.1 illustrating a meniscus confined between two parallel plates, and in figure 5.7.1 illustrating the interface of a drop attached to a horizontal plane.

The line where two fluids meet a solid surface is the contact line. In the case of a two-dimensional or axisymmetric interface, the contact line is represented by the contact point, which is the trace of the contact line in the xy or a meridional plane, marked by a circle in figures 5.5.1, 5.6.1, and 5.7.1.

Contact angle

The angle subtended between (a) the line that is normal to the contact line and tangential to the solid surface, and (b) the line that is normal to the contact line and tangential to the interface, measured on the side of fluid labeled 2, as illustrated in figure 5.4.2, is called the contact angle. The static contact angle is a physical constant determined by the prevailing physical conditions

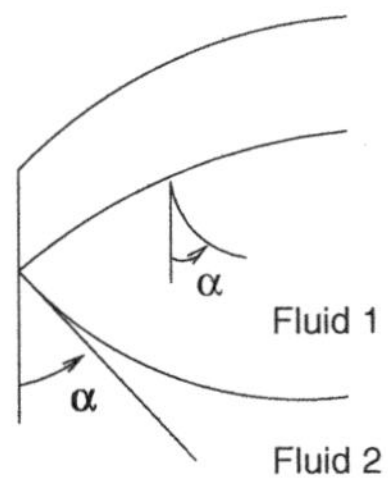

Figure 5.4.2 An interface ending at a contact line. The contact angle, α, is measured on the side of the fluid labeled 2.

and physical properties of the solid and fluids. If fluid 1 wets the solid better than fluid 2, then the contact angle is smaller than $\pi/2$. If fluid 2 wets the solid better than fluid 1, the contact angle is larger than $\pi/2$ but less than the maximum possible value of π.

If the side-wall illustrated in figure 5.4.2 is vertical and the static contact angle is equal to $\pi/2$, the interface remains flat all the way up to the contact line. Under more general conditions, the interface assumes a curved shape with nonzero curvature established to satisfy a physical requirement on the contact angle.

5.4.2 The Laplace-Young equation

To derive the equation governing the shape of a two-dimensional curved interface, we substitute the pressure distributions (5.4.1) in the interfacial condition (4.4.11), finding

$$-\rho_2\, g\, y_I + c_2 + \rho_1\, g\, y_I - c_1 = \gamma\, \kappa. \tag{5.4.5}$$

Rearranging, we obtain the Laplace-Young equation governing the shape of a two-dimensional interface in hydrostatics,

$$\kappa = -\frac{\Delta\rho\, g}{\gamma}\, y_I + B, \tag{5.4.6}$$

where $\Delta\rho \equiv \rho_2 - \rho_1$, and

$$B \equiv \frac{c_2 - c_1}{\gamma} \tag{5.4.7}$$

is a constant with dimensions of inverse length determined by enforcing an appropriate boundary condition or global constraint.

The Laplace-Young equation (5.4.6) requires that the curvature of an interface is a linear function of the elevation from a reference state. An obvious solution arises by assuming that y_I is constant, and then setting $B = \Delta\rho\, g y_I/\gamma$ to find $\kappa = 0$. However, the flat shape of the interface computed in this manner will not necessarily conform with the boundary condition on the static contact angle, and the obvious solution will not generally be admissible. The shape of the interface must be found so that (5.4.6) and a prescribed boundary condition on the contact angle are both satisfied.

Implicit in (5.4.6) is the assumption that the acceleration of gravity points in the negative direction of the y axis. The general expression for an arbitrary orientation is

$$\kappa = \frac{\Delta\rho}{\gamma}\,\mathbf{g}\cdot\mathbf{x}_I + B, \tag{5.4.8}$$

where $\mathbf{x}_I$ is the position of the interface.

Capillary length

Assuming that the fluids are stably stratified, $\rho_2 > \rho_1$ or $\Delta\rho > 0$, we introduce the capillary length

$$\ell = \Big(\frac{\gamma}{\Delta\rho\, g}\Big)^{1/2}, \tag{5.4.9}$$

and recast equation (5.4.6) into the simpler form

$$\kappa = -\frac{y_I}{\ell^2} + B. \tag{5.4.10}$$

For an air-water interface at 20° Celsius, $\gamma = 73$ dynes/cm $= 73\times 10^{-3}$ kg/sec^2, ρ_1=0.0 kg/m^3, ρ_2=1000.0 kg/m^3, yielding a capillary length of 2.72 mm.

5.4.3 Three-dimensional interfaces

The equations derived in this section for two-dimensional interfaces also apply for interfaces with three-dimensional shapes, provided that the curvature in the xy plane is replaced by twice the mean curvature, $2\kappa_m$. For example, the counterpart of equation (5.4.10) for a three-dimensional interface is

$$2\,\kappa_m = -\frac{y_I}{\ell^2} + B. \tag{5.4.11}$$

Numerical solutions of the Laplace-Young equation will be presented in the remainder of this chapter.

When gravitational effects are insignificant, the first term on the right-hand side of (5.4.11) can be discarded, and the interface has a constant mean curvature. A thin soap film attached to a wire frame has constant mean curvature.

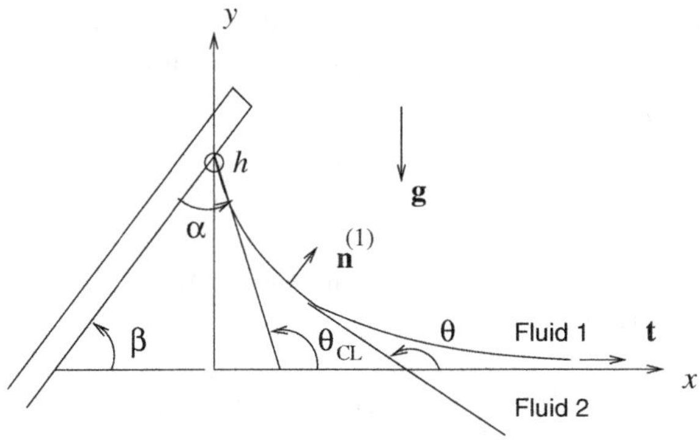

Figure 5.5.1 Illustration of a semi-infinite interface attached to an inclined plate. Far from the plate, the interface becomes horizontal.

Problem

5.4.1 *Pressure in a layer*

Derive expressions for the pressure distribution across a horizontal liquid layer of thickness h sandwiched between two semi-infinite fluids.

5.4.2 *Constant mean curvature*

Prepare a list of five geometrical shapes with constant mean curvature.

5.5 A semi-infinite interface attached to an inclined plate

We begin the study of two-dimensional interfacial shapes by considering a semi-infinite interface attached to a flat plate that is inclined at an angle β with respect to the horizontal plane, as illustrated in figure 5.5.1. Far from the plate, as x tends to infinity, the interface tends to become horizontal. The contact angle subtended between the inclined plate and the tangent to the interface at the contact point is required to have a prescribed value, α.

It is convenient to set the origin of the y axis at the position of the flat interface far from the plate and describe the interface by the equation $y_I = f(x)$. As x tends to infinity, the function $f(x)$ vanishes, yielding a flat interface. Since the interfacial curvature tends to zero far from the plate, the constant B on the right-hand side of the Laplace-Young equation (5.4.10) must be zero, yielding the simpler form

$$\kappa = -\frac{f}{\ell^2}. \tag{5.5.1}$$

Substituting in (5.5.1) the expressions for the curvature given in (4.3.19) and (4.3.21), we derive a differential equation governing the shape of the interface,

$$\frac{\mathrm{d}}{\mathrm{d}x}\Big(\frac{1}{(1+f'^2)^{1/2}}\Big) = \frac{\mathrm{d}|\cos\theta|}{\mathrm{d}x} = -\frac{ff'}{\ell^2} = -\frac{1}{2}\frac{(f^2)'}{\ell^2}, \tag{5.5.2}$$

where a prime denotes a derivative with respect to x. Integrating once with respect to x, we obtain

$$\frac{1}{(1+f'^2)^{1/2}} = |\cos\theta| = -\frac{1}{2}\frac{f^2}{\ell^2} + C, \tag{5.5.3}$$

where C is an integration constant. Demanding that f vanishes as x tends to infinity and correspondingly θ tends to π, we find $C = 1$.

Capillary rise

At the contact line located at $x = 0$, the slope angle θ defined by the equation $\tan\theta = f'$ takes the value

$$\theta_{CL} = \alpha + \beta. \tag{5.5.4}$$

Evaluating (5.5.3) at $x = 0$, setting $C = 1$ and rearranging, we obtain an expression for the positive or negative capillary rise $h \equiv f(0)$,

$$\frac{h^2}{2\,\ell^2} = 1 - |\cos(\alpha+\beta)|, \tag{5.5.5}$$

which shows that the maximum possible value of $|h|$ occurs when $\alpha + \beta$ is a multiple of $\pi/2$, and is equal to $\sqrt{2}\ell$.

Numerical formulation

To compute the shape of the interface, we set the left-hand side of (5.5.3) equal to the right-hand side with $C = 1$, and rearrange to obtain the first-order ordinary differential equation

$$\frac{\mathrm{d}f}{\mathrm{d}x} = \pm\Big(\frac{4}{(2-\hat{f}^2)^2} - 1\Big)^{1/2} = \pm\hat{f}\,\frac{(4-\hat{f}^2)^{1/2}}{2-\hat{f}^2}, \tag{5.5.6}$$

where $\hat{f} \equiv f/\ell$ is a dimensionless function. The plus or minus sign on the right-hand side must be selected according to the expected interface shape.

The preceding analysis assumes that the interface has a monotonic shape, which is true if θ_{CL} lies in the range $(\pi/2, 3\pi/2)$. Outside this range, the capillary rise is given by equation (5.5.5) with the minus sign replaced by a plus sign on the right-hand side.

When the shape of the interface is non-monotonic, the interface becomes vertical at a point, the function $f(x)$ is multi-valued, and the integration of (5.5.6) requires special attention. To bypass this subtlety, we regard x along the interface as a function of the independent variable f, and recast (5.5.6) into the form

$$\frac{\mathrm{d}x}{\mathrm{d}f} = \pm \frac{2 - \hat{f}^2}{\hat{f}\,(4 - \hat{f}^2)^{1/2}}. \tag{5.5.7}$$

The solution of (5.5.7) must be found for $|f| < |h|$, where $|h|$ is the capillary rise computed from equation (5.5.5).

5.5.1 Numerical method

A numerical solution of (5.5.7) may be computed according to the following steps:

1. Compute the angle θ_{CL} from equation (5.5.4).

2. Compute the capillary rise h using the formulas

$$\frac{h}{\sqrt{2}\,\ell} = \begin{cases} +(1 + |\cos(\theta_{CL})|)^{1/2} & \text{if} \quad 0 < \theta_{CL} < \pi/2, \\ +(1 - |\cos(\theta_{CL})|)^{1/2} & \text{if} \quad \pi/2 < \theta_{CL} < \pi, \\ -(1 - |\cos(\theta_{CL})|)^{1/2} & \text{if} \quad \pi < \theta_{CL} < 3\pi/2, \\ -(1 + |\cos(\theta_{CL})|)^{1/2} & \text{if} \quad 3\pi/2 < \theta_{CL} < 2\pi. \end{cases} \tag{5.5.8}$$

3. Integrate the differential equation (5.5.7) from $f = h$ to 0 with initial condition $x(f = h) = 0$ using, for example, the explicit Euler method or the modified Euler method discussed in section 1.5. If h is negative, we use a negative spatial step.

To implement the explicit Euler method, we select a small positive or negative integration step, $\Delta f = h/N$, where N defines the level of numerical discretization, evaluate equation (5.5.7) at the point f and approximate the derivative on the left-hand side with the finite difference $[x(f+\Delta f) - x(f)]/\Delta f$. Rearranging, we obtain

$$x(f + \Delta f) = x(f) + \Delta f\, \frac{2 - \hat{f}^2}{\hat{f}\,(4 - \hat{f}^2)^{1/2}}. \tag{5.5.9}$$

The repetitive application of this formula starting from $f = h$ where $x = 0$ generates a sequence of points distributed over the interface.

To implement the modified Euler method, we replace formula (5.5.9) with the slightly more involved formula

$$x(f + \Delta f) = x(f) + \Delta f\, \frac{1}{2} \Big(\frac{2 - \hat{f}^2}{\hat{f}\,(4 - \hat{f}^2)^{1/2}} + \frac{2 - \hat{f}^2_{tmp}}{\hat{f}_{tmp}\,(4 - \hat{f}^2_{tmp})^{1/2}} \Big), \tag{5.5.10}$$

where $f_{tmp} = f + \Delta f$. The numerical method is implemented in the MATLAB code *men_2d_plate* residing in directory *03_hydrostat* of FDLIB. The program scans tilting angles while holding the contact angle constant and displays the interfacial profile in animation, as follows:

```
%---
% data
%---

gac = 1.0;    % acceleration of gravity
rhop = 1.0;    % pool density
rhoa = 0.0;    % ambient fluid density
gamma = 1.0;    % surface tension
beta = 0.01*pi;    % plate inclination
alpha = 0.01*pi;    % contact angle
ndiv = 2*64;    % interface divisions

%---
% prepare
%---

drho = rhop-rhoa;

if(drho<0)
  disp 'The density of the pool must be higher than'
  disp 'the density of the ambient fluid'
  return
end

capls = gamma/(gac*abs(drho));
capl = sqrt(capls);
dbeta = 0.05;

Irepeat = 1;

%========
while (Irepeat==1)
%========

%--------------------------------
% angle theta at the contact line
%--------------------------------

thcl = alpha+beta;
cst = cos(thcl);

%----------------------------------------
% compute the magnitude of meniscus rise h
%----------------------------------------
```

```
if((thcl>0.5*pi)&(thcl<1.5*pi))
  h = sqrt(2.0*capls*(1.0-abs(cst)));
else
  h = sqrt(2.0*capls*(1.0+abs(cst)));
end

%---
% sign of meniscus rise
%---

if(thcl<pi)
  h = abs(h);    % meniscus goes up
else
  h =-abs(h);    % meniscus goes down
end

%---
% plate position at y=0
%---

xpl = -h/tan(beta);

%-------------------------------------------
% integrate meniscus equation dx/df = G(f)
% from f=h to f=0 using the modified Euler
% method with constant step
%-------------------------------------------

df = h/ndiv;    % note that df can be positive or negative
dfh = 0.5*df;

y(1) = h;
x(1) = 0.0;    % starting point

for i=1:ndiv-1
  fred = y(i)/capl;
  freds = fred*fred;
  xp = (2.0-freds)/sqrt(4.0-freds)/fred;
  xsv = x(i);    % save
  xpsv = xp;    % save
  y(i+1) = y(i)-df;
  x(i+1) = x(i)+xp*df;
  fred = y(i+1)/capl;
  freds = fred*fred;
  xp = (2.0-freds)/(fred*sqrt(4.0-freds));
  x(i+1) = xsv+(xpsv+xp)*dfh;
end
```

```
%---
% plotting
%---

plot(x,y)
hold on
patch([xpl, 0, x, x(ndiv), (-2-h)/tan(beta)] ...
  ,[0, h, y,-2, -2],'y');
plot([-1 3 3 -1 -1],[-2 -2 2 2 -2])
plot([-10 10],[-10*tan(beta)+h, 10*tan(beta)+h] ...
  ,'r','LineWidth',3);
plot([xpl, 10],[0, 0],'c--','LineWidth',1);
hold off
axis equal
xlabel('x','fontsize',15); ylabel('y','fontsize',15)
axis([-1 3 -2 2])
pause(0.1)

%---
% tilt the plate
%---

beta = beta+dbeta;

if(beta>0.99*pi|beta<0.01*pi)
  dbeta = -dbeta;
end

%====
end
%====
```

The graphics generated by the code for two plate inclination angles and a fixed contact angle is shown in figure 5.5.2.

Computer Problem

5.5.1 *Semi-infinite meniscus*

Run the code *men_2d_plate* to generate a family of shapes corresponding to a fixed value of the plate inclination angle, β, and various contact angles, α. Generate another family of shapes corresponding to a fixed value of the contact angle and various plate inclination angles. Discuss the behavior of the capillary rise in each case.

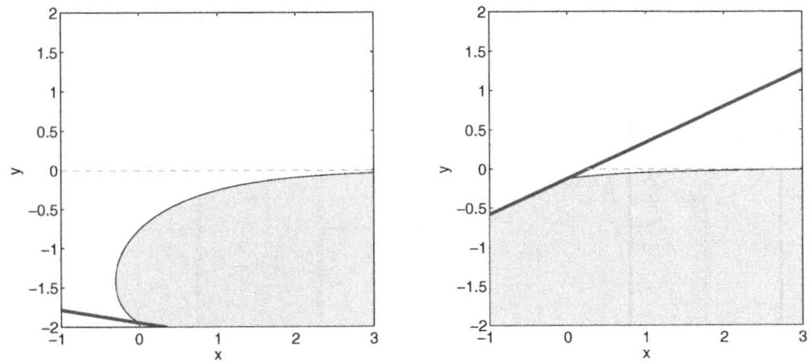

Figure 5.5.2 A semi-infinite meniscus attached to an inclined plate generated by the FDLIB code *men_2d_plate*. The plate inclination angles are different, but the contact angle is the same in both cases.

5.6 A meniscus between two parallel plates

Next, we consider a two-dimensional interface subtended between two parallel vertical plates separated by the distance $2b$, as illustrated in figure 5.6.1(a). The two contact points are at the same elevation, and the interface is symmetric with respect to the mid-plane located at $x = 0$. It is convenient to set the origin of the Cartesian axes at the interface midway between the plates and describe the position of the interface by the equation $y_I = f(x)$. Outside and far from the plates, the interface assumes a horizontal shape located at $y_I = -h$, where h is the positive or negative capillary rise of the meniscus midway between the plates. Our objective is to compute h along with the unknown shape of the meniscus by solving the Laplace-Young equation (5.4.10).

Our choice of Cartesian axes requires that $y_I = 0$ at $x = 0$. Equation (5.4.10) then gives

$$\kappa(0) = B. \tag{5.6.1}$$

Because the interface is symmetric with respect to $x = 0$, $\mathrm{d}f/\mathrm{d}x = 0$ at $x = 0$, the expression for the curvature (4.3.19) produces

$$\kappa(0) = -f''(0), \tag{5.6.2}$$

where a prime denotes a derivative with respect to x. Combining equations (5.6.1) and (5.6.2), we find

$$B = -f''(0). \tag{5.6.3}$$

Now evaluating (5.4.10) at a point outside and far from the plates where the

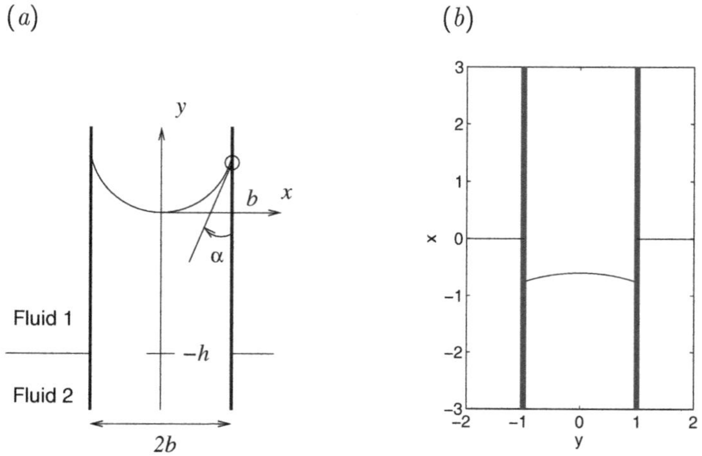

Figure 5.6.1 (*a*) Illustration of a meniscus subtended between two parallel vertical plates for contact angle α less than $\pi/2$. When α is greater than $\pi/2$, the meniscus submerges and the capillary rise h is negative. (*b*) A meniscus shape produced by the FDLIB code *men_2d*.

curvature of the interface tends to vanish and y_I tends to $-h$, we obtain

$$B = -\frac{h}{\ell^2}. \tag{5.6.4}$$

Combining equations (5.6.2) – (5.6.4), we find

$$h = -\ell^2\,\kappa(0) = \ell^2\,f''(0), \tag{5.6.5}$$

which shows that the capillary rise is determined by the curvature of the interface at the mid-plane.

Governing equations

To complete the mathematical formulation, we substitute the first expression for the curvature given in (4.3.19) and the value of B given in (5.6.4) in the Laplace-Young equation (5.4.10), and rearrange to derive a second-order ordinary differential equation involving the capillary rise as an unknown,

$$f'' = \frac{1}{\ell^2}\,(f+h)\,(1+f'^2)^{3/2}. \tag{5.6.6}$$

The solution must be found in the interval $0 < x < b$ subject to the boundary conditions

$$f(0) = 0, \qquad f'(0) = 0, \qquad f'(b) = \cot\alpha. \tag{5.6.7}$$

The third condition specifies the prescribed value of the contact angle. If $\alpha = \pi/2$, all three boundary (5.6.7) are homogeneous and the obvious solution describes a flat, non-elevated and non-submerged interface, $h = 0$ and $f = 0$.

Canonical form

To compute the meniscus shape under general conditions, we recast the second-order differential equation (5.6.6) involving the unspecified parameter h into the canonical form of a system of three first-order differential equations. The word "canonical" derives from the Greek work κανονικος which means *normal*. This is done by introducing the three new variables

$$y_1 \equiv f, \qquad y_2 \equiv f', \qquad y_3 \equiv h, \tag{5.6.8}$$

where $y_2' = f''$. Given these definitions, equation (5.6.6) is resolved into three first-order component equations,

$$\frac{\mathrm{d}y_1}{\mathrm{d}x} = q_2, \qquad \frac{\mathrm{d}y_2}{\mathrm{d}x} = \frac{1}{\ell^2}\,(y_1 + y_3)\,(1 + y_2^2)^{3/2}, \qquad \frac{\mathrm{d}y_3}{\mathrm{d}x} = 0. \tag{5.6.9}$$

The third equation simply states that $y_3 \equiv h$ is a constant. In terms of the new variables, the boundary conditions (5.6.7) become

$$y_1(0) = 0, \qquad y_2(0) = 0, \qquad y_2(b) = \cot\alpha. \tag{5.6.10}$$

If the value of $y_3(0) = h$ were known, we would be able to integrate the system (5.6.9) from $x = 0$ to b using, for example, the explicit Euler or the modified Euler method discussed in sections 1.5 and 5.5.

Explicit Euler method

To implement the explicit Euler method, we recast the system (5.6.9) into the general symbolic form

$$\begin{aligned} \frac{\mathrm{d}y_1}{\mathrm{d}x} &= f_1(y_1, y_2, y_3, x), \qquad \frac{\mathrm{d}y_2}{\mathrm{d}x} = f_2(y_1, y_2, y_3, x), \\ \frac{\mathrm{d}y_3}{\mathrm{d}x} &= f_3(y_1, y_2, y_3, x), \end{aligned} \tag{5.6.11}$$

where

$$\begin{aligned} f_1(y_1, y_2, y_3, x) &\equiv y_2, \\ f_2(y_1, y_2, y_3, x) &\equiv \frac{1}{\ell^2}\,(y_1 + y_3)\,(1 + y_2^2)^{3/2}, \\ f_3(y_1, y_2, y_3, x) &\equiv 0 \end{aligned} \tag{5.6.12}$$

are the phase-space velocities.

Next, we evaluate equations (5.6.11) at a point, x, select a small spatial step, Δx, and approximate the derivatives on the left-hand sides with finite differences writing, for example, $dy_1/dx = (y_1(x + \Delta x) - y_1(x))/\Delta x$. Solving for $y_1(x + \Delta x)$ and repeating for the second and third equation, we find

$$
\begin{aligned}
y_1(x + \Delta x) &= y_1(x) + \Delta x\, f_1(y_1(x), y_2(x), y_3(x), x), \\
y_2(x + \Delta x) &= y_2(x) + \Delta x\, f_2(y_1(x), y_2(x), y_3(x), x), \\
y_3(x + \Delta x) &= y_3(x) + \Delta x\, f_3(y_1(x), y_2(x), y_3(x), x).
\end{aligned}
\tag{5.6.13}
$$

In vector notation,

$$
\mathbf{y}(x + \Delta x) = \mathbf{y}(x) + \Delta x\, \mathbf{f}(\mathbf{x}(x), x), \tag{5.6.14}
$$

where $\mathbf{y} \equiv (y_1, y_2, y_3)$ is the solution vector, and $\mathbf{f} \equiv (f_1, f_2, f_3)$ is the corresponding phase-space velocity vector. The repetitive application of formula (5.6.14) starting from $x = 0$ allows us to generate a sequence of points distributed over the meniscus.

Modified Euler method

To implement the modified Euler method, we replace formula (5.6.14) with a predictor and a corrector formula,

$$
\begin{aligned}
\mathbf{y}^{tmp} &= \mathbf{y}(x) + \Delta x\, \mathbf{f}(\mathbf{y}(x), x), \\
\mathbf{y}(x + \Delta x) &= \mathbf{x}(x) + \Delta x\, \frac{1}{2}\, [\mathbf{f}(\mathbf{y}(x), x) + \mathbf{f}(\mathbf{y}^{tmp}, x + \Delta x)],
\end{aligned}
\tag{5.6.15}
$$

where the superscript *tmp* denotes a preliminary value computed by the explicit Euler method. The first equation generates a provisional value; the second equation advances the solution using the initial and provisional value. The method is implemented in the following MATLAB code *men_2d_ode* residing in directory *men_2d* inside directory *03_hydrostat* of FDLIB:

```
function [x,y1,y2] = men_2d_ode (npts,capls,b,h)

%-----------------------------------------
% Integrate the ODEs by the modified Euler method
% Integration interval:  (0, b)
% Initial condition y1(0) = 0, y2(0) = 0
%-----------------------------------

%----------------------
% prepare and initialize
%----------------------

dx = b/npts;    % uniform x step
x(1) = 0.0; y1(1) = 0.0; y2(1) = 0.0;
```

```
%----------------------------------------
% integrate by the modified Euler method
%----------------------------------------

for i=1:npts

  if(i==1)
    y1p = 0.0   % value at mid-plane
    y2p = h/capls   % value at mid-plane
  else
    y1p = y2(i);
    y2p = (y1(i)+h)*sqrt( (1.0+y2(i)*y2(i))^3 )/capls;
  end

  y1sv = y1(i); y2sv = y2(i);    % save
  y1psv = y1p; y2psv = y2p;
  x(i+1) = x(i) + dx;
  y1(i+1) = y1(i) + y1p*dx;
  y2(i+1) = y2(i) + y2p*dx;
  y1p = y2(i+1);   % second velocity evaluation
  y2p = (y1(i+1)+h)*sqrt((1.0+y2(i+1)^2)^3 )/capls;
  y1(i+1) = y1sv + 0.5*(y1psv+y1p)*dx;
  y2(i+1) = y2sv + 0.5*(y2psv+y2p)*dx;

end

%-----
% done
%-----

return
```

5.6.1 The shooting method

Because the value of h is *a priori* unknown, the starting vector $\mathbf{q}(0)$ is not available and the solution of (5.6.9) must be found using an iterative method. The shooting method prescribes the obvious: guess the value of $q_3(0) = h$, compute the solution of (5.6.9), and then check whether the third condition in (5.6.10) is fulfilled; if not, repeat the computation with an improved guess.

To improve the guess in a systematic fashion that guarantees rapid convergence, we note that the value of $x_2(b)$ computed by solving equations (5.6.9) is a function of the guessed value, $x_3(0) = h$. To signify this dependence, we extend the list of arguments of x_2, writing $x_2(b; h)$. The third boundary condition in (5.6.10) requires $q_2(b; h) - \cot\alpha = 0$, which means that h is a root of the objective function

$$Q(h) \equiv x_2(b; h) - \cot\alpha. \tag{5.6.16}$$

The problem has been reduced to computing the solution of the algebraic equation $Q(h) = 0$, where the left-hand side is evaluated by integrating equations (5.6.9) with a specified value of h.

Secant updates

The secant method provides us with a simple algorithm for solving the nonlinear algebraic equation $Q(h) = 0$ according to the following steps:

1. First, we select a value for h that approximates the root, called $h^{(1)}$, and compute $Q(h^{(1)})$ by integrating system (5.6.9).

2. Second, we select another value for h called $h^{(2)}$, and compute $Q(h^{(2)})$ by integrating (5.6.9).

3. Third, we approximate the graph of the function $Q(h)$ with a straight line passing through the points computed in steps 1 and 2. The slope of the approximating straight line is

$$s^{(2)} = \frac{Q(h^{(2)}) - Q(h^{(1)})}{h^{(2)} - h^{(1)}}. \tag{5.6.17}$$

4. Next, we identify the improved value $h^{(3)}$ with the root of the linear function describing the approximating straight line. Elementary algebra shows that the root is given by

$$h^{(3)} = h^{(2)} - \frac{Q(h^{(2)})}{s^{(2)}}. \tag{5.6.18}$$

5. Finally, we repeat the computation with the pairs $h^{(2)}$ and $h^{(3)}$ until convergence.

A reasonable choice for $h^{(1)}$ can be obtained by assuming that the meniscus has a circular shape of radius R, which is positive when the interface is concave upward and negative when the interface is concave downward. Using elementary trigonometry, we find that the prescribed boundary condition on the contact angle will be satisfied when $\cos\alpha = b/R$. Rearranging, we derive the approximation $\kappa \simeq -1/R = -\cos\alpha/b$. Combining equations (5.6.2) and (5.6.5), we obtain a desired educated guess,

$$h \simeq \cos\alpha \, \frac{\ell^2}{b}. \tag{5.6.19}$$

Equation (5.6.19) reveals that the maximum possible value of $|h|$ for a circular interface is l^2/b.

The method is implemented in the following MATLAB function *men_2d* residing in directory *03_hydrostat* of FDLIB:

```
function [Iflag,x,y1,hmen] = men_2d ...
...
    (b,gac,gamma,rhop,rhoa ...
    ,alpha,npts,epsilon,maxiter,tol)

%-------------------------------------------
% Hydrostatic shape of a 2D meniscus between
% two vertical parallel plates computed
% by the shooting method
%-------------------------------------------

Iflag=0;   % flag for success

%-----
% prepare
%----

drho = rhop-rhoa;   % density difference

% square of the capillary length:
capls = gamma/(gac*abs(drho));

%-----------------------------------------
% initial guess for h
%-----------------------------------------

h(1) = capls*cos(alpha)/b;

if(abs(alpha-0.5*pi)<0.0000001)
  cota = 0.0;
else
  cota = 1.0/tan(alpha);
end

%---
% compute the first solution of the odes
% to start-up the secant method
%---

Ic=1;   % counter
[x,y1,y2] = men_2d_ode(npts,capls,b,h(Ic));
error(Ic) = y2(npts+1)-cota;
err = abs(error(Ic));

%------------------------
% second start-up solution
%------------------------

Ic=2;
```

```
h(2) = h(1)+epsilon;
[x,y1,y2] = men_2d_ode(npts,capls,b,h(Ic));
error(Ic) = y2(npts+1)-cota;
err = abs(error(Ic));

%----------------------------------------
% iterate using the secant method
% until convergence
%----------------------------------------

for iter=1:maxiter

  Ic = Ic+1;

%---
% secant updating
%---

  Icb = Ic-2; Ica = Ic-1;
  dedh = (error(Ica)-error(Icb))/(h(Ica)-h(Icb));
  h(Ic) = h(Ica)-error(Ica)/dedh;

  [x,y1,y2] = men_2d_ode(npts,capls,b,h(Ic));

  error(Ic) = y2(npts+1)-cota;
  err = abs(error(Ic));

  if(err<tol)
    break
  end

%---
end
%---

if(iter==maxiter)
  disp('men_2d:  ODE solver failed')
  Iflag=1;
  return
end

hmen=h(Ic);

%---
% done
%---

return
```

Results of a computation for a contact angle higher than $\pi/2$ are shown in figure 5.6.1(*b*). In this case, the meniscus submerges below the level of the liquid outside the plates.

Computer Problem

5.6.1 *Meniscus between plates*

(*a*) Run the code *men_2d* to generate a family of shapes corresponding to a fixed value of the plate separation and various contact angles. (*b*) Produce another family of shapes corresponding to a fixed contact angle and various plate separations. Discuss the behavior of the capillary rise in each case.

5.7 A two-dimensional drop on a horizontal or inclined plane

Proceeding in order of ascending difficulty, we consider the shape of a two-dimensional liquid drop hanging underneath or resting on a horizontal or inclined plane. Our analysis also applies for a gas bubble.

5.7.1 Drop on a horizontal plane

First, we consider a two-dimensional drop of a fluid labeled 2 resting above or hanging below a horizontal plane. The drop is surrounded by a stationary ambient fluid labeled 1, as illustrated in figure 5.7.1. The resting drop shown in frame (*a*) is called a *sessile* drop, and the hanging drop shown in frame (*b*) is called a *pendant* drop.

Our objective is to compute the shape of the interface for a specified value of the contact angle, α, and drop area, A_D. It is convenient to introduce Cartesian axes with origin at the extreme point of the interface, and the x axis pointing normal to the interface and into the ambient fluid, as depicted in figure 5.7.1 for a sessile or pendant drop.

Parametric representation

One difference between the problem presently considered and those discussed previously in this chapter, is that, neither the range of x nor the range of y is known over the span of the drop interface at the outset. To circumvent this difficulty, we describe the shape of the interface in parametric form in terms of the slope angle ψ defined in figure 5.7.1, increasing from zero at the origin to α at the contact point.

Our objective is to compute the two scalar functions of ψ such that the x and y coordinates of a point at the interface are described by the functions

$$x = X(\psi), \qquad y = Y(\psi), \tag{5.7.1}$$

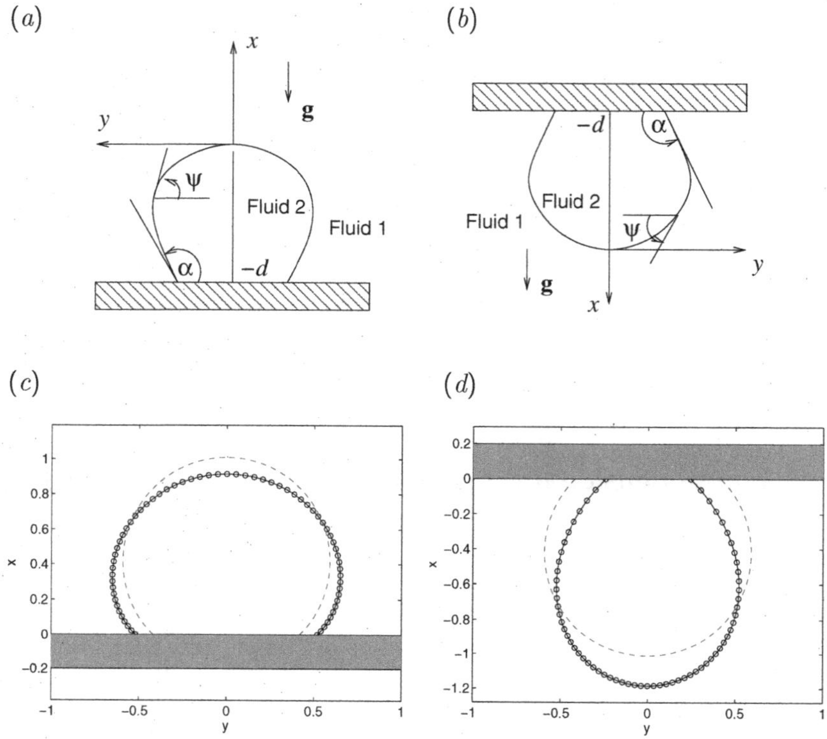

Figure 5.7.1 Illustration of a two-dimensional (*a*) sessile liquid drop resting on a horizontal plane, and (*b*) pendant liquid drop hanging below a horizontal plane. (*c, d*) Shapes of a sessile and a pendant drop computed by code *drop_2d*. The circular symbols represent data arising from the numerical integration of the differential equations; the dashed lines show the approximate circular shape arising for small drops or large surface tension.

for $0 < \psi < \alpha$. To compute $X(\psi)$ and $Y(\psi)$, we require two ordinary differential equations and a suitable number of boundary conditions or global constraints.

Ordinary differential equations

The first differential equation is the definition of the chosen parameter ψ in terms of the interface slope,

$$\cot\psi = -\frac{dY}{dX}. \tag{5.7.2}$$

One important benefit of the adopted parametrization is that the boundary condition for the contact angle at the contact point is satisfied automatically and can be removed from further discussion. The second differential equation must arise from the Laplace-Young equation determining the jump in pressure across the interface,

$$\kappa = -s_1 \frac{\Delta \rho \, g}{\gamma} X + B, \tag{5.7.3}$$

where $\Delta\rho = \rho_2 - \rho_1$. The coefficient s_1 is equal to 1 for a sessile drop and -1 for a pendant drop reflecting the orientation of the gravity with respect to the positive direction of the x axis. In terms of the square of the capillary length, $\ell^2 \equiv \gamma/(|\Delta\rho|\, g)$, equation (5.7.3) takes the compact form

$$\kappa = -s_1 \, s_2 \, \frac{X}{\ell^2} + B, \tag{5.7.4}$$

where the coefficient s_2 is equal to 1 when $\rho_2 > \rho_1$, or -1 when $\rho_2 < \rho_1$. Evaluating equation (5.7.4) at the origin, we find that the constant B is equal to the unknown curvature of the interface at the plane of symmetry located at $y = 0$.

Now we introduce expression (4.3.19) for the curvature, set $f' = \mathrm{d}Y/\mathrm{d}X$, and use (5.7.2) to find

$$\kappa = -\frac{1}{\cot\psi}\frac{\mathrm{d}}{\mathrm{d}X}\Big(\frac{1}{(1+\cot^2\psi)^{1/2}}\Big) = -\frac{1}{\cot\psi}\frac{\mathrm{d}\sin\psi}{\mathrm{d}X} = -\sin\psi\,\frac{\mathrm{d}\psi}{\mathrm{d}X}. \tag{5.7.5}$$

Substituting the last expression in (5.7.5) in the left-hand side of (5.7.4), we obtain

$$\sin\psi\,\frac{\mathrm{d}\psi}{\mathrm{d}X} = s_1\, s_2\, \frac{X}{\ell^2} - B, \tag{5.7.6}$$

which can be rearranged to give the desired parametric dependence

$$\frac{\mathrm{d}X}{\mathrm{d}\psi} = \frac{\sin\psi}{Q}, \tag{5.7.7}$$

where

$$Q \equiv s_1\, s_2\, \frac{X}{\ell^2} - B. \tag{5.7.8}$$

To derive a corresponding parametric dependence for Y, we recast (5.7.2) into the form

$$\frac{\mathrm{d}Y}{\mathrm{d}\psi} = -\cot\psi\,\frac{\mathrm{d}X}{\mathrm{d}\psi}. \tag{5.7.9}$$

Substituting (5.7.7) in (5.7.9), we find

$$\frac{\mathrm{d}Y}{\mathrm{d}\psi} = -\frac{\cos\psi}{Q}. \tag{5.7.10}$$

Equations (5.7.7) and (5.7.10) provide us with the desired system of two first-order differential equations involving an unspecified parameter, B. The boundary conditions require

$$X(0) = 0, \qquad Y(0) = 0, \tag{5.7.11}$$

and the constraint on the drop area requires

$$2 \int_{-d}^{0} Y \, \mathrm{d}x = A_D, \tag{5.7.12}$$

where $x = -d$ describes the position of the plane.

The shooting method

Because the value of the constant B is *a priori* unknown, the solution must be found using an iterative method. The shooting method combined with the secant method for improving the guess provides us with an efficient algorithm. The numerical procedure involves the following steps:

1. We begin by guessing a value for B.
2. Next, we integrate the system of equations (5.7.7) and (5.7.10).
3. Third, we compute the integral on the right-hand side of (5.7.12) using the trapezoidal rule, and then evaluate the objective function

$$Q \equiv 2 \int_{-d}^{0} y \, \mathrm{d}x - A_D. \tag{5.7.13}$$

4. Fourth, we improve the guess for B with the goal of driving the objective function Q to zero using, for example, the secant method discussed in section 5.6.

Since the constant B is equal to the unknown curvature of the interface at the mid-plane located at $y = 0$, a reasonable guess can be obtained by assuming that the interface has the shape of a section of a circle, and then computing the radius of the circle, a, according to specified values of the contact angle and drop area. Using elementary trigonometry, we find

$$a = \left(\frac{2 A_D}{2\alpha - \sin 2\alpha} \right)^{1/2}. \tag{5.7.14}$$

Fourth-order Runge-Kutta method

The fourth-order Runge-Kutta method (RK4) is an improvement of the modified Euler method discussed earlier in this section, involving three exploratory steps and a final step. The following MATLAB function *drop_2d_ode* residing in directory *drop_2d* inside subdirectory *03_hydrostat* of FDLIB implements the method to solve the system of equations (5.7.7) and (5.7.10) and compute the drop area:

```
function [x,y,area] = drop_2d_ode ...
  ...
  (npts ...
  ,capls ...
  ,Isp ...
  ,dpsi ...
  ,B ...
  )

%---------------------------------------------
% Integrate ODEs by RK4 with uniform
% step size for the angle psi
%---------------------------------------------

%--------
% prepare
%--------

dpsih = 0.5*dpsi;

%----------------
% top of the drop
%----------------

psi = 0.0; x(1) = 0.0; y(1) = 0.0;

%----------
% integrate
%----------

for i=1:npts

  if(i==1)
    xp = 0.0; yp = 1.0/B;
  else
    Q = Isp*x(i)/capls-B;
    xp = sin(psi)/Q;
    yp =-cos(psi)/Q;
  end
  xp1 = xp; yp1 = yp;

  psi = psi +dpsih;
  x(i+1) = x(i)+xp*dpsih;
  y(i+1) = y(i)+yp*dpsih;

  Q = Isp*x(i+1)/capls-B;
  xp = sin(psi)/Q;
  yp =-cos(psi)/Q;
  xp2 = xp; yp2 = yp;
```

```
    x(i+1) = x(i)+xp*dpsih;
    y(i+1) = y(i)+yp*dpsih;

    Q = Isp*x(i+1)/capls-B;
    xp = sin(psi)/Q;
    yp =-cos(psi)/Q;
    xp3 = xp; yp3 = yp;

    psi = psi +dpsih;
    x(i+1) = x(i)+xp*dpsi;
    y(i+1) = y(i)+yp*dpsi;

    Q = Isp*x(i+1)/capls-B;
    xp = sin(psi)/Q;
    yp =-cos(psi)/Q;
    xp4 = xp; yp4 = yp;

    x(i+1) = x(i) + (xp1+2*xp2+2*xp3+xp4)*dpsi/6.0;
    y(i+1) = y(i) + (yp1+2*yp2+2*yp3+yp4)*dpsi/6.0;

end

%----------------------------------------
% compute the area of the integrated shape
% by the trapezoidal rule
%----------------------------------------

area = 0.0;
for i=1:npts
  area= area+(y(i+1)+y(i))*abs(x(i+1)-x(i));
end

area = 0.5D0*area;   % to account for trapezoidal weights

%---
% double the area to get the full shape
%---

area = 2.0*area

%-----
% done
%-----

return
```

The following MATLAB function *drop_2d* residing in directory *03_hydrostat* of FDLIB implements the secant method:

```
function [a,Bfinal,x,y] = drop_2d ...
  ...
  (Jsp,gac,gamma,rhod,rhoa,area...
  ,alpha,npts,epsilon,maxiter,tol)

%--------------------------------------------
% Hydrostatic shape of a two-dimensional
% sessile drop resting on a horizontal plate,
% or pendant drop hanging underneath
% a horizontal plate, for a specified
% specified area and contact angle
%--------------------------------------------

%--------
% prepare
%--------

Iflag=0;   % signals failure

drho = rhod-rhoa;   % density difference
capls = gamma/(gac*abs(drho));   % square of the
                                 % capillary length

Isp = 1;   % Isp is an orientation index

if(drho<0)
  Isp = -Isp;
end
if(Jsp==-1)
  Isp = -Isp;
end

%---------------------------------
% to start, assume that the drop shape
% is a truncated circle
% and compute the circle radius "a"
% in terms of the drop area and contact angle
%---------------------------------

a = sqrt(area/(alpha-0.5*sin(2.0*alpha)));
B(1) = 1.0/a;

%---
% compute the initial solution of the odes
% to start-up the secant method
%---

dpsi = alpha/npts;
Ic = 1;   % counter
```

```
[x,y,area_sh] = drop_2d_ode ...
  ...
  (npts ...
  ,capls ...
  ,Isp ...
  ,dpsi ...
  ,B(Ic) ...
  );

error(Ic) = area_sh - area;
err = abs(error(Ic));

%-------------------------
% second start-up solution
%-------------------------

Ic=2;
B(2) = B(1)+epsilon;

[x,y,area_sh] = drop_2d_ode ...
  ...
  (npts ...
  ,capls ...
  ,Isp ...
  ,dpsi ...
  ,B(Ic) ...
  );

error(Ic) = area_sh - area;
err = abs(error(Ic));

%----------------------------------------
% iterate on B using the secant method
%----------------------------------------

for iter=1:maxiter

  Ic = Ic+1;

%---
% secant updating
%---

  Icb = Ic-2;
  Ica = Ic-1;
  dedc = (error(Ica)-error(Icb))/(B(Ica)-B(Icb));
  B(Ic) = B(Ica)-error(Ica)/dedc;
```

```
    [x,y,area_sh] = drop_2d_ode ...
      ...
      (npts ...
      ,capls ...
      ,Isp ...
      ,dpsi ...
      ,B(Ic) ...
      );

    error(Ic) = area_sh - area;
    err = abs(error(Ic));
    if(err<tol) break; end

  %---
  end
  %---

  if(iter==maxiter)
    disp('drop_2d:  ODE solver failed')
    Iflag=1;
    return
  end

  Bfinal = B(Ic);

  %---
  % done
  %---

  return
```

Drop shapes computed using this code are shown in figure 5.7.1(*c, d*). Gravity squeezes the sessile drop toward the wall and pulls the pendant drop away from the wall.

5.7.2 A drop on an inclined plane

Next, we consider the more challenging case of a two-dimensional drop resting above or hanging underneath an inclined plane, as shown in figure 5.7.2. Note that the origin of the x axis is fixed on the plane. In the case of the sessile drop depicted in figure 5.7.2(*a*), the contact angle at the first contact point, α_1, is larger than the contact angle at the second contact point, α_2. In the case of the pendant drop shown in figure 5.7.2(*b*), α_1, is smaller than α_2.

In the inclined system of coordinates depicted in figure 5.7.2(*a, b*), the Cartesian components of the acceleration of gravity are

$$g_x = -g \cos\beta \qquad g_y = -g \sin\beta, \tag{5.7.15}$$

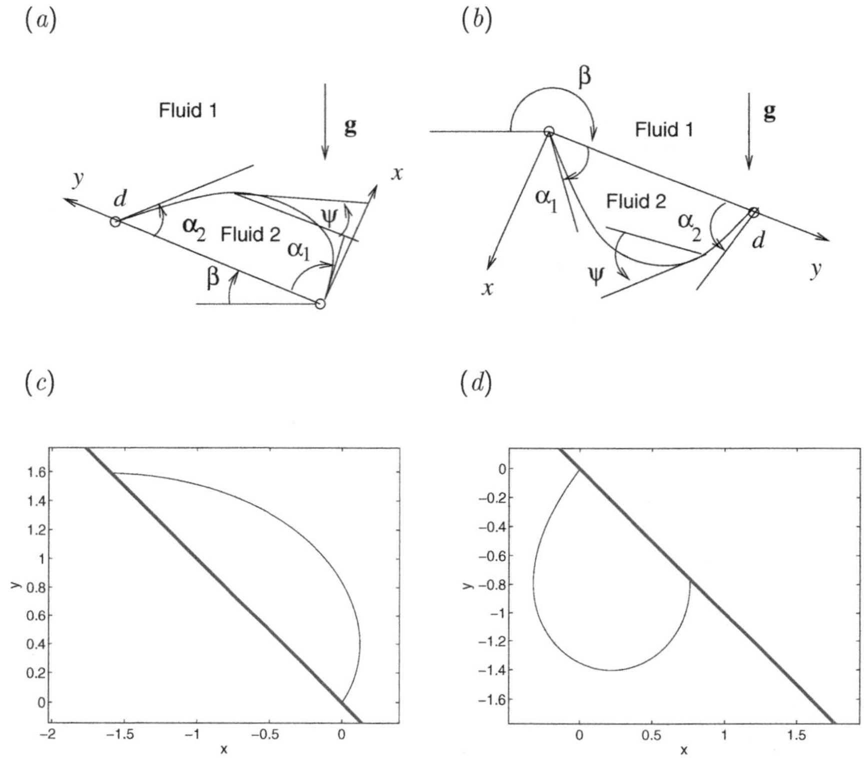

Figure 5.7.2 Illustration of a two-dimensional liquid drop, (*a*) resting above, and (*b*) hanging underneath an inclined plane. (*c, d*) Interfacial shapes produce by the FDLIB code *drop_2di*.

where β is the plane inclination angle ranging from 0 to 2π. The pressure distribution in the ambient fluid and inside the drop is given by

$$p^{(1)}(x,y) = -\rho_1\, g\,(x\,\cos\beta + y\,\sin\beta) + c_1,$$
$$p^{(2)}(x,y) = -\rho_2\, g\,(x\,\cos\beta + y\,\sin\beta) + c_2, \qquad (5.7.16)$$

where c_1 and c_2 are two constants. Substituting these expressions in the interfacial balance (4.4.11), we find

$$-\rho_2\, g\,(X\,\cos\beta + Y\,\sin\beta) + c_2 + \rho_1\, g\,(X\,\cos\beta + Y\,\sin\beta) + c_1 = \gamma\,\kappa, \qquad (5.7.17)$$

where (X, Y) are the interface coordinates. Rearranging, we obtain the Laplace-Young equation

$$\kappa = -s_1\,\frac{\Delta\rho\, g}{\gamma}(X\,\cos\beta + Y\,\sin\beta) + C, \qquad (5.7.18)$$

where $\Delta\rho \equiv \rho_2 - \rho_1$, and $C \equiv (c_2 - c_1)/\gamma$. Physically, the constant C is the curvature of the interface at the first contact point.

The interface will be described parametrically in terms of the slope angle, ψ, varying from $-\alpha_1$ at the first contact point, to α_2 at the second contact point, as shown in figure 5.7.2. Regarding X and Y as functions of ψ and working as in the case of a drop above or below a horizontal plane, we derive the differential equations

$$\frac{\mathrm{d}X}{\mathrm{d}\psi} = \frac{\sin\psi}{Q}, \qquad \frac{\mathrm{d}Y}{\mathrm{d}\psi} = -\frac{\cos\psi}{Q}, \tag{5.7.19}$$

where

$$Q \equiv s\,\frac{X\cos\beta + Y\sin\beta}{\ell^2} - C, \tag{5.7.20}$$

the coefficient s is equal to 1 when $\rho_2 > \rho_1$, or -1 when $\rho_2 < \rho_1$. The first contact line boundary condition sets the origin of the inclined Cartesian axes,

$$X(-\alpha_1) = 0, \qquad Y(-\alpha_1) = 0. \tag{5.7.21}$$

The second contact line boundary conditions requires

$$X(\alpha_2) = 0. \tag{5.7.22}$$

The constraint on the drop area requires

$$\int_0^d X\,\mathrm{d}y = A_D, \tag{5.7.23}$$

where $y = d$ marks the position of the second contact line.

Specified contact angle

Given the drop area, A_D, we may specify the first contact angle, α_1, and compute the second contact angle, α_2, and the constant C to satisfy conditions (5.7.22) and (5.7.23). A shooting method can be implemented for this purpose according to the following steps:

1. We begin by guessing values for C and α_2.

2. Next, we integrate the system of equations (5.7.19) with initial conditions provided in (5.7.21).

3. Third, we compute the integral on the right-hand side of (5.7.23) using the trapezoidal rule.

4. Fourth, we evaluate the two components of an objective function,

$$F(1) = X(\alpha_2), \qquad F(2) \equiv \int_0^d X\,\mathrm{d}y - A_D. \tag{5.7.24}$$

5. Now we improve the values of C and α_2 to drive the two components of the objective function, $F(1)$ and $F(2)$ to zero, and return to step 2.

The following MATLAB function *drop_2di_ode* residing in directory *drop_2d* inside directory *03_hydrostat* of FDLIB implements the method to solve the system of equations (5.7.19) and compute the objective function, **F**:

```
function F = drop_2di_ode(solution)

%-----------------------------------------------
% Two-dimensional drop on an inclined plane.
% Integrate ODEs by RK4 with uniform
% step size for the slope angle psi
% dpsi:  increments in psi
% C: shooting parameter
%-----------------------------------------------

global area capls Isp npts beta alpha1 alpha2 x y

%--------
% dispense variables
%--------

C = solution(1); alpha2 = solution(2);

%--------
% prepare
%--------

csb = cos(beta); snb = sin(beta);
dpsi = (alpha1+alpha2)/npts;
dpsih = 0.5*dpsi;

%----------------
% first contact point
%----------------

psi = -alpha1; x(1) = 0.0; y(1) = 0.0;

%----------
% integrate
%----------

for i=1:npts

  Q = Isp*(csb*x(i)+snb*y(i))/capls-C;
  xp = sin(psi)/Q; yp =-cos(psi)/Q;
  xp1 = xp; yp1 = yp;
```

```
   psi = psi +dpsih;
   x(i+1) = x(i)+xp*dpsih;
   y(i+1) = y(i)+yp*dpsih;

   Q = Isp*(csb*x(i+1)+snb*y(i+1))/capls-C;
   xp = sin(psi)/Q; yp =-cos(psi)/Q;
   xp2 = xp; yp2 = yp;

   x(i+1) = x(i)+xp*dpsih;
   y(i+1) = y(i)+yp*dpsih;

   Q = Isp*(csb*x(i+1)+snb*y(i+1))/capls-C;
   xp = sin(psi)/Q; yp =-cos(psi)/Q;
   xp3 = xp; yp3 = yp;

   psi = psi +dpsih;
   x(i+1) = x(i)+xp*dpsi;
   y(i+1) = y(i)+yp*dpsi;

   Q = Isp*(csb*x(i+1)+snb*y(i+1))/capls-C;
   xp = sin(psi)/Q; yp =-cos(psi)/Q;
   xp4 = xp; yp4 = yp;

   x(i+1) = x(i) + (xp1+2*xp2+2*xp3+xp4)*dpsi/6.0;
   y(i+1) = y(i) + (yp1+2*yp2+2*yp3+yp4)*dpsi/6.0;

end

G(1) = x(npts+1);

%----------------------------------------
% compute the area of the integrated shape
% by the trapezoidal rule
%----------------------------------------

ar = 0.0;
for i=1:npts
  ar= ar+(x(i+1)+x(i))*(y(i+1)-y(i));
end
ar = 0.5*ar;   % to account for trapezoidal weights

F(2) = ar-area;

%-----
% done
%-----

return
```

The improvement in the values of C and α_2 in step 5 of the algorithm can be done using Newton's method for solving a system of two nonlinear equations.[1] To simplify the implementation, we will use instead the internal MATLAB function *fsolve*.

The procedure is implemented in the following MATLAB code *drop_2di* residing in directory *drop_2d* inside directory *03_hydrostat* of FDLIB:

```
%-------------------------------------------------------
% Hydrostatic shape of a two-dimensional
% sessile drop resting on an inclined plane
% or a pendant drop hanging underneath a inclined plane,
% for a specified area and first contact angle
%-------------------------------------------------------

global area capls Isp npts beta alpha1 alpha2 x y

gac = 1.0;    % acceleration of gravity
rhod = 1.0;    % density of the drop
rhoa = 0.0;    % density of the ambient fluid
area = 1.0;    % drop volume
beta = 0.125*pi;    % inclination angle
alpha1 = 0.35*pi;    % first contact angle
npts = 64;    % number of interfacial markers

%---
% prepare
%---

drho = rhod-rhoa;    % density difference
Isp = 1.0;      % Isp is an orientation index
if(drho<0) Isp = -Isp; end
csb = cos(beta); snb = sin(beta);
ROT = [csb, snb; -snb, csb];    % rotation matrix for graphics

%---
% initial guess for a circular interface
%---

a = sqrt(area/(alpha1-0.5*sin(2.0*alpha1)));
Bcrc = 1.0/a;
solution = [Bcrc alpha1];

%---
% loop over surface tension and animate
%---
```

[1] POZRIKIDIS, C., 2008, *Numerical Computation in Science and Engineering*, Second Edition, Oxford University Press.

```
for repeat=1:100

  gamma = 10.0-repeat*0.10;
  capls = gamma/(gac*abs(drho));    % square of the cap length
  solution = fsolve(@drop_2di_ode, solution)
  alpha2=solution(2)
  xplot = -y; yplot = x;

  for i=1:npts+1
    xx = [xplot(i), yplot(i)];
    xx = ROT*xx';
    xplot(i) = xx(1);
    yplot(i) = xx(2);
  end

  plot(xplot,yplot,'-')

  hold on
  plot(csb*[0.2,-2.5],snb*[-0.2,2.5],'r','LineWidth',3)
  axis equal
  hold off
  pause(0.1)

  if(alpha2<0.05*pi) break; end
  if(alpha2>0.95*pi) break; end

end
```

The shape of a sessile and a pendant drop computed using this code are shown in figure 5.7.2(*c, d*).

Specified contact points

Consider a physical experiment where a droplet is placed on a horizontal plane and assumes a symmetric equilibrium shape corresponding to a specified static contact angle at both contact points, α. The distance between the two contact points is denoted by d. The plane is then rotated and the drop deforms while both contact points remain pinned at the plane. Instead, the first and second contact angles, α_1 and α_2, deviate from the reference value α in response to the changing orientation of the acceleration of gravity vector with respect to the direction of the plane. Physically, the contact points will remain stationary only if the two contact angles are confined inside a contact angle hysteresis window bounded by the advancing and receding contact line.

For each plane inclination angle, the three unknown, C, α_1, and α_2, must be found as part of the solution to satisfy (a) the requirement on the drop area expressed by (5.7.23), and (*b*) the two contact point conditions,

$$X(\alpha_2) = 0, \qquad Y(\alpha_2) = d. \tag{5.7.25}$$

A shooting method can be implemented according to the following steps:

1. First, we guess values for C, α_1, and α_2.
2. Second, we integrate the system of equations (5.7.19) with initial conditions provided in (5.7.21).
3. Third, we compute the integral on the right-hand side of (5.7.23) using the trapezoidal rule.
4. Fourth, we evaluate the three components of the objective function

$$F(1) = X(\alpha_2), \qquad F(2) = Y(\alpha_2) - d, \qquad F(3) \equiv \int_0^d X\,\mathrm{d}y - A_D. \tag{5.7.26}$$

5. Now we improve the values of C, α_1, and α_2, with the goal of driving the three components of the objective function, $F(1)$, $F(2)$, and $F(3)$ to zero, and return to step 2.

The following MATLAB function *drop_2di1_ode* residing in directory *drop_2d* inside directory *03_hydrostat* of FDLIB solves the system of equations (5.7.19) and computes the three components of the objective function, **F**:

```
function F = drop_2di1_ode(solution)

%---------------------------------------------
% Two-dimensional drop on an inclined plane.
%
% Integrate ODEs by RK4 with uniform
% step size for the slope angle psi
%
% SYMBOLS:
% -------
%
% dpsi:  increments in psi
% C: shooting parameter
%---------------------------------------------

global area capls Isp npts beta alpha1 alpha2 d x y

%--------
% dispense variables
%--------
```

```
C = solution(1);
alpha1 = solution(2);
alpha2 = solution(3);

%--------
% prepare
%--------

csb = cos(beta); snb = sin(beta);
dpsi = (alpha1+alpha2)/npts;
dpsih = 0.5*dpsi;

%----------------
% first contact point
%----------------

psi = -alpha1; x(1) = 0.0; y(1) = 0.0;

%----------
% integrate
%----------

for i=1:npts

  Q = Isp*(csb*x(i)+snb*y(i))/capls-C;
  xp = sin(psi)/Q; yp =-cos(psi)/Q;
  xp1 = xp; yp1 = yp;

  psi = psi +dpsih;
  x(i+1) = x(i)+xp*dpsih;
  y(i+1) = y(i)+yp*dpsih;

  Q = Isp*(csb*x(i+1)+snb*y(i+1))/capls-C;
  xp = sin(psi)/Q; yp =-cos(psi)/Q;
  xp2 = xp; yp2 = yp;

  x(i+1) = x(i)+xp*dpsih;
  y(i+1) = y(i)+yp*dpsih;

  Q = Isp*(csb*x(i+1)+snb*y(i+1))/capls-C;
  xp = sin(psi)/Q; yp =-cos(psi)/Q;
  xp3 = xp; yp3 = yp;

  psi = psi +dpsih;
  x(i+1) = x(i)+xp*dpsi;
  y(i+1) = y(i)+yp*dpsi;

  Q = Isp*(csb*x(i+1)+snb*y(i+1))/capls-C;
```

```
   xp = sin(psi)/Q; yp = -cos(psi)/Q;
   xp4 = xp; yp4 = yp;

   x(i+1) = x(i) + (xp1+2*xp2+2*xp3+xp4)*dpsi/6.0;
   y(i+1) = y(i) + (yp1+2*yp2+2*yp3+yp4)*dpsi/6.0;

end

F(1) = x(npts+1); F(2) = y(npts+1)-d;

%----------------------------------------
% compute the are of the integrated shape
% by the trapezoidal rule
%----------------------------------------

ar = 0.0;
for i=1:npts
  ar = ar+(x(i+1)+x(i))*(y(i+1)-y(i));
end
ar = 0.5*ar; % to account for trapezoidal weights

F(3) = ar-area;

%-----
% done
%-----

return
```

The shooting method is implemented in the following MATLAB code *drop 2di1* residing in directory *drop_2d* inside directory *03_hydrostat* of FDLIB:

```
%------------------------------------------------------
% Hydrostatic shape of a two-dimensional
% sessile drop resting on an inclined plane
% or a pendant drop hanging underneath a inclined plane,
% for a specified area and first contact points
%------------------------------------------------------

global area capls Isp npts beta alpha1 alpha2 d x y

Jsp = -1;   % pendant
Jsp = 1;    % sessile

gac = 1.0;    % acceleration of gravity
gamma = 2.0;    % surface tension
rhod = 1.0;    % density of the drop
rhoa = 0.0;    % density of the ambient fluid
area = 1.0;    % drop area
```

```
alpha = 0.75*pi;   % contact angle
npts = 32;   % number of interfacial markers

epsilon = 0.01;   % for the shooting method
maxiter = 16;
tol = 0.000000001;

%---
% compute the drop shape
% on a horizontal plane
%---

[rcrc,B,x,y] = drop_2d ...
   ...
   (Jsp,gac,gamma,rhod,rhoa,area ...
   ,alpha,npts,epsilon,maxiter,tol ...
   );

%---
% position of the contact line
%---

d = 2*y(npts+1);

%---
% prepare to rotate
%---

drho = rhod-rhoa;   % density difference
Isp = 1.0;   % Isp is an orientation index
if(drho<0) Isp = -1.0; end
capls = gamma/(gac*abs(drho));   % square of the capillary length
npts = 2*npts;
C = B-shiftx/capls;
alpha1 = alpha;
alpha2 = alpha;
options = optimset('TolFun',0.0000001,'MaxFunEvals',200000);

%------------
% start rotating
%------------

for repeat=1:1024

  beta = 0.002*(repeat-1.0)*pi;
  csb = cos(beta); snb = sin(beta);
  ROT = [csb, snb; -snb, csb];   % rotation matrix for graphics
```

```
  solution = [C alpha1 alpha2];
  solution = fsolve(@drop_2di1_ode, solution,options)

  xplot = -y;

  for i=1:npts+1
    xx = [xplot(i), yplot(i)];
    xx = ROT*xx';
    xplot(i) = xx(1);
    yplot(i) = xx(2);
  end

  plot(xplot,yplot,'-')
  hold on
  plot(csb*[0.5,-2.0],snb*[-0.5,2.0],'r','LineWidth',3)
  axis equal
  xlabel('x','fontsize',15)
  ylabel('y','fontsize',15)
  hold off
  pause(0.1)

end
```

Running the code animates interfacial profiles on the rotated plane.

Problem

5.7.1 *Two-dimensional drop on a horizontal plane*

(*a*) Derive formula (5.7.14).

(*b*) Derive a relation between the constant B introduced in (5.7.3) and the constant C introduced in (5.7.18) for a horizontal plane, $\beta = 0$.

Computer Problems

5.7.2 *Two-dimensional drop on a horizontal plane*

Run the program *drop_2d* to generate a family of shapes corresponding to a fixed value of the drop area and various contact angles. Produce another family of shapes corresponding to a fixed value of the contact angle and various drop areas. Discuss the computed interfacial shapes.

5.7.3 *Two-dimensional drop on an inclined plane*

Write a code that compute interfacial shapes for a fixed drop area and contact points, and unspecified contact lines.

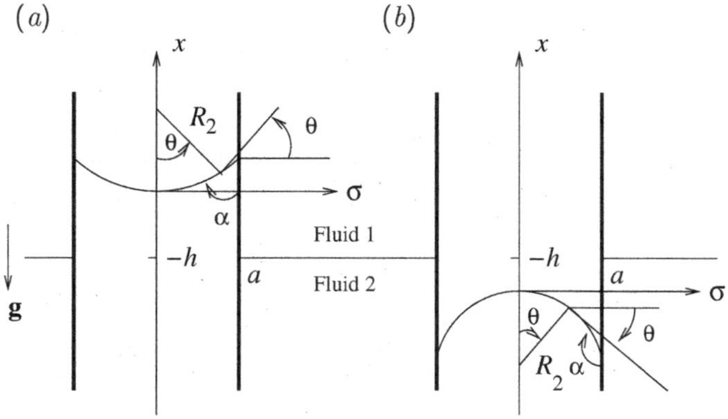

Figure 5.8.1 Illustration of an axisymmetric meniscus inside a vertical circular tube for (a) contact angle α less than $\pi/2$, and (b) α greater than $\pi/2$.

5.8 Axisymmetric meniscus inside a tube

To compute the shape of axisymmetric interfaces, we work as in the case of two-dimensional interfaces discussed previously in this chapter. Additional considerations include possible subtleties in the computation of the mean curvature, and a more pronounced numerical sensitivity to the parameters of the computation. The new features will be illustrated by discussing the axisymmetric versions of the two-dimensional problems considered in sections 5.6 and 5.7.

In this section, we consider an axisymmetric meniscus inside a vertical cylindrical tube of radius a, as illustrated in figure 5.8.1, which is the counterpart of the two-dimensional meniscus depicted in figure 5.6.1. In the cylindrical polar coordinates defined in figure 5.8.1, the interface is located at $x_I = f(\sigma)$. Outside the tube, the interface assumes a flat horizontal shape with vanishing curvature located at $x = -h$.

Laplace-Young equation

Working as in section 5.4 but using the interfacial condition (4.4.12) in place of (4.4.11), we derive the three-dimensional version of the Laplace-Young equation (5.4.6),

$$2\kappa_m = -\frac{\Delta\rho\, g}{\gamma}\, x_I + B, \tag{5.8.1}$$

where κ_m is the mean curvature. The constant B is defined in equation (5.4.7) and is given in terms of the capillary rise in equation (5.6.4).

The mean curvature is given by formula (4.3.43), and the second principal curvature is given by formulas (4.3.45). In the case of a raised meniscus depicted in figure 5.8.1(a), the angle θ is positive and the second principal radius of curvature is negative, $R_2 < 0$. In the case of a submerged meniscus depicted in figure 5.8.1(b), θ is negative and $R_2 > 0$.

Substituting the expression for the mean curvature in the left-hand side of the Laplace-Young equation (5.8.1) and rearranging, we derive the second-order differential equation

$$f'' = (1 + f'^2)[-\frac{f'}{\sigma} + (1 + f'^2)^{1/2}(\frac{f}{\ell^2} - B)\,], \tag{5.8.2}$$

where $\ell = (\gamma/(\Delta\rho\, g))^{1/2}$ is the capillary length. An equivalent first-order system is

$$\frac{\mathrm{d}f}{\mathrm{d}\sigma} = q, \qquad \frac{\mathrm{d}q}{\mathrm{d}\sigma} = (1 + q^2)\,[-\frac{q}{\sigma} + (1 + q^2)^{1/2}(\frac{f}{\ell^2} - B)\,]. \tag{5.8.3}$$

The first equation defines the slope function q, and the second equation implements the Laplace-Young equation.

Parametric representation

It is expedient to describe the shape of the interface in parametric form in terms of the slope angle θ varying from 0 at the center of the tube to $\pi/2 - \alpha$ at the inner wall of the tube, where α is the contact angle, as shown in figure 5.8.1. The axial and radial position of a point at the interface are described by the functions

$$x = X(\theta), \qquad \sigma = \Sigma(\theta). \tag{5.8.4}$$

In terms of the angle θ, the mean curvature is

$$2\,\kappa_m = \kappa_1 + \kappa_2 = \frac{1}{f'}\frac{\mathrm{d}}{\mathrm{d}\sigma}\Big(\frac{1}{(1 + f'^2)^{1/2}}\Big) - \frac{\sin\theta}{\sigma} \tag{5.8.5}$$

or

$$2\,\kappa_m = \frac{1}{\tan\theta}\frac{\mathrm{d}}{\mathrm{d}\sigma}\Big(\frac{1}{(1 + \tan\theta^2)^{1/2}}\Big) - \frac{\sin\theta}{\sigma} = \frac{1}{\tan\theta}\frac{\mathrm{d}\cos\theta}{\mathrm{d}\sigma} - \frac{\sin\theta}{\sigma} = -\cos\theta\,\frac{\mathrm{d}\theta}{\mathrm{d}\sigma} - \frac{\sin\theta}{\sigma}. \tag{5.8.6}$$

Substituting the right-hand side of (5.8.6) in the left-hand side of the Laplace-Young equation (5.8.1) and rearranging, we find

$$\frac{\mathrm{d}\Sigma}{\mathrm{d}\theta} = \frac{\cos\theta}{Q}, \tag{5.8.7}$$

where

$$Q \equiv \frac{X+h}{\ell^2} - \frac{\sin\theta}{\Sigma}. \tag{5.8.8}$$

Equation (5.8.7) governs the parametric representation of the radial position, σ, in terms of the slope angle, θ. To derive a corresponding equation for the axial position X, we combine the definition $f' = \tan\theta = \mathrm{d}X/\mathrm{d}\Sigma$ with (5.8.7), and obtain

$$\frac{\mathrm{d}X}{\mathrm{d}\theta} = \frac{\sin\theta}{Q}. \tag{5.8.9}$$

The boundary conditions require

$$\Sigma = 0 \qquad \text{and} \qquad X = 0 \qquad \text{at } \theta = 0, \tag{5.8.10}$$

and

$$\Sigma = a \qquad \text{at } \theta = \frac{\pi}{2} - \alpha. \tag{5.8.11}$$

An apparent difficulty arises when we attempt to evaluate the function Q defined in equation (5.8.8) at $\theta = 0$ corresponding to $\Sigma = 0$, as the second fraction on the right-hand side is undefined. However, using the l'Hôpital rule, we find that, as θ tends to zero, this ratio reduces to $\mathrm{d}\theta/\mathrm{d}\Sigma$. Substituting this asymptotic limit in (5.8.8) and the result in (5.8.7) and (5.8.9), we derive the regularized initial conditions

$$\left(\frac{\mathrm{d}\Sigma}{\mathrm{d}\theta}\right)_{\theta=0} = \frac{2\,\ell^2}{h}, \qquad \left(\frac{\mathrm{d}X}{\mathrm{d}\theta}\right)_{\theta=0} = 0, \tag{5.8.12}$$

which are used to start up the integration.

Solution by iteration

Since the value of the capillary rise h is *a priori* unknown, the solution must be found by iteration. The shooting method involves the following steps:

- We begin by guessing a value for h.
- Second, we compute the solution of (5.8.7) and (5.8.9) subject to the initial conditions (5.8.10).
- Next, we check whether (5.8.11) is satisfied. If not, we improve the guess using, for example, the secant method discussed in section 5.6.

A reasonable initial guess emerges by assuming that the meniscus has a spherical shape consistent with the prescribed contact angle. Using elementary trigonometry, we find

$$h \simeq 2\cos\alpha\,\frac{\ell^2}{a}. \tag{5.8.13}$$

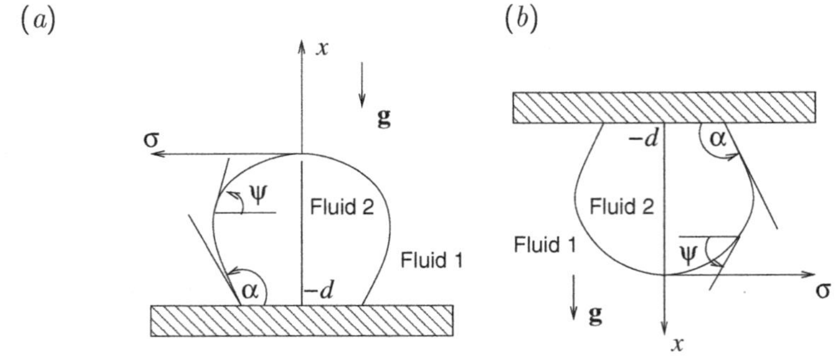

Figure 5.9.1 Illustration of (a) an axisymmetric sessile liquid drop resting on a horizontal plane, and (b) an axisymmetric pendant liquid drop hanging under a horizontal plate.

Note that, when $\alpha > \pi/2$, the predicted rise is negative.

The numerical method is implemented in code *men_ax* residing in directory *03_hydrostat* of FDLIB, incorporating a minor modification of the code *men_2d* listed in section 5.6.

Computer Problem

5.8.1 *Axisymmetric meniscus*

Run the code *men_ax* to generate a family of shapes corresponding to a fixed value of the tube radius and various contact angles. Produce another family of shapes corresponding to a fixed value of the contact angle and various tube radii. Discuss the behavior of the capillary rise in each case.

5.9 Axisymmetric drop on a horizontal plane

Next, we consider an axisymmetric drop of a fluid labeled 2 resting above or hanging underneath a horizontal plane. The drop is surrounded by an ambient fluid labeled 1, as illustrated in figure 5.9.1. Our objective is to compute the shape of the interface for a specified value of the contact angle, α, and drop volume, V_D. The analysis also applies for a gas bubble.

Working as in the case of the two-dimensional drop discussed in section 5.7, we describe the interface parametrically in terms of the slope angle ψ defined in figure 5.9.1, as

$$x = X(\psi), \qquad \sigma = \Sigma(\psi), \tag{5.9.1}$$

where

$$\cot\psi = -\frac{\mathrm{d}\Sigma}{\mathrm{d}X}. \tag{5.9.2}$$

The counterparts of equations (5.7.7) and (5.7.10) are

$$\frac{\mathrm{d}X}{\mathrm{d}\psi} = \frac{\sin\psi}{Q}, \qquad \frac{\mathrm{d}\Sigma}{\mathrm{d}\psi} = -\frac{\cos\psi}{Q}, \tag{5.9.3}$$

where

$$Q \equiv \frac{\sin\psi}{\Sigma} + s_1\, s_2\, \frac{X}{\ell^2} - B, \tag{5.9.4}$$

and $\ell^2 = \gamma/(\Delta\rho\, g)$ is the square of the capillary length. The boundary conditions require

$$X(0) = 0, \qquad \Sigma(0) = 0, \tag{5.9.5}$$

and the constraint on the drop volume V_D takes the form

$$\pi \int_{-d}^{0} \Sigma^2\, \mathrm{d}x = V_D, \tag{5.9.6}$$

where $x = -d$ describes the position of the plane. At the axis of symmetry located at $\sigma = 0$, equations (5.9.3) are replaced by the regularized forms

$$\left(\frac{\mathrm{d}\Sigma}{\mathrm{d}\psi}\right)_{\psi=0} = \frac{2}{B}, \qquad \left(\frac{\mathrm{d}X}{\mathrm{d}\psi}\right)_{\psi=0} = 0, \tag{5.9.7}$$

arising from Taylor series expansions.

The solution of equations (5.9.3) can be found using the shooting method discussed in section 5.7 for the corresponding problem in two dimensions. In the present case, the constant B is equal to twice the mean curvature of the interface at the x axis. A reasonable guess for B can be found by assuming that the interface is a section of a sphere, and then computing the radius of the sphere, a, to satisfy the constraints on the contact angle and drop volume. Using elementary trigonometry, we find

$$a = \left(\frac{3V_D/\pi}{2 + \cos^3\alpha - 3\cos\alpha}\right)^{1/3}, \tag{5.9.8}$$

and then set $B = 2/a$.

The numerical method is implemented in the code *drop_ax* residing in directory *03_hydrostat* of FDLIB, incorporating a minor modification of the code *drop_2d* listed in section 5.7.

5.9.1 Solution space

The numerical method described in this section fails when the drop develops re-entrant shapes near the base. When this occurs, the functions $X(\psi)$ and $\Sigma(\psi)$ cease to be single valued. To address this difficulty, we integrate the parametric differential equations from $\psi = 0$ up to a specified maximum value $\psi_{max} < \alpha$, and then continue the integration regarding the radial distance a function of the axial position, $\sigma = w(x)$.

To perform the integration with respect to x, we substitute the expression for the mean curvature given in (4.3.39) in the Laplace-Young equation, and find

$$-\frac{w''}{(1+w'^2)^{3/2}} + \frac{1}{\sigma}\,\frac{1}{\sqrt{1+w'^2}} = -\frac{\Delta\rho\, g}{\gamma}\, x + B, \tag{5.9.9}$$

where a prime denotes a derivative with respect to x. Rearranging, we derive the second-order differential equation,

$$w'' = (1+w'^2)\,\Big[\frac{1}{w} + \sqrt{1+w'^2}\,(s_1\, s_2 \frac{x}{\ell^2} - B)\Big], \tag{5.9.10}$$

which can be resolved into a system of two first-order equations,

$$\begin{aligned} \frac{\mathrm{d}w}{\mathrm{d}x} &= q, \\ \frac{\mathrm{d}q}{\mathrm{d}x} &= (1+q^2)\,\Big[\frac{1}{w} + \sqrt{1+q'^2}\,\Big(s_1\, s_2 \frac{x}{\ell^2} - B\Big)\Big]. \end{aligned} \tag{5.9.11}$$

Initial conditions are provided by the values computed at the end of the integration with respect to ψ.

The following MATLAB function *drop_ax1_ode* residing in directory *drop_ax* inside subdirectory *03_hydrostat* of FDLIB integrates the differential equations:

```
function [x,s,Ntot,vlm,slope] = drop_ax1_ode ...
...
(npts1 ...      % number of steps using the psi parametrization
,capls ...      % square of the capillary length
,Isp ...
,psi_max ...
,B ...
,ratio ...      % grading of psi integration intervals
,Next ...      % determines the x position of the drop base
,npts2 ...      % number of steps using the x parametrization
)

%-----------------------------------------
% Integrate ODEs using RK4
% Return the interfacial profile (x, s)
```

```
% total number of divisions (Ntot)
% volume of the drop (vlm)
% slope at the base (slope)
%---------------------------------------------

%==============
% first section
%==============

%---------
% set the step size vector dpsi
% so that it increases geometrically
% by the factor "ratio"
%---------

if(ratio==1)
  alp = 1.0;
  factor = 1.0/npts1;
else
  texp = 1/(npts1-1);
  alp = ratio^texp;
  factor = (1.0-alp)/(1.0-alp^npts1);
end

dpsi(1)=psi_max*factor;
for i=2:npts1
  dpsi(i) = dpsi(i-1)*alp;
end

%---------------
% top of the drop
%---------------

psi = 0.0; x(1) = 0.0;
s(1) = 0.0;

%----------
% integrate
%----------

for i=1:npts1

  dpsih(i) = 0.5*dpsi(i);

  if(i==1)
    xp = 0.0;
    sp = 2.0/B;
  else
    Q = sin(psi)/s(i)+Isp*x(i)/capls-B;
```

```
    xp = sin(psi)/Q;
    sp =-cos(psi)/Q;
  end

  xp1 = xp;
  sp1 = sp;

  psi = psi +dpsih(i);
  x(i+1) = x(i)+xp*dpsih(i);
  s(i+1) = s(i)+sp*dpsih(i);

  Q = sin(psi)/s(i+1)+Isp*x(i+1)/capls-B;
  xp = sin(psi)/Q;
  sp =-cos(psi)/Q;
  xp2 = xp;
  sp2 = sp;

  x(i+1) = x(i)+xp*dpsih(i);
  s(i+1) = s(i)+sp*dpsih(i);

  Q = sin(psi)/s(i+1)+Isp*x(i+1)/capls-B;
  xp = sin(psi)/Q;
  sp =-cos(psi)/Q;
  xp3 = xp;
  sp3 = sp;

  psi = psi +dpsih(i);
  x(i+1) = x(i)+xp*dpsi(i);
  s(i+1) = s(i)+sp*dpsi(i);

  Q = sin(psi)/s(i+1)+Isp*x(i+1)/capls-B;
  xp = sin(psi)/Q;
  sp =-cos(psi)/Q;
  xp4 = xp;
  sp4 = sp;

  x(i+1) = x(i) + (xp1+2*xp2+2*xp3+xp4)*dpsi(i)/6.0;
  s(i+1) = s(i) + (sp1+2*sp2+2*sp3+sp4)*dpsi(i)/6.0;

end

Ntot=npts1;

%==============
% continue the integration with a uniform step "dx"
% using the x parametrization up to x = Next*x(npts1+1)
%==============

dx = Next*x(npts1+1)/npts2;
```

```
%---
% initial slope
%---

Q = sin(psi)/s(npts1+1)+Isp*x(npts1+1)/capls-B;
xp = sin(psi)/Q;
sp =-cos(psi)/Q;
q(npts1+1) = xp/sp;   % dx/dsigma

%---
% integrate
%---

for i=npts1+1:npts1+npts2+2

  dsdx = q(i);
  tmp = 1+q(i)*q(i);
  dqdx = tmp*( 1/s(i)+sqrt(tmp)*(Isp*x(i)/capls-B) );
  dsdx1 = dsdx;
  dqdx1 = dqdx;

  x(i+1) = x(i)+0.5*dx;
  s(i+1) = s(i)+0.5*dx*dsdx;
  q(i+1) = q(i)+0.5*dx*dqdx;

  dsdx = q(i+1);
  tmp = 1+q(i+1)*q(i+1);
  dqdx = tmp*( 1/s(i+1)+sqrt(tmp)*(Isp*x(i+1)/capls-B) );
  dsdx2 = dsdx;
  dqdx2 = dqdx;

  x(i+1) = x(i)+0.5*dx;
  s(i+1) = s(i)+0.5*dx*dsdx;
  q(i+1) = q(i)+0.5*dx*dqdx;

  dsdx = q(i+1);
  tmp = 1+q(i+1)*q(i+1);
  dqdx = tmp*( 1/s(i+1)+sqrt(tmp)*(Isp*x(i+1)/capls-B) );
  dsdx3 = dsdx;
  dqdx3 = dqdx;

  x(i+1) = x(i)+dx;
  s(i+1) = s(i)+dx*dsdx;
  q(i+1) = q(i)+dx*dqdx;

  dsdx = q(i+1);
  tmp = 1+q(i+1)*q(i+1);
  dqdx = tmp*( 1/s(i+1)+sqrt(tmp)*(Isp*x(i+1)/capls-B) );
```

```
  dsdx4 = dsdx;
  dqdx4 = dqdx;

  s(i+1) = s(i)+dx*(dsdx1+2.0*dsdx2+2.0*dsdx3+dsdx4)/6.0;
  q(i+1) = q(i)+dx*(dqdx1+2.0*dqdx2+2.0*dqdx3+dqdx4)/6.0;

  Ntot = Ntot+1;
  if(s(i+1)<0) break; end
  end

slope = q(i+1);

%-------------------------------------------
% compute the volume of the integrated shape
% by the trapezoidal rule
%-------------------------------------------

vlm = 0.0;
for i=1:Ntot
  vlm = vlm+(s(i+1)*s(i+1)+s(i)*s(i))*abs(x(i+1)-x(i));
end
vlm = 0.5*vlm;   % to account for trapezoidal weights
vlm = pi*vlm;

%-----
% done
%-----

return
```

Families of shapes parametrized by the constant B can be generated by specifying the capillary length, and then computing the parameter $Next$ to ensure a specified drop volume. The parameter B implicitly determines the contact angle, and the parameter $Next$ explicitly determines the location of the drop base. Finding the proper value of $Next$ can be done using the secant method implemented in the following MATLAB code:

```
%----
% Solution branches of a sessile or pendant drop
%----

a = 1.0;   % drop radius
Isp = -1;   % pendant drop
psi_max = 0.25*pi;
npts1 = 24; npts2 = 48;
ratio = 0.9;
maxiter = 10;   % for secant iterations
tol = 0.0000001;   % for secant iterations
```

```
%---
% prepare
%---

volume = 1.0*4*pi*a*a*a/3;

%---
% plot the wall
%---

xwall(1) = -1.8; ywall(1) = 0;
xwall(2) = 1.8; ywall(2) = 0;

%---
% family parameters
%---

capls = 2.0;
Nloop = 2*2*2*2*2*2*128;
Bmin = 1.00; Bmax = 4.0;
DB = 0.0020;
Nplot = 32;   % will plot after Nplot shapes

%---
% prepare to loop
%---

B = Bmin;
Next(1) = 2.0;
Iplot = Nplot;

%============
for Iloop=1:Nloop   % loop over B
%============

  B = B+DB;
  Ic = 1;

  [ x,s,Ntot,vlm,slope ] = drop_ax1_ode ...
  ...
  (npts1 ...
  ,capls ...
  ,Isp ...
  ,psi_max ...
  ,B ...
  ,ratio ...
  ,Next(1) ...
  ,npts2 ...
  );
```

```
  error(Ic)=vlm-volume;
  Ic = 2;
  Next(Ic) = Next(Ic-1)+0.10;

  [x,s,Ntot,vlm,slope] = drop_ax1_ode ...
  ...
  (npts1 ...
  ,capls ...
  ,Isp ...
  ,psi_max ...
  ,B ...
  ,ratio ...
  ,Next(2) ...
  ,npts2 ...
  );

  error(Ic)=vlm-volume;

%----------------------------------------
% iterate on Next using the secant method
%----------------------------------------

  for iter=1:maxiter

    Ic = Ic+1;

%---
% secant updating
%---

    Icb = Ic-2;
    Ica = Ic-1;
    dedc = (error(Ica)-error(Icb))/(Next(Ica)-Next(Icb));
    Next(Ic) = Next(Ica)-error(Ica)/dedc;

    [x,s,Ntot,vlm,slope] = drop_ax1_ode ...
    ...
    (npts1 ...
    ,capls ...
    ,Isp ...
    ,psi_max ...
    ,B ...
    ,ratio ...
    ,Next(Ic) ...
    ,npts2 ...
    );

    error(Ic) = vlm-volume;
```

```
    err = abs(error(Ic));

    if(err<tol) break; end

%----
  end
%----

  Next(1) = Next(Ic);
  xplot(Iloop) = a*B;
  yplot(Iloop) = atan(slope)/pi+0.5;

%---
% plotting and such
%---

  hold on
  xlabel('y/a','fontsize',15)
  ylabel('x/a','fontsize',15)
  set(gca,'fontsize',15)
  axis equal
  box

  if(Isp==1)
    patch([xwall xwall(2) xwall(1)], ...
    [ywall ywall(2)-0.2 ywall(1)-0.2],'g')
  else
    patch([xwall xwall(2) xwall(1)], ...
    [ywall ywall(2)+0.2 ywall(1)+0.2],'g')
  end

  plot(xwall,ywall,'k')

  if(Iplot==Nplot)
    x=x-x(Ntot+1);    % shift the profile
    plot( s,Isp*x,'r-')
    plot(-s,Isp*x,'r-')
    Iplot = 0;
  end

  Iplot = Iplot+1;

  if(B>Bmax) break; end

%============
end
%============
```

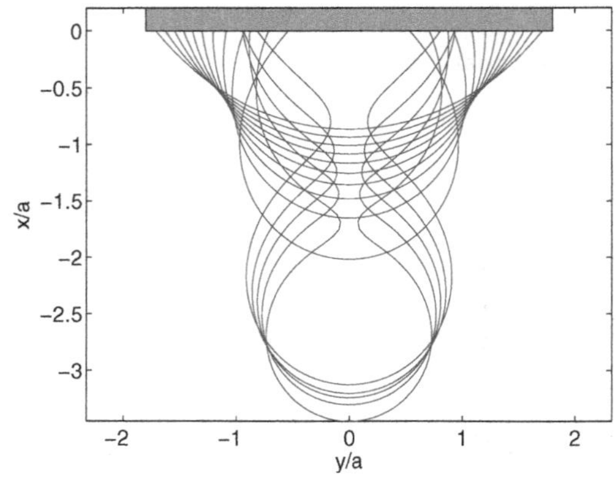

Figure 5.9.2 A family of axisymmetric drop shapes corresponding to a fixed capillary length and varying contact angle.

A family of shapes produced by the code is shown in figure 5.9.2. We observe interesting compact, light-bulb, and hourglass interfacial contours.

Problem

5.9.1 *Drop on a plane*

(*a*) Derive the regularized expressions (5.9.7) departing from equations (5.9.3) and (5.9.4). (*b*) Derive formula (5.9.8).

Computer Problem

5.9.2 *Axisymmetric drop*

Run the code *drop_ax* to generate a family of shapes corresponding to a fixed value drop volume and various contact angles. Produce another family of shapes corresponding to a fixed contact angle and various drop volumes.

5.10 A sphere straddling an interface

In section 5.2.1, we discussed the equilibrium position of a spherical particle floating on a flat interface. The flat interfacial shape is established under special conditions or when the capillary length is much smaller than the particle size. A more sophisticated analysis is required under general circumstances.

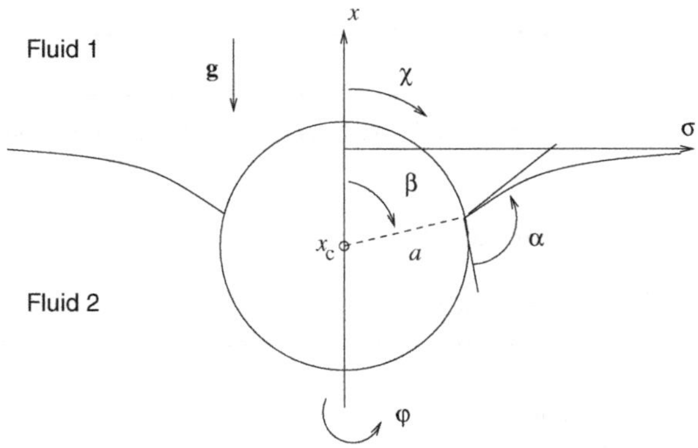

Figure 5.10.1 Illustration of a spherical particle straddling a curved interface between two fluids.

Figure 5.10.1 illustrates a spherical particle of radius a straddling a curved axisymmetric interface. The origin of the x is set at the position of the flat interface far from the particle. The center of the particle is located at $x = x_c$, and the circular contact line is located at the axial position $x = x_{cl} = x_c + a\cos\beta$, and radial position $\sigma = \sigma_{cl} = a\sin\beta$, where the angle β varies in the range $[0, \pi]$. In the case of a flat interface, $x_{cl} = 0$ and $x_c = -a\cos\beta$.

The shape of the meniscus is described by the function $x = f(\sigma)$ satisfying the first-order differential equations (5.8.3). Requiring that the mean curvature vanishes as σ tends to infinity, we find that the constant B is zero, yielding

$$\frac{\mathrm{d}f}{\mathrm{d}\sigma} = q,$$
$$\frac{\mathrm{d}q}{\mathrm{d}\sigma} = (1+q^2)\Big[-\frac{q}{\sigma} + (1+q^2)^{1/2}\frac{f}{\ell^2}\Big], \tag{5.10.1}$$

where $\ell^2 = \gamma/(\Delta\rho g)$ is the square of the capillary length, and $\Delta\rho = \rho_2 - \rho_1$ is the density difference between the lower and upper fluids. The Bond number is defined in the terms of the capillary length as $Bo = \Delta\rho g a^2/\gamma = (a/\ell)^2$.

The contact line condition requires

$$f(\sigma_{cl}) = x_{cl}, \qquad q(\sigma_{cl}) = \tan(\alpha - \beta), \tag{5.10.2}$$

and the far-field condition requires

$$f(\infty) = 0. \tag{5.10.3}$$

The following MATLAB function *flsphere_ode* residing in directory *flsphere* inside directory *03_hydrostat* of FDLIB integrates the differential equations using the fourth-order Runge-Kutta method:

```
function [f,s,q]=flsphere_ode(ndiv,smax,capls,xcl,scl,slope);

%=======================================================
% solve ODEs for a semi-infinite axisymmetric meniscus
% using RK4
%
% scl:  sigma of contact line
%
% will integrate from scl to smax
%=======================================================

dsg = (smax-scl)/ndiv;
dsgh = 0.5*dsg;

s(1) = scl;   % starting point
f(1) = xcl;
q(1) = slope;

%---
% integrate
%---

for i=1:ndiv

  fp = q(i);
  tmp = 1+q(i)*q(i);
  tmg = sqrt(tmp)/capls;
  qp = tmp*(-fp/s(i) + f(i)*tmg );
  fp1 = fp;
  qp1 = qp;

  s(i+1) = s(i)+ dsgh;
  f(i+1) = f(i)+fp*dsgh;
  q(i+1) = q(i)+qp*dsgh;
  fp2 = fp;
  qp2 = qp;

  fp = q(i+1);
  tmp = 1+q(i+1)*q(i+1);
  tmg = sqrt(tmp)/capls;
  qp = tmp*(-fp/s(i+1) + f(i+1)*tmg);
  fp3 = fp;
  qp3 = qp;

  f(i+1) = f(i)+fp*dsgh;
  q(i+1) = q(i)+qp*dsgh;

  fp = q(i+1);
  tmp = 1+q(i+1)*q(i+1);
```

```
  tmg = sqrt(tmp)/capls;
  qp = tmp*(-fp/s(i+1) + f(i+1)*tmg);

  s(i+1) = s(i)+ dsg;
  f(i+1) = f(i)+fp*dsg;
  q(i+1) = q(i)+qp*dsg;

  fp = q(i+1);
  tmp = 1+q(i+1)*q(i+1);
  tmg = sqrt(tmp)/capls;
  qp = tmp*(-fp/s(i+1) + f(i+1)*tmg);
  fp4 = fp;
  qp4 = qp;

  f(i+1) = f(i) + (fp1+2*fp2+2*fp3+fp4)*dsg/6.0;
  q(i+1) = q(i) + (qp1+2*qp2+2*qp3+qp4)*dsg/6.0;

end

%---
% done
%---

return
```

Force on the sphere

The pressure in the upper and lower fluids is given by

$$p^{(1)}(x) = -\rho_1\, g\, x + P_0, \qquad\qquad p^{(2)}(x) = -\rho_2\, g\, x + P_0, \tag{5.10.4}$$

where P_0 is an unspecified constant. By symmetry, the y and z components of the hydrostatic force exerted on the sphere are zero. The x component of the force is given by

$$F_x^h = -\iint p\, n_x\, \mathrm{d}S, \tag{5.10.5}$$

where $n_x = \cos\chi$ is the x component of the unit vector normal to the sphere, and the meridional angle χ is defined in figure 5.10.1. Writing $\mathrm{d}S = (\sigma \mathrm{d}\varphi)(a \mathrm{d}\chi)$, $\sigma = a \sin\chi$, and thus $\mathrm{d}S = a^2 \sin\chi\, \mathrm{d}\chi \mathrm{d}\varphi$, and integrating with respect to the meridional angle, φ, we find

$$F_x^h = 2\pi a^2 \,\Big(\int_0^\beta p^{(1)}(\chi) \cos\chi\, \mathrm{d}\cos\chi + \int_\beta^\pi p^{(2)}(\chi) \cos\chi\, \mathrm{d}\cos\chi\Big). \tag{5.10.6}$$

Next, we substitute the expressions for the pressure, set $x = x_c + a\cos\chi$, and simplify to find

$$F_x^h = 2\pi a^2 \Big(\int_0^\beta [-\rho_1\, g\,(x_c + a\cos\chi)] \cos\chi \, \mathrm{d}\cos\chi + \int_\beta^\pi [-\rho_2\, g\,(x_c + a\cos\chi)] \cos\chi \, \mathrm{d}\cos\chi \Big). \qquad (5.10.7)$$

As expected on physical grounds, the constant P_0 does not make a net contribution to the force. Carrying out the integration, we find

$$F_x^h = 2\pi g a^2 \Big(-\rho_1\, x_c \frac{\cos^2\beta - 1}{2} - \rho_1\, a \frac{\cos^3\beta - 1}{3} - \rho_2\, x_c \frac{1 - \cos^2\beta}{2} + \rho_2\, a \frac{1 + \cos^3\beta}{3} \Big). \qquad (5.10.8)$$

Rearranging, we find

$$F_x^h = \pi g a^2 \Big(x_c\, \Delta\rho\, (\cos^2\beta - 1) + \frac{2}{3}\, a\, \Delta\rho\, \cos^3\beta + \frac{4}{3}\, a\, \bar{\rho} \Big), \qquad (5.10.9)$$

where $\bar{\rho} = (\rho_1 + \rho_2)/2$ is the mean fluid density. When $\rho_1 = \rho_2 = \rho$, only the last term survives, yielding the buoyancy force on a submerged sphere.

Surface tension pulls the sphere tangentially to the interface at the contact line. Integrating the tension along the circular contact line, we derive the resultant x component of the capillary force,

$$F_x^{cap} = \gamma\,(2\pi\sigma_{cl})\,\sin(\alpha - \beta) = 2\pi a \gamma\, \sin\beta\, \sin(\alpha - \beta). \qquad (5.10.10)$$

Force equilibrium requires

$$F_x^h + F_x^{cap} = W, \qquad (5.10.11)$$

where W is the weight of the sphere. Setting $W = 4\pi a^3 \rho_P g/3$ and simplifying, we derive the force equilibrium equation

$$\frac{x_c}{a}\, \sin^2\beta = \frac{2}{3}\, \cos^3\beta + \frac{2}{3}\, \kappa + 2\, \frac{\ell^2}{a^2}\, \sin\beta\, \sin(\alpha - \beta), \qquad (5.10.12)$$

where

$$\kappa = 2\, \frac{\bar{\rho} - \rho_P}{\Delta\rho}, \qquad (5.10.13)$$

and ρ_p is the particle density. The limit $\kappa = -1$ or 1 corresponds to $\rho_p = \rho_2$ or ρ_1, where the sphere is neutrally buoyant in the lower or upper fluid.

In the case of a flat interface, $x_c = -a \cos\beta$, the expression for the hydrodynamic force simplifies to

$$F_x^{h,flat} = \pi g a^3 \Delta\rho \cos\beta \,(1 - \frac{1}{3} \cos^2\beta) + \frac{2}{3}\pi g a^3 \,(\rho_2 + \rho_1). \qquad (5.10.14)$$

The contact angle is equal to the contact line aperture, $\alpha = \beta$, the capillary force vanishes, and the trigonometric equation (5.10.12) simplifies to a cubic equation for $\cos\beta$,

$$\cos^3\beta - 3\cos\beta - 2\kappa = 0. \qquad (5.10.15)$$

A solution for $\cos\beta$ in the admissible range $[-1, 1]$ exists only when $|\kappa| < 1$.

Algorithm

When $|\kappa| < 1$, equation (5.10.12) and the accompanying Laplace-Young equation admit a solution for a limited range of sufficiently large capillary lengths. A numerical method for computing hydrostatic shapes can be implemented according to the following steps:

1. We begin by considering a flat interface, solve the cubic equation (5.10.15) for $\cos\beta$, and recover β.
2. Second, we choose a value for the contact angle, $\alpha \simeq \beta$.
3. Next, we solve (5.10.12) for x_c to satisfy the particle force balance.
4. We now integrate the system of differential equations (5.10.1) from $\sigma = \sigma_{cl}$ up to a specified distance, $\sigma = \sigma_{max}$, and check whether $\sigma_{max} = 0$. If not, we adjust β to make it so. The adjustment can be done using the secant or Newton's method.
5. We change the contact angle, α, and return to step 3 to obtain a new configuration.

This procedure ensures that a good initial guess is available for the shooting method, obtained by parameter continuation. The numerical method is implemented in the following MATLAB function *flsphere* residing in directory *03_hydrostat* of FDLIB:

```
function [s,x,q,beta,alpha,xc,xcl,scl ...
              ,al_scan,xc_scan] ...
     ...
     = flsphere(a,capl,rho1,rho2,kappa,alphain,smax,ndiv)

%-----------
% parameters
%-----------
```

```
eps = 0.0000001;    % for Newton's method
tol = 0.0000001;    % for Newton's method

%---
% will scan contact angle space with step dal
% that depends on the capillary length
%---

dal = -0.005;

if(capl<1.5) dal = -0.002; end
if(capl<0.9) dal = -0.001; end
if(capl<0.8) dal = -0.0005; end

Nmax = 2*abs(floor(pi/dal))

%----
% prepare
%---

Drho = rho2-rho1;
Brho = 0.5*(rho1+rho2);
rhop = Brho-0.5*kappa*Drho;

%----
% flat interface
%---

C(1) = 1; C(2) = 0; C(3) = -3; C(4) = -4*(Brho-rhop)/Drho;
cosbeta = roots(C);
if(abs(cosbeta(1))<1)
  beta = acos(cosbeta(1));
elseif(abs(cosbeta(2))<1)
  beta = acos(cosbeta(2));
elseif(abs(cosbeta(3))<1)
  beta = acos(cosbeta(3));
end
alpha = beta;
xc = -cos(beta);
Ido = 1; Iflag = 0; Icount = 0;

%---
while(Ido==1) % over contact angles
%---

Icount = Icount+1;

for iter=1:50
```

```
%---
% solve for beta using Newton's method
%---

  cs = cos(beta);
  sn = sin(beta);
  amb = alpha-beta;
  xc = (2/3*cs*cs*cs +4*(Brho-rhop)/(3*Drho) ...
      +2*capls/a^2*sn*sin(amb))/(1-cs*cs);
  xcl = xc+a*cs;
  scl = a*sn;
  slope = tan(amb);
  [x,s,q] = flsphere_ode(ndiv,smax,capls,xcl,scl,slope);
  error = x(ndiv+1);
  f(abs(error)<tol) break; end
  beta = beta+eps;
  cs = cos(beta);

  sn = sin(beta);
  amb = alpha-beta;
  xc = (2/3*cs*cs*cs +4*(Brho-rhop)/(3*Drho) ...
    +2*capls/a^2*sn*sin(amb))/(sn*sn);
  xcl = xc+a*cs;
  scl = a*sn;

  slope = tan(amb);
  [x1,s1,q1] = flsphere_ode(ndiv,smax,capls,xcl,scl,slope);
  error1 = x1(ndiv+1);
  beta = beta-eps;   % reset
  der = (error1-error)/eps;
  beta = beta-error/der;

end

if(abs(error)>tol)
  disp "iterations did not converge"
  return
end

%---
% plot the interface profile
%---

figure(2)
plot(s,x,'r');
hold on
plot(-s,x,'r');
axis equal
xlabel('y/a','fontsize',15)
```

```
ylabel('x/a','fontsize',15)

%---
% plot the particle contour
%---

ncrc=64;
for i=1:ncrc+1
  tht = (i-1)*2*pi/ncrc;
  xcrc(i) = xc+a*cos(tht);
  scrc(i) = a*sin(tht);
end

plot(scrc,xcrc);
hold off
pause(0.001)

%---
% end of plotting
%---

if(Iflag==1) break; end

al_scan(Icount) = alpha/pi;
xc_scan(Icount) = xc/a;
alpha = alpha+dal;

%-------
% change alpha scanning direction or lock
%-------

if(alpha<0.01*pi)
  dal = abs(dal);
  alpha = alpha+dal;
elseif(alpha>0.99*pi)
  dal = -abs(dal);
  alpha = alpha+dal;
elseif(abs(alpha-alphain)<1.0*abs(dal))   % lock on alphain
  alpha = alphain+0.00;
  Iflag = 1;
end

if(Icount>Nmax) break; end

%---
end
%---

return
```

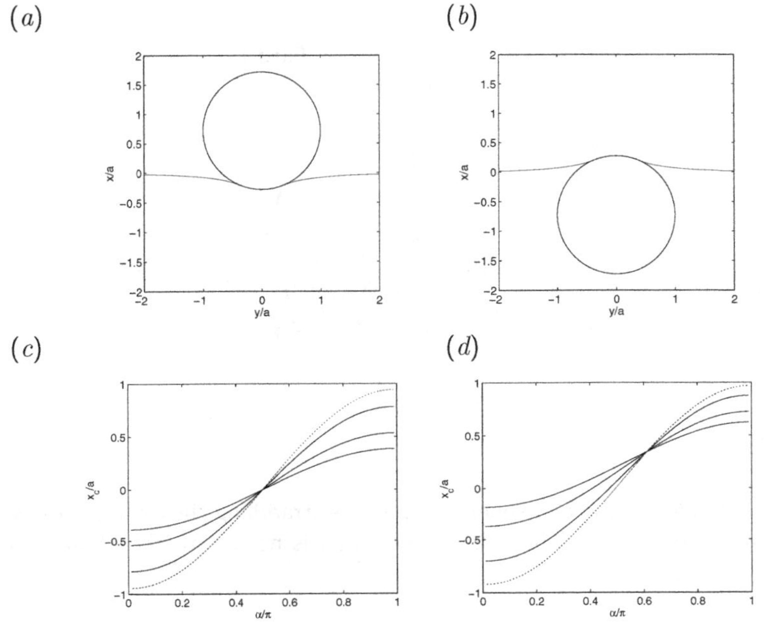

Figure 5.10.2 Axisymmetric interfacial profiles attached to a floating sphere for $\ell/a = 1$ and (a) $\kappa = 0.5$ and $\alpha/\pi = 0.98$, or (b) $\kappa = -0.5$ and $\alpha/\pi = 0.0242$. Particle center position for (c) $\kappa = 0$ and (d) $\kappa = 0.5$; the solid, dashed, dotted, and dot-dashed lines correspond to $\ell/a = 0.7$, 1.0, 2.0, and 5.0 (dashed line).

Nondimensionalizing lengths by the sphere radius, a, we find that the meniscus profile depends on the ratio ℓ/a, the density difference ratio, κ, and the contact angle, α. Two interfacial shapes computed with truncation level $\sigma_{max} = 6a$ are shown in figure 5.10.2(a, b). In (a), a heavy hydrophobic particle is held afloat, and in (b), a light hydrophilic particle is held captive.

Figure 5.10.2(c, d) shows graphs of the particle center position plotted against the contact angle for $\kappa = 0$ and 0.5. When $\kappa = 0$ and the contact angle is $\alpha = \pi/2$, the interface is flat and the particle is divided equally between the upper and lower fluid, independent of the capillary length. When $\kappa = 0.5$ the interface is flat at a certain contact angle, independent of the capillary length. The particle center position for $\kappa = -0.5$ is the mirror image of that for $\kappa = 0.5$, subject to a reflection of the contact angle. As ℓ/a becomes larger, the particle center becomes independent of κ, given by $x_c = -a \cos\alpha$.

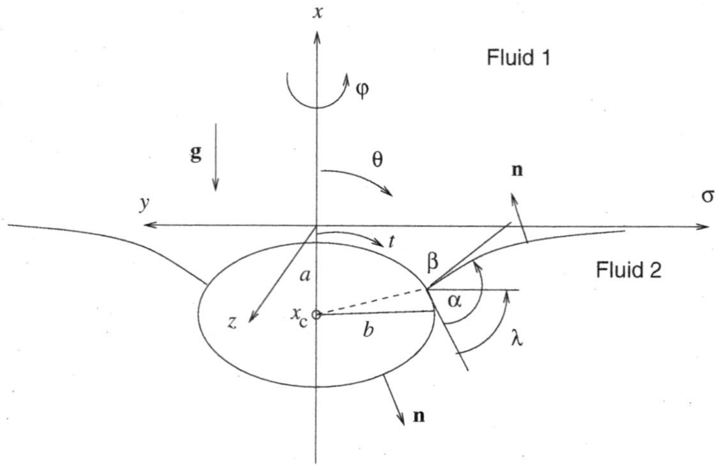

Figure 5.10.3 Illustration of a spheroidal particle straddling the interface between two immiscible fluids. The axis of revolution is normal to the planar undisturbed interface.

5.10.1 Spheroidal particle

As a generalization, we consider the equilibrium position of a spheroidal floating particle whose axis revolution is parallel to the acceleration of gravity, as depicted in figure 5.10.3. The particle center is located at $x = x_c$, and the elliptical particle contour in a meridional plane is described by the equations

$$x = x_c + a\,\cos t, \qquad \sigma = b\,\sin t, \tag{5.10.16}$$

where a, b are the particle semi-axes, and the parameter t varies from 0 at the top of the particle to π at the bottom of the particle. The origin of the x axis is set at the position of the flat interface far from the particle. The circular contact line is located at the axial position $x = x_{cl} = x_c + a\cos\beta$ and radial position $\sigma = \sigma_{cl} = b\sin\beta$, where the parameter β varies in the range $[0, \pi]$. In the case of a flat interface, $x_{cl} = 0$ and $x_c = -a\cos\beta$. The shape of the meniscus is governed by the equations previously described for a spherical particle.

The vertical component of the force exerted on the particle is given by (5.10.5). Introducing the arc length around the particle contour, l, measured in the direction of t, setting $\mathrm{d}S = 2\pi\sigma\mathrm{d}l$ and $n_x = \mathrm{d}\sigma/\mathrm{d}l$, performing the integration around the axis of revolution, and recalling that $t = \beta$ marks the location of the contact line, we find

$$F_x^h = -\pi b^2 \int_0^\pi p\,\sin 2t\,\mathrm{d}t = -\pi b^2 \Big(\int_0^\beta p^{(1)}\,\sin 2t\,\mathrm{d}t + \int_\beta^\pi p^{(2)}\,\sin 2t\,\mathrm{d}t \Big). \tag{5.10.17}$$

Next, we substitute the expressions for the pressure, set $x = x_c + a\cos t$, and find

$$F_x^h = -2\pi b^2 g\,[\rho_1 \int_0^\beta (x_c + a\cos t)\cos t \,\mathrm{d}\cos t + \rho_2 \int_\beta^\pi (x_c + a\cos t)\cos t\,\mathrm{d}\cos t\,]. \tag{5.10.18}$$

Carrying out the integration, we find

$$F_x^h = 2\pi g b^2 \Big(\rho_1\, x_c\, \frac{\sin^2\beta}{2} - \rho_1\, a\, \frac{\cos^3\beta - 1}{3} \\ - \rho_2\, x_c\, \frac{\sin^2\beta}{2} + \rho_2\, a\, \frac{1+\cos^3\beta}{3} \Big). \tag{5.10.19}$$

Rearranging, we find

$$F_x = \pi g b^2 \Big(- x_c\,\Delta\rho\,\sin^2\beta + \frac{2}{3}\, a\,\Delta\rho\,\cos^3\beta + \frac{4}{3}\, a\,\bar\rho \Big). \tag{5.10.20}$$

The x component of the capillary force is given by

$$F_x^{cap} = \gamma\,(2\pi\sigma_{cl})\,\sin(\alpha-\lambda) = 2\pi b\gamma\,\sin\beta\,\sin(\alpha-\lambda), \tag{5.10.21}$$

where λ is the angle subtended between the tangential vector and the σ axis at the contact line, as shown in figure 5.10.3, given by

$$\lambda = \arctan(\frac{b}{a}\,\tan\beta). \tag{5.10.22}$$

Substituting the expression $W = 4\pi b^2 a \rho_p g/3$ in the force equilibrium equation (5.10.11) and simplifying, we find

$$\frac{x_c}{a}\,\sin^2\beta = \frac{2}{3}\,\cos^3\beta + \frac{2}{3}\,\kappa + 2\,\frac{\ell^2}{ab}\,\sin\beta\,\sin(\alpha-\lambda). \tag{5.10.23}$$

In the case of a flat interface, $x_c = -a\cos\beta$, the expression for the hydrostatic force simplifies to

$$F_x^{h,flat} = \pi g b^2 a\,\Delta\rho\,\cos\beta\,(1 - \frac{1}{3}\,\cos^2\beta) + \frac{4}{3}\pi g b^2\, a\,\bar\rho, \tag{5.10.24}$$

the contact angle is equal to the contact line aperture angle, $\alpha = \lambda$, the capillary force vanishes, and the trigonometric equation (5.10.23) simplifies to (5.10.15).

The particle position and meniscus shape can be found using a modification of the method previously discussed for the spherical particle (problem 5.10.3).

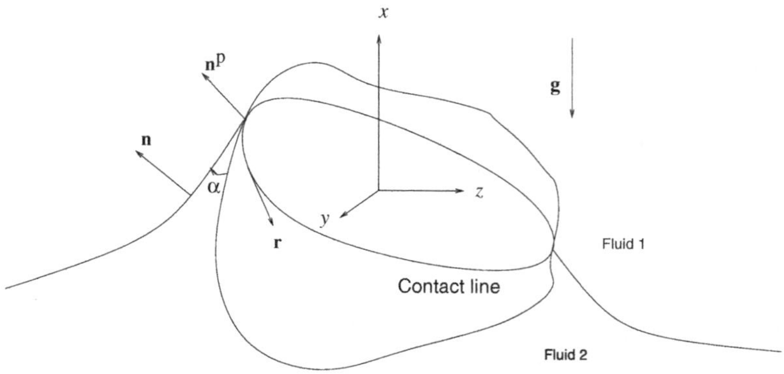

Figure 5.11.1 Illustration of a three-dimensional particle straddling the interface between two immiscible fluids.

Problems

5.10.1 *Flat interface*

Prove that a solution of (5.10.15) for $\cos\beta$ in the admissible range $[-1, 1]$ exists only when $|\kappa| < 1$.

Computer Problems

5.10.2 *Floating position*

Prepare graphs similar to those shown in figure 5.10.2(c, d) for $\kappa = 0.1$ and 0.9. Discuss the physical interpretation of the results.

5.10.3 *Spheroidal particle*

Modify the code given in the text to compute the floating position of a spheroidal particle. Prepare graphs similar to those shown in figure 5.10.2 for particle aspect ratio $a/b = 0.5$ and 0.1.

5.11 A three-dimensional meniscus

Having discussed axisymmetric configurations, we now consider a small particle with arbitrary three-dimensional shape straddling the interface between two quiescent immiscible fluids, as shown in figure 5.11.1. The x axis points against the acceleration of gravity and passes through the designated particle center. The upper fluid is designated with the index 1, and the lower fluid is designated with the index 2. The fluids are stably stratified, $\rho_2 > \rho_1$ or $\Delta\rho \equiv \rho_2 - \rho_1 > 0$.

The interface meets the particle around a contact line and becomes horizontal far from the contact line.

If $\mathbf{n}^p$ is the unit vector normal to the particle, and $\mathbf{n}$ is the unit vector normal to the interface, the contact angle α is given by

$$\alpha = \arccos(\mathbf{n}^p \cdot \mathbf{n}), \tag{5.11.1}$$

where $0 < \alpha < \pi$, and the left-hand side is evaluated at the contact line.

Laplace-Young equation

The shape of the interface is described by the function $x = \zeta(y, z)$. Far from the particle, as y and z tend to infinity, the function ζ decays to zero, yielding a flat interface. The Laplace-Young equation requires

$$2\kappa_m = -\frac{\zeta}{\ell^2}, \tag{5.11.2}$$

where $\ell^2 = \gamma/(\Delta\rho\, g)$ is the square of the capillary length, γ is the surface tension, g is the magnitude of the acceleration of gravity, and κ_m is the mean curvature reckoned to be positive when the interface is downward concave. Substituting in the Laplace-Young equation the expression for the mean curvature given in (4.3.31), we derive a nonlinear partial differential equation,

$$\nabla^2\zeta + \zeta_z^2\,\zeta_{yy} - 2\,\zeta_y\,\zeta_z\,\zeta_{yz} + \zeta_y^2\,\zeta_{zz} - (1 + |\nabla\zeta|^2)^{3/2}\,\frac{\zeta}{\ell^2} = 0, \tag{5.11.3}$$

where ∇ is the gradient and ∇^2 is the Laplacian operator in the yz plane; a subscript denotes a derivative with respect to the corresponding variable.

5.11.1 Elliptic coordinates

We will assume that the projection of the contact line in the yz plane is an ellipse, as shown in figure 5.11.2(a). To solve equation (5.11.3) in the exterior of the ellipse, we introduce elliptic coordinates, (u, φ), defined by the conformal mapping function $y + \mathrm{i}\,z = A\,\sinh(u + \mathrm{i}\,\varphi)$, where i is the imaginary unit, and A is a real constant. Decomposing the mapping function into its real and imaginary parts, we find

$$y = A\,\sinh u \cos\varphi, \qquad z = A\,\cosh u \sin\varphi. \tag{5.11.4}$$

The variable u ranges from a value u_0 around the ellipse, to infinity far from the ellipse. The variable φ varies in the range $[0, 2\pi]$. As u tends to infinity, the contour lines of constant u tend to become circles, as shown in figure 5.11.2(a). The metric of the transformation is

$$h = A\,|\cosh(u + \mathrm{i}\,\varphi)| = A\,\sqrt{\cosh^2 u - \sin^2\varphi}. \tag{5.11.5}$$

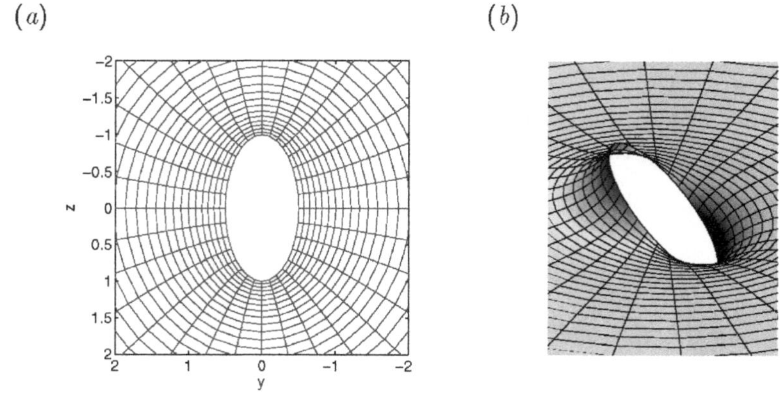

Figure 5.11.2 (*a*) Illustration of grid lines based on orthogonal elliptic coordinates. (*b*) Hydrostatic shape of a meniscus computed in these coordinates.

The magnitude of the gradient and Laplacian are given by

$$|\nabla\zeta| = \frac{1}{h}\,|\widehat{\nabla}\zeta|, \qquad \nabla^2\zeta = \frac{1}{h^2}\,\widehat{\nabla}^2\zeta, \tag{5.11.6}$$

where a hat denotes differentiation with respect to the elliptic coordinates.

The Laplace-Young equation (5.11.3) takes the form

$$\frac{1}{h^2}\,\widehat{\nabla}^2\zeta + \zeta_z^2\,\zeta_{yy} - 2\,\zeta_y\,\zeta_z\,\zeta_{yz} + \zeta_y^2\,\zeta_{zz} - (1+|\nabla\zeta|^2)^{3/2}\,\frac{\zeta}{\ell^2} = 0. \tag{5.11.7}$$

To fit the elliptical shape, we set

$$b = A\,\sinh u_0, \qquad c = A\,\cosh u_0, \tag{5.11.8}$$

where b and c are the ellipse semi-axes along the y and z axes. Solving for A and u_0, we find

$$A = \frac{b}{\sinh u_0}, \qquad \tanh u_0 = \frac{b}{c}. \tag{5.11.9}$$

Note that the elliptic coordinates (u, φ) are orthogonal in the yz plane but not over the three-dimensional interface. Moreover, φ differs from the meridional angle around the vertical x axis.

5.11.2 Finite-difference method

Equation (5.11.7) is now regarded as a nonlinear Poisson-like equation forced by an *a priori* unknown source term. The solution can be found numerically using a finite-difference method in elliptic coordinates with evenly spaced grid lines. The second-order finite-difference equation corresponding to (5.11.7) at the (i,j) interior grid point in the $u\varphi$ plane is

$$\frac{1}{h_{i,j}^2}\Big(\frac{\zeta_{i-1,j}-2\,\zeta_{i,j}-\zeta_{i+1,j}}{\Delta u^2}+\frac{\zeta_{i,j-1}-2\,\zeta_{i,j}-\zeta_{i,j+1}}{\Delta\varphi^2}\Big) \tag{5.11.10}$$
$$+\Big(\,\zeta_z^2\,\zeta_{yy}-2\,\zeta_y\,\zeta_z\,\zeta_{yz}+\zeta_y^2\,\zeta_{zz}\Big)_{i,j}-(1+|\nabla\zeta|^2)_{i,j}^{3/2}\,\frac{\zeta_{i,j}}{\ell^2}=0.$$

To implement Gauss-Seidel iterations, we rearrange to obtain

$$\zeta_{i,j}=\frac{1}{G}\Big[\frac{1}{h_{i,j}^2}\Big(\frac{\zeta_{i-1,j}+\zeta_{i+1,j}}{\Delta u^2}+\frac{\zeta_{i,j-1}+\zeta_{i,j+1}}{\Delta\varphi^2}\Big)$$
$$+\Big(\,\zeta_z^2\,\zeta_{yy}-2\,\zeta_y\,\zeta_z\,\zeta_{yz}+\zeta_y^2\,\zeta_{zz}\Big)_{i,j}\Big], \tag{5.11.11}$$

where

$$G=\Big[\frac{2}{h_{i,j}^2}\big(\frac{1}{\Delta u^2}+\frac{1}{\Delta\varphi^2}\big)+\frac{(1+|\nabla\zeta|^2)_{i,j}^{3/2}}{\ell^2}\Big]. \tag{5.11.12}$$

The iterations proceed by guessing grid values, and then replacing them with the right-hand side of (5.11.11).

The first derivatives with respect to y and z can be computed from derivatives with respect to u and φ by solving a system of 2×2 equations arising from the chain rule,

$$\begin{bmatrix}\partial y/\partial u & \partial y/\partial\varphi\\ \partial z/\partial u & \partial z/\partial\varphi\end{bmatrix}\begin{bmatrix}\partial\zeta/\partial y\\ \partial\zeta/\partial z\end{bmatrix}=\begin{bmatrix}\partial\zeta/\partial u\\ \partial\zeta/\partial\varphi\end{bmatrix}. \tag{5.11.13}$$

The derivatives with respect to u and φ on the right-hand side are computed by numerical differentiation. The second derivatives can be computed by a similar method.

The method is implemented in the following MATLAB code *men_3d* residing inside directory *03_hydrostat* of FDLIB:

```
%=================
% meniscus in the exterior of an ellipse
%=================

b = 0.5;
c = 1.0;
capl = 0.75;      % capillary length
xcline = 0.5;      % height of the contact line

tol = 0.000001;      % iteration tolerance
Niter = 1000;
```

```
%----------
% divisions
%----------

Nu = 32; Nphi = 32;

%---
% prepare
%---

Dphi = 2*pi/Nphi;
capls = capl*capl;
Dphi2 = 2.0*Dphi;
u0 = atanh(b/c);
snhu0 = sinh(u0);
cshu0 = cosh(u0);
A = b/snhu0;
umax = log(32.0*b/A);
Du = (umax-u0)/Nu;
Du2 = 2.0*Du;

%---
% grid
%---

for i=1:Nu+1
  u(i) = u0+(i-1)*Du;
  snhu(i) = sinh(u(i));
  cshu(i) = cosh(u(i));
  for j=1:Nphi+1
    phi(j) = (j-1)*Dphi;
    csphi(j) = cos(phi(j));
    snphi(j) = sin(phi(j));
    y(i,j) = A*snhu(i)*csphi(j);
    z(i,j) = A*cshu(i)*snphi(j);
    h(i,j) = A*sqrt(cshu(i)^2-snphi(j)^2);
  end
end

%---
% initialize
%---

for i=1:Nu+1
  for j=1:Nphi+2
    x(i,j) = 0.0;
  end
end
```

```
%---
% boundary conditions
%---

for j=1:Nphi+1
  x(1,j) = xcline*y(1,j);
  x(Nu+1,j) = 0.0;
end

x(1,Nphi+2) = x(1,2);
x(Nu+1,Nphi+2) = 0.0;

%---
% iterations
%---

for iterations=1:Niter

%---
% compute the first derivatives:  dx/dy and dx/dz
%---

% interior nodes:

for i=2:Nu
  for j=2:Nphi+1
    MAT(1,1) = A*cshu(i)*csphi(j);
    MAT(1,2) = A*snhu(i)*snphi(j);
    MAT(2,1) = -A*snhu(i)*snphi(j);
    MAT(2,2) = A*cshu(i)*csphi(j);
    RHS(1) = (x(i+1,j)-x(i-1,j))/Du2;
    RHS(2) = (x(i,j+1)-x(i,j-1))/Dphi2;
    SOL = RHS/MAT';
    dxdy(i,j) = SOL(1);
    dxdz(i,j) = SOL(2);
  end
  dxdy(i,1) = dxdy(i,Nphi+1);
  dxdz(i,1) = dxdz(i,Nphi+1);
  dxdy(i,Nphi+2) = dxdy(i,2);
  dxdz(i,Nphi+2) = dxdz(i,2);
end

% boundary nodes by one-sided differences:

for j=1:Nphi+2
  dxdy(1,j) = 2.0*dxdy(2,j)-dxdy(3,j);
  dxdy(Nu+1,j) = 2.0*dxdy(Nu,j)-dxdy(Nu-1,j);
  dxdz(1,j) = 2.0*dxdz(2,j)-dxdz(3,j);
```

```
  dxdz(Nu+1,j) = 2.0*dxdz(Nu,j)-dxdz(Nu-1,j);
end

%---
% compute the second derivatives
%---

for i=2:Nu
  for j=2:Nphi+1
    MAT(1,1) = A*cshu(i)*csphi(j);
    MAT(1,2) = A*snhu(i)*snphi(j);
    MAT(2,1) = -A*snhu(i)*snphi(j);
    MAT(2,2) = A*cshu(i)*csphi(j);
    RHS(1) = (dxdy(i+1,j)-dxdy(i-1,j))/Du2;
    RHS(2) = (dxdy(i,j+1)-dxdy(i,j-1))/Dphi2;
      SOL = RHS/MAT';
    dxdyy(i,j) = SOL(1);
    dxdyz(i,j) = SOL(2);
    RHS(1) = (dxdz(i+1,j)-dxdz(i-1,j))/Du2;
    RHS(2) = (dxdy(i,j+1)-dxdy(i,j-1))/Dphi2;
      SOL = RHS/MAT';
    dxdyy(i,j) = SOL(1);
    dxdyz(i,j) = SOL(2);
    RHS(1) = (dxdz(i+1,j)-dxdz(i-1,j))/Du2;
    RHS(2) = (dxdz(i,j+1)-dxdz(i,j-1))/Dphi2;
      SOL = RHS/MAT';
    dxdzy(i,j) = SOL(1);
    dxdzz(i,j) = SOL(2);
  end
end

%------
% scan the grid points
%------

errr = 0.0;

for i=2:Nu
  for j=2:Nphi+1
    H = h(i,j); HS = H*H;
    tmp = 1.0+dxdy(i,j)^2+dxdz(i,j)^2;
    G = 2.0*(1.0/Du^2+1.0/Dphi^2)/HS + tmp^(3/2)/capls;
    xnew = (x(i+1,j)+x(i-1,j))/(HS*Du^2) ...
      +(x(i,j+1)+x(i,j-1))/(HS*Dphi^2);
    xnew = xnew + dxdz(i,j)^2 * dxdyy(i,j);
    xnew = xnew - 2.0*dxdy(i,j)*dxdz(i,j)*dxdyz(i,j);
    xnew = xnew + dxdy(i,j)^2 * dxdzz(i,j);
    xnew=xnew/G;
    corr = abs(xnew-x(i,j));
```

```
    x(i,j) = xnew;
    if(corr>errr) errr = corr; end
  end
end

for i=2:Nu+1
  x(i,1) = x(i,Nphi+1);
  x(i,Nphi+2) = x(i,2);
end

if(errr<tol) break; end

%--
end      % of iterations
%--

if(errr>tol)
  disp('the iterations did not converge')
  errr
  return
end

%---
% plotting
%---

hold on;
for i=1:Nu
  for j=1:Nphi
    patch([y(i,j), y(i,j+1), y(i+1,j+1), y(i+1,j)], ...
      [z(i,j), z(i,j+1), z(i+1,j+1), z(i+1,j)], ...
      [x(i,j), x(i,j+1), x(i+1,j+1), x(i+1,j)], ...
      [x(i,j), x(i,j+1), x(i+1,j+1), x(i+1,j)]);
  end
end

xlabel('y','fontsize',15);
ylabel('z','fontsize',15);
zlabel('x','fontsize',15);

axis square
axis([-1 1 -1 1 -1 1])
view(162, 12)
```

A solution subject to the boundary condition implemented in the code is shown in figure 5.11.2(*b*).

5.11.3 Capillary force and torque

Surface tension pulls the particle around the contact line in a direction that is tangential to the interface and lies in a plane that is normal to the contact line. The resultant capillary force and torque with respect to an arbitrary point, $\mathbf{x}_0$, are given by

$$\mathbf{F}^{cap} = \gamma \oint \mathbf{r} \times \mathbf{n} \, dl, \qquad \mathbf{T}^{cap} = \gamma \oint (\mathbf{x} - \mathbf{x}_0) \times (\mathbf{r} \times \mathbf{n}) \, dl, \quad (5.11.14)$$

where $\mathbf{n}$ is the unit vector normal to the interface given by

$$\mathbf{n} = \frac{1}{|\frac{\partial \mathbf{x}}{\partial u} \times \frac{\partial \mathbf{x}}{\partial \varphi}|} \frac{\partial \mathbf{x}}{\partial u} \times \frac{\partial \mathbf{x}}{\partial \varphi}, \quad (5.11.15)$$

$\mathbf{r}$ is the unit vector tangential to the contact line, the integration is performed around the contact line, and l is the arc length around the contact line. Using a vector identity, we may express the torque as

$$\mathbf{T}^{Cap} = \gamma \oint \Big[\mathbf{r} \, [(\mathbf{x} - \mathbf{x}_0) \cdot \mathbf{n}] - \mathbf{n} \, [\, (\mathbf{x} - \mathbf{x}_0) \cdot \mathbf{r}] \Big] \, dl. \quad (5.11.16)$$

After the computation of the meniscus has been concluded, the capillary force and torque can be obtained using the following module:

```
%---
% compute dxdu, dxdphi,
% the surface normal and tension vector
% around the contact line
%---

x0=0.0; y0=0.0; z0=0.0;

for j=1:Nphi+1

  dxdu(j) = (-x(3,j)+4.0*x(2,j)-3.0*x(1,j))/Du2;
  dydu(j) = A*cshu0*csphi(j);
  dzdu(j) = A*snhu0*snphi(j);
  dmdu(j) = sqrt(dxdu(j)^2+dydu(j)^2+dzdu(j)^2);

  if(j==1)
    dxdphi(j) = (x(1,2)-x(1,Nphi))/Dphi2;
  else
    dxdphi(j) = (x(1,j+1)-x(1,j-1))/Dphi2;
  end

  dydphi(j) =-A*snhu0*snphi(j);
  dzdphi(j) = A*cshu0*csphi(j);
  dmdphi(j) = sqrt(dxdphi(j)^2+dydphi(j)^2+dzdphi(j)^2);
```

```
    vnx(j) = dydu(j)*dzdphi(j)-dzdu(j)*dydphi(j);
    vny(j) = dzdu(j)*dxdphi(j)-dxdu(j)*dzdphi(j);
    vnz(j) = dxdu(j)*dydphi(j)-dydu(j)*dxdphi(j);
    vnm(j) = sqrt(vnx(j)^2+ vny(j)^2+vnz(j)^2);
    vnx(j) = vnx(j)/vnm(j); vny(j) = vny(j)/vnm(j);
    vnz(j) = vnz(j)/vnm(j);
    tngx(j) = dydphi(j)*vnz(j)-dzdphi(j)*vny(j);
    tngy(j) = dzdphi(j)*vnx(j)-dxdphi(j)*vnz(j);
    tngz(j) = dxdphi(j)*vny(j)-dydphi(j)*vnx(j);
    crsx(j) = (y(1,j)-y0)*tngz(j)-(z(1,j)-z0)*tngy(j);
    crsy(j) = (z(1,j)-z0)*tngx(j)-(x(1,j)-x0)*tngz(j);
    crsz(j) = (x(1,j)-x0)*tngy(j)-(y(1,j)-y0)*tngx(j);

end

%---
% force and torque
%---

forcex = 0.0; forcey = 0.0; forcez = 0.0;
torqux = 0.0; torquy = 0.0; torquz = 0.0;

for j=1:Nphi
    forcex = forcex+tngx(j);
    forcey = forcey+tngy(j);
    forcez = forcez+tngz(j);
    torqux = torqux+crsx(j);
    torquy = torquy+crsy(j);
    torquz = torquz+crsz(j);
end

forcex = forcex*Dphi;
forcey = forcey*Dphi;
forcez = forcez*Dphi;
torqux = torqux*Dphi;
torquy = torquy*Dphi;
torquz = torquz*Dphi;
```

Problem

5.11.1 *Convergence of iterations*

Consider a meniscus originating from a horizontal elliptical contact line. Study the shape of the meniscus for several contact line elevations and capillary lengths, and investigate the convergence of the Gauss-Seidel iterations.

6 Equation of Motion and Vorticity Transport

Fluid flow is established in response to an external action mediated by boundary motion, by the application of a surface force, or by the presence of a body force. The evolution of a transient flow and the structure of a steady flow established after an initial start-up period are governed by two fundamental principles of thermodynamics and classical mechanics: mass conservation, and Newton's second law for the motion of a fluid parcel. The implementation of Newton's law of motion in continuum mechanics leads to Cauchy's equation of motion, providing us with an expression for the point particle acceleration in terms of stresses, and to the vorticity transport equation governing the point particle rotation. The derivation and interpretation of these equations in general and specific terms, and their solution for simple flow configurations will be discussed in this chapter.

6.1 Newton's second law of motion for a fluid parcel

Newton's second law of motion requires that the rate of change of the linear momentum of a fluid parcel, $\mathbf{M}_p$, is equal to the sum of the forces exerted on the parcel, including the body force due to gravity given in equation (5.1.1) and the surface force given in equation (5.1.2),

$$\frac{\mathrm{d}\mathbf{M}_p}{\mathrm{d}t} = \mathbf{F}^S + \mathbf{F}^B = \iiint_{Parcel} \mathbf{f}\,\mathrm{d}S + \iiint_{Parcel} \rho\,\mathbf{g}\,\mathrm{d}V, \qquad (6.1.1)$$

C. Pozrikidis, *Fluid Dynamics: Theory, Computation, and Numerical Simulation*,
DOI: 10.1007/978-0-387-95871-2_6,

where $\mathbf{f}$ is the hydrodynamic traction exerted on the parcel surface. Expressing the traction in terms of the stress tensor, as shown in (4.2.9), we find

$$\frac{\mathrm{d}\mathbf{M}_p}{\mathrm{d}t} = \iiint_{Parcel} \mathbf{n} \cdot \boldsymbol{\sigma}\, \mathrm{d}S + \iiint_{Parcel} \rho\, \mathbf{g}\, \mathrm{d}V, \tag{6.1.2}$$

where the unit normal vector, $\mathbf{n}$, points into the parcel exterior. Our next task is to relate the rate of change of the parcel momentum to the fluid density and velocity.

6.1.1 Rate of change of linear momentum

An expression for the linear momentum arises by subdividing the parcel into elementary sub-parcels with volume $\mathrm{d}V_p$ and mass $\mathrm{d}m_P = \rho\, \mathrm{d}V$, and summing the contributions by integration to obtain

$$\mathbf{M}_p = \iiint_{Parcel} \mathbf{u}\, \mathrm{d}m_p = \iiint_{Parcel} \mathbf{u}\, \rho\, \mathrm{d}V, \tag{6.1.3}$$

where $\mathbf{u}$ is the fluid velocity. The rate of change of linear momentum is given by

$$\frac{\mathrm{d}\mathbf{M}_p}{\mathrm{d}t} = \frac{\mathrm{d}}{\mathrm{d}t} \iiint_{Parcel} \mathbf{u}\, \mathrm{d}m_p = \frac{\mathrm{d}}{\mathrm{d}t} \iiint_{Parcel} \mathbf{u}\, \rho\, \mathrm{d}V, \tag{6.1.4}$$

where the time derivative is taken for a fixed parcel identity. Because the integral is computed over the volume of the parcel which is not stationary but changes in time, interchanging the time differentiation and volume integration on the right-hand side of (6.1.4) is permissible only if the time derivative is replaced by the material derivative, $\mathrm{D}/\mathrm{D}t$, under the integral sign, yielding

$$\frac{\mathrm{d}\mathbf{M}_p}{\mathrm{d}t} = \iiint_{Parcel} \frac{\mathrm{D}\,(\mathbf{u}\, \mathrm{d}m_p)}{\mathrm{D}t} = \iiint_{Parcel} \Big(\frac{\mathrm{D}\mathbf{u}}{\mathrm{D}t}\, \mathrm{d}m_p + \frac{\mathrm{D}\, \mathrm{d}m_p}{\mathrm{D}t}\, \mathbf{u} \Big). \tag{6.1.5}$$

Mass conservation requires that the material derivative of the elementary mass, $\mathrm{d}m_p$ vanish, yielding the simplified expression

$$\frac{\mathrm{d}\mathbf{M}_p}{\mathrm{d}t} = \iiint_{Parcel} \frac{\mathrm{D}\mathbf{u}}{\mathrm{D}t}\, \mathrm{d}m_p = \iiint_{Parcel} \frac{\mathrm{D}\mathbf{u}}{\mathrm{D}t}\, \rho\, \mathrm{d}V, \tag{6.1.6}$$

where $\mathrm{D}\mathbf{u}/\mathrm{D}t$ is the point particle acceleration.

6.1.2 Equation of parcel motion

Substituting the right-hand side of (6.1.6) in the left-hand side of (6.1.2), we obtain the desired equation of parcel motion,

$$\iiint_{Parcel} \frac{\mathrm{D}\mathbf{u}}{\mathrm{D}t}\, \rho\, \mathrm{d}V = \iint_{Parcel} \mathbf{n} \cdot \boldsymbol{\sigma}\, \mathrm{d}S + \iiint_{Parcel} \rho\, \mathbf{g}\, \mathrm{d}V, \tag{6.1.7}$$

involving the point particle acceleration, the stress tensor, and the body force. Explicitly, the x, y, and z components of (6.1.7) are

$$\begin{aligned}
\iiint_{Parcel} \frac{\mathrm{D}u_x}{\mathrm{D}t}\,\rho\,\mathrm{d}V &= \iint_{Parcel} (n_x\sigma_{xx} + n_y\sigma_{yx} + n_z\sigma_{zx})\,\mathrm{d}S + \iiint_{Parcel} \rho g_x\,\mathrm{d}V, \\
\iiint_{Parcel} \frac{\mathrm{D}u_y}{\mathrm{D}t}\,\rho\,\mathrm{d}V &= \iint_{Parcel} (n_x\sigma_{xy} + n_y\sigma_{yy} + n_z\sigma_{zy})\,\mathrm{d}S + \iiint_{Parcel} \rho g_y\,\mathrm{d}V, \\
\iiint_{Parcel} \frac{\mathrm{D}u_z}{\mathrm{D}t}\,\rho\,\mathrm{d}V &= \iint_{Parcel} (n_x\sigma_{xz} + n_y\sigma_{yz} + n_z\sigma_{zz})\,\mathrm{d}S + \iiint_{Parcel} \rho g_z\,\mathrm{d}V.
\end{aligned} \tag{6.1.8}$$

Equations (6.1.8) are valid irrespective of whether the fluid is compressible or incompressible.

6.1.3 *Two-dimensional flow*

The counterpart of the equation of parcel motion (6.1.7) for two-dimensional flow in the xy plane is

$$\iint_{Parcel} \frac{\mathrm{D}\mathbf{u}}{\mathrm{D}t}\,\rho\,\mathrm{d}A = \oint_{Parcel} \mathbf{n}\cdot\boldsymbol{\sigma}\,\mathrm{d}l + \iint_{Parcel} \rho\,\mathbf{g}\,\mathrm{d}A, \tag{6.1.9}$$

where $\mathrm{d}A$ is a differential area in the domain of flow, and $\mathrm{d}l$ is the differential arc length along the contour of a parcel in the xy plane. Explicitly, the x and y components of (6.1.9) are

$$\begin{aligned}
\iint_{Parcel} \frac{\mathrm{D}u_x}{\mathrm{D}t}\,\rho\,\mathrm{d}A &= \oint_{Parcel} (n_x\sigma_{xx} + n_y\sigma_{yx})\,\mathrm{d}l + \iint_{Parcel} \rho\,g_x\,\mathrm{d}A, \\
\iint_{Parcel} \frac{\mathrm{D}u_y}{\mathrm{D}t}\,\rho\,\mathrm{d}A &= \oint_{Parcel} (n_x\sigma_{xy} + n_y\sigma_{yy})\,\mathrm{d}l + \iint_{Parcel} \rho\,g_y\,\mathrm{d}A.
\end{aligned} \tag{6.1.10}$$

We note again that these equations are valid irrespective of whether the fluid is compressible or incompressible.

A rectangular parcel

As an application, we consider the motion of a fluid parcel with an instantaneous rectangular shape whose sides are parallel to the x and y axes, as depicted in figure 6.1.1. The parcel will remain rectangular only if the fluid exhibits rigid-body motion. Under more general conditions, the parcel will deform to obtain the warped shape drawn with the dashed line in figure 6.1.1. However, parcel deformation does not prevent us from evaluating the integrals in (6.1.10) over the instantaneous rectangular shape.

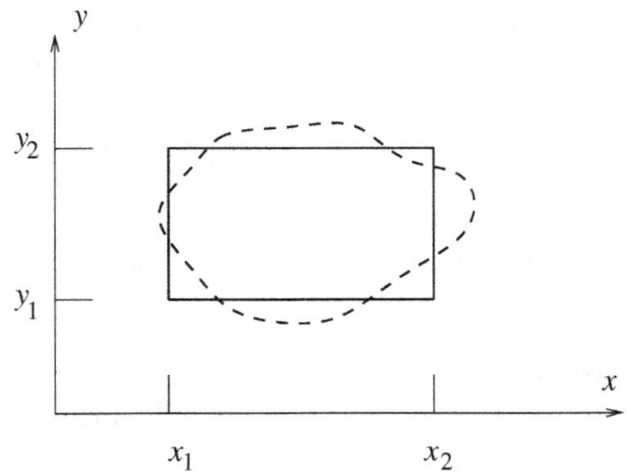

Figure 6.1.1 Illustration of a fluid parcel with a rectangular instantaneous shape drawn with the solid line in a two-dimensional flow. Even though the parcel generally deforms to obtain a warped shape, drawn with the dashed line, Newton's second law of motion can be applied in its integral form over the instantaneous parcel shape.

For simplicity, we assume that the density of the fluid is uniform and the acceleration of gravity is constant over the parcel volume. We note that the unit normal vector is parallel to the x or y axis over each side, and find that equations (6.1.10) assume the simpler forms

$$\int_{x_1}^{x_2}\int_{y_1}^{y_2}(\frac{\mathrm{D}u_x}{\mathrm{D}t} - g_x)\,\rho\,\mathrm{d}y\,\mathrm{d}x = \int_{y_1}^{y_2}[(\sigma_{xx})_{x=x_2} - (\sigma_{xx})_{x=x_1}]\,\mathrm{d}y + \int_{x_1}^{x_2}[(\sigma_{yx})_{y=y_2} - (\sigma_{yx})_{y=y_1}]\,\mathrm{d}x, \tag{6.1.11}$$

and

$$\int_{x_1}^{x_2}\int_{y_1}^{y_2}(\frac{\mathrm{D}u_y}{\mathrm{D}t} - g_y)\,\rho\,\mathrm{d}y\,\mathrm{d}x = \int_{y_1}^{y_2}[(\sigma_{xy})_{x=x_2} - (\sigma_{xy})_{x=x_1}]\,\mathrm{d}y + \int_{x_1}^{x_2}[(\sigma_{yy})_{y=y_2} - (\sigma_{yy})_{y=y_1}]\,\mathrm{d}x. \tag{6.1.12}$$

The first integral on the right-hand side of (6.1.11) involves normal stresses exerted on the vertical sides, and the second integral involves shear stresses exerted on the horizontal sides; the converse is true for (6.1.12).

Steady unidirectional flow

In the case of steady unidirectional flow along the x axis, point particles move along the x axis with constant velocity and vanishing acceleration,

$\mathrm{Du}/\mathrm{D}t = 0$. Setting the left-hand side of the equation of parcel motion (6.1.9) to zero, we obtain a balance between the hydrodynamic and the body force,

$$\oint_{Parcel} \mathbf{n} \cdot \boldsymbol{\sigma} \, \mathrm{d}l + \iint_{Parcel} \rho \, \mathbf{g} \, \mathrm{d}A = \mathbf{0}. \tag{6.1.13}$$

Restricting our attention to Newtonian fluids, we use the constitutive equation (4.5.4), and make two observations:

- In the absence of axial and transverse stretching, $\partial u_x/\partial x = 0$ and $\partial u_y/\partial y = 0$, the normal stresses σ_{xx} and σ_{yy} are equal to the negative of the pressure, $\sigma_{xx} = \sigma_{yy} = -p$.
- The shear stresses $\sigma_{xy} = \sigma_{yx}$ are independent of streamwise position, x, but may depend on the lateral position, y.

Subject to these simplifications, the balance equations (6.1.11) and (6.1.12) reduce to

$$\int_{y_1}^{y_2} [p_{x=x_2} - p_{x=x_1}] \, \mathrm{d}y - [(\sigma_{yx})_{y=y_2} - (\sigma_{yx})_{y=y_1}] \, \Delta x = \rho \, g_x \, \Delta x \, \Delta y,$$

$$\int_{x_1}^{x_2} [p_{y=y_2} - p_{y=y_1}] \, \mathrm{d}x = \rho \, g_y \, \Delta x \, \Delta y, \tag{6.1.14}$$

where $\Delta x \equiv x_2 - x_1$ and $\Delta y = y_2 - y_1$. The second equation is satisfied when

$$\frac{p_{x,y=y_2} - p_{x,y=y_1}}{\Delta y} = \rho \, g_y, \tag{6.1.15}$$

reflecting the hydrostatic pressure variation. The first equation in (6.1.14) is satisfied when

$$\frac{p_{x=x_2,y} - p_{x=x_1,y}}{\Delta x} = \rho \, g_x - G, \tag{6.1.16}$$

and

$$\frac{(\sigma_{yx})_{x,y=y_2} - (\sigma_{yx})_{x,y=y_1}}{\Delta y} = -G, \tag{6.1.17}$$

where G is a free parameter called the modified pressure gradient.

Shear- pressure- and gravity-driven flow

Physically, G is determined by the mechanism driving the flow. Three modular flow configurations can be identified:

1. When $G = 0$, equation (6.1.16) shows that the pressure variation in the x direction is hydrostatic, and equation (6.1.17) shows that the shear stress σ_{yx} is constant, independent of y. This is the case of shear-driven flow.

2. When the streamwise pressure drop is zero, $p_{x=x_2,y} = p_{x=x_1,y}$, equation (6.1.16) requires $G = \rho\, g_x$ and equation (6.1.17) shows that the difference in the shear stress, $(\sigma_{yx})_{x,y=y_2} - (\sigma_{yx})_{x,y=y_1}$, is equal to $-\rho\, g_x\, \Delta y$. This is the case of gravity-driven flow.

3. When the flow is horizontal, $g_x = 0$, equation (6.1.16) shows that G is the negative of the streamwise pressure gradient, and equation (6.1.17) shows that the difference in the shear stress, $(\sigma_{yx})_{x,y=y_2} - (\sigma_{yx})_{x,y=y_1}$ is equal to $-G\, \Delta y$. This is the case of pressure-driven flow.

Problem

6.1.1 *Body force in terms of a surface integral*

Show that the body force expressed by the second integral on the right-hand side of (6.1.7) can be expressed as a surface integral in the form

$$\iint_{Parcel} \rho\, (\mathbf{g} \cdot \mathbf{x})\, \mathbf{n}\, \mathrm{d}S. \tag{6.1.18}$$

Hint: Use the Gauss divergence theorem (2.6.29).

6.2 Integral momentum balance

Consider the integrand of the rate of change of momentum on the left-hand side of equation (6.1.7). Using the rules of product differentiation and the continuity equation (2.8.5), we write

$$\rho\, \frac{\mathrm{D}\mathbf{u}}{\mathrm{D}t} = \frac{\mathrm{D}(\rho\, \mathbf{u})}{\mathrm{D}t} - \mathbf{u}\, \frac{\mathrm{D}\rho}{\mathrm{D}t} = \frac{\mathrm{D}(\rho\, \mathbf{u})}{\mathrm{D}t} + (\rho\, \mathbf{u})\, (\nabla \cdot \mathbf{u}), \tag{6.2.1}$$

where

$$\nabla \cdot \mathbf{u} \equiv \frac{\partial u_x}{\partial x} + \frac{\partial u_y}{\partial y} + \frac{\partial u_z}{\partial z} \tag{6.2.2}$$

is the divergence of the velocity. If the fluid is incompressible, the second term on the right-hand side of (6.2.1) does not appear.

The x component of the vectorial expression (6.2.1) can be manipulated to give

$$\rho\, \frac{\mathrm{D}u_x}{\mathrm{D}t} = \frac{\mathrm{D}(\rho\, u_x)}{\mathrm{D}t} + (\rho\, u_x)\, (\nabla \cdot \mathbf{u}) = \frac{\partial(\rho\, u_x)}{\partial t} + \mathbf{u} \cdot \nabla(\rho\, u_x) + (\rho\, u_x)\, (\nabla \cdot \mathbf{u})$$

$$= \frac{\partial(\rho\, u_x)}{\partial t} + u_x\, \frac{\partial(\rho\, u_x)}{\partial x} + u_y\, \frac{\partial(\rho\, u_x)}{\partial y} + u_z\, \frac{\partial(\rho\, u_x)}{\partial z} + (\rho\, u_x)\, (\nabla \cdot \mathbf{u}), \tag{6.2.3}$$

where the time derivative $\partial/\partial t$ is taken keeping the spatial position fixed. Combining the last four terms in the last expression, we find

$$\rho \frac{\mathrm{D}u_x}{\mathrm{D}t} = \frac{\partial(\rho\, u_x)}{\partial t} + \frac{\partial(\rho\, u_x u_x)}{\partial x} + \frac{\partial(\rho\, u_y u_x)}{\partial y} + \frac{\partial(\rho\, u_z u_x)}{\partial z}. \tag{6.2.4}$$

Working in a similar fashion with the y and z components of (6.2.1), we derive the corresponding expressions

$$\rho \frac{\mathrm{D}u_y}{\mathrm{D}t} = \frac{\partial(\rho\, u_y)}{\partial t} + \frac{\partial(\rho\, u_x u_y)}{\partial x} + \frac{\partial(\rho\, u_y u_y)}{\partial y} + \frac{\partial(\rho\, u_z u_y)}{\partial z} \tag{6.2.5}$$

and

$$\rho \frac{\mathrm{D}u_z}{\mathrm{D}t} = \frac{\partial(\rho\, u_z)}{\partial t} + \frac{\partial(\rho\, u_x u_z)}{\partial x} + \frac{\partial(\rho\, u_y u_z)}{\partial y} + \frac{\partial(\rho\, u_z u_z)}{\partial z}. \tag{6.2.6}$$

These expressions hold for incompressible as well as compressible fluids.

Momentum tensor

To collect equations (6.2.4)–(6.2.6) into a unified form, we introduce the momentum tensor, M_{ij}, defined as

$$M_{ij} \equiv \rho\, u_i u_j, \tag{6.2.7}$$

where the indices i and j range over x, y, and z or, correspondingly, 1, 2, and 3. It is evident from the definition (6.2.7) that the tensor $\mathbf{M}$ is symmetric,

$$M_{ij} = M_{ji}. \tag{6.2.8}$$

Next, we introduce the divergence of the momentum tensor defined as a vector whose ith component is given by

$$\frac{\partial M_{ji}}{\partial x_j} = \frac{\partial M_{ij}}{\partial x_j}, \tag{6.2.9}$$

where summation is implied over the repeated index j. For example, the x component of the divergence of $\mathbf{M}$ is

$$\frac{\partial M_{jx}}{\partial x_j} = \frac{\partial M_{xx}}{\partial x} + \frac{\partial M_{yx}}{\partial y} + \frac{\partial M_{zx}}{\partial z}. \tag{6.2.10}$$

Subject to these definitions, equations (6.2.4) – (6.2.6) are collected into the form

$$\rho \frac{\mathrm{D}u_i}{\mathrm{D}t} = \frac{\partial(\rho\, u_i)}{\partial t} + \frac{\partial M_{ji}}{\partial x_j}, \tag{6.2.11}$$

where $i = x, y$, or z. The corresponding vector form is

$$\rho \frac{\mathrm{D}\,\mathbf{u}}{\mathrm{D}t} = \frac{\partial(\rho\,\mathbf{u})}{\partial t} + \nabla \cdot \mathbf{M}. \qquad (6.2.12)$$

The right-hand sides of equations (6.2.11) and (6.2.12) involve Eulerian derivatives; that is, derivatives with respect to time and spatial coordinates.

Equation of parcel motion

Now substituting (6.2.12) in the left-hand side of the equation of parcel motion (6.1.7), we derive the alternative form

$$\iiint_{Parcel} \Big(\frac{\partial(\rho\,\mathbf{u})}{\partial t} + \nabla \cdot \mathbf{M} \Big)\, \mathrm{d}V = \iint_{Parcel} \mathbf{n} \cdot \boldsymbol{\sigma}\, \mathrm{d}S + \iiint_{Parcel} \rho\, \mathbf{g}\, \mathrm{d}V. \qquad (6.2.13)$$

We can use the Gauss divergence theorem stated in equation (2.6.29) to convert the volume integral of the divergence of the momentum tensor into a surface integral over the parcel volume, obtaining

$$\iiint_{Parcel} \frac{\partial(\rho\,\mathbf{u})}{\partial t}\, \mathrm{d}V + \iint_{Parcel} \mathbf{n} \cdot \mathbf{M}\, \mathrm{d}S = \iint_{Parcel} \mathbf{n} \cdot \boldsymbol{\sigma}\, \mathrm{d}S + \iiint_{Parcel} \rho\, \mathbf{g}\, \mathrm{d}V, \qquad (6.2.14)$$

where the unit normal vector, $\mathbf{n}$, points outward from the parcel. In index notation,

$$\iiint_{Parcel} \frac{\partial(\rho\, u_i)}{\partial t}\, \mathrm{d}V + \iint_{Parcel} n_j M_{ji}\, \mathrm{d}S = \iint_{Parcel} n_j \sigma_{ji}\, \mathrm{d}S + \iiint_{Parcel} \rho\, g_i\, \mathrm{d}V, \qquad (6.2.15)$$

where summation is implied over the repeated index j.

Control volumes and integral momentum balance

It is important to remember that equation (6.2.15) originates from Newton's second law of motion for a fluid parcel. However, in the process of expressing the material derivative in terms of Eulerian derivatives taken with respect to time and position in space, the parcel has lost its significance as a material body of fluid and has become relevant only insofar as to define the volume it occupies in space.

To signify the new interpretation, we rewrite the integral momentum balance (6.2.15) in identical form, except that the volume of integration is now called a control volume (CV). Using the definition of the momentum tensor shown in (6.2.7), we express the integral momentum balance in the form

$$\iiint_{CV} \frac{\partial(\rho\, u_i)}{\partial t}\, \mathrm{d}V + \iint_{CV} (n_j u_j)\, \rho u_i\, \mathrm{d}S = \iint_{CV} n_j \sigma_{ji}\, \mathrm{d}S + \iiint_{CV} \rho\, g_i\, \mathrm{d}V. \tag{6.2.16}$$

The four integrals on the left- and right-hand sides admit the following interpretation:

1. The first integral is the rate of change of the ith component of momentum of the fluid residing inside the control volume. At steady state, this term vanishes.

2. The scalar $n_j u_j$ of the second integrand on the left-hand side is the component of the fluid velocity normal to the boundary of the control volume. The corresponding integral expresses the rate of convective transport of the ith component of the fluid momentum across the boundary of the control volume.

3. The first integral on the right-hand side is the ith component of the surface force exerted on the boundary of the control volume.

4. The second integral on the right-hand side is the ith component of the body force exerted on the control volume.

Equation (6.2.16) is sometimes called an integral momentum balance. It is important to bear in mind that an integral momentum can only be written in Cartesian coordinates. In vector notation, we obtain

$$\iiint_{CV} \frac{\partial(\rho \mathbf{u})}{\partial t}\, \mathrm{d}V + \iint_{CV} \rho\, \mathbf{n} \cdot (\mathbf{u}\mathbf{u})\, \mathrm{d}S = \iint_{CV} \mathbf{n} \cdot \boldsymbol{\sigma}\, \mathrm{d}S + \iiint_{CV} \rho\, \mathbf{g}\, \mathrm{d}V. \tag{6.2.17}$$

Grouping two integrals, we find

$$\iiint_{CV} \frac{\partial(\rho \mathbf{u})}{\partial t}\, \mathrm{d}V = \iint_{CV} \mathbf{n} \cdot (\boldsymbol{\sigma} - \rho\, \mathbf{u}\mathbf{u})\, \mathrm{d}S + \iiint_{CV} \rho\, \mathbf{g}\, \mathrm{d}V, \tag{6.2.18}$$

where $\boldsymbol{\sigma} - \rho\, \mathbf{u}\mathbf{u}$ is the stress-momentum tensor.

6.2.1 Flow through a sudden enlargement

To illustrate the usefulness of the integral momentum balance in engineering analysis, we consider steady flow through a duct with a sudden enlargement, as illustrated in figure 6.2.1.[1] We begin by introducing a control volume identified

[1] The solution of practical engineering problems by the use of integral mass, momentum, and energy balances is discussed in a classical text by BIRD, R. B., STEWART, W. E. & LIGHTFOOT, E. N., 2006, *Transport Phenomena*, Second Edition, Wiley.

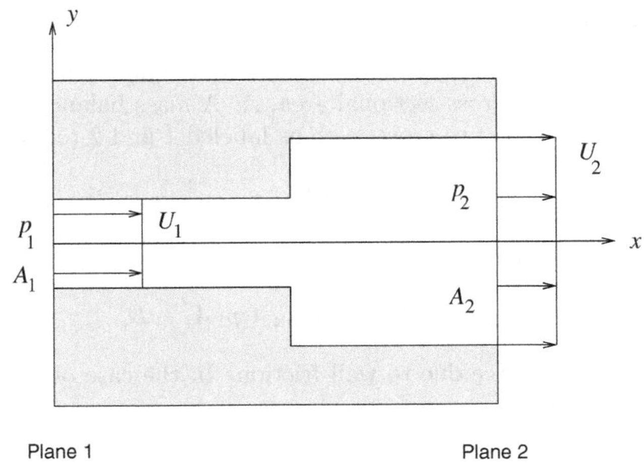

Figure 6.2.1 Simplified model of flow through a duct with a sudden enlargement. An integral momentum balance allows us to compute the pressure rise, $P_2 - P_1$, in terms of the inlet and outlet cross-sectional areas, A_1 and A_2.

with the section of the duct confined between the vertical planes labeled 1 and 2, and assume that the density of the fluid is uniform and the velocity profile is flat at the inlet and outlet.

Neglecting the shear stress at the walls, approximating the normal stress at the inlet and outlet with the negative of the pressure, assuming that the pressure at the washer-shaped area is equal to the inlet pressure, and neglecting the effects of gravity, we find that the integral momentum balance (6.2.16) at steady state simplifies to

$$\rho U_2^2 A_2 - \rho U_1^2 A_1 = -p_2 A_2 + p_1 A_1 + P_1 (A_2 - A_1), \tag{6.2.19}$$

where A_1 and A_2 are the cross-sectional areas at the inlet and outlet. The three terms on the right-hand side of (6.2.19) are approximations to the first integral on the right-hand side of (6.2.16) for the outlet, washer-shaped area, and inlet. Mass conservation requires

$$U_1 A_1 = U_2 A_2. \tag{6.2.20}$$

Solving for U_1 and substituting the result in (6.2.19), we derive an expression for the pressure drop,

$$P_2 - P_1 = \left(\frac{A_2}{A_1} - 1 \right) \rho U_2^2, \tag{6.2.21}$$

which predicts a rise in pressure, in agreement with laboratory observation.

6.2.2 *Isentropic flow through a conduit*

In a second application, we consider the flow of a compressible fluid through a conduit with variable cross-sectional area, A. A mass balance over a control volume confined between two cross-sections labeled 1 and 2 requires

$$\rho_1 U_1 A_1 = \rho_2 U_2 A_2. \tag{6.2.22}$$

The counterpart of equation (6.2.19) is

$$\rho_2 U_2^2 A_2 - \rho_1 U_1^2 A_1 = -p_2 A_2 + p_1 A_1 - D_x, \tag{6.2.23}$$

where D_x is the drag force due to wall friction. In the case of isentropic flow, we use (4.5.35) to find

$$\frac{p_1}{\rho_1^k} = \frac{p_2}{\rho_2^k} \tag{6.2.24}$$

where $k \equiv c_p/c_v$ is the ratio of two heat capacities. The last three equations can be used to compute p_2, V_2, and D_x, from knowledge of p_1, ρ_1, and V_1.

Energetics

Energy conservation under adiabatic conditions requires

$$h_1 + \frac{1}{2} U_1^2 + g\, y_1 = h_2 + \frac{1}{2} U_2^2 + g\, y_2, \tag{6.2.25}$$

where h is the specific enthalpy and y is the vertical positions of the conduit centerline. In terms of the heat capacity under constant pressure, $h = c_p\, T$, we find

$$c_p\, T_1 + \frac{1}{2} U_1^2 + g\, y_1 = c_p\, T_2 + \frac{1}{2} U_2^2 + g\, y_2. \tag{6.2.26}$$

This equation relates the velocity to the temperature to the elevation at two stations.

Stagnation-point temperature

At a stagnation point, $U_2 = 0$. Setting $y_1 = y_2$, we find

$$T_2 - T_1 = \frac{U_1^2}{2c_p} = \frac{(k-1)\, U_1^2}{2kR} = \frac{(k-1)\, U_1^2\, T_1}{2c^2}, \tag{6.2.27}$$

where c is the speed of sound given in (4.5.38). Rearranging, we derive an expression for the stagnation-point temperature,

$$\frac{T_2 - T_1}{T_1} = M^2\, \frac{k-1}{2}, \tag{6.2.28}$$

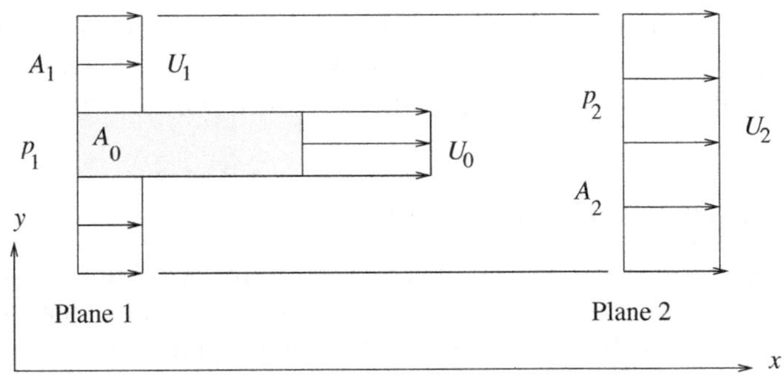

Figure 6.2.2 Schematic illustration of an ejector pump. The pressure rise, $p_2 - p_1$, can be estimated by performing an integral momentum balance.

where $M = V/c$ is the Mach number. Using the equation of state for isentropic conditions, we find

$$\frac{p_2}{p_1} = \left(\frac{k-1}{2} M^2 + 1\right)^{k/(k-1)}. \tag{6.2.29}$$

For sufficiently small Mach numbers,

$$\frac{p_2}{p_1} \simeq 1 + \frac{k}{2} M^2. \tag{6.2.30}$$

Problem

6.2.1 *Operation of an ejector pump*

A schematic illustration of an ejector pump is shown in figure 6.2.2. At plane 1, two fluid streams merge: one with uniform velocity U_1 over a cross-sectional area A_1, and the second with uniform velocity U_0 over a cross sectional area A_0. At plane 2, the velocity profile is uniform over the cross-sectional area $A_2 = A_0 + A_1$. The pressure is assumed to be uniform over the cross-section of the inlet and outlet, respectively, equal to p_1 and p_2. The fluid density is assumed to be uniform throughout the flow. Derive an expression for the rise in pressure, $p_2 - p_1$, in terms of ρ, U_0, U_1, A_0, and A_1, similar to that shown in equation (6.2.21).

6.3 Cauchy's equation of motion

Equation (6.1.7) contains two volume integrals and one surface integral over the boundary of a fluid parcel. If we could manage to convert the surface integral into a volume integral, we would be able to collect all integrands into one common expression. Since the volume of the parcel is arbitrary, the unified integrand would have to be identically zero, providing us with a differential equation.

6.3.1 *Hydrodynamic volume force*

Transforming the surface integral of the traction into a volume integral can be done using once again the Gauss divergence theorem stated in equation (2.6.25). Identifying the vector $\mathbf{h}$ with each one of the three columns of the stress tensor, we obtain

$$\iint_{Parcel} \mathbf{n} \cdot \boldsymbol{\sigma} \, dS = \iiint_{Parcel} \nabla \cdot \boldsymbol{\sigma} \, dV. \tag{6.3.1}$$

In index notation, the ith component of this equation is

$$\iint_{Parcel} n_j \, \sigma_{ji} \, dS = \iiint_{Parcel} \frac{\partial \sigma_{ji}}{\partial x_j} \, dV, \tag{6.3.2}$$

where summation of the repeated index j is implied.

The divergence of the stress tensor under the integral sign on the right-hand side of (6.3.1) or (6.3.2) is a vector denoted by

$$\boldsymbol{\Sigma} \equiv \nabla \cdot \boldsymbol{\sigma}, \tag{6.3.3}$$

with Cartesian components

$$\Sigma_x = \frac{\partial \sigma_{xx}}{\partial x} + \frac{\partial \sigma_{yx}}{\partial y} + \frac{\partial \sigma_{zx}}{\partial z}, \qquad \Sigma_y = \frac{\partial \sigma_{xy}}{\partial x} + \frac{\partial \sigma_{yy}}{\partial y} + \frac{\partial \sigma_{zy}}{\partial z},$$

$$\Sigma_z = \frac{\partial \sigma_{xz}}{\partial x} + \frac{\partial \sigma_{yz}}{\partial y} + \frac{\partial \sigma_{zz}}{\partial z}. \tag{6.3.4}$$

Physically, $\boldsymbol{\Sigma}$ is the hydrodynamic force per differential volume of fluid; in contrast, the traction $\mathbf{f}$ is the hydrodynamic force per differential surface area in the fluid.

6.3.2 *Force on an infinitesimal parcel*

To confirm identity (6.3.2), we consider a small fluid parcel with the shape of a rectangular parallelepiped centered at the origin, as illustrated in figure 5.1.1(*b*). The six flat sides of the parcel are perpendicular to the x, y, or z axis, the lengths of the three edges are, respectively, equal to Δx, Δy, and Δz, and the volume of the parcel is equal to $\Delta V = \Delta x \, \Delta y \, \Delta z$.

Consider the surface integral on the left-hand side of equation (6.3.2). Over the sides that are perpendicular to the x axis, located at $x = \pm\frac{\Delta x}{2}$ and designated as the first and second side, the unit normal vector is parallel to the x

axis; over the first side $n_x = 1$, and over the second side $n_x = -1$. Because the size of the parcel is small, the stresses over each side can be approximated with corresponding values at the side center-points.

Subject to this approximation, the surface integral on the left-hand side over the first and second side take the forms

$$\sigma_{xi}(x = \frac{\Delta x}{2}, y = 0, z = 0)\, \Delta y\, \Delta z, \tag{6.3.5}$$

and

$$-\sigma_{xi}(x = -\frac{\Delta x}{2}, y = 0, z = 0)\, \Delta y\, \Delta z, \tag{6.3.6}$$

where the parentheses enclose the coordinates of the evaluation point. Adding these two contributions and factoring out the common product $\Delta y \Delta z$ expressing the surface area, we obtain

$$\Big[\sigma_{xi}(x = \frac{\Delta x}{2}, y = 0, z = 0) - \sigma_{xi}(x = -\frac{\Delta x}{2}, y = 0, z = 0)\Big]\, \Delta y\, \Delta z. \tag{6.3.7}$$

Next, we observe that, in the limit as Δx tends to zero, the ratio of the differences

$$\begin{aligned} &\frac{\sigma_{xi}(x = \frac{\Delta x}{2}, y = 0, z = 0) - \sigma_{xi}(x = -\frac{\Delta x}{2}, y = 0, z = 0)}{\frac{\Delta x}{2} - (-\frac{\Delta x}{2})} \\ &\quad = \frac{\sigma_{xi}(x = \frac{\Delta x}{2}, y = 0, z = 0) - \sigma_{xi}(x = -\frac{\Delta x}{2}, y = 0, z = 0)}{\Delta x} \end{aligned} \tag{6.3.8}$$

tends to the partial derivative $\partial \sigma_{xi}/\partial x$ evaluated at the origin. Correspondingly, the difference (6.3.7) reduces to

$$\frac{\partial \sigma_{xi}}{\partial x}\, \Delta x\, \Delta y\, \Delta z = \frac{\partial \sigma_{xi}}{\partial x}\, \Delta V, \tag{6.3.9}$$

where the derivatives are evaluated at the origin.

Working in a similar fashion with pairs of sides that are perpendicular to the y or z axis, and summing the three contributions, we find that the left-hand side of (6.3.2) takes the approximate form

$$\Big(\frac{\partial \sigma_{xi}}{\partial x} + \frac{\partial \sigma_{yi}}{\partial y} + \frac{\partial \sigma_{zi}}{\partial z}\Big)\, \Delta V, \tag{6.3.10}$$

where the quantity enclosed by the parentheses is evaluated at the origin. Expression (6.3.10) is an approximation to the volume integral on the right-hand side of (6.3.2).

6.3.3 The equation of motion

Substituting (6.3.1) in (6.1.7), consolidating various terms, and noting that because the volume of integration is arbitrary the combined integrand must vanish, we obtain Cauchy's differential equation governing the motion of an incompressible or compressible fluid,

$$\rho \frac{\mathrm{D}\mathbf{u}}{\mathrm{D}t} = \nabla \cdot \boldsymbol{\sigma} + \rho \, \mathbf{g}. \tag{6.3.11}$$

In index notation,

$$\rho \frac{\mathrm{D}u_i}{\mathrm{D}t} = \frac{\partial \sigma_{ji}}{\partial x_j} + \rho \, g_i, \tag{6.3.12}$$

where summation of the repeated index j is implied on the right had side, and the index i is free to vary over x, y, and z.

In terms of the point particle acceleration, $\mathbf{a}$, and the hydrodynamic volume force defined in equation (6.3.3), Cauchy's equation of motion takes the simple form

$$\rho \, \mathbf{a} = \boldsymbol{\Sigma} + \rho \, \mathbf{g} \tag{6.3.13}$$

for incompressible as well as compressible fluids.

Eulerian form

Using equations (2.8.9) and (6.2.12), we derive two alternative forms of (6.3.11) involving Eulerian derivatives,

$$\rho \, (\frac{\partial \, \mathbf{u}}{\partial t} + \mathbf{u} \cdot \nabla \mathbf{u}) = \nabla \cdot \boldsymbol{\sigma} + \rho \, \mathbf{g}, \tag{6.3.14}$$

and

$$\frac{\partial \, (\rho \, \mathbf{u})}{\partial t} + \nabla \cdot (\rho \, \mathbf{u} \, \mathbf{u}) = \nabla \cdot \boldsymbol{\sigma} + \rho \, \mathbf{g}, \tag{6.3.15}$$

applicable for incompressible as well as compressible fluids.

Explicitly, the three scalar components of (6.3.14) are given in table 6.3.1. The terms enclosed by the parentheses on the left-hand sides are the Cartesian components of the point particle acceleration. The right-hand sides include the Cartesian components of the volume force due to the hydrodynamic stresses and the components of the body force.

$$\rho\,(\frac{\partial u_x}{\partial t} + u_x\frac{\partial u_x}{\partial x} + u_y\frac{\partial u_x}{\partial y} + u_z\frac{\partial u_x}{\partial z}) = \frac{\partial \sigma_{xx}}{\partial x} + \frac{\partial \sigma_{yx}}{\partial y} + \frac{\partial \sigma_{zx}}{\partial z} + \rho\, g_x$$

$$\rho\,(\frac{\partial u_y}{\partial t} + u_x\frac{\partial u_y}{\partial x} + u_y\frac{\partial u_y}{\partial y} + u_z\frac{\partial u_y}{\partial z}) = \frac{\partial \sigma_{xy}}{\partial x} + \frac{\partial \sigma_{yy}}{\partial y} + \frac{\partial \sigma_{zy}}{\partial z} + \rho\, g_y$$

$$\rho\,(\frac{\partial u_z}{\partial t} + u_x\frac{\partial u_z}{\partial x} + u_y\frac{\partial u_z}{\partial y} + u_z\frac{\partial u_z}{\partial z}) = \frac{\partial \sigma_{xz}}{\partial x} + \frac{\partial \sigma_{yz}}{\partial y} + \frac{\partial \sigma_{zz}}{\partial z} + \rho\, g_z$$

Table 6.3.1 The Cartesian components of the equation of motion involving the point particle momentum, the hydrodynamic volume force, and the body force.

6.3.4 Evolution equations

Given the instantaneous velocity and stress fields, we can evaluate the right-hand sides of (6.3.11) and (6.3.14), as well as the second term on the left-hand side of (6.3.14), and thereby compute the rates of change $\mathrm{D}\mathbf{u}/\mathrm{D}t$ and $\partial\mathbf{u}/\partial t$. This observation suggests that the equation of motion (6.3.11) is, in fact, an evolution equation for the point particle velocity, whereas equation (6.3.14) is an evolution equation for the velocity at a fixed point in the flow.

A similar evolution equation for the density was derived in chapter 2 on the basis of the continuity equation, as shown in (2.7.18). The evolution equations for the density and velocity originate from fundamental physical laws: mass conservation, and Newton's second law of motion for a deformable medium.

6.3.5 Cylindrical polar coordinates

In the cylindrical polar coordinates illustrated in figure 1.3.2, the hydrodynamic volume force defined in equation (6.3.3) is expressed by

$$\boldsymbol{\Sigma} = \Sigma_x\,\mathbf{e}_x + \Sigma_\sigma\,\mathbf{e}_\sigma + \Sigma_\varphi\,\mathbf{e}_\varphi. \tag{6.3.16}$$

Using the rules of coordinate transformation and the chain rule of differentiation, we derive the expressions shown in table 6.3.2(a).

Equation of motion

The cylindrical polar components of the equation of motion are

$$\rho\, a_x = \Sigma_x + \rho\, g_x, \qquad \rho\, a_\sigma = \Sigma_\sigma + \rho\, g_\sigma, \qquad \rho\, a_\varphi = \Sigma_\varphi + \rho\, g_\varphi, \tag{6.3.17}$$

(a)

$$\Sigma_x = \frac{\partial \sigma_{xx}}{\partial x} + \frac{1}{\sigma}\frac{\partial(\sigma \sigma_{\sigma x})}{\partial \sigma} + \frac{1}{\sigma}\frac{\partial \sigma_{\varphi x}}{\partial \varphi}$$

$$\Sigma_\sigma = \frac{\partial \sigma_{x\sigma}}{\partial x} + \frac{1}{\sigma}\frac{\partial(\sigma \sigma_{\sigma\sigma})}{\partial \sigma} + \frac{1}{\sigma}\frac{\partial \sigma_{\varphi\sigma}}{\partial \varphi} - \frac{1}{\sigma}\sigma_{\varphi\varphi}$$

$$\Sigma_\varphi = \frac{\partial \sigma_{x\varphi}}{\partial x} + \frac{1}{\sigma^2}\frac{\partial(\sigma^2 \sigma_{\varphi\sigma})}{\partial \sigma} + \frac{1}{\sigma}\frac{\partial \sigma_{\varphi\varphi}}{\partial \varphi}$$

(b)

$$\Sigma_r = \frac{1}{r^2}\frac{\partial(r^2\,\sigma_{rr})}{\partial r} + \frac{1}{r\sin\theta}\frac{\partial(\sigma_{r\theta}\,\sin\theta)}{\partial\theta} + \frac{1}{r\sin\theta}\frac{\partial \sigma_{\varphi r}}{\partial \varphi} - \frac{\sigma_{\theta\theta} + \sigma_{\varphi\varphi}}{r}$$

$$\Sigma_\theta = \frac{1}{r^2}\frac{\partial(r^2\,\sigma_{r\theta})}{\partial r} + \frac{1}{r\sin\theta}\frac{\partial(\sigma_{\theta\theta}\,\sin\theta)}{\partial\theta} + \frac{1}{r\sin\theta}\frac{\partial \sigma_{\varphi\theta}}{\partial \varphi} + \frac{\sigma_{r\theta} - \sigma_{\varphi\varphi}\,\cot\theta}{r}$$

$$\Sigma_\varphi = \frac{1}{r^2}\frac{\partial(r^2\,\sigma_{r\varphi})}{\partial r} + \frac{1}{r}\frac{\partial \sigma_{\theta\varphi}}{\partial\theta} + \frac{1}{r\sin\theta}\frac{\partial \sigma_{\varphi\varphi}}{\partial \varphi} + \frac{\sigma_{r\varphi} + 2\,\sigma_{\theta\varphi}\,\cot\theta}{r}$$

(c)

$$\Sigma_r = \frac{1}{r}\frac{\partial(r\sigma_{rr})}{\partial r} + \frac{1}{r}\frac{\partial \sigma_{\theta r}}{\partial \theta} - \frac{1}{r}\sigma_{\theta\theta}$$

$$\Sigma_\theta = \frac{1}{r^2}\frac{\partial(r^2\sigma_{r\theta})}{\partial r} + \frac{1}{r}\frac{\partial \sigma_{\theta\theta}}{\partial \theta}$$

Table 6.3.2 Components of the hydrodynamic volume force in (a) cylindrical, (b) spherical, and (c) plane polar coordinates.

where a_x, a_σ, and a_φ are the cylindrical polar components of the point particle acceleration given in equations (2.8.14). Using the alternative expressions (2.8.15), we find

$$\rho\,\frac{\mathrm{D}u_x}{\mathrm{D}t} = \Sigma_x + \rho\, g_x, \qquad \rho\,\frac{\mathrm{D}u_\sigma}{\mathrm{D}t} = \rho\,\frac{u_\varphi^2}{\sigma} + \Sigma_\sigma + \rho\, g_\sigma,$$

$$\rho\,\frac{\mathrm{D}u_\varphi}{\mathrm{D}t} = -\rho\,\frac{u_\sigma u_\varphi}{\sigma} + \Sigma_\varphi + \rho\, g_\varphi. \tag{6.3.18}$$

These equations apply for incompressible as well as compressible fluids.

Centrifugal force

The first term on the right-hand side of the second equation in (6.3.18), $\rho\, u_\varphi^2/\sigma$, expresses an effective volume force in the radial (σ) direction due to fluid motion in the meridional (φ) direction, known as the centrifugal force. A centrifugal force arises in the flow generated by the rotation of a solid circular cylinder about its axis in a viscous liquid, as will be discussed in section 7.5.

Coriolis force

The negative of the first term on the right-hand side of third equation in (6.3.18), $\rho\, u_\sigma u_\varphi/\sigma$, expresses an effective force in the meridional (φ) direction, arising when flow occurs in both the σ and φ directions, known as the Coriolis force. The Coriolis force arises in the flow due to the spinning of a circular disk immersed in a fluid.

6.3.6 Spherical polar coordinates

In the spherical polar coordinates depicted in figure 1.3.3, the hydrodynamic volume force defined in equation (6.3.3) is expressed by

$$\mathbf{\Sigma} = \Sigma_r\, \mathbf{e}_r + \Sigma_\theta\, \mathbf{e}_\theta + \Sigma_\varphi\, \mathbf{e}_\varphi. \tag{6.3.19}$$

Using the rules of coordinate transformation and the chain rule of differentiation, we derive the expressions shown in table 6.3.2(*b*).

The spherical polar components of the equation of motion are

$$\rho\, a_r = \Sigma_r + \rho\, g_r, \qquad \rho\, a_\theta = \Sigma_\theta + \rho\, g_\theta, \qquad \rho\, a_\varphi = \Sigma_\varphi + \rho\, g_\varphi, \tag{6.3.20}$$

where a_r, a_θ, and a_φ are the spherical polar components of the point particle acceleration given by expressions (2.8.17).

6.3.7 Plane polar coordinates

In the plane polar coordinates depicted in figure 1.3.4, the hydrodynamic volume force defined in equation (6.3.3) is expressed by

$$\mathbf{\Sigma} = \Sigma_r\, \mathbf{e}_r + \Sigma_\theta\, \mathbf{e}_\theta. \tag{6.3.21}$$

Using the coordinate transformation rules and the chain rule of differentiation, we derive the expressions shown in table 6.3.2(*c*).

The plane polar components of the equation of motion are

$$\rho\, a_r = \Sigma_r + \rho\, g_r, \qquad \rho\, a_\theta = \Sigma_\theta + \rho\, g_\theta, \tag{6.3.22}$$

where a_r and a_θ are the plane polar components of the point particle acceleration given in equations (2.8.19). Alternative expressions are

$$\rho\, \frac{\mathrm{D}u_r}{\mathrm{D}t} = \rho\, \frac{u_\theta^2}{r} + \Sigma_r + \rho\, g_r, \qquad \rho\, \frac{\mathrm{D}u_\theta}{\mathrm{D}t} = -\rho\, \frac{u_r u_\theta}{r} + \Sigma_\theta + \rho\, g_\theta, \tag{6.3.23}$$

involving, respectively, the centrifugal force and the negative of the Coriolis force on the right-hand sides.

6.3.8 Vortex force

Returning to equation (6.3.14), we replace the second term in the parentheses on the left-hand side with the right-hand side of (2.8.25), and find

$$\rho\,(\frac{\partial \mathbf{u}}{\partial t} + \frac{1}{2}\nabla u^2 + \boldsymbol{\omega} \times \mathbf{u}) = \nabla \cdot \boldsymbol{\sigma} + \rho\, \mathbf{g}, \tag{6.3.24}$$

where

$$u^2 \equiv u_x^2 + u_y^2 + u_z^2 \tag{6.3.25}$$

is the square of the magnitude of the velocity. The third term on the left-hand side of (6.3.24) represents a vortex force appearing when the vorticity vector is non-parallel to the velocity vector; otherwise, their cross product is identically zero. In a Beltrami flow, the vorticity vector is parallel to the velocity vector at every point.

6.3.9 Summary of governing equation

In summary, the flow of an incompressible or compressible fluid is governed by the continuity equation (2.7.10),

$$\frac{\partial \rho}{\partial t} + \nabla \cdot (\rho\, \mathbf{u}) = 0, \tag{6.3.26}$$

and Cauchy's equation of motion expressed by (6.3.14) or (6.3.15). The Cauchy stress tensor is defined in terms of the velocity and the pressure by means of a constitutive equation, as discussed in chapter 4. The five unknowns include the three velocity components, the density, and the pressure.

The continuity equation and the three components of the equation of motion provide us with four equations. In the case of incompressible fluids, a fifth equation is provided by the idealized incompressibility condition, $\mathrm{D}\rho/\mathrm{D}t = 0$. In the case of compressible fluids, a fifth equation relating the density to the pressure is provided by thermodynamics, as shown in (4.5.35) for isentropic flow.

6.3.10 Accelerating frame of reference

The equation of motion applies to a stationary frame of reference where Newton's second law of motion applies. Suppose that the Cartesian axes translate with velocity $\mathbf{V}(t)$. The point particle acceleration in the stationary frame is

$$\mathbf{a}^{stat} = \mathbf{a} + \frac{\mathrm{d}\mathbf{V}}{\mathrm{d}t}. \tag{6.3.27}$$

Substituting this expression in the equation of motion (6.3.13) and rearranging, we find

$$\rho\,\mathbf{a} = \mathbf{\Sigma} + \rho\,\mathbf{g} - \rho\,\frac{\mathrm{d}\mathbf{V}}{\mathrm{d}t}. \tag{6.3.28}$$

The last term on the right-hand side represents a fictitious inertial acceleration force.

A more general equation can be written to describe fluid motion in a frame of reference that undergoes steady or unsteady translation and rotation.[2]

Problems

6.3.1 *Beltrami flow*

Explain why a two-dimensional or an axisymmetric flow cannot be a Beltrami flow.

6.3.2 *Free fall*

A bucket of fluid is moving in free gravitational fall. Write the equation of motion in a frame of reference attached to the bucket.

6.4 Euler's and Bernoulli's equations

Euler's equation derives from the equation of motion (6.3.11) by substituting the simplest possible constitutive equation for the stress tensor corresponding to an ideal fluid expressed by (4.5.15). Considering the individual components of the volume force $\mathbf{\Sigma}$ given in (6.3.4), we find

$$\mathbf{\Sigma} \equiv \nabla \cdot \boldsymbol{\sigma} = -\nabla p = -\frac{\partial p}{\partial x}\,\mathbf{e}_x - \frac{\partial p}{\partial y}\,\mathbf{e}_y - \frac{\partial p}{\partial z}\,\mathbf{e}_z. \tag{6.4.1}$$

The equation of motion (6.3.11) then reduces to Euler's equation

$$\rho\,\frac{\mathrm{D}\mathbf{u}}{\mathrm{D}t} = -\nabla p + \rho\,\mathbf{g}. \tag{6.4.2}$$

The associated Eulerian form is

$$\rho\,(\frac{\partial \mathbf{u}}{\partial t} + \mathbf{u} \cdot \nabla \mathbf{u}) = -\nabla p + \rho\,\mathbf{g}. \tag{6.4.3}$$

Explicitly, the three Cartesian components of (6.4.3) are

[2] Pozrikidis, C., 1997, *Introduction to Theoretical and Computational Fluid Dynamics*, Oxford University Press.

$$\rho\left(\frac{\partial u_x}{\partial t} + u_x\frac{\partial u_x}{\partial x} + u_y\frac{\partial u_x}{\partial y} + u_z\frac{\partial u_x}{\partial z}\right) = -\frac{\partial p}{\partial x} + \rho\, g_x,$$

$$\rho\left(\frac{\partial u_y}{\partial t} + u_x\frac{\partial u_y}{\partial x} + u_y\frac{\partial u_y}{\partial y} + u_z\frac{\partial u_y}{\partial z}\right) = -\frac{\partial p}{\partial y} + \rho\, g_y, \qquad (6.4.4)$$

$$\rho\left(\frac{\partial u_z}{\partial t} + u_x\frac{\partial u_z}{\partial x} + u_y\frac{\partial u_z}{\partial y} + u_z\frac{\partial u_z}{\partial z}\right) = -\frac{\partial p}{\partial z} + \rho\, g_z.$$

These equations apply for incompressible as well as compressible flow.

Polar coordinates

The cylindrical, spherical, and plane polar components of Euler's equation follow readily from equations (6.3.17), (6.3.20), and (6.3.22), using the constitutive equations given in tables 4.7.1–3 for vanishing fluid viscosity.

Vortex force

Using the expression for the point particle acceleration shown on the left-hand side of equation (6.3.24), we derive an alternative form of Euler's equation in terms of the vortex force,

$$\rho\left(\frac{\partial \mathbf{u}}{\partial t} + \frac{1}{2}\nabla u^2 - \mathbf{u}\times\boldsymbol{\omega}\right) = -\nabla p + \rho\,\mathbf{g}, \qquad (6.4.5)$$

where $u^2 \equiv u_x^2 + u_y^2 + u_z^2$ is the square of the magnitude of the velocity.

6.4.1 Boundary conditions

Euler's equation is a first-order differential equation for the velocity and pressure in the spatial domain. To compute a solution, we require one scalar boundary condition or two scalar continuity or jump conditions for the velocity or pressure over each boundary.

Over an impermeable surface, we require the no-penetration condition. Over a free surface, we require that the pressure is equal to the ambient pressure increased or decreased by an amount that is equal to the product of the surface tension and twice the local mean curvature. Over a fluid interface, we require a kinematic and a dynamic continuity or jump condition. The kinematic condition requires that the normal component of the fluid velocity is continuous across the interface. The dynamic condition requires that the pressure undergoes a discontinuity by an amount that is equal to the product of the surface tension and twice the local mean curvature.

6.4.2 Irrotational flow

The third term on the left-hand side of (6.4.5) vanishes in the case of irrotational flow. Expressing the velocity in terms of the gradient of the velocity potential, ϕ, as shown in equation (3.2.3) and more explicitly in equations (3.2.13), we find that Euler's equation (6.4.5) takes the simple form

$$\rho\,(\frac{\partial\,\nabla\phi}{\partial t} + \frac{1}{2}\,\nabla u^2) = -\nabla p + \rho\,\mathbf{g}. \tag{6.4.6}$$

The order of time and space differentiation in the gradient of the potential can be switched on the left-hand side.

The acceleration of gravity can be expressed as the gradient of the scalar $s \equiv \mathbf{g} \cdot \mathbf{x}$,

$$\mathbf{g} = \nabla s = \nabla(\mathbf{g} \cdot \mathbf{x}) = \nabla(g_x\,x + g_y\,y + g_z\,z). \tag{6.4.7}$$

Substituting this expression in (6.4.6), assuming that the density is uniform throughout the domain of flow and collecting all terms under the gradient, we find

$$\nabla\Big(\frac{\partial\,\phi}{\partial t} + \frac{1}{2}\,u^2 + \frac{p}{\rho} - \mathbf{g} \cdot \mathbf{x}\Big) = \mathbf{0}. \tag{6.4.8}$$

Since the spatial derivatives of the scalar quantity enclosed by the parentheses on the left-hand side of (6.4.8) are zero, the quantity must be independent of position although it may change in time. Euler's equation for irrotational flow then reduces to Bernoulli's equation describing the irrotational flow of a uniform-density fluid,

$$\frac{\partial\,\phi}{\partial t} + \frac{1}{2}\,u^2 + \frac{p}{\rho} - \mathbf{g} \cdot \mathbf{x} = c(t), \tag{6.4.9}$$

where $c(t)$ is an unspecified function of time.

Evolution of the velocity potential

Bernoulli's equation (6.4.9) can be regarded as an evolution equation for the harmonic potential. Given the instantaneous velocity and pressure, we can evaluate the second, third, and fourth terms on the left-hand side, compute the time derivative $\partial\phi/\partial t$, and advance the potential over a small period of elapsed time. The last term, $c(t)$, causes the potential to increase or decrease uniformly by the same rate throughout the domain of flow. However, because the velocity is computed by taking derivatives of the potential with respect to the spatial coordinates, this uniform change is inconsequential to the velocity.

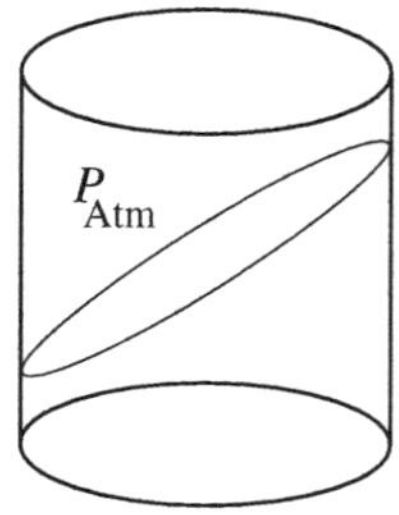

Figure 6.4.1 Irrotational motion due to the sloshing of a fluid in a container. Bernoulli's equation provides us with an evolution equation for the potential following the motion of point particles distributed over the free surface.

Lagrangian form

When a flow is bounded by a free surface where the pressure is prescribed on one side, it is beneficial to convert the Eulerian time derivative $\partial\phi/\partial t$ on the left-hand side of (6.4.9) to the material derivative, $\mathrm{D}\phi/\mathrm{D}t$. Expressing the velocity as the gradient of the potential, $\mathbf{u} = \nabla\phi$, we find

$$\frac{\mathrm{D}\phi}{\mathrm{D}t} \equiv \frac{\partial\phi}{\partial t} + \mathbf{u}\cdot\nabla\phi = \frac{\partial\phi}{\partial t} + \mathbf{u}\cdot\mathbf{u} = \frac{\partial\phi}{\partial t} + u^2, \tag{6.4.10}$$

where $u = |\mathbf{u}|$ is the magnitude of the velocity. Combining equations (6.4.9) and (6.4.10), we find

$$\frac{\mathrm{D}\,\phi}{\mathrm{D}t} = \frac{1}{2}\,u^2 - \frac{p}{\rho} + \mathbf{g}\cdot\mathbf{x} - c(t), \tag{6.4.11}$$

which provides us with the rate of change of the potential following a point particle.

Fluid sloshing in a tank

As an application, we consider the sloshing of a fluid inside a tank, as illustrated in figure 6.4.1. The pressure at the free surface on the side of the liquid, denoted by p_{fs}, is related to the ambient pressure, P_{atm}, by the dynamic boundary condition

$$p_{fs} = P_{atm} + \gamma\,2\,\kappa_m, \tag{6.4.12}$$

where γ is the surface tension, and κ_m is the mean curvature of the free surface. Applying equation (6.4.11) at a point on the free surface and using (6.4.12), we derive an expression for the rate of change of the potential following a point particle,

$$\frac{\mathrm{D}\phi}{\mathrm{D}t} = \frac{1}{2}\,u^2 - \frac{P_{atm} + \gamma\,2\,\kappa_m}{\rho} + \mathbf{g}\cdot\mathbf{x} - c(t). \tag{6.4.13}$$

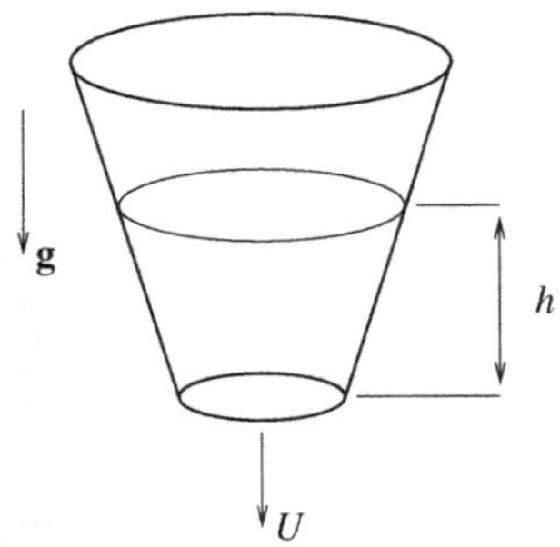

Figure 6.4.2 Illustration of the gravitational drainage of a fluid from a tank. The exit velocity can be computed from Bernoulli's equation for irrotational flow, and is given by Torricelli's law.

Integrating this equation in time following the motion of interfacial point particles provides us with a boundary condition for the potential over the free surface.

6.4.3 Steady irrotational flow

At steady state, the time derivative of the potential on the left-hand side of (6.4.9) vanishes, yielding the best known version of Bernoulli's equation,

$$\frac{1}{2}u^2 + \frac{p}{\rho} - \mathbf{g}\cdot\mathbf{x} = c(t). \tag{6.4.14}$$

The time-dependent function $c(t)$ on the right-hand side accounts for a possible uniform change in the pressure throughout the domain of flow. The three terms on the left-hand side of (6.4.14) express, respectively, the kinetic energy, the potential energy due to the pressure, and the potential energy due to the body force, all three per unit mass of the fluid. Bernoulli's equation requires that the sum of the three energies is the same at every point in the flow.

Torricelli's law

Bernoulli's equation allows us to perform approximate engineering analysis of a broad class of internal and external high-speed flows. As an example, we consider the gravitational drainage of a fluid from a tank, as illustrated in figure 6.4.2. If the rate of drainage is sufficiently slow, the flow can be assumed to be quasi-steady. This means that the magnitude of the time derivative is small compared to the rest of the terms in the unsteady Bernoulli equation (6.4.9), and the steady version of Bernoulli's equation (6.4.14) can be employed.

To compute the velocity at the point of drainage, U, we evaluate the left-hand side of (6.4.14) first at the free surface and then at the point of drainage,

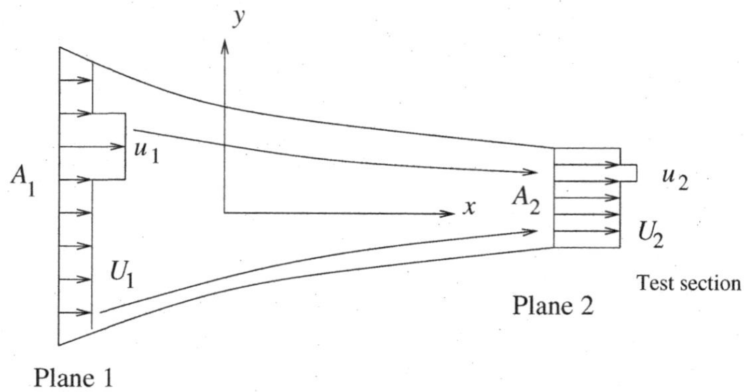

Figure 6.4.3 Illustration of flow in a wind or water tunnel with a contraction that dampens small perturbations.

and equate the two expressions thus obtained. Neglecting the velocity at the free surface as small, and setting the pressure at the free surface and at the point of drainage equal to the ambient atmospheric pressure, we derive Torricelli's law

$$U = \sqrt{2gh}, \tag{6.4.15}$$

which also describes the velocity of a rigid body in free gravitational fall.

Decay of perturbations in a wind or water tunnel

Wind and water tunnels are used extensively in studies of high-speed flow. To obtain a desirable uniform velocity profile, the tunnel is designed with a smooth contraction upstream a test section where measurement or observation takes place, as illustrated in figure 6.4.3.

Consider a small perturbation of the otherwise flat upstream velocity profile at plane 1, as illustrated in figure 6.4.3. The pressure is nearly uniform over any cross-section along the contraction. Applying Bernoulli's equation (6.4.14) for the fluid outside or inside the perturbed region and neglecting the effects of gravity, we find

$$\frac{1}{2}U_1^2 + \frac{P_1}{\rho} = \frac{1}{2}U_2^2 + \frac{P_2}{\rho}, \qquad \frac{1}{2}u_1^2 + \frac{P_1}{\rho} = \frac{1}{2}u_2^2 + \frac{P_2}{\rho}. \tag{6.4.16}$$

Combining these equations to eliminate the pressure and rearranging, we find

$$\frac{U_2 - u_2}{U_1 - u_1} = \frac{U_1 + u_1}{U_2 + u_2}. \tag{6.4.17}$$

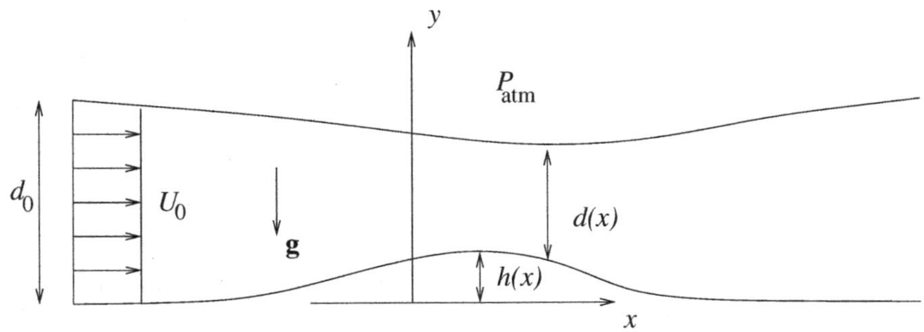

Figure 6.4.4 Irrotational free-surface flow of a horizontal stream over a hump.

Because the perturbation has been assumed small, the actual velocities, u_1 and u_2, can be replaced with the unperturbed velocities U_1 and U_2 in the numerator and denominator of the fraction on the right-hand side of (6.4.17), yielding

$$\frac{U_2 - u_2}{U_1 - u_1} = \frac{U_1}{U_2}. \tag{6.4.18}$$

Combining the approximate mass balance $U_1 A_1 = U_2 A_2$ with equation (6.4.18) and rearranging, we obtain

$$\frac{(U_2 - u_2)/U_2}{(U_1 - u_1)/U_1} = \left(\frac{A_2}{A_1}\right)^2, \tag{6.4.19}$$

which shows that the relative amplitude of the perturbation decays like the square of the contraction ratio, A_2/A_1, confirming that the contraction promotes a uniform velocity profile.

Flow of a horizontal stream over a hump

In the third example, we consider steady two-dimensional irrotational flow of a horizontal stream over a gently sloped hump, called the Venturi flume, as illustrated in figure 6.4.4. The free surface is located at $y = h(x) + d(x)$, where $h(x)$ is the height of the hump and $d(x)$ is the depth of the stream. As x tends to infinity on either side, $h(x)$ tends to zero. The profile of the streamwise velocity is assumed to be uniform across the stream; thus, $u_x = u(x)$.

Applying Bernoulli's equation (6.4.14) first at a point at the free surface located far upstream and then at an arbitrary point at the free surface, neglecting the y component of the free-surface velocity and the pressure drop across the free surface due to surface tension, and noting that the gravitational acceleration vector is given by $\mathbf{g} = (0, -g)$, we find

$$\frac{1}{2} U_0^2 + \frac{P_{atm}}{\rho} + g\, d_0 = \frac{1}{2} u^2(x) + \frac{P_{atm}}{\rho} + g\,[h(x) + d(x)], \tag{6.4.20}$$

where U_0 is the upstream velocity, and $d_0 \equiv d(x = -\infty)$ is the upstream depth. Combining the mass conservation equation $U_0 d_0 = u(x) d(x)$ with equation (6.4.20) to eliminate $u(x)$ and rearranging the resulting expression, we derive a cubic equation for the reduced layer depth $\hat{d}(x) \equiv d(x)/d_0$,

$$\hat{d}^3(x) + \hat{d}^2(x)\,[\hat{h}(x) - 1 - \frac{1}{2}\,Fr^2] + \frac{1}{2}\,Fr^2 = 0, \tag{6.4.21}$$

where $\hat{h}(x) \equiv h(x)/d_0$ is the reduced height of the hump. We have introduced the dimensionless ratio

$$Fr \equiv \frac{U_0}{\sqrt{g\,d_0}}, \tag{6.4.22}$$

expressing the relative magnitude of inertial and gravitational forces, called the Froude number.

In practice, the Venturi flume is used to deduce the flow rate from measurements of the deflection of the free surface from the horizontal position. As the Froude number tends to zero, gravitational forces dominate and equation (6.4.21) has the obvious solution $\hat{d}(x) = 1 - \hat{h}(x)$, which shows that the free surface tends to become flat. As the Froude number tends to infinity, inertial forces dominate and equation (6.4.21) has an obvious solution, $\hat{d}(x) = 1$, which shows that the depth of the stream remains constant and the free surface follows the topography of the hump. Intermediate values of the Froude number yield free surface profiles with a downward deflection (problem 6.4.5).

6.4.4 Steady rotational flow

We return to Euler's equation (6.4.5) and consider a rotational flow at steady state. The time derivative on the left-hand side vanishes, yielding

$$\frac{1}{2}\,\nabla u^2 - \mathbf{u} \times \boldsymbol{\omega} = -\frac{1}{\rho}\,\nabla p + \mathbf{g}, \tag{6.4.23}$$

where $u^2 \equiv u_x^2 + u_y^2 + u_z^2$ is the square of the magnitude of the velocity. The x component of equation (6.4.23) reads

$$\frac{1}{2}\,\frac{\partial u^2}{\partial x} - u_y\,\omega_z + u_z\,\omega_y = -\frac{1}{\rho}\,\frac{\partial p}{\partial x} + g_x. \tag{6.4.24}$$

Next, we put the origin of the Cartesian axes at a point in the fluid, identify the streamline that passes through that point, and orient the x axis tangentially to the streamline and thus parallel to the local velocity, as shown in figure 6.4.5. By design, the y and z velocity components are zero at the origin; consequently, the second and third terms on the left-hand side of equation (6.4.24) vanish. Taking advantage of these simplifications, we derive the reduced form

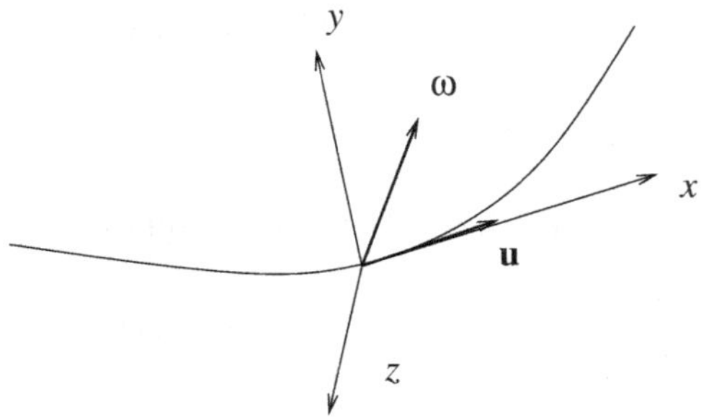

Figure 6.4.5 A system of Cartesian coordinates with the x axis tangential to a streamline is used to derive Bernoulli's equation for steady rotational flow, equation (6.4.26).

$$\frac{\partial}{\partial x}\left(\frac{1}{2}u^2 + \frac{p}{\rho} - g_x x\right) = 0, \tag{6.4.25}$$

where the left-hand side is evaluated at the origin.

Equation (6.4.25) states that the rate of change of the quantity enclosed by the parentheses on the left-hand side with respect to distance along the streamline is zero. Consequently, the quantity enclosed by the parentheses must remain constant along the streamline,

$$\frac{1}{2}u^2 + \frac{p}{\rho} - \mathbf{g}\cdot\mathbf{x} = f(x, y, z), \tag{6.4.26}$$

where the function $f(x, y, z)$ remains constant along a streamline. In two-dimensional or axisymmetric flow, $f(x, y, z)$ is a function of the stream function, ψ, which, by definition, is constant along a streamline.

6.4.5 Flow with uniform vorticity

The velocity field of a two-dimensional flow with uniform vorticity in the xy plane, $\omega_z = \Omega$, can be decomposed into (a) the velocity field of a simple two-dimensional flow with uniform vorticity Ω, denoted by $\mathbf{v}$, and (b) a potential flow expressed by a harmonic potential ϕ,

$$\mathbf{u} = \mathbf{v} + \nabla\phi. \tag{6.4.27}$$

Two examples of simple flows are simple shear flow with velocity

$$\mathbf{v} = (-\Omega\, y, 0), \tag{6.4.28}$$

and flow expressing rigid-body rotation with velocity

$$\mathbf{v} = \frac{\Omega}{2}\,(-y, x). \tag{6.4.29}$$

Using this decomposition, we derive Bernoulli's equation

$$\frac{\partial \phi}{\partial t} + \frac{1}{2}\,u^2 + \frac{p}{\rho} - \mathbf{g}\cdot\mathbf{x} + \Omega\,\psi = c(t), \tag{6.4.30}$$

where ψ is the stream function and $u = |\mathbf{u}|$ is the magnitude of the velocity (problem 6.4.2).

Problems

6.4.1 *Flow through a sudden enlargement*

Consider flow through a sudden enlargement depicted in figure 6.2.1. Use Bernoulli's equation to compute the rise in pressure $P_2 - P_1$. Compare your answer to that shown in equation (6.2.21) obtained by an approximate integral momentum balance.

6.4.2 *Flow with uniform vorticity*

(*a*) Derive Bernoulli's equation (6.4.30) for two-dimensional flow. (*b*) Derive a similar equation for axisymmetric flow where the meridional vorticity is proportional to the distance from the axis of symmetry, σ.

6.4.3 *Flow due to an unsteady point source or point vortex*

(*a*) Discuss whether the flow due to a two- or three-dimensional point source with time-dependent strength satisfies the Euler equation for inviscid flow. (*b*) Repeat (*a*) for a two-dimensional point vortex.

6.4.4 *Force on a sphere in accelerating potential flow*

Consider unsteady irrotational flow past a sphere that is held stationary in an accelerating stream with velocity $V_x(t)$. The velocity potential and Cartesian components of the velocity are given in equations (3.6.11) and (3.6.12). Use Bernoulli's equation (6.4.9) to evaluate the pressure, and then compute the force exerted on the sphere by evaluating the surface integral

$$\mathbf{F} = -\iint_{Sphere} p\,\mathbf{n}\,\mathrm{d}S, \tag{6.4.31}$$

where $\mathbf{n}$ is the unit vector normal to the sphere pointing into the fluid. Based on this result, compute the force exerted on a sphere held stationary in a non-accelerating steady flow. Discuss the physical relevance of the assumption of irrotational flow.

Computer Problem

6.4.5 *Flow over a hump*

Consider the flow of a horizontal stream over a hump, as illustrated in figure 6.4.4. The height of the hump is described by the parabolic shape function $h(x) = h_0\,[1 - (x/a)^2]$, for $-a \le x \le a$, where a is the half-length of the hump and h_0 is the maximum height. Substituting this profile in (6.4.21), we find

$$\hat{d}^3(x) + \hat{d}^2(x)\,\Big(\frac{h_0}{d_0}(1-\hat{x}^2) - 1 - \frac{1}{2}\,Fr^2\Big) + \frac{1}{2}\,Fr^2 = 0, \tag{6.4.32}$$

for $-1 \le \hat{x} \le 1$, where $\hat{x} \equiv x/a$ is the reduced distance from the mid-point. Compute and plot the reduced layer depth $\hat{d}$ against $\hat{x}$ for $h_0/d_0 = 0.01$, 0.05, and 0.10, and $Fr = 0.01$, 0.1, 10, and 100. Discuss the free surface shapes.

6.5 The Navier-Stokes equation

The Navier-Stokes equation follows from the equation of motion (6.3.11) by substituting the constitutive equation for the stress tensor for an incompressible Newtonian fluid given in (4.5.3). For a fluid with uniform viscosity, the hydrodynamic volume force is

$$\boldsymbol{\Sigma} \equiv \nabla\cdot\boldsymbol{\sigma} = \nabla\cdot(-p\,\mathbf{I} + \mu\,2\,\mathbf{E}) = -\nabla p + \mu\,2\,\nabla\cdot\mathbf{E}, \tag{6.5.1}$$

where $\mathbf{I}$ is the identity matrix. Working in index notation, we find that the ith component of twice the divergence of the rate of deformation tensor, $\mathbf{E}$, on the right-hand side is

$$\begin{aligned} 2\,\frac{\partial E_{ji}}{\partial x_j} = 2\,\frac{\partial}{\partial x_j}\Big[\frac{1}{2}\big(\frac{\partial u_i}{\partial x_j} + \frac{\partial u_j}{\partial x_i}\big)\Big] &= \frac{\partial^2 u_i}{\partial x_j \partial x_j} + \frac{\partial^2 u_j}{\partial x_j \partial x_i} \\ &= \frac{\partial^2 u_i}{\partial x_j \partial x_j} + \frac{\partial}{\partial x_i}\Big(\frac{\partial u_j}{\partial x_j}\Big), \end{aligned} \tag{6.5.2}$$

where summation is implied over the repeated index j. Because the fluid has been assumed incompressible, the divergence of the velocity in the last term enclosed by the parentheses is zero. The penultimate term is the Laplacian of the ith component of the velocity,

$$\frac{\partial^2 u_i}{\partial x_j \partial x_j} = \frac{\partial^2 u_i}{\partial x^2} + \frac{\partial^2 u_i}{\partial y^2} + \frac{\partial^2 u_i}{\partial z^2} \equiv \nabla^2 u_i. \tag{6.5.3}$$

Using these results to simplify expression (6.5.1), we find that the hydrodynamic volume force is given by

$$\boldsymbol{\Sigma} \equiv \nabla\cdot\boldsymbol{\sigma} = -\nabla p + \mu\,\nabla^2\mathbf{u}. \tag{6.5.4}$$

$$\rho\,(\frac{\partial u_x}{\partial t} + u_x \frac{\partial u_x}{\partial x} + u_y \frac{\partial u_x}{\partial y} + u_z \frac{\partial u_x}{\partial z})$$
$$= -\frac{\partial p}{\partial x} + \mu\,(\frac{\partial^2 u_x}{\partial x^2} + \frac{\partial^2 u_x}{\partial y^2} + \frac{\partial^2 u_x}{\partial z^2}) + \rho\, g_x$$

$$\rho\,(\frac{\partial u_y}{\partial t} + u_x \frac{\partial u_y}{\partial x} + u_y \frac{\partial u_y}{\partial y} + u_z \frac{\partial u_y}{\partial z})$$
$$= -\frac{\partial p}{\partial y} + \mu\,(\frac{\partial^2 u_y}{\partial x^2} + \frac{\partial^2 u_y}{\partial y^2} + \frac{\partial^2 u_y}{\partial z^2}) + \rho\, g_y$$

$$\rho\,(\frac{\partial u_z}{\partial t} + u_x \frac{\partial u_z}{\partial x} + u_y \frac{\partial u_z}{\partial y} + u_z \frac{\partial u_z}{\partial z})$$
$$= -\frac{\partial p}{\partial z} + \mu\,(\frac{\partial^2 u_z}{\partial x^2} + \frac{\partial^2 u_z}{\partial y^2} + \frac{\partial^2 u_z}{\partial z^2}) + \rho\, g_z$$

Table 6.5.1 Eulerian form of the three Cartesian components of the Navier-Stokes equation.

Correspondingly, the equation of motion (6.3.11) reduces to the Navier-Stokes equation,

$$\rho\,\frac{\mathrm{D}\mathbf{u}}{\mathrm{D}t} = -\nabla p + \mu\,\nabla^2 \mathbf{u} + \rho\,\mathbf{g}, \tag{6.5.5}$$

which is distinguished from Euler's equation (6.4.2) by the presence of the viscous force represented by the product of the viscosity and the Laplacian of the velocity on the right-hand side.

The Eulerian form of the Navier-Stokes equation involving time and space derivatives is

$$\rho\,(\frac{\partial \mathbf{u}}{\partial t} + \mathbf{u} \cdot \nabla \mathbf{u}\,) = -\nabla p + \mu\,\nabla^2 \mathbf{u} + \rho\,\mathbf{g}. \tag{6.5.6}$$

The three Cartesian scalar components of this vectorial equation are shown in table 6.5.1.

6.5.1 Pressure and viscous forces

The negative of the pressure gradient on the right-hand side of (6.5.6) represents the pressure force. The Laplacian of the velocity multiplied by the viscosity on the right-hand side of (6.5.6) represents the viscous force. Working in index notation under the assumption that the fluid is incompressible and thus the velocity field is solenoidal, $\nabla \cdot \mathbf{u} = 0$, we find that the Laplacian of the velocity is equal to the negative of the curl of the vorticity (problem 6.5.1),

$$\nabla^2 \mathbf{u} = -\nabla \times \boldsymbol{\omega}. \tag{6.5.7}$$

An important consequence of this identity is that, if the flow is irrotational, or if the vorticity is constant, or if the vorticity field is irrotational, then the viscous force vanishes even though the fluid is not inviscid. In that case, the Navier-Stokes equation reduces to Euler's equation, which can be integrated to yield Bernoulli's equation (6.4.9) for irrotational flow or equation (6.4.26) for steady rotational flow.

6.5.2 A radially expanding or contracting bubble

The flow generated by the radial expansion or contraction of a spherical bubble with time-dependent radius, $a(t)$, provides us with an example of an irrotational flow with nonzero viscous stresses but vanishing viscous forces. The velocity can be represented by a three-dimensional point source with time dependent strength, $m(t)$, placed at the bubble center. In spherical polar coordinates with the origin at the bubble center, the velocity potential and radial component of the velocity are given, respectively, by

$$\phi(r,t) = -\frac{m(t)}{4\pi}\,\frac{1}{r}, \qquad u_r(r,t) = \frac{\partial \phi}{\partial r} = \frac{m(t)}{4\pi}\,\frac{1}{r^2}, \tag{6.5.8}$$

where $m(t)$ is the strength of the point source. The no-penetration condition at the surface of the bubble requires

$$\frac{\mathrm{d}a}{\mathrm{d}t} = u_r(r=a), \tag{6.5.9}$$

which can be rearranged to give an expression for the strength of the point source in terms of the bubble radius,

$$m(t) = 4\pi a^2(t)\,\frac{\mathrm{d}a}{\mathrm{d}t}. \tag{6.5.10}$$

Substituting (6.5.10) in equations (6.5.8), we find

$$\phi(r) = -a^2(t)\,\frac{\mathrm{d}a}{\mathrm{d}t}\,\frac{1}{r} = -\frac{1}{3}\,\frac{\mathrm{d}a^3}{\mathrm{d}t}\,\frac{1}{r}, \qquad u_r(r) = \frac{\partial \phi}{\partial r} = \frac{1}{3}\,\frac{\mathrm{d}a^3}{\mathrm{d}t}\,\frac{1}{r^2}. \tag{6.5.11}$$

Now referring to Bernoulli's equation (6.4.9) for unsteady irrotational flow, we compute the first and second terms on the left-hand side,

$$\frac{\partial \phi}{\partial t} = -\frac{1}{3}\,\frac{\mathrm{d}^2 a^3}{\mathrm{d}t^2}\,\frac{1}{r}, \qquad u^2 = u_r^2(r) = \frac{1}{9}\left(\frac{\mathrm{d}a^3}{\mathrm{d}t}\right)^2 \frac{1}{r^4}. \tag{6.5.12}$$

Substituting these expressions in Bernoulli's equation (6.4.9) and solving for the pressure, we find

$$\frac{p(r)}{\rho} = \frac{1}{3}\,\frac{\mathrm{d}^2 a^3}{\mathrm{d}t^2}\,\frac{1}{r} - \frac{1}{18}\left(\frac{\mathrm{d}a^3}{\mathrm{d}t}\right)^2 \frac{1}{r^4} + c(t) + \mathbf{g}\cdot\mathbf{x}. \tag{6.5.13}$$

Far from the bubble, the first and second terms on the right-hand side vanish, and the pressure assumes the linear and possibly time-dependent distribution $p_\infty(\mathbf{x}, t) = \rho\,[c(t) + \mathbf{g} \cdot \mathbf{x}]$.

The normal stress, σ_{rr}, undergoes a jump across the bubble surface, determined by the mean curvature and surface tension, γ. Using the simplified version of the interfacial condition (4.3.25) for an interface with constant surface tension, we find

$$\sigma_{rr}(r = a) + p_B(t) = \gamma\, 2\kappa_m = \gamma\, \frac{2}{a}, \tag{6.5.14}$$

where $p_B(t)$ is the pressure in the bubble interior and $\kappa_m = 1/a$ is the mean curvature of the interface. Substituting the second equation in (6.5.11) in the Newtonian constitutive equation

$$\sigma_{rr} = -p + 2\mu\, \frac{\partial u_r}{\partial r}, \tag{6.5.15}$$

and the result in (6.5.14), we find

$$p(r = a) = p_B(t) - 4\mu\, \frac{\mathrm{d}a}{\mathrm{d}t}\, \frac{1}{a} - \gamma\, \frac{2}{a}. \tag{6.5.16}$$

Finally, we apply (6.5.13) at the bubble surface, evaluate the pressure from (6.5.16), neglect hydrostatic variations over the diameter of the bubble, and rearrange the resulting expression to obtain the generalized Rayleigh equation

$$\rho a\, \frac{\mathrm{d}^2 a}{\mathrm{d}t^2} + \frac{3}{2}\, \rho \left(\frac{\mathrm{d}a}{\mathrm{d}t}\right)^2 + 4\mu\, \frac{\mathrm{d}a}{\mathrm{d}t}\, \frac{1}{a} + \gamma \frac{2}{a} = p_B(t) - p_\infty(\mathbf{x}_B, t), \tag{6.5.17}$$

where $\mathbf{x}_B$ is the location of the bubble center.

The second-order nonlinear ordinary differential equation (6.5.17) governs the evolution of the bubble radius. To compute the solution, we require the initial bubble radius, the initial rate of expansion $\mathrm{d}a/\mathrm{d}t$, the bubble pressure, and the liquid pressure at infinity. The bubble pressure may be further related to the bubble volume by means of an appropriate equation of state provided by thermodynamics.

6.5.3 Boundary conditions

The Navier-Stokes equation is a second-order differential equation for the velocity with respect to the spatial coordinates. To compute a solution, we require one scalar boundary condition for each component of the velocity or traction over each boundary.

Over an impermeable solid surface, we require the no-penetration and the no-slip or slip boundary condition discussed in sections 2.10 and 4.8.

Over a free surface with constant surface tension, we require that the tangential component of the traction vanishes, while the normal component is equal to the negative of the ambient pressure increased or decreased by the capillary pressure defined as the product of the surface tension and twice the local mean curvature, as discussed in section 4.3.

Over a fluid interface, we require kinematic and dynamic continuity or jump conditions. The kinematic condition requires that all velocity components are continuous across the interface. The dynamic condition requires that the normal component of the traction undergoes a discontinuity by an amount that is equal to the capillary pressure, while the tangential component of the traction undergoes a discontinuity that is determined by the Marangoni traction due to variations in surface tension, as discussed in section 4.3.

6.5.4 Polar coordinates

The cylindrical polar components of the hydrodynamic volume force for a Newtonian fluid arise by substituting the constitutive relations shown in table 4.7.1 in the expressions shown in table 4.3.1(a). After a fair amount of algebra, we derive the expressions shown in table 6.5.2(a). The cylindrical polar components of the Navier-Stokes equation arise by substituting these relations in the right-hand sides of (6.3.17).

The spherical polar components of the hydrodynamic volume force arise by substituting the constitutive relations given in table 4.7.2 in the expressions shown in table 4.3.1(b). After a fair amount of algebra, we derive the expressions shown in table 6.5.2(b). The Laplacian operator ∇^2 in spherical polar coordinates is defined in equation (3.2.21). The spherical polar components of the Navier-Stokes equation arise by substituting these expressions in the right-hand sides of (6.3.20).

The plane polar components of the hydrodynamic volume force arise by substituting the constitutive relations shown in table 4.7.3 in the expressions shown in table 4.3.1(c). After a fair amount of algebra, we derive the expressions shown in table 6.5.2(c). The plane polar components of the Navier-Stokes equation arise by substituting these expressions in the right-hand sides of (6.3.23).

Problems

6.5.1 *Viscous force*

Prove identity (6.5.7). *Hint:* Set the vorticity equal to the curl of the velocity; express the curl of the vorticity in index notation; and then use identity (2.3.9).

(a)

$$\Sigma_x = -\frac{\partial p}{\partial x} + \mu \Big(\frac{\partial^2 u_x}{\partial x^2} + \frac{1}{\sigma}\frac{\partial}{\partial \sigma}\Big(\sigma \frac{\partial u_x}{\partial \sigma}\Big) + \frac{1}{\sigma^2}\frac{\partial^2 u_x}{\partial \varphi^2} \Big)$$

$$\Sigma_\sigma = -\frac{\partial p}{\partial \sigma} + \mu \Big(\frac{\partial^2 u_\sigma}{\partial x^2} + \frac{\partial}{\partial \sigma}\Big(\frac{1}{\sigma}\frac{\partial(\sigma u_\sigma)}{\partial \sigma}\Big) + \frac{1}{\sigma^2}\frac{\partial^2 u_\sigma}{\partial \varphi^2} - \frac{2}{\sigma^2}\frac{\partial u_\varphi}{\partial \varphi} \Big)$$

$$\Sigma_\varphi = -\frac{1}{\sigma}\frac{\partial p}{\partial \varphi} + \mu \Big(\frac{\partial^2 u_\varphi}{\partial x^2} + \frac{\partial}{\partial \sigma}\Big(\frac{1}{\sigma}\frac{\partial(\sigma\, u_\varphi)}{\partial \sigma}\Big) + \frac{1}{\sigma^2}\frac{\partial^2 u_\varphi}{\partial \varphi^2} + \frac{2}{\sigma^2}\frac{\partial u_\sigma}{\partial \varphi} \Big)$$

(b)

$$\Sigma_r = -\frac{\partial p}{\partial r} + \mu \Big(\nabla^2 u_r - \frac{2}{r^2} u_r - \frac{2}{r^2}\frac{\partial u_\theta}{\partial \theta} - \frac{2}{r^2} u_\theta \cot\theta - \frac{2}{r^2 \sin\theta}\frac{\partial u_\varphi}{\partial \varphi} \Big)$$

$$\Sigma_\theta = -\frac{1}{r}\frac{\partial p}{\partial \theta} + \mu \Big(\nabla^2 u_\theta + \frac{2}{r^2}\frac{\partial u_r}{\partial \theta} - \frac{u_\theta}{r^2 \sin^2\theta} - \frac{2\cos\theta}{r^2 \sin^2\theta}\frac{\partial u_\varphi}{\partial \varphi} \Big)$$

$$\Sigma_\varphi = -\frac{1}{r\,\sin\theta}\frac{\partial p}{\partial \varphi} + \mu \Big(\nabla^2 u_\varphi - \frac{u_\varphi}{r^2 \sin^2\theta} + \frac{2}{r^2 \sin\theta}\frac{\partial u_r}{\partial \varphi} + \frac{2\cos\theta}{r^2 \sin^2\theta}\frac{\partial u_\theta}{\partial \varphi} \Big)$$

(c)

$$\Sigma_r = -\frac{\partial p}{\partial r} + \mu \Big(\frac{\partial}{\partial r}\Big(\frac{1}{r}\frac{\partial(r\, u_r)}{\partial r}\Big) + \frac{1}{r^2}\frac{\partial^2 u_r}{\partial \theta^2} - \frac{2}{r^2}\frac{\partial u_\theta}{\partial \varphi} \Big)$$

$$\Sigma_\theta = -\frac{1}{r}\frac{\partial p}{\partial \theta} + \mu \Big(\frac{\partial}{\partial r}\Big(\frac{1}{r}\frac{\partial(r\, u_\theta)}{\partial r}\Big) + \frac{1}{r^2}\frac{\partial^2 u_\theta}{\partial \theta^2} + \frac{2}{r^2}\frac{\partial u_r}{\partial \theta} \Big)$$

Table 6.5.2 Components of the hydrodynamic volume force for a Newtonian fluid in (a) cylindrical, (b) spherical, and (c) plane polar coordinates. The Laplacian operator ∇^2 in spherical polar coordinates is defined in equation (3.2.21).

6.5.2 *Steady flow*

Consider a flow at steady state. Explain why it is not generally possible to specify an arbitrary solenoidal velocity field that satisfies the boundary conditions, and then compute the pressure by solving the Navier-Stokes equation (6.5.6).

Hint: Consider the conditions for the equation $\nabla p = \mathbf{F}$ to have a solution, where $\mathbf{F}$ is a vector field. Recall that the curl of the gradient of a function vanishes, as shown in equation (6.6.12).

6.5.3 *Expansion of a bubble*

Show that, when the right-hand side of (6.5.17) vanishes and viscous stresses and surface tension are insignificant, an exact solution of equation (6.5.17) is

$$\frac{a(t)}{a(t=0)} = \Big(1 + \frac{5}{2}\,\frac{t}{a(t=0)}\,\Big(\frac{\mathrm{d}a}{\mathrm{d}t}\Big)_{t=0}\Big)^{2/5}. \tag{6.5.18}$$

6.6 Vorticity transport

In section 6.3, we interpreted the equation of motion as an evolution equation determining either the rate of change of the velocity of a moving point particle, or the rate of change of the fluid velocity at a fixed point in the flow. Descendant evolution equations governing the rate of change of the spatial derivatives of the velocity comprising the velocity gradient tensor and its symmetric and skew-symmetric components comprising the rate of deformation and vorticity tensors can be derived by straightforward differentiation.

Of particular interest is the evolution of the skew-symmetric part of the velocity gradient tensor related to the vorticity as shown in (2.3.13). The availability of an evolution equation for the vorticity allows us to study the rate of change of the angular velocity of a small fluid parcel as it translates and deforms while it is convected in a flow.

6.6.1 Two-dimensional flow

We begin by considering the evolution of the strength of the vorticity in a two-dimensional flow in the xy plane, defined in terms of the velocity as

$$\omega_z = \frac{\partial u_y}{\partial x} - \frac{\partial u_x}{\partial y}. \tag{6.6.1}$$

To derive an evolution equation for ω_z, we divide both sides of the equation of motion (6.3.14) by the density so as to remove it from the left-hand side, and then take the y derivative of the x component of the resulting equation and subtract it from the x derivative of the corresponding y component. The result is the vorticity transport equation

$$\frac{\partial \omega_z}{\partial t} + \frac{\partial}{\partial x}(u_x \frac{\partial u_y}{\partial x} + u_y \frac{\partial u_y}{\partial y}) - \frac{\partial}{\partial y}(u_x \frac{\partial u_x}{\partial x} + u_y \frac{\partial u_x}{\partial y})$$
$$= \frac{\partial}{\partial x}(\frac{1}{\rho}\,\Sigma_y) - \frac{\partial}{\partial y}(\frac{1}{\rho}\,\Sigma_x). \tag{6.6.2}$$

Expanding out the derivatives on the left-hand side and using the continuity equation, $\partial u_x/\partial x + \partial u_y/\partial y = 0$, we obtain the simpler form

$$\frac{\mathrm{D}\omega_z}{\mathrm{D}t} \equiv \frac{\partial \omega_z}{\partial t} + u_x \frac{\partial \omega_z}{\partial x} + u_y \frac{\partial \omega_z}{\partial y} = \frac{\partial}{\partial x}(\frac{1}{\rho}\,\Sigma_y) - \frac{\partial}{\partial y}(\frac{1}{\rho}\,\Sigma_x). \tag{6.6.3}$$

The left-hand side of (6.6.3) expresses the material derivative of the vorticity which, according to equation (2.3.7), is equal to twice the rate of change of the angular velocity of a small fluid parcel.

Next, we expand the derivatives on the right-hand side of (6.6.3) and express the hydrodynamic volume force in terms of the stresses using the definitions

$$\Sigma_x \equiv \frac{\partial \sigma_{xx}}{\partial x} + \frac{\partial \sigma_{yx}}{\partial y}, \qquad \Sigma_y \equiv \frac{\partial \sigma_{xy}}{\partial x} + \frac{\partial \sigma_{yy}}{\partial y}. \tag{6.6.4}$$

The result is the general form of the vorticity transport equation for incompressible fluids,

$$\begin{aligned} \frac{\mathrm{D}\omega_z}{\mathrm{D}t} &= \frac{1}{\rho^2}\,(-\Sigma_y \frac{\partial \rho}{\partial x} + \Sigma_x \frac{\partial \rho}{\partial y}) \\ &\quad + \frac{1}{\rho}\Big(\frac{\partial^2 \sigma_{xy}}{\partial x^2} - \frac{\partial^2 \sigma_{xy}}{\partial y^2} + \frac{\partial^2 (\sigma_{yy} - \sigma_{xx})}{\partial x\, \partial y}\Big). \end{aligned} \tag{6.6.5}$$

Baroclinic production of vorticity

The term enclosed by the first set of parentheses on the right-hand side of (6.6.5) expresses production of vorticity due to density inhomogeneity, known as baroclinic production. To illustrate the physical mechanism responsible for this term, we consider a vertical column of fluid whose density increases upward in the direction of the y axis, $\partial\rho/\partial y > 0$, as depicted in figure 6.6.1. The x component of the hydrodynamic volume force, Σ_x, causes the column to accelerate in the positive direction of the x axis. Because the density and thus the inertia of the fluid increases with height, the top accelerates less than the bottom, and the column buckles backwards exhibiting counterclockwise rotation expressed by the second term inside the first set of parentheses on the right-hand side of (6.6.5).

A similar interpretation is possible in the case of a horizontal layer whose density increases in the direction of the x axis, $\partial\rho/\partial x > 0$, rotating under the influence of a vertical volume force, Σ_y. The associated baroclinic production of vorticity is represented the first term inside the first set of parentheses on the right-hand side of (6.6.5).

Negligible viscous forces

When viscous forces are insignificant, the shear stress virtually vanishes, the normal stress is equal to the negative of the pressure, and the hydrodynamic volume force is equal to the negative of the pressure gradient. Consequently, the term enclosed by the large parentheses on the right-hand side of (6.6.5) makes a negligible contribution.

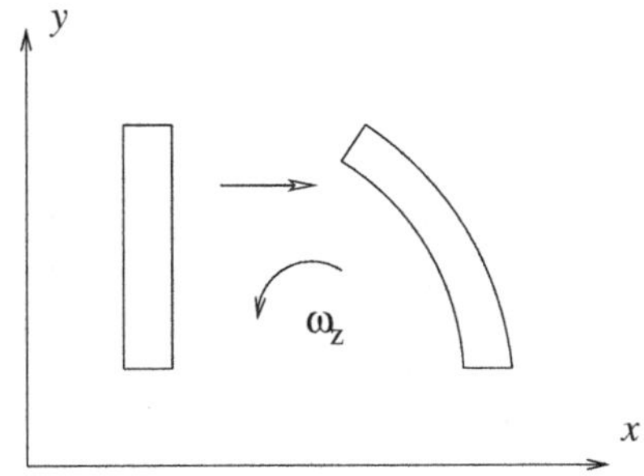

Figure 6.6.1 Vorticity is produced when a column of fluid that is heavier at the top buckles in acceleration under the influence of a volume force.

Substituting $\Sigma_x = -\partial p/\partial x$ and $\Sigma_y = -\partial p/\partial y$, we find that the term expressing baroclinic production obtains a simple form, yielding the vorticity transport equation

$$\rho^2 \frac{\mathrm{D}\omega_z}{\mathrm{D}t} = \frac{\partial p}{\partial y}\frac{\partial \rho}{\partial x} - \frac{\partial p}{\partial x}\frac{\partial \rho}{\partial y} = (\nabla\rho \times \nabla p)\cdot \mathbf{e}_z, \tag{6.6.6}$$

where $\mathbf{e}_z$ is the unit vector along the z axis that is perpendicular to the xy plane of the flow.

In the case of a uniform-density fluid, $\nabla\rho = \mathbf{0}$, equation (6.6.6) predicts

$$\frac{\mathrm{D}\omega_z}{\mathrm{D}t} = 0, \tag{6.6.7}$$

which shows that a small fluid parcel rotates with constant angular velocity as it moves around the domain of flow. The physical origin of this remarkably simple result can be traced back to conservation of angular momentum in the absence of a shearing stress imparting a torque.

Incompressible Newtonian fluids

Next, we consider the evolution of vorticity in an incompressible Newtonian fluid with uniform density and viscosity. Substituting the constitutive equation for the stress tensor (4.5.4) in the right-hand side of (6.6.5) and simplifying the resulting expression using the continuity equation, we derive the vorticity transport equation

$$\frac{\mathrm{D}\omega_z}{\mathrm{D}t} = \frac{\partial \omega_z}{\partial t} + u_x\frac{\partial \omega_z}{\partial x} + u_y\frac{\partial \omega_z}{\partial y} = \nu\,(\frac{\partial^2 \omega_z}{\partial x^2} + \frac{\partial^2 \omega_z}{\partial y^2}) \equiv \nu\,\nabla^2\omega_z, \tag{6.6.8}$$

where $\nu \equiv \mu/\rho$ is a physical constant with dimensions of length squared divided by time, called the kinematic viscosity of the fluid.

Temperature ^{o}C	Water cm^2/sec	Air $\text{cm}^2\ \text{sec}$
20	1.004×10^{-2}	15.05×10^{-2}
40	0.658×10^{-2}	18.86×10^{-2}
80	0.365×10^{-2}	20.88×10^{-2}

Table 6.6.1 The kinematic viscosity of water and air at three temperatures.

Table 6.6.1 displays the kinematic viscosity of water and air at three temperatures. Note that the kinematic viscosity of air is higher than that of water by two or three orders of magnitude. In contrast, the viscosity of water is higher than that of air by one or two orders of magnitude. Strangely, air is kinematically more viscous than water.

The right-hand side of (6.6.8) expresses diffusion of vorticity in the xy plane. Like temperature or concentration of a species, vorticity spreads from regions of highly rotational flow to regions of irrotational flow; that is, from regions where small spherical parcels exhibit intense rotation to regions of weakly rotational or irrotational motion. The actual mechanism by which this occurs will be exemplified in chapters 7 and 10 with reference to unsteady and boundary-layer flows.

6.6.2 Axisymmetric flow

Next, we consider an axisymmetric flow without swirling motion and refer to cylindrical polar coordinates, (x, σ, φ), depicted in figure 6.6.2. Working as previously for two-dimensional flow, we derive the vorticity transport equation for an incompressible Newtonian fluid with uniform density and viscosity,

$$\frac{\mathrm{D}}{\mathrm{D}t}\left(\frac{\omega_\varphi}{\sigma}\right) = \frac{\nu}{\sigma^2}\, E^2(\sigma\, \omega_\varphi). \tag{6.6.9}$$

The second-order linear differential operator E^2 on the right-hand side, defined in equations (2.9.14) and (2.9.17), is the counterpart of the Laplacian operator for two-dimensional flow shown in (6.6.8).

Vortex stretching

When viscous forces are negligible, the right-hand side of (6.6.9) is zero and the resulting vorticity transport equation takes the form

$$\frac{\mathrm{D}}{\mathrm{D}t}\left(\frac{\omega_\varphi}{\sigma}\right) = 0. \tag{6.6.10}$$

This equation requires that the meridional component of the vorticity of a point particle, ω_φ, is proportional to the distance of the point particle from the axis of

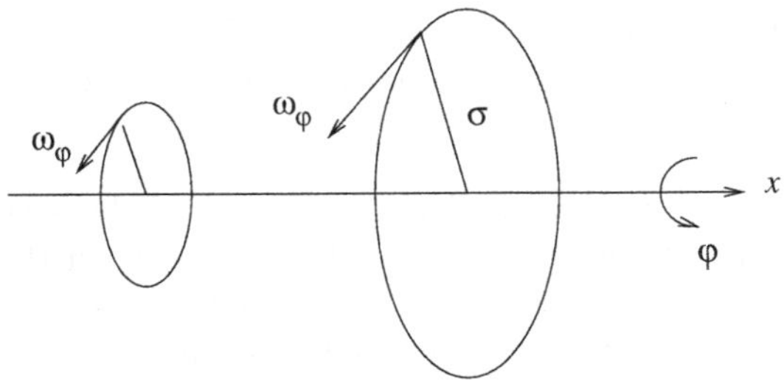

Figure 6.6.2 The vorticity of a point particle in axisymmetric flow increases as the particle moves farther away from the axis of symmetry due to vortex stretching.

symmetry, σ, so that the ratio between the two is constant in time and equal to the initial value, as illustrated schematically in figure 6.6.2. This fundamental evolution law expresses a physical process known as vortex stretching. The significance of vortex stretching will be further discussed in chapter 11 in the context of vortex flow.

6.6.3 Three-dimensional flow

Generalizing the preceding discussion, we now derive an evolution equation for the vorticity vector in a three-dimensional incompressible Newtonian flow. The density and viscosity are assumed to be uniform throughout the domain of flow. Our point of departure is the Navier-Stokes equation (6.4.2). Using the expression for the point particle acceleration shown on the left-hand side of equation (6.3.24), we find the following alternative form of the Navier-Stokes equation in terms of the vortex force,

$$\rho\,(\frac{\partial \mathbf{u}}{\partial t} + \frac{1}{2}\,\nabla u^2 + \boldsymbol{\omega} \times \mathbf{u}) = -\nabla p + \mu\,\nabla^2 \mathbf{u} + \rho\,\mathbf{g}, \qquad (6.6.11)$$

where $u^2 \equiv u_x^2 + u_y^2 + u_z^2$ is the square of the magnitude of the velocity.

To derive an evolution equation for the vorticity, we take the curl of both sides of equation (6.6.11). A vector identity states that the curl of the gradient of a smooth scalar function of position, f, is identically zero,

$$\nabla \times \nabla f = \mathbf{0}. \qquad (6.6.12)$$

To prove this identity, we work in index notation and express the ith component of the left-hand side as

$$\epsilon_{ijk}\frac{\partial}{\partial x_j}\left(\frac{\partial f}{\partial x_k}\right) = \epsilon_{ijk}\frac{\partial^2 f}{\partial x_j\,\partial x_k} = -\epsilon_{ikj}\frac{\partial^2 f}{\partial x_k\,\partial x_j}. \qquad (6.6.13)$$

The symmetry of the second derivative on the right-hand side combined with the inherent antisymmetry of the alternating tensor ensures that the right-hand side is identically zero.

Using identity (6.6.12), we find that the curl of the second term on the left-hand side of (6.6.11) and the curl of the first term on the right-hand side of (6.6.11) are both zero. Invoking the definition $\boldsymbol{\omega} = \nabla \times \mathbf{u}$, we obtain the vorticity transport equation for three-dimensional flow,

$$\frac{\partial \boldsymbol{\omega}}{\partial t} + \nabla \times (\boldsymbol{\omega} \times \mathbf{u}) = \nu \nabla^2 \boldsymbol{\omega}, \tag{6.6.14}$$

where $\nu \equiv \mu/\rho$ is the kinematic viscosity.

The second term on the left-hand side of (6.6.14) can be manipulated to obtain a physical interpretation. In index notation, the ith component of the vector expressed by this term is

$$\epsilon_{ijk} \frac{\partial}{\partial x_j}(\epsilon_{klm}\, \omega_l\, u_m) = \epsilon_{ijk}\, \epsilon_{klm} \frac{\partial(\omega_l\, u_m)}{\partial x_j} = \epsilon_{ijk}\, \epsilon_{lmk} \frac{\partial(\omega_l\, u_m)}{\partial x_j}. \tag{6.6.15}$$

Using property (2.3.9), we recast the right-hand side into the form

$$\begin{aligned}(\delta_{il}\, \delta_{jm} - \delta_{im}\, \delta_{jl}) \frac{\partial(\omega_l\, u_m)}{\partial x_j} &= \frac{\partial(\omega_i\, u_j)}{\partial x_j} - \frac{\partial(\omega_j\, u_i)}{\partial x_j} \\ &= u_j \frac{\partial \omega_i}{\partial x_j} + \omega_i \frac{\partial u_j}{\partial x_j} - u_i \frac{\partial \omega_j}{\partial x_j} - \omega_j \frac{\partial u_i}{\partial x_j}.\end{aligned} \tag{6.6.16}$$

An identity states that the divergence of the curl of a smooth vector field is zero. A consequence of this identity is that the vorticity field is solenoidal. Because the fluid has been assumed incompressible, the velocity field is also solenoidal. Consequently, the second and third terms on the right-hand side of (6.6.16) are zero.

Substituting the result back in equation (6.6.14), we derive the targeted vorticity transport equation

$$\frac{\mathrm{D}\omega_i}{\mathrm{D}t} = \frac{\partial \omega_i}{\partial t} + u_j \frac{\partial \omega_i}{\partial x_j} = \omega_j \frac{\partial u_i}{\partial x_j} + \nu \nabla^2 \omega_i. \tag{6.6.17}$$

In vector notation,

$$\frac{\mathrm{D}\boldsymbol{\omega}}{\mathrm{D}t} = \frac{\partial \boldsymbol{\omega}}{\partial t} + \mathbf{u} \cdot \nabla \boldsymbol{\omega} = \boldsymbol{\omega} \cdot \nabla \mathbf{u} + \nu \nabla^2 \boldsymbol{\omega}. \tag{6.6.18}$$

The left-hand side of (6.6.18) is the material derivative of the vorticity expressing the rate of change of the vorticity following the motion of point particles.

Vorticity rotation and stretching

To understand the nature of the first term on the right-hand side of (6.6.18), we consider a small material vector, $d\mathbf{X}$, and label the first point as A, and the last point as B. Using a Taylor series expansion, we find that the difference in the velocity across the end points is

$$\mathbf{u}^B - \mathbf{u}^A = d\mathbf{X} \cdot \nabla \mathbf{u}. \tag{6.6.19}$$

Comparing this expression with the first term on the right-hand side of (6.6.18), we find that the vorticity vector behaves like a material vector convected by the flow. This means that the vorticity vector rotates and stretches or compresses under the influence of the local flow.

In the case of two-dimensional flow, because the vorticity vector is normal to the plane of the flow, neither rotation, nor stretching, nor compression can take place. In the case of axisymmetric flow, because the vorticity vector points in the meridional direction, rotation is prohibited, but stretching or compression can take place, as discussed in section 6.6.2.

Persistence of irrotational motion in inviscid flow

One important consequence of (6.6.18) is that, if the vorticity of a point particle is zero at the initial instant, it will remain zero at all times. Thus, volumes of rotational fluid remain rotational, volumes of irrotational fluid remain irrotational, and the interface between them remains sharp and well-defined at all times.

Source of vorticity in viscous flow

In practice, because a fluid flow is always established from the state of rest, the initial vorticity distribution is zero. Since the right-hand side of the vorticity transport equation (6.6.18) vanishes throughout the fluid, the initial rate of production of vorticity is also zero, and this seemingly suggests that the flow will remain irrotational at all times. However, this is not true, as vorticity, like heat, enters the fluid by diffusion across the boundaries. The precise mechanism by which this occurs will be discussed in chapters 7 and 10.

Problems

6.6.1 *Reduction to two-dimensional flow*

Show that the vorticity transport equation (6.6.18) reproduces the transport equation (6.6.8) for the strength of the vorticity, ω_z, of a two-dimensional flow in the xy plane.

6.6.2 *Convection of vorticity*

Prove the identity

$$\omega_j \, \frac{\partial u_i}{\partial x_j} = \omega_j \, \frac{\partial u_j}{\partial x_i}. \tag{6.6.20}$$

This identify allows us to express the first term on the left-hand side of (6.6.18) in the alternative form

$$\boldsymbol{\omega} \cdot \nabla \mathbf{u} = \nabla \mathbf{u} \cdot \boldsymbol{\omega}. \tag{6.6.21}$$

Hint: Begin with the identity $\boldsymbol{\omega} \times \boldsymbol{\omega} = \boldsymbol{\omega} \times \nabla \times \mathbf{u} = \mathbf{0}$, and then work in index notation using identity (2.3.9).

6.7 Dynamic similitude and the Reynolds number

Consider a uniform (streaming) flow along the x axis with velocity U_1 past a stationary body with designated size L_1, as illustrated in figure 6.7.1(a). Also consider another streaming flow along the x axis with velocity U_2 past a second body that arises by shrinking or expanding the first body by a certain factor, as illustrated in figure 6.7.1(b). If the surface of the first body is described by the equation

$$f(x, y, z) = 0, \tag{6.7.1}$$

then the surface of the second body is described by the equation

$$f(\frac{x}{\alpha}, \frac{y}{\alpha}, \frac{z}{\alpha}) = 0, \tag{6.7.2}$$

where $\alpha \equiv L_2/L_1$ is a scaling factor.

For example, if the first body is a sphere with diameter L_1 centered at the point (x_c, y_c, z_c), then

$$f(x, y, z) = (x - x_c)^2 + (y - y_c)^2 + (z - z_c)^2 - \frac{1}{4} \, L_1^2 \tag{6.7.3}$$

and

$$f(\frac{x}{\alpha}, \frac{y}{\alpha}, \frac{z}{\alpha}) = (\frac{x}{\alpha} - x_c)^2 + (\frac{y}{\alpha} - y_c)^2 + (\frac{z}{\alpha} - z_c)^2 - \frac{1}{4} \, L_1^2. \tag{6.7.4}$$

Simplifying, we find

$$f(\frac{x}{\alpha}, \frac{y}{\alpha}, \frac{z}{\alpha}) = \frac{1}{\alpha^2} \left[(x - \alpha \, x_c)^2 + (y - \alpha \, y_c)^2 + (z - \alpha \, z_c)^2 - \frac{1}{4} \, L_2^2 \right]. \tag{6.7.5}$$

Setting the right-hand side of (6.7.5) to zero, we obtain the equation of a sphere with diameter L_2 centered at the point $(\alpha x_c, \alpha y_c, \alpha z_0)$.

Let ρ_1 and μ_1 be the density and viscosity of the first fluid, and ρ_2 and μ_2 be the density and viscosity of the second fluid. Both fluids are assumed to be incompressible and Newtonian.

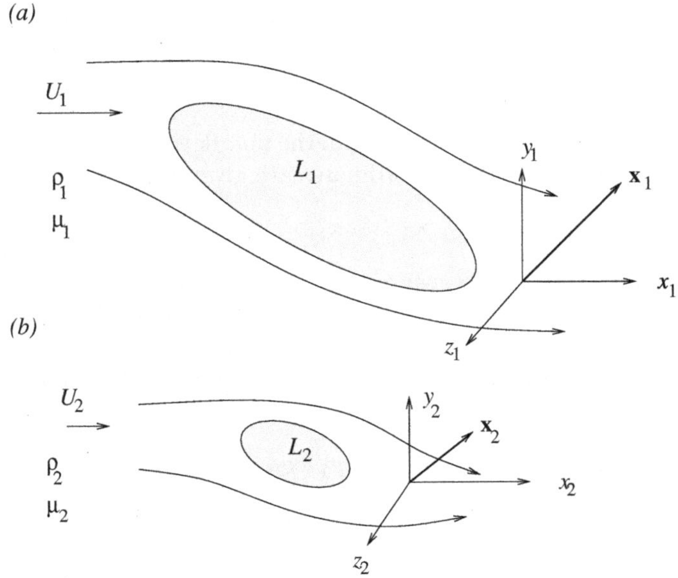

Figure 6.7.1 Illustration of flow in two similar domains. If the Reynolds numbers are equal, as shown in equation (6.7.6), the velocity and pressure field of the second flow may be deduced from the velocity and pressure of the first flow, and *vice versa*, by rescaling.

Reynolds number

We will show that, when the values of the four control and physical parameters defining the first flow, L_1, U_1, ρ_1, μ_1, and the corresponding values of the four control and physical parameters defining the second flow, L_2, U_2, ρ_2, μ_2, are related by the equations

$$\frac{\rho_1 U_1 L_1}{\mu_1} = \frac{\rho_2 U_2 L_2}{\mu_2}, \tag{6.7.6}$$

the structure of the second flow can be inferred from the structure of the first flow, and *vice versa*, by carrying out a simple computation described as rescaling. The left-hand side of (6.7.6) is the Reynolds number of the first flow, and the right-hand side of (6.7.6) is the Reynolds number of the second flow.

Rescaling

To deduce the structure of the second flow from the structure of the first flow, we introduce the dynamic pressure established due to the flow, defined as the pressure deviation from the hydrostatic distribution,

$$p_1^D \equiv p_1 - \rho_1 \, \mathbf{g} \cdot \mathbf{x}, \qquad p_2^D \equiv p_2 - \rho_2 \, \mathbf{g} \cdot \mathbf{x}. \tag{6.7.7}$$

The superscript D stands for "dynamic" or, more accurately, "hydrodynamic." In the absence of an imposed flow, the pressure assumes the hydrostatic distribution and the dynamic pressure vanishes throughout the domain of flow.

Now we consider an arbitrary point in the first flow, $\mathbf{x}_1$, and a corresponding point in the second flow whose coordinates are given by

$$\mathbf{x}_2 = \alpha\, \mathbf{x}_1. \tag{6.7.8}$$

Equations (6.7.1) and (6.7.2) ensure that, if $\mathbf{x}_1$ lies at the surface of the body in the first flow, then $\mathbf{x}_2$ will lie at the surface of the body in the second flow. We will show that, when relation (6.7.6) is fulfilled, the velocity and dynamic pressure at the second point in the second flow are related to those at the first point in the first flow by the equations

$$\mathbf{u}_2(\mathbf{x}_2) = \delta\, \mathbf{u}_1(\mathbf{x}_2), \qquad p_2^D(\mathbf{x}_2) = \beta\, \delta^2\, p_1^D(\mathbf{x}_2). \tag{6.7.9}$$

We have introduced the ratio of the velocities of the incident flow, δ, the viscosity ratio λ, and the density ratio β,

$$\delta \equiv \frac{U_2}{U_1}, \qquad \lambda \equiv \frac{\mu_2}{\mu_1}, \qquad \beta \equiv \frac{\rho_2}{\rho_1}. \tag{6.7.10}$$

The equality of the Reynolds numbers expressed by (6.7.6) requires

$$\beta\, \delta\, \alpha = \lambda. \tag{6.7.11}$$

Relations (6.7.9) are also valid for unsteady flow, provided that the velocity field of the first flow at the designated origin of time is related to the velocity of the second flow at the designated origin of time by the first of equations (6.7.9), and the comparison is made at times t_1 and t_2 related by

$$t_2 = \frac{\delta t_1}{\alpha}. \tag{6.7.12}$$

6.7.1 Dimensional analysis

To prove relations (6.7.9), we consider the Navier-Stokes equation (6.5.6) and the continuity equation $\nabla \cdot \mathbf{u} = 0$ governing the structure and dynamics of the two flows with appropriate physical constants corresponding to the two fluids, subject to appropriate far-field and boundary conditions, and work as follows.

First flow

Considering the first flow, we introduce the dimensionless independent variables

$$\hat{x}_1 = \frac{x}{L_1}, \qquad \hat{y}_1 = \frac{y}{L_1}, \qquad \hat{z}_1 = \frac{z}{L_1}, \qquad \hat{t}_1 = \frac{t\, U_1}{L_1}, \tag{6.7.13}$$

and the dimensionless dependent variables

$$\hat{u}_{x_1} = \frac{u_{x1}}{U_1}, \qquad \hat{u}_{y_1} = \frac{u_{y1}}{U_1}, \qquad \hat{u}_{z_1} = \frac{u_{z1}}{U_1}, \qquad \hat{p}_1^D = \frac{p_1^D}{\rho_1 U_1^2}. \tag{6.7.14}$$

Solving for the dimensional variables in terms of their dimensionless counterparts denoted by the hats, and substituting the result in the Navier-Stokes and continuity equation, we find

$$\frac{\partial\, \hat{\mathbf{u}}_1}{\partial \hat{t}_1} + \hat{\mathbf{u}}_1 \cdot \hat{\nabla}_1 \hat{\mathbf{u}}_1 = -\hat{\nabla}_1\, \hat{p}_1^D + \frac{1}{Re_1}\, \hat{\nabla}_1^2\, \hat{\mathbf{u}}_1 \tag{6.7.15}$$

and

$$\hat{\nabla}_1 \cdot \hat{\mathbf{u}}_1 = 0, \tag{6.7.16}$$

where $Re_1 \equiv \rho_1 U_1 L_1/\mu_1$ is the Reynolds number of the first flow shown on the left-hand side of (6.7.6). We have introduced the dimensionless gradient

$$\hat{\nabla}_1 \equiv (\frac{\partial}{\partial \hat{x}_1}, \quad \frac{\partial}{\partial \hat{y}_1}, \quad \frac{\partial}{\partial \hat{z}_1}), \tag{6.7.17}$$

and associated Laplacian operator

$$\hat{\nabla}_1^2 \equiv \frac{\hat{\partial}^2}{\partial \hat{x}_1^2} + \frac{\hat{\partial}^2}{\partial \hat{y}_1^2} + \frac{\hat{\partial}^2}{\partial \hat{z}_1^2}. \tag{6.7.18}$$

The far-field condition requires that, far from the body, the dimensionless velocity $\hat{u}_{x_1}$ tends to unity, whereas $\hat{u}_{y_1}$ and $\hat{u}_{z_1}$ vanish. The no-slip and no-penetration boundary conditions require that the velocity vanishes at points (x, y, z) that satisfy equation (6.7.1) or, equivalently, points $(\hat{x}_1, \hat{y}_1, \hat{z}_1)$ that satisfy

$$f(L_1\hat{x}_1, L_1\hat{y}_1, L_1\hat{z}_1) = 0. \tag{6.7.19}$$

Second flow

Next, we consider the second flow and introduce the dimensionless independent variables

$$\hat{x}_2 = \frac{x}{L_2}, \qquad \hat{y}_2 = \frac{y}{L_2}, \qquad \hat{z}_2 = \frac{z}{L_2}, \qquad \hat{t}_2 = \frac{t\, U_2}{L_2}, \tag{6.7.20}$$

and the dimensionless dependent variables

$$\hat{u}_{x_2} = \frac{u_{x2}}{U_2}, \qquad \hat{u}_{y_2} = \frac{u_{y2}}{U_2}, \qquad \hat{u}_{z_2} = \frac{u_{z2}}{U_2}, \qquad \hat{p}_2^D = \frac{p_2^D}{\rho_2 U_2^2}. \tag{6.7.21}$$

Solving for the dimensional variables in terms of their dimensionless counterparts, and substituting the result in the Navier-Stokes and continuity equation, we find

$$\frac{\partial\, \hat{\mathbf{u}}_2}{\partial \hat{t}_2} + \hat{\mathbf{u}}_2 \cdot \hat{\nabla}_2 \hat{\mathbf{u}}_2 = -\hat{\nabla}_2\, \hat{p}_2^D + \frac{1}{Re_2}\, \hat{\nabla}_2^2\, \hat{\mathbf{u}}_2 \tag{6.7.22}$$

and

$$\hat{\nabla}_2 \cdot \hat{\mathbf{u}}_2 = 0, \tag{6.7.23}$$

where $Re_2 \equiv \rho_2 U_2 L_2 / \mu_2$ is the Reynolds number shown on the right-hand side of (6.7.6). We have introduced the dimensionless gradient

$$\hat{\nabla}_2 \equiv (\frac{\partial}{\partial \hat{x}_2},\quad \frac{\partial}{\partial \hat{y}_2},\quad \frac{\partial}{\partial \hat{z}_2}), \tag{6.7.24}$$

and associated Laplacian operator

$$\hat{\nabla}_2^2 \equiv \frac{\hat{\partial}^2}{\partial \hat{x}_2^2} + \frac{\hat{\partial}^2}{\partial \hat{y}_2^2} + \frac{\hat{\partial}^2}{\partial \hat{z}_2^2}. \tag{6.7.25}$$

The far-field condition requires that, far from the body, the dimensionless velocity $\hat{u}_{x_2}$ tends to unity, whereas $\hat{u}_{y_2}$ and $\hat{u}_{z_2}$ vanish. The no-slip and no-penetration boundary conditions require that the velocity vanishes at points (x, y, z) that satisfy equation (6.7.2) or, equivalently, points $(\hat{x}_2, \hat{y}_2, \hat{z}_2)$ that satisfy

$$f(L_1 \hat{x}_2, L_1 \hat{y}_2, L_1 \hat{z}_2) = 0. \tag{6.7.26}$$

Comparison

To this end, we compare one by one the equations and boundary conditions governing the two flows in the dimensionless variables indicated by a hat, and draw four important conclusions:

1. The Navier-Stokes equation (6.7.15) is identical to the Navier-Stokes equation (6.7.22), provided that the two Reynolds numbers are equal, as stated in (6.7.6).

2. The continuity equation (6.7.16) is identical to the continuity equation (6.7.23).

3. The far-field conditions are identical; both dimensionless velocities designated by hats tend to $[1, 0, 0]$ far from the body.

4. The boundary conditions on the first body described by (6.7.19) are identical to the boundary conditions on the second body described by (6.7.26).

These results demonstrate that, when the Reynolds numbers of the two flows are equal, values of the dimensionless dependent variables in the two flows at corresponding dimensionless times and corresponding dimensionless positions are the same. Setting $\hat{p}_1^D(\hat{\mathbf{x}}_1) = \hat{p}_2^D(\hat{\mathbf{x}}_2)$ and using the definitions (6.7.14) and (6.7.21), we derive the second relation in (6.7.9) subject to (6.7.10).

Problem

6.7.1 *Reynolds number*

Compute the Reynolds number of (a) an ant crawling, (b) a person running, (c) a car moving at 100 km per hour, and (d) an elephant running across a plain at maximum speed.

6.8 Structure of a flow as a function of the Reynolds number

Let us consider flow in a domain with characteristic length L, identify an appropriate characteristic velocity, U, and compute the Reynolds number

$$Re \equiv \frac{\rho L U}{\mu} = \frac{LU}{\nu}, \tag{6.8.1}$$

where $\nu \equiv \mu/\rho$ is the kinematic viscosity. Next, we introduce the dimensionless independent variables

$$\hat{x} = \frac{x}{L}, \qquad \hat{y} = \frac{y}{L}, \qquad \hat{z} = \frac{z}{L}, \qquad \hat{t} = \frac{t\,U}{L}, \tag{6.8.2}$$

and the dimensionless dependent variables

$$\hat{u}_x = \frac{u_x}{U}, \qquad \hat{u}_y = \frac{u_y}{U}, \qquad \hat{u}_z = \frac{u_z}{U}, \qquad \hat{p}^D = \frac{p^D}{\rho U^2}, \tag{6.8.3}$$

where the superscript D signifies the dynamic pressure incorporating the effect of gravity. Solving for the dimensional variables in terms of their dimensionless counterparts and substituting the result in the Navier-Stokes and continuity equation, we find

$$\frac{\partial \hat{\mathbf{u}}}{\partial \hat{t}} + \hat{\mathbf{u}} \cdot \hat{\nabla}\hat{\mathbf{u}} = -\hat{\nabla}\hat{p}^D + \frac{1}{Re}\,\hat{\nabla}^2\hat{\mathbf{u}}, \tag{6.8.4}$$

and

$$\hat{\nabla} \cdot \hat{\mathbf{u}} = 0. \tag{6.8.5}$$

These dimensionless forms reveals that, given the boundary shape, the structure of a flow depends on L, U, ρ, and μ collectively through the dimensionless Reynolds number in the sense of the dynamic similitude expressed by equations (6.7.9) and (6.7.10).

Velocity and length scales

The choice of characteristic velocity, U, and length scale, L, is not always apparent. Subtleties arise when the domain of flow contains intrinsic or a multitude of length scales. Examples include the flow of a particle suspension and the flow established spontaneously due to an instability in the absence of external forcing.

For the successful choice of L and U, all terms in the dimensionless Navier-Stokes equation (6.8.4), with the possible exception of the pressure term, are of order unity. The Reynolds number then expresses the relative importance of the inertial forces, assumed to scale with $\rho U^2/L$, and viscous forces, assumed to scale with $\mu U/L^2$. Their ratio is the Reynolds number defined in (6.8.1).

6.8.1 Stokes flow

If the Reynolds number is small, viscous forces dominate and the left-hand side of the dimensionless Navier-Stokes equation (6.8.4) makes a negligible contribution. Although the dimensionless pressure gradient also appears to make a negligible contribution, this is only a mathematical illusion.

To see this, we observe that the dimensionless pressure arose from the arbitrary scaling shown in the equation in (6.8.3), which can be contrasted with the physical scaling of the position vector and velocity in terms of the unambiguous length and velocity scales, L and U. As a consequence, the dimensionless pressure gradient may become singular as the Reynolds number tends to zero, requiring an alternative scaling. To prevent this failure, we retain the pressure gradient in the dimensionless form of the Navier-Stokes equation independent of the Reynolds number.

Reverting to dimensional variables, we find that the Navier-Stokes equation reduces to the Stokes equation

$$\mathbf{0} = -\nabla p + \mu \nabla^2 \mathbf{u} + \rho \mathbf{g}, \tag{6.8.6}$$

describing steady or unsteady creeping flow with negligible inertial forces. Cursory inspection now reveals that the pressure scales with $\mu U/L$ instead of ρU^2. The analysis and computation of creeping flow will be the exclusive topic of our discussion in chapter 9.

6.8.2 Flow at high Reynolds numbers

Inspecting (6.8.4), we find that, when the Reynolds number is high, viscous forces are negligible and can be neglected provided that the velocity does not change rapidly over small distances across fluid layers that are thin compared to the global size of the boundaries; otherwise, the standard scaling with respect to U and L may not be valid. Such thin layers typically occur along flow boundaries or interfaces between two adjacent streams of the same or different fluids. In chapter 10, we will see that viscous forces are significant inside these layers, even though the bulk of the flow may occur at high Reynolds numbers.

6.8.3 *Laminar and turbulent flow*

When the Reynolds number exceeds a certain threshold, an unsteady small-scale motion characterized by rapid fluctuations in the velocity and vorticity fields is spontaneously established. In practice, turbulent motion is often superposed on a steady or unsteady macroscopic or large-scale flow that evolves at a slower rate. Flows below the critical Reynolds number are called laminar to indicate that the streamlines are smooth, whereas flows above the critical Reynolds number are called turbulent to indicate that the streamlines are highly convoluted.

The transition from laminar to turbulent motion may occur by several mechanisms, including the amplification of internal waves. The critical Reynolds number where transition occurs can be estimated theoretically by carrying out a stability analysis, as will be discussed in chapter 10. The dynamics of turbulent motion can be studied by several methods, including statistical analysis, nonlinear dynamical systems theory, and vortex dynamics, as will be discussed in chapter 10.

Problems

6.8.1 *Characteristic scales*

Identify the characteristic velocity scale U, length scale L, and Reynolds number of (a) simple shear flow past a stationary body, (b) flow due to the settling of a particle in the atmosphere, and (c) flow due to a breaking wave in the ocean.

6.9 Dimensionless numbers in fluid dynamics

We have demonstrated that two geometrically related flows occurring at the same Reynolds number are similar, in that one can be deduced from the other by rescaling. The proof was carried out for a flow that is bounded by a solid surface over which the no-slip and no-penetration boundary conditions are required. A time-independent velocity field was imposed in the far-field boundary condition in lieu of a driving mechanism.

If the driving mechanism is time-dependent or the flow is bounded by fluid interfaces and free surfaces, additional conditions for dynamic similitude requiring the equality of further dimensionless numbers are necessary. These dimensionless numbers enter the problem formulation either through the governing equations or through boundary and interfacial conditions.

Frequency number for a time-dependent flow

Let us consider an externally forced time-dependent flow and identify a velocity scale, U, a length scale, L, and a time scale, T. In the case of periodic flow with angular frequency ω due, for example, to an oscillating pressure gradient, T can be identified with the period, $T = 2\pi/\omega$. The relative importance of inertial and viscous forces in the equation of motion is expressed by the dimensionless frequency parameter

$$\beta \equiv \frac{L^2}{\nu T}, \tag{6.9.1}$$

where ν is the kinematic viscosity of the fluid. In the case of an intrinsically time-dependent flow, $T = L/U$, and the frequency parameter reduces to the Reynolds number.

Froude number

Consider flow in the ocean due to the propagation of water waves. The relative importance of inertial and gravitational forces is determined by the Froude number,

$$Fr \equiv \frac{U}{\sqrt{gL}}, \tag{6.9.2}$$

where g is the magnitude of the acceleration of gravity. In the case of flow over a hump discussed in section 6.4, the Froude number takes the specific form shown in equation (6.4.22).

Bond number

The relative importance of gravitational forces and surface tension in a fluid bounded by a free surface or fluid interface is determined by the Bond number,

$$Bo \equiv \frac{\rho g L^2}{\gamma}, \tag{6.9.3}$$

where γ is the surface tension (problem 6.9.1).

Weber number

The relative importance of inertial forces and surface tension in a flow that is bounded by a free surface or fluid interface is determined by the Weber number,

$$We \equiv \frac{\rho U^2 L}{\gamma}. \tag{6.9.4}$$

For example, the Weber number determines the deformation and nature of the flow around a gas bubble rising or convected at high speed through an ambient liquid.

Capillary number

The relative importance of viscous forces and surface tension in a fluid bounded by a free surface or fluid interface is determined by the capillary number,

$$Ca \equiv \frac{\mu U}{\gamma}. \tag{6.9.5}$$

For example, the capillary number determines the deformation and thus the nature of the flow around a liquid droplet immersed in a shear flow.

Problems

6.9.1 *Bond number in hydrostatics*

Explain how the Bond number emerges from the scaling of the Laplace-Young equation (5.4.6) in hydrostatics.

6.9.2 *Ratio of two numbers*

What is the ratio between the Weber number and the capillary number?

7 Channel, Tube, and Film Flow

In previous chapters, we have derived differential equations governing the motion of an incompressible Newtonian fluid by requiring mass conservation and enforcing Newton's second law for the motion of fluid parcels. The governing equations are accompanied by initial, boundary, and interfacial conditions. In this chapter, we proceed to derive analytical and semi-analytical solutions for an important class of steady and unsteady flows with rectilinear and circular streamlines. The engineering significance of these flows combined with their ability to demonstrate the salient mechanisms of momentum and vorticity transport at steady and unsteady state justify an in-depth study.

7.1 Steady flow in a two-dimensional channel

We begin by considering flow in a two-dimensional channel confined between two parallel plane walls separated by the distance $2h$ and inclined at an angle θ_0 with respect to the horizontal plane, as illustrated in figure 7.1.1. In the inclined system of Cartesian coordinates depicted in this figure, the x axis is parallel to the walls and the y axis is perpendicular to the walls. The components of the acceleration of gravity vector are

$$g_x = g \sin\theta_0, \qquad g_y = -g\cos\theta_0, \tag{7.1.1}$$

where g is the magnitude of the acceleration of gravity. The lower wall translates parallel to itself with constant velocity U_1, and the upper wall translates parallel to itself with constant velocity U_2. The motion of the fluid is governed by the

C. Pozrikidis, *Fluid Dynamics: Theory, Computation, and Numerical Simulation*,
DOI: 10.1007/978-0-387-95871-2_7,

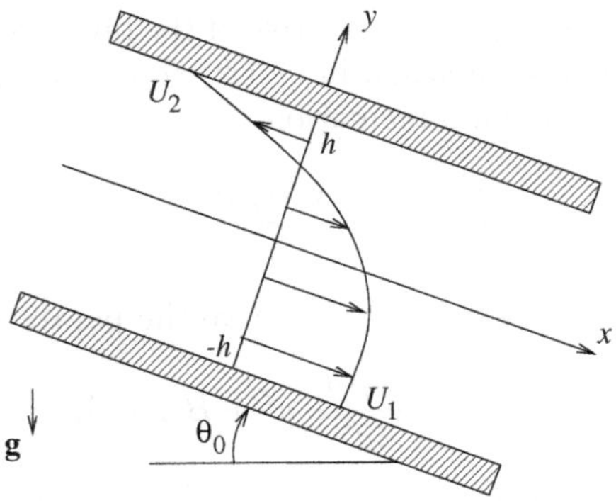

Figure 7.1.1 Illustration of steady flow in a two-dimensional channel confined between two parallel walls.

Navier-Stokes equation (6.5.5) whose Cartesian components are displayed in table 6.5.1.

Unidirectional and fully-developed flow

Our analysis will be based on the assumption of steady unidirectional flow requiring that the y and z velocity components vanish, $u_y = 0$ and $u_z = 0$, and the x component is constant in time, $\partial u_x/\partial t = 0$. This assumption precludes the occurrence of turbulent motion where small-scale three-dimensional fluctuations arise, as discussed in chapter 10. The continuity equation for two-dimensional flow, $\partial u_x/\partial x + \partial u_y/\partial y = 0$, requires $\partial u_x/\partial x = 0$, stating that the flow is fully developed. Thus, the axial velocity, u_x, is a function of position across the channel, y, alone.

Pressure and velocity profile

Simplifying the x and y components of the equation of motion shown in table 6.5.1 by discarding terms that are identically zero, we find

$$0 = -\frac{\partial p}{\partial x} + \mu\,\frac{\mathrm{d}^2 u_x}{\mathrm{d}y^2} + \rho\, g_x, \qquad 0 = -\frac{\partial p}{\partial y} + \rho\, g_y. \tag{7.1.2}$$

The no-slip boundary condition at the two walls requires

$$u_x = U_1 \text{ at } y = -h, \qquad u_x = U_2 \text{ at } y = h. \tag{7.1.3}$$

The second equation in (7.1.2) determines the pressure distribution in hydrostatics. It is convenient to screen out the hydrostatic variation by expressing the pressure in the form

$$p = P_0 - \chi x + \rho\, g_y y, \tag{7.1.4}$$

where P_0 is a constant and χ is the negative of the pressure gradient (derivative) along the x axis. The first equation in (7.1.2) thus takes the form of a second-order linear ordinary differential equation,

$$\frac{d^2 u_x}{dy^2} = -\frac{\chi + \rho\, g_x}{\mu}. \tag{7.1.5}$$

Integrating twice with respect to y, we derive the parabolic velocity profile

$$u_x(y) = -\frac{\chi + \rho\, g_x}{2\mu}\, y^2 + B\, y + A, \tag{7.1.6}$$

where A and B are two constants determined by the no-slip boundary conditions (7.1.3). Solving for these constants and substituting the result back in (7.1.6), we find

$$u_x(y) = U_1 + \frac{y+h}{2h}\,(U_2 - U_1) + \frac{\chi + \rho\, g_x}{2\mu}\,(h^2 - y^2). \tag{7.1.7}$$

Function *chan_2d* in directory *04_various* of FDLIB evaluates the velocity profile given in (7.1.7).

Shear- pressure- and gravity-driven flow

It is instructive to identify three special cases of the general flow expressed by equation (7.1.7):

1. When $\chi + \rho g_x = 0$, the last term on the right-hand side of (7.1.7) disappears and the flow is driven by boundary motion in a shear-driven mode; this is the plane Couette flow.

2. When $U_1 = 0$, $U_2 = 0$, and the channel is horizontal, $g_x = 0$, the flow is driven by an imposed pressure gradient along the x axis equal to the negative of the constant χ, in a pressure-driven mode; this is the Hagen or plane Poiseuille flow, also called the Hagen-Poiseuille flow.

3. When $U_1 = 0$, $U_2 = 0$, and in the absence of pressure variation along the x axis at any y position, $\chi = 0$, we obtain gravity-driven flow.

Shear stress

The shear stress, σ_{xy}, arises by differentiating the velocity profile (7.1.7) with respect to y, finding

$$\sigma_{xy} = \mu\,\frac{du_x}{dy} = \mu\,\frac{U_2 - U_1}{2h} - (\chi + \rho\, g_x)\, y. \tag{7.1.8}$$

In Couette flow, the shear stress is a constant determined by the velocities of the walls and the distance between the walls. In gravity- and pressure-driven flow, the shear stress varies linearly across the channel.

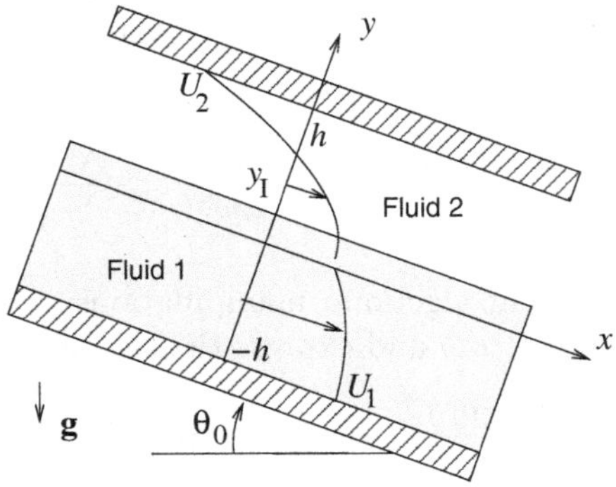

Figure 7.1.2 Illustration of steady two-layer flow in a channel confined between two parallel walls.

Flow rate

The flow rate along the x axis is found by integrating the velocity profile across the channel, obtaining

$$Q = \int_{-h}^{h} u_x(y)\,\mathrm{d}y = (U_1 + U_2)\,h + 2\,\frac{\chi + \rho\,g_x}{3\,\mu}\,h^3. \tag{7.1.9}$$

As an application, we consider flow a rectangular cavity that is closed at the left and right ends, so that the flow rate is zero, $Q = 0$. Physically, the flow is driven by the translation of two moving belts identified with the lower and upper walls. Equation (7.1.9) shows that a pressure gradient is established for the right-hand side to vanish.

7.1.1 Two-layer flow

Next, we consider the flow of two superimposed layers with generally different viscosities and densities, as illustrated in figure 7.1.2. In the inclined system of coordinates depicted in the figure, the interface is located at $y = y_I$. The lower layer is labeled as fluid 1, and the upper layer is labeled as fluid 2.

Working as in the case of single-fluid flow, we derive the pressure distribution inside each layer corresponding to (7.1.4),

$$p^{(1)} = P_0 - \chi\,x + \rho_1 g_y\,(y - y_I), \qquad p^{(2)} = P_0 - \chi\,x + \rho_2 g_y\,(y - y_I). \tag{7.1.10}$$

Note that the negative of the pressure gradient, χ, must be the same in both fluids; otherwise, the interfacial condition requiring continuity of normal stress, which amounts to continuity of pressure, cannot be satisfied.

The velocity profile across each layer is governed by the counterparts of equation (7.1.5) for each fluid,

$$\frac{d^2 u_x^{(1)}}{dy^2} = -\frac{\chi + \rho_1 g_x}{\mu_1}, \qquad \frac{d^2 u_x^{(2)}}{dy^2} = -\frac{\chi + \rho_2 g_x}{\mu_2}. \tag{7.1.11}$$

To facilitate impending algebraic manipulations, we integrate equations (7.1.11) twice with respect to y and express the solution in the form

$$\begin{aligned} u_x^{(1)}(y) &= -\frac{\chi + \rho_1 g_x}{2\,\mu_1}(y - y_I)^2 + \xi_1 (y - y_I) + u_I, \\ u_x^{(2)}(y) &= -\frac{\chi + \rho_2 g_x}{\mu_2}(y - y_I)^2 + \xi_2 (y - y_I) + u_I, \end{aligned} \tag{7.1.12}$$

where u_I is the common interfacial velocity and ξ_1, ξ_2 are the interfacial shear rates expressing the slope of the velocity profile on either side of the interface,

$$\xi_1 \equiv \left(\frac{du_x^{(1)}}{dy}\right)_{y=y_I}, \qquad \xi_2 \equiv \left(\frac{du_x^{(2)}}{dy}\right)_{y=y_I}. \tag{7.1.13}$$

To compute the three unknowns, ξ_1, ξ_2, and u_I, we enforce the no-slip boundary condition at the lower and upper walls,

$$u_x^{(1)}(y = -h) = U_1, \qquad u_x^{(2)}(y = h) = U_2, \tag{7.1.14}$$

and also require continuity of shear stress across the interface,

$$\mu_1\,\xi_1 = \mu_2\,\xi_2. \tag{7.1.15}$$

After a fair amount of algebra, we find

$$u_I = \frac{1}{2\mu_1}\,\frac{h_1 h_2}{\lambda + r}\,[(1 + r)\,\chi + \rho_1 g_x\,(1 + \beta\,r)] + \frac{r\,U_1 + \lambda\,U_2}{r + \lambda} \tag{7.1.16}$$

and

$$\xi_1 = -\frac{\chi + \rho_1\,g_x}{2\mu_1}\,h_1 + \frac{u_I - U_1}{h_1}, \qquad \xi_2 = \frac{\chi + \rho_2\,g_x}{2\mu_2}\,h_2 - \frac{u_I - U_2}{h_2}, \tag{7.1.17}$$

where $h_1 = h + y_I$ and $h_2 = h - y_I$ are the lower and upper layer thicknesses satisfying $h_1 + h_2 = 2h$. We have introduced the viscosity, density, and thickness ratios,

$$\lambda \equiv \frac{\mu_2}{\mu_1}, \qquad \beta \equiv \frac{\rho_2}{\rho_1}, \qquad r \equiv \frac{h_2}{h_1}. \tag{7.1.18}$$

Program *chan_2d_2l* in directory *04_various* of FDLIB evaluates the interfacial velocity, shear rates, velocity profile across the two layers, and the corresponding flow rates.

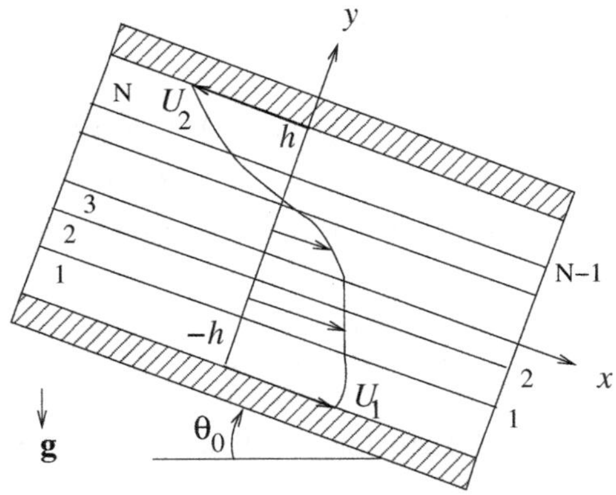

Figure 7.1.3 Illustration of steady multi-layer flow in a channel confined between two parallel walls. The integers printed on the right are interfacial labels.

7.1.2 Multi-layer flow

We now generalize the two-layer flow by considering an arbitrary number of N superimposed layers, as illustrated in figure 7.1.3. The bottom layer is labeled 1, and the top layer is labeled N. The $N-1$ interfaces separating the layers are located at $y = y_I^{(i)}$, where $i = 1, \dots, N-1$. The velocity profile across the ith layer is governed by the generalized version of equation (7.1.5),

$$\frac{\mathrm{d}^2 u_x^{(i)}}{\mathrm{d}y^2} = -\frac{\chi + \rho_i g_x}{\mu_i}, \tag{7.1.19}$$

where $i = 1, \dots, N$.

Integrating (7.1.19) twice with respect to y, we derive the parabolic profile

$$u_x^{(i)}(y) = -\frac{\chi + \rho_i g_x}{2\mu_i}\, y^2 + B^{(i)}\, y + A^{(i)}, \tag{7.1.20}$$

where $A^{(i)}$ and $B^{(i)}$ are unknown constants to be determined by requiring (a) the no-slip boundary condition at the lower and upper walls, (b) continuity of velocity at the interfaces expressed by

$$u_x^{(i)}(y = y_I^{(i)}) = u_x^{(i+1)}(y = y_I^{(i)}), \tag{7.1.21}$$

for $i = 1, \dots, N-1$, and (c) continuity of shear stress at the interfaces expressed by

$$\mu_i \left(\frac{\mathrm{d}u_x}{\mathrm{d}y}\right)^{(i)}_{y=y_I^{(i)}} = \mu_{i+1} \left(\frac{\mathrm{d}u_x}{\mathrm{d}y}\right)^{(i+1)}_{y=y_I^{(i)}}, \tag{7.1.22}$$

for $i = 1, \ldots, N-1$. Substituting the profile (7.1.20) in (7.1.22) and solving for $B^{(i)}$, we derive the recursion relation

$$B^{(i)} = \frac{\chi + \rho_i g_x}{\mu_i}\, y_I^{(i)} + \frac{\mu_{i+1}}{\mu_i}\, \Big(B^{(i+1)} - \frac{\chi + \rho_{i+1} g_x}{\mu_{i+1}}\, y_I^{(i)}\Big), \tag{7.1.23}$$

where $i = 1, \ldots, N-1$. It is convenient to introduce the shear rate at the upper wall,

$$\alpha \equiv \Big(\frac{\mathrm{d}u_x}{\mathrm{d}y}\Big)^{(N)}_{y=h}. \tag{7.1.24}$$

Differentiating the profile (7.1.20) for $i = N$ with respect to y and evaluating the derivative at $y = h$, we find

$$B^{(N)} = \alpha + \frac{\chi + \rho_N g_x}{\mu_N}\, h. \tag{7.1.25}$$

If we knew the value of α, we would be able to compute the coefficient $B^{(N)}$ from (7.1.25), and then evaluate the rest of the coefficients $B^{(i)}$ using the recursion relation (7.1.23), for $i = N-1, \ldots, 1$. Once this has been accomplished, we would be able to compute $A^{(1)}$ to satisfy the no-slip boundary condition at the bottom wall using the equation

$$u_x^{(1)}(y = -h) = -\frac{\chi + \rho_1\, g_x}{2\mu_1}\, h^2 - B^{(1)}\, h + A^{(1)} = U_1, \tag{7.1.26}$$

and then evaluate the rest of the coefficients $A^{(i)}$ by requiring continuity of velocity across each interface expressed by (7.1.21). At the end, the no-slip boundary condition at the upper wall would also be satisfied. Unfortunately, the value of α is *a priori* unknown and must be computed as part of the solution.

Numerical method

An expedient method of producing the value of α and simultaneously computing the unknown coefficients of the velocity profiles can be devised based on the no-slip boundary condition at the upper wall. We begin by expressing this condition in the form

$$f(\alpha) \equiv u_x^{(N)}(y = h) - U_2 = 0 \tag{7.1.27}$$

or

$$f(\alpha) = -\frac{\chi + \rho_N g_x}{2\,\mu_N}\, h^2 + B^{(N)}\, h + A^{(N)} - U_2 = 0, \tag{7.1.28}$$

where $f(\alpha)$ is an objective function. A key observation is that $f(\alpha)$ is a linear function of α, and may thus be expressed in the form

$$f(\alpha) = C\alpha + D, \tag{7.1.29}$$

where

$$C = f(1) - f(0), \qquad D = f(0). \tag{7.1.30}$$

The linear dependence becomes evident by observing that, if we assign a certain value to α, we can use the procedure described in the paragraph following equation (7.1.25) to evaluate the coefficients of the velocity profiles across each layer, and then compute the left-hand side of (7.1.28) by linear algebraic manipulations.

Combining the last two equations, we find that the desired value of α satisfying $f(\alpha) = 0$ is given by

$$\alpha = -\frac{D}{C} = \frac{f(0)}{f(1) - f(0)}. \tag{7.1.31}$$

The algorithm involves evaluating $f(0)$ and $f(1)$, and then using (7.1.31) to obtain α. The method is implemented in the function *chan_2d_ml* residing in directory *04_various* of FDLIB. The MATLAB implementation is:

```
function [u,Q] = chan_2d_ml ...

  (NLR,h,thick,visc,den,gac,U1,U2,pg,thet0,y ...
  ,yI,uI,xi1,xi2)

%-----------------------------------------
% Multi-layer flow through a channel confined
% between two parallel plane walls located at y = +-h %
% Layer numbered 1 is adjacent to the the lower wall,
% Layer numbered NLR is adjacent to the the upper wall.
% The velocity in the ith layer is given by:
%
% u(i) = af(i)+bf(i)*y-gf(i)*y**2
%
% SYMBOLS:
% --------
%
% NLR: Number of layers
% thick:  layer thicknesses
% yI(i):  y-position of the ith interface
% uI(i):  velocity of the ith interface
% xi1(i):  shear rate on lower side of the ith interface
% xi2(i):  shear rate on upper side of the ith interface
% Q(i):  flow rate of the ith layer
% th0:  plane inclination angle
% gac:  magnitude of the acceleration of gravity
% pg:  pressure gradient
%-----------------------------------------
```

```
%-----------------------------
% set the interface positions
%-----------------------------

yI(1) = thick(1);

for i=2:NLR
  yI(i) = yI(i-1)+thick(i);
end

%-----------------------------------------
% The velocity in the ith layer is given by:
% u(i) = af(i)+bf(i)*y-gf(i)*y*y
% compute gf(i)
%-----------------------------------------

fc = gac*sin(thet0);

for i=1:NLR
  gf(i) = 0.5*(pg+fc*den(i))/visc(i);
end

%--------------------------------------------------
% In the first pass,
% set the shear rate at upper wall "uw_shear" to zero
% and compute the residual of the upper wall boundary
% condition:  error0
%
% In the second pass,
% set the shear rate at upper wall "uw_shear" to -1
% and compute the residual of the upper wall boundary
% condition:  error1
%
% In the third pass,
% set the proper shear rate:
% uw_shear = error0/(error1-error0)
%-------------------

uw_shear = 0.0;

for Ipass=1:3

%---
% specified upper wall shear stress
%---

bf(NLR) = 2.0*gf(NLR)*yi(NLR)-uw_shear;
```

```
%---
% compute bf(i) by recursion starting from the top
% and requiring continuity of shear stress
% across the interfaces
%---

for i=NLR-1:-1:1
  bf(i) = 2.0*gf(i)*yi(i) + visc(i+1)/visc(i) ...
    *(bf(i+1)-2.0*gf(i+1)*yi(i));
end

%---
% compute af(1) to satisfy the non-slip condition
% at the lower wall
%---

af(1) = U1+(bf(1)+gf(1)*h)*h;

%---
% compute af(i), i=2,..., NLR by recursion
% starting from the bottom
% and requiring continuity of velocity
%---

for i=1:NLR-1
  af(i+1) = af(i)+( (bf(i)-bf(i+1)) ...
    -(gf(i)-gf(i+1))*yI(i))*yI(i);
end

%---
% error in the no-slip condition on the upper wall
% error = u(y=h)-U2
%---

if(Ipass==1)
  error0 = af(NLR)+(bf(NLR)-gf(NLR)*h)*h-U2;
  uw_shear = -1.0;
elseif(Ipass==2)
  error1 = af(NLR)+(bf(NLR)-gf(NLR)*h)*h-U2;
  uw_shear = error0/(error1-error0);
end

%---
end   % of Ipass
%---

%---
% compute interfacial velocities and shear rates
%---
```

```
for i=1:NLR-1
  ui(i) = af(i)+(bf(i)-gf(i)*yI(i))*yI(i);
  xi1(i) = bf(i) -2.0*gf(i) *yI(i);
  xi2(i) = bf(i+1)-2.0*gf(i+1)*yI(i);
end

%----
% compute the flow rates
% the ith layer is confined between i-1 and i interfaces
%----

Q(1) = af(1)*yI(1)+bf(1)*yI(1)*yI(1)/2.0 ...
     -gf(1)*yI(1)*yI(1)*yI(1)/3.0;

for i=2:NLR
  Q(i) = af(i)*(yI(i)-yI(i-1)) ...
    +bf(i)*(yI(i)*yI(i)-yI(i-1)*yI(i-1))/2.0 ...
    -gf(i)*(yI(i)*yI(i)-yI(i-1)*yI(i-1)*yI(i-1))/3.0;
end

%-------------------------
% compute the velocity at y
%-------------------------

% identify the host layer:

for i=1,N
  if(y<yI(i)) break; end
end
i=i-1
if(i>NLR) i=NLR; end

%---
% evaluate the velocity
%---

u = af(i)+(bf(i)-gf(i)*y)*y;

%-----
% done
%-----

return
```

7.1.3 *Power-law fluids*

Next, we derive the counterpart of the velocity profile (7.1.7) for a power-law fluid whose viscosity is a function of the shear rate according to the constitutive equation (4.6.1). The x component of the equation of motion for steady unidirectional flow requires

$$0 = \chi + \frac{\mathrm{d}\,\sigma_{xy}}{\mathrm{d}y} + \rho\, g_x, \tag{7.1.32}$$

where χ is the negative of the streamwise pressure gradient. Integrating with respect to y, we find that the shear stress is a linear function of position,

$$\sigma_{xy} = A - (\chi + \rho\, g_x)\, y, \tag{7.1.33}$$

where A is a constant. Substituting the expression for the shear stress given in (4.6.1), we find

$$\mu_0 \left| \frac{du_x}{dy} \right|^{n-1} \frac{du_x}{dy} = A - (\chi + \rho\, g_x)\, y, \tag{7.1.34}$$

where A is a constant.

When $\mathrm{d}u_x/\mathrm{d}y > 0$, we find

$$\mu_0 \left(\frac{\mathrm{d}u_x}{\mathrm{d}y} \right)^n = A - (\chi + \rho\, g_x)\, y, \tag{7.1.35}$$

yielding

$$\frac{\mathrm{d}u_x}{\mathrm{d}y} = \left(\frac{A - (\chi + \rho\, g_x)\, y}{\mu_0} \right)^{1/n}. \tag{7.1.36}$$

When $\mathrm{d}u_x/\mathrm{d}y < 0$, we find

$$\frac{\mathrm{d}u_x}{\mathrm{d}y} = -\left(\frac{-A + (\chi + \rho\, g_x)\, y}{\mu_0} \right)^{1/n}, \tag{7.1.37}$$

where the positive nth root is implied.

Assuming that $\chi + \rho\, g_x \neq 0$, we integrate the shear rate with respect to y and derive the velocity profile

$$u_x(y) = -\frac{n\, \mu_0}{(n+1)\, (\chi + \rho\, g_x)} \left(\frac{A - (\chi + \rho\, g_x)\, y}{\mu_0} \right)^{(n+1)/n} + B, \tag{7.1.38}$$

where B is a new constant. The constants A and B are determined by the boundary conditions.

In the case of pressure- and gravity-driven flow where $U_1 = U_2 = 0$ and $\chi + \rho\, g_x > 0$, the velocity profile is symmetric with respect to the centerline, $y = 0$. Requiring $\mathrm{d}u_x/\mathrm{d}y = 0$ at $y = 0$, we find $A = 0$. Requiring $u_x = 0$ at $y = -h$, we find

$$B = \frac{n\mu_0}{(n+1)(\chi + \rho\, g_x)} \left(\frac{(\chi + \rho\, g_x)\, h}{\mu_0} \right)^{(n+1)/n}. \tag{7.1.39}$$

Thus, the velocity profile is given by

$$u_x(y) = -\frac{n}{(n+1)} \left(\frac{\chi + \rho\, g_x}{\mu_0} \right)^{1/n} \left(h^{(n+1)/n} - (-y)^{(n+1)/n} \right) \tag{7.1.40}$$

in the lower half of the channel, $-h < y < 0$, and by

$$u_x(y) = -\frac{n}{(n+1)} \left(\frac{\chi + \rho\, g_x}{\mu_0} \right)^{1/n} \left(h^{(n+1)/n} - y^{(n+1)/n} \right) \tag{7.1.41}$$

in the upper half of the channel, $0 < y < h$. When $n = 1$, we recover the expressions derived previously in this section for Newtonian flow.

In the case of Couette (shear-driven) flow, $\chi + \rho\, g_x = 0$, the velocity profile is identical to the linear profile of a Newtonian fluid, even though the shear stress is determined by the power-law exponent, n.

Problems

7.1.1 *Reduction to single-layer flow*

(*a*) Verify that, when $\lambda = 1$ and $\beta = 1$, expressions (7.1.16) and (7.1.17) are consistent with the velocity profile (7.1.7) for single-fluid flow.

(*b*) Confirm that, when the densities and viscosities of all layers are identical, the coefficients $A^{(i)}$ and $B^{(i)}$ introduced in (7.1.20) are all equal to those corresponding to single-fluid flow.

7.1.2 *Integral momentum balance*

Verify that the shear stress shown in (7.1.8) satisfies the integral momentum balance over the rectangular fluid parcel drawn with the solid line in figure 6.1.1.

Computer Problems

7.1.3 *Multi-layer flow*

Use program *chan_2d_ml* to compute and plot the velocity profile of a three-layer pressure-driven flow of your choice. Investigate and discuss the effect of the viscosity of each layer.

7.1.4 *Non-Newtonian flow*

Prepare graphs of the velocity profile expressed by (7.1.40) and (7.1.41) for $n = 1$, 2, 4, 8, and 16. Discuss effect of the power-law exponent, n.

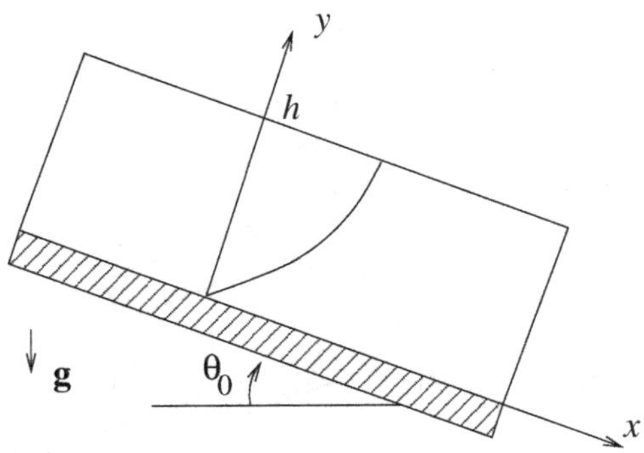

Figure 7.2.1 Illustration of gravity-driven flow of a liquid film down an inclined plane.

7.2 Steady film flow down an inclined plane

Gravity-driven flow of a liquid film down an inclined surface is encountered in a broad range of technological applications. Examples include photographic film and paper, electronics, and magnetic recording media coating. In the simplest configuration, a liquid film of thickness h flows down a plane that is inclined by the angle θ_0 with respect to the horizontal plane, as illustrated in figure 7.2.1.

Pressure and velocity profile

The motion of the fluid in steady unidirectional flow is governed by the simplified equations (7.1.2). The no-slip boundary condition requires that the velocity is zero at the plane located at $y = 0$, and the free-surface condition requires that the shear stress is zero at the free surface located at $y = h$,

$$u_x = 0 \quad \text{at } y = 0, \qquad \frac{du_x}{dy} = 0 \quad \text{at } y = h. \tag{7.2.1}$$

The pressure distribution is given by equation (7.1.4) with $\chi = 0$, yielding $p(y) = P_0 + \rho\, g_y y$, where $g_y = -g \cos\theta_0$. Setting the pressure at the free surface equal to the ambient atmospheric pressure, we obtain $P_{atm} = P_0 + \rho\, g_y h$ or $P_0 = P_{atm} - \rho\, g_y h$, and then derive the final expression

$$p(y) = P_{atm} + \rho\, g \cos\theta_0\, (h - y). \tag{7.2.2}$$

Working as in section 7.1 for channel flow, we obtain the Nusselt streamwise velocity profile

$$u_x(y) = \frac{\rho\, g_x}{2\mu}\, y\,(2h - y), \tag{7.2.3}$$

where $g_x = g \sin\theta_0$. This semi-parabolic profile is half the complete parabolic profile in pressure- and gravity-driven flow through a two-dimensional channel

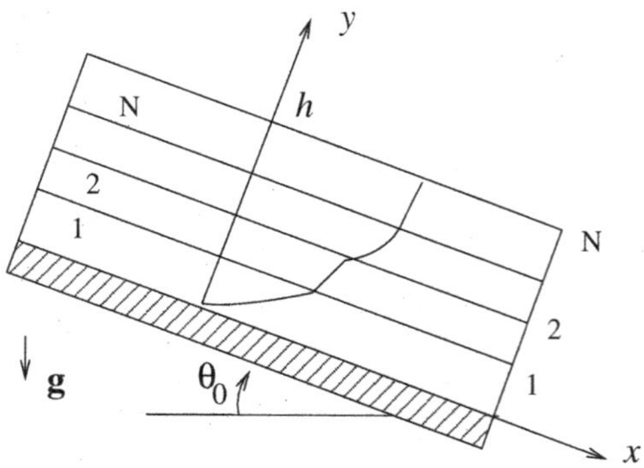

Figure 7.2.2 Illustration of gravity-driven multi-film flow down an inclined plane.

with width $2h$, where the free surface is located at the centerline. The shear stress varies linearly from a certain value at the wall to the required value of zero at the free surface.

Flow rate

The flow rate arises by integrating the velocity profile across the film, obtaining

$$Q \equiv \int_0^h u_x(y)\,\mathrm{d}y = \frac{g\,\rho\,h^3}{3\,\mu} = \frac{2}{3}\,h\,u_x(h), \tag{7.2.4}$$

where $u_x(h)$ is the maximum velocity occurring at the free surface. The mean velocity, $u_{mean} \equiv Q/h$, is equal to two thirds the free surface velocity.

Function *film* in directory *04_various* of FDLIB evaluates the velocity profile and flow rate given in (7.2.3) and (7.2.4).

7.2.1 Multi-film flow

Next, we consider the gravity-driven flow of an arbitrary number of N superimposed films down an inclined plane, as illustrated in figure 7.2.2. In photographic film manufacturing, as many as thirteen films may flow down an inclined plane to be deposited onto a moving support.

The bottom film is labeled 1 and the top film is labeled N. The $N-1$ interfaces separating the films are located at $y = y_I^{(i)}$, for $i = 1, \ldots, N-1$, and the free surface is located at $y = y_I^{(N)}$. The velocity profile across the ith film is governed by the simplified equations of motion (7.1.2) with $\chi = 0$. Integrating the first equation twice with respect to y, we derive the parabolic profile

$$u_x^{(i)}(y) = -\frac{\rho_i g_x}{2\mu_i}\,y^2 + B^{(i)}\,y + A^{(i)}, \tag{7.2.5}$$

where $A^{(i)}$ and $B^{(i)}$ are unknown constants determined by enforcing (a) the no-slip boundary condition at the plane, (b) continuity of velocity at the interfaces expressed by equation (7.1.20), (c) continuity of shear stress at the interfaces expressed by equation (7.1.21), and (d) the condition of zero shear stress at the free surface. With reference to equation (7.1.24), $\alpha = 0$. Knowledge of the value of the shear stress at the free surface allows us to evaluate the coefficients $B^{(i)}$ and $A^{(i)}$, working as described in the paragraph following equation (7.1.24). The numerical method is implemented in program *films_2d* located in directory *04_various* of FDLIB.

Wall shear stress

One interesting feature of the multi-film flow is that the wall shear stress and velocity profile across the first film that is adjacent to the wall are independent of the viscosities of the rest of the films. To see this, we write the velocity profile (7.2.5) for $i = 1$ and require the no-slip boundary condition to find $A^{(1)} = 0$. To compute the coefficient $B^{(1)}$, we perform a force balance over a section of the composite film confined between the planes $x = x_1$ and x_2. The balance requires that the x component of the force exerted by the shear stress at the wall and at the free surface counterbalance the x component of the weight of the fluid residing inside the control volume. We note that the shear stress is zero at the free surface and obtain

$$\sigma_{xy}^{(1)}(y=0) = \sum_{i=1}^{N} \rho_i g_x \, (y_I^{(i)} - y_I^{(i-1)}), \tag{7.2.6}$$

with the understanding that $y_I^{(0)} = 0$ and $y_I^{(N)} = h$. Using the profile (7.2.5) for $i = 1$, we find

$$\sigma_{xy}^{(1)}(y=0) = \mu_1 \left(\frac{\partial u_x^{(1)}}{\partial y} \right)_{y=0} = \mu_1 \, B^{(1)}. \tag{7.2.7}$$

Setting the right-hand sides of the last two equations equal, solving for $B^{(1)}$ and substituting the result in the profile (7.2.5) for $i = 1$ proves the stated independence of the first velocity profile and wall shear stress on the viscosity of the overlying fluids.

7.2.2 Power-law fluids

Next, we derive the counterpart of the velocity profile (7.2.3) for a power-law fluid whose viscosity is a function of the shear rate, as shown in equation (4.6.1). The x velocity increases from zero at the wall because of the no-slip boundary condition, to a maximum at the free surface. Thus, $du_x/dy > 0$. Working as in section 7.1.7, we find

$$\frac{du_x}{dy} = \left(\frac{A - (\chi + \rho g_x) \, y}{\mu_0} \right)^{1/n}, \tag{7.2.8}$$

where A is a constant. Integrating with respect to y, we find

$$u_x(y) = -\frac{n\,\mu_0}{(n+1)\,(\chi+\rho g_x)}\,\Big(\frac{A-(\chi+\rho\,g_x)\,y}{\mu_0}\Big)^{(n+1)/n} + B, \qquad (7.2.9)$$

where B is a new constant. The constants A and B are found by enforcing the boundary conditions (7.2.1). Applying the second condition and using (7.2.8), we find

$$A = (\chi + \rho\, g_x)\, h. \qquad (7.2.10)$$

The first condition leads to

$$B = \frac{n\,\mu_0}{(n+1)\,(\chi+\rho g_x)}\,\Big(\frac{(\chi+\rho\,g_x)h}{\mu_0}\Big)^{(n+1)/n}. \qquad (7.2.11)$$

Substituting these expressions in (7.2.9), we obtain the velocity profile

$$u_x(y) = -\frac{n}{(n+1)}\,\Big(\frac{\chi+\rho\,g_x}{\mu_0}\Big)^{1/n}\Big(h^{(n+1)/n} - (h-y)^{(n+1)/n}\Big). \qquad (7.2.12)$$

When $n = 1$, we recover the parabolic profile for Newtonian flow.

Problems

7.2.1 *Multi-film flow*

Confirm that the wall shear stress is independent of the viscosity of the films that are not adjacent to the wall on the basis of (*a*) the recursion relation (7.1.23) for gravity-driven flow, and (*b*) the free-surface condition expressed by (7.1.25) with $\alpha = 0$.

Computer Problems

7.2.2 *Multi-film flow*

Use program *films_2d* in directory *04_various* of FDLIB to compute and plot the velocity profile of a three-film flow of your choice. Investigate and discuss the effect of the film densities.

7.2.3 *Non-Newtonian flow*

Prepare a graph of the velocity profile expressed by (7.2.12) for $n = 1, 2, 4, 8$, and 16. Discuss the effect of the power-law exponent, n.

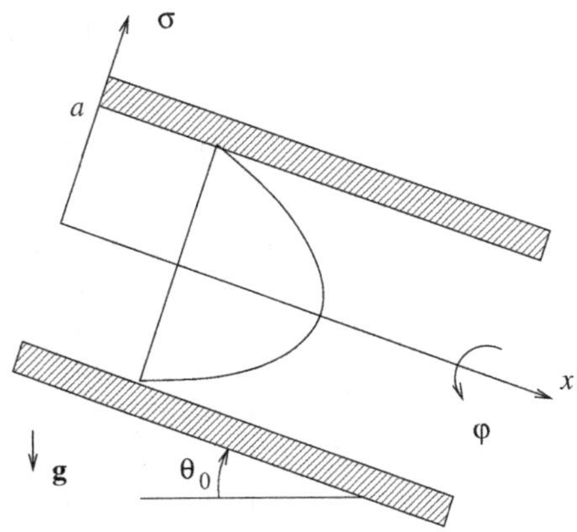

Figure 7.3.1 Illustration of steady unidirectional flow through a circular tube.

7.3 Steady flow through a circular tube

Figure 7.3.1 illustrates unidirectional flow through an inclined cylindrical tube of radius a. To derive the velocity profile, we introduce cylindrical polar coordinates, (x, σ, φ), where the x axis coincides with the tube centerline. The radial and meridional velocity components vanish in steady unidirectional flow, $u_\sigma = 0$ and $u_\varphi = 0$, and the x component depends only on the distance from the tube centerline, $u_x(\sigma)$.

Pressure and velocity profile

It is convenient to screen out the hydrostatic pressure variation normal to the x axis by expressing the pressure in the form

$$p = P_0 - \chi x + \rho\,(g_y y + g_z z), \tag{7.3.1}$$

where P_0 is a constant, and χ is the negative of the streamwise pressure gradient.

Inspecting the individual terms on the left-hand side of the x component of the equation of motion displayed in the first equation in (6.3.17), we find that the left-hand side is identically zero. Using the expression for the x component of the hydrodynamic volume force given in the first entry of table 6.5.2, and rearranging, we derive the simplified equation of motion

$$\frac{1}{\sigma}\frac{\mathrm{d}}{\mathrm{d}\sigma}\Big(\sigma\frac{\mathrm{d}u_x}{\mathrm{d}\sigma}\Big) = -\frac{\chi + \rho\, g_x}{\mu}, \tag{7.3.2}$$

which is the counterpart of equation (7.1.5) for two-dimensional channel flow. The solution must satisfy the no-slip boundary condition $u_x = 0$ at $\sigma = a$, and a regularity condition ensuring that u_x is finite at the centerline, $\sigma = 0$.

Two straightforward integrations of (7.3.2) with respect to σ provide us with the velocity profile

$$u_x(\sigma) = -\frac{\chi + \rho\, g_x}{4\mu}\,\sigma^2 + B\,\ln\sigma + A, \tag{7.3.3}$$

where A and B are two constants. For the velocity to be finite at the centerline, the constant B must be zero. To evaluate the constant A, we enforce the no-slip condition at the tube surface and derive the parabolic Poiseuille flow profile

$$u_x(\sigma) = \frac{\chi + \rho\, g_x}{4\mu}\,(a^2 - \sigma^2). \tag{7.3.4}$$

Maximum velocity occurs at the tube centerline, $\sigma = 0$,

$$u_{max} = u_x(0) = \frac{\chi + \rho\, g_x}{4\,\mu}\,a^2. \tag{7.3.5}$$

Doubling the tube radius while holding the pressure gradient constant quadruples the maximum velocity.

Program *tube_crc* in directory *04_various* of FDLIB evaluates the velocity profile described by (7.3.4).

Shear stress

The shear stress, $\sigma_{\sigma x}$, arises by differentiating the velocity profile (7.3.3) with respect to σ, finding

$$\sigma_{\sigma x} = \mu\,\frac{\mathrm{d}u_x}{\mathrm{d}\sigma} = -\frac{\chi + \rho\, g_x}{2}\,\sigma. \tag{7.3.6}$$

Thus, the shear stress is proportional to the distance from the centerline, σ, and reaches a maximum at the tube wall.

Integral momentum balance

It is instructive to verify that the derived expression for the shear stress satisfies an integral momentum balance over an annular control volume confined by two parallel planes located at $x = x_1$ and x_2, and two cylindrical surfaces located at $\sigma = \sigma_1$ and σ_2. Because the flow is steady, the rate of change of the x momentum of the fluid inside the control volume is zero. Balancing the normal force exerted on the planar sides, the shear force exerted at the cylindrical sides, and the body force, and observing that the normal viscous stress is zero while the normal stress is equal to the negative of the pressure over the planar sides, we find

$$\begin{aligned} 2\,\pi \int_{\sigma_1}^{\sigma_2} [-p_{(x=x_2)} + p_{(x=x_1)}]\,\sigma\,\mathrm{d}\sigma - \sigma_{\sigma x}\Big|_{\sigma=\sigma_1} 2\pi\sigma_1\,(x_2 - x_1) \\ +\sigma_{\sigma x}\Big|_{\sigma=\sigma_2} 2\pi\sigma_2\,(x_2 - x_1) + \rho\, g_x\,\pi\,(\sigma_2^2 - \sigma_1^2)\,(x_2 - x_1) = 0. \end{aligned} \tag{7.3.7}$$

Now using (7.3.1) to express the pressure difference in terms of the negative of the pressure gradient, χ, we find

$$\pi\,\chi\,(\sigma_2^2-\sigma_1^2)\,(x_2-x_1)-\sigma_{\sigma x}\Big|_{\sigma=\sigma_1} 2\pi\,\sigma_1\,(x_2-x_1)$$

$$+\sigma_{\sigma x}\Big|_{\sigma=\sigma_2} 2\,\pi\,\sigma_2\,(x_2-x_1)+\rho g_x\,\pi\,(\sigma_2^2-\sigma_1^2)\,(x_2-x_1)=0. \qquad (7.3.8)$$

Setting $\sigma_1=0$ and solving for $\sigma_{\sigma x}|_{\sigma=\sigma_2}$, we obtain precisely expression (7.3.6) evaluated at $\sigma=\sigma_2$.

Flow rate and the Poiseuille law

The flow rate along the tube arises by integrating the velocity profile over the tube cross-section, finding

$$Q\equiv 2\pi\int_0^a u_x(\sigma)\,\sigma\,\mathrm{d}\sigma=\frac{\chi+\rho\,g_x}{8\mu}\,\pi a^4=\frac{1}{2}\,u_x(0)\,(\pi a^2), \qquad (7.3.9)$$

where $u_x(0)$ is the maximum velocity occurring at the centerline. Equation (7.3.9) expresses the Poiseuille law first established by laboratory observation of blood flow. The data suggested that the flow rate through a circular tube subject to a constant pressure drop depends on the fourth power of the tube diameter.

The mean velocity is defined as the ratio of the flow rate to the tube cross-sectional area,

$$u_{mean}\equiv\frac{Q}{\pi a^2}=\frac{1}{2}\,u_x(0). \qquad (7.3.10)$$

Thus, the mean velocity is equal to half the maximum velocity occurring at the centerline.

Tube bifurcation

As an application, we consider a blood vessel of radius a, bifurcating into two vessels of equal radius a_1, as shown in figure 7.3.2. The flow through each tube is assumed to satisfy the equations of Poiseuille flow. It has been theorized that blood vessels in the human circulation bifurcate such that the wall shear stress is constant and the endothelial cells lining capillary walls are exposed to the same hydrodynamic stimulus. Using expression (7.3.6) for the shear stress, we find

$$(\chi+\rho g_x)\,a=(\chi_1+\rho g_x)\,a_1. \qquad (7.3.11)$$

Mass conservation requires $Q=2\,Q_1$. Using Poiseuille's law, we find

$$(\chi+\rho g_x)\,a^4=(\chi_1+\rho g_x)\,a_1^4. \qquad (7.3.12)$$

Combining the last two equations, we find

$$\frac{a_1}{a}=\frac{1}{2^{1/3}}\simeq 0.794. \qquad (7.3.13)$$

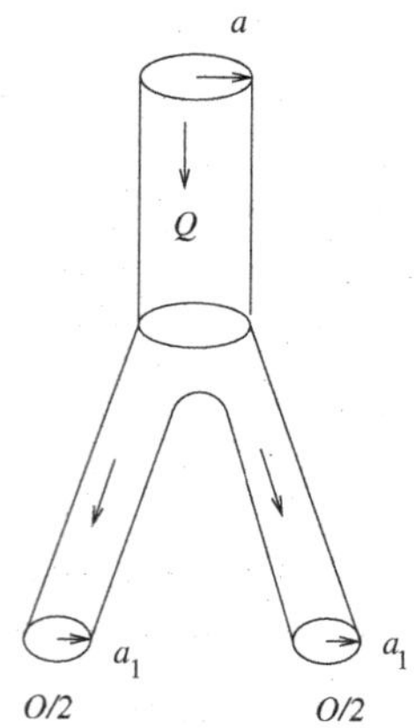

Figure 7.3.2 Bifurcation of a tube with radius a into two descendant tubes with equal radius a_1.

7.3.1 Multi-layer tube flow

Figure 7.3.3 illustrates axisymmetric multi-layer pressure- and gravity-driven flow through a circular tube. For the interfaces to remain concentric, the tube must be vertical or else the fluid densities must be matched; otherwise, hydrostatic pressure variations will cause a non-axisymmetric displacement.

Program *tube_crc_ml* in directory *04_various* of FDLIB computes the velocity profile. The numerical method is similar to that for multi-layer film flow down an inclined plane discussed in section 7.2.

7.3.2 Flow due to a translating sector

We have considered pressure- and gravity-driven flow through a circular tube. Suppose now that the axial pressure gradient is zero, the tube is horizontal, and the flow is driven exclusively by boundary motion. If the whole surface of the tube translates parallel to itself with constant velocity, the fluid will also translate with the same uniform velocity in a plug-flow mode. However, if only a sector of the tube confined inside $-\alpha \le \varphi \le \alpha$ translates with velocity U, and the remainder of the tube surface is stationary, a distributed velocity field will be established, as illustrated in figure 7.3.4(a).

Assuming that the flow is unidirectional and fully-developed, we set the radial and meridional velocity components to zero, $u_\sigma = 0$ and $u_\varphi = 0$, and regard the axial component, u_x, as a function of distance from the tube centerline, σ, and meridional angle, φ. The no-slip boundary condition at the surface of the cylinder requires $u_x(\sigma = a, \varphi) = U$ for $-\alpha < \varphi < \alpha$, and $u_x(\sigma = a, \varphi) = 0$ otherwise.

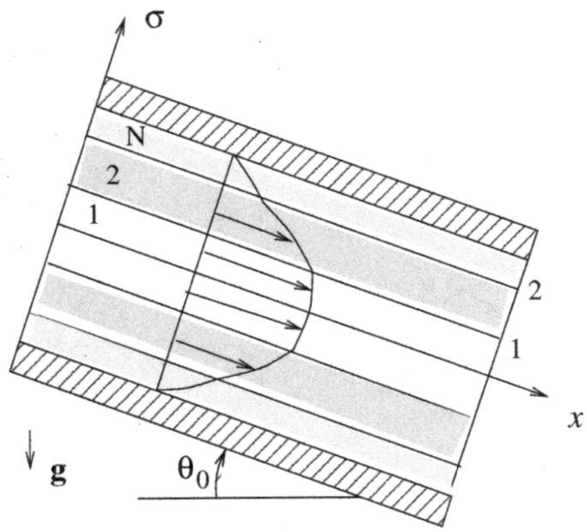

Figure 7.3.3 Illustration of multi-layer unidirectional flow through a circular tube. The layer labels are shown on the left, and the interface labels are shown on the right.

Consideration of the individual terms on the left-hand side of the x component of the equation of motion given in (6.3.17) reveals that the left-hand side is identically zero. Using the expression for the x component of the hydrodynamic volume force shown in table 6.5.2(a) with a constant pressure, we derive Laplace's equation

$$\nabla^2 u_x = \frac{1}{\sigma}\,\frac{\partial}{\partial \sigma}\Big(\sigma\,\frac{\partial u_x}{\partial \sigma}\Big) + \frac{1}{\sigma^2}\,\frac{\partial^2 u_x}{\partial \varphi^2} = 0, \tag{7.3.14}$$

where ∇^2 is the Laplacian operator in the yz plane expressed in cylindrical polar coordinates.

The solution can be found using a powerful method for solving Laplace's equation in two dimensions in the interior or exterior of a circle, subject to specified boundary conditions for the unknown function, expressed by the Poisson integral formula. For the problem under consideration, we find

$$u_x(\sigma,\varphi) = \frac{U}{\pi}\Big[\arctan\Big(\frac{a+\sigma}{a-\sigma}\tan\frac{\alpha-\varphi}{2}\Big) + \arctan\Big(\frac{a+\sigma}{a-\sigma}\tan\frac{\alpha+\varphi}{2}\Big)\Big]. \tag{7.3.15}$$

Inspecting the right-hand side, we confirm that, as the semi-angle of the translating sector α tends to π, the fluid tends to translate with uniform velocity in a plug-flow mode (problem 7.3.2).

Program *tube_crc_sec* in directory *04_various* of FDLIB evaluates evaluates the velocity profile given in (7.3.15).

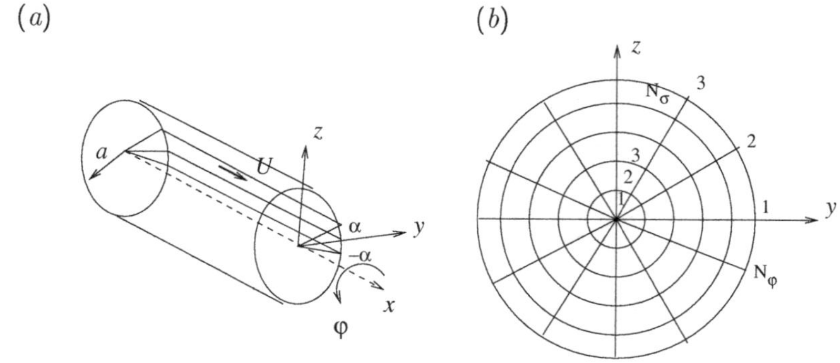

Figure 7.3.4 (a) Illustration of unidirectional flow through a circular tube due to the translation of a sector with semi-angle α. (b) Discretization of the tube cross-section for the purpose of computing the flow rate using the mid-point rule.

Flow rate

The flow rate through the tube arises by integrating the velocity distribution over the tube cross-section,

$$Q \equiv \int_0^{2\pi} \int_0^a u_x(\sigma, \varphi)\, \sigma \, \mathrm{d}\sigma \, \mathrm{d}\varphi. \tag{7.3.16}$$

The integral on the right-hand side cannot be found in analytical form. To obtain a numerical approximation, we divide the integration domain with respect to φ and σ into N_φ and N_σ evenly spaced intervals of equal size $\Delta\varphi = 2\pi/N_\varphi$ and $\Delta\sigma = a/N_\sigma$ defining elemental cross-sectional areas, as shown in figure 7.3.4(b), and approximate the velocity over each elemental area with the value at the center-point. Replacing the double integral in (7.3.16) with a double sum, we find

$$Q \simeq \Delta\varphi\, \Delta\sigma \sum_{i=1}^{N_\varphi} \sum_{j=1}^{N_\sigma} u_x(\sigma_j, \varphi_i)\, \sigma_j, \tag{7.3.17}$$

where $\varphi_i = (i - \frac{1}{2})\, \Delta\varphi$ and $\sigma_j = (j - \frac{1}{2})\, \Delta\sigma$. Equation (7.3.17) implements the mid-point rule for two-dimensional integration in a plane.

Problems

7.3.1 *Flow in a tube due to a translating sector*

Demonstrate that, as the translating sector semi-angle α tends to π, the right-hand side of (7.3.15) tends to the wall velocity U everywhere inside the tube.

7.3.2 *Axisymmetric film flow*

A liquid film drains due to gravity downward on the exterior surface of a vertical circular rod of radius a. Show that, in cylindrical polar coordinates with the x axis coaxial with the rod pointing upward, the velocity profile across the film is given by

$$u_x = \frac{g}{4\,\nu}\Big(\sigma^2 - a^2 - 2\,(a+h)^2 \ln\frac{\sigma}{a}\Big), \tag{7.3.18}$$

where h is the film thickness.

Computer Problems

7.3.3 *Flow in a circular tube due to a translating sector*

Prepare a graph of the dimensionless mean velocity, $\hat{u}_{mean} \equiv Q/(\pi a^2 U)$, plotted against the reduced semi-angle, α/π, using the numerical approximation implemented in (7.3.17). Deduce the slope of the graph at the origin, $\alpha = 0$. Your results should be accurate to the third significant figure.

7.3.4 *Multi-layer flow through a tube*

(*a*) Outline and explain the numerical procedure implemented in the code. *tube_crc_ml*.

(*b*) Plot the velocity profile for a three-layer configuration of your choice. Investigate and discuss the effect of the layer viscosities.

7.4 Steady flow through an annular tube

The velocity profile of unidirectional flow through an annular tube confined between two coaxial cylinders with radii R_1 and R_2 can be derived by a straightforward generalization of the analysis presented in section 7.3 for a circular tube. A new feature is that the two cylinders are allowed to translate parallel to themselves along the x axis with respective velocities equal to U_1 and U_2, as illustrated in figure 7.4.1.

Velocity profile

Integrating equation (7.3.2) with respect to σ, we find that the axial component of the velocity is given by (7.3.3). To evaluate the constants A and B, we enforce the no-slip conditions $u_x(\sigma = R_1) = U_1$ and $u_x(\sigma = R_2) = U_2$, and derive the velocity profile

$$\begin{aligned} u_x(\sigma) &= U_2 + (U_1 - U_2)\,\frac{\ln(R_2/\sigma)}{\ln(R_2/R_1)} \\ &\quad + \frac{\chi + \rho\, g_x}{4\,\mu}\Big(R_2^2 - \sigma^2 - (R_2^2 - R_1^2)\,\frac{\ln(R_2/\sigma)}{\ln(R_2/R_1)}\Big). \end{aligned} \tag{7.4.1}$$

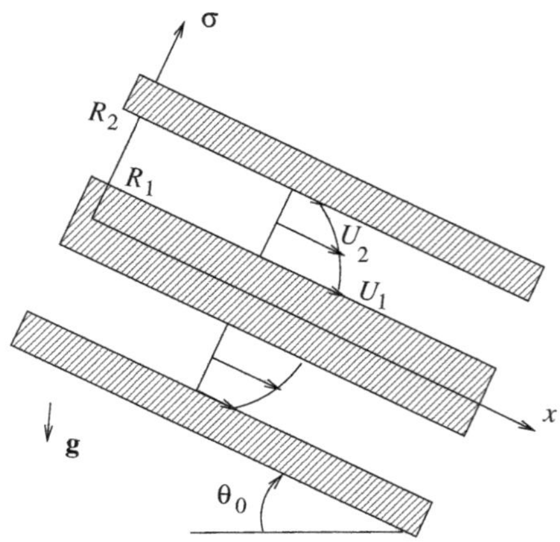

Figure 7.4.1 Illustration of unidirectional flow through an annular tube with translating walls.

The flow rate along the tube arises by integrating the velocity profile over the annular cross-section confined between the two concentric cylinders, obtaining

$$Q \equiv 2\pi \int_{R_1}^{R_2} u_x(\sigma)\,\sigma\,\mathrm{d}\sigma = \pi\,(U_2\,R_2^2 - U_1\,R_1^2) - \frac{\pi}{2}\,(U_2 - U_1)\,\frac{R_2^2 - R_1^2}{\ln(R_2/R_1)}$$

$$+\pi\,\frac{\chi + \rho\,g_x}{8\,\mu}\,(R_2^2 - R_1^2)\,(R_2^2 + R_1^2 - \frac{R_2^2 - R_1^2}{\ln(R_2/R_1)}). \qquad (7.4.2)$$

Program *tube_ann* in directory *04_various* of FDLIB evaluates the velocity profile and the flow rate given in (7.4.1) and (7.4.2).

Small gaps

If the clearance of the channel is small compared to the inner cylinder radius, $R_2 - R_1 < R_1$, the curvature of the cylindrical surfaces is insignificant. Equations (7.4.1) and (7.4.2) then reduce to equations (7.1.7) and (7.1.9) for unidirectional flow in a channel with parallel-sided walls separated by the distance $2h = R_2 - R_1$, where $y = \sigma - \frac{1}{2}(R_1 + R_2)$ is the radial distance of a point from the mean tube surface. To demonstrate this reduction, we introduce the small dimensionless number

$$\epsilon = \frac{R_2 - R_1}{R_1} = \frac{2h}{R_1} \ll 1, \qquad (7.4.3)$$

and write

$$\frac{R_2}{R_1} = 1 + \epsilon, \qquad \ln\frac{R_2}{R_1} = \ln(1+\epsilon) \simeq \epsilon - \frac{1}{2}\,\epsilon^2. \qquad (7.4.4)$$

Using the definition of h, we find

$$y = \sigma - (R_1 + h) = \sigma - (R_2 - h). \tag{7.4.5}$$

Next, we write

$$\ln\frac{R_2}{\sigma} = -\ln\frac{\sigma}{R_2} = -\ln\Big(\frac{\sigma}{R_1}\frac{R_1}{R_2}\Big) = -\ln\frac{\sigma}{R_1} - \ln\frac{R_1}{R_2} = -\ln\frac{\sigma}{R_1} + \ln\frac{R_2}{R_1}, \tag{7.4.6}$$

and

$$\ln\frac{\sigma}{R_1} = \ln\frac{y + R_1 + h}{R_1} = \ln\Big(1 + \frac{y+h}{R_1}\Big) \simeq \eta - \frac{1}{2}\,\eta^2, \tag{7.4.7}$$

where

$$\eta = \frac{y+h}{R_1}. \tag{7.4.8}$$

Combining these results, we find

$$\begin{aligned} \frac{\ln(R_2/\sigma)}{\ln(R_2/R_1)} &= -\frac{\ln(\sigma/R_1)}{\ln(R_2/R_1)} + 1 \simeq -\frac{\eta - \frac{1}{2}\eta^2}{\epsilon - \frac{1}{2}\,\epsilon^2} + 1 \simeq -\frac{\eta\,(1 - \frac{1}{2}\,\eta)}{\epsilon\,(1 - \frac{1}{2}\,\epsilon)} + 1 \\ &\simeq -\frac{\eta}{\epsilon}\Big(1 - \frac{1}{2}\,\eta\Big)\,\Big(1 + \frac{1}{2}\,\epsilon\Big) + 1 \simeq 1 - \frac{\eta}{\epsilon}\Big(1 - \frac{1}{2}\,\eta + \frac{1}{2}\,\epsilon\Big). \end{aligned} \tag{7.4.9}$$

We have omitted cubic terms in ϵ and η. Note that the ratio η/ϵ is on the order of unity. Continuing, we find

$$R_2^2 - R_1^2 = (R_2 + R_1)(R_2 - R_1) = (2R_1 + 2h)\,2h = R_1^2\,(2+\epsilon)\,\epsilon, \tag{7.4.10}$$

and

$$R_2^2 - \sigma^2 = (R_2 + \sigma)(R_2 - \sigma) = R_1^2\,(2 + \epsilon + \eta)(\epsilon - \eta). \tag{7.4.11}$$

Substituting (7.4.9)–(7.4.11) in (7.4.1), we find

$$\begin{aligned} u_x(\sigma) &\simeq U_2 + (U_1 - U_2)\,\Big(1 - \frac{\eta}{\epsilon}\Big) \\ &+ \frac{\chi + \rho\, g_x}{4\,\mu}\, R_1^2\,\Big(\,(2 + \epsilon + \eta)(\epsilon - \eta) - (2+\epsilon)\big[\epsilon - \eta\,\Big(1 - \frac{1}{2}\,\eta + \frac{1}{2}\,\epsilon\Big)\big]\Big) \end{aligned} \tag{7.4.12}$$

or

$$\begin{aligned} u_x(\sigma) &\simeq U_2 + (U_1 - U_2)\,\frac{\epsilon - \eta}{\epsilon} \\ &+ \frac{\chi + \rho\, g_x}{4\,\mu}\, R_1^2\,\Big(\,2\,(\epsilon - \eta) + \epsilon^2 - \eta^2 - (2+\epsilon)\big[\epsilon - \eta + \frac{\eta}{2}\,(\eta - \epsilon)\big]\,\Big). \end{aligned} \tag{7.4.13}$$

Simplifying and omitting cubic terms in ϵ and η, we find

$$u_x(\sigma) \simeq U_2 + (U_1 - U_2)\,\frac{h-y}{2h} + \frac{\chi + \rho\, g_x}{4\,\mu}\, R_1^2 \Big(2\,(\epsilon - \eta) + \epsilon^2 - \eta^2 - 2\,(\epsilon - \eta) - \eta\,(\eta - \epsilon) - \epsilon\,(\epsilon - \eta) \Big), \tag{7.4.14}$$

and then

$$u_x(\sigma) \simeq U_2 + (U_1 - U_2)\,\frac{h-y}{2h} + \frac{\chi + \rho\, g_x}{4\,\mu}\, R_1^2\, 2\,\eta\,(\epsilon - \eta). \tag{7.4.15}$$

After simplifications, we find

$$u_x(\sigma) \simeq U_2 + (U_1 - U_2)\,\frac{h-y}{2h} + \frac{\chi + \rho\, g_x}{2\mu}\,(h+y)\,(h-y), \tag{7.4.16}$$

which is precisely the profile for channel flow given in (7.1.7).

To recover the expression for the flow rate in channel flow, we substitute in (7.4.2)

$$R_2^2 = R_1^2\,(1+\epsilon)^2, \qquad \ln\frac{R_2}{R_1} = \ln(1+\epsilon) \simeq \epsilon - \frac{1}{2}\,\epsilon^2 + \frac{1}{3}\,\epsilon^3, \tag{7.4.17}$$

and find

$$Q \simeq \pi\, R_1^2\,(U_2\,(1+\epsilon)^2 - U_1) - \frac{\pi}{2}\,(U_2 - U_1)\,R_1^2\,\frac{\epsilon\,(2+\epsilon)}{\epsilon(1-\frac{1}{2}\,\epsilon)} + \pi\,\frac{\chi + \rho\, g_x}{8\,\mu}\,R_1^4\,\epsilon\,(2+\epsilon)\Big[(1+\epsilon)^2 + 1 - \frac{\epsilon\,(2+\epsilon)}{\epsilon(1-\frac{1}{2}\epsilon + \frac{1}{3}\epsilon^2)}\Big], \tag{7.4.18}$$

or

$$Q \simeq \pi\, R_1^2\,(U_2\,(1+\epsilon)^2 - U_1) - \frac{\pi}{2}\,(U_2 - U_1)\,R_1^2\,(2+\epsilon)(1+\frac{1}{2}\,\epsilon) + \pi\,\frac{\chi + \rho\, g_x}{8\,\mu}\,R_1^4\,\epsilon\,(2+\epsilon)\Big[(1+\epsilon)^2 + 1 - (2+\epsilon)(1+\frac{1}{2}\epsilon - \frac{1}{12}\epsilon^2)\Big], \tag{7.4.19}$$

which simplifies to

$$\begin{aligned} Q &\simeq \pi\, R_1^2\,(U_2 + U_1)\,\epsilon + \pi\,\frac{\chi + \rho\, g_x}{8\mu}\,R_1^4\,\frac{4}{3}\,\epsilon^3 \\ &= 2\pi R_1\,\Big((U_2 + U_1)\,h + \frac{\chi + \rho\, g_x}{\mu}\,\frac{2}{3}\,h^3\Big). \end{aligned} \tag{7.4.20}$$

The expression inside the large parentheses is the flow rate of channel flow.

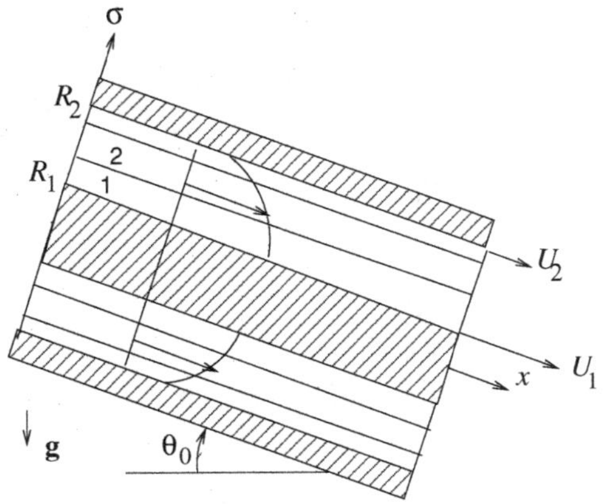

Figure 7.4.2 Illustration of multi-layer unidirectional flow through an annular tube with translating walls.

7.4.1 Multi-layer annular flow

Figure 7.4.2 illustrates axisymmetric multi-layer pressure-, gravity-, and shear-driven flow through am annular tube. For the interfaces to remain concentric, the tube must be vertical or else the fluid densities must be identical. Otherwise, hydrostatic pressure variations will cause a non-axisymmetric displacement.

Program *tube_ann_ml* in directory *04_various* of FDLIB computes the composite velocity profile. The algorithm is similar to that discussed in section 7.1 for multi-layer channel flow.

Computer Problem

7.4.1 *Multi-layer annular flow*

Repeat the two parts of problem 7.3.4 for multi-layer annular flow.

7.5 Steady flow in channels and tubes

Thus far, we have considered channel and tube flow where the streamwise velocity, u_x, is a function of lateral position, y, distance from the tube centerline, σ, and possibly meridional angle, φ. In this section, we consider flow in a channel or tube with arbitrary cross-section where the streamwise velocity is a function of two spatial coordinates determining the position over the cross-section, $u_x(y, z)$. The assumption of unidirectional motion allows us to simplify the x, y, and z components of the equation of motion displayed in table 6.5.1, obtaining

$$
\begin{aligned}
0 &= -\frac{\partial p}{\partial x} + \mu\,(\frac{\partial^2 u_x}{\partial y^2} + \frac{\partial^2 u_x}{\partial z^2}) + \rho\, g_x, \\
0 &= -\frac{\partial p}{\partial y} + \rho\, g_y, \\
0 &= -\frac{\partial p}{\partial z}.
\end{aligned}
\tag{7.5.1}
$$

Without loss of generality, we have assumed that the acceleration of gravity lies in the xy plane, $g_z = 0$.

The y and z components in (7.5.1) are satisfied by the pressure distribution (7.1.4). To satisfy the x component, we require that the streamwise velocity is a solution of the second-order partial differential equation

$$
\nabla^2 u_x = \frac{\partial^2 u_x}{\partial y^2} + \frac{\partial^2 u_x}{\partial z^2} = -\frac{\chi + \rho\, g_x}{\mu}, \tag{7.5.2}
$$

where

$$
\nabla^2 \equiv \frac{\partial^2}{\partial y^2} + \frac{\partial^2}{\partial z^2} \tag{7.5.3}
$$

is the Laplacian operator in the yz plane. Equation (7.5.2) is a Poisson equation with a constant right-hand side generalizing the ordinary differential equations (7.1.5) and (7.3.2) for two-dimensional and axisymmetric flow. In the remainder of this section, we derive exact solutions of (7.5.2) for selected boundary geometries.

7.5.1 Elliptical tube

First, we consider flow through a cylindrical tube with elliptical cross-section whose contour in the yz plane is described by the equation

$$
f(y, z) = \frac{y^2}{a^2} + \frac{z^2}{b^2} - 1 = 0, \tag{7.5.4}
$$

where a and b are the tube semi-axes, as illustrated in figure 7.5.1(a). The coordinates of a point on the elliptical contour can be identified by a parameter η taking values in the range $[0, 2\pi)$, as

$$
y = a\,\cos\eta, \qquad z = b\,\sin\eta. \tag{7.5.5}
$$

The no-slip boundary condition requires $u_x(y, z) = 0$ for (y, z) pairs generated by (7.5.5) and thus satisfying $f(y, z) = 0$ according to (7.5.4).

Because $f(y, z)$ is a quadratic function of y and z, its Laplacian defined in (7.5.3) is constant. Motivated by this observation, we express the velocity in the

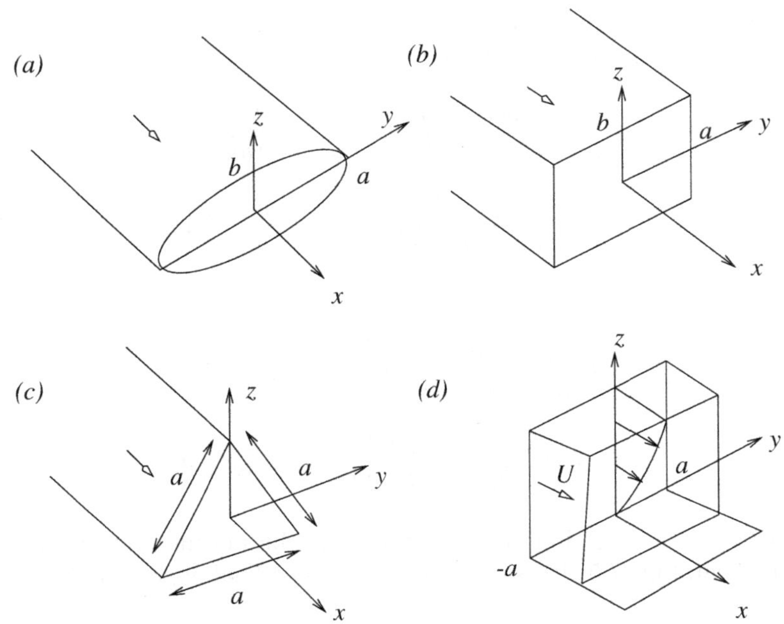

Figure 7.5.1 Illustration of steady flow through a tube with (a) elliptical, (b) rectangular, and (c) equilateral triangle cross-section. (d) Flow between two parallel plates moving over a plane wall regarded as a painted surface.

form $u_x(y, z) = c\, f(y, z)$ to guarantee the satisfaction of the no-slip boundary condition, and adjust the constant c to satisfy the Poisson equation (7.5.2). The result is the quadratic velocity distribution

$$u_x(y, z) = \frac{\chi + \rho\, g_x}{2\mu} \frac{a^2 b^2}{a^2 + b^2} \left(1 - \frac{y^2}{a^2} - \frac{z^2}{b^2}\right). \tag{7.5.6}$$

Maximum velocity occurs at the tube centerline located at $y = 0$ and $z = 0$.

Flow rate

Taking advantage of the symmetry of the velocity profile with respect to the xy and xz planes, we express the axial flow rate as an integral of the velocity over the first quadrant,

$$Q = \iint_{Ellipse} u_x(y, z)\, \mathrm{d}y\, \mathrm{d}z = 4 \int_0^a \int_0^{z_{max}} u_x(y, z)\, \mathrm{d}z\, \mathrm{d}y, \tag{7.5.7}$$

where z_{max} is computed by solving equation (7.5.4) for z, yielding

$$z_{max} = b\,\sqrt{1 - y^2/a^2}. \tag{7.5.8}$$

Substituting the velocity profile (7.5.6) in the integrand and carrying out the integration with respect to z, we find

$$Q = 2\,\frac{\chi + \rho\, g_x}{\mu}\,\frac{a^2 b^2}{a^2+b^2}\int_0^a (1 - \frac{y^2}{a^2} - \frac{z_{max}^2}{3\,b^2})\, z_{max}\, \mathrm{d}y$$
$$= 4\,\frac{\chi + \rho\, g_x}{3\mu}\,\frac{a^2 b^3}{a^2+b^2}\int_0^a (1 - \frac{y^2}{a^2})^{3/2}\, \mathrm{d}y, \qquad (7.5.9)$$

and then

$$Q = 4\,\frac{\chi + \rho\, g_x}{3\mu}\,\frac{a^3 b^3}{a^2+b^2}\int_0^1 (1 - \hat{y}^2)^{3/2}\, \mathrm{d}\hat{y}$$
$$= 4\,\frac{\chi + \rho\, g_x}{3\mu}\,\frac{a^3 b^3}{a^2+b^2}\int_0^{\pi/2} \cos^4\xi\, \mathrm{d}\xi, \qquad (7.5.10)$$

where $\hat{y} \equiv y/a \equiv \sin\xi$. Referring to standard tables of integrals, we find that the last integral in (7.5.10) is equal to $3\pi/16$. The final expression for the flow rate is

$$Q = \frac{\chi + \rho\, g_x}{4\mu}\,\pi\,\frac{a^3 b^3}{a^2+b^2}. \qquad (7.5.11)$$

As the second semi-axis, b, tends to become equal to the first semi-axis, a, the cross-section of the tube tends to become circular and the expressions for the velocity distribution and flow rate reduce to those shown in equations (7.3.4) and (7.3.9) for Poiseuille flow.

Program *tube_ell* in directory *04_various* of FDLIB evaluates the velocity profile and flow rate given by (7.5.6) and (7.5.11).

7.5.2 *Rectangular tube*

Next, we consider flow through a tube whose cross-section is a rectangle with side-lengths equal to $2a$ and $2b$, as illustrated in figure 7.5.1(b). A standard method of solving the Poisson equation (7.5.2) involves expressing the velocity as the sum of (a) a particular solution that satisfies the Poisson equation (7.5.2) but not all boundary conditions, denoted as $u_x^p(y,z)$, and (b) a homogeneous solution, $u_x^h(y,z)$, that satisfies Laplace's equation $\nabla^2 u_x^h = 0$ subject to boundary conditions that arise by requiring that the total velocity, $u_x = u_x^p + u_x^h$, satisfies the no-slip boundary condition along the four walls.

A convenient particular solution is

$$u_x^p(y,z) = \frac{\chi + \rho\, g_x}{2\,\mu}\,(b^2 - z^2). \qquad (7.5.12)$$

Requiring $u_x^p + u_x^h = 0$ along the four walls, we derive boundary conditions for the homogeneous solution,

$$u_x^h(y = \pm a, z) = -u_x^p(y = \pm a, z) = -\frac{\chi + \rho\, g_x}{2\,\mu}\,(b^2 - z^2), \qquad (7.5.13)$$

and

$$u_x^h(y, z = \pm b) = -u_x^p(y, z = \pm b) = 0. \qquad (7.5.14)$$

Fourier expansion

The homogeneous solution can be expanded in a Fourier series with respect to z,

$$u_x^h(y, z) = \sum_{n=1}^{\infty} f_n(y)\,\cos(\alpha_n \frac{z}{b}). \qquad (7.5.15)$$

To satisfy the boundary conditions (7.5.14), we set $\alpha_n = (n - \frac{1}{2})\,\pi$. The functions $f_n(y)$ will be computed so that the right-hand side of (7.5.15) satisfies Laplace's equation $\nabla^2 u_x^h = 0$, and conforms with the boundary conditions (7.5.13). Taking the Laplacian of (7.5.15), we find

$$\nabla^2 u_x^h(y, z) = \sum_{n=1}^{\infty} [\,\frac{d^2 f_n(y)}{dy^2} - \frac{\alpha_n^2}{b^2}\, f_n(y)\,]\,\cos(\alpha_n \frac{z}{b}). \qquad (7.5.16)$$

For the infinite sum on the right-hand side to be zero for any value of y and z, each term enclosed by the square brackets must be zero,

$$\frac{d^2 f_n(y)}{dy^2} - \frac{\alpha_n^2}{b^2}\, f_n(y) = 0. \qquad (7.5.17)$$

The general solution of this equation is

$$f_n(y) = A_n\,\cosh(\alpha_n\,\frac{y}{b}) + B_n\,\sinh(\alpha_n\,\frac{y}{b}), \qquad (7.5.18)$$

where A_n and B_n are constant coefficients. Discarding the hyperbolic sines to ensure that the functions $f_n(y)$ are even and the velocity profile is symmetric with respect to $y = 0$, and substituting the result in (7.5.15), we find

$$u_x^h(y, z) = \sum_{n=1}^{\infty} A_n\,\cosh(\alpha_n\,\frac{y}{b})\,\cos(\alpha_n \frac{z}{b}). \qquad (7.5.19)$$

To satisfy the remaining boundary conditions (7.5.13), we require

$$\sum_{n=1}^{\infty} A_n\,\cosh(\alpha_n\,\frac{a}{b})\,\cos(\alpha_n \frac{z}{b}) = -\frac{\chi + \rho\, g_x}{2\,\mu}\,(b^2 - z^2). \qquad (7.5.20)$$

The left-hand side of (7.5.20) represents the Fourier series of the quadratic function with respect to z on the right-hand side.

The solution follows readily using the trigonometric identity

$$\int_{-b}^{b} \cos(\alpha_n \frac{z}{b}) \cos(\alpha_m \frac{z}{b})\, \mathrm{d}z = \begin{cases} b & \text{if } n = m, \\ 0 & \text{if } n \neq m. \end{cases} \tag{7.5.21}$$

Multiplying both sides of (7.5.20) by $\cos(\alpha_m\, z/b)$, integrating with respect to z from $-b$ to b, using (7.5.21) to compute the integrals on the left-hand side, and then renaming m to n, we find

$$A_n = -\frac{\chi + \rho\, g_x}{2\,\mu\, b\, \cosh(\alpha_n\, \frac{a}{b})} \int_{-b}^{b} \cos(\alpha_n \frac{z}{b})\, (b^2 - z^2)\, \mathrm{d}z \tag{7.5.22}$$

$$= -\frac{\chi + \rho\, g_x}{2\mu\, \cosh(\alpha_n\, \frac{a}{b})}\, b^2 \int_{-\alpha_n}^{\alpha_n} (1 - \frac{\eta^2}{\alpha_n^2})\, \cos\eta\, \mathrm{d}\eta = -\frac{\chi + \rho\, g_x}{2\mu\, \cosh(\alpha_n\, \frac{a}{b})}\, \frac{4b^2}{\alpha_n^3}\, (-1)^n,$$

where $\eta \equiv \alpha_m\, z/b$ is an intermediate variable. Substituting expression (7.5.22) in (7.5.19) and adding the homogeneous to the particular solution expressed by (7.5.12), we obtain the velocity distribution in the form of an infinite Fourier series,

$$u_x(y,z) = \frac{\chi + \rho\, g_x}{2\,\mu}\, b^2\, [1 - \frac{z^2}{b^2} + 4 \sum_{n=1}^{\infty} \frac{(-1)^n}{\alpha_n^3} \frac{\cosh(\alpha_n\, \frac{y}{b})}{\cosh(\alpha_n\, \frac{a}{b})} \cos(\alpha_n \frac{z}{b})]. \tag{7.5.23}$$

Flow rate

The flow rate arises by integrating the velocity distribution over the tube cross-section,

$$Q \equiv \int_{-a}^{a} \int_{-b}^{b} u_x(y,z)\, \mathrm{d}y\, \mathrm{d}z. \tag{7.5.24}$$

Substituting expression (7.5.23) and carrying out the integration, we find

$$Q = \frac{\chi + \rho\, g_x}{3\,\mu}\, 4\, a\, b^3 \left[1 - 6\, \frac{b}{a} \sum_{n=1}^{\infty} \frac{1}{\alpha_n^5} \tanh(\alpha_n \frac{a}{b}) \right]. \tag{7.5.25}$$

Program *tube_rec* in directory *04_various* of FDLIB evaluates the velocity profile, the flow rate, and the maximum velocity occurring at the tube centerline.

7.5.3 Triangular tube

In the third case-study, we consider pressure- or gravity-driven flow through a tube whose cross-section is an equilateral triangle with side-length a, as illustrated in figure 7.5.1(c). The origin of the Cartesian axis is placed at the centroid of the triangle, and the y axis is parallel to the lower side of the triangle.

The following velocity distribution satisfies the Poisson equation (7.5.2) and is consistent with the no-slip boundary condition along the three sides,

$$u_x(y,z) = \frac{\chi + \rho\, g_x}{36\,\mu\, a}\,(2\sqrt{3}\,z + a)\,(\sqrt{3}\,z + 3\,y - a)\,(\sqrt{3}\,z - 3\,y - a). \tag{7.5.26}$$

Integrating the velocity over the cross-section, we derive an expression for the flow rate

$$Q = \frac{\sqrt{3}}{320}\,\frac{\chi + \rho\, g}{\mu}\,a^4. \tag{7.5.27}$$

7.5.4 Semi-infinite rectangular channel

Next, we consider flow between two semi-infinite parallel plates sliding with velocity U against a stationary flat surface, as illustrated in figure 7.5.1(d). The motion of the plates generates a unidirectional flow along the x axis, regarded as a model of the flow occurring between two hairs of an idealized two-dimensional paint brush moving over a flat surface.

In the absence of a pressure gradient and significant gravitational forces, the x component of the velocity satisfies Laplace's equation $\nabla^2 u_x = 0$, arising from (7.5.2) by setting the right-hand side to zero. The no-slip boundary condition requires that the velocity is zero over the flat surface located at $z = 0$, and takes a constant value, U, over the sliding plates located at $y = \pm a$. Far from the painted surface, the fluid moves with the plate velocity U in a plug-flow mode.

Fourier expansion

To compute the solution, we expand the streamwise velocity in a Fourier series as discussed earlier in this section for flow through a tube with rectangular cross-section. Requiring that u_x tends to U as z moves to infinity, we obtain the counterpart of (7.5.19),

$$u_x(y,z) = U + \sum_{n=1}^{\infty} A_n \exp(-\alpha_n \frac{z}{a})\, \cos(\alpha_n \frac{y}{a}), \tag{7.5.28}$$

where $\alpha_n = (n - \frac{1}{2})\,\pi$. Expansion (7.5.28) satisfies the no-slip condition at the side walls. To also satisfy the no-slip condition at the painted surface, we require

$$u_x(y, z=0) = U + \sum_{n=1}^{\infty} A_n \cos(\alpha_n \frac{y}{a}) = 0. \tag{7.5.29}$$

Using the counterpart of the trigonometric identity (7.5.21)

$$\int_{-a}^{a} \cos(\alpha_n \frac{y}{a}) \cos(\alpha_m \frac{y}{a}) \, dy = \begin{cases} a & \text{if } n = m, \\ 0 & \text{if } n \neq m, \end{cases} \tag{7.5.30}$$

and working as previously for flow in a rectangular channel, we find

$$A_n = -\frac{U}{a} \int_{-a}^{a} \cos(\alpha_n \frac{y}{a}) \, dy = -\frac{2U}{\alpha_n} \sin \alpha_n = (-1)^n \frac{2U}{\alpha_n}. \tag{7.5.31}$$

Substituting this expression in (7.5.28), we obtain the desired velocity distribution

$$u_x(y, z) = U \, [\, 1 + 2 \sum_{n=1}^{\infty} \frac{(-1)^n}{\alpha_n} \exp(-\alpha_n \frac{z}{a}) \cos(\alpha_n \frac{y}{a}) \,]. \tag{7.5.32}$$

As the two plates travel over the stationary surface, they dispense an amount of fluid with flow rate

$$\begin{aligned} Q &= \int_0^{\infty} \int_{-a}^{a} [\, U - u_x(y, z) \,] \, dy \, dz \\ &= -2U \sum_{n=1}^{\infty} \frac{(-1)^n}{\alpha_n} \int_0^{\infty} \int_{-a}^{a} \exp(-\alpha_n \frac{z}{a}) \cos(\alpha_n \frac{y}{a}) \, dy \, dz \\ &= 4Ua^2 \sum_{n=1}^{\infty} \frac{1}{\alpha_n^3} = 1.085 \, U \, a^2. \end{aligned} \tag{7.5.33}$$

Program *chan_brush* in directory *04_various* of FDLIB evaluates the velocity field and flow rate given by equations (7.5.32) and (7.5.33).

Problems

7.5.1 *Area and perimeter of an ellipse*

Show that the area of an ellipse with semi-axes a and b is equal to πab.

7.5.2 *Flow through a triangular tube*

Compute the ratio of the flow rate through a circular or triangular tube with the same cross-sectional areas, subject to the same pressure gradient.

Computer Problem

7.5.3 *An elliptical hose*

Using the parametrization (7.5.5), we find that the differential arc length along the contour of an ellipse is given by

$$\mathrm{d}l \equiv (\mathrm{d}x^2 + \mathrm{d}y^2)^{1/2} = (a^2 \, \sin^2 \eta + b^2 \, \cos^2 \eta)^{1/2} \, \mathrm{d}\eta \tag{7.5.34}$$

or

$$\mathrm{d}l = \Big(a^2 - (a^2 - b^2) \, \cos^2 \eta \Big)^{1/2} \mathrm{d}\eta = a \, \Big(1 - k^2 \, \cos^2 \eta \Big)^{1/2} \mathrm{d}\eta, \tag{7.5.35}$$

where $k^2 \equiv 1 - b^2/a^2$. For convenience and without loss of generality, we have assumed $b \le a$. The arc length of the ellipse is given by

$$L = \oint \mathrm{d}l = 4\,a \int_0^{\pi/2} (1 - k^2 \, \cos^2 \eta)^{1/2} \, \mathrm{d}\eta = 4\,a \int_0^{\pi/2} (1 - k^2 \, \sin^2 \eta)^{1/2} \, \mathrm{d}\eta. \tag{7.5.36}$$

When $a = b$, we find $k^2 = 0$ and obtain the well-known result for a circle, $L = 2\pi a$. The integral on the right-hand side of (7.5.36) is the complete elliptic integral of the second kind.

In section 11.5, we will see that the complete elliptic integral can be evaluated using an efficient iterative method. As an alternative, we use the mid-point rule to approximate the integral with a sum,

$$L \simeq 4\,a \, \frac{\pi}{2\,N} \sum_{i=1}^{N} (1 - k^2 \, \sin^2 \eta_i)^{1/2}, \tag{7.5.37}$$

where N is a specified discretization level, and $\eta_i = (i - \frac{1}{2}) \frac{\pi}{2N}$.

(*a*) Prepare a graph of the reduced arc length L/a against the axes ratio b/a, where $0 < b/a < 1$. The value of N in (7.5.37) should be large enough for the arc length to be accurate at the third decimal place.

(*b*) A gardener delivers water through a circular hose made of a flexible but inextensible material. By pinching the end of the hose, she is able to produce elliptical shapes with variable cross-sections, while maintaining the cross-sectional perimeter of the hose constant. Prepare a plot of the delivered flow rate against the aspect ratio of the cross-section for a certain pressure gradient, and discuss your results.

7.6 Steady swirling flow

We have considered steady unidirectional flows with rectilinear streamlines. Because point particles travel with constant velocity along straight paths, inertial

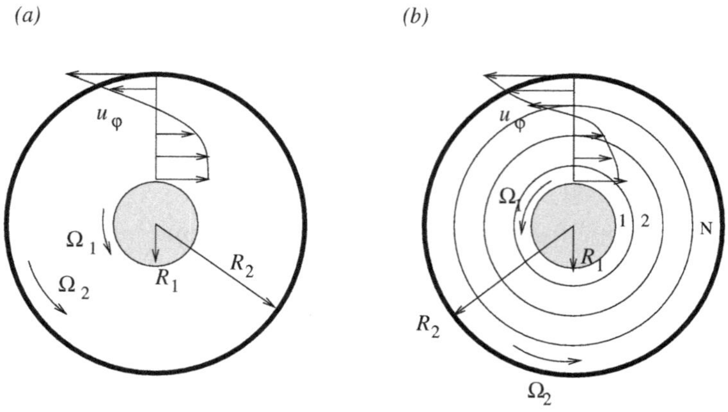

Figure 7.6.1 Illustration of swirling flow between two rotating concentric cylinders: (a) single-fluid, and (b) multi-layer flow.

forces are identically zero and the flow is governed by a linear equation of motion expressing a balance between the pressure gradient, the viscous force, and the body force. In this section, we turn our attention to swirling flow with circular streamlines where centrifugal forces arise.

7.6.1 Annular flow

We begin by considering steady swirling flow between two infinite concentric cylinders with radii R_1 and R_2 rotating around their common axis with angular velocities Ω_1 and Ω_2, as illustrated in figure 7.6.1(a). The induced circular Couette flow is the counterpart of the plane Couette flow with rectilinear streamlines discussed in section 7.1.

It is natural to introduce cylindrical polar coordinates with the x axis coaxial with the cylinders. Our analysis will be based on two key assumptions: the axial and radial components of the velocity vanish, $u_x = 0$ and $u_\sigma = 0$, and the meridional component, u_φ, is independent of the meridional angle φ, $\partial u_\varphi / \partial \varphi = 0$. Consequently, $\mathrm{D}u_\varphi/\mathrm{D}t = 0$, and u_φ depends only on distance from the x axis, σ, where $\mathrm{D}/\mathrm{D}t$ is the material derivative. The no-slip boundary condition at the cylinder surfaces requires

$$u_\varphi = R_1\,\Omega_1 \qquad \text{at } \sigma = R_1,$$
$$u_\varphi = R_2\,\Omega_2 \qquad \text{at } \sigma = R_2. \tag{7.6.1}$$

For flow with circular streamlines, the cylindrical polar components of the equation of motion displayed in equations (6.3.18) take the simplified forms

$$
\begin{aligned}
0 &= -\frac{\partial p}{\partial x} + \rho\, g_x, \\
0 &= \rho\, \frac{u_\varphi^2}{\sigma} - \frac{\partial p}{\partial \sigma} + \rho\, g_\sigma, \\
0 &= -\frac{1}{\sigma}\frac{\partial p}{\partial \varphi} + \mu\, \frac{\mathrm{d}}{\mathrm{d}\sigma}\Big(\frac{1}{\sigma}\,\frac{\mathrm{d}(\sigma u_\varphi)}{\mathrm{d}\sigma}\Big) + \rho\, g_\varphi.
\end{aligned}
\tag{7.6.2}
$$

To remove the hydrostatic pressure variation, we introduce the dynamic pressure defined as

$$p^D \equiv p - \rho\,(g_x\, x + g_y\, y + g_z\, z), \tag{7.6.3}$$

and reduce the three differential equations (7.6.2) to two ordinary differential equations,

$$\frac{\mathrm{d}p^D}{\mathrm{d}\sigma} = \rho\, \frac{u_\varphi^2}{\sigma}, \qquad \frac{\mathrm{d}}{\mathrm{d}\sigma}\Big(\frac{1}{\sigma}\,\frac{\mathrm{d}\,(\sigma u_\varphi)}{\mathrm{d}\sigma}\Big) = 0. \tag{7.6.4}$$

Straightforward integration of the second equation with respect to σ subject to the boundary conditions (7.6.1) provides us with the velocity profile

$$u_\varphi(\sigma) = \frac{\Omega_2 - \alpha\,\Omega_1}{1-\alpha}\,\sigma - \frac{\Omega_2 - \Omega_1}{1-\alpha}\,\frac{R_1^2}{\sigma}, \tag{7.6.5}$$

where

$$\alpha \equiv \Big(\frac{R_1}{R_2}\Big)^2 < 1. \tag{7.6.6}$$

The first term on the right-hand side of (7.6.5) expresses rigid-body rotation. The second term expresses swirling motion due to a point vortex situated at the cylinder axis. The pressure distribution follows by substituting (7.6.5) in the first of equations (7.6.4) and carrying out the integration with respect to σ. The shear stress exerted on the inner cylinder is

$$\sigma_{\sigma\varphi} = \mu\,\Big(\frac{\mathrm{d}u_\varphi}{\mathrm{d}\sigma}\Big)_{\sigma=R_1}. \tag{7.6.7}$$

The torque exerted on either cylinder is

$$T \equiv 2\pi\sigma\,\sigma_{\sigma\varphi} = -4\pi\mu\, R_1^2\,\frac{\Omega_2 - \Omega_1}{1-\alpha}. \tag{7.6.8}$$

Program *tube_ann_sw* in directory *04_various* of FDLIB evaluates the velocity profile given in (7.6.5).

Rigid-body rotation

Equation (7.6.5) confirms that when, $\Omega_1 = \Omega_2 = \Omega$, or when the inner cylinder is absent, $R_1 = 0$, the fluid rotates as a rigid body with angular velocity Ω. In this case, the pressure distribution is readily found to be

$$p^D = P_0 + \frac{1}{2}\,\rho\,\Omega^2 \sigma^2, \tag{7.6.9}$$

where P_0 is a constant. Thus, rigid body rotation is associated with a quadratic pressure distribution with respect to the radial position established to counteract the inertial centrifugal force.

Small gaps

When the clearance of the channel is small compared to the inner cylinder radius, $R_2 - R_1 < R_1$, the circular Couette flow reduces to plane Couette in a channel with parallel-sided walls studied in section 7.6.1. To demonstrate this reduction, we substitute $\Omega_1 = U_1/R_1$ and $\Omega_2 = U_2/R_2$ in (7.6.5) and find

$$u_\varphi(\sigma) = \frac{U_2/R_2 - \alpha\, U_1/R_1}{1-\alpha}\,\sigma - \frac{U_2/R_2 - U_1/R_1}{1-\alpha}\,\frac{R_1^2}{\sigma}. \tag{7.6.10}$$

Next, we define the gap, $2h = R_2 - R_1$, and the scaled distance from the inner cylinder, η, and write

$$R_2 = R_1 + 2h, \qquad \frac{R_2}{R_1} = 1 + \epsilon, \qquad \epsilon = \frac{2h}{R_1}, \qquad \frac{\sigma}{R_1} = \eta + 1, \qquad \eta = \frac{y+h}{R_1}, \tag{7.6.11}$$

where ϵ is a small parameter. The velocity profile becomes

$$u_\varphi(\sigma) = \frac{1}{R_2}\,\frac{U_2 - \alpha\,(1+\epsilon)\,U_1}{1-\alpha}\,\sigma - \frac{1}{R_2}\,\frac{U_2 - (1+\epsilon)\,U_1}{1-\alpha}\,\frac{R_1^2}{\sigma}, \tag{7.6.12}$$

which can be rearranged into

$$u_\varphi(\sigma) = \Big([U_2 - \alpha\,(1+\epsilon)\,U_1]\,\frac{\sigma}{R_1} - [U_2 - (1+\epsilon)\,U_1]\,\frac{R_1}{\sigma} \Big)\,\frac{R_1}{R_2}\,\frac{1}{1-\alpha}. \tag{7.6.13}$$

By definition,

$$\alpha = \frac{1}{(1+\epsilon)^2} \simeq \frac{1}{1+2\epsilon} \simeq 1 - 2\epsilon, \tag{7.6.14}$$

and

$$1 - \alpha = 1 - \frac{1}{(1+\epsilon)^2} = \frac{(1+\epsilon)^2 - 1}{(1+\epsilon)^2} \simeq \frac{2\,\epsilon}{(1+\epsilon)^2}, \tag{7.6.15}$$

and thus,

$$u_\varphi(\sigma) \simeq \Big([U_2 - (1-2\,\epsilon)\,(1+\epsilon)\,U_1]\,(1+\eta) - [U_2 - (1+\epsilon)\,U_1]\,(1-\eta) \Big)\,\frac{1+\epsilon}{2\,\epsilon}. \tag{7.6.16}$$

Linearizing the product $(1-2\,\epsilon)\,(1+\epsilon)$, we find

$$u_\varphi(\sigma) \simeq \Big([U_2 - (1-\epsilon)\,U_1]\,(1+\eta) - [U_2 - (1+\epsilon)\,U_1]\,(1-\eta) \Big)\,\frac{1+\epsilon}{2\,\epsilon}. \tag{7.6.17}$$

Linearizing the expression inside the large parentheses, we obtain

$$u_\varphi(\sigma) \simeq \Big((U_2 - U_1)\,2\,\eta + 2\,\epsilon\,U_1 \Big)\,\frac{1}{2\,\epsilon} = (U_2 - U_1)\,\frac{\eta}{\epsilon} + U_1, \tag{7.6.18}$$

which is the linear velocity profile of plane Couette flow.

7.6.2 Multi-layer flow

Program *tube_ann_sw_ml* in directory *04_various* of FDLIB computes the velocity profile of multi-layer swirling annular flow, as illustrated in figure 7.6.1(*b*). The algorithm is similar to that discussed in section 7.1 for multi-layer channel flow.

Problems

7.6.1 *Free surface of a rotating liquid.*

Am amount of liquid is placed inside a vertical cylindrical container that is closed at the bottom and rotates about the x axis pointing against the direction of gravity, with angular velocity Ω. Using equation (7.6.9), we find that the pressure distribution in the liquid is given by

$$p = P_0 + \frac{1}{2}\,\rho\,\Omega^2 + \rho\,g_x x. \tag{7.6.19}$$

(*a*) Show that, when surface tension is insignificant, the shape of the free surface is described by the equation

$$x = \frac{\Omega^2\sigma^2}{2g} + c, \tag{7.6.20}$$

where c is a constant and g is the magnitude of the acceleration of gravity.

(*b*) Derive a differential equation governing the shape of the free surface in the presence of surface tension.

Computer Problem

7.6.2 *Multi-layer swirling flow*

(*a*) Outline and explain the numerical procedure implemented in the code *tube_ann_sw_ml*.

(*b*) Plot and discuss the velocity profile of a three-layer swirling flow of your choice.

7.7 Transient channel flow

Having investigated the structure of steady unidirectional flows with rectilinear or circular streamlines, we now turn our attention to corresponding unsteady flows produced by sudden or oscillatory boundary motion or tilting, and by the application of a time-dependent pressure gradient. Our analysis will continue to be based on the assumption of unidirectional flow with rectilinear or circular streamlines.

7.7.1 *Couette flow*

We begin by considering flow in a two-dimensional channel with parallel walls separated by the distance d, where the lower wall is suddenly set in motion parallel to itself along the x axis with constant velocity U, while the upper wall is held stationary, as illustrated in figure 7.7.1. Working as in section 7.1, we find that, in the absence of a pressure gradient along the x axis, the pressure assumes the hydrostatic distribution. The x component of the equation of motion provides us with a partial differential equation for the x component of the velocity,

$$\rho \frac{\partial u_x}{\partial t} = \mu \frac{\partial^2 u_x}{\partial y^2}, \tag{7.7.1}$$

subject to the initial condition

$$u_x = 0 \quad \text{for } 0 < y \le d \ \text{ at } t = 0, \tag{7.7.2}$$

and the no-slip boundary conditions

$$u_x(y=0) = U, \qquad u_x(y=d) = 0, \qquad \text{for } t \ge 0. \tag{7.7.3}$$

At steady state, the solution is given by the linear Couette velocity profile

$$u_x^c(y) = U\,(1 - \frac{y}{d}), \tag{7.7.4}$$

as discussed in section 7.1.

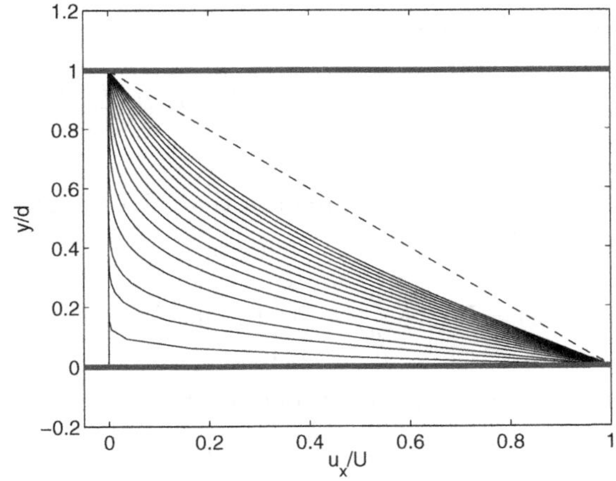

Figure 7.7.1 Transient Couette flow in a channel with parallel walls due to the impulsive translation of the lower wall, at dimensionless times $\hat{t} \equiv t\,\nu/d^2 =$ 0.001, 0.005, 0.010, 0.020, 0.030, ..., 0.14, and 0.15. The straight line is the linear profile established at long times.

Fourier expansion

To compute the solution of (7.7.1), we consider the deviation of the transient velocity profile from the linear profile established at steady state. Expanding the difference in a Fourier series with respect to y, we find

$$u_x(y,t) = u_x^C(y) + \sum_{n=1}^{\infty} A_n(t)\,\sin\frac{n\pi y}{d}. \tag{7.7.5}$$

The arguments of the trigonometric functions on the right-hand side have been designed to satisfy the boundary conditions (7.7.3) at all times. To compute the functions $A_n(t)$, we substitute expansion (7.7.5) in the governing equation (7.7.1) and carry out the differentiation with respect to t and x to find

$$\sum_{n=1}^{\infty} \Big[\rho\,\frac{\mathrm{d}A_n(t)}{\mathrm{d}t} + \mu\left(\frac{n\pi}{d}\right)^2 A_n(t)\Big]\,\sin\frac{n\pi y}{d} = 0. \tag{7.7.6}$$

To ensure the satisfaction of this equation for any value of y, we set the expression enclosed by the square brackets to zero and obtain a first-order linear differential equation for $A_n(t)$ whose solution is

$$A_n(t) = b_n \exp(-\frac{n^2\pi^2\nu}{d^2}\,t), \tag{7.7.7}$$

where $\nu = \mu/\rho$ is the kinematic viscosity, and b_n is a constant. Substituting expression (7.7.7) in (7.7.5), we find

$$u_x(y,t) = u_x^c(y) + \sum_{n=1}^{\infty} b_n \exp(-\frac{n^2\pi^2\nu}{d^2}\, t) \, \sin\frac{n\pi y}{d}. \qquad (7.7.8)$$

To compute the constants b_n, we require that (7.7.8) satisfies the initial condition (7.7.2) for $y > 0$,

$$u_x(y, t=0) = U\,(1 - \frac{y}{d}) + \sum_{n=1}^{\infty} b_n \, \sin\frac{n\pi y}{d} = 0. \qquad (7.7.9)$$

The solution follows readily using the trigonometric identity

$$\int_0^d \sin\frac{n\pi y}{d}\, \sin\frac{m\pi y}{d}\, dy = \begin{cases} 0 & \text{if } n = 0 \text{ or } m = 0, \\ \frac{d}{2} & \text{if } n = m \neq 0, \\ 0 & \text{if } n \neq m. \end{cases} \qquad (7.7.10)$$

Multiplying (7.7.9) by $\sin(m\pi y/d)$, integrating with respect to y from 0 to d, using identity (7.7.10), and then switching m to n, we find

$$b_n = -\frac{2U}{d} \int_0^d (1 - \frac{y}{d}) \, \sin\frac{n\pi y}{d}\, dy = -\frac{2U}{n\pi}. \qquad (7.7.11)$$

The velocity profile (7.7.8) takes the final form

$$u_x(y,t) = U\,[\,1 - \frac{y}{d} - \frac{2}{\pi} \sum_{n=1}^{\infty} \frac{1}{n}\, \exp(-\frac{n^2\pi^2\nu}{d^2}\, t) \, \sin\frac{n\pi y}{d}\,]. \qquad (7.7.12)$$

Program *chan_2d_imp* in directory *04_various* of FDLIB evaluates the velocity profile given in (7.7.12). A sequence of developing profiles of the scaled velocity $\hat{u} \equiv u_x/U$ illustrating the approach to steady state is shown in figure 7.7.1.

Approach to steady state

At long times, the summed terms on the right-hand side of (7.7.12) decay at an exponential rate, and the linear profile of the steady Couette flow expressed by the first two terms inside the angular brackets on the right-hand side dominates. The elapsed time where steady flow has virtually been established, t_s, can be estimated by setting the magnitude of the argument of the exponential term on the right-hand side of (7.7.12) equal to unity for the lowest possible value $n = 1$, obtaining

$$t_s \simeq \frac{d^2}{\pi^2 \nu}. \qquad (7.7.13)$$

Apart from the factor π^2 in the denominator, this estimate could have been deduced on the basis of dimensional analysis at the outset.

Diffusion of vorticity

The only non-vanishing vorticity component in unidirectional flow is $\omega_z = -\partial u_x/\partial y$. Differentiating (7.7.1) with respect to y, we find that the evolution of the vorticity is governed by equation (7.7.1), provided that u_x is replaced by ω_z on both sides. Figure 7.7.1 illustrates that, as soon as the lower wall starts translating, a thin layer of highly rotational flow is established. The vorticity then diffuses away from the wall to occupy the clearance of the channel. At steady state, the vorticity takes the uniform value U/d. This process exemplifies how vorticity enters a fluid across a boundary during a period of transient flow.

7.7.2 Impulsive motion of a plate in a semi-infinite fluid

A detailed analysis of the transient Couette flow at short times can be carried out by neglecting the presence of the upper stationary wall and concentrating on the behavior of the flow near the lower moving wall. In principle, this can be done by setting $d = \infty$ in the initial condition (7.7.2) and boundary conditions (7.7.3) to obtain a semi-infinite flow. However, cursory inspection reveals that the Fourier series solution (7.7.12) is no longer useful in this limit.

Similarity solution

The absence of a length scale in semi-infinite flow suggests that the time and space dependencies must combine into a unified variable which can be nondimensionalized by the kinematic viscosity. Recalling that the kinematic viscosity has units of length squared over time, we introduce the dimensionless variable

$$\eta \equiv \frac{y}{\sqrt{\nu t}}, \tag{7.7.14}$$

ranging from zero to infinity, and express the velocity in the form

$$u_x(y,t) = U f(\eta), \tag{7.7.15}$$

where $f(\eta)$ is a function of a single variable. This functional form implies that the fluid velocity as seen by an observer who finds herself at the position $y = \sqrt{\nu t}$, and is thus traveling upward with velocity $\mathrm{d}y/\mathrm{d}t = \sqrt{\nu/(4t)}$, remains constant in time. The initial and boundary conditions are satisfied provided that the function $f(\eta)$ satisfies the boundary condition $f(\eta = 0) = 1$ and the far-field condition $f(\eta = \infty) = 0$.

Substituting the similarity solution (7.7.15) in the governing equation (7.7.1), we find

$$\rho \frac{\partial u_x}{\partial t} = \rho U \frac{\mathrm{d}f}{\mathrm{d}\eta}\frac{\partial \eta}{\partial t} = \mu U \frac{\partial^2 u_x}{\partial y^2} = \mu U \frac{\partial}{\partial y}\left(\frac{\partial \eta}{\partial y}\frac{\mathrm{d}f}{\mathrm{d}\eta}\right). \tag{7.7.16}$$

Rearranging, we find

$$\frac{1}{\nu}\frac{\mathrm{d}f}{\mathrm{d}\eta}\frac{\partial\eta}{\partial t} = \frac{\partial}{\partial y}\Big(\frac{1}{\sqrt{\nu t}}\frac{\mathrm{d}f}{\mathrm{d}\eta}\Big) = \frac{1}{\sqrt{\nu t}}\frac{\mathrm{d}^2 f}{\mathrm{d}\eta^2}\frac{\partial\eta}{\partial y} = \frac{1}{\nu t}\frac{\mathrm{d}^2 f}{\mathrm{d}\eta^2}. \tag{7.7.17}$$

We note that $\partial\eta/\partial t = -y/(2\nu^{1/2}t^{3/2})$, and derive a second-order nonlinear ordinary differential equation,

$$\frac{\mathrm{d}^2 f}{\mathrm{d}\eta^2} = -\frac{1}{2}\,\eta\,\frac{\mathrm{d}f}{\mathrm{d}\eta}. \tag{7.7.18}$$

To compute the solution, we recast (7.7.18) into the form

$$\frac{\mathrm{d}}{\mathrm{d}\eta}\Big(\ln\frac{\mathrm{d}f}{d\eta}\Big) = -\frac{1}{2}\,\eta. \tag{7.7.19}$$

Integrating, we find

$$\frac{\mathrm{d}f}{\mathrm{d}\eta} = A\,\exp\Big(-\frac{\eta^2}{4}\Big), \tag{7.7.20}$$

where A is an integration constant. Carrying out a second integration, we find

$$f(\eta) = A\int_0^{\eta}\exp\Big(-\frac{q^2}{4}\Big)\,\mathrm{d}q + B = 2\,A\int_0^{\eta/2}\exp(-v^2)\,\mathrm{d}v + B, \tag{7.7.21}$$

where B is a new integration constant, q is an auxiliary integration variable, and $v \equiv q/2$.

The integral on the right-hand side of (7.7.21) is not available in analytical form. To formalize the solution, we introduce the error function defined as

$$\mathrm{erf}(w) \equiv \frac{2}{\sqrt{\pi}}\int_0^{w}\exp(-v^2)\,\mathrm{d}v, \tag{7.7.22}$$

and the complementary error function defined as

$$\mathrm{erfc}(w) = 1 - \mathrm{erf}(w). \tag{7.7.23}$$

Graphs of these functions are shown in figure 7.7.2. As w tends to infinity, the error function tends to the asymptotic value of unity; correspondingly, the complementary error function tends to zero.

The frequent occurrence of the error function in various branches of mathematical physics has motivated its tabulation and representation in terms of various approximations. Program *error_f* in directory *99_spec_fnc* inside directory *01_num_meth* of FDLIB evaluates the error function and the complementary error function using accurate algebraic approximations.

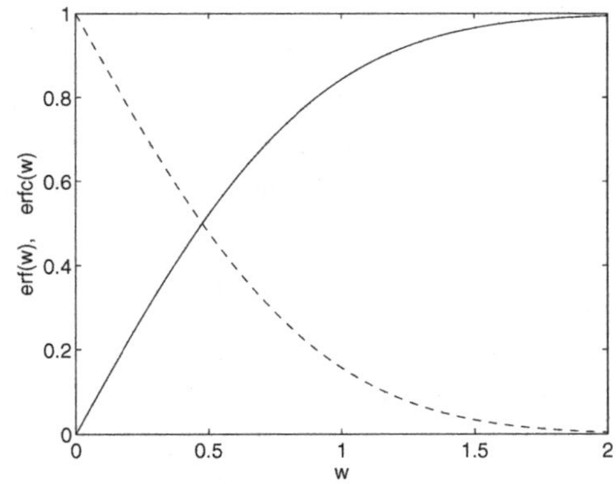

Figure 7.7.2 Graphs of the error function (solid line) and complementary error function (dashed line) defined in equations (7.7.21) and (7.7.22), computed using algebraic approximations.

Returning to (7.7.21), we express the right-hand side in terms of the error function and require $f(\eta = 0) = 1$, and $f(\eta = \infty) = 0$ to obtain $A = -1/\sqrt{\pi}$ and $B = 1$, yielding the velocity profile

$$u_x(y, t) = U f(\eta) = U \, \mathrm{erfc}(\frac{\eta}{2}). \tag{7.7.24}$$

The accompanying vorticity profile is

$$\omega_z(y, t) = -\frac{\partial u_x}{\partial y} = -U \frac{\mathrm{d}f}{\mathrm{d}\eta} \frac{\partial \eta}{\partial y} = \frac{U}{\sqrt{\pi \nu t}} \exp(-\frac{y^2}{4\nu t}). \tag{7.7.25}$$

This expression explicitly illustrates the singular nature of the vorticity at the initial instant and the spreading of the vorticity away from the moving wall due to viscous diffusion at later times.

The wall shear stress is given by

$$\sigma_{xy}(y = 0, t) = \mu \left(\frac{\partial u_x}{\partial u} \right)_{y=0} = -\frac{\mu U}{\sqrt{\pi \nu t}}. \tag{7.7.26}$$

This expression illustrates that an unphysical singularity occurs at the initial instant, as soon as the plate is set in motion. In reality, a plate cannot start moving with constant velocity in an impulsive fashion. Instead, the plate velocity must gradually increase from zero to the final value over a non-infinitesimal period of time.

7.7.3 *Pressure- and gravity-driven flow*

As a second case study, we consider flow in a channel with stationary walls where the motion is due to the sudden application of a constant pressure gradient or sudden tilting of the walls. The x component of the equation of motion provides us with a partial differential equation for the x component of the velocity,

$$\rho \frac{\partial u_x}{\partial t} = \chi + \mu \frac{\partial^2 u_x}{\partial y^2} + \rho\, g_x. \tag{7.7.27}$$

The pressure field is given by equation (7.1.4), where χ is a constant. The velocity profile satisfies the initial condition

$$u_x = 0 \qquad \text{for } 0 < y \le d \;\; \text{at } t = 0, \tag{7.7.28}$$

and the no-slip boundary condition

$$u_x(y=0) = 0, \qquad u_x(y=d) = 0, \qquad \text{for } t \ge 0. \tag{7.7.29}$$

At steady state, we obtain the Hagen parabolic velocity profile

$$u_x^h(y) = \frac{\chi + \rho\, g_x}{2\mu}\, y\,(d-y), \tag{7.7.30}$$

as discussed in section 7.1.

Fourier series

To compute the unsteady flow, we consider the deviation of the transient velocity from the parabolic profile. Expanding the difference in a Fourier series with respect to y, we find

$$u_x(y,t) = u_x^h(y) + \sum_{n=1}^{\infty} A_n(t)\, \sin\frac{n\pi y}{d}. \tag{7.7.31}$$

The arguments of the trigonometric functions on the right-hand side have been designed to satisfy the wall boundary conditions at all times. To compute the functions $A_n(t)$, we substitute the Fourier expansion in the governing equation (7.7.27) and carry out the differentiation with respect to t and x to find

$$\sum_{n=1}^{\infty} \Big[\rho \frac{\mathrm{d}A_n(t)}{\mathrm{d}t} + \mu \left(\frac{n\pi}{d}\right)^2 A_n(t) \Big] \sin\frac{n\pi y}{d} = 0. \tag{7.7.32}$$

To ensure the satisfaction of this equation for any value of y, we set the expression enclosed by the square brackets to zero and obtain a first-order linear differential equation for $A_n(t)$ whose solution is

$$A_n(t) = b_n \exp\left(-\frac{n^2\pi^2\nu}{d^2}\, t\right), \tag{7.7.33}$$

where $\nu = \mu/\rho$ is the kinematic viscosity, and b_n is a constant. Substituting this expression in (7.7.31), we find

$$u_x(y,t) = u_x^h(y) + \sum_{n=1}^{\infty} b_n \exp(-\frac{n^2\pi^2\nu}{d^2}\, t) \, \sin \frac{n\pi y}{d}. \tag{7.7.34}$$

To compute the constants b_n, we require that (7.7.34) satisfies the initial condition (7.7.28),

$$u_x(y, t=0) = \frac{\chi + \rho\, g_x}{2\mu}\, y\,(d-y) + \sum_{n=1}^{\infty} b_n \, \sin \frac{n\pi y}{d} = 0, \tag{7.7.35}$$

for $y > 0$.

The solution follows readily from the trigonometric identity

$$\int_0^d \sin \frac{n\pi y}{d} \, \sin \frac{m\pi y}{d} \, dy = \begin{cases} 0 & \text{if } n = 0 \text{ or } m = 0, \\ \frac{d}{2} & \text{if } n = m \neq 0, \\ 0 & \text{if } n \neq m. \end{cases} \tag{7.7.36}$$

Multiplying the middle and right-hand sides of (7.7.35) by $\sin(m\pi y/d)$, integrating with respect to y from 0 to d, and using identity (7.7.36), we find

$$\frac{\chi + \rho\, g_x}{2\mu} \int_0^d y\,(d-y) \, \sin \frac{m\pi y}{d} \, dy + b_m \, \frac{d}{2} = 0, \tag{7.7.37}$$

or

$$b_m = -\frac{\chi + \rho\, g_x}{\mu d} \int_0^d y\,(d-y) \, \sin \frac{m\pi y}{d} \, dy. \tag{7.7.38}$$

Manipulating the integral, we find

$$\begin{aligned} b_m &= \frac{\chi + \rho\, g_x}{\mu m \pi} \int_0^d y\,(d-y) \, \mathrm{d} \cos \frac{m\pi y}{d} \qquad (7.7.39) \\ &= \frac{\chi + \rho\, g_x}{\mu m \pi} \Big(\Big[y\,(d-y) \, \cos \frac{m\pi y}{d} \Big]_0^d - \int_0^d \cos \frac{m\pi y}{d} \, (d - 2y) \, dy \Big), \end{aligned}$$

and then

$$b_m = 2\, \frac{\chi + \rho\, g_x}{\mu m \pi} \int_0^d \cos \frac{m\pi y}{d} \, y \, dy = 2\, d\, \frac{\chi + \rho\, g_x}{\mu m^2 \pi^2} \int_0^d y \, \mathrm{d} \sin \frac{m\pi y}{d}$$

$$= -2\, d\, \frac{\chi + \rho\, g_x}{\mu m^2 \pi^2} \int_0^d \sin \frac{m\pi y}{d} \, dy = 2\, d^2 \, \frac{\chi + \rho\, g_x}{\mu m^3 \pi^3} \Big[\cos \frac{m\pi y}{d} \Big]_0^d. \tag{7.7.40}$$

The term

$$\Big[\cos \frac{m\pi y}{d} \Big]_0^d = \cos \frac{m\pi d}{d} - 1 \tag{7.7.41}$$

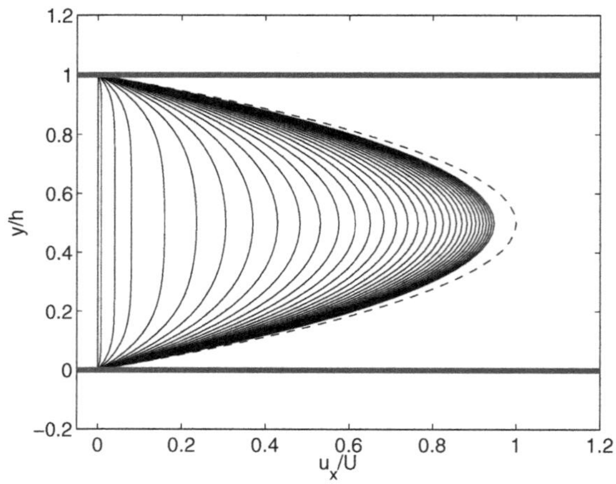

Figure 7.7.3 Transient flow in a channel with parallel walls separated by the distance d, due to the sudden application of a constant pressure gradient or sudden tilting, at dimensionless times $\hat{t} \equiv t\,\nu/d^2 = 0.001$, 0.005, 0.010, 0.020, 0.030, ... The dashed parabolic line corresponds to the Hagen flow established at long times.

is equal to 2 if m is odd, and 0 if m is even. Substituting the expression for b_m in (7.7.35) yields the transient velocity profile

$$u_x(y,t) = \frac{\chi + \rho\, g_x}{2\mu}\big[\, y\,(d-y) - \frac{8\,d^2}{\pi^3} \sum_{n=1,3,\ldots}^{\infty} \frac{1}{n^3}\, \exp(-\frac{n^2\pi^2\nu}{d^2}\, t)\, \sin\frac{n\pi y}{d}\big]. \tag{7.7.42}$$

Program *chan_2d_trans* in directory *04_various* of FDLIB evaluates the velocity profile at a specified time. A sequence of developing profiles of the scaled velocity u_x/U is shown in figure 7.7.3, where U is the maximum velocity of the steady Hagen flow established at long times.

Problem

7.7.1 *Flow due to a constant shear stress*

Show that the velocity field due to the application of a constant shear stress of magnitude τ at the planar boundary of a semi-infinite fluid occupying the upper half-space, $y \geq 0$, is given by

$$u(y,t) = \frac{\tau}{\mu}\sqrt{\nu t}\,\Big[\, \frac{2}{\sqrt{\pi}}\, \exp(-\frac{\eta^2}{4}) - \eta\, \mathrm{erfc}(\frac{\eta}{2})\,\Big], \tag{7.7.43}$$

where $\eta \equiv y/\sqrt{\nu t}$. Discuss the asymptotic behavior at long times.

7.8 Oscillatory channel flow

In section 7.7, we discussed transient Couette flow and transient pressure- or gravity-driven flow. In all cases, vorticity enters the flow through the boundaries and then diffuses to occupy the domain of flow. In this section, we consider corresponding problems for oscillatory flow. An important new feature is that vorticity of alternating sign enters the flow across the boundaries, and cancellation by diffusion prevents the establishment of significant motion far from the boundaries.

7.8.1 Oscillatory Couette flow

Suppose that the lower wall of a two-dimensional channel oscillates parallel to itself with velocity $U\cos(\Omega t)$, while the upper wall is stationary, thereby generating an unsteady unidirectional flow, as illustrated in figure 7.8.1. The velocity field is governed by the unsteady diffusion equation (7.7.1) subject to the boundary conditions

$$u_x(y=0) = U\ \cos(\Omega t), \qquad u_x(y=d) = 0, \tag{7.8.1}$$

for $t \geq 0$. Anticipating that the flow will be periodic in time due to the linearity of the governing equations, we express the velocity in the form

$$u_x(y,t) = U\,[\,f_c(y)\ \cos(\Omega t) + f_s(y)\ \sin(\Omega t)\,]. \tag{7.8.2}$$

To compute the functions $f_c(y)$ and $f_s(y)$, we substitute (7.8.2) in (7.7.1), carry out the time differentiation, collect the coefficients of the sines and cosines, and then set the compiled expressions to zero to obtain two coupled linear ordinary differential equations,

$$\frac{\mathrm{d}^2 f_s}{\mathrm{d}y^2} = -\frac{\Omega}{\nu}\, f_c, \qquad \frac{\mathrm{d}^2 f_c}{\mathrm{d}y^2} = \frac{\Omega}{\nu}\, f_s, \tag{7.8.3}$$

where $\nu \equiv \mu/\rho$ is the kinematic viscosity.

To expedite the solution, we introduce the imaginary unit i defined such that $\mathrm{i}^2 = -1$, formulate the complex function

$$f(y) \equiv f_c(y) + \mathrm{i}\, f_s(y), \tag{7.8.4}$$

and combine the two equations in (7.8.3) into a single complex form,

$$\frac{\mathrm{d}^2 f}{\mathrm{d}y^2} = -\frac{\mathrm{i}\,\Omega}{\nu}\, f. \tag{7.8.5}$$

The real and imaginary parts of the complex equation (7.8.5) are equal, respectively, to the second and first equation in (7.8.3). Now using the Euler decomposition

$$\exp(-\mathrm{i}\,\Omega\, t) = \cos(\Omega\, t) - \mathrm{i}\ \sin(\Omega\, t), \tag{7.8.6}$$

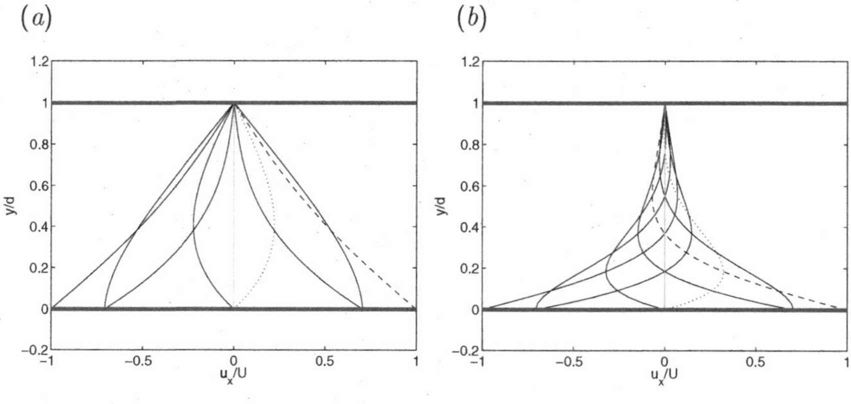

Figure 7.8.1 Velocity profiles of oscillatory Couette flow in a two-dimensional channel at phase angles $\Omega t/\pi = 0$ (dashed lines), 0.25, 0.50 (dotted lines), 0.75, 1.0, 1.25, 1.50, 1.75, and 2.0, for Womersley number (*a*) $N_W = 2$, and (*b*) 6. In the first case, the flow is nearly quasi-steady.

we find that u_x is given by the real part of the complex function

$$w(y,t) \equiv U\, f(y)\, \exp(-\mathrm{i}\,\Omega\, t). \tag{7.8.7}$$

The imaginary part of $w(y,t)$ also satisfies the governing equation (7.7.1). The solution of (7.8.5) is readily found to be

$$f(y) = A\, \exp\Big(y\,\sqrt{-\mathrm{i}\Omega/\nu}\,\Big) + B\, \exp\Big(-y\,\sqrt{-\mathrm{i}\Omega/\nu}\,\Big), \tag{7.8.8}$$

where A and B are two complex constants, with the understanding that, if β is a positive or negative real number, then

$$\begin{aligned}\exp(\beta\sqrt{-\mathrm{i}}) &= \exp[\frac{\beta}{\sqrt{2}}\,(-1+\mathrm{i})] = \exp(-\frac{\beta}{\sqrt{2}})\ \exp(\mathrm{i}\,\frac{\beta}{\sqrt{2}}) \\ &= \exp(-\frac{\beta}{\sqrt{2}})\ \Big[\cos(\frac{\beta}{\sqrt{2}}) + \mathrm{i}\,\sin(\frac{\beta}{\sqrt{2}})\Big].\end{aligned} \tag{7.8.9}$$

The no-slip boundary conditions (7.8.1) require $f(0) = 1$ and $f(d) = 0$. Using these equations to evaluate A and B, we find

$$\begin{aligned}A &= \frac{\exp(-d\,\sqrt{-\mathrm{i}\Omega/\nu}\,)}{\exp(-d\,\sqrt{-\mathrm{i}\Omega/\nu}\,) - \exp(d\,\sqrt{-\mathrm{i}\Omega/\nu}\,)} = \frac{1}{1-\exp(2\,d\,\sqrt{-\mathrm{i}\Omega/\nu}\,)}, \\ B &= 1 - A.\end{aligned} \tag{7.8.10}$$

Substituting these expressions in (7.8.7) and rearranging, we derive the complex function

$$f(y) = \frac{\exp(y\sqrt{-\mathrm{i}\Omega/\nu}\,) - \exp[(2\,d - y)\sqrt{-\mathrm{i}\Omega/\nu}\,]}{1 - \exp(2\,d\sqrt{-\mathrm{i}\Omega/\nu}\,)}, \tag{7.8.11}$$

whose real and imaginary parts provide us with the velocity according to equations (7.8.4) and (7.8.2).

Program *chan_2d_osc* in directory *04_various* of FDLIB evaluates the velocity profile for a specified angular frequency and reduced temporal phase angle, $\Omega\, t$.

Womersley number

The argument of the exponential term on the right-hand side of (7.8.11) suggests that the structure of the flow is determined by the dimensionless Womersley number

$$N_W \equiv d\sqrt{\frac{\Omega}{\nu}}, \tag{7.8.12}$$

expressing the reduced angular frequency. Figure 7.8.1 shows profiles of the scaled velocity, u_x/U, at times $\Omega t/\pi = 0$, 0.25, 0.50, 0.75, 1.0, 1.25, 1.50, 1.75, and 2.0, for $N_W = 2$ and 6. When the Womersley number is small, the flow evolves in quasi-steady fashion and the velocity profile is almost linear at any time. As the Womersley number increases, the motion is confined inside a boundary layer close to the moving wall, and the rest of the fluid is virtually stationary.

7.8.2 Rayleigh's oscillating plate

To investigate the behavior at high frequencies or large channel widths, we consider the structure of the flow at large values of the Womersley number. In this limit, the second term in the numerator and the second term in the denominator on the right-hand side of (7.8.11) become exponentially small, leaving the simplified form

$$f(y) \simeq \exp(y\sqrt{-\mathrm{i}\Omega/\nu}\,) \tag{7.8.13}$$

$$= \exp\Big(-y\sqrt{\Omega/(2\,\nu)}\Big)\,\Big[\cos\Big(y\sqrt{\Omega/(2\,\nu)}\Big) + \mathrm{i}\,\sin\Big(y\sqrt{\Omega/(2\,\nu)}\Big)\Big].$$

The associated velocity profile is

$$u_x(y,t) = U\,\exp(-y\sqrt{\Omega/(2\,\nu)}\,)\,\Big[\,\cos(\Omega t)\,\cos\Big(y\sqrt{\Omega/(2\,\nu)}\Big) + \sin(\Omega t)\,\sin\Big(y\sqrt{\Omega/(2\,\nu)}\Big)\Big]. \tag{7.8.14}$$

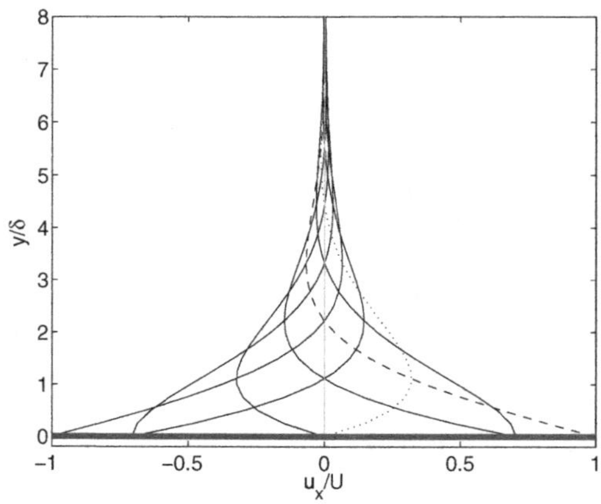

Figure 7.8.2 Velocity profiles due to the in-plane oscillation of a plate in a semi-infinite viscous fluid, where δ is the thickness of the Stokes boundary layer defined in (7.8.16).

Combining the two terms, we find

$$u_x(y,t) = U \exp\Big(- y \sqrt{\Omega/(2\,\nu)}\Big) \cos\Big(\Omega\, t - y\sqrt{\Omega/(2\,\nu)}\Big). \qquad (7.8.15)$$

Program *plate_osc* in directory *04_various* of FDLIB evaluates the velocity profile at a specified time. A sequence of profiles is shown in figure 7.8.2.

Stokes boundary layer

Expression (7.8.15) shows that the magnitude of the velocity decays exponentially with respect to distance normal to the oscillating plate. The region where substantial fluid motion occurs is identified as the Stokes boundary layer. A rough measure of the thickness of the boundary layer, δ, is the distance where the magnitude of the argument of the exponential term on the right-hand side of (7.8.15) is equal to unity,

$$\delta = \sqrt{\frac{2\nu}{\Omega}}. \qquad (7.8.16)$$

Apart from the factor of two in the numerator, this estimate could have been deduced on the basis of dimensional analysis at the outset.

In terms of the Stokes layer thickness, the velocity profile (7.8.15) takes the form

$$u_x(y,t) = U \exp(-y/\delta) \cos(\Omega t - y/\delta). \qquad (7.8.17)$$

The shear stress exerted on an oscillating plate is

$$\sigma_{xy}(y=0,t) = \mu \left(\frac{\partial u}{\partial y}\right)_{y=0} = U\,\sqrt{\mu\rho\,\Omega}\,\cos(\Omega t - \frac{3\pi}{4}), \tag{7.8.18}$$

or

$$\sigma_{xy}(y=0,t) = \frac{\mu U}{\delta}\,\sqrt{2}\,\cos(\Omega\,t - \frac{3\pi}{4}). \tag{7.8.19}$$

This expression reveals that the wall shear stress lags behind the wall velocity by the phase angle $3\pi/4$, independent of the angular frequency.

7.8.3 Pulsating pressure-driven flow

As a last case study, we consider pulsating flow in a channel due to an oscillatory pressure gradient. The pressure field is given in equation (7.1.4), where $\chi = \bar{\chi}\,\sin(\Omega t)$. Thus, the pressure gradient varies harmonically in time with angular frequency Ω and amplitude $\bar{\chi}$. Working as in the case of oscillatory Couette flow, we express the velocity in the form

$$u_x(y,t) = -\frac{\bar{\chi}}{\rho\,\Omega}\Big[\,f_c(y)\,\cos(\Omega t) + f_s(y)\,\sin(\Omega t)\,\Big], \tag{7.8.20}$$

and introduce the complex function

$$f(y) = f_c(y) + \mathrm{i}\,f_s(y), \tag{7.8.21}$$

where i is the imaginary unit. The no-slip boundary condition requires $f(0) = 0$ and $f(d) = 0$. The solution is found to be

$$f(y) = 1 - \frac{\cosh[(\frac{2y}{d} - 1)\,\frac{d}{2}\,\sqrt{-\mathrm{i}\Omega/\nu}]}{\cosh(\frac{d}{2}\,\sqrt{-\mathrm{i}\Omega/\nu})}. \tag{7.8.22}$$

The hyperbolic cosine of a complex argument on the right-hand side is computed by decomposing it into real exponentials and using the definition

$$\cosh z \equiv \frac{\exp(z) + \exp(-z)}{2}, \tag{7.8.23}$$

in conjunction with (7.8.9).

Program *chan_2d_wom* in directory *04_various* of FDLIB evaluates the velocity profile using the preceding expressions. Figure 7.8.3 shows time sequences of profiles of the scaled velocity u_x/U at a low and a moderate value of the Womersley number, where U is the maximum velocity of the steady Hagen flow. At low frequencies, the motion is quasi-steady and the velocity profile is nearly parabolic throughout each cycle. At high frequencies, the flow consists of a core oscillating in a plug-flow mode, and two Stokes boundary layers, one attached to each wall (problem 7.8.2). One interesting feature of the motion is

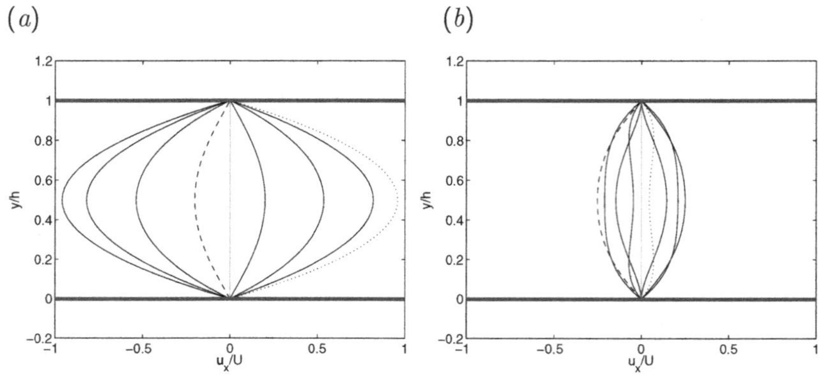

Figure 7.8.3 Velocity profiles of pulsating pressure-driven flow in a two-dimensional channel confined between two parallel walls at phase angles $\Omega\, t/\pi$ = 0 (dashed lines), 0.25, 0.50 (dotted lines), 0.75, 1.0, 1.25, 1.50, 1.75, and 2.0, for Womersley number (*a*) $N_W = \sqrt{2}$, and (*b*) $\sqrt{30}$.

that, under certain conditions, the amplitude of the velocity at the edges of the boundary layers may exceed that in the central core.

Problems

7.8.1 *Shear stress on an oscillating plate*

(*a*) Derive expression (7.8.18).

(*b*) We have noted that the wall shear stress lags behind the wall velocity by the phase angle $3\pi/4$, independent of the angular frequency. Does a similar independence arise in the case of channel flow where the velocity profile is given by (7.8.2), (7.8.4), and (7.8.11)?

7.8.2 *Womersley flow*

Show that, at high values of the Womersley number, equation (7.8.22) takes the approximate form

$$f(y) = 1 - \exp(y\,\sqrt{-\mathrm{i}\Omega/\nu}) - \exp\Big(\,(d-y)\,\sqrt{-\mathrm{i}\Omega/\nu}\,\Big). \tag{7.8.24}$$

This expression confirms the development of a plug-flow core and two Stokes boundary layers, one attached to each wall.

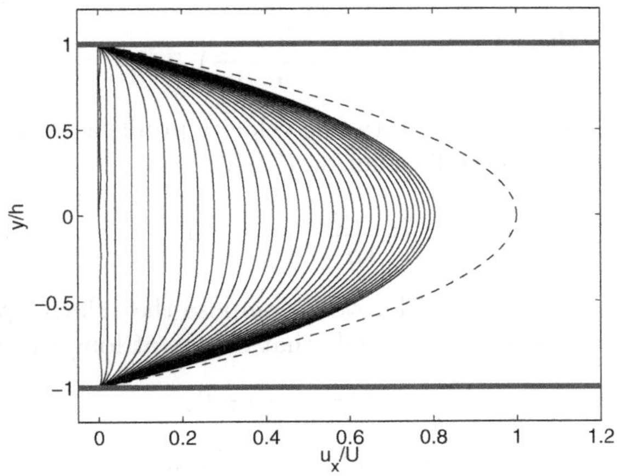

Figure 7.9.1 Transient flow in a circular tube of radius a due to the sudden application of a constant pressure gradient or sudden tilting, at dimensionless times $\hat{t} \equiv t\,\nu/a^2 = 0.001$, 0.005, 0.010, 0.020, 0.030, $\ldots$. The dashed parabolic line corresponds to the Poiseuille flow established at long times.

7.9 Transient and oscillatory flow in a circular tube

The study of transient and oscillatory channel flow in sections 7.7 and 7.8 has revealed the physical mechanisms by which momentum is transmitted and vorticity diffuses away from planar boundaries in unsteady flow. In this section, we consider corresponding mechanisms for curved boundaries by studying the dynamics of unsteady unidirectional and swirling tube flow.

7.9.1 *Transient Poiseuille flow*

First, we consider transient unidirectional flow through a cylindrical tube of radius a due to the sudden application of a pressure gradient or tilting, as illustrated in figure 7.9.1. The motion is described in cylindrical polar coordinates, (x, σ, φ), where the x axis is parallel to the tube centerline. The radial and meridional velocity components are identically zero, $u_\sigma = 0$ and $u_\varphi = 0$, while the axial component is a function of distance from the x axis and time, $u_x(\sigma, t)$. The pressure field is given by equation (7.3.1) with a constant axial pressure gradient, χ. The no-slip boundary condition requires that the velocity vanishes at the tube surface at all times, $u_x(\sigma = a, t) = 0$. At the initial instant, the fluid is quiescent, $u_x(\sigma, t = 0) = 0$.

Simplifying the x component of the equation of motion expressed in cylindrical polar coordinates, we obtain a partial differential equation for the axial velocity,

$$\rho \frac{\partial u_x}{\partial t} = \chi + \mu \left(\frac{\partial^2 u_x}{\partial \sigma^2} + \frac{1}{\sigma} \frac{\partial u_x}{\partial \sigma} \right) + \rho g_x. \tag{7.9.1}$$

When the flow reaches a steady state, the velocity profile assumes the parabolic form described by the Poiseuille solution (7.3.4),

$$u_x^p(\sigma) = \frac{\chi + \rho g_x}{4 \mu} (a^2 - \sigma^2). \tag{7.9.2}$$

To compute the solution of (7.9.1), we decompose the velocity profile into the Poiseuille profile and a transient profile that decays at long times,

$$u_x(\sigma, t) = u_x^p(\sigma) + v_x(\sigma, t). \tag{7.9.3}$$

Substituting (7.9.3) in (7.9.1), we find that the transient profile, $v_x(\sigma, t)$, satisfies the homogeneous equation

$$\frac{\partial v_x}{\partial t} = \nu \left(\frac{\partial^2 v_x}{\partial \sigma^2} + \frac{1}{\sigma} \frac{\partial v_x}{\partial \sigma} \right), \tag{7.9.4}$$

where $\nu \equiv \mu/\rho$ is the kinematic viscosity.

Separation of variables

Equation (7.9.4) can be solved by separation of variables implemented by the expansion

$$v_x(\sigma, t) = \sum_{n=1}^{\infty} c_n \, \phi_n(\sigma) \, \psi_n(t), \tag{7.9.5}$$

where $\phi_n(\sigma)$, and $\psi_n(t)$ are unknown functions of their respective arguments, and c_n are constant coefficients. Substituting this expansion in (7.9.4), carrying out the differentiation and rearranging, we obtain

$$\sum_{n=1}^{\infty} c_n \, \phi_n(\sigma) \, \psi_n(t) \left[\frac{1}{\psi_n} \frac{\mathrm{d}\psi_n}{\mathrm{d}t} - \frac{\nu}{\phi_n} \left(\frac{\mathrm{d}^2 \phi_n}{\mathrm{d}\sigma^2} + \frac{1}{\sigma} \frac{\mathrm{d}\phi_n}{\mathrm{d}\sigma} \right) \right] = 0. \tag{7.9.6}$$

For this equality to hold true for any value of σ and t, the expression enclosed by the square brackets on the right-hand side must be identically zero,

$$\frac{1}{\psi_n} \frac{\mathrm{d}\psi_n}{\mathrm{d}t} = \frac{\nu}{\phi_n} \left(\frac{\mathrm{d}^2 \phi_n}{\mathrm{d}\sigma^2} + \frac{1}{\sigma} \frac{\mathrm{d}\phi_n}{\mathrm{d}\sigma} \right) = -\nu \, b_n^2, \tag{7.9.7}$$

where b_n^2 is a positive constant. The right-hand side of (7.9.7) was designed to facilitate forthcoming algebraic manipulations. Rearranging the two equalities in (7.9.7) we derive two ordinary differential equations,

$$\frac{\mathrm{d}\psi_n}{\mathrm{d}t} = -\nu \, b_n^2 \, \psi_n, \qquad \frac{\mathrm{d}^2 \phi_n}{\mathrm{d}\sigma^2} + \frac{1}{\sigma} \frac{\mathrm{d}\phi_n}{\mathrm{d}\sigma} + b_n^2 \phi_n = 0. \tag{7.9.8}$$

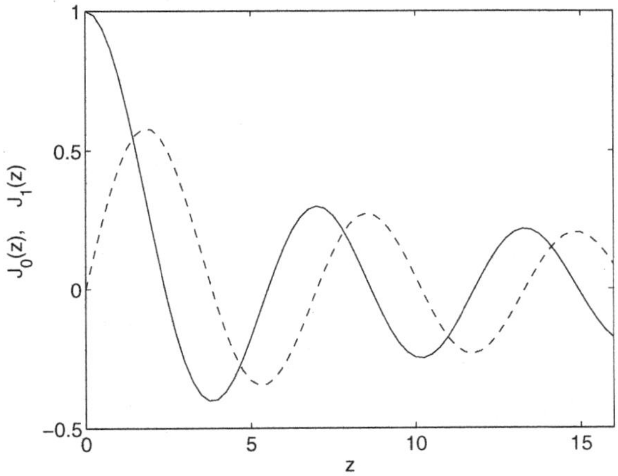

Figure 7.9.2 Graphs of the Bessel functions $J_0(z)$ and $J_1(z)$ drawn, respectively, with the solid and dashed line, necessary for the computation of transient tube flow.

The solution of the first equation is readily found to be

$$\psi_n(t) = \exp(-\nu\, b_n^2\, t), \tag{7.9.9}$$

which reveals that the transient flow decays exponentially in time.

Bessel functions

To derive the solution of the second equation in (7.9.8), we introduce the Bessel function of the first kind, $J_0(z)$, satisfying the zeroth-order Bessel equation

$$z\,\frac{\mathrm{d}}{\mathrm{d}z}\Big(z\,\frac{\mathrm{d}J_0(z)}{\mathrm{d}z}\Big) + z^2\, J_0(z) = z^2\,\frac{\mathrm{d}^2 J_0(z)}{\mathrm{d}z^2} + z\,\frac{\mathrm{d}J_0(z)}{\mathrm{d}z} + z^2\, J_0(z) = 0. \tag{7.9.10}$$

The graph of $J_0(z)$ is shown with the solid line in figure 7.9.2. The frequent occurrence of the Bessel functions in various branches of mathematical physics has motivated its tabulation and representation in terms of infinite series and algebraic approximations. Program *bess_J0* in directory *99_spec_fnc* inside directory *01_num_meth* of FDLIB evaluates J_0 using an accurate algebraic approximation.

Replacing z in Bessel's equation (7.9.10) with $b_n\sigma$ and simplifying, we find that the function

$$\phi_n(\sigma) = J_0(b_n\sigma) \tag{7.9.11}$$

satisfies the second equation in (7.9.8), and is thus the desired solution. To satisfy the no-slip boundary condition, we require

$$\phi_n(a) = J_0(b_n a) = 0, \tag{7.9.12}$$

which shows that $b_n a$ is a zero of $J_0(z)$, denoted by α_n. The first seven zeros are

$$\begin{aligned} &\alpha_1 = 2.4044448, \quad \alpha_2 = 5.5201, \quad \alpha_3 = 8.6537, \\ &\alpha_4 = 11.7915, \quad \alpha_5 = 14.9309, \quad \alpha_6 = 18.0711, \\ &\alpha_7 = 21.2116, \end{aligned} \tag{7.9.13}$$

to shown accuracy. Accordingly, we set

$$b_n = \frac{\alpha_n}{a}. \tag{7.9.14}$$

Substituting (7.9.14) in (7.9.11) and (7.9.9), and the result in (7.9.5), we derive the expansion

$$v_x(\sigma, t) = \sum_{n=1}^{\infty} c_n \, J_0(\frac{\alpha_n \, \sigma}{a}) \, \exp(-\frac{\alpha_n^2 \, \nu \, t}{a^2}). \tag{7.9.15}$$

This representation clearly satisfies the requirement that the transient flow decays at long times. To also satisfy the initial condition $u_x(\sigma, t = 0) = 0$, we evaluate the decomposition (7.9.3) at $t = 0$ and use (7.9.2) and (7.9.15) to obtain

$$\sum_{n=1}^{\infty} c_n \, J_0(\frac{\alpha_n \, \sigma}{a}) = -\frac{\chi + \rho \, g_x}{4 \, \mu} \, (a^2 - \sigma^2). \tag{7.9.16}$$

The coefficients c_n must be such that this representation holds true for any value of σ.

Orthogonality of the zeroth-order Bessel functions

To extract the coefficients c_n from (7.9.16), we use the following orthogonality property of the Bessel functions,

$$\int_0^1 z \, J_0(\alpha_n z) \, J_0(\alpha_m z) \, dz = \begin{cases} \frac{1}{2} \, J_1^2(\alpha_n) & \text{if } n = m, \\ 0 & \text{if } n \neq m, \end{cases} \tag{7.9.17}$$

where $J_1(z)$ is the Bessel function of the first kind. By definition, $J_1(z)$ satisfies the differential equation

$$z^2 \, \frac{\mathrm{d}^2 J_1(z)}{\mathrm{d}z^2} + z \, \frac{\mathrm{d}J_1(z)}{\mathrm{d}z} + (z^2 - 1) \, J_1(z) = 0, \tag{7.9.18}$$

and is required to be finite at every value of z between, and including, zero and infinity. The graph of $J_1(z)$ is shown with the dashed line in figure 7.9.2.

Replacing z in identity (7.9.17) with σ/a and rearranging, we find

$$\int_0^a \sigma\, J_0(\frac{\alpha_n\, \sigma}{a})\, J_0(\frac{\alpha_m\, \sigma}{a})\, \mathrm{d}\sigma = \begin{cases} \frac{1}{2}\, a^2\, J_1^2(\alpha_n) & \text{if } n = m, \\ 0 & \text{if } n \neq m. \end{cases} \tag{7.9.19}$$

Inspired by this identity, we multiply both sides of (7.9.16) by $\sigma\, J_0(\alpha_m\, \sigma/a)$, integrate with respect to σ from 0 to a, use the orthogonality property (7.9.19), and then switch m to n to find

$$\begin{aligned} c_n &= -\frac{\chi + \rho\, g_x}{2\,\mu\, a^2\, J_1^2(\alpha_n)} \int_0^a J_0(\frac{\alpha_n\, \sigma}{a})\, (a^2 - \sigma^2)\, \sigma\, \mathrm{d}\sigma \\ &= -\frac{\chi + \rho\, g_x}{2\,\mu\, J_1^2(\alpha_n)}\, a^2 \int_0^1 J_0(\alpha_n\, v)\, (1 - v^2)\, v\, \mathrm{d}v, \end{aligned} \tag{7.9.20}$$

where $v \equiv \sigma/a$. To compute the last integral, we replace z in Bessel's equation (7.9.10) with $\alpha_n v$, and simplify to obtain

$$\alpha_n^2\, v\, J_0(\alpha_n\, v) = -\frac{\mathrm{d}}{\mathrm{d}v}\Big(v\, \frac{\mathrm{d}J_0(\alpha_n v)}{\mathrm{d}v}\Big). \tag{7.9.21}$$

Note that the left- and thus the right-hand side of (7.9.21) vanishes when $v = 0$ or 1. Integrating (7.9.21) with respect to v from 0 to 1, we find

$$\alpha_n^2 \int_0^1 J_0(\alpha_n\, v)\, v\, \mathrm{d}v = -[v\, \frac{\mathrm{d}J_0(\alpha_n v)}{\mathrm{d}v})]_{v=1} + [v\, \frac{\mathrm{d}J_0(\alpha_n v)}{\mathrm{d}v})]_{v=0}. \tag{7.9.22}$$

Using the identity $\mathrm{d}J_0(z)/\mathrm{d}z = -J_1(z)$, we find

$$\int_0^1 J_0(\alpha_n\, v)\, v\, \mathrm{d}v = \frac{1}{\alpha_n}\, J_1(\alpha_n). \tag{7.9.23}$$

A similar computation yields

$$\begin{aligned} \alpha_n^2 \int_0^1 J_0(\alpha_n\, v)\, v^3\, \mathrm{d}v &= -\int_0^1 v^2\, \frac{\mathrm{d}}{\mathrm{d}v}(v\, \frac{\mathrm{d}J_0(\alpha_n v)}{\mathrm{d}v})\, \mathrm{d}v \\ &= -(v^3\, \frac{\mathrm{d}J_0(\alpha_n v)}{\mathrm{d}v})_{v=1} + (v^3\, \frac{\mathrm{d}J_0(\alpha_n v)}{\mathrm{d}v})_{v=0} + 2\int_0^1 v^2\, \frac{\mathrm{d}J_0(\alpha_n v)}{\mathrm{d}v}\, \mathrm{d}v \\ &= \alpha_n\, J_1(\alpha_n) + 2\,[v^2\, J_0(\alpha_n v)]_{v=1} - 2\,[v^2\, J_0(\alpha_n v)]_{v=0} \\ &\quad -4\int_0^1 v\, J_0(\alpha_n v)\, \mathrm{d}v = (\alpha_n - \frac{4}{\alpha_n})\, J_1(\alpha_n). \end{aligned} \tag{7.9.24}$$

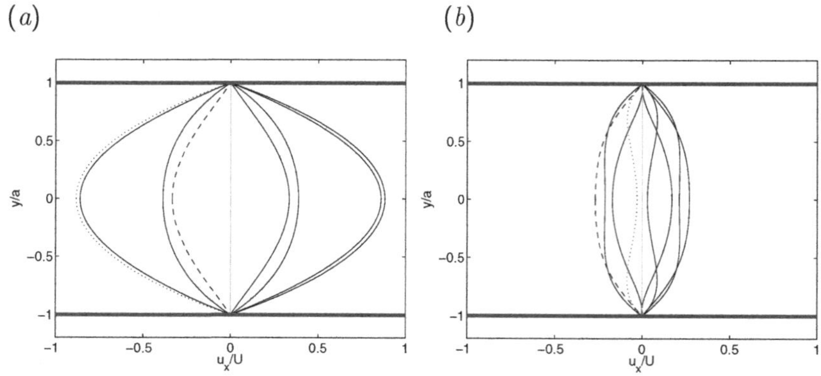

Figure 7.9.3 Velocity profiles of oscillatory pressure-driven flow in a circular tube at phase angles $\Omega t/\pi = 0$ (dashed lines), 0.25, 0.50 (dotted lines), 0.75, 1.0, 1.25, 1.50, 1.75, and 2.0, for Womersley number (*a*) $N_W = 2$, and (*b*) 6.

Substituting (7.9.23) and (7.9.24) in (7.9.20), we obtain the desired result

$$c_n = -\frac{\chi + \rho\, g_x}{\mu\, J_1(\alpha_n)\, \alpha_n^3}\, 2\, a^2. \tag{7.9.25}$$

Finally, we substitute (7.9.25) in (7.9.15) and then in (7.9.3), and thus derive the velocity profile

$$u_x(\sigma, t) = \frac{\chi + \rho\, g_x}{4\,\mu}\, [a^2 - \sigma^2 - 8\, a^2 \sum_{n=1}^{\infty} \frac{1}{\alpha_n^3}\, \frac{J_0(\alpha_n\, \sigma/a)}{J_1(\alpha_n)}\, \exp(-\frac{\alpha_n^2\, \nu\, t}{a^2})\,]. \tag{7.9.26}$$

Program *tube_crc_trans* in directory *04_various* of FDLIB computes the velocity described by (7.9.26). A sequence of profiles evolving toward the parabolic distribution established at steady state is illustrated in figure 7.9.1.

7.9.2 *Pulsating pressure-driven flow*

Next, we consider pulsating flow through a circular tube due to an oscillatory pressure gradient, as illustrated in figure 7.9.3. The no-slip boundary condition requires that the velocity is zero over the tube wall. The pressure field is given by equation (7.3.1), where $\chi = \bar{\chi}\, \sin(\Omega t)$. Thus, the pressure gradient varies harmonically in time with angular frequency Ω and amplitude $\bar{\chi}$.

Working as in section 7.8.3 for pulsating channel flow, we express the velocity in the form

$$u_x(\sigma, t) = -\frac{\bar{\chi}}{\rho\,\Omega}\, [f_c(\sigma)\, \cos(\Omega t) + f_s(\sigma)\, \sin(\Omega t)], \tag{7.9.27}$$

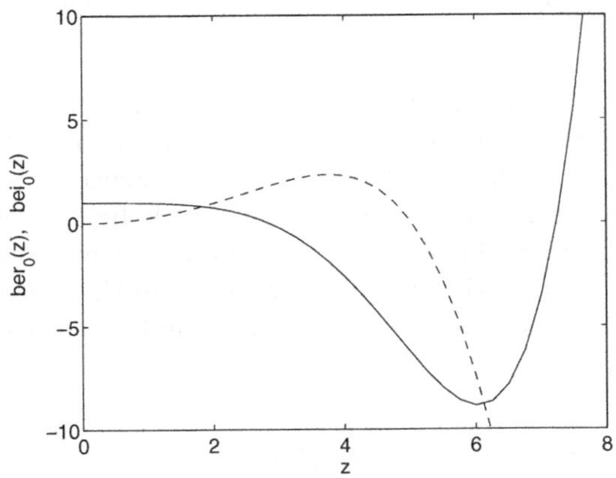

Figure 7.9.4 Graphs of the Kelvin functions, $\text{ber}_0(z)$ and $\text{bei}_0(z)$ plotted, respectively, with the solid and dashed line, arising in the computation of pulsating tube flow.

introduce the complex function

$$f(\sigma) = f_c(\sigma) + \mathrm{i}\, f_s(\sigma), \tag{7.9.28}$$

and require the no-slip boundary condition $f(a) = 0$, where i is the imaginary unit, $\mathrm{i}^2 = -1$. The solution is

$$f(\sigma) = 1 - \frac{J_0(\sigma\sqrt{-\mathrm{i}\Omega/\nu})}{J_0(a\sqrt{-\mathrm{i}\Omega/\nu})}, \tag{7.9.29}$$

with the understanding that, if β is a positive real number, then

$$J_0(\beta\sqrt{-\mathrm{i}}) = \text{ber}_0(\beta) + \mathrm{i}\,\text{bei}_0(\beta), \tag{7.9.30}$$

where ber_0 and bei_i are the zeroth-order Kelvin functions (problem 7.9.1). Graphs of these functions are displayed in figure 7.9.4.

Program *ber_bei_0* in directory *99_spec_fnc* inside directory *01_num_meth* of FDLIB evaluates the Kelvin functions using algebraic approximations.

Womersley number

The functional form of the arguments of the Bessel functions on the right-hand side of (7.9.29) suggests that the structure of the flow is determined by the reduced angular frequency expressed by the Womersley number

$$N_W \equiv \frac{1}{2}\, a \sqrt{\frac{\Omega}{\nu}}, \tag{7.9.31}$$

which is the counterpart of the Womersley number for channel flow defined in (7.8.12).

Program *tube_crc_wom* in directory *04_various* of FDLIB evaluates the velocity profile using the expressions derived in this section. Profiles at two frequencies are shown in figure 7.9.3. At low frequencies, the flow is nearly quasi-steady and the velocity profile is nearly parabolic. At high frequencies, the flow consists of a central core oscillating in a plug-flow mode, and a Stokes boundary layer attached to the cylindrical wall. As in the case of channel flow, the amplitude of the velocity at the edge of the boundary layer may exceed that in the central core.

7.9.3 Transient circular Couette flow

The meridional velocity of an unsteady swirling flow with circular streamlines satisfies the linear partial differential equation

$$\frac{\partial u_\varphi}{\partial t} = \nu \,(\frac{\partial^2 u_\varphi}{\partial \sigma^2} + \frac{1}{\sigma}\frac{\partial u_\varphi}{\partial \sigma} - \frac{u_\varphi}{\sigma^2}), \tag{7.9.32}$$

where $\nu \equiv \mu/\rho$ is the kinematic viscosity.

Consider transient flow inside a hollow circular cylinder of inner radius a filled with a fluid. At the origin of time, the cylinder starts rotating suddenly around its axis with constant angular velocity, Ω. Working as in the case of axial flow, we derive the transient velocity profile

$$u_\varphi(\sigma, t) = \Omega \Big[\sigma + 2\,a \sum_{n=1}^{\infty} \frac{1}{\alpha_n} \frac{J_1(\alpha_n \sigma / a)}{J_0(\alpha_n)} \exp(-\frac{\alpha_n^2\,\nu\, t}{a^2}) \Big], \tag{7.9.33}$$

where α_n are the positive zeros of the first-order Bessel function, J_1. At long times, the summed terms on the right-hand side of (7.9.33) vanish, leaving a velocity profile that expresses rigid-body rotation.

Program *tube_crc_sw* in directory *04_various* of FDLIB evaluates the velocity profile of transient swirling flow described by equation (7.9.33).

7.9.4 More on Bessel functions

We have seen that Bessel functions arise in the computation of flow in an axisymmetric domain. In concluding this section, we state an orthogonality property that is useful in evaluating the coefficients of Bessel expansions; an example is shown in (7.9.14).

By definition, the pth-order Bessel function of the first kind, $J_p(z)$, satisfies the pth-order Bessel equation

$$z \frac{d}{dz}\Big(z \frac{dJ_p(z)}{dz} \Big) + (z^2 - p^2)\, J_p(z)$$
$$= z^2 \frac{d^2 J_p(z)}{dz^2} + z \frac{dJ_p(z)}{dz} + (z^2 - p^2)\, J_p(z) = 0, \qquad (7.9.34)$$

and is required to be finite at every value of z between, and including, zero and infinity. The particular cases $p = 0$ and 1 are shown in equations (7.9.10) and (7.9.18). The orthogonality property of the Bessel functions states that, if α_n and α_m are two zeros of $J_p(z)$, then

$$2 \int_0^1 z\, J_p(\alpha_n z)\, J_p(\alpha_m z)\, dz = \begin{cases} J_{p+1}^2(\alpha_n) = J_{p-1}^2(\alpha_n) = J_p'^2(\alpha_n) & \text{if } n = m, \\ 0 & \text{if } n \neq m. \end{cases} \qquad (7.9.35)$$

The Bessel function orthogonality property is analogous to the Fourier orthogonality property for trigonometric functions, as shown, for example, in equations (7.5.21).

Problems

7.9.1 *Pulsating pressure-driven flow*

Derive the velocity profile described by equations (7.9.27)–(7.9.29).

7.9.2 *Transient circular Couette flow*

Derive the velocity profile given in equation (7.9.33).

7.9.3 *Transient swirling flow*

Consider transient flow outside a solid circular cylinder of radius a immersed in an infinite liquid. At the origin of time, the cylinder starts rotating suddenly around its axis with constant angular velocity Ω. Derive the transient velocity profile.

Computer Problem

7.9.4 *Transient circular Couette flow*

Compute and plot profiles of the transient circular Couette flow at a sequence of times. Discuss the effect of truncation of the infinite sum on the right-hand side of (7.9.33).

Finite-Difference Methods 8

In previous chapters, we have discussed the equations governing the structure of a steady flow and the evolution of an unsteady flow, and derived selected solutions for elementary flow configurations by analytical and simple numerical methods. To generate solutions for arbitrary boundary geometries and flow conditions, it is necessary to develop further general-purpose numerical methods. In this chapter, we discuss the choice of governing equations whose solution is sought, and the implementation of finite-difference methods for incompressible Newtonian flow. The discourse will introduce us to the main challenges encountered in computational fluid dynamics (CFD).

8.1 Choice of governing equations

General-purpose methods for computing the flow of an incompressible fluid can be classified into two categories distinguished by the choice of governing equations.

In the first class of methods, the flow is described in terms of primary variables including the velocity and the pressure. The structure of the velocity and pressure field in a steady flow, and the evolution of the velocity and pressure field in an unsteady flow, are computed by solving the Navier-Stokes equation and the continuity equation, subject to appropriate boundary conditions, initial conditions, and possibly supplemental constraints.

In the second class of methods, the flow is computed on the basis of the vorticity transport equation. The numerical procedure involves two stages: first,

C. Pozrikidis, *Fluid Dynamics: Theory, Computation, and Numerical Simulation*,
DOI: 10.1007/978-0-387-95871-2_8,

we compute the structure or evolution of the vorticity field using the vorticity transport equation discussed in section 6.6; then we obtain the simultaneous structure or evolution of the velocity field by inverting the equation defining the vorticity,

$$\boldsymbol{\omega} = \nabla \times \mathbf{u}, \tag{8.1.1}$$

subject to constraints imposed by the continuity equation and boundary conditions. Inverting (8.1.1) involves solving for $\mathbf{u}$ in terms of $\boldsymbol{\omega}$. Descendant methods are distinguished by the particular procedure used to recover the velocity field from a specified vorticity distribution.

The strengths and weaknesses of the aforementioned two classes of methods will become apparent as we describe their implementation. One appealing feature of the second class of methods based on the vorticity transport equation is the lack of need to solve for the pressure. Bypassing the computation of the pressure is desirable when boundary conditions for the pressure are not directly available and must be derived from the governing equations. Disadvantages include the need to derive boundary conditions for the vorticity.

Problem

8.1.1 *Inversion of the vorticity*

Show that, if $\mathbf{u}$ is a solenoidal velocity field corresponding to a certain vorticity field $\boldsymbol{\omega}$, then the velocity field $\mathbf{v} = \mathbf{u} + \nabla f$ corresponds to the same vorticity field, where f is an arbitrary smooth scalar function. Explain why, for the velocity field $\mathbf{v}$ to be solenoidal, $\nabla \cdot \mathbf{v} = 0$, the function f must be harmonic; that is, it must satisfy Laplace's equation $\nabla^2 f = 0$.

8.2 Unidirectional flow; velocity/pressure formulation

We begin developing finite-difference methods by discussing the velocity/pressure formulation for unidirectional flow in a channel confined between two parallel walls located at $y = 0$ and $y = h$, as illustrated in figure 8.2.1. The lower and upper walls translate parallel to themselves along the x axis with generally time-dependent velocities, $U_1(t)$ and $U_2(t)$.

In practice, channel flow occurs under two complementary conditions reflecting the physical mechanism driving the flow. In the first case, the flow rate along the channel $Q(t)$ is specified, and the streamwise pressure gradient $\partial p(t)/\partial x$ is computed as part of the solution. In the second case, the pressure gradient is specified, and the flow rate is computed as part of the solution. In this section and in section 8.3, we consider the case of flow due to a specified, possibly time-dependent, pressure gradient. In section 8.4, we consider the complementary case of flow subject to a specified flow rate.

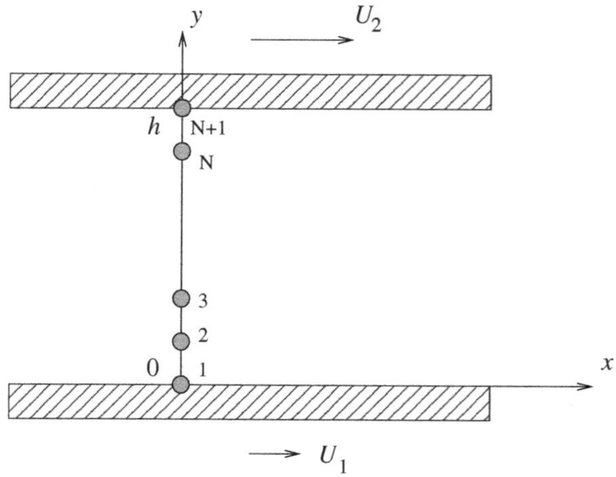

Figure 8.2.1 Illustration of a one-dimensional finite-difference grid used to compute the velocity profile in unidirectional channel flow.

8.2.1 Governing equations

To set up the mathematical formulation, we consider the x component of the equation of motion. In the case of unidirectional flow, we obtain the simplified form

$$\frac{\partial u_x}{\partial t} = -\frac{1}{\rho}\frac{\partial p}{\partial x} + \nu \frac{\partial^2 u_x}{\partial y^2} + g_x, \tag{8.2.1}$$

where ρ is the fluid density, ν is the kinematic viscosity, and g_x is the x component of the acceleration of gravity. Equation (8.2.1) is to be solved subject to a specified initial condition and to the possibly time-dependent velocity boundary conditions

$$u_x(y=0) = U_1(t), \qquad u_x(y=h) = U_2(t), \tag{8.2.2}$$

requiring no-slip at the walls.

8.2.2 Explicit finite-difference method

To implement the finite-difference method, we divide the cross-section of the channel extending over $0 \le y \le h$ into N intervals defined by $N+1$ grid points, as shown in figure 8.2.1. For convenience, we denote the x component of the velocity at the i grid point as $u_i(t) \equiv u_x(y_i, t)$.

Next, we evaluate both sides of (8.2.1) at the ith interior grid point at time t, where $i = 2, \ldots, N$, and approximate the time derivative on the left-hand side with a first-order forward finite difference and the second derivative on the right-hand side with a second-order centered finite difference. The result is the finite-difference equation (FDE)

$$\frac{u_i(t+\Delta t) - u_i(t)}{\Delta t} = -\frac{1}{\rho}\frac{\partial p}{\partial x}(t) + \nu\,\frac{u_{i-1}(t) - 2\,u_i(t) + u_{i+1}(t)}{\Delta y^2} + g_x. \quad (8.2.3)$$

Solving for $u_i(t+\Delta t)$ on the left-hand side, we find

$$u_i(t+\Delta t) = \alpha\,u_{i-1}(t) + (1-2\,\alpha)\,u_i(t) + \alpha\,u_{i+1}(t) + \Delta t\,(-\frac{1}{\rho}\frac{\partial p}{\partial x}(t) + g_x), \quad (8.2.4)$$

for $i = 2, \ldots, N$. We have introduced the dimensionless ratio

$$\alpha \equiv \frac{\nu \Delta t}{\Delta y^2}, \quad (8.2.5)$$

called the numerical diffusion number. Equation (8.2.4) allows us to update the velocity at the interior grid points explicitly, starting from the specified initial condition, subject to the boundary conditions

$$u_1(t) = U_1(t), \qquad u_{N+1}(t) = U_2(t). \quad (8.2.6)$$

The following MATLAB code *channel_ftcs* residing in directory *channel* inside directory *11_fdm* of FDLIB performs the animation of the evolving velocity profile:

```
%----
% parameters
%----

h = 1.0;
mu = 0.6; rho = 0.5;
N = 32;
dpdx = -2.0; gx = 0.4;
U1 = 0.0; U2 = 0.0;
al = 0.51;        % alpha

nstep = 20000;       % number of steps
nu = mu/rho;      % kinematic viscosity

%---
% prepare
%---

Dy = h/N;
Dt = al*Dy*Dy/nu;
```

```
%---
% grid and initial condition
%---

for i=1:N+1
  y(i) = (i-1)*Dy;
  u(i) = 0;
end
u(1) = U1;
u(N+1) = U2;

t = 0.0;

%---
% time stepping
%---

for step=1:nstep

  t = t+Dt;
  unew(1) = U1;
  for i=2:N
    unew(i) = al*u(i-1)+(1-2*al)*u(i)+al*u(i+1) ...
      + Dt*(-dpdx/rho+gx);
  end
  unew(N+1) = U2;
  u=unew;

  if(step==1)
    Handle1 = plot(u,y,'o-');
    set(Handle1, 'erasemode', 'xor');
    set(gca,'fontsize',15)
    axis([0 0.5 0 h])
    xlabel('u','fontsize',15)
    ylabel('y','fontsize',15)
  else
    set(Handle1,'XData',u,'YData',y);
    pause(0.02)
    drawnow
  end

end     % of time stepping
```

Snapshots of an evolving profile are shown in figure 8.2.2 for two values of α.

Numerical instability

Numerical experimentation reveals and theoretical analysis confirms that the explicit method of updating the velocity based on equation (8.2.4) is free of

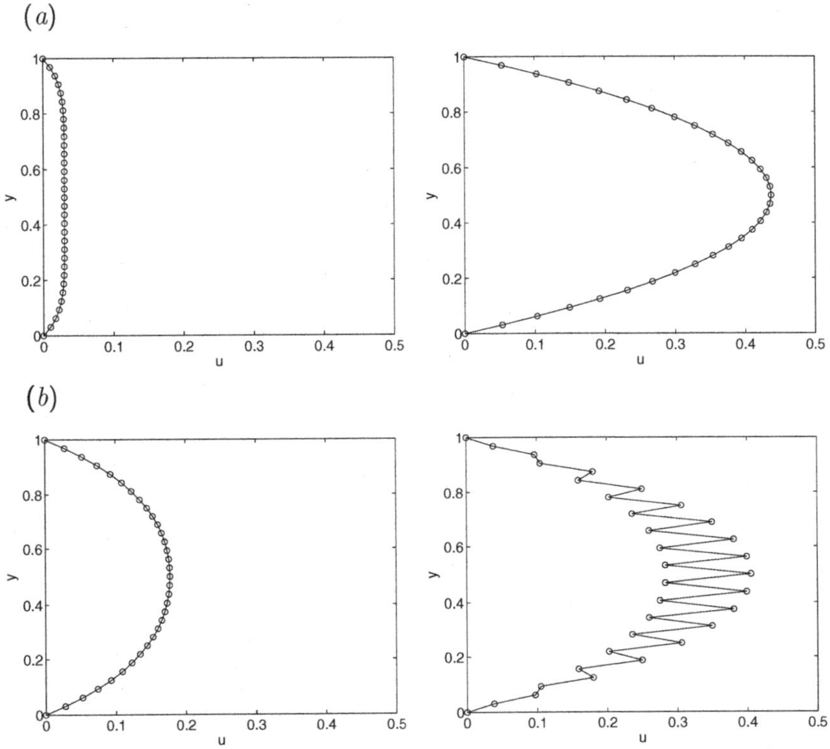

Figure 8.2.2 Snapshots of an evolving profile for unidirectional flow in a channel computed by an explicit finite difference method for (a) $\alpha = 0.40$ and (b) 0.51. The oscillations in the second case is a manifestation of numerical instability.

oscillations only when the time step Δt is small enough so that the dimensionless numerical diffusion number α defined in (8.2.5) is less than 0.5.[1] For larger time steps, the velocity profile develops unphysical growing numerical oscillations unrelated to the physics of the motion. This behavior is illustrated in the results shown in figure 8.2.2. Accordingly, the explicit finite-difference method is conditionally stable.

8.2.3 *Implicit finite-difference method*

To avoid the restriction on the time step for numerical stability, we implement an implicit finite-difference method. Evaluating (8.2.1) at the ith interior grid point at time $t + \Delta t$, where $i = 2, \ldots, N$, and then approximating the time

[1] Pozrikidis, C., 2008, *Numerical Computation in Science and Engineering*, Second Edition, Oxford University Press.

derivative on the left-hand side with a first-order backward finite difference and the second derivative on the right-hand side with a second-order centered finite difference, we obtain

$$\frac{u_i(t+\Delta t)-u_i(t)}{\Delta t} = -\frac{1}{\rho}\frac{\partial p}{\partial x}(t+\Delta t) \tag{8.2.7}$$

$$+\nu\,\frac{u_{i+1}(t+\Delta t)-2\,u_i(t+\Delta t)+u_{i-1}(t+\Delta t)}{\Delta y^2}+g_x.$$

Rearranging (8.2.7), we find

$$-\alpha\,u_{i-1}(t+\Delta t)+(1+2\,\alpha)\,u_i(t+\Delta t)-\alpha\,u_{i+1}(t+\Delta t)$$

$$= u_i(t)-\frac{\Delta t}{\rho}\frac{\partial p}{\partial x}(t+\Delta t)+\Delta t\,g_x, \tag{8.2.8}$$

where α is the numerical diffusion number defined in (8.2.5). Equation (8.2.8) allows us to compute the velocity at the interior grid points at the time level $t+\Delta t$ in an implicit fashion, which involves simultaneously solving for all unknown grid values, subject to the boundary conditions

$$u_1(t+\Delta t)=U_1(t+\Delta t), \qquad u_{N+1}(t+\Delta t)=U_2(t+\Delta t). \tag{8.2.9}$$

To formalize the implicit algorithm, we write equation (8.2.8) for $i=2,\ldots,N$, and enforce the boundary conditions to obtain a system of $N-1$ linear equations for the velocity at the $N-1$ interior grid points at time $t+\Delta t$,

$$\mathbf{A}\cdot\mathbf{u}(t+\Delta t)=\mathbf{u}(t)+\mathbf{b}. \tag{8.2.10}$$

We have introduced the tridiagonal coefficient matrix

$$\mathbf{A}=\begin{bmatrix} 1+2\alpha & -\alpha & 0 & 0 & 0 & \cdots & 0 & 0 & 0 \\ -\alpha & 1+2\alpha & -\alpha & 0 & 0 & \cdots & 0 & 0 & 0 \\ 0 & -\alpha & 1+2\alpha & -\alpha & 0 & \cdots & 0 & 0 & 0 \\ \cdots & \cdots & \cdots & \cdots & \cdots & \cdots & \cdots & \cdots & \cdots \\ 0 & 0 & 0 & 0 & 0 & \cdots & -\alpha & 1+2\alpha & -\alpha \\ 0 & 0 & 0 & 0 & 0 & \cdots & 0 & -\alpha & 1+2\alpha \end{bmatrix}, \tag{8.2.11}$$

the vector of unknown velocities

$$\mathbf{u}(t+\Delta t)=\begin{bmatrix} u_2(t+\Delta t) \\ u_3(t+\Delta t) \\ \vdots \\ u_{N-1}(t+\Delta t) \\ u_N(t+\Delta t) \end{bmatrix}, \tag{8.2.12}$$

and the known vectors

$$\mathbf{u}(t) = \begin{bmatrix} u_2(t) \\ u_3(t) \\ \vdots \\ u_{N-1}(t) \\ u_N(t) \end{bmatrix}, \qquad \mathbf{b} = \begin{bmatrix} -\frac{\Delta t}{\rho}\frac{\partial p}{\partial x}(t+\Delta t) + \Delta t\, g_x + \alpha\, U_1(t+\Delta t) \\ -\frac{\Delta t}{\rho}\frac{\partial p}{\partial x}(t+\Delta t) + \Delta t\, g_x \\ \vdots \\ -\frac{\Delta t}{\rho}\frac{\partial p}{\partial x}(t+\Delta t) + \Delta t\, g_x \\ -\frac{\Delta t}{\rho}\frac{\partial p}{\partial x}(t+\Delta t) + \Delta t\, g_x + \alpha\, U_2(t+\Delta t) \end{bmatrix}. \tag{8.2.13}$$

The numerical method involves solving the linear system (8.2.10) at the time instant t to produce the velocity profile at the next time instant $t + \Delta t$, beginning from a specified initial state.

Thomas algorithm

The tridiagonal structure of the matrix $\mathbf{A}$ displayed in (8.2.11) allows us to compute the solution efficiently using the legendary Thomas algorithm. We formalize the numerical method in a general framework in terms of a $K \times K$ linear system for an unknown vector $\mathbf{x}$,

$$\mathbf{D} \cdot \mathbf{x} = \mathbf{s}. \tag{8.2.14}$$

The vector $\mathbf{s}$ is assumed to be known, and the matrix $\mathbf{D}$ has the tridiagonal form

$$\mathbf{D} = \begin{bmatrix} a_1 & b_1 & 0 & 0 & 0 & \cdots & 0 & 0 & 0 \\ c_2 & a_2 & b_2 & 0 & 0 & \cdots & 0 & 0 & 0 \\ 0 & c_3 & a_3 & b_3 & 0 & \cdots & 0 & 0 & 0 \\ \cdots & \cdots & \cdots & \cdots & \cdots & \cdots & \cdots & \cdots & \cdots \\ 0 & 0 & 0 & 0 & 0 & \cdots & c_{K-1} & a_{K-1} & b_{K-1} \\ 0 & 0 & 0 & 0 & 0 & \cdots & 0 & c_K & a_K \end{bmatrix}. \tag{8.2.15}$$

Thomas's algorithm proceeds in two stages. At the first stage, the tridiagonal system (8.2.14) is transformed into the bidiagonal system

$$\mathbf{D}' \cdot \mathbf{x} = \mathbf{y}, \tag{8.2.16}$$

involving the bidiagonal coefficient matrix

$$\mathbf{D}' = \begin{bmatrix} 1 & d_1 & 0 & 0 & 0 & \cdots & 0 & 0 & 0 \\ 0 & 1 & d_2 & 0 & 0 & \cdots & 0 & 0 & 0 \\ 0 & 0 & 1 & d_3 & 0 & \cdots & 0 & 0 & 0 \\ \cdots & \cdots & \cdots & \cdots & \cdots & \cdots & \cdots & \cdots & \cdots \\ 0 & 0 & 0 & 0 & 0 & \cdots & 0 & 1 & d_{K-1} \\ 0 & 0 & 0 & 0 & 0 & \cdots & 0 & 0 & 1 \end{bmatrix}. \tag{8.2.17}$$

Reduction to bidiagonal :

$$\begin{bmatrix} d_1 \\ y_1 \end{bmatrix} = \frac{1}{a_1} \begin{bmatrix} b_1 \\ s_1 \end{bmatrix}$$

Do $i = 1, K-1$

$$\begin{bmatrix} d_{i+1} \\ y_{i+1} \end{bmatrix} = \frac{1}{a_{i+1} - c_{i+1} d_i} \begin{bmatrix} b_{i+1} \\ s_{i+1} - c_{i+1} y_i \end{bmatrix}$$

End Do

Backward substitution :

$x_K = y_K$

Do $i = K-1, 1$ (step $= -1$)

$x_i = y_i - d_i \, x_{i+1}$

End Do

Table 8.2.1 Thomas algorithm for solving a tridiagonal system of equations.

At the second stage, the bidiagonal system (8.2.16) is solved by backward substitution, which involves solving the last equation for the last unknown, x_K, and then moving upward to compute the rest of the unknowns in a sequential fashion. The combined algorithm is shown in table 8.2.1 and implemented in the following MATLAB function:

```
function x = thomas (n,a,b,c,s)

%=====================================================
% Thomas algorithm for a tridiagonal system
%
% n:  system size
% a,b,c:  diagonal, superdiagonal,
%         and subdiagonal elements
% s:  right-hand side
%=====================================================

%------------------------------
% reduction to upper bidiagonal
%------------------------------
```

```
d(1) = b(1)/a(1);
y(1) = s(1)/a(1);

for i=1:n-1
  i1 = i+1;
  den = a(i1)-c(i1)*d(i);
  d(i1) = b(i1)/den;
  y(i1) = (s(i1)-c(i1)*y(i))/den;
end

%------------------
% back substitution
%------------------

x(n) = y(n);

for i=n-1:-1:1
  x(i) = y(i)-d(i)*x(i+1);
end

%-----
% done
%-----

return;
```

The Thomas algorithm is a special implementation of the inclusive method of Gauss elimination discussed in section 3.4 for general linear systems of equations. The key idea is to bypass idle multiplications by zeros.

Finite-difference code

The following MATLAB code *channel_btcs* residing in directory *channel* inside directory *11_fdm* of FDLIB performs the time integration of an initial velocity profile using the implicit method:

```
%-----
% parameters
%-----

h = 1.0; mu = 0.6; rho = 0.5; N = 32; dpdx = -2.0;
gx = 0.4;
U1 = 0.1; U2 = -0.5;
al = 0.40;      % alpha
nstep = 20000;      % number of steps
nu = mu/rho;
Dy = h/N
Dt = al*Dy*Dy/nu;
```

```
%---
% grid and initial condition
%---

for i=1:N+1
  y(i) = (i-1)*Dy;
  u(i) = 0.0;
end
u(1) = U1;
u(N+1) = U2;

%---
% tridiagonal projection matrix
%---

for i=1:N-1
  atr(i) = 1.0+2*al;
  btr(i) = -al;
  ctr(i) = -al;
end

%---
% time stepping
%---

t=0.0;

for step=1:nstep

  for i=1:N-1     % right-hand side
    s(i) = u(i+1)+Dt*(-dpdx/rho+gx);
  end

  t = t+Dt;
  u(1) = U1;
  u(N+1)= U2;
  s(1) = s(1) + al*u(1);
  s(N-1) = s(N-1) + al*u(N+1);

  sol = thomas (N-1,atr,btr,ctr,s);

  for i=2:N
    u(i) = sol(i-1);
  end

  if(step==1)
    Handle1 = plot(u,y,'o-');
    set(Handle1, 'erasemode', 'xor');
    set(gca,'fontsize',15)
```

```
    axis([min(U1,U2) max(U1,U2)+2.0 0 h])
    xlabel('u','fontsize',15)
    ylabel('y','fontsize',15)
  else
    set(Handle1,'XData',u,'YData',y);
    pause(0.2)
    drawnow
  end

end     % of time stepping
```

Numerical stability

Numerical experimentation reveals that the implicit method of updating the velocity based on equation (8.2.10) is free of numerical oscillations irrespective of the size of the time step, Δt. Accordingly, the implicit finite-difference method is unconditionally stable and thus highly desirable.

8.2.4 Steady state

To obtain the velocity profile at steady state, we return to equation (8.2.10) and set $\mathbf{u}(t + \Delta t) = \mathbf{u}(t) = \mathbf{u}$ to find

$$(\mathbf{A} - \mathbf{I}) \cdot \mathbf{u} = \mathbf{b}, \tag{8.2.18}$$

where $\mathbf{I}$ is the unit matrix. Dividing the individual equations encapsulated in (8.2.18) by α, we obtain the simpler form

$$\mathbf{C} \cdot \mathbf{u} = \mathbf{d}, \tag{8.2.19}$$

involving the tridiagonal coefficient matrix

$$\mathbf{C} = \begin{bmatrix} 2 & -1 & 0 & 0 & 0 & \cdots & 0 & 0 & 0 \\ -1 & 2 & -1 & 0 & 0 & \cdots & 0 & 0 & 0 \\ 0 & -1 & 2 & -1 & 0 & \cdots & 0 & 0 & 0 \\ \cdots & \cdots & \cdots & \cdots & \cdots & \cdots & \cdots & \cdots & \cdots \\ 0 & 0 & 0 & 0 & 0 & \cdots & -1 & 2 & -1 \\ 0 & 0 & 0 & 0 & 0 & \cdots & 0 & -1 & 2 \end{bmatrix}, \tag{8.2.20}$$

the vector of unknown velocities at steady state

$$\mathbf{u} = \begin{bmatrix} u_2 \\ u_3 \\ \vdots \\ u_{N-1} \\ u_N \end{bmatrix}, \tag{8.2.21}$$

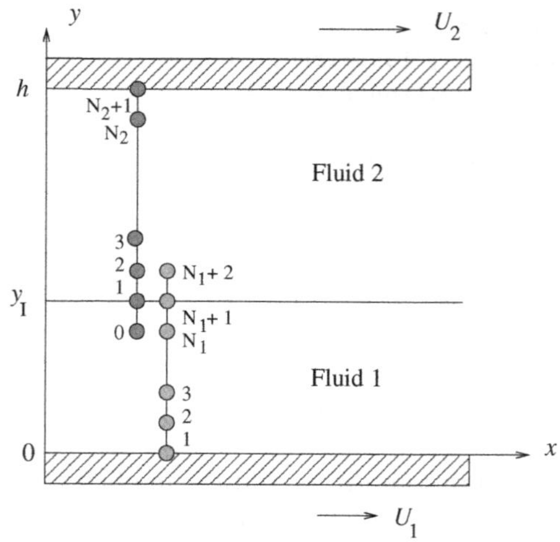

Figure 8.2.2 Illustration of a composite finite-difference grid with phantom nodes used to compute the velocity profile in unidirectional two-fluid channel flow.

and the known vector

$$\mathbf{d} = \begin{bmatrix} -\frac{1}{\mu}\frac{\partial p}{\partial x} + \frac{\Delta y^2}{\nu}\, g_x + U_1 \\ -\frac{1}{\mu}\frac{\partial p}{\partial x} + \frac{\Delta x^2}{\nu}\, g_x \\ \vdots \\ -\frac{1}{\mu}\frac{\partial p}{\partial x} + \frac{\Delta y^2}{\nu}\, g_x \\ -\frac{1}{\mu}\frac{\partial p}{\partial x} + \frac{\Delta y^2}{\nu}\, g_x + U_2 \end{bmatrix}. \qquad (8.2.22)$$

To compute the velocity profile at steady state, we merely solve the system of linear algebraic equations (8.2.19) using a numerical method.

8.2.5 *Two-layer flow*

Next, we consider the flow of two superimposed layers in a channel, as illustrated in figure 8.2.2. The lower layer is labeled 1, and the upper layer is labeled 2. The fluids are separated by a flat interface located at $y = h_1$, where $h_1 < h$ is the lower layer thickness, and h is the channel width. The thickness of the upper layer is $h_2 = h - h_1$.

Interfacial conditions

At the interface, we require three conditions: continuity of velocity, continuity of shear stress, and continuity of normal stress. To satisfy the third condition, we require that the streamwise pressure gradient, $\partial p/\partial x$, is the same inside both layers. Continuity of velocity and shear stress at the interface require

$$u_x^{(1)}(y = h_1) = u_x^{(2)}(y = h_1), \tag{8.2.23}$$

and

$$\mu_1 \left(\frac{\partial u_x^{(1)}}{\partial y}\right)_{y=h_1} = \mu_2 \left(\frac{\partial u_x^{(2)}}{\partial y}\right)_{y=h_1}. \tag{8.2.24}$$

Using the equation of motion (8.2.1), we find that, if (8.2.23) is true at the initial instant, it will also be true at all times provided that

$$-\frac{1}{\rho_1}\frac{\partial p}{\partial x} + \nu_1 \left(\frac{\partial^2 u_x^{(1)}}{\partial y^2}\right)_{y=h_1} = -\frac{1}{\rho_2}\frac{\partial p}{\partial x} + \nu_2 \left(\frac{\partial^2 u_x^{(2)}}{\partial y^2}\right)_{y=h_1}. \tag{8.2.25}$$

Finite-difference implementation

We begin developing the finite-difference method by dividing the lower layer into N_1 evenly spaced intervals defined by the $N_1 + 1$ grid points $y_i^{(1)}$, where $i = 1, \ldots, N_1 + 1$, and the upper layer into N_2 evenly spaced intervals defined by the $N_2 + 1$ grid points $y_i^{(2)}$, where $i = 1, \ldots, N_2 + 1$, as shown in figure 8.2.2.

For reasons that will soon become apparent, we extend the domain of definition of each layer into the adjacent layer by one artificial grid point labeled $N_1 + 2$ for the lower layer or 0 for the upper layer. To simplify the notation, we denote

$$u_i^{(1)} \equiv u_x^{(1)}(y_i^{(1)}), \qquad u_i^{(2)} \equiv u_x^{(2)}(y_i^{(2)}). \tag{8.2.26}$$

Approximating the derivatives in (8.2.24) and (8.2.25) with centered finite differences, we derive two equations relating the values of the velocity at the extended nodes, $u_{N_1+2}^{(1)}$ and $u_0^{(2)}$,

$$\mu_1 \frac{u_{N_1+2}^{(1)} - u_{N_1}^{(1)}}{2\Delta y_1} = \mu_2 \frac{u_2^{(2)} - u_0^{(2)}}{2\Delta y_2} \tag{8.2.27}$$

and

$$-\frac{1}{\rho_1}\frac{\partial p}{\partial x} + \nu_1 \frac{u_{N_1+2}^{(1)} - 2\,u_{N_1+1}^{(1)} + u_{N_1}^{(1)}}{\Delta y_1^2} = -\frac{1}{\rho_2}\frac{\partial p}{\partial x} + \nu_2 \frac{u_2^{(2)} - 2\,u_1^{(2)} + u_0^{(2)}}{\Delta y_2^2}, \tag{8.2.28}$$

where $\Delta y_1 \equiv h_1/N_1$ and $\Delta y_2 \equiv h_2/N_2$. Setting $u^{(1)}_{N_1+1} = u^{(2)}_1$ and introducing the ratios

$$\lambda \equiv \frac{\mu_2}{\mu_1}, \qquad \delta \equiv \frac{\rho_2}{\rho_1}, \qquad \gamma \equiv \frac{\nu_2}{\nu_1} = \frac{\lambda}{\delta}, \qquad \beta \equiv \frac{\Delta y_2}{\Delta y_1}, \tag{8.2.29}$$

we recast equations (8.2.27) and (8.2.28) into a system of two linear equations for the velocity at the extended nodes,

$$\beta\, u^{(1)}_{N_1+2} + \lambda\, u^{(2)}_0 = \beta\, u^{(1)}_{N_1} + \lambda\, u^{(2)}_2, \tag{8.2.30}$$

and

$$\beta^2\, u^{(1)}_{N_1+2} - \gamma\, u^{(2)}_0 = 2\,(\beta^2 - \gamma)\, u^{(1)}_{N_1+1} - \beta^2 u^{(1)}_{N_1} + \gamma\, u^{(2)}_2 + \frac{\Delta y_2^2}{\mu_1}\,(1 - \frac{1}{\delta})\frac{\partial p}{\partial x}. \tag{8.2.31}$$

In matrix notation,

$$\begin{bmatrix} \beta & \lambda \\ -\beta^2 & \gamma \end{bmatrix} \cdot \begin{bmatrix} u^{(1)}_{N_1+2} \\ u^{(2)}_0 \end{bmatrix} \tag{8.2.32}$$
$$= \begin{bmatrix} \beta\, u^{(1)}_{N_1} + \lambda\, u^{(2)}_2, \\ -2\,(\beta^2 - \gamma)\, u^{(1)}_{N_1+1} + \beta^2 u^{(1)}_{N_1} - \gamma u^{(2)}_2 - \frac{\Delta y_2^2}{\mu_1}\,(1 - \frac{1}{\delta})\frac{\partial p}{\partial x} \end{bmatrix}.$$

Solving for the velocity at the lower extended node, we find

$$u^{(1)}_{N_1+2} = a_1\, u^{(1)}_{N_1} + a_2\, u^{(1)}_{N_1+1} + a_3\, u^{(2)}_2 + a_4\, \frac{\partial p}{\partial x}, \tag{8.2.33}$$

where

$$a_1 = \frac{\gamma - \beta\lambda}{\gamma + \beta\lambda}, \qquad a_2 = 2\lambda \frac{\beta^2 - \gamma}{\beta(\gamma + \beta\lambda)}, \qquad a_3 = \frac{2\,\gamma\lambda}{\beta(\gamma + \beta\lambda)},$$

$$a_4 = \frac{\lambda}{\beta(\gamma + \beta\lambda)}\, \frac{\Delta y_2^2}{\mu_1} \left(1 - \frac{1}{\delta}\right). \tag{8.2.34}$$

When the physical properties of the layers are matched, $\lambda = \gamma = \delta = 1$, and the lower and upper grid sizes are equal, $\beta = 1$, then $u^{(1)}_{N_1+2} = u^{(2)}_2$ by equation (8.2.33), and $u^{(2)}_0 = u^{(1)}_{N_1}$ by equation (8.2.30), as required.

Explicit integration

Working as in the case of single-fluid flow, we derive the explicit finite-difference equation

$$u_i^{(1)}(t+\Delta t) = \alpha_1\, u_{i+1}^{(1)}(t) + (1-2\,\alpha_1)\, u_i^{(1)}(t) + \alpha_1\, u_{i-1}^{(1)}(t) \\ -\Delta t\left(-\frac{1}{\rho_1}\frac{\partial p}{\partial x}(t) + g_x\right) \tag{8.2.35}$$

for the lower layer, and the corresponding equation

$$u_i^{(2)}(t+\Delta t) = \alpha_2\, u_{i+1}^{(2)}(t) + (1-2\,\alpha_2)\, u_i^{(2)}(t) + \alpha_2\, u_{i-1}^{(2)}(t) \\ -\Delta t\left(-\frac{1}{\rho_2}\frac{\partial p}{\partial x}(t) + g_x\right) \tag{8.2.36}$$

for the upper layer, where

$$\alpha_1 \equiv \frac{\Delta t\,\nu_1}{\Delta y_1^2}, \qquad \alpha_2 \equiv \frac{\Delta t\,\nu_2}{\Delta y_2^2} \tag{8.2.37}$$

are the numerical diffusivities for the lower and upper layer. The numerical procedure involves the following steps:

1. First, we initialize the nodal velocities.

2. Second, we compute the velocity at the extended node $u_{N_1+2}^{(1)}$ from equation (8.2.33).

3. Third, we use equation (8.2.36) to update the velocity at the grid points of the lower layer, $i = 2, \ldots, N_1+1$.

4. Fourth, we set $u_1^{(2)} = u_{N_1+1}^{(1)}$.

5. Fifth, we use equation (8.2.37) to update the velocity at the internal grid points of the upper layer, for $i = 2, \ldots, N_2$.

6. Next, we use the boundary conditions to update the velocity at the lower and upper wall.

7. Now we return to step 2 and repeat the computation for another step.

The method is implemented in the following MATLAB code *two_layers* residing in directory *channel* inside directory *11_fdm* of FDLIB, performing the animation of the developing velocity profile:

```
%---
% parameters
%---
```

```
h = 1.0; h1 = 0.25;
N1 = 4; N2 = 32;
mu1 = 1.0; mu2 = 2.0;
rho1 = 1.5; rho2 = 1.0;
dpdx = -1.0; gx = 0.2;

alpha = 0.4;

%---
% prepare
%---

h2 = h-h1;
Dy1 = h1/N1;
Dy2 = h2/N2;

lambda = mu2/mu1;
delta = rho2/rho1;
gamma = lambda/delta;
beta = Dy2/Dy1;

tmp = gamma+beta*lambda;
a1 = (gamma-beta*lambda)/tmp;
a2 = 2*lambda*(beta*beta-gamma)/(beta*tmp);
a3 = 2*gamma*lambda/(beta*tmp);
a4 = lambda*Dy2*Dy2*(1-1/delta)/(beta*mu1*tmp);

Dt1 = rho1*alpha*Dy1*Dy1/mu1;
Dt2 = rho2*alpha*Dy2*Dy2/mu2;

Dt = min(Dt1,Dt2)

al1 = Dt*mu1/(Dy1*Dy1*rho1);
al2 = Dt*mu2/(Dy2*Dy2*rho2);

%---
% initialize
%---

for i=1:N1+1
  u1(i) = 0.0;
  y1(i) = (i-1)*Dy1;
end

for i=1:N2+1
  u2(i) = 0.0;
  y2(i) = (i-1)*Dy2+h1;
end
```

```
%---
% time stepping
%---

for step=1:1000

  u1(N1+2) = a1*u1(N1)+a2*u1(N1+1)+a3*u2(2)+a4*dpdx;

  unew1(1) = 0.0;

  for i=2:N1+1
    unew1(i) = al1*u1(i+1)+(1-2*al1)*u1(i)+al1*u1(i-1)...
      +Dt*(-dpdx/rho1+gx);
  end

  unew2(1) = unew1(N1+1);

  for i=2:N2
    unew2(i) = al2*u2(i+1)+(1-2*al2)*u2(i)+al2*u2(i-1)...
      +Dt*(-dpdx/rho2+gx);
  end

  unew2(N2+1)=0.0;

  u1 = unew1;
  u2 = unew2;

%---
 % animation
%---

  if(step==1)
    handle1 = plot(u1,y1,'o-',u2,y2,'o-');
    set(gca,'fontsize',15)
    axis([0 0.1 0 h])
    xlabel('u','fontsize',15)
    ylabel('y','fontsize',15)
  else
    set(handle1,'XData',[u1, u2],'YData',[y1, y2],'Marker','o')
    drawnow
    pause(0.01)
  end

%---
end     % of time stepping
%---
```

Snapshots of an evolving profile are shown in figure 8.2.3 for two values of α defined as the minimum of α_1 and α_2.

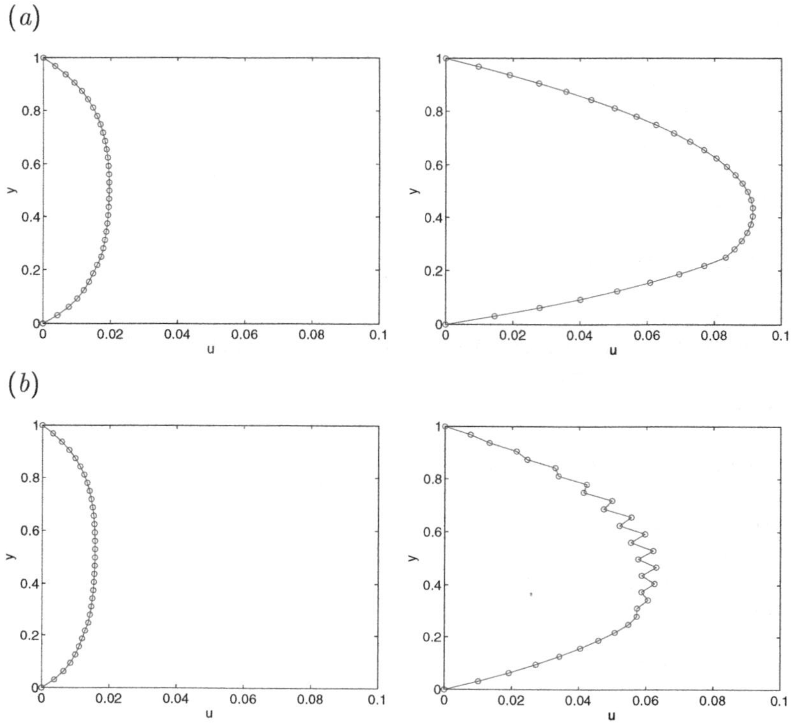

Figure 8.2.3 Snapshots of an evolving profile of two-layer flow in a channel computed by an explicit finite difference method for (a) $\alpha = 0.4$ and (b) 0.5, where α is the minimum of α_1 and α_2. A numerical instability arises in the second case.

Problems

8.2.1 *Steady state*

Derive the system (8.2.18) departing from the explicit finite-difference formula (8.2.3).

8.2.2 *Two-layer channel flow*

Derive a system of finite-difference equations governing the velocity profile of the two-layer channel flow at steady state.

8.2.3 *Flow in a circular tube*

Develop an explicit finite-difference method based on the velocity/pressure formulation for computing the velocity profile developing inside a tube with a circular cross-section due to a suddenly imposed constant pressure gradient.

Computer Problems

8.2.4 *Thomas algorithm*

Use the MATLAB function *thomas* listed in the text to solve a system of equations of your choice. Verify the accuracy of the solution by confirming that it satisfies the chosen system of equations.

8.2.5 *Single-fluid channel flow*

(*a*) Write a program that computes the evolution of the velocity profile in a channel with stationary walls due to a sinusoidal pressure gradient based on the explicit finite-difference method discussed in the text. Run the program for fluid properties and flow conditions of your choice, and for several time step sizes corresponding to numerical diffusion number α that is larger and lower than 0.5. Discuss the performance of the numerical method.

(*b*) Repeat (*a*) for the implicit finite-difference method discussed in the text.

8.2.6 *Two-layer channel flow*

Develop an implicit finite-difference method for computing the evolution of the two-layer flow. Implement the method in a program that computes the evolution of the velocity profile of two-layer flow in a channel with stationary walls due to the sudden application of a constant pressure gradient. Run the program for fluid properties and flow conditions of your choice, and for several time step sizes. Discuss the performance of the numerical method.

8.3 Unidirectional flow; velocity/vorticity formulation

In the case of unsteady unidirectional flow along the x axis, the vorticity transport equation reduces to the unsteady diffusion equation for the z component of the vorticity, ω_z,

$$\frac{\partial \omega_z}{\partial t} = \nu \, \frac{\partial^2 \omega_z}{\partial y^2}, \tag{8.3.1}$$

where ν is the kinematic viscosity. Invoking the definition of the vorticity stated in (8.1.1), we find

$$\omega_z = -\frac{\partial u_x}{\partial y}. \tag{8.3.2}$$

Integrating (8.3.2) with respect to y from the lower wall up to an arbitrary point, we obtain an integral representation for the velocity in terms of the vorticity,

$$u_x(y) = U_1 - \int_0^y \omega_z(y')\,\mathrm{d}y'. \tag{8.3.3}$$

Without loss of generality, we have chosen to satisfy the boundary condition at the lower wall. It remains to ensure that the no-slip boundary condition is also satisfied at the upper wall.

The numerical method involves computing the evolution of the vorticity from a specified initial state using (8.3.1), and simultaneously recovering the evolution of the velocity field based on equation (8.3.2) or its integrated version (8.3.3). Since the velocity does not appear in equation (8.3.1), the two steps are decoupled.

8.3.1 Boundary conditions for the vorticity

Because the unsteady diffusion equation (8.3.1) is a second-order differential equation with respect to y, two boundary conditions for the vorticity are required, one at each end of the solution domain located at $y = 0$ and h. These boundary conditions must be such that the integral condition

$$\int_0^h \omega_z(y')\, \mathrm{d}y' = U_1 - U_2 \tag{8.3.4}$$

is fulfilled so that the right-hand side of (8.3.3) satisfies the no-slip boundary condition $u_x(y = h) = U_2$, and either the flow rate through the channel has a specified value $Q(t)$, or the streamwise pressure gradient has a specified value $\partial p(t)/\partial x$.

We consider flow subject to a specified pressure gradient and recast the x component of the equation of motion for unidirectional flow into the form

$$\frac{\partial u_x}{\partial t} = -\frac{1}{\rho}\frac{\partial p}{\partial x} + \nu\,\frac{\partial^2 u_x}{\partial y^2} + g_x = -\frac{1}{\rho}\frac{\partial p}{\partial x} - \nu\,\frac{\partial \omega_z}{\partial y} + g_x. \tag{8.3.5}$$

Evaluating (8.3.5) at the lower and upper wall and rearranging, we obtain boundary conditions for the slope of the vorticity,

$$\begin{aligned} \left(\frac{\partial \omega_z}{\partial y}\right)_{y=0} &= -\frac{1}{\nu}\frac{\mathrm{d}U_1}{\mathrm{d}t} - \frac{1}{\mu}\frac{\partial p}{\partial x} + \frac{1}{\nu}\,g_x, \\ \left(\frac{\partial \omega_z}{\partial y}\right)_{y=h} &= -\frac{1}{\nu}\frac{\mathrm{d}U_2}{\mathrm{d}t} - \frac{1}{\mu}\frac{\partial p}{\partial x} + \frac{1}{\nu}\,g_x. \end{aligned} \tag{8.3.6}$$

These equations provide us with Neumann boundary conditions on either end of the solution domain.

Special attention must be paid to the case of impulsive motion. If a wall suddenly moves in an impulsive fashion with the velocity changing from one value to another in an infinitesimal period of time, the corresponding time derivative on the right-hand side of one or both of (8.3.6) develops an infinite spike described by the Dirac delta function discussed in chapter 11. This singular behavior is too demanding to be handled by the numerical method.

Next, we investigate whether the vorticity boundary conditions (8.3.6) ensure the satisfaction of the integral constraint (8.3.4), which is necessary for the satisfaction of the no-slip boundary condition at the upper wall. Integrating (8.3.1) with respect to y from 0 to h, interchanging the order of the integration and time differentiation on the left-hand side, and using (8.3.6) to simplify the right-hand side, we find

$$\frac{\mathrm{d}}{\mathrm{d}t}\int_0^h \omega_z(y')\,\mathrm{d}y' = \frac{\mathrm{d}}{\mathrm{d}t}(U_1 - U_2). \tag{8.3.7}$$

Time-integration of (8.3.7) reproduces (8.3.4) up to a time-independent constant determined by the initial state. Thus, if (8.3.4) is satisfied at the initial instant, it will also be satisfied at all subsequent times.

8.3.2 Alternative set of equations

In an alternative approach, we take the derivative of (8.3.2) with respect to y to derive the second-order equation

$$\frac{\partial^2 u_x}{\partial y^2} = -\frac{\partial \omega_z}{\partial y} \equiv -q, \tag{8.3.8}$$

where $q \equiv \partial\omega_z/\partial y$ is the slope of the vorticity. To compute the velocity, we integrate the second-order equation (8.3.8) with respect to y using the velocity boundary conditions $u_x(y=0) = U_1$ and $u_x(y=h) = U_2$.

The important benefit stemming from the use of (8.3.8) instead of (8.3.2), is that, in order to compute the velocity, the slope of the vorticity q instead of the vorticity itself is required. An evolution equation for q arises by differentiating both sides of (8.3.1) with respect to y, finding

$$\frac{\partial q}{\partial t} = \nu\,\frac{\partial^2 q}{\partial y^2}. \tag{8.3.9}$$

Boundary conditions are provided by equations (8.3.6).

In summary, the numerical procedure involves integrating in time equation (8.3.9) from an initial state subject to the derived boundary conditions (8.3.6), while simultaneously computing the velocity by solving the second-order equation (8.3.8) subject to the velocity boundary conditions $u_x(y=0) = U_1$ and $u_x(y=h) = U_2$.

Explicit finite-difference method

To implement a finite-difference method, we divide the flow domain $0 \leq y \leq h$ into N intervals separated by $N+1$ grid points, as shown in figure 8.2.1, and evaluate equation (8.3.9) at time t at the interior nodes corresponding to $i = 2, \ldots, N$.

Approximating the time derivative on the left-hand side with a first-order finite difference and the y derivative on the left-hand side with a second-order finite difference, we obtain

$$\frac{q_i(t+\Delta t) - q_i(t)}{\Delta t} = \nu \, \frac{q_{i-1}(t) - 2\, q_i(t) + q_{i+1}(t)}{\Delta y^2}, \tag{8.3.10}$$

where we have defined $q_i \equiv q(y_i)$. Solving for $q_i(t+\Delta t)$ on the left-hand side, we find

$$q_i(t+\Delta t) = \alpha\, q_{i-1}(t) + (1 - 2\,\alpha)\, q_i(t) + \alpha\, q_{i+1}(t), \tag{8.3.11}$$

where $\alpha \equiv \nu\, \Delta t/\Delta y^2$ is the numerical diffusion number. Equation (8.3.11) allows us to explicitly update the values of q at the grid points subject to boundary conditions for q_1 and q_{N+1} given by the right-hand sides of equations (8.3.6).

The centered-difference discretization of equation (8.3.8) leads us to the linear system (8.2.19), where the coefficient matrix $\mathbf{C}$ is given in (8.2.20) and the constant vector on the right-hand side is given by

$$\mathbf{d} = \begin{bmatrix} \Delta y^2\, q_2 + U_1 \\ \Delta y^2\, q_3 \\ \vdots \\ \Delta y^2\, q_{N-1} \\ \Delta y^2\, q_N + U_2 \end{bmatrix}. \tag{8.3.12}$$

The linear system can be solved efficiently using the Thomas algorithm.

8.3.3 Comparison with the velocity/pressure formulation

Comparing the velocity/vorticity formulation discussed in this section with the velocity/pressure formulation discussed in section 8.2, we find that the latter is significantly simpler in conception and implementation. While this is undoubtedly true in the case of unidirectional flow presently considered, we will see later in this chapter that the vorticity-velocity formulation is more competitive in the more general case of two- and three-dimensional flow.

Problems

8.3.1 *Steady flow*

Discuss the implementation of the velocity/vorticity formulation for steady channel flow due to a specified pressure gradient.

8.3.2 *Two-layer flow*

Develop a velocity/vorticity formulation for two-layer channel flow discussed in section 8.2.

8.3.3 *Flow in a circular tube*

Develop an explicit method based on the velocity/vorticity formulation for computing the velocity profile developing inside a circular tube due to a suddenly imposed constant pressure gradient.

Computer Problem

8.3.4 *Explicit finite-difference method.*

Write a program that computes the evolution of the velocity profile in a channel with stationary walls due to the sudden application of a constant pressure gradient, based on the explicit finite-difference method discussed in the text. Run the program for fluid properties and flow conditions of your choice and several time step sizes corresponding to a numerical diffusion number α that is larger and lower than 0.5. Discuss the performance of the numerical method.

8.4 Unidirectional flow; stream function/vorticity formulation

In sections 8.2 and 8.3, we discussed methods for computing the evolution of the velocity profile in channel flow subject to a specified pressure gradient. In this section, we consider the complementary case of flow subject to a specified flow rate and develop a numerical method stemming from the velocity/vorticity formulation.

For reasons that will soon become apparent, we introduce the stream function, ψ, satisfying the equation $u_x = \partial\psi/\partial y$. In the case of unidirectional flow, u_x and thus ψ is a function of position, y, and time, t. The flow rate across a line that begins and ends at two parallel planes located at $y = y_1$ and y_2 is equal to the difference in the corresponding values of the stream function, $Q_{12} = \psi(y_2) - \psi(y_1)$. The flow rate through the entire channel is $Q = \psi(y = h) - \psi(y = 0)$.

Using equation (8.3.2), we find that the nonzero component of the vorticity is related to the stream function by the equation

$$\omega_z = -\frac{\partial^2 \psi}{\partial y^2}. \tag{8.4.1}$$

The numerical method involves computing the evolution of the vorticity from a specified initial state using the vorticity transport equation (8.3.1), while simultaneously recovering the evolution of the stream function using the one-dimensional Poisson equation (8.4.1). Since the differential equations (8.3.1) and (8.4.1) are of second order with respect to y, two boundary conditions for the vorticity and two boundary conditions for the stream function are required, one at each end of the solution domain, $y = 0$ and h.

Since adding an arbitrary constant to the stream function does not affect the velocity, the base level of the stream function is unspecified. Accordingly, we may stipulate $\psi(y = 0) = 0$, finding $\psi(y = h) = Q(t)$. It is now evident that, by introducing the stream function, we have considerably facilitated the implementation of the condition on the flow rate.

8.4.1 Boundary conditions for the vorticity

The boundary conditions for the vorticity must involve the specified wall velocities U_1 and U_2 by way of the no-slip boundary condition. To illustrate the implementation of this condition, we divide the flow domain, $0 \le y \le h$, into N intervals defined by $N + 1$ grid points, as shown in figure 8.2.1, and evaluate (8.4.1) at time t at the boundary nodes corresponding to $i = 1$ and $N + 1$. Approximating the y derivative on the right-hand side with a combination of finite differences, we obtain

$$\omega_1 = -\frac{(\frac{\partial \psi}{\partial y})_{\frac{1}{2}(y_1+y_2)} - (\frac{\partial \psi}{\partial y})_{y_1}}{\frac{1}{2}\,\Delta y} = -2\,\frac{\frac{\psi_2-\psi_1}{\Delta y} - U_1}{\Delta y}$$
$$= 2\,\frac{\psi_1 - \psi_2}{\Delta y^2} + 2\,\frac{U_1}{\Delta y}, \tag{8.4.2}$$

and

$$\omega_{N+1} = -\frac{(\frac{\partial \psi}{\partial y})_{y_{N+1}} - (\frac{\partial \psi}{\partial y})_{\frac{1}{2}(y_N+y_{N+1})}}{\frac{1}{2}\,\Delta y} = -2\,\frac{U_2 - \frac{\psi_{N+1}-\psi_N}{\Delta y}}{\Delta y}$$
$$= 2\,\frac{\psi_{N+1} - \psi_N}{\Delta y^2} - 2\,\frac{U_2}{\Delta y}, \tag{8.4.3}$$

where we have denoted $\omega_i \equiv \omega_z(y_i)$ and $\psi_i \equiv \psi(y_i)$.

It is somewhat alarming to observe that the no-slip condition is implemented indirectly in terms of the vorticity. Specifically, it is not clear that solving (8.4.1) for the stream function and subsequently differentiating it to produce the velocity generates a velocity field that is consistent with the prescribed boundary velocity. However, a thorough analysis of the numerical method reveals that this will be true, except under unusual circumstances associated with singular boundary conditions involving discontinuous functions.

8.4.2 A semi-implicit method

Proceeding with the finite-difference method, we apply equation (8.3.1) at the interior nodes corresponding to $i = 2, \ldots, N$ at time t. Approximating the time derivative on the left-hand side with a first-order finite difference and the y derivative on the right-hand side with a second-order second finite difference, we obtain

$$\frac{\omega_i(t+\Delta t) - \omega_i(t)}{\Delta t} = \nu \, \frac{\omega_{i-1}(t) - 2\,\omega_i(t) + \omega_{i+1}(t)}{\Delta y^2}. \tag{8.4.4}$$

Solving for $\omega_i(t + \Delta t)$, we find

$$\omega_i(t+\Delta t) = \alpha\,\omega_{i+1}(t) + (1-2\,\alpha)\,\omega_i(t) + \alpha\,\omega_{i-1}(t), \tag{8.4.5}$$

where $\alpha \equiv \nu\Delta t/\Delta y^2$ is the numerical diffusion number. Equation (8.4.5) allows us to explicitly update the values of the vorticity at the interior grid points, subject to boundary conditions for ω_1 and ω_{N+1} given by the right-hand sides of (8.4.2) and (8.4.3); the stream function at time t is assumed to be known.

The implicit discretization of equation (8.4.1) leads to the linear system

$$\mathbf{C} \cdot \boldsymbol{\psi}(t + \Delta t) = \mathbf{d}(t + \Delta t), \tag{8.4.6}$$

where the coefficient matrix $\mathbf{C}$ is given in (8.2.20), the vector $\boldsymbol{\psi}$ is defined as

$$\boldsymbol{\psi} = \begin{bmatrix} \psi_2 \\ \psi_3 \\ \vdots \\ \psi_{N-1} \\ \psi_N \end{bmatrix}, \tag{8.4.7}$$

and the vector on the right-hand side of (8.4.6) is given by

$$\mathbf{d} = \begin{bmatrix} \Delta y^2\,\omega_2 \\ \Delta y^2\,\omega_3 \\ \vdots \\ \Delta y^2\,\omega_{N-1} \\ \Delta y^2\,\omega_N + Q \end{bmatrix}. \tag{8.4.8}$$

The system (8.4.6) can be solved efficiently using the Thomas algorithm.

The numerical method involves the following steps:

1. First, we assign initial values to the stream function and vorticity at all nodes.
2. Second, we compute the vorticity at the boundary nodes using (8.4.2) and (8.4.3).
3. Third, we update the vorticity at the internal nodes using (8.4.5).
4. Fourth, we update the stream function at the internal nodes by solving the linear system (8.4.6).
5. We may now return to step 2 and repeat the calculation for another step.

The velocity profile arises by numerically differentiating the stream function with respect to y.

Problems

8.4.1 *Steady flow*

Develop a finite-difference method based on the stream function / vorticity formulation for computing the velocity profile of steady channel flow subject to a specified flow rate.

8.4.2 *Two-layer flow*

Develop a finite-difference method based on the stream function / vorticity formulation for unsteady two-layer channel flow discussed in section 8.2.

8.4.3 *Flow inside a circular tube*

Develop a finite-difference method based on the stream function/vorticity formulation for computing the velocity profile developing inside a circular tube, subject to a specified flow rate.

Computer Problem

8.4.4 *Explicit finite-difference method*

Write a program that computes the evolution of the velocity profile in a channel with stationary walls using an explicit finite-difference method. The flow rate increases gradually toward a steady value according to the equation

$$Q(t) = Q_0\,[1 - \exp(-\frac{\beta \nu t}{h^2})], \tag{8.4.9}$$

where Q_0 is the constant flow rate prevailing of long times, and β is a dimensionless constant. Run the program for fluid properties and flow conditions of your choice, and for several sizes of the time step corresponding to α larger and lower than 0.5. Discuss the performance of the numerical method.

8.5 Two-dimensional flow; stream function/vorticity formulation

Having discussed finite-difference methods for unidirectional flow, we now turn our attention to the more general case of two-dimensional flow where further issues concerning the satisfaction of the continuity equation and choice of boundary conditions arise. We begin in this section by considering the vorticity/stream function formulation developed as an extension of the corresponding formulation for unidirectional flow discussed in section 8.4.

Texts on computer language programming traditionally introduce elementary programming procedures by explaining the structure of a program entitled "world" which prints the important message "Hello World." Texts on computational fluid dynamics explain numerical methods by discussing the prototypical example of flow in a two-dimensional cavity driven by a moving lid, known as the "driven-cavity flow." We will follow this time-honored tradition.

8.5.1 *Flow in a cavity*

Consider flow in a cavity driven by the motion of a lid translating parallel to itself with a generally time-dependent velocity $U(t)$, as illustrated in figure 8.5.1. We begin by introducing the stream function, ψ, defined from the differential relations $u_x = \partial\psi/\partial y$ and $u_y = -\partial\psi/\partial x$, where u_x and u_y are the x and y velocities. The no-penetration condition requires that the component of the velocity normal to each one of the four walls is zero. In terms of the stream function,

$$\psi = 0 \qquad \text{over all walls,} \tag{8.5.1}$$

so that the tangential derivative of the stream function, being proportional to the normal component of the velocity, is zero. The zero on the right-hand side of (8.5.1) can be replaced by an arbitrary constant without a consequence on the numerical solution or physical structure of the flow.

The no-slip boundary condition requires that the tangential component of the velocity is zero at the bottom, left, and right walls, and equal to $U(t)$ on the upper wall. In terms of the stream function,

$$\frac{\partial\psi}{\partial y} = 0 \qquad \text{at the bottom,} \qquad \frac{\partial\psi}{\partial x} = 0 \qquad \text{at the sides,}$$

$$\frac{\partial\psi}{\partial y} = U(t) \qquad \text{at the top.} \tag{8.5.2}$$

We can derive simplified expressions for the boundary values of the non-vanishing component of the vorticity, $\omega_z \equiv -\partial u_x/\partial y + \partial u_y/\partial x$, in terms of the stream function, based on the boundary conditions for the velocity. For

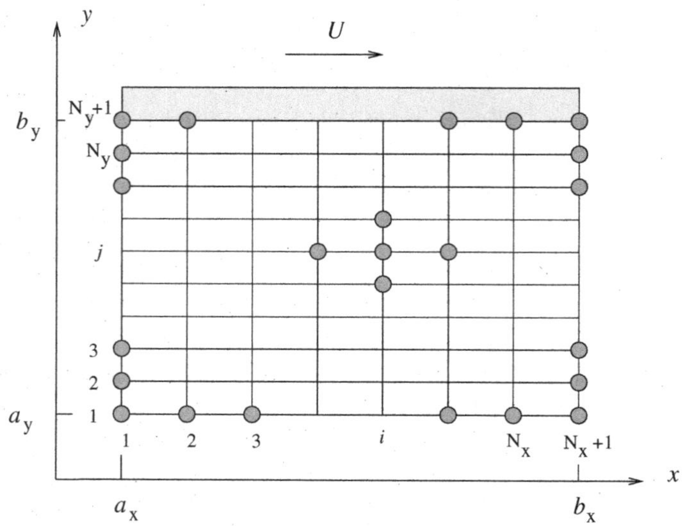

Figure 8.5.1 Illustration of a finite-difference grid used to compute flow in a cavity driven by a moving lid.

example, recalling that $u_y = 0$ and thus $\partial u_y/\partial x = 0$ over the bottom wall, we find $\omega_z = -\partial u_x/\partial y = -\partial^2\psi/\partial y^2$. Working in this fashion, we find

$$\omega_z = -\frac{\partial^2\psi}{\partial y^2} \quad \text{at the top and bottom,}$$
$$\omega_z = -\frac{\partial^2\psi}{\partial x^2} \quad \text{at the sides,} \tag{8.5.3}$$

which are simplified versions of the more general expression for the vorticity in terms of the stream function, $\omega_z = -(\partial^2\psi/\partial x^2 + \partial^2\psi/\partial y^2)$.

8.5.2 Finite-difference grid

To prepare the ground for the implementation of the finite-difference method, we cover the rectangular solution domain with a uniform two-dimensional Cartesian grid consisting of $N_x + 1$ uniformly spaced vertical lines and $N_y + 1$ uniformly spaced horizontal lines, as shown in figure 8.5.1. Parallel grid lines are separated by the intervals Δx or Δy, defining the grid size. The intersections of grid lines define grid points or nodes labeled with a pair of integers, (i, j), where $i = 1, \ldots, N_x + 1$ and $j = 1, \ldots, N_y + 1$. The vertical side walls correspond to $i = 1$ and $N_x + 1$, and the bottom and top walls correspond to $j = 1$ and $N_y + 1$.

The goal of the finite-difference method is to produce the values of flow variables at the grid points. For simplicity, we denote

$$\omega_{i,j} \equiv \omega_z(x_i, y_j), \qquad \psi_{i,j} \equiv \psi(x_i, y_j). \tag{8.5.4}$$

Similar notation is used for other variables.

8.5.3 Unsteady flow

Following the general protocol of methods based on the vorticity transport equation, we compute the evolution of the flow by advancing the vorticity field using the vorticity transport equation for two-dimensional flow written in the form of an evolution equation for the vorticity,

$$\frac{\partial \omega_z}{\partial t} = -u_x \frac{\partial \omega_z}{\partial x} - u_y \frac{\partial \omega_z}{\partial y} + \nu \left(\frac{\partial^2 \omega_z}{\partial x^2} + \frac{\partial^2 \omega_z}{\partial y^2} \right), \tag{8.5.5}$$

subject to appropriate derived boundary conditions for the vorticity, while simultaneously obtaining the evolution of the stream function by solving the Poisson equation

$$\nabla^2 \psi \equiv \frac{\partial^2 \psi}{\partial x^2} + \frac{\partial^2 \psi}{\partial y^2} = -\omega_z, \tag{8.5.6}$$

subject to specified boundary conditions for the stream function.

A simple method for computing the evolution of the flow when the lid suddenly starts translating involves the following steps:

1. At the initial instant, we set the stream function and velocity at all interior and boundary grid nodes to zero. Then we set the x component of the velocity at the grid nodes along the lid equal to $U(t = 0)$.

2. At the second step, we differentiate the velocity to produce the vorticity using the definition $\omega_z \equiv -\partial u_x/\partial y + \partial u_y/\partial x$. For the interior grid points, we use centered differences, as shown in (2.5.19), to obtain

$$\omega_{i,j} = -\frac{(u_x)_{i,j+1} - (u_x)_{i,j-1}}{2\Delta y} + \frac{(u_y)_{i+1,j} - (u_y)_{i-1,j}}{2\Delta x}. \tag{8.5.7}$$

   For the top wall, we use backward differences to obtain

$$\omega_{i,N_y+1} \simeq -\left(\frac{\partial u_x}{\partial y}\right)_{i,N_y+1} \simeq \frac{-3U + 4\,(u_x)_{i,N_y} - (u_x)_{i,N_y-1}}{2\Delta y}, \tag{8.5.8}$$

   involving values at interior grid points. For the bottom and side walls, we use forward or backward differences to obtain

$$\omega_{i,1} \simeq -\left(\frac{\partial u_x}{\partial y}\right)_{i,1} \simeq \frac{-4\,(u_x)_{i,2} + (u_x)_{i,3}}{2\,\Delta y}, \tag{8.5.9}$$

and

$$\omega_{1,j} \simeq (\frac{\partial u_y}{\partial x})_{1,j} \simeq \frac{4\,(u_y)_{2,j} - (u_y)_{3,j}}{2\,\Delta x},$$

$$\omega_{N_x+1,j} \simeq (\frac{\partial u_y}{\partial x})_{N_x+1,j} \simeq \frac{-4\,(u_y)_{N_x,j} + (u_y)_{N_x-1,j}}{2\,\Delta x}, \tag{8.5.10}$$

3. Now we integrate in time equation (8.5.5) to obtain the vorticity at the interior grid points at time $t + \Delta t$. Using a fully explicit method, we set

$$\omega_{i,j}(t + \Delta t) = \omega_{i,j}(t) + G_{(i,j)}(t), \tag{8.5.11}$$

where $G_{(i,j)}(t)$ is the right-hand side of (8.5.5) evaluated at the (i, j) grid point at time t. To evaluate $G_{(i,j)}(t)$, we approximate the first spatial derivatives and the Laplacian of the vorticity using centered differences. For example, the Laplacian of the vorticity can be approximated with the finite-difference formula shown in equation (3.3.14), written for ω_z.

4. Next, we solve the Poisson equation (8.5.6) for the stream function subject to the boundary condition $\psi = 1$, using a slightly generalized version of the finite-difference method for Laplace's equation discussed in section 3.3. Approximating the second derivatives with centered differences, we obtain the counterpart of equation (3.3.15),

$$\psi_{i+1,j} - 2\,(1 + \beta)\,\psi_{i,j} + \psi_{i-1,j} + \beta\,\psi_{i,j+1} + \beta\,\psi_{i,j-1} = -\Delta x^2\,\omega_{i,j}, \tag{8.5.12}$$

where $\beta \equiv (\Delta x/\Delta y)^2$.

5. Finally, we differentiate the stream function to compute the components of the velocity at time $t + \Delta t$ at the interior grid points.

Having completed one time step, we return to step 2 and repeat the computation for another time step.

8.5.4 Steady flow

To compute the steady flow, we follow a somewhat different approach. In this case, the left-hand side of (8.5.5) vanishes, yielding a differential relation between the velocity and the vorticity. Solving for the Laplacian of the vorticity, we find

$$\frac{\partial^2 \omega_z}{\partial x^2} + \frac{\partial^2 \omega_z}{\partial y^2} = \frac{1}{\nu}\,(u_x \frac{\partial \omega_z}{\partial x} + u_y \frac{\partial \omega_z}{\partial y}), \tag{8.5.13}$$

which can be regarded as a Poisson equation for the vorticity forced by the *a priori* unknown source function on the right-hand side. Computing the flow in terms of the stream function and vorticity involves simultaneously solving equations (8.5.6) and (8.5.13) according to the following steps:

1. First, we guess the distribution of the stream function and associated vorticity distribution.

2. Second, we solve the Poisson equation (8.5.6) for the stream function subject to the no-penetration boundary condition, $\psi = 0$.

3. Third, we compute the right-hand side of (8.5.13).

4. Fourth, we derive boundary conditions for the vorticity using the stream function obtained in step 2.

5. Fifth, we solve the Poisson equation (8.5.13) for the vorticity.

6. Now we check whether the vorticity computed in step 5 agrees with that assigned in step 1 within a specified tolerance. If not, we replace the latter by the former and return to step 2.

The method is implemented according to the following steps:

1. First, we assign values to the stream function at all $(N_x + 1) \times (N_y + 1)$ interior and boundary grid points, and to the vorticity at all $N_x \times N_y$ interior grid points. A simple choice is to set them all to zero.

2. Second, we solve the Poisson equation (8.5.6) subject to the boundary condition $\psi = 0$ using an iterative method. To perform the iterations, we approximate the second derivatives with centered differences and obtain equation (8.5.12), expressed in the form

$$R_{i,j} \equiv \psi_{i+1,j} - 2\,(1+\beta)\psi_{i,j} + \psi_{i-1,j} + \beta\psi_{i,j+1} + \beta\psi_{i,j-1} + \Delta x^2\, \omega_{i,j} = 0, \tag{8.5.14}$$

   where $R_{i,j}$ is defined as the residual. The iterative method involves computing a time-like sequence of grid values parametrized by the index l, computed using the formula

$$\psi_{i,j}^{(l+1)} = \psi_{i,j}^{(l)} + \frac{\mathcal{R}}{2(1+\beta)} R_{i,j}^{(l)}, \tag{8.5.15}$$

   for $l = 1, 2, \ldots$, where $\mathcal{R}$ is a specified relaxation factor used to control the iterations.

3. In the next step, we compute the vorticity at the boundary grid points taking into consideration the velocity boundary conditions. Considering the grid points on the lid, we expand the stream function in a Taylor

series with respect to y about a top grid point. Evaluating the expansion at the grid point immediately below, we obtain

$$\psi_{i,N_y} \simeq \psi_{i,N_y+1} - \Delta y \, (\frac{\partial \psi}{\partial y})_{i,N_y+1} + \frac{\Delta y^2}{2} \, (\frac{\partial^2 \psi}{\partial y^2})_{i,N_y+1}. \quad (8.5.16)$$

Setting $(\partial\psi/\partial y)_{i,N_y+1} = U$ and $\omega_{i,N_y+1} = -(\partial^2\psi/\partial y^2)_{i,N_y+1}$, as discussed in the paragraph following equation (8.5.2), and solving for ω_{i,N_y+1}, we find

$$\omega_{i,N_y+1} = 2 \, \frac{\psi_{i,N_y+1} - \psi_{i,N_y}}{\Delta y^2} - 2 \, \frac{U}{\Delta y}. \quad (8.5.17)$$

Working in a similar fashion, we derive analogous expressions for the bottom, left, and right walls,

$$\omega_{i,1} = 2 \, \frac{\psi_{i,1} - \psi_{i,2}}{\Delta y^2}, \qquad \omega_{1,j} = 2 \, \frac{\psi_{1,j} - \psi_{2,j}}{\Delta y^2},$$
$$\omega_{N_x+1,j} = 2 \, \frac{\psi_{N_x+1,j} - \psi_{N_x,j}}{\Delta y^2}. \quad (8.5.18)$$

More accurate expressions can be derived using higher-order expansions.

4. In the next step, we differentiate the stream function to produce the velocity at the interior grid points subject to the no-penetration condition, $\psi = 0$.

5. Now we differentiate the vorticity to produce the x and y derivatives at the interior grid points subject to boundary values computed in step 3.

6. Next, we compute the right-hand side of (8.5.13) at the interior grid points.

7. Finally, we solve equation (8.5.13) by iteration, as discussed in step 2. The counterparts of equations (8.5.14) and (8.5.15) are

$$R_{i,j} \equiv \omega_{i+1,j} - 2\,(1+\beta)\,\omega_{i,j} + \omega_{i-1,j} + \beta\,\omega_{i,j+1} + \beta\,\omega_{i,j-1}$$
$$-\Delta x^2 \, N_{i,j} = 0, \quad (8.5.19)$$

   and

$$\omega_{i,j}^{(l+1)} = \omega_{i,j}^{(l)} + \frac{\mathcal{R}'}{2(1+\beta)} \, R_{i,j}^{(l)}, \quad (8.5.20)$$

   where $N_{i,j}$ is the right-hand side of (8.5.13) evaluated at the (i,j) grid point, and $\mathcal{R}'$ is a new relaxation factor.

The method is implemented in the following MATLAB code *cvt_sv* residing in directory *11_fdm* of FDLIB:

```
%============
% code cvt_sv
%============

%-----
% parameters
%-----

Ulid = 1.0;       % lid velocity
Lx = 2.0; Ly = 1.0;     % cavity dimensions
Nx = 32; Ny = 16;     % grid size
visc = 0.01;     % viscosity
rho = 1.0;     % density
relax = 0.5;     % relaxation parameter
Niteri = 5;     % number of inner iterations
Niterg = 100;     % number of global iterations
vort_init = 0.0;     % initial vorticity

%-------
% prepare
%--------

Dx = Lx/Nx; Dy = Ly/Ny;
Dx2 = 2.0*Dx; Dy2 = 2.0*Dy; Dxs = Dx*Dx; Dys = Dy*Dy;
beta = Dxs/Dys; beta1 = 2.0*(1.0+beta);
nu = visc/rho;     % kinematic viscosity

%------------------------------------------
% generate the grid
% initialize stream function and vorticity
%------------------------------------------

for i=1:Nx+1
  for j=1:Ny+1
    x(i,j) = (i-1.0)*Dx;
    y(i,j) = (j-1.0)*Dy;
    psi(i,j) = 0.0;     % stream function
    vort(i,j) = -vort_init;
  end
end

%------------------
% global iterations
%------------------

for iter=1:Niterg

  save = vort;
```

```
%----------------------------------------
% Jacobi updating of the stream function
% at the interior nodes
%----------------------------------------

  for iteri=1:Niteri
    for j=2:Ny
      for i=2:Nx
        res = (psi(i+1,j)+psi(i-1,j)+ beta*psi(i,j+1) ...
        +beta*psi(i,j-1)+Dxs*vort(i,j))/beta1-psi(i,j);
        psi(i,j) = psi(i,j) + relax*res;
      end
    end
  end

%-------------------------------------
% Compute the vorticity at boundary grid points
% using the velocity boundary conditions
% Lower-order boundary conditions are commented out
%-------------------------------------

%---
% top and bottom walls
%---

for i=2:Nx
% vort(i,1) = 2.0*(psi(i,1)-psi(i,2))/Dys;
% vort(i,Ny+1) = 2.0*(psi(i,Ny+1)-psi(i,Ny))/Dys-2.0*Ulid/Dys;
  vort(i,1) = (7.0*psi(i,1)-8.0*psi(i,2)+psi(i,3))/(2.0*Dys);
  vort(i,Ny+1) = (7.0*psi(i,Ny+1)-8.0*psi(i,Ny) ...
              +psi(i,Ny-1))/(2.0*Dys)-3.0*Ulid/Dy;
end

%---
% left and right walls
%---

for j=2:Ny
% vort(1,j) = 2.0*(psi(1,j)-psi(2,j) )/Dxs;
% vort(Nx+1,j) = 2.0*(psi(Nx+1,j)-psi(Nx,j))/Dxs;
  vort(1,j) = (7.0*psi(1,j)-8.0*psi(2,j)+psi(3,j))/(2.0*Dxs);
  vort(Nx+1,j) = (7.0*psi(Nx+1,j)-8.0*psi(Nx,j) ...
  +psi(Nx-1,j))/(2.0*Dxs);
end

%--------------------------------
% Compute the velocity at the interior
% grid points by central differences
%--------------------------------
```

```
for j=2:Ny
  for i=2:Nx
    ux(i,j) = (psi(i,j+1)-psi(i,j-1))/Dy2;
    uy(i,j) = - (psi(i+1,j)-psi(i-1,j))/Dx2;
  end
end

%--------------------------------------------
% iterate Poisson's equation for the vorticity
%--------------------------------------------

for iteri=1:Niteri
  for j=2:Ny
      for i=2:Nx
      source(i,j) = ux(i,j)*(vort(i+1,j)-vort(i-1,j))/Dx2...
          + uy(i,j)*(vort(i,j+1)-vort(i,j-1))/Dy2;
      source(i,j) = -source(i,j)/nu;
      res = (vort(i+1,j)+vort(i-1,j) + beta*vort(i,j+1) ...
          +beta*vort(i,j-1)+Dxs*source(i,j))/beta1-vort(i,j);
      vort(i,j) = vort(i,j) + relax*res;
    end
  end
end     % of iteri

%------------------
% monitor the error
%------------------

cormax = 0.0;

for i=1:Nx+1
  for j=1:Ny+1
    res = abs(vort(i,j)-save(i,j));
    if(res>cormax)
      cormax = res;
    end
  end
end

if(cormax<tol)
  break
end

end     % of iterg
%--

%============
% graphics
%============
```

```
for i=1:Nx+1      % set up plotting vectors
  xgr(i) = Dx*(i-1);
end
for j=1:Ny+1
  ygr(j) = Dy*(j-1);
end

surf(20*xgr,20*ygr,vort')
xlabel('x','fontsize',15)
ylabel('y','fontsize',15)
zlabel('\omega','fontsize',15)
set(gca,'fontsize',15)
axis([0 Lx 0 Ly -10 10])
axis equal

figure
contour(xgr,ygr,psi',32)
xlabel('x','fontsize',15)
ylabel('y','fontsize',15)
zlabel('\psi','fontsize',15)
axis([0 Lx 0 Ly])
axis equal
```

Vorticity and stream function contour plots for a cavity with aspect ratio $L_x/L_y = 2$, and Reynolds number $Re = UL_x/\nu = 1$ and 100 are shown in figure 8.5.2. Stream function contours are streamlines and particle paths in a two-dimensional flow.

The vorticity is singular at the upper two cavity corners due to the discontinuous boundary velocity, but this does not deter the performance of the numerical method. A local analysis shows that the shear stress diverges at these corners, and an infinite force is required to slide the lid as a result of the sharp-corner idealization. As the Reynolds number increases, the center of the eddy developing inside the cavity is shifted toward the right wall due to the fluid inertia. Regions of recirculating flow develop at the bottom two corners requiring increased spatial resolution.

8.5.5 Summary

In summary, the vorticity/stream function formulation is distinguished by the following features:

1. Expressing the velocity in terms of the stream function ensures the automatic satisfaction of the continuity equation.

2. The formulation bypasses the computation of the pressure. If we had to solve for the pressure, we would have to derive appropriate boundary conditions, as will be discussed later in this chapter.

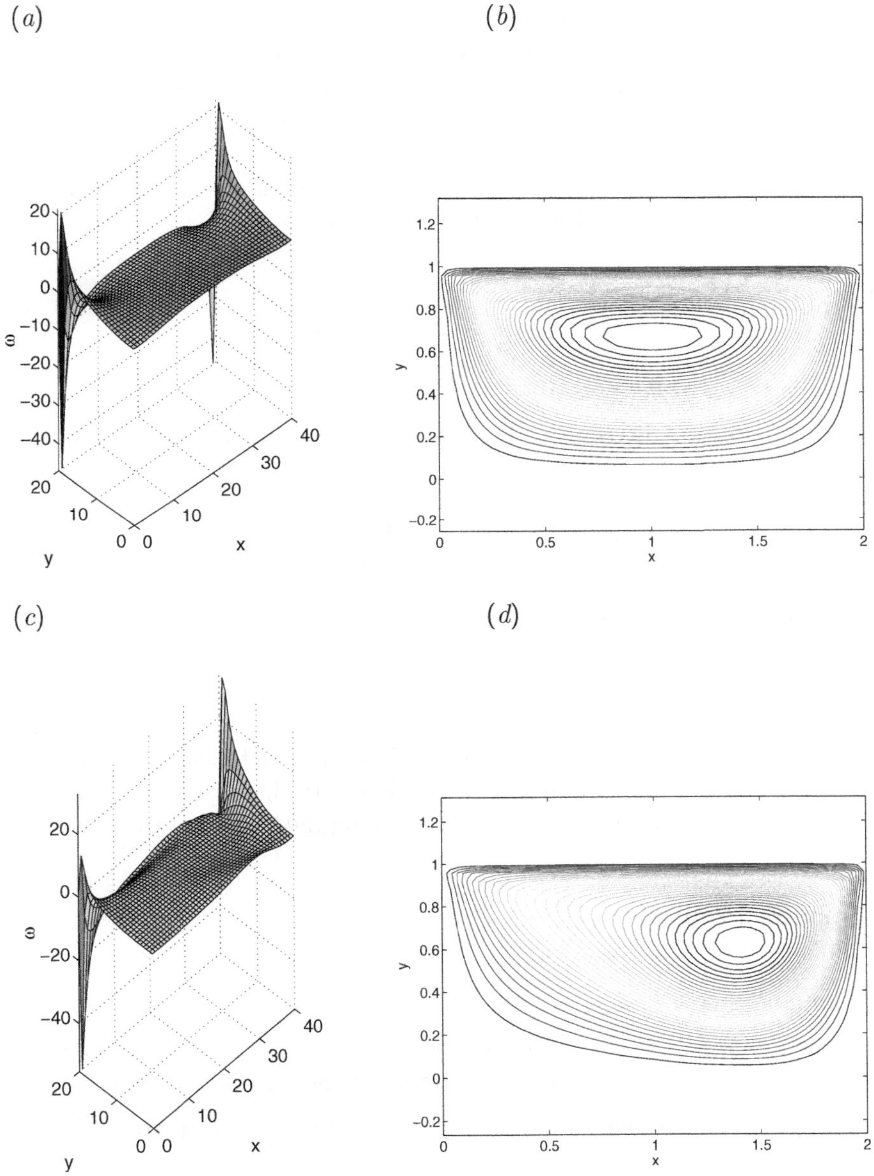

Figure 8.5.2 Vorticity and stream function contour plots for flow in a rectangular cavity with aspect ratio $L_x/L_y = 2$, at Reynolds number (a, b) $Re = UL_x/\nu$ = 1 and (c, d) 100.

3. Enforcing the no-penetration and no-slip boundary condition is done sequentially rather than simultaneously. The no-penetration condition is enforced when solving for the stream function, and the no-slip condition is enforced when deriving boundary conditions for the vorticity.

Similar simplifications occur when solving for the Stokes stream function describing axisymmetric flow without swirling motion.

Problems

8.5.1 *Computation of the pressure*

Show that the pressure distribution in an incompressible fluid satisfies Poisson's equation

$$\nabla^2 p = 2\,\rho\,\Big(\frac{\partial^2\psi}{\partial x^2}\,\frac{\partial^2\psi}{\partial y^2} - \Big(\frac{\partial^2\psi}{\partial x \partial y}\Big)^2\Big). \tag{8.5.21}$$

Hint: Take the divergence of the Navier-Stokes equation and use the continuity equation.

8.5.2 *Axisymmetric flow*

Develop a finite-difference method based on the stream function / vorticity formulation for steady flow in an axisymmetric cavity depressed on a circular cylinder. The flow is driven by a sleeve sliding along the surface of the cylinder.

Computer Problem

8.5.3 *Steady flow in a cavity*

(*a*) Run the program *cvt_sv* and prepare velocity vector plots for the two flows illustrated in figure 8.5.2. Discuss the structure of the streamline pattern.

(*b*) Investigate the performance of the numerical method for a square cavity at high Reynolds numbers.

(*c*) Duplicate the results shown in figure 8.5.2 for a slender cavity with aspect ratio $L_x/L_y = 8$. Discuss the structure of the flow.

(*d*) Implement a stopping check so that the computations terminate when the vorticity field has been computed with a specified accuracy.

(*e*) Resolve and discuss the structure of the eddies developing near the lower two corners.

8.6 Velocity/pressure formulation

Although the stream function/vorticity formulation discussed in section 8.5 is simple and efficient, its extension to three dimensions and its generalization to flow in the presence of interfaces are cumbersome. To handle arbitrary flow configurations, we develop a direct formulation in primary variables including the velocity and the pressure.

To compute the evolution of an unsteady flow, we require an evolution equation for the velocity and another evolution equation for the pressure. The former is provided by the Navier-Stokes equation stated as

$$\frac{\partial \mathbf{u}}{\partial t} = \mathbf{N}(\mathbf{u}) - \frac{1}{\rho}\nabla p + \nu\,\mathbf{L}(\mathbf{u}). \tag{8.6.1}$$

We have introduced the nonlinear convective operator, $\mathbf{N}$, and the linear diffusive operator $\mathbf{L}$, defined as

$$\mathbf{N}(\mathbf{u}) \equiv -\mathbf{u}\cdot\nabla\mathbf{u}, \qquad \mathbf{L}(\mathbf{u}) \equiv \nabla^2\mathbf{u}. \tag{8.6.2}$$

If the fluid were compressible, the continuity equation would provide us with an evolution equation for the density, as shown in equations (2.7.10) and (2.7.11). An evolution equation for the pressure could then be obtained by introducing an equation of state relating the density to the local pressure and temperature.

An explicit evolution equation for the pressure developing in a incompressible fluid is not available. Instead, the continuity equation requires the kinematic constraint

$$\nabla\cdot\mathbf{u} = 0, \tag{8.6.3}$$

which demands that the pressure field evolves so that the rate of expansion, $\nabla\cdot\mathbf{u}$, is zero throughout the domain of flow at all times. To translate this requirement into a mathematical restriction, we take the divergence of the Navier-Stokes equation (8.6.1), interchange the divergence with the time derivative on the left-hand side, and thus derive an evolution equation for the rate of expansion,

$$\frac{\partial\,\nabla\cdot\mathbf{u}}{\partial t} = \nabla\cdot\mathbf{N}(\mathbf{u}) - \frac{1}{\rho}\nabla^2 p + \nu\,\nabla\cdot\mathbf{L}(\mathbf{u}). \tag{8.6.4}$$

Note that the divergence operator and the nonlinear operator $\mathbf{N}$ on the right-hand side do not commute, $\nabla\cdot\mathbf{N}(\mathbf{u}) \neq \mathbf{N}(\nabla\cdot\mathbf{u})$. For simplicity, we have assumed that the density and viscosity are uniform throughout the domain of flow.

Equation (8.6.3) requires that the left-hand side of (8.6.4) vanishes at all times, which will be true provided that the pressure satisfies the pressure Poisson equation (PPE),

$$\nabla^2 p = \rho \nabla \cdot \mathbf{N}(\mathbf{u}) + \mu \nabla \cdot \mathbf{L}(\mathbf{u}). \tag{8.6.5}$$

It can be argued that, since the divergence and the linear operator $\mathbf{L}$ commute, $\nabla \cdot \mathbf{L}(\mathbf{u}) = \mathbf{L}(\nabla \cdot \mathbf{u})$, the last term on the right-hand side of (8.6.5) should be set to zero, yielding the simplified pressure Poisson equation (SPPE),

$$\nabla^2 p = \rho \nabla \cdot \mathbf{N}(\mathbf{u}). \tag{8.6.6}$$

However, in practice, the magnitude of the last term on the right-hand side of (8.6.5) is nonzero due to numerical error associated with the approximation of partial derivatives with finite differences. It turns out that the complete absence of this term may be detrimental to the performance of the numerical method by fostering the growth of small oscillations. To prevent the onset of these oscillations, the PPE is preferred over its simplified counterpart.

8.6.1 Alternative system of governing equations

The preceding discussion suggests a numerical procedure for computing the evolution of an unsteady flow based on equations (8.6.1) and (8.6.5) or (8.6.6): compute the evolution of the velocity using (8.6.1), and simultaneously obtain the evolution of the pressure by solving the Poisson equation (8.6.5) or (8.6.6). The procedure is analogous to that employed in the stream function / vorticity formulation discussed in section 8.5. One important difference is that, by employing the stream function, the satisfaction of the continuity equation (8.6.3) is guaranteed, independent of the magnitude of the numerical error.

To examine whether the velocity/pressure formulation ensures the satisfaction of (8.6.3), we substitute (8.6.5) in the right-hand side of the pressure Poison equation (8.6.4) and obtain the expected result

$$\frac{\partial \nabla \cdot \mathbf{u}}{\partial t} = 0, \tag{8.6.7}$$

which ensures that, if the rate of expansion vanishes at the initial instant by a sensible choice of the initial condition, it will also vanish at all times.

Substituting (8.6.6) in the right-hand side of the simplified pressure Poison equation (8.6.4), we obtain an unsteady diffusion equation for the rate of expansion,

$$\frac{\partial \nabla \cdot \mathbf{u}}{\partial t} = \nu \nabla \cdot \mathbf{L}(\mathbf{u}), \tag{8.6.8}$$

which ensures that, if the rate of expansion vanishes at the initial instant by an appropriate choice of an initial condition, it will also vanish at all times provided that the boundary values of the rate of expansion also vanish at all times. The additional condition underlines the importance of accurately satisfying mass conservation at the boundaries and justifies why (8.6.5) is preferred over its simplified counterpart (8.6.6).

8.6.2 Pressure boundary conditions

To solve the pressure Poisson equation, we require a boundary condition derived from specified boundary conditions for the velocity. The pressure boundary condition emerges by evaluating the Navier-Stokes equation (8.6.1) at the boundaries, and then taking the inner product of both sides with the unit vector normal to the boundaries pointing outward, $\mathbf{n}$. The result is the Neumann boundary condition

$$\mathbf{n} \cdot \nabla p = \rho\, \mathbf{n} \cdot \Big(-\frac{\partial \mathbf{u}}{\partial t} + \mathbf{N}(\mathbf{u}) + \nu\, \mathbf{L}(\mathbf{u}) \Big). \tag{8.6.9}$$

The left-hand side expresses the derivative of the pressure normal to the boundaries, expressing the rate of change of the pressure with respect to distance normal to the boundaries. The right-hand side is then simplified by implementing the no-slip and no-penetration boundary conditions.

For example, in the case of two-dimensional flow over a horizontal stationary wall located at $y = 0$, we require $u_x = 0$ and $u_y = 0$ at $y = 0$, and obtain

$$\mathbf{n} \cdot \nabla p = \frac{\partial p}{\partial y} = \mu\, \frac{\partial^2 u_y}{\partial y^2}. \tag{8.6.10}$$

The left-hand side is the normal derivative of the pressure, and the right-hand side is the negative of the normal derivative of the vorticity multiplied by the fluid viscosity.

8.6.3 Compatibility condition for the pressure

The Poisson equation governing the pressure distribution in an incompressible fluid is analogous to the Poisson equation governing the steady-state temperature distribution in a conductive medium identified with the domain of flow, subject to a homogeneous heat production term expressed by the right-hand side. The boundary condition (8.6.9) specifies the boundary distribution of the flux in terms of the instantaneous velocity. Physical reasoning suggests that a steady distribution will exist only if the total rate of heat production is balanced by the total rate of heat removal across the boundaries, so that heat neither accumulates to elevate the temperature nor is depleted to lower the temperature.

In the case of two-dimensional flow, the mathematical expression of this requirement takes the form of a compatibility condition stating that the areal integral of the right-hand side of (8.6.5) or (8.6.6) over the domain of flow must be equal to the line integral of the right-hand side of (8.6.9) or (8.6.10) over the boundaries. If this condition is not fulfilled, a solution for the pressure cannot be found. In the case of three-dimensional flow, the compatibility condition requires that the volume integral of the right-hand side of (8.6.5) or (8.6.6) over the domain of flow should be equal to the surface integral of the right-hand side of (8.6.9) or (8.6.10) over the boundaries,; otherwise, a solution for the pressure cannot be found. In numerical practice, this compatibility condition is enforced implicitly or explicitly depending on the chosen implementation of the numerical method.

Problem

8.6.1 *Pressure boundary condition*

Derive the pressure boundary condition (8.6.10).

8.7 Operator splitting and solenoidal projection

In practice, the velocity/pressure formulation is implemented in a way that expedites the numerical solution and reduces the cost of the computations. For illustration, we discuss the computation of an evolving two-dimensional flow. The extension to three-dimension flow is straightforward in principle and implementation.

In the most popular implementation of the velocity/pressure formulation, the Navier-Stokes equation (8.6.1) is decomposed into two constituent equations,

$$\frac{\partial \mathbf{u}}{\partial t} = \mathbf{N}(\mathbf{u}) + \nu\, \mathbf{L}(\mathbf{u}), \tag{8.7.1}$$

and

$$\frac{\partial \mathbf{u}}{\partial t} = -\frac{1}{\rho}\,\nabla p, \tag{8.7.2}$$

where the operators $\mathbf{N}$ and $\mathbf{L}$ are defined in equations (8.6.2). The right-hand sides of equations (8.7.1) and (8.7.2) arise by splitting the full Navier-Stokes operator on the right-hand side of (8.6.1) into two parts, subject to the following interpretation.

Consider the change in the velocity field over a small time interval, Δt, following the current time, t. The decomposition (8.7.1) and (8.7.2) is inspired by the idea of updating the velocity in two sequential steps, where the first update is due to inertia and viscosity, and the second update is due to the pressure gradient. Time is reset to $t + \Delta t$ after the completion of the second step. We will see that this decomposition significantly simplifies the implementation of the numerical method by allowing the convection–diffusion and pressure gradient steps to be handled independently using appropriate numerical methods.

However, certain issues arise. First, the boundary condition for the velocity to be used for integrating (8.7.1) cannot be the same as the specified physical boundary condition, otherwise the second step mediated by (8.7.2) will cause a departure. Second, the boundary condition for the pressure may no longer be computed from (8.6.9), and should be derived instead using equation (8.7.2).

The second observation suggests that p in equation (8.7.2) may no longer be considered the hydrodynamic pressure, p, and should be regarded a fictitious pressure whose role is to ensure that the velocity field is solenoidal at the end of the second step. To make this distinction clear, we replace (8.7.2) with the equation

$$\frac{\partial \mathbf{u}}{\partial t} = -\frac{1}{\rho} \nabla \chi, \tag{8.7.3}$$

where χ is a projection function. Equation (8.7.3) receives the velocity field delivered by the convection–diffusion equation (8.7.1), which is not necessarily solenoidal, and removes the non-solenoidal component in a process that can be described as projection into the space of solenoidal functions.

The choice of boundary conditions for the projection function χ has been the subject of extensive discussion. It can be shown that the homogeneous Neumann boundary condition requiring that the derivative of the projection function χ with respect to distance normal to a boundary vanishes, is appropriate. The associated boundary conditions for the velocity will be discussed later in this section.

Next, we discuss the implementation of numerical methods for carrying out the convection–diffusion and projection steps expressed by equations (8.7.1) and (8.7.3).

8.7.1 Convection–diffusion step

To prevent numerical instabilities, we perform the convection–diffusion step expressed by (8.7.1) using an implicit finite difference method. This means that updating the velocity requires solving linear systems of algebraic equations.

Evaluating the x and y components of equation (8.7.1) at the (i, j) grid point at time $t + \Delta t$, and approximating the time derivatives with backward differences and the spatial derivatives with differences of our choice, we derive a system of equations for the unknown velocity vector comprised of the x and y velocity components at the grid points at time $t + \Delta t$. The size of the velocity vector is equal to twice the number of grid points. For a 32×32 grid, we obtain a velocity vector with nearly 2000 unknowns and an equal number of equations whose solution incurs significant computational cost.

As an alternative, we split the operator on the right-hand side of (8.7.1) into two spatial constituents expressing convection–diffusion in the x or y direction, given by

$$\frac{\partial \mathbf{u}}{\partial t} = -u_x \frac{\partial \mathbf{u}}{\partial x} + \nu \frac{\partial^2 \mathbf{u}}{\partial x^2}, \tag{8.7.4}$$

and

$$\frac{\partial \mathbf{u}}{\partial t} = -u_y \frac{\partial \mathbf{u}}{\partial y} + \nu \frac{\partial^2 \mathbf{u}}{\partial y^2}. \tag{8.7.5}$$

We then advance the velocity over the time interval Δt in a sequential fashion based on this decomposition.

To achieve second-order accuracy, we discretize equation (8.7.4) using the Crank-Nicolson method. The implementation involves evaluating (8.7.4) at the (i,j) grid point at time $t + \frac{1}{2}\Delta t$, approximating the time and space derivatives with central differences, and averaging the space derivatives over the time levels t and $t + \Delta t$,

$$\frac{\mathbf{u}^*_{i,j} - \mathbf{u}_{i,j}(t)}{\Delta t} = -\frac{1}{2}(u_x)_{i,j}(t)\,[(\frac{\mathbf{u}_{i+1,j} - \mathbf{u}_{i-1,j}}{2\Delta x})(t) + \frac{\mathbf{u}^*_{i+1,j} - \mathbf{u}^*_{i-1,j}}{2\Delta x})]$$

$$+\frac{1}{2}\nu\,[(\frac{\mathbf{u}_{i+1,j} - 2\,\mathbf{u}_{i,j} + \mathbf{u}_{i-1,j}}{\Delta x^2})(t) + \frac{\mathbf{u}^*_{i+1,j} - 2\,\mathbf{u}^*_{i,j} + \mathbf{u}^*_{i-1,j}}{\Delta x^2}]. \tag{8.7.6}$$

An asterisk designates the first intermediate velocity field. To simplify the notation, we define

$$\mathbf{u}^n_{i,j} \equiv \mathbf{u}_{i,j}(t), \tag{8.7.7}$$

where the superscript n denotes the nth time level corresponding to t. Rearranging equation (8.7.6), we derive the finite-difference equation

$$\begin{aligned} &-(c_x + 2\alpha_x)\,\mathbf{u}^*_{i-1,j} + 4(1+\alpha_x)\,\mathbf{u}^*_{i,j} + (c_x - 2\alpha_x)\,\mathbf{u}^*_{i+1,j} \\ &\quad = (c_x + 2\alpha_x)\,\mathbf{u}^n_{i-1,j} + 4(1-\alpha_x)\,\mathbf{u}^n_{i,j} - (c_x - 2\alpha_x)\,\mathbf{u}^n_{i+1,j}, \end{aligned} \tag{8.7.8}$$

involving the local x convection number

$$c_x \equiv \frac{(u_x)^n_{i,j}\,\Delta t}{\Delta x}, \tag{8.7.9}$$

and the x diffusion number,

$$\alpha_x \equiv \frac{\nu\,\Delta t}{\Delta x^2}. \tag{8.7.10}$$

The right-hand side of (8.7.8) can be computed in terms of the velocity at the grid points at the nth time level, which is available.

Evaluating (8.7.8) at grid points that lie along y grid lines corresponding to fixed values of j, we obtain tridiagonal systems of equations for the x and y components of the first intermediate velocity. The paramount advantage of the method of directional splitting is that these tridiagonal systems can be solved efficiently using the Thomas algorithm discussed in section 8.2.5, subject to boundary conditions to be shortly discussed.

An analogous discretization of (8.7.5) yields

$$-(c_y + 2\alpha_y)\,\mathbf{u}^{**}_{i,j-1} + 4(1+\alpha_y)\,\mathbf{u}^{**}_{i,j} + (c_y - 2\alpha_y)\,\mathbf{u}^{**}_{i,j+1}$$
$$= (c_y + 2\alpha_y)\,\mathbf{u}^{*}_{i,j-1} + 4(1-\alpha_y)\,\mathbf{u}^{*}_{i,j} - (c_y - 2\alpha_y)\,\mathbf{u}^{*}_{i,j+1}, \tag{8.7.11}$$

where

$$c_y \equiv \frac{(u_y)^n_{i,j}\,\Delta t}{\Delta y} \tag{8.7.12}$$

is the local y convection number and

$$\alpha_y \equiv \frac{\nu\,\Delta t}{\Delta y^2} \tag{8.7.13}$$

is the y diffusion number. A double asterisk in (8.7.11) designates the second intermediate velocity field.

The right-hand side of (8.7.11) can be computed in terms of the first intermediate velocity delivered by equation (8.7.8). Evaluating (8.7.11) at grid points that lie along x grid lines corresponding to fixed values of i, we obtain tridiagonal systems of equations for the x and y components of the second intermediate velocity. The solution can be found using the Thomas algorithm subject to boundary conditions to be shortly derived.

8.7.2 Projection step

Next, we advance the velocity field using the projection step (8.7.3), where the projection function is computed to satisfy the continuity equation at the end of this step. Evaluating (8.7.4) at the (i,j) grid point and approximating the time derivative with a finite difference, we find

$$\frac{\mathbf{u}_{i,j}(t+\Delta t) - \mathbf{u}^{**}_{i,j}}{\Delta t} = -\frac{1}{\rho}\,(\nabla\chi)^n_{i,j}, \tag{8.7.14}$$

which can be rearranged to give

$$\mathbf{u}_{i,j}(t+\Delta t) = \mathbf{u}^{**}_{i,j} - \frac{\Delta t}{\rho}\,(\nabla\chi)^n_{i,j}. \tag{8.7.15}$$

The gradient on the right-hand side of (8.7.15) can be approximated by centered, forward, or backward differences.

Now we consider the numerical discretization of the continuity equation, $\nabla \cdot \mathbf{u} = 0$. Using centered differences, we approximate the rate of expansion at the (i, j) grid point with a discrete form,

$$D_{i,j} \equiv (\nabla \cdot \mathbf{u})_{i,j} \simeq \frac{(u_x)_{i+1,j} - (u_x)_{i-1,j}}{2\Delta x} + \frac{(u_y)_{i,j+1} - (u_y)_{i,j-1}}{2\Delta y}. \quad (8.7.16)$$

Evaluating (8.7.16) at the $n+1$ time level corresponding to time $t+\Delta t$, requiring that the left-hand side is zero, and using (8.7.15) to express $\mathbf{u}(t + \Delta t)$ on the right-hand side in terms of the second intermediate velocity denoted by the double asterisk and the projection function, we find

$$\frac{\rho}{\Delta t} (\nabla \cdot \mathbf{u}^{**})_{i,j} = \frac{(\frac{\partial \chi}{\partial x})_{i+1,j} - (\frac{\partial \chi}{\partial x})_{i-1,j}}{2\Delta x} + \frac{(\frac{\partial \chi}{\partial y})_{i,j+1} - (\frac{\partial \chi}{\partial y})_{i,j-1}}{2\Delta y}. \quad (8.7.17)$$

The right-hand side of (8.7.17) is recognized as the discrete divergence of the gradient of the projection function χ.

For grid points that are not adjacent to a wall, we approximate the partial derivatives of the right-hand side of (8.7.17) with centered differences and simplify to obtain

$$\frac{\rho}{\Delta t} (\nabla \cdot \mathbf{u}^{**})_{i,j} = \frac{\chi_{i-2,j} - 2\,\chi_{i,j} + \chi_{i+2,j}}{4\,\Delta x^2} + \frac{\chi_{i,j-2} - 2\,\chi_{i,j} + \chi_{i,j+2}}{4\,\Delta y^2}. \quad (8.7.18)$$

The right-hand side of (8.7.18) is recognized as the finite-difference approximation of the Laplacian of χ, computed with spatial intervals equal to $2\,\Delta x$ and $2\,\Delta y$.

For points that are adjacent to a wall, we derive corresponding formulas incorporating the homogeneous Neumann boundary condition. For example, applying (8.7.17) at the near-corner point $i = 2$ and $j = 2$, and setting $(\partial\chi/\partial y)_{2,1} = 0$ and $(\partial\chi/\partial x)_{1,2} = 0$, we obtain

$$\frac{\rho}{\Delta t} (\nabla \cdot \mathbf{u}^{**})_{2,2} = \frac{\chi_{4,2} - \chi_{2,2}}{4\,\Delta x^2} + \frac{\chi_{2,4} - \chi_{2,2}}{4\,\Delta y^2}. \quad (8.7.19)$$

Returning to (8.7.17), we capriciously reduce the intervals of the centered spatial differences to Δx and Δy, and derive the alternative expression

$$\frac{\rho}{\Delta t} (\nabla \cdot \mathbf{u}^{**})_{i,j} = \frac{\chi_{i+1,j} - 2\chi_{i,j} + \chi_{i-1,j}}{\Delta x^2} + \frac{\chi_{i,j+1} - 2\chi_{i,j} + \chi_{i,j-1}}{\Delta y^2}, \quad (8.7.20)$$

which is applicable at all interior grid points. This finite-difference equation could have been derived directly from (8.7.14), by taking the divergence of both sides and then approximating the emerging Laplacian of χ on the right-hand side with the five-point formula, as shown in (8.7.20).

Evaluating (8.7.18) or (8.7.20) at the interior grid points and their counterparts for the wall-adjacent points, and introducing boundary conditions for χ, we derive a system of linear equations for the grid values of χ, which is the counterpart of the linear system descending from the pressure Poisson equation discussed in section 8.6. Having computed the grid values of the projection function, we return to equation (8.7.15) and perform the final step, advancing the velocity to the $n+1$ time level corresponding to time $t + \Delta t$.

Because the coefficient matrix of the linear system associated with (8.7.18) or (8.7.20) is independent of time, we may either compute the matrix inverse at the outset and then solve the system at each step by simple matrix-vector multiplication, or employ efficient custom-made iterative solution algorithms.

8.7.3 Boundary conditions for the intermediate velocity

Next, we address the issue of boundary conditions for the intermediate velocities denoted by a single or double asterisk. The choice of these boundary conditions stems from a key observation: because of the homogeneous Neumann condition chosen for the projection function, the projection step introduces a tangential but not a normal component of the boundary velocity. Accordingly, the boundary conditions for the intermediate velocity should be such that the tangential velocity introduced in the projection step brings the total velocity to the specified physical value at the end of a complete step. In practice, this is done by estimating the magnitude of the intermediate slip velocity, and then improving the guess by iteration, as will be explained later in this section.

8.7.4 Flow in a cavity

The implementation of the numerical method involves further considerations that are best illustrated with reference to the familiar problem of flow in a two-dimensional cavity driven by a translating lid.

Homogeneous Neumann boundary condition for the projection function

Consider the numerical implementation of the condition of zero normal derivative of the projection function at the boundaries of the cavity illustrated in figure 8.5.1. Requiring $\partial\chi/\partial y = 0$ at the bottom and top walls and approximating the first derivative with a second-order forward or backward finite differences, we find

$$\left(\frac{\partial \chi}{\partial y}\right)_{i,1} \simeq \frac{-3\,\chi_{i,1} + 4\,\chi_{i,2} - \chi_{i,3}}{2\Delta y} = 0, \qquad (8.7.21)$$

and

$$\left(\frac{\partial \chi}{\partial y}\right)_{i,N_y+1} \simeq \frac{\chi_{i,N_y-1} - 4\,\chi_{i,N_y} + 3\,\chi_{i,N_y+1}}{2\Delta y} = 0. \tag{8.7.22}$$

Requiring $\partial\chi/\partial x = 0$ at the left and right walls and approximating the first derivative with a second-order forward or backward finite-difference formula, we find

$$\left(\frac{\partial \chi}{\partial x}\right)_{1,j} \simeq \frac{-3\,\chi_{1,j} + 4\,\chi_{2,j} - \chi_{3,j}}{2\Delta x} = 0, \tag{8.7.23}$$

and

$$\left(\frac{\partial \chi}{\partial x}\right)_{N_x+1,j} \simeq \frac{\chi_{N_x-1,j} - 4\,\chi_{N_x,j} + 3\,\chi_{N_x+1,j}}{2\Delta x} = 0. \tag{8.7.24}$$

These difference equations complement those arising from the discretization of the Poisson equation.

Compatibility condition for system (8.7.18)

The linear system descending from (8.7.18), accompanied by the homogeneous Neumann boundary conditions is singular, which means that it has either no solution or an infinite number of solutions depending on the right-hand side. If multiple solutions exist, any particular solution can be shifted by an arbitrary constant vector with equal elements. Correspondingly, the value of the projection function at the grid points can be shifted by a physically irrelevant constant. Reference to (8.7.15) ensures that this constant has no effect on the structure of the flow.

When the discrete divergence of the second intermediate velocity is computed using (8.7.16), the discrete form of the compatibility condition discussed at the end of section 8.6 is fulfilled and the linear system has a multiplicity of solutions. A solution can be found by assigning an arbitrary value to one of the unknowns, discarding one equation, and solving the rest of the equations for the remaining unknowns. Unfortunately, the numerical solution computed in this manner is contaminated by artificial oscillations described as odd-even coupling.

Compatibility condition for system (8.7.20)

The linear system descending from (8.7.20) accompanied by the homogeneous Neumann boundary conditions is also singular, reflecting the arbitrary level of the projection function. Unfortunately, when the discrete divergence of the second intermediate velocity field is computed using (8.7.16), the discrete form of the compatibility condition is not satisfied. Consequently, one equation of the linear system cannot be satisfied to machine precision.

Resisting the temptation to fudge the computation by discarding one arbitrary equation, we add a small term to the right-hand side of (8.7.20) and then adjust the magnitude of this term to satisfy the compatibility condition of the modified system of equations obtained in this fashion. If $\mathbf{A} \cdot \mathbf{x} = \mathbf{b}$ is the linear system corresponding to (8.7.20), then the modified system is

$$\mathbf{A} \cdot \mathbf{x} = \mathbf{b} + \epsilon\, \mathbf{c}, \tag{8.7.25}$$

where ϵ is an *a priori* unknown constant, and $\mathbf{c}$ is a constant vector that emerges by replacing the left-hand side of (8.7.17) with an arbitrary value, while retaining the linear equations implementing the homogeneous Neumann boundary conditions. Our objective is to adjust the value of the constant ϵ so that the system (8.7.25) has an infinite number of solutions.

In one approach, we work as follows:

1. First, we set the value of the last component of $\mathbf{x}$ to zero, discard the last equation of $\mathbf{A} \cdot \mathbf{x} = \mathbf{b}$, solve the remaining equations, and call the solution $\mathbf{x}^{(1)}$. Then we evaluate the difference between the left-hand side and the right-hand of the last equation, $r^{(1)}$.

2. Second, we set the value of the last component of $\mathbf{x}$ to zero, discard the last equation of $\mathbf{A} \cdot \mathbf{x} = \mathbf{c}$, solve the remaining equations, and call the solution $\mathbf{x}^{ref}$. Then evaluate the difference between the left-hand side and the right-hand of the last equation, r^{ref}.

3. The desired solution is

$$\mathbf{x} = \mathbf{x}^{(1)} + \epsilon\, \mathbf{x}^{ref}, \tag{8.7.26}$$

   where $\epsilon = -r^{(1)}/r^{ref}$.

A more rigorous approach involves removing from the right-hand side, $\mathbf{b}$, its projection on the eigenvector corresponding to the null eigenvalue of the transpose matrix, $\mathbf{A}^T$.

Boundary conditions for the intermediate velocity

The boundary conditions for the intermediate velocity must be such that the right-hand side of (8.7.15) is consistent with the specified physical boundary conditions at time $t + \Delta t$. Requiring that the left-hand side of (8.7.15) is zero over a stationary boundary, we obtain the boundary condition

$$\mathbf{u}^{**}_{i,j} = \frac{\Delta t}{\rho}\, (\nabla \chi)^n_{i,j}. \tag{8.7.27}$$

Because the projection function was required to satisfy the homogeneous Neumann boundary condition, the right-hand side of (8.7.27) has only a tangential component expressing a numerical wall slip.

An apparent difficulty in computing the tangential component of the intermediate velocity is that the right-hand side of (8.7.27) is not available during the convection–diffusion step. To circumvent this difficulty, we may approximate the projection function with that at the previous step, proceed with the projection step, and then improve the approximation by repeating the convection–diffusion step until the slip velocity has fallen below a sufficiently small threshold.

The following MATLAB code *cvt_pm* residing inside directory *11_fdm* of FDLIB performs the time integration using the projection method discussed in this section and animates the evolving velocity vector field:

```
%============
% code cvt_pm
%============

close all
clear all

%------------------------
% settings and parameters
%------------------------

Lx = 1.0;     % cavity length
Ly = 0.5;     % cavity depth
Nx = 32;      % grid size
Ny = 16;      % grid size
Dt = 0.001;   % time step
visc = 0.1;   % viscosity
dens = 1.0;   % density
lidU = 1.0;   % lid velocity
Nstep = 100;  % number of steps

%-----------------------
% parameters for solving
% the Poisson equation
%-----------------------

itermax = 50000;
tol = 0.000001;
relax = 0.2;

qleft = 0.0;  % Neumann BC
qright = 0.0;
qbot = 0.0;
qtop = 0.0;

Ishift = 1;

%--------------------
% parameters for slip
% velocity iterations
%--------------------

slipN = 50;  % max iterations
sliptol = 0.00001; % tolerance
sliprel = 1.0; % relaxation

%--------
% prepare
%--------

nu=visc/dens; % kinem visc
Dx  = Lx/Nx;
Dy  = Ly/Ny;
Dx2 = 2.0*Dx;
Dy2 = 2.0*Dy;
Dxs = Dx*Dx;
Dys = Dy*Dy;

Dtor = Dt/dens;

%------------
% lid velocity
%------------

for i=1:Nx+1
  Ulid(i) = lidU;
end

%---------------------
% define grid lines
% initialize time,
% velocity (U, V)
```

```
% and the projection
% function (T)
%--------------------

time = 0.0;

for j=1:Ny+1
 for i=1:Nx+1
   X(i,j) = (i-1.0)*Dx;
   Y(i,j) = (j-1.0)*Dy;
   U(i,j) = 0.0; % x vel
   V(i,j) = 0.0; % y vel
   T(i,j) = 0.0;  % projn fnc
 end
end

for i=1:Nx+1
   U(i,Ny+1) = Ulid(i);
end

%------------------
% naive velocity
% boundary conditions
%--------------------

for i=1:Nx+1
  BCxt(i) = Ulid(i);  % top
  BCxb(i) = 0.0;      % bottom
end

for j=1:Ny+1
  BCyl(j) = 0.0; % left wall
  BCyr(j) = 0.0; % right wall
end

%-------------
% time stepping
%-------------

%==============
for step=1:Nstep
%==============

%---
% animation
%---

if(step==1)
  Handle1 = quiver(X,Y,U,V);
  set(Handle1, ...
     'erasemode','xor');
  axis ([0, Lx, 0, Ly])
  axis equal
  set(gca,'fontsize',15)
  xlabel('x','fontsize',15)
  ylabel('y','fontsize',15)
  hold on
  plot([0,Lx,Lx,0,0] ...
      ,[0,0,Ly,Ly,0]);
else
  set(Handle1,'UData' ...
     ,U,'VData',V);
  pause(0.01)
  drawnow
end

%-------------------------
% initialize
% the intermediate velocity
%------------------------

for j=1:Ny+1
  for i=1:Nx+1
   UU(i,j) = U(i,j);
   VV(i,j) = V(i,j);
  end
end

%===========================
% Perform inner iterations
% for the projection function
% to satisfy the no-slip
% boundary condition
%===========================

for inner=1:slipN

%---
% zero the tridiagonal matrix
%---

 for i=1:Nx-1
   atr(i) = 0.0;
   btr(i) = 0.0;
   ctr(i) = 0.0;
 end
```

```
%-----------------------------
% Integrate the conv-diff
% equation in the x direction
% using the Crank-Nicolson
% method. Advance the velocity
% from u^n to u*
%-----------------------------

  AL = nu*Dt/Dxs;

%--------------
  for j=2:Ny   % run over rows
%----

  for i=2:Nx
    % cell Re:
    RC = U(i,j)*Dt/Dx;
    C1 =   RC + 2.0*AL;
    C2 =          4.0*(1.0-AL);
    C3 =  -RC + 2.0*AL;
    ctr(i-1) =  -RC - 2.0*AL;
    atr(i-1) =   4.0*(1.0+AL);
    btr(i-1) =   RC - 2.0*AL;
    Blu(i-1) = C1*U(i-1,j) ...
     + C2*U(i,j) ...
     + C3*U(i+1,j);  % RHS
    Blv(i-1) = C1*V(i-1,j) ...
     + C2*V(i,j) ... % RHS
     + C3*V(i+1,j);
  end

  Blv(1)     = Blv(1) ...
    - ctr(1)*BCyl(j);
  Blv(Nx-1) = Blv(Nx-1) ...
    - btr(Nx-1)*BCyr(j);

  Xsu = thomas ...
    (Nx-1,atr,btr,ctr,Blu);  % x
  Xsv = thomas  ...
    (Nx-1,atr,btr,ctr,Blv);  % y

  for k=1:Nx-1
   UU(k+1,j) = Xsu(k);
   VV(k+1,j) = Xsv(k);
  end

%---
   end  % of run over rows
%----------

  for i=1:Nx-1     % reset
    atr(i) = 0.0;
    btr(i) = 0.0;
    ctr(i) = 0.0;
  end

%-----------------------------
% integrate the conv-diffusion
% equation in the y direction
% using the Crank-Nicolson
% method. Advance the velocity
% from u* to u**
%-----------------------------

 AL = nu*Dt/Dys;

%---
 for i=2:Nx % over columns
%---

  for j=2:Ny  % bottom to top
   RC = V(i,j)*Dt/Dy; % cell Re
   C1 =  RC +2.0*AL;
   C2 =        4.0*(1.0-AL);
   C3 = -RC +2.0*AL;
   ctr(j-1) = -RC -2.0*AL;
   atr(j-1) =       4.0*(1.0+AL);
   btr(j-1) =  RC -2.0*AL;
   Blu(j-1) = C1*UU(i,j-1) ...
    + C2*UU(i,j) ...
    + C3*UU(i,j+1); % RHS
   Blv(j-1) = C1*VV(i,j-1) ...
    + C2*VV(i,j) ...
    + C3*VV(i,j+1); % RHS
  end

  % Boundary conditions:

  Blu(1)    = Blu(1) ...
    -ctr(1)*BCxb(i);
  Blu(Ny-1) = Blu(Ny-1) ...
    -btr(Ny-1)*BCxt(i);

  Xsu = thomas(Ny-1 ...
    ,atr,btr,ctr,Blu); % x
  Xsv = thomas(Ny-1 ...
    ,atr,btr,ctr,Blv); % y
```

```
  for k=1:Ny-1
    UU(i,k+1) = Xsu(k);
    VV(i,k+1) = Xsv(k);
  end

%---
  end   % of run over columns
%---

%------------------------
% Compute the intermediate
% compressibility
% by centered differences
%
% Divus = Div u**
%------------------------

% initialize

  for j=1:Ny+1
    for i=1:Nx+1
      Divus(i,j) = 0.0;
    end
  end

% interior nodes

  for i=2:Nx
    for j=2:Ny
      DuDx = (UU(i+1,j) ...
             -UU(i-1,j))/Dx2;
      DvDy = (VV(i,j+1) ...
             -VV(i,j-1))/Dy2;
      Divus(i,j) = DuDx+DvDy;
     end
   end

% left wall

  for j=1:Ny+1
   DuDx = (-3.0*UU(1,j) ...
      +4.0*UU(2,j) ...
      -UU(3,j))/Dx2;
   DvDy = 0.0;
   Divus(1,j) = DuDx+DvDy;
  end

% save for corners

  save11l = Divus(1,j);
  save12l = Divus(1,Ny+1);

% bottom wall

  for i=1:Nx+1
    DuDx = 0.0;
    DvDy = (-3.0*VV(i,1) ...
      +4.0*VV(i,2)-VV(i,3))/Dy2;
    Divus(i,1) = DuDx+DvDy;
  end

% save for corners

  save11b = Divus(1,1);
  save21b = Divus(Nx+1,1);

% right wall

  for j=1:Ny+1
     DuDx = (3.0*UU(Nx+1,j) ...
     -4.0*UU(Nx,j) ...
     +UU(Nx-1,j))/Dx2;
     DvDy = 0.0;
     Divus(Nx+1,j) = DuDx+DvDy;
  end

% save for corners

  save21r = Divus(Nx+1,1);
  save22r = Divus(Nx+1,Ny+1);

% top wall

   for i=1:Nx+1
%    DuDx = (Ulid(i+1) ...
%       -Ulid(i-1))/Dx2;
     DuDx = 0.0;
     DvDy = (3.0*VV(i,Ny+1) ...
      -4.0*VV(i,Ny) ...
      +VV(i,Ny-1))/Dy2;
     Divus(i,Ny+1) = DuDx+DvDy;
   end

% save for corners

   save12t = Divus(1,   Ny+1);
   save22t = Divus(Nx+1,Ny+1);
```

```
% corners by averaging

   Divus(1,      1) ...
   = 0.5*(save11l+save11b);
   Divus(1,    Ny+1) ...
   = 0.5*(save12l+save12t);
   Divus(Nx+1,    1) ...
   = 0.5*(save21b+save21r);
   Divus(Nx+1,Ny+1) ...
   = 0.5*(save22r+save22t);

%-----------------------
% Solve for the proj fnc
% by Gauss-Siedel (GS)
%-----------------------

%---
% source term
%-----

 for i=1:Nx+1
  for j=1:Ny+1
    source(i,j) ...
    = -Divus(i,j)/Dtor;
  end
 end

 [T,iter,Iflag] = pois_gs_nnnn ...
 ...
 (Nx,Ny,Dx,Dy,source ...
 ,itermax,tol,relax ...
 ,qleft,qright,qbot ...
 ,qtop,T,Ishift);

  if(Iflag==0)
   disp "Poisson solver"
   disp "did not converge"
   break
  end

%--------------------------
% Project the velocity
% at all nodes
% except at the corner nodes
%--------------------------

%---
% interior nodes
%----

for i=2:Nx
 for j=2:Ny
  DTDx = (T(i+1,j)-T(i-1,j))/Dx2;
  DTDy = (T(i,j+1)-T(i,j-1))/Dy2;
  UU(i,j) = UU(i,j) - Dtor*DTDx;
  VV(i,j) = VV(i,j) - Dtor*DTDy;
 end
end

%-------------
% Lower boundary
%
% Use forward differences with
% special treatment of the
% near-corner nodes
%-------------

for i=2:Nx
 DTDy = 0.0;
 if(i==2)
   DTDx = (-3.0*T(2,1) ...
   +4.0*T(3,1)-T(4,1))/Dx2;
 elseif(i==Nx)
   DTDx =-(-3.0*T(Nx,1) ...
   +4.0*T(Nx-1,1) ...
       -T(Nx-2,1))/Dx2;
 else
   DTDx = (T(i+1,1) ...
          -T(i-1,1))/Dx2;
 end
 UU(i,1) = BCxb(i)-Dtor*DTDx;
 VV(i,1) =        -Dtor*DTDy;
end

%-------------
% Upper boundary
%
% Use backward differences
% special treatment for the
% near-corner nodes
%-------------

for i=2:Nx
 DTDy = 0.0;
 if(i==2)
   DTDx = (-3.0*T(2,Ny+1) ...
   +4.0*T(3,Ny+1) ...
   -T(4,Ny+1))/Dx2;
```

```
 elseif(i==Nx)
   DTDx =-(-3.0*T(Nx,Ny+1) ...
   +4.0*T(Nx-1,Ny+1) ...
   -T(Nx-2,Ny+1))/Dx2;
 else
   DTDx = (T(i+1,Ny+1) ...
          -T(i-1,Ny+1))/Dx2;
 end
 UU(i,Ny+1) = BCxt(i)-Dtor*DTDx;
 VV(i,Ny+1) =         -Dtor*DTDy;
end

%------------
% Left boundary
%
% Use forward differences
% special treatment for the
% near-corner nodes
%------------

for j=2:Ny
 DTDx = 0.0;
 if(j==2)
   DTDy = (-3.0*T(1,2) ...
   +4.0*T(1,3)-T(1,4))/Dy2;
 elseif(j==Ny)
   DTDy =-(-3.0*T(1,Ny) ...
   +4.0*T(1,Ny-1) ...
       -T(1,Ny-2))/Dy2;
 else
   DTDy = (T(1,j+1)...
          -T(1,j-1))/Dy2;
 end
 UU(1,j) =          -Dtor*DTDx;
 VV(1,j) = BCyl(j)-Dtor*DTDy;
end

%-------------
% Right boundary
%
% Use backward differences
% special treatment for the
%  near-corner nodes
%-------------

for j=2:Ny
  DTDx = 0.0;
  if(j==2)
   DTDy = (-3.0*T(Nx+1,2) ...
          +4.0*T(Nx+1,3) ...
              -T(Nx+1,4))/Dy2;
  elseif(j==Ny)
   DTDy =-(-3.0*T(Nx+1,Ny) ...
   +4.0*T(Nx+1,Ny-1) ...
    -T(Nx+1,Ny-2))/Dy2;
  else
    DTDy = (T(Nx+1,j+1) ...
      -T(Nx+1,j-1))/Dy2;
  end
  UU(Nx+1,j) =         -Dtor*DTDx;
  VV(Nx+1,j) = BCyr(j)-Dtor*DTDy;
end

%------------------------------
% Compute the wall slip velocity
% and modify the BC
% for the intermediate
% (star) velocities
%------------------------------

slipmax = 0.0;

% top and bottom:

for i=1:Nx+1
  cor = UU(i,Ny+1)-Ulid(i);
  if(abs(cor)>slipmax)
    slipmax = cor;
  end
  BCxt(i) = BCxt(i)-sliprel*cor;
  cor = UU(i,1);
  if(abs(cor)>slipmax)
    slipmax = cor;
  end
  BCxb(i) = BCxb(i)-sliprel*cor;
end

% left and right:

for j=1:Ny+1
 cor = VV(1,j);
 if(abs(cor)>slipmax)
  slipmax = cor;
 end
 BCyl(j) = BCyl(j)-sliprel*cor;
 corr = VV(Nx+1,j);
 if(abs(cor)>slipmax)
  slipmax = cor;
```

```
 end
 BCyr(j) = BCyr(j)-sliprel*cor;
end

slipmax
if(slipmax<sliptol) break; end

end % of inner iterations

%-------------------
% Update velocity
% to the final value
%-------------------

  for j=1:Ny+1
    for i=1:Nx+1
      U(i,j) = UU(i,j);
      V(i,j) = VV(i,j);
    end
  end

%-----------
% Reset time
%-----------

  time = time + Dt

%===========
 end % of time stepping
%===========

figure
mesh(X,Y,U);
set(gca,'fontsize',15)
xlabel('x','fontsize',15)
ylabel('y','fontsize',15)
zlabel('u','fontsize',15)

figure
mesh(X,Y,V);
xlabel('x','fontsize',15)
ylabel('y','fontsize',15)
zlabel('V','fontsize',15)

figure
mesh(X,Y,T);
xlabel('x','fontsize',15)
ylabel('y','fontsize',15)
zlabel('T','fontsize',15)
```

The code calls the MATLAB function *thomas* listed in section 8.2 to solve tridiagonal systems of equations using the Thomas algorithm. The code also calls the following MATLAB function *pois_gs_nnnn* to solve the Poisson equation for the projection function, subject to the Neumann boundary condition along the four sides:

```
function [f,iter,Iflag] = pois_gs_nnnn ...
...
    (Nx,Ny,Dx,Dy,g,itermax,tol,relax,qleft ...
      ,qright,qbot,qtop,f,Ishift)

%-----------------------------------------
% Solution of Poisson's equation
% in a rectangular domain
% with the uniform Neumann boundary condition
% along the four sides:
%
% bottom:  df/dy = qbot
% top:  df/dy = -qtop
% left:  df/dx = qleft
% right:  df/dx = -qright
%
% The solution is found by
```

```
% point Gauss-Seidel iterations
%----------------------------------------

%--------
% prepare
%--------

Dx2 = 2.0*Dx; Dy2 = 2.0*Dy;
Dxs = Dx*Dx; Dys = Dy*Dy;
beta = Dxs/Dys; beta1 = 2.0*(beta+1.0);
Iflag = 0;     % convergence flag, 1 indicates convergence

%-----------------------
% Gauss-Seidel iterations
%-----------------------

for iter=1:itermax

%-----------------------
% update nodes row-by-row
%-----------------------

fsv = f;
cormax = 0.0;

%---
% interior nodes
%---

for j=2:Ny
  for i=2:Nx
    res = (f(i+1,j)+f(i-1,j)+beta*(f(i,j+1)+f(i,j-1)) ...
    + Dxs*g(i,j))/beta1-f(i,j);
    f(i,j) = f(i,j) + relax*res;
  end
end

%-------------
% left boundary
%-------------

i=1;
for j=2:Ny
  res = (2*f(i+1,j)-Dx2*qleft+beta*(f(i,j+1)+f(i,j-1)) ...
    +Dxs*g(i,j))/beta1 - f(i,j);
  f(i,j) = f(i,j) + relax*res;
  end
end
```

```
% corner points:

j=1;
res = (2*f(i+1,j)-Dx2*qleft ...
    +beta*(f(i,j+1)+f(i,j+1)-Dy2*qbot) ...
    +Dxs*g(i,j))/beta1 - f(i,j);
f(i,j) = fsave(i,j) + relax*res;

j=Ny+1;
res = (2*f(i+1,j)-Dx2*qleft ...
    +beta*(f(i,j-1)+f(i,j-1)-Dy2*qtop) ...
    +Dxs*g(i,j))/beta1 - f(i,j);
f(i,j) = fsave(i,j) + relax*res;

%---------------
% right boundary
%---------------

i=Nx+1;
for j=2:Ny
  res = (2*f(i-1,j)+-Dx2*qright+beta*(f(i,j+1)+f(i,j-1)) ...
    +Dxs*g(i,j))/beta1 -f(i,j);
  f(i,j) = fsave(i,j) + relax*res;
end

% corner points:

j=1;
res = (2*f(i-1,j)-Dx2*qright ...
    +beta*(f(i,j+1)+f(i,j+1)-Dy2*qbot) ...
    +Dxs*g(i,j))/beta1 - f(i,j);
f(i,j) = fsave(i,j) + relax*res;

j=Ny+1;
res = (2*f(i-1,j)-Dx2*qright)+beta*(2*f(i,j-1)-Dy2*qtop) ...
    +Dxs*g(i,j))/beta1 - f(i,j);
f(i,j) = fsave(i,j) + relax*res;

%----------------
% bottom boundary
%----------------

j=1;
for i=2:Nx
  res = (f(i+1,j)+f(i-1,j)+beta*(2*f(i,j+1)-Dy2*qbot) ...
      +Dxs*g(i,j))/beta1 - f(i,j);
  f(i,j) = fsave(i,j) + relax*res;
end
```

```
%------------
% top boundary
%------------

j=Ny+1;
for i=2:Nx
  res = (f(i+1,j)+f(i-1,j)+beta*(2*f(i,j-1)-Dy2*qtop) ...
      +Dxs*g(i,j))/beta1 - f(i,j);
  f(i,j) = fsave(i,j) + relax*res;
end

%------
% shift
%------

if(Ishift==1)
  shift = f(Nx/2,Ny/2);
  for i=1:Nx+1
    for j=1:Ny+1
      f(i,j) = f(i,j)-shift;
    end
  end
end

%------------------
% maximum correction
%------------------

cormax = 0;

for i=1:Nx+1
  for j=1:Ny+1
    cor = abs(f(i,j)-fsv(i,j));
    if(abs(cor)>cormax) cormax = cor; end
  end
end

%-----
% stopping check
%-----

if(cormax<tol)
  Iflag = 1;
  break
end

%---
end     % of iterations
%---
```

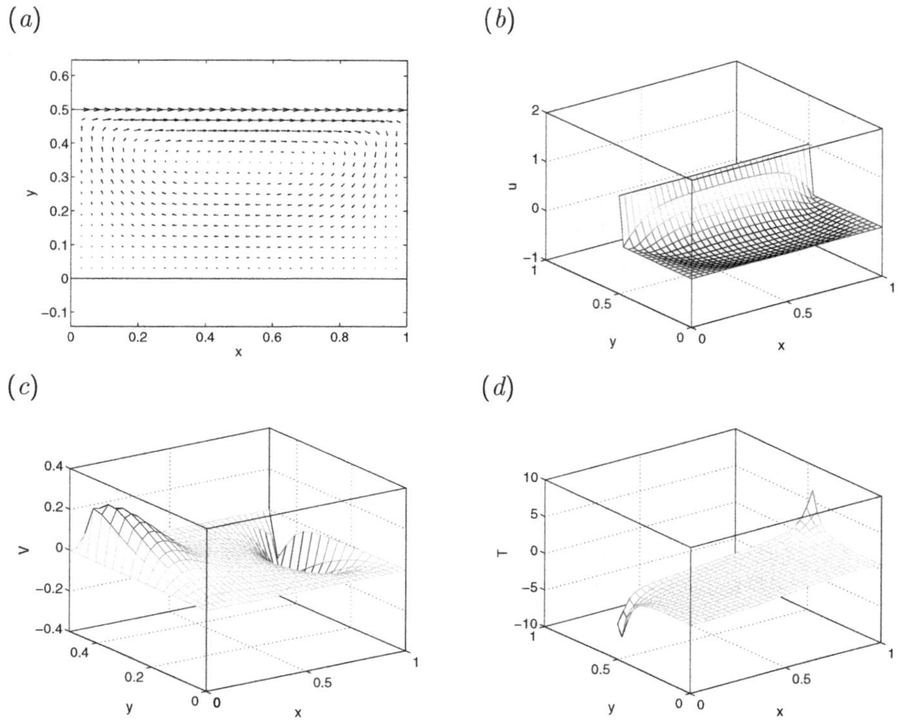

Figure 8.7.1 Flow in a rectangular cavity computed using a projection method. The graphs show (a) the velocity vector field, (b) the x velocity component, (c) the y velocity component, and (d) the projection function at steady state.

```
%-----
% done
%-----

return
```

The graphics produced by the code for the parameter values implemented in the code is shown in figure 8.7.1. In the early stages of the motion, the flow is similar to that generated by the impulsive translation of a plate in a semi-infinite fluid. At later times, a fully developed recirculating flow is established.

8.7.5 Computation of the pressure

Two methods are available for extracting the pressure field, if desired. The first method involves combining equations (8.7.8), (8.7.11), and (8.7.15)–or any other appropriate set of equations–to produce a relationship between $\mathbf{u}^n$ and $\mathbf{u}^{(n+1)}$. Requiring that, in the limit as Δt tends to zero, this relationship reduces to a spatially discretized version of the Navier-Stokes equation, we derive an expression for an effective pressure. If the boundary conditions satisfied by the effective pressure are consistent with the Neumann boundary condition satisfied by the physical pressure, then the former is accepted as an approximation to the latter.

The second method involves substituting the computed velocity field in the Navier-Stokes equation and solving the resulting equation for the pressure subject to the Neumann boundary condition, as discussed in section 8.6.2.

Problem

8.7.1 *Singular system for the projection function*

Show that equation (8.7.26) provides us with the solution of (8.7.25) subject to the homogeneous Neumann boundary condition.

Computer Problem

8.7.2 *Developing flow in a cavity*

Run the code *cvt_pm* located in directory *11_fdm* of FDLIB for a cavity with aspect ratio $L_x/L_y = 8$ and other parameter values of your choice. Prepare velocity vector fields and discuss the structure of the flow and the performance of the numerical method.

8.8 Staggered grids

The derivation of boundary conditions for the pressure can be bypassed by using a staggered grid consisting of two superposed displaced grids where the pressure or the velocity is defined. We illustrate the methodology with reference to steady two-dimensional Stokes flow in a rectangular cavity driven by a moving lid. The motion of the fluid is governed by the equations of Stokes flow, including the continuity equation and the Stokes equation,

$$\nabla \cdot \mathbf{u} = 0, \qquad -\nabla p + \mu \nabla^2 \mathbf{u} = \mathbf{0}, \tag{8.8.1}$$

where p is the dynamic pressure incorporating hydrostatic variations.

Pressure and velocity nodes

The staggered grid consists of two interwoven grids parametrized by two index pairs, (i, j) and (i', j'), as illustrated in figure 8.8.1 where the primed

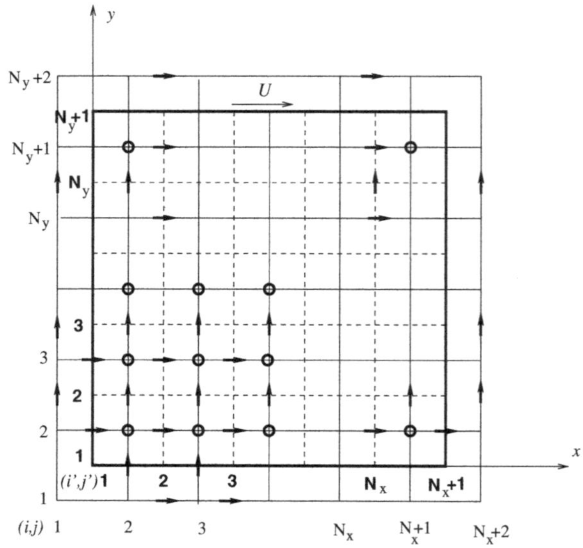

Figure 8.8.1 Illustration of a staggered grid for computing two-dimensional flow in a rectangular cavity. The pressure is defined at nodes indicated by the circles, the x velocity component is defined at nodes indicated by the horizontal arrows, and the y velocity component is defined at nodes indicated by the vertical arrows.

indices are printed in bold. The grid lines of the primary grid are represented by the solid lines, and the grid lines of the secondary grid are represented by the broken lines. The secondary grid conforms with the physical boundaries of the flow.

Discrete values of the pressure are assigned to the primary nodes, (i, j), defined by the intersection of the solid lines, shown as circles. Discrete values of the x component of the velocity are defined at the intersection of horizontal primary grid lines and vertical secondary grid lines, (i', j), shown as horizontal arrows. Discrete values of the y component of the velocity are defined at the intersection of vertical primary grid lines and horizontal secondary grid lines, (i, j'), shown as vertical arrows.

Finite-difference equations

A distinguishing feature of the staggered grid method is that the governing equations are enforced at different nodes. For convenience, we denote u_x by u and u_y by v. Enforcing the continuity equation at the (i, j) pressure node and approximating the partial derivative with central differences, we obtain

$$\frac{u_{i,j} - u_{i-1,j}}{\Delta x} + \frac{v_{i,j} - u_{i,j-1}}{\Delta y} = 0, \qquad (8.8.2)$$

for $i = 2, \ldots, N_x + 1$ and $j = 2, \ldots, N_y + 1$. Applying the x component of the Stokes equation at the (i, j) x-velocity node and introducing similar difference approximations, we obtain

$$\frac{p_{i+1,j} - p_{i,j}}{\Delta x} = \mu\,(\frac{u_{i-1,j} - 2u_{i,j} + u_{i+1,j}}{\Delta x^2} + \frac{u_{i,j-1} - 2u_{i,j} + u_{i,j+1}}{\Delta y^2}), \qquad (8.8.3)$$

for $i = 2, \ldots, N_x$ and $j = 2, \ldots, N_y + 1$. Applying the y component of the Stokes equation at the (i, j), y-velocity nodes and working in a similar fashion, we obtain

$$\frac{p_{i,j+1} - p_{i,j}}{\Delta y} = \mu\,(\frac{v_{i-1,j} - 2v_{i,j} + v_{i+1,j}}{\Delta x^2} + \frac{v_{i,j-1} - 2v_{i,j} + v_{i,j+1}}{\Delta y^2}), \qquad (8.8.4)$$

for $i = 2, \ldots, N + 1$ and $j = 2, \ldots, N$.

Boundary conditions

Unphysical x velocity components are required at the horizontal lines $j = 1$ and $j = N_y + 2$ for the computation of the second y derivative of u near the top and bottom boundaries. Corresponding unphysical y velocity components are required at the vertical levels $i = 1$ and $N_x + 2$ for the computation of the second x derivative of v near the left and right boundaries. These exterior velocities are computed by extrapolation to satisfy the boundary conditions at the physical levels $i' = 1$, $i' = N_x + 1$, $j' = 1$, and $j' = N_y + 1$.

For example, approximating u with a parabola near the lid located at $y = a_y$, and enforcing the no-slip boundary condition $u = U$, we find

$$u = U + A\,(y - a_y)^2 + B\,(y - a_y), \qquad (8.8.5)$$

where U is the lid velocity, and A, B are unknown coefficients. Applying this expression at three neighboring grid levels, we obtain,

$$u_{i,N_y} = U + A\,\frac{9}{4}\,h^2 - B\,\frac{3}{2}\,h, \qquad u_{i,N_y+1} = U + A\,\frac{1}{4}\,h^2 - B\,\frac{1}{2}\,h,$$
$$u_{i,N_y+2} = U + A\,\frac{1}{4}\,h^2 + B\,\frac{1}{2}\,h, \qquad (8.8.6)$$

where $h = \Delta y$. Eliminating A and B, we find

$$u_{i,N_y+2} = \frac{1}{3}\,(8\,U - 6\,u_{i,N_y+1} + u_{i,N_y}). \qquad (8.8.7)$$

Similar expressions are derived for the other external velocities.

The preceding difference equations provide us with a complete system of linear algebraic equations for the nodal velocities and pressures. The system can be compiled and solved at once, as illustrated in the following MATLAB code *cvt_stag* residing in directory *11_fdm* of FDLIB:

```
%==========
% steady Stokes flow in a rectangular cavity
% occupying 0<x<ax, 0<y<ay
% computed on a staggered Cartesian grid
%======

ax = 1.0; ay = 0.75;

Nx = 24;    % x divisions
Ny = 16;    % y divisions
mu = 1.0;    % viscosity
Ulid = 1;    % lid velocity

%---
% prepare
%---

Dx = ax/Nx; Dy = ay/Ny;
Dxs = Dx*Dx; Dys = Dy*Dy;

%---
% initialize
%---

mats = Nx*Ny+Nx*(Ny-1)+Ny*(Nx-1);    % matrix size

u = zeros(Nx+1,Ny+1);    % impulse matrices
v = zeros(Nx+1,Ny+1);
p = zeros(Nx+1,Ny+1);

%=======
% compile the coefficient matrix
%=======

Jc = 0;    % counter of impulses

for ipass=1:3    % impulse for u,v,p

if(ipass==1)
  klim = Nx; llim = Ny+1;    % u vel
elseif(ipass==2)
  klim = Nx+1; llim = Ny;    % v vel
elseif(ipass==3)
  klim = Nx+1; llim = Ny+1;    % pressure
end

%---
for l=2:llim    % scan
  for k=2:klim
```

```
  Jc=Jc+1;
%---

  if(ipass==1);
    u(k,l) = 1.0;
  elseif(ipass==2);
    v(k,l) = 1.0;
  elseif(ipass==3);
    p(k,l) = 1.0;
  end

%---
% boundary conditions
%---
  for ii=2:Nx
    u(ii,1)= (-6*u(ii,2)+u(ii,3))/3.0;
    u(ii,Ny+2) = (-6*u(ii,Ny+1)+u(ii,Ny))/3.0;
  end
  for jj=2:Ny
    v(1,jj) = (-6*v(2,jj)+v(3,jj))/3.0;
    v(Nx+2,jj) = (-6*v(Nx+1,jj)+v(Nx,jj))/3.0;
  end
%---

  Ic=0;   % counter of equations

  for j=2:Ny+1
    for i=2:Nx   % x Stokes
      Ic = Ic+1;
      MAT(Ic,Jc) = -(p(i+1,j)-p(i,j))/Dx ...
        +mu*(u(i-1,j)-2*u(i,j)+u(i+1,j))/Dxs ...
        +mu*(u(i,j-1)-2*u(i,j)+u(i,j+1))/Dys;
      MAT(Ic,Jc) = MAT(Ic,Jc)*Dxs;
    end
  end

  for j=2:Ny
    for i=2:Nx+1   % y Stokes
      Ic = Ic+1;
      MAT(Ic,Jc) = -(p(i,j+1)-p(i,j))/Dy ...
        +mu*(v(i-1,j)-2*v(i,j)+v(i+1,j))/Dxs ...
        +mu*(v(i,j-1)-2*v(i,j)+v(i,j+1))/Dys;
      MAT(Ic,Jc) = MAT(Ic,Jc)*Dxs;
    end
  end
  for j=2:Ny+1
    for i=2:Nx+1   % continuity
      Ic = Ic+1;
        MAT(Ic,Jc) = (u(i,j)-u(i-1,j))/Dx+(v(i,j)-v(i,j-1))/Dy;
```

```
        MAT(Ic,Jc) = MAT(Ic,Jc)*Dx;
      end
    end

    u = zeros(Nx+1,Ny+1);    % reset
    v = zeros(Nx+1,Ny+1);
    p = zeros(Nx+1,Ny+1);

    end   % of for k
    end   % of for l
  end   % of ipass

%====
% set the right-hand side
%===

for i=1:mats
  RHS(i) = 0.0;
end

Ic = (Ny-2)*(Nx-1);
for i=2:Nx   % x Stokes
  Ic = Ic+1;
  RHS(Ic) = RHS(Ic)-8*mu*Ulid*Dxs/3/Dys;
end

%====
% set p=0 at the last node
% and find the solution
%====

SOL = RHS(1:mats-1)/MAT(1:mats-1,1:mats-1)';
SOL(mats) = 0.0;

%====
% distribute the solution
%====

Ic = 0;
for j=2:Ny+1
  for i=2:Nx
    Ic = Ic+1;
    uvel(i,j) = SOL(Ic);
  end
end
for j=2:Ny
  for i=2:Nx+1
    Ic=Ic+1;
    vvel(i,j) = SOL(Ic);
```

```
  end
end
for j=2:Ny+1
  for i=2:Nx+1
    Ic = Ic+1;
    pressure(i,j) = SOL(Ic);
  end
end

%=====
% interpolate the velocity field at the pressure nodes
% for plotting purposes
%=====

for j=2:Ny+1
  uvel(1,j) = 0;
  uvel(Nx+1,j) = 0;
  for i=2:Nx+1
    uint(i,j) = 0.5*(uvel(i,j)+uvel(i-1,j));
  end
end

for i=2:Nx+1
  vvel(i,1) = 0;
  vvel(i,Ny+1) = 0;
  for j=2:Ny+1
    vint(i,j) = 0.5*(vvel(i,j)+vvel(i,j-1));
  end
end

%====
% prepare a velocity vector plot
%====

plot([0 ax ax 0 0],[0 0 ay ay 0])
hold on
axis([-0.1*ax 1.1*ax, -0.1*ax 1.1*ax])
xlabel('x');ylabel('y')

for j=2:Ny+1
  ylevel = 0.5*Dy+(j-2)*Dy;
  for i=2:Nx+1
    xlevel = 0.5*Dx+(i-2)*Dx;
    xarrow = 0.5*Dx*uint(i,j)/Ulid;
    yarrow = 0.5*Dx*vint(i,j)/Ulid;
    arrow = arrow_cp(xlevel,ylevel,xarrow,yarrow);
    plot(arrow(:,1),arrow(:,2));
  end
end
```

A velocity vector field produced by the code is shown in figure 8.8.2.

The implementation of the numerical procedure features the following details:

- The linear system is compiled so that the first block of equations encapsulates the x component of the Stokes equation, the second block encapsulates the y component of the Stokes equation, and the third block encapsulates the continuity equation.
- The first block of unknowns encapsulates the x velocity components, the second block encapsulates the y velocity components, and the third block encapsulates the pressure.
- To generate the coefficient matrix of the linear system, we sequentially set one nodal value of the x velocity component, y velocity component, or pressure to one, while holding all other values to zero. The corresponding column of the coefficient matrix is given by the residual of the governing equations scanned in the aforementioned order.
- Since the pressure field is defined up to an arbitrary constant, the pressure is set arbitrarily to zero at the last pressure node. One equation expressing the discrete implementation of the continuity equation is then discarded to balance the number of equations to the number of unknowns.
- The code calls the custom-made MATLAB function *arrow_cp* near the end to draw nice-looking arrows.

By using a staggered grid, we have circumvented the explicit derivation of boundary conditions for the pressure. Happily, a careful analysis shows that implicit in the numerical formulation is the Neumann boundary condition that arises by projecting the equation of motion normal to the boundaries.

Unfortunately, the staggered-grid method becomes considerably more involved and prohibitively expensive when applied to grids defined in general curvilinear coordinates.

Problem

8.8.1 *Coefficient matrix*

(*a*) Present a pictorial depiction of the coefficient matrix arising from the difference equations, and identify large non-zero blocks.

(*b*) Verify that the coefficient matrix is singular.

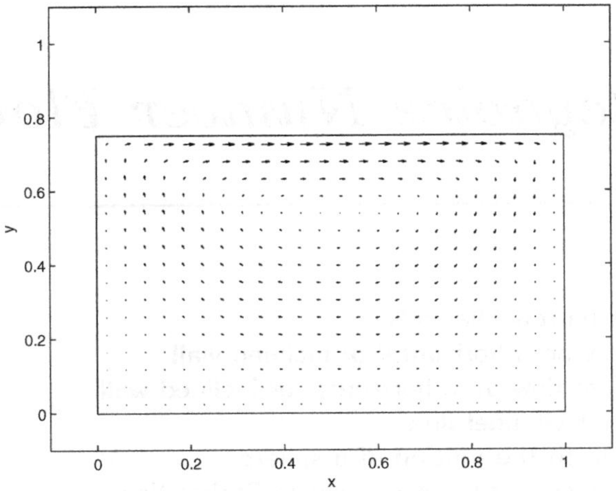

Figure 8.8.2 Velocity vector field of Stokes flow in a rectangular cavity produced by code *cvt_stag*.

Computer Problem

8.8.2 *Pressure distribution*

Modify the graphics portion of code *cvt_stag* to visualize the pressure field. Present and discuss contour plots of the pressure field for flow in a square cavity.

Low Reynolds Number Flow 9

Newton's second law of motion requires that the rate of change of momentum of a fluid parcel must be balanced by the body force exerted over the parcel volume and by the surface force exerted on the parcel boundary. Under certain conditions, the rate of change of momentum of the parcel is small compared to the body and surface force, and can be neglected without introducing serious error. The flow is then governed by a balance between the volume force and the surface force. Physically, this occurs when the fluid viscosity is high, when the fluid density is small, when the velocity changes rapidly over a small distance yielding a sharp spatial gradient, or when the convection velocity of a fluid parcel is sufficiently small. The formal requirement for fluid inertia to be negligible is that a properly defined Reynolds number is sufficiently small. How small it should be, depends on the particular problem under consideration. In this chapter, we consider a family of flows occurring at small Reynolds numbers and discuss the solution of simplified systems of governing equations that arise by dropping the inertial terms in the equation of motion. The simplification will allow us to address a multitude of physical and engineering problems and derive solutions by a host of analytical and numerical methods.

9.1 Flow in narrow channels

We begin by considering steady, nearly unidirectional flow in a two-dimensional channel confined between a gently sloped upper wall and a perfectly flat lower wall as illustrated in figure 9.1.1. The flow is driven by the translation of the upper wall parallel to itself with velocity V, and possibly by a pressure drop imposed across the length of the channel extending between $x = 0$ and L.

C. Pozrikidis, *Fluid Dynamics: Theory, Computation, and Numerical Simulation*,
DOI: 10.1007/978-0-387-95871-2_9,

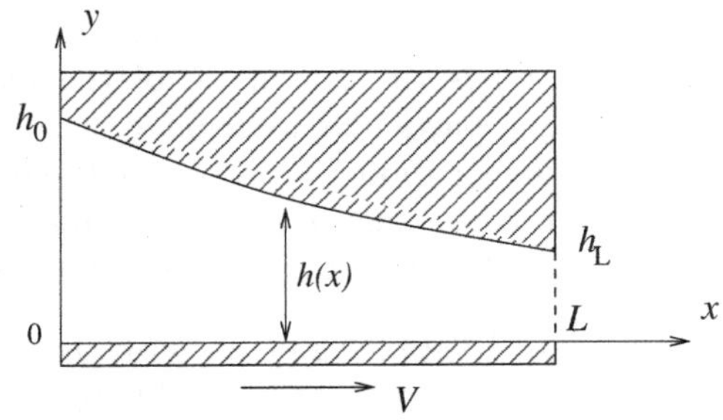

Figure 9.1.1 Lubrication flow in a slider bearing with a flat lower wall and a curved upper wall.

In the Cartesian system of coordinates depicted in figure 9.1.1, the lower wall is located at $y = 0$ and the upper wall is located at $y = h(x)$. If the channel width is constant, we obtain steady unidirectional flow in a channel with parallel-sided walls discussed in section 7.1. Our objective is to compute the velocity and pressure field, and then evaluate the force exerted on the lower wall for an arbitrary upper wall profile.

9.1.1 Governing equations

The motion of the fluid is governed by the continuity equation

$$\frac{\partial u_x}{\partial x} + \frac{\partial u_y}{\partial y} = 0, \tag{9.1.1}$$

and by the steady version of the Navier-Stokes equation for an incompressible fluid whose x and y scalar components read

$$\rho\,(u_x \frac{\partial u_x}{\partial x} + u_y \frac{\partial u_x}{\partial y}) = -\frac{\partial p}{\partial x} + \mu\,(\frac{\partial^2 u_x}{\partial x^2} + \frac{\partial^2 u_x}{\partial y^2}) + \rho\, g_x, \tag{9.1.2}$$

and

$$\rho\,(u_x \frac{\partial u_y}{\partial x} + u_y \frac{\partial u_y}{\partial y}) = -\frac{\partial p}{\partial y} + \mu\,(\frac{\partial^2 u_y}{\partial x^2} + \frac{\partial^2 u_y}{\partial y^2}) + \rho\, g_y. \tag{9.1.3}$$

Our first task is to show that, under certain conditions, several terms in equations (9.1.2) and (9.1.3) are negligible compared to other terms and can be discarded yielding a simplified system of governing equations known as the equations of lubrication flow.

9.1.2 Scaling

We begin the analysis by considering the term $\partial u_x/\partial x$ on the left-hand side of (9.1.2). If the upper wall were perfectly flat and parallel to the lower wall, the flow would be unidirectional and this term would be identically zero. More generally, suppose that $U_x(x_1)$ is the maximum of the magnitude of the x component of the velocity at a particular location $x = x_1$, and $U_x(x_2)$ is the corresponding maximum at the location $x = x_2$, where the maximum is defined with respect to y. Common sense suggests that the magnitude of the term $\partial u_x/\partial x$ will be comparable to the ratio $[U_x(x_2) - U_x(x_1)]/(x_2 - x_1)$.

To this end, we identify the distance $x_2 - x_1$ with the length necessary for the difference $U_x(x_2) - U_x(x_1)$ to become comparable to $U_x(x_1)$, and scale the partial derivative $\partial u_x/\partial x$ with the ratio $U_x(x_1)/(x_2 - x_1)$. If the upper wall were perfectly flat and parallel to the lower wall, the distance $x_2 - x_1$ would be infinite and this ratio would vanish.

Similar arguments can be made to show that the term $\partial u_x/\partial y$ evaluated at $x = x_1$, scales with $U_x(x_1)/h(x_1)$, the term $\partial^2 u_x/\partial x^2$ evaluated at $x = x_1$ scales with $U_x(x_1)/(x_2 - x_1)^2$, and the term $\partial^2 u_x/\partial y^2$ evaluated at $x = x_1$ scales with $U_x(x_1)/h^2(x_1)$.

Continuity equation

Next, we consider the continuity equation (9.1.1), and scale the partial derivative $\partial u_y/\partial y$ evaluated at $x = x_1$ with $U_y(x_1)/h(x_1)$, where $U_y(x_1)$ is the maximum of the magnitude of the y component of the velocity at $x = x_1$. The continuity equation requires that the magnitude of $\partial u_y/\partial y$ is comparable to the magnitude of $\partial u_x/\partial x$, which was found to be of order $U_x(x_1)/(x_2 - x_1)$. This will be true only if $U_y(x_1)$ scales with $U_x(x_1)\, h(x_1)/(x_2 - x_1)$.

Relative magnitudes

Using these arguments, we find that the ratio of the magnitude of the first term on the left-hand side of (9.1.2) to the magnitude of the penultimate term on the right-hand side is

$$\frac{\rho\, U_x(x_1)\, \frac{U_x(x_1)}{x_2 - x_1}}{\mu\, \frac{U_x(x_1)}{h^2(x_1)}} = \frac{\rho\, h(x_1)\, U_x(x_1)}{\mu} \, \frac{h(x_1)}{x_2 - x_1}. \tag{9.1.4}$$

The ratio of the magnitude of the second term on the left-hand side of (9.1.2) to the magnitude of the penultimate term on the right-hand side is

$$\frac{\rho\, h(x_1)\, \frac{U_x(x_1)}{x_2 - x_1}\, \frac{U_x(x_1)}{h(x_1)}}{\mu\, \frac{U_x(x_1)}{h^2(x_1)}} = \frac{\rho\, h(x_1)\, U_x(x_1)}{\mu} \, \frac{h(x_1)}{x_2 - x_1}. \tag{9.1.5}$$

The ratio of the magnitude of the second term on the right-hand side of (9.1.2) to the magnitude of the penultimate term on the right-hand side is

$$\frac{\mu\, \frac{U_x(x_1)}{(x_2 - x_1)^2}}{\mu\, \frac{U_x(x_1)}{h^2(x_1)}} = \Big(\frac{h(x_1)}{x_2 - x_1}\Big)^2. \tag{9.1.6}$$

9.1.3 Equations of lubrication flow

The first fraction on the right-hand sides of equations (9.1.4) and (9.1.5) is the local Reynolds number of the flow defined with respect to the local channel width. If the magnitude of the local Reynolds number and the magnitude of the ratio $h(x_1)/(x_2-x_1)$ are such that the right-hand sides of equations (9.1.4) – (9.1.6) are much smaller than unity, then the first and penultimate terms on the right-hand side of (9.1.2) will dominate the x component of the equation of motion leading to the simplified form

$$0 = -\frac{\partial p}{\partial x} + \mu\,\frac{\partial^2 u_x}{\partial y^2} + \rho\, g_x, \tag{9.1.7}$$

which describes a locally unidirectional flow.

Pressure field

Next, we consider the pressure field. Equation (9.1.7) requires that the magnitude of the term $-\partial p/\partial x + \rho\, g_x$ is comparable to the magnitude of $\mu \partial^2 u_x/\partial y^2$, which scales with $\mu\, U_x(x_1)/h^2(x_1)$. Consideration of the individual terms on both sides of the y component of the equation of motion (9.1.3) shows that the term $-\partial p/\partial y + \rho\, g_y$ scales with $\mu\, U_x(x_1)/[(x_2-x_1)h(x_1)]$. It is evident then that, when the ratio $h(x_1)/(x_2-x_1)$ is small, non-hydrostatic pressure variations are negligible along the y axis. Accordingly, the y component of the equation of motion (9.1.3) reduces to

$$0 = -\frac{\partial p}{\partial y} + \rho\, g_y. \tag{9.1.8}$$

Differentiating this equation once with respect to x, we find that the axial pressure gradient, $\partial p/\partial x$, is independent of the lateral position, y, and is only a function of the axial position, x.

Physically, the flow can be assumed to be locally unidirectional and parallel to the x axis. At every x station, the upper and lower walls appear to be parallel, separated by a distance that is equal to the local channel width, $h(x)$.

9.1.4 Lubrication in a slider bearing

As an application, we consider flow in the lubrication zone of a slider bearing illustrated in figure 9.1.2. The lower wall is horizontal, and the upper wall is plane but gently sloped by an angle α. The pressure is held constant at both ends of the lubrication zone extending from $x = 0$ to L. If the fluid at one end of the lubrication zone is exposed to the atmosphere, then the pressure at that end will be equal to the pressure of the atmosphere. Hydrostatic pressure variations in the y direction are negligibly small.

Reviewing the scaling arguments discussed in sections 9.1.2 and 9.1.3, we identify the distance $x_2 - x_1$ with the length of the lubrication zone, L. A

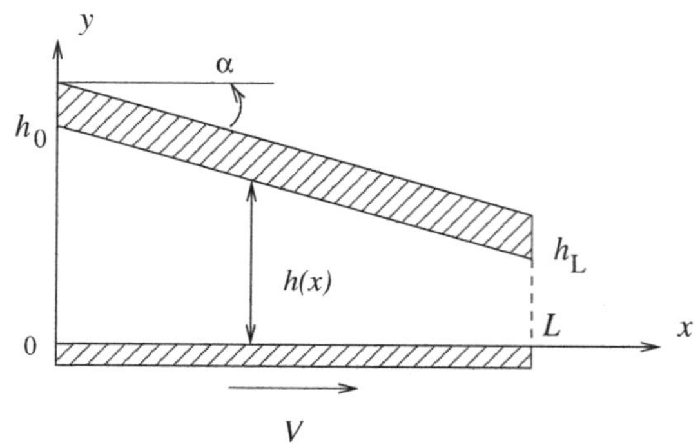

Figure 9.1.2 Lubrication flow in a slider bearing with a planar upper surface.

special case arises when the two walls are parallel, as the distance $x_2 - x_1$ becomes infinite and no error is introduced by dropping the inertial terms in the equation of motion.

Velocity and flow rate

To simplify the analysis, we introduce the approximation $\tan\alpha \simeq \alpha$, and express the local clearance of the channel as

$$h(x) = h_0 - \alpha\, x, \tag{9.1.9}$$

where $h_0 \equiv h(x = 0)$ is the clearance of the channel at the beginning of the lubrication zone. Next, we use the velocity profile derived in section 7.1 for flow in a two-dimensional channel with parallel-sided walls, and find that the solution of (9.1.7) is

$$u_x(x, y) = V\Big(1 - \frac{y}{h(x)}\Big) - \frac{1}{2\mu}\, y\,(\,h(x) - y\,)\,\frac{\mathrm{d}p}{\mathrm{d}x}(x). \tag{9.1.10}$$

The local axial pressure gradient $(\mathrm{d}p/\mathrm{d}x)(x)$ is an unknown that must be computed as part of the solution. The local flow rate along the channel corresponding to the velocity profile (9.1.10) is

$$Q(x) \equiv \int_0^{h(x)} u_x(x, y)\,\mathrm{d}y = \frac{1}{2}\,V h(x) - \frac{h^3(x)}{12\mu}\,\frac{\mathrm{d}p}{\mathrm{d}x}(x). \tag{9.1.11}$$

A key observation is that, because in a stationary frame of reference the flow is steady, the flow rate, $Q(x)$, is not only constant in time, but also independent of the streamwise position, x. To verify this statement, we perform a mass balance over a control area confined between two cross-sections of the channel separated by an infinitesimal distance, and note that fluid neither enters nor escapes from the control volume through the bottom or top.

Computation of the pressure

Solving equation (9.1.11) for the axial pressure gradient and substituting $h(x) = h_0 - \alpha x$ for the channel width, we find

$$\frac{\mathrm{d}p}{\mathrm{d}x} = \frac{6\mu V}{(h_0 - \alpha x)^2} - \frac{12\mu Q}{(h_0 - \alpha x)^3}. \tag{9.1.12}$$

Integrating the pressure gradient with respect to x should produce a specified pressure drop across the lubrication zone. This mathematical condition reflects the physical environment under which the lubrication flow takes place. Performing the integration, we find

$$-\Delta p \equiv p(x = L) - p(x = 0) = \int_0^L \frac{\mathrm{d}p}{\mathrm{d}x}\,\mathrm{d}x$$
$$= \int_0^L \Big(\frac{6\mu V}{(h_0 - \alpha x)^2} - \frac{12\mu Q}{(h_0 - \alpha x)^3} \Big)\,\mathrm{d}x, \tag{9.1.13}$$

where Δp is the pressure drop. Expression (9.1.13) relates the flow rate, Q, to the pressure at the two ends of the lubrication zone.

Vanishing pressure drop

If the end pressures are equal, $\Delta p = 0$. Carrying out the integration on the right-hand side of (9.1.13), solving for Q and simplifying, we derive the expression

$$Q = V\frac{h_0\,h_L}{h_0 + h_L}, \tag{9.1.14}$$

where $h_L = h_0 - \alpha L$ is the channel width at the end of the lubrication zone. Note that, if either h_0 or h_L vanishes, in which case the channel is closed at one end, the flow rate is zero, as expected.

Substituting the right-hand side of (9.1.14) for the flow rate in the right-hand side of (9.1.13) and carrying out the integration, we derive the desired pressure distribution,

$$p(x) = p(x = 0) + \int_0^x \frac{\mathrm{d}p}{\mathrm{d}x}\,\mathrm{d}x = p(x = 0) + \frac{6\mu V}{h_0 + h_L}\,\frac{\alpha x(L - x)}{(h_0 - \alpha x)^2}. \tag{9.1.15}$$

It is reassuring to confirm that the second term on the right-hand side of (9.1.15) vanishes when $x = L$, yielding the specified zero pressure drop.

Lift force

The y component of the hydrodynamic force exerted on the upper sloped surface can be approximated with the integral

$$F_y = \int_0^L p(x)\,\mathrm{d}x. \tag{9.1.16}$$

Substituting the pressure distribution (9.1.15) in the integrand and carrying out the integration, we find

$$F_y = p(x=0)\,L + \frac{6\mu V L^2}{h_0^2}\kappa^2 \left(\ln\frac{h_0}{h_L} - 2\,\frac{h_0 - h_L}{h_0 + h_L}\right). \tag{9.1.17}$$

We have introduced the geometrical factor

$$\kappa \equiv \frac{h_0}{\alpha L}, \tag{9.1.18}$$

taking values in the range $(-\infty, 0)$ or $(1, \infty)$. A negative value corresponds to an upper wall sloping upward, whereas a positive value corresponds to an upper wall sloping downward, as shown in figure 9.1.2. Values of κ in the range $[0, 1]$ are prohibited by the requirement that the upper wall does not slope downward so much as to touch the lower wall before the end of the lubrication zone. When $\kappa = 1$, the two walls meet at $x = L$, and the channel is closed at the right end.

Lift coefficient

We have derived expressions for the flow rate, pressure distribution, and normal force given, respectively, by equations (9.1.14), (9.1.15), and (9.1.17). These derivations conclude the main goal of our analysis. To interpret the results in physical terms, we restate expression (9.1.17) in the form

$$F_y = p(x=0)\,L + \frac{6\mu V L^2}{h_0^2}\,G(\kappa), \tag{9.1.19}$$

where the function $G(\kappa)$ expressing the hydrodynamic lift or load force is given by

$$G(\kappa) \equiv \kappa^2\,\ln\frac{\kappa}{\kappa - 1} - \frac{2\,\kappa^2}{2\kappa - 1}. \tag{9.1.20}$$

Figure 9.1.3 presents a graph of $G(\kappa)$ in its domain of definition. The results show that the lubrication force is positive when the lower wall moves towards the minimum gap, and negative when the lower wall moves towards the maximum gap. In the first case, the lift force will be able to balance the weight of an overlying object whose lower surface is represented by the inclined plane, provided that κ is sufficiently close to unity. In the second case, the lubrication force pulls the object toward the plane wall closing the gap and choking the flow.

9.1.5 Flow in a wavy channel

The preceding analysis can be extended in a straightforward fashion to arbitrary channel geometries. A fundamental assumption is that the conditions for the

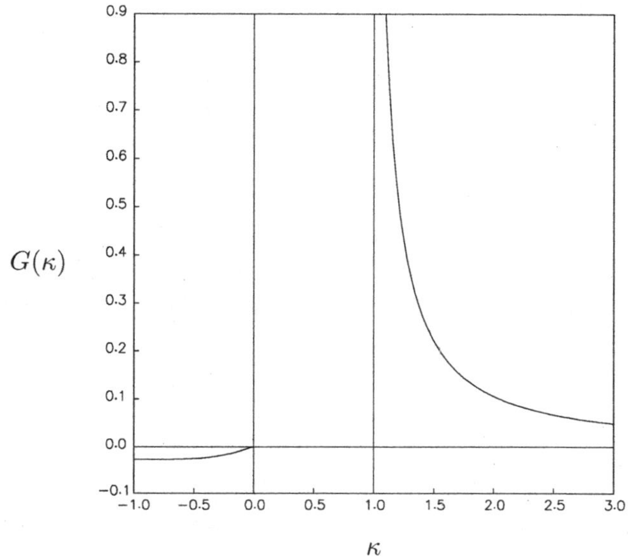

Figure 9.1.3 Graph of the function $G(\kappa)$ expressing the hydrodynamic lift force exerted on the inclined plane shown in figure 9.1.1.

right-hand sides of (9.1.4) – (9.1.6) to be small are satisfied. This means that the local Reynolds number and the ratio $h(x_1)/(x_2 - x_1)$ must both be small.

As an example, we consider flow in a periodic channel confined between a plane and a wavy wall, as illustrated in figure 9.1.4. The local channel width is described by

$$h(x) = h_0 + a \cos kx, \tag{9.1.21}$$

where h_0 is the average channel width, L is the period, $k = 2\pi/L$ is the wave number, and a is the amplitude of the corrugations. The streamwise velocity and flow rate are given by equations (9.1.10) and (9.1.11), where $h(x)$ given in (9.1.21). The counterpart of equation (9.1.12) is

$$\frac{\mathrm{d}p}{\mathrm{d}x} = \frac{6\mu V}{(h_0 + a\cos kx)^2} - \frac{12\mu Q}{(h_0 + a\cos kx)^3}. \tag{9.1.22}$$

The negative of the pressure drop over one period is

$$-\Delta p \equiv p(x = L) - p(x = 0) = \int_0^L \frac{\mathrm{d}p}{\mathrm{d}x}\,\mathrm{d}x, \tag{9.1.23}$$

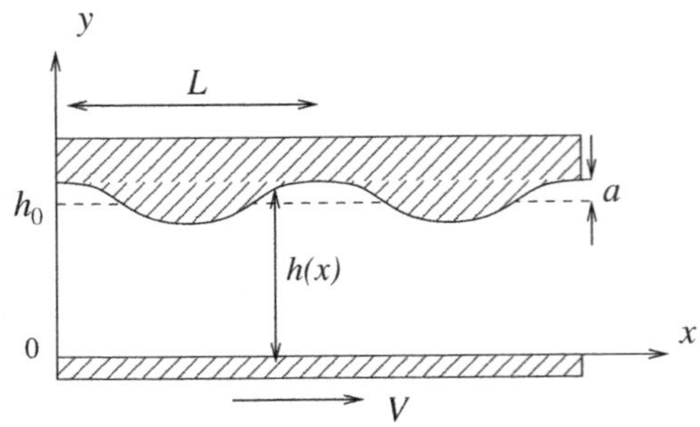

Figure 9.1.4 Lubrication flow in a furrowed channel confined between a translating plane wall and a stationary wavy wall.

yielding

$$\Delta p = -\int_0^L \left(\frac{6\mu V}{(h_0 + a\cos kx)^2} - \frac{12\mu Q}{(h_0 + a\cos kx)^3} \right) \mathrm{d}x. \tag{9.1.24}$$

Solving for Q, we find

$$Q = \frac{\frac{\Delta p}{6\mu} + V \int_0^L \frac{\mathrm{d}x}{(h_0 + a\cos kx)^2}}{2\int_0^L \frac{\mathrm{d}x}{(h_0 + a\cos kx)^3}}, \tag{9.1.25}$$

which provides us with an expression for the flow rate in terms of the specified wall velocity and pressure drop.

The definite integrals on the right-hand side of (9.1.25) are best computed by numerical methods. Interestingly, because the integrands are periodic, the best results are obtained by using the simplest algorithm of numerical integration expressed by the trapezoidal rule. The method involves dividing the integration domain $(0, L)$ into N intervals of equal length $\Delta x = L/N$, and introducing the approximation

$$\int_0^L \frac{dx}{(h_0 + a\cos kx)^2} \simeq \Delta x \sum_{i=1}^N \frac{1}{(h_0 + a\cos kx_i)^2}, \tag{9.1.26}$$

where $x_i = (i-1)\,\Delta x$ are the interval end points. A similar approximation is made for the integral in the denominator of (9.1.25). The Euler-Maclaurin theorem ensures that, as the number of divisions N becomes larger, the difference between the left-and right-hand side of (9.1.26), defined as the numerical error, decreases faster than any power of $1/N$, and this allows for rapid convergence.

Once the flow rate has been found, the result can be substituted in the right-hand side of (9.1.24), and the expression thus obtained can be integrated by analytical or numerical methods to yield the pressure distribution along the channel. The velocity then follows from (9.1.10).

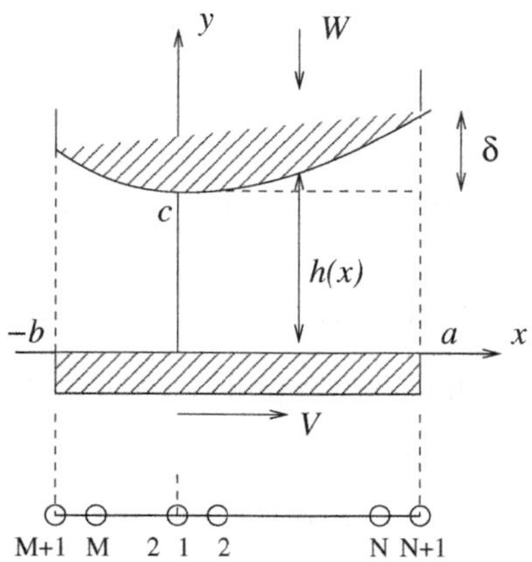

Figure 9.1.5 A two-dimensional body presses against a sliding wall under the influence of its weight. The lubrication force developing between the body and the wall lifts the body away from the wall.

9.1.6 Dynamic lifting

A generalization of the lubrication analysis allows us to develop a method for simulating the motion of a two-dimensional body pressing against a horizontal wall that translates along the x axis with velocity V, as illustrated in figure 9.1.5. The clearance between the body and the translating wall is occupied by a lubricating fluid extending from $x = -b$ to a. This configuration serves as a two-dimensional model of the flow between a piston ring pressing against the cylinder of a combustion chamber in an internal combustion engine.

The lubrication flow generates a pressure field, and the associated lift force causes the body to move along the y axis with velocity $V_B = \mathrm{d}c/\mathrm{d}t$, where c is the minimum film thickness defined as the minimum clearance between the body and the translating wall. Our objective is to compute the evolution of the function $c(t)$ from a specified initial state. Hydrostatic pressure variations in the y direction are assumed to be negligibly small.

We begin by writing a mass balance for the lubricating fluid requiring

$$\frac{\mathrm{d}c}{\mathrm{d}t} = -\frac{\partial Q}{\partial x}. \tag{9.1.27}$$

Evaluating the flow rate from the right-hand side of (9.1.11), we derive a partial-differential equation for the pressure,

$$\frac{\partial}{\partial x}\Big(\frac{h^3}{\mu}\frac{\partial p}{\partial x}\Big) = 6\,V\frac{\partial h}{\partial x} + 12\,\frac{\mathrm{d}c}{\mathrm{d}t}, \tag{9.1.28}$$

which is to be solved subject to the specified, and possibly time-dependent, boundary conditions $p(x = -b) = P_1(t)$ and $p(x = a) = P_2(t)$. The vertical velocity of the body, $V_B = \mathrm{d}c/\mathrm{d}t$, is determined by a balance between the weight of the body, W, and the lift force due to the pressure,

$$W = \int_{-b}^{a} p(x,t)\, \mathrm{d}x. \tag{9.1.29}$$

The solution can be computed using the following algorithm:

1. First, we solve the following simplified version of equation (9.1.28),

$$\frac{\partial}{\partial x}\Big(\frac{h^3}{\mu}\frac{\partial p}{\partial x}\Big) = 6\, V\, \frac{\partial h}{\partial x}, \tag{9.1.30}$$

   subject to the required boundary conditions

$$p(x = -b) = P_1(t), \qquad\qquad p(x = a) = P_2(t), \tag{9.1.31}$$

   and call the solution $p_1(x)$.

2. Second, we solve the following simplified version of equation (9.1.28),

$$\frac{\partial}{\partial x}\Big(\frac{h^3}{\mu}\frac{\partial p}{\partial x}\Big) = 12, \tag{9.1.32}$$

   subject to the homogeneous boundary conditions

$$p(x = -b) = 0, \qquad\qquad p(x = a) = 0, \tag{9.1.33}$$

   and call the solution $p_2(x)$.

3. Third, we set

$$p(x) = p_1(x) + \frac{\mathrm{d}c}{\mathrm{d}t}\, p_2(x). \tag{9.1.34}$$

4. Fourth, we substitute the pressure distribution (9.1.34) in the force balance (9.1.29) and carry out the integration to compute $\mathrm{d}c/\mathrm{d}t$.

5. Having evaluated $\mathrm{d}c/\mathrm{d}t$, we update the minimum clearance, c.

6. We return to step 1 and repeat.

Straightforward substitution shows that the pressure distribution (9.1.34) satisfies the governing equation (9.1.28) as well as the pressure boundary condition specified at either end.

Code *bear_2d* residing in directory *05_lub* of FDLIB implements the numerical method. The geometry of the body is defined by the lengths a, b, and δ defined in figure 9.1.5. In the finite-difference implementation, the left part of the lubrication zone, $-b < x < 0$, is divided into M intervals, and the right part of the lubrication zone, $0 < x < a$, is divided into N intervals, as depicted at the bottom of figure 9.1.5. The derivatives with respect to x in equations (9.1.30) and (9.1.32) are approximated with second-order finite differences, yielding tridiagonal systems of algebraic equations for the pressure at the nodes. The solution is found using the Thomas algorithm.

Problems

9.1.1 *Lubrication in a slider bearing*

Confirm that the function $G(\kappa)$ defined in (9.1.20) satisfies the symmetry property $H(\kappa - 0.5) = -H(0.5 - \kappa)$, and explain the underlying physical reason.

9.1.2 *Flow in a symmetric channel*

Consider pressure-drive flow in a symmetric channel with stationary wavy walls located at $y = h(x) = \pm(h_0 + a \cos kx)$. Derive an expression for the flow rate in terms of the pressure drop over one period.

Computer Problems

9.1.3 *Flow in a wavy channel*

(*a*) Consider flow in a channel confined between a plane wall and a wavy wall, as discussed in the text, driven by an imposed pressure drop; in this case, $V = 0$. Prepare a graph of the dimensionless flow rate $\widehat{Q} \equiv 12\mu LQ/(h_0^3\, \Delta P)$ against the scaled amplitude a/h_0, and discuss its functional form. The integral on the right-hand side of (9.1.25) should be computed using the trapezoidal rule, as shown in equation (9.1.26).

(*b*) Repeat (*a*) for flow driven by boundary motion, corresponding to $\Delta p = 0$. Prepare a graph of the dimensionless flow rate $\widehat{Q} \equiv 2Q/(V\, h_0)$ against the reduced amplitude a/h_0, and discuss its functional form.

9.1.4 *Dynamical simulation of the lifting of a body due to lubrication*

Run the program *bear_2d* for a set of conditions of your choice. Prepare a plot of the minimum film thickness as a function of time, and discuss the results of the simulations.

9.2 Film flow on a horizontal or inclined wall

One distinguishing feature of the flow in narrow channels discussed in section 9.1 is that the velocity profile can be approximated locally with the parabolic profile corresponding to unidirectional flow in a channel confined between two parallel plane walls. An analogous simplification is possible in the case of a

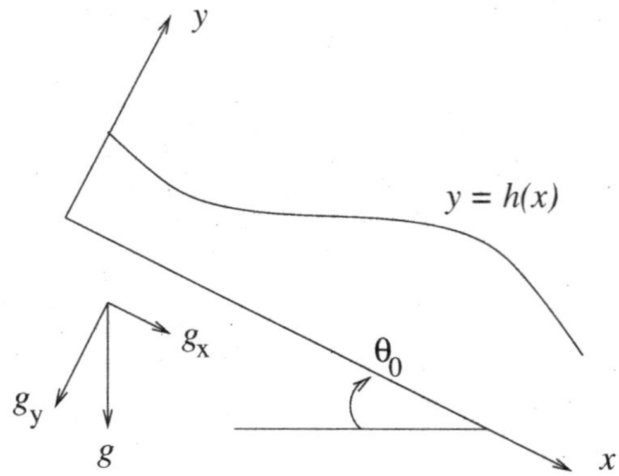

Figure 9.2.1 Illustration of a viscous film flowing down an inclined plane wall. The motion of the fluid is governed by the simplified equations of lubrication flow.

liquid film bounded by a free surface where the shear stress is required to vanish, as discussed in this section, as well as in the case of a liquid layer bounded by a fluid interface where the shear stress is continuous on either side, as discussed in section 9.3.

9.2.1 Thin-film flow

We consider the unsteady flow of a liquid film over a horizontal wall or down an inclined plane wall, as illustrated in figure 9.2.1. In the inclined system of coordinates depicted in this figure, the wall is located at $y = 0$ and the free surface is located at $y = h(x, t)$. The components of the acceleration of gravity are given by

$$g_x = g \sin\theta_0, \qquad g_y = -g \cos\theta_0, \tag{9.2.1}$$

where g is the magnitude of the acceleration of gravity and θ_0 is the inclination angle of the wall; if the wall is horizontal, $\theta_0 = 0$; if the wall is vertical, $\theta_0 = \pi/2$. The pressure above the film is assumed to be uniform and equal to the atmospheric pressure, P_{atm}. Our objective is to derive a differential equation governing the evolution of the film thickness, $h(x, t)$, from a specified initial profile.

Governing equations

Repeating the scaling arguments discussed in section 9.1 for lubrication flow, we find that, if the free surface is gently sloped, the magnitude of the streamwise derivative $\partial h/\partial x$ is uniformly small, and the x and y components of the equation of motion reduce to the equations of lubrication flow stated in (9.1.7) and (9.1.8). Integrating equation (9.1.8) with respect to y from an arbitrary location up to the location of the free surface, we find

$$p(x,y,t) = p_{fs} - \rho\, g_y\, (h - y), \tag{9.2.2}$$

where $p_{fs} = p(x, y = h(t), t)$ is the pressure on the side of the liquid at the free surface.

The simple form of the lubrication equation (9.1.7) allows us to pretend that the flow is locally unidirectional and parallel to the x axis. Approximating the velocity profile across the uneven film with the Nusselt parabolic profile across a flat film bounded by a planar free surface given in equation (7.2.3), we find

$$u_x(x,y,t) = (-\frac{\partial p}{\partial x} + \rho\, g_x)\,\frac{1}{2\mu}\, y\,(2h - y). \tag{9.2.3}$$

Note that the x dependence enters the problem formulation in a parametric fashion by way of the pressure gradient. The term enclosed by the first set of parentheses on the right-hand side of (9.2.3) incorporates the effective body force due to gravity and the streamwise pressure gradient due to surface tension, as will be shortly discussed. The flow rate across a plane normal to the x axis is

$$Q(x,t) \equiv \int_o^{h(x,t)} u_x(x,y,t)\, dy = (-\frac{\partial p}{\partial x} + \rho\, g_x)\,\frac{1}{3\mu}\, h^3(x,t). \tag{9.2.4}$$

Further analysis involves two key steps.

Evaluation of the pressure gradient

First, we observe that the normal stress undergoes a jump across the interface due to the surface tension, γ, according to the interfacial condition (4.3.4). In the case of locally unidirectional flow, the normal stress can be approximated with the negative of the pressure, yielding the pressure jump condition

$$p_{FS} = P_{atm} + \gamma\, \kappa. \tag{9.2.5}$$

Moreover, the curvature of the mildly sloped free surface can be approximated with the linear form

$$\kappa(x) \simeq -\frac{\partial^2 h}{\partial x^2}. \tag{9.2.6}$$

Substituting (9.2.6) in (9.2.5) and the result in (9.2.2), we obtain an expression for the pressure distribution inside the film in terms of the film thickness,

$$p(x,y,t) \simeq P_{atm} - \rho\, g_y\, (h - y) - \gamma\,\frac{\partial^2 h}{\partial x^2}. \tag{9.2.7}$$

Substituting further (9.2.7) in the right-hand side of (9.2.4), we derive an expression for the flow rate in terms of the film thickness,

$$Q(x,t) = \frac{\rho}{3\mu}\, h^3 (\, g_x + g_y\, \frac{\partial h}{\partial x} + \frac{\gamma}{\rho}\, \frac{\partial^3 h}{\partial x^3}\,). \tag{9.2.8}$$

Mass balance

Next, we perform a mass balance over a control area confined by (*a*) the wall, (*b*) two parallel planes that are normal to the plane and are separated by an infinitesimal distance, (*c*) the corresponding section of the evolving free surface. The mass balances requires that the rate of mass accumulation in the control volume is equal to the difference between the flow rates into and out from the control volume. In differential form,

$$\frac{\partial h}{\partial t} = -\frac{\partial Q}{\partial x}. \tag{9.2.9}$$

If $\partial Q/\partial x$ is negative, the rate of change $\partial h/\partial t$ is positive.

Evolution equation for the film thickness

To complete the mathematical formulation, we substitute the right-hand side of (9.2.8) in the right-hand side of (9.2.9) to obtain a first-order partial differential equation in time governing the evolution of the film thickness,

$$\frac{\partial h}{\partial t} + \frac{\rho}{3\mu}\,\frac{\partial}{\partial x}\Big(\, h^3\,(g_x + g_y\,\frac{\partial h}{\partial x} + \frac{\gamma}{\rho}\,\frac{\partial^3 h}{\partial x^3})\Big) = 0. \tag{9.2.10}$$

The products of the film thickness, h, and its spatial derivatives renders equation (9.2.10) highly nonlinear. To compute the solution, an initial condition for $h(x, t = 0)$ at the origin of computational time must be specified.

Carrying out the differentiation with respect to x on the left-hand side, we find that the fourth derivative, $\partial^4 h/\partial x^4$, emerges from the term involving the surface tension. Since (9.2.10) is a fourth-order equation, two boundary conditions at each end of the solution domain involving h, $\partial h/\partial x$, or $\partial^2 h/\partial x^2$, are required. If the film is periodic, the boundary conditions are replaced by periodicity conditions for the film thickness and its spatial derivatives.

In the absence of surface tension, $\gamma = 0$, we obtain a second-order differential equation. In that case, only one boundary condition at each end of the solution domain, involving h or $\partial h/\partial x$, is required. If the film is periodic, the boundary conditions are replaced by periodicity conditions.

At steady state, $\partial h/\partial t = 0$, the shape of the free surface is described by the nonlinear ordinary differential equation

$$\frac{\mathrm{d}}{\mathrm{d}x}\Big(\, h^3 (g_x + g_y \frac{\mathrm{d}h}{\mathrm{d}x} + \frac{\gamma}{\rho}\frac{\mathrm{d}^3 h}{\mathrm{d}x^3})\Big) = 0, \tag{9.2.11}$$

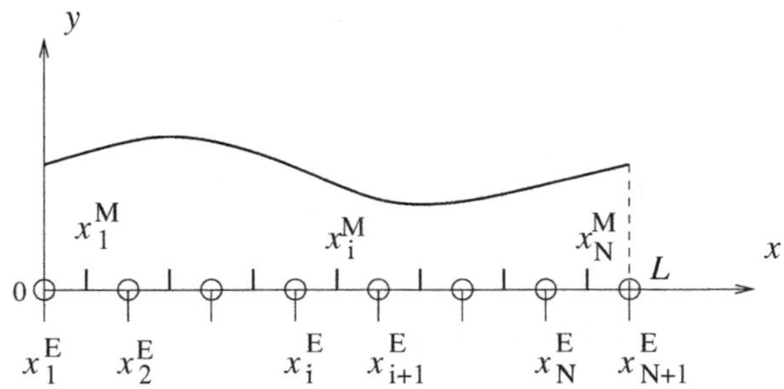

Figure 9.2.2 A finite-volume grid used to compute the evolution of a periodic film.

which admits the flat-film Nusselt solution derived in section 7.2 for constant film thickness.

9.2.2 Numerical methods

Several numerical methods are available for solving the differential equation (9.2.10) including finite-difference, finite-volume, finite-element, and spectral methods. The finite-difference method will be discussed in section 9.3 in the context of multi-layer film flow. To illustrate an alternative, in this section we discuss the implementation of an entry-level finite-volume method.

Finite-volume method

We want to compute the evolution of a film that is, and remains, spatially periodic with period L. To implement the finite-volume method, we divide the computational domain extending from $x = 0$ to L into N intervals of equal length $\Delta x = L/N$, also called finite volumes, as illustrated in figure 9.2.2. The end points of the ith interval are denoted by $x_i^E = (i-1)\Delta x$, and the mid-point is denoted by $x_i^M = (i-0.5)\Delta x$, where $i = 1, \ldots, N$. Values of the film-thickness corresponding to x_i^E and x_i^M are denoted, respectively, by h_i^E and h_i^M.

A distinguishing feature of the finite-volume method is that the governing equation (9.2.10) is integrated over the length of each element to reduce the order of the highest derivative. Considering the ith element, we write

$$\int_{x_i^E}^{x_{i+1}^E} \frac{\partial h}{\partial t}\, \mathrm{d}x + \frac{\rho}{3\mu} \Big(h^3 (g_x + g_y \frac{\partial h}{\partial x} + \frac{\gamma}{\rho} \frac{\partial^3 h}{\partial x^3}) \Big)_{x_{i+1}^E}$$

$$- \frac{\rho}{3\mu} \Big(h^3 \; (g_x + g_y \frac{\partial h}{\partial x} + \frac{\gamma}{\rho} \frac{\partial^3 h}{\partial x^3}) \Big)_{x_i^E} = 0. \qquad (9.2.12)$$

Using the mid-point rule to approximate the integral on the left-hand side of (9.2.12), we obtain

$$\int_{x_i^E}^{x_{i+1}^E} \frac{\partial h}{\partial t}\, \mathrm{d}x \simeq \Big(\frac{\partial h}{\partial t}\Big)_{x_i^M} \Delta x. \tag{9.2.13}$$

Approximating further the time derivative on the right-hand side of (9.2.13) with a first-order forward-difference, we obtain

$$\Big(\frac{\partial h}{\partial t}\Big)_{x_i^M} \simeq \Big(\frac{h(t+\Delta t)-h(t)}{\Delta t}\Big)_{x_i^M}. \tag{9.2.14}$$

Next, we substitute (9.2.14) in (9.2.13) and the result in (9.2.12). Solving the emerging expression for $h(x_i^M, t+\Delta t)$, we find

$$h(x_i^M, t+\Delta t) = h(x_i^M, t) - \frac{\Delta t}{\Delta x}\,\frac{\rho}{3\mu}\,\Big(h^3(g_x + g_y\frac{\partial h}{\partial x} + \frac{\gamma}{\rho}\frac{\partial^3 h}{\partial x^3})\Big)_{x_{i+1}^E}$$
$$+\frac{\Delta t}{\Delta x}\,\frac{\rho}{3\mu}\,\Big(h^3(g_x + g_y\frac{\partial h}{\partial x} + \frac{\gamma}{\rho}\frac{\partial^3 h}{\partial x^3})\Big)_{x_i^E}, \tag{9.2.15}$$

where the right-hand side is evaluated at t.

Finally, we express the values of the function h and its spatial derivatives at the element end points in terms of values at the mid-points using a combination of averaging and finite-difference approximations. For example, we write

$$h(x_{i+1}^E) \simeq \frac{1}{2}[h(x_i^M) + h(x_{i+1}^M)],$$
$$\frac{\partial h}{\partial x}(x_{i+1}^E) \simeq \frac{h(x_{i+1}^M) - h(x_i^M)}{\Delta x}, \tag{9.2.16}$$
$$\frac{\partial^3 h}{\partial x^3}(x_{i+1}^E) \simeq \frac{h(x_{i+2}^M) - 3\,h(x_{i+1}^M) + 3\,h(x_i^M) - h(x_{i-1}^M)}{\Delta x^3}.$$

The periodicity condition requires

$$h(x_{-1}^M) = h(x_{N-1}^M), \qquad h(x_0^M) = h(x_N^M),$$
$$h(x_{N+1}^M) = h(x_1^M), \qquad h(x_{N+2}^M) = h(x_2^M), \tag{9.2.17}$$

and this allows the evaluation of the right-hand sides of equation (9.2.16) at the ends of the computational domain spanning one period.

The numerical procedure for simulating the evolution of the film from a given initial condition involves the following steps:

1. First, we specify the values of h_i^M at the origin of computational time.

2. Second, we choose a time step, Δt.

3. Third, we evaluate expressions (9.2.16) and (9.2.17) at the element end points.

4. Fourth, we compute the right-hand side of (9.2.15) and thereby obtain the updated value of the film thickness at the mid-points.

5. We may now return to step 3 and repeat the calculation for another step.

The method requires a small time step to prevent the onset of numerical instability. Implicit methods where the second and third terms on the right-hand side of (9.2.15) are evaluated at time $t + \Delta t$ overcome this limitation at the expense of having to solve a system of nonlinear equations.

Computer Problem

9.2.1 *Finite-volume method*

Write a computer code that uses the finite-volume method described in the text to simulate the evolution of a periodic film resting on a horizontal wall or flowing down an inclined plane. Compute the evolution of a film with an initially sinusoidal free surface resting on a horizontal wall. Investigate by numerical experimentation and discuss the effect of surface tension.

9.3 Multi-film flow on a horizontal or inclined wall

Next, we consider the more general case of an arbitrary number of superimposed layers flowing down an inclined plane, as illustrated in figure 9.3.1. For the flow configuration to be stable, the density of the films must be constant or decrease with distance from the wall; otherwise, a gravitational instability due to unstable density stratification, known as the Rayleigh-Taylor instability, will arise. Multi-layer flow is encountered in the manufacturing of photographic films during the process of slide coating.

To generalize our discussion, we allow the inclined plane to exhibit periodic corrugations around a mean value. In the inclined system of coordinates defined in figure 9.3.1, the wall is described by the function $y = y_0(x)$, and the interfaces are described by the functions $y = y_i(x, t)$, where $i = 1, \ldots, N$, and the Nth interface is the free surface. The ith film is confined between the interfaces labeled i and $i - 1$, with the understanding that the zeroth interface represents the wall.

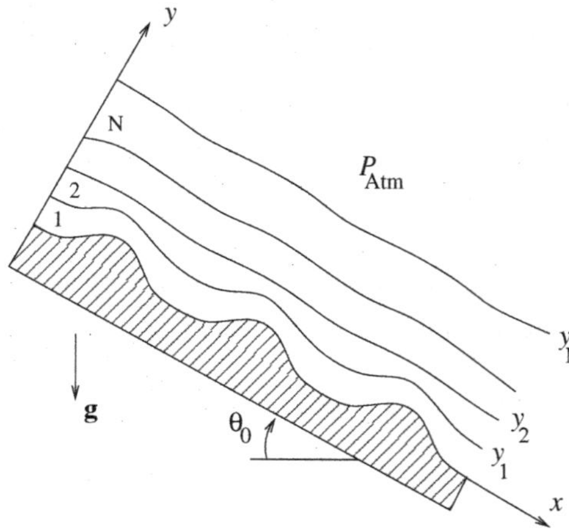

Figure 9.3.1 Multi-film flow down an inclined wall with periodic corrugations.

When the wall and interfaces are nearly flat, $|\partial y_i/\partial x| < 1$, for $i = 0, \ldots N$, the flow inside each layer is locally nearly unidirectional along the x axis. The x and y components of the equation of motion then simplify to

$$\frac{\partial p^{(i)}}{\partial x} = \mu_i \, \frac{\partial^2 u_x^{(i)}}{\partial y^2} + \rho_i \, g_x, \qquad \frac{\partial p^{(i)}}{\partial y} = \rho_i \, g_y, \tag{9.3.1}$$

where μ_i and ρ_i is the viscosity and density of the ith layer. The no-slip boundary condition at the wall requires

$$u_x^{(1)}(y = y_0) = 0. \tag{9.3.2}$$

Interfacial conditions

Continuity of velocity across the interfaces requires

$$u_x^{(i)}(y = y_i) = u_x^{(i+1)}(y = y_i), \tag{9.3.3}$$

for $i = 1, \ldots, N-1$. Continuity of shear stress across the interfaces requires

$$\mu_i \left(\frac{\partial u_x^{(i)}}{\partial y} \right)_{y=y_i} = \mu_{i+1} \left(\frac{\partial u_x^{(i+1)}}{\partial y} \right)_{y=y_i}, \tag{9.3.4}$$

with the understanding that $\mu_{N+1} = 0$, which ensures that the shear stress vanishes at the free surface. Finally, a balance of the normal stresses on either side of the ith interface involving the interfacial tension, γ_i, requires

$$(p^{(i)})_{y=y_i} = (p^{(i+1)})_{y=y_i} + \gamma_i \, \kappa_i, \tag{9.3.5}$$

where κ_i is the curvature of the ith interface or free surface, with the understanding that $p^{(N+1)} = P_{atm}$. The statement of the problem is now complete, and we proceed to formulate the solution.

Pressure gradient

Our first task is to compute the pressure gradient on either side of each interface. We begin by integrating the second equation in (9.3.1) with respect to y from the ith interface up to an arbitrary point, and use the interfacial condition (9.3.5) to find

$$p^{(i)}(x,y) = p^{(i+1)}(x, y = y_i) + \gamma_i\, \kappa_i + \rho_i\, g_y\, [\, y - y_i(x)\,], \tag{9.3.6}$$

for $i = 1, \ldots, N$, with the understanding that $p^{(N+1)} = P_{atm}$.

Next, we differentiate equation (9.3.6) with respect to x, and use the chain rule to write

$$\frac{\partial p^{(i)}}{\partial x} = \frac{\partial p^{(i+1)}}{\partial x} + \frac{\partial p^{(i+1)}}{\partial y}\, \frac{\partial y_i}{\partial x} + \gamma_i\, \frac{\partial \kappa_i}{\partial x} - \rho_i\, g_y\, \frac{\partial y_i}{\partial x}. \tag{9.3.7}$$

Evaluating $\partial p^{(i+1)}/\partial y$ from the second equation in (9.3.1), we obtain

$$\frac{\partial p^{(i)}}{\partial x} = \frac{\partial p^{(i+1)}}{\partial x} + \gamma_i\, \frac{\partial \kappa_i}{\partial x} + (\rho_{i+1} - \rho_i)\, g_y\, \frac{\partial y_i}{\partial x}, \tag{9.3.8}$$

with the understanding that $\rho_{N+1} = 0$. Writing this equation for the interfaces numbered $i, i+1, \ldots, N$, and adding corresponding sides, we obtain

$$\frac{\partial p^{(i)}}{\partial x} = \sum_{j=i}^{N} \Big[\gamma_j\, \frac{\partial \kappa_j}{\partial x} + (\rho_{j+1} - \rho_j)\, g_y\, \frac{\partial y_j}{\partial x} \Big]. \tag{9.3.9}$$

Finally, we introduce the familiar approximation for the curvature of the jth interface, $\kappa_j \simeq -\partial^2 y_j/\partial x^2$, and obtain

$$\frac{\partial p^{(i)}}{\partial x} = \sum_{j=i}^{N} \Big[-\gamma_j\, \frac{\partial^3 y_j}{\partial x^3} + (\rho_{j+1} - \rho_j)\, g_y\, \frac{\partial y_j}{\partial x} \Big]. \tag{9.3.10}$$

The right-hand side of (9.3.10) can be evaluated from knowledge of the instantaneous interfacial profiles.

Velocity profiles

The velocity profile inside the ith layer arises by integrating the first equation in (9.3.1) with respect to y, obtaining

$$u_x^{(i)} = A_i(x) + B_i(x)\, y - G_i(x) y^2, \tag{9.3.11}$$

where we have defined

$$G_i(x) \equiv \frac{1}{2\mu_i}\Big(-\frac{\partial p^{(i)}}{\partial x} + \rho_i\, g_x \Big). \tag{9.3.12}$$

To compute the unknown functions $A_i(x)$ and $B_i(x)$, we use the wall condition to find

$$A_1(x) + B_1(x)\, y_0 - G_1(x)\, y_0^2 = 0, \tag{9.3.13}$$

and

$$A_i(x) + B_i(x)\, y_i - G_i(x)\, y_i^2 = A_{i+1}(x) + B_{i+1}(x)\, y_i - G_{i+1}(x)\, y_i^2, \tag{9.3.14}$$

for $i = 1, \ldots, N-1$, and the interfacial condition to find

$$B_N(x) = 2\, G_N(x)\, y_N \tag{9.3.15}$$

and

$$\mu_i\, [B_i(x) - 2\, G_i(x)\, y_i] = \mu_{i+1}\, [B_{i+1}(x) - 2\, G_{i+1}(x)\, y_i\,], \tag{9.3.16}$$

for $i = N-1, N-2, \ldots 1$. Rearranging (9.3.16), we find

$$B_i(x) = 2\, G_i(x)\, y_i + \frac{\mu_{i+1}}{\mu_i}\, [B_{i+1}(x) - 2\, G_{i+1}(x)\, y_i\,]. \tag{9.3.17}$$

Note that the coefficients A_i are computed recursively from bottom to top, whereas the coefficients B_i are computed recursively from top to bottom.

A straightforward substitution allows us to replace the recursion relation (9.3.17) with the explicit formula

$$B_i(x) = 2\, G_i(x)\, y_i + 2 \sum_{k=i+1}^{N} \frac{\mu_k}{\mu_i}\, G_k(x)\, (y_k - y_{k-1}), \tag{9.3.18}$$

for $i = 1, \ldots, N-1$. Equations (9.3.15) and (9.3.18) provide us with expressions for evaluating the coefficients $B_i(x)$. Once these expressions are available, $A_1(x)$ follows from (9.3.13) and the rest of the coefficients $A_i(x)$ follow from (9.3.14).

9.3.1 Evolution equations

The counterpart of the mass balance (9.2.9) for the ith film is

$$\frac{\partial h_i}{\partial t} = \frac{\partial y_i}{\partial t} - \frac{\partial y_{i-1}}{\partial t} = -\frac{\partial Q_i}{\partial x}, \tag{9.3.19}$$

where h_i is the film thickness, and

$$Q_i(x,t) \equiv \int_y^{y_{i-1}} u^{(i)}\,\mathrm{d}y \tag{9.3.20}$$

$$= A_i(x)\,(y_i - y_{i-1}) + \frac{1}{2}\,B_i(x)\,(y_i^2 - y_{i-1}^2) - \frac{1}{3}\,G_i(x)\,(y_i^3 - y_{i-1}^3)$$

is the local flow rate. Writing equation (9.3.19) for $i = 1,\ldots,N$, and combining the expressions thus obtained, we derive the evolution equations

$$\frac{\partial y_i}{\partial t} = -\sum_{j=1}^{i} \frac{\partial Q_j}{\partial x}, \tag{9.3.21}$$

for $i = 1,\ldots,N$. Substituting equation (9.3.10) in (9.3.12), the result in (9.3.21), and the outcome in (9.3.21), we derive a system of fourth-order nonlinear partial differential equations governing the evolution of the interfaces and free surface.

One layer

In the case of one layer, $N = 1$, and a plane wall corresponding to $y_0 = 0$, equations (9.3.15) and (9.3.13) yield $B_1(x) = 2\,G_1(x)\,y_1$ and $A_1(x) = 0$. The evolution equation (9.3.21) then reduces to (9.2.10) for single-film flow.

Two layers

In the case of two layers , $N = 2$, and a plane wall corresponding to $y_0 = 0$, we find

$$\begin{aligned} B_2(x) &= 2\,G_2(x)\,y_2, \\ B_1(x) &= 2\,G_1(x)\,y_1 + 2\,\lambda_1\,G_2(x)\,(y_2 - y_1), \\ A_1(x) &= 0, \\ A_2(x) &= G_1(x)\,y_1^2 + G_2(x)\,y_1\,[2\,\lambda_1\,(y_2 - y_1) - 2\,y_2 + y_1], \end{aligned} \tag{9.3.22}$$

where $\lambda_1 \equiv \mu_2/\mu_1$. Using (9.3.10) and (9.3.12), we obtain

$$G_2(x) = \frac{\rho_2}{2\mu_2}\Big(g_x + g_y\,\frac{\partial y_2}{\partial x} + \frac{\gamma_2}{\rho_2}\,\frac{\partial^3 y_2}{\partial x^3} \Big), \tag{9.3.23}$$

and

$$\begin{aligned} G_1(x) = \frac{\rho_2}{2\mu_2}\,\lambda_1\,\Big(&\frac{1}{\beta_1}\,g_x + g_y\,(\frac{1}{\beta_1} - 1)\,\frac{\partial y_1}{\partial x} + g_y\,\frac{\partial y_2}{\partial x} \\ &+ \frac{\gamma_1}{\rho_2}\,\frac{\partial^3 y_1}{\partial x^3} + \frac{\gamma_2}{\rho_2}\,\frac{\partial^3 y_2}{\partial x^3} \Big), \end{aligned} \tag{9.3.24}$$

where $\beta_1 \equiv \rho_2/\rho_1$. The evolution of the interfaces is governed by the coupled differential equations

$$\frac{\partial y_1}{\partial t} = -\frac{\partial Q_1}{\partial x}, \qquad \frac{\partial y_2}{\partial t} = -\frac{\partial Q_1}{\partial x} - \frac{\partial Q_2}{\partial x}, \tag{9.3.25}$$

subject to a specified initial condition.

9.3.2 Numerical methods

The solution of equation (9.3.21) for $i = 1, \dots N$ can be found by a standard finite-difference method. In the numerical implementation, one period of the flow is discretized into a uniform grid, and the spatial derivatives in the differential equations are approximated with finite differences.

The following MATLAB function *films_pde* residing in directory *films* inside directory *05_lub* of FDLIB evaluates the rates of evolution, $\mathrm{d}y_i/\mathrm{d}t$, for an arbitrary number of layers:

```
function dydt = films_pde ...
  (NLR,NSG,Dx,wall,y,rho,mu,gamma,gx,gy,ICU)

%-----------------------------------------
% Compute the rate of change of y position
% of the interfaces:
% dy_i/dt, for i=1,...,NLR
%
% NLR: number of layers
% NSG: number of segments
% Dx:  x finite-difference interval
% wall:  wall profile
% y:  position of interfaces
% ICU=1 for central differences
% ICU=2 for backward differences
%-----------------------------------------

%---
% prepare
%---

Dx2 = 2.0*Dx;

%---
% compute derivatives
%---

for i=1:NLR   % run over interfaces

%---
% first derivative
%---

y1(1,i) = (y(2,i)-y(NSG,i))/Dx2;
for k=2:NSG
  y1(k,i) = (y(k+1,i)-y(k-1,i))/Dx2;
end
y1(NSG+1,i) = y1(1,i);
```

```
%---
% second derivative
%---

y2(1,i) = (y1(2,i)-y1(NSG,i))/Dx2;
for k=2:NSG
  y2(k,i) = (y1(k+1,i)-y1(k-1,i))/Dx2;
end
y2(NSG+1,i) = y2(1,i);

%---
% third derivative
%---

y3(1,i) = (y2(2,i)-y2(NSG,i))/Dx2;
for k=2:NSG
  y3(k,i) = (y2(k+1,i)-y2(k-1,i))/Dx2;
end
y3(NSG+1,i) = y3(1,i);

%---
end
%---

%--------------------------------------
% Compute the effective pressure drop G
% using equations (9.3.10) and (9.3.12)
%--------------------------------------

for i=1:NLR   % over interfaces
  for k=1:NSG   % over points
    dpdx = 0.0;
    for j=i:NLR
      dpdx = dpdx - gamma(j)*y3(k,j) ...
      + gy*(rho(j+1)-rho(j))*y1(k,j);
    end
    G(k,i) = 0.50*(-dpdx + rho(i)*gx)/mu(i);
  end
end

%----------------------------
% Compute the coefficients B
%----------------------------

for k=1:NSG
  B(k,NLR) = 2.0*G(k,NLR)*y(k,NLR);
end
```

```
for i=NLR-1:-1:1
  for k=1:NSG
    B(k,i) = 2.0*G(k,i)*y(k,i) ...
      +mu(i+1)/mu(i)*(B(k,i+1)-2.0*G(k,i+1)*y(k,i));
  end
end

%----------------------------
% Compute the coefficients A
%----------------------------

for k=1:NSG
  A(k,1) = wall(k)*(wall(k)*G(k,1)-B(k,1));
end

for i=1:NLR-1
  for k=1:NSG
    A(k,i+1) = A(k,i)+(B(k,i)-B(k,i+1))*y(k,i) ...
      -(G(k,i)-G(k,i+1))*y(k,i)*y(k,i);
  end
end

%--------------------------
% Compute the flow rates Q
%--------------------------

for k=1:NSG
  Q(k,1) = A(k,1)*(y(k,1)-wall(k)) ...
    +0.5*B(k,1)*(y(k,1)*y(k,1)-wall(k)*wall(k));
    -G(k,1)*(y(k,1)^3-wall(k)^3)/3.0;
end
Q(NSG+1,1) = Q(1,1);

for i=2:NLR
  for k=1:NSG
    Q(k,i) = A(k,i)*(y(k,i)-y(k,i-1)) ...
    +0.5*B(k,i)*(y(k,i)*y(k,i)-y(k,i-1)*y(k,i-1)) ...
    -G(k,i)*(y(k,i)^3-y(k,i-1)^3)/3.0;
  end
  Q(NSG+1,i) = Q(1,i);
end

%---
% compute dQ/dx
%
% by central differences (ICU = 1)
% or upwind differences (ICU ne 1)
%---
```

```
%---
if(ICU==1)   % central differences
%---

for i=1:NLR
  dQdx(1,i) = (Q(2,i)-Q(NSG,i))/Dx2;
  for k=2:NSG
    dQdx(k,i) = (Q(k+1,i)-Q(k-1,i))/Dx2;
  end
end

%---
else   % backward differences
%---

for i=1:NLR
  dQdx(1,i) = (Q(1,i)-Q(NSG,i))/Dx;
  for k=2:NSG
    dQdx(k,i) = (Q(k,i)-Q(k-1,i))/Dx;
  end
end

%---
end
%---

%---
% finally compute dydt
%---

for i=1:NLR
  for k=1:NSG
  dydt(k,i) = 0.0;
  for j=1:i
    dydt(k,i) = dydt(k,i)-dQdx(k,j);
  end
  dydt(NSG+1,i) = dydt(1,i);
  end
end

%---
% Done
%---

return
```

To implement an explicit finite-difference method, we approximate the time derivative on the left-hand side of (9.3.21) at a grid point with a forward difference using a sufficiently small time step, Δt, and then advance the position of the interfaces. The method is implemented in following code *films* residing in directory 05_*lub* of FDLIB:

```
NLR = 2;    % number of layers
L = 2.0;    % wave length
Dt = 0.01;    % time step
Nstep = 128*2*128;    % number of steps
th0 = 0.125*pi;    % inclination angle
gac = 1.0;    % acceleration of gravity
NSG = 2*16;  % number of divisions (segments)
ICU = 2;    % backward differences
ICU = 1;    % central differences

mu(1) = 1.0; rho(1) = 1.0; gamma(1) = 0.000;    % first layer
rho(2) = 0.5; mu(2) = 1.0; gamma(2) = 0.00;    % second layer

%---
% prepare
%---

rho(NLR+1) = 0.0;
cs0 = cos(th0); sn0 = sin(th0);
gx = gac*sn0; gy = -gac*cs0;

ROT = [cs0,sn0;-sn0,cs0]; % rotation matrix

%---
% wall and initial profiles
%---

Dx = L/NSG;

for k=1:NSG+1
  x(k) = (k-1)*Dx;
  arg = 2*pi*x(k)/L;
  wall(k) = 0.0;
  y(k,1) = 0.2+0.1*cos(arg);
  if(NLR>=2)
    y(k,2) = 0.4+0.1*cos(arg);
  end
  if(NLR>=3)
    y(k,3) = 0.6+0.1*cos(arg);
  end
end

%-----
% time stepping
%-----
```

```
for step=1:Nstep

  dydt = films_pde ...
  ...
    (NLR,NSG,Dx,wall,y,rho,mu,gamma,gx,gy,ICU);

  y = y+dydt*Dt;

%-----
% plotting
%-----

  for k=1:NSG+1
    xx = ROT*[x(k) wall(k)]';
    xplot0(k) = xx(1)/L; yplot0(k) = xx(2)/L;
    xx = ROT*[x(k) y(k,1)]';
    xplot1(k) = xx(1)/L; yplot1(k) = xx(2)/L;
    if(NLR>=2)
    xx = ROT*[x(k) y(k,2)]';
      xplot2(k) = xx(1)/L; yplot2(k) = xx(2)/L;
    end
    if(NLR>=3)
      xx = ROT*[x(k) y(k,3)]';
      xplot3(k) = xx(1)/L; yplot3(k) = xx(2)/L;
    end
  end

  if(step==1)
    Handle0 = plot(xplot0,yplot0,'k');
    hold on
    Handle1 = plot(xplot1,yplot1,'r.-');
    if(NLR>=2)
      Handle2 = plot(xplot2,yplot2,'r.-');
  end
    if(NLR>=3)
      Handle3 = plot(xplot3,yplot3,'r.-');
    end
    xlabel('x/L','fontsize',15)
    ylabel('y/L','fontsize',15)
  end

  set(Handle0,'XData',xplot0,'YData',yplot0)
  set(Handle1,'XData',xplot1,'YData',yplot1)
  if(NLR>=2)
    set(Handle2,'XData',xplot2,'YData',yplot2)
  end
  if(NLR>=3)
    set(Handle3,'XData',xplot3,'YData',yplot3)
  end
```

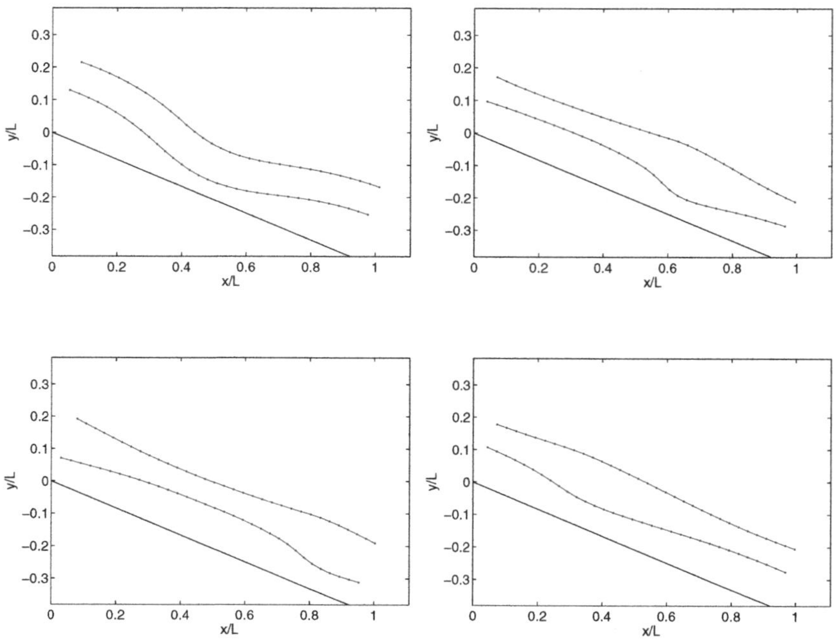

Figure 9.3.2 Snapshots of an evolving two-layer film down an inclined plane produced by code *films* of FDLIB.

```
    drawnow

  end
```

A sequence of evolving profiles for a two-layer flow is shown in figure 9.3.2.

Problem

9.3.1 *Two-layer flow*

Consider a two-layer film flow with fluids of equal viscosity and density in the absence of interfacial tension, $\lambda_1 = 1$, $\beta_1 = 1$, and $\gamma_1 = 0$. Show that the second evolution equation in (9.3.25) reduces to (9.2.10) for single-film flow. Discuss the significance of the first evolution equation in (9.3.25).

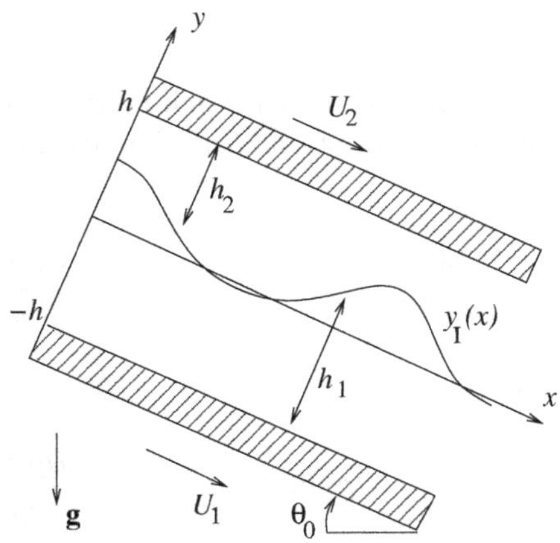

Figure 9.4.1 Illustration of flow of two layers in a channel confined between two parallel plane walls driven by gravity, boundary motion, or an imposed pressure gradient.

Computer Problem

9.3.2 *Multi-film flow*

Run the code *films* for a two-layer configuration of your choice on a horizontal wall and then on an inclined wall. Investigate and discuss the significance of the interfacial tensions.

9.4 Two-layer channel flow

Multi-layer channel flows are encountered in polymer co-extrusion and in the manufacturing of composite and laminated materials. In this section, we consider the flow of two superposed layers in a channel confined between two parallel plane walls separated by the distance $2h$, as illustrated in figure 9.4.1. The lower layer is labeled 1, and the upper layer is labeled 2. The flow is driven partly by the parallel translation of the lower and upper walls with respective velocities U_1 and U_2, partly by a pressure gradient imposed along the x axis, and partly by the gravitational body force.

The velocity profile corresponding to steady unidirectional flow with a flat interface was discussed in section 7.1.1. Our present goal is to derive an evolution equation for the layer thickness when the interface is gently sloped, also accounting for the presence of an insoluble surfactant.

Equations of lubrication flow

We begin by assuming that the flow in each layer is governed by the simplified x and y components of the equation of motion arising from the assumption of lubrication flow,

$$0 = -\frac{\partial p^{(1)}}{\partial x} + \mu_1 \frac{\partial^2 u_x^{(1)}}{\partial y^2} + \rho_1 \, g_x, \qquad 0 = -\frac{\partial p^{(1)}}{\partial y} + \rho_1 \, g_y \tag{9.4.1}$$

for the lower layer, and

$$0 = -\frac{\partial p^{(2)}}{\partial x} + \mu_2 \frac{\partial^2 u_x^{(2)}}{\partial y^2} + \rho_2 \, g_x, \qquad 0 = -\frac{\partial p^{(2)}}{\partial y} + \rho_2 \, g_y \tag{9.4.2}$$

for the upper layer. The components of the acceleration of gravity are given in (9.2.1). The y component of the equation of motion states that the pressure in each layer changes in the y direction only because of gravity.

Velocity profiles

Next, we describe the position of the interface by the equation $y = y_I(x, t)$. Integrating the first equations in (9.4.1) and (9.4.2) twice with respect to y while treating the streamwise pressure gradient as a constant, we obtain the parabolic velocity profile

$$u_x^{(1)}(y, t) = -\frac{1}{2\mu_1} \left(-\frac{\partial p^{(1)}}{\partial x} + \rho_1 \, g_x \right) (y - y_I)^2 + \xi_1 \, (y - y_I) + u_I \tag{9.4.3}$$

in the lower fluid, $-h < y < y_I(x)$, and

$$u_x^{(2)}(y, t) = -\frac{1}{2\mu_2} \left(-\frac{\partial p^{(2)}}{\partial x} + k\rho_2 \, g_x \right) (y - y_I)^2 + \xi_2 \, (y - y_I) + u_I \tag{9.4.4}$$

in the upper fluid, $y_I(x) < y < h$, where u_I is the *a priori* unknown streamwise velocity at the position of the interface, and

$$\xi_1 \equiv \left(\frac{\partial u_x^{(1)}}{\partial y} \right)_{y=y_I}, \qquad \xi_2 \equiv \left(\frac{\partial u_x^{(2)}}{\partial y} \right)_{y=y_I}, \tag{9.4.5}$$

are the *a priori* unknown shear rates evaluated on either side of the interface. If the interface were flat and the surface tension constant, u_I, ξ_1 and ξ_2 would be given by expressions (7.1.16) and (7.1.17).

Shear rate and interface velocity

To compute the interfacial shear rates, we enforce the no-slip boundary condition at the upper and lower wall, $u_x^{(1)}(y = -h) = U_1$ and $u_x^{(2)}(y = h) = U_2$.

Evaluating the velocity from (9.4.3) and (9.4.4) and solving for the shear rates, we find

$$\xi_1 = -\frac{h_1}{2\mu_1}\Big(-\frac{\partial p^{(1)}}{\partial x} + \rho_1\, g_x \Big) + \frac{u_I - U_1}{h_1},$$

$$\xi_2 = \frac{h_2}{2\mu_2}\Big(-\frac{\partial p^{(2)}}{\partial x} + \rho_2\, g_x \Big) - \frac{u_I - U_2}{h_2}, \tag{9.4.6}$$

where

$$h_1(x,t) = h + y_I(x,t), \qquad h_2(x,t) = h - y_I(x,t) \tag{9.4.7}$$

are the local and instantaneous layer thicknesses. In the case of a flat interface with uniform surface tension, the streamwise pressure gradients are equal and expressions (9.4.6) reduce to those given in (7.1.17) with $\partial p^{(1)}/\partial x = \partial p^{(2)}/\partial x = -\chi$.

Next, we substitute the right-hand sides of equations (9.4.6) in the tangential interfacial stress balance

$$\mu_1\, \xi_1 = \mu_2\, \xi_2 + \frac{\partial \gamma}{\partial x}, \tag{9.4.8}$$

and solve for the interfacial velocity,

$$u_I = \frac{1}{\delta + \lambda}\Big[\frac{h_1 h_2}{2\mu_1}\Big(-\frac{\partial p^{(1)}}{\partial x} - \delta\, \frac{\partial p^{(2)}}{\partial x} + \rho_1\, g_x\,(1 + \beta\,\delta) \Big) + \delta\, U_1 + \lambda\, U_2 + \frac{h_2}{\mu_1}\frac{\partial \gamma}{\partial x} \Big], \tag{9.4.9}$$

where γ is the surface tension,

$$\lambda \equiv \frac{\mu_2}{\mu_1}, \qquad \beta \equiv \frac{\rho_2}{\rho_1}, \qquad \delta \equiv \frac{h_2}{h_1}, \tag{9.4.10}$$

and $h_1(1+\delta) = 2h$. In the case of a flat interface with uniform surface tension, the streamwise pressure gradients are equal and expression (9.4.9) reduces to that given in (7.1.16) with $\partial p^{(1)}/\partial x = \partial p^{(2)}/\partial x = -\chi$.

Streamwise pressure field

Next, we note that the pressure undergoes a jump across the interface due to the surface tension, γ, and write

$$p^{(1)}(x, y = y_I) = p^{(2)}(x, y = y_I) + \gamma\, \kappa, \tag{9.4.11}$$

where κ is the interfacial curvature. Adopting the familiar approximation $\kappa(x) \simeq -\partial^2 h/\partial x^2$ and differentiating (9.4.11) with respect to x using the chain rule, we find

$$\frac{\partial p^{(1)}}{\partial x} + \frac{\partial p^{(1)}}{\partial y}\,\frac{\partial y_I}{\partial x} = \frac{\partial p^{(2)}}{\partial x} + \frac{\partial p^{(2)}}{\partial y}\,\frac{\partial y_I}{\partial x} - \gamma\, \frac{\partial^3 h}{\partial x^3}, \tag{9.4.12}$$

where both sides are evaluated at the interface. Using the second equations in (9.4.1) and (9.4.2) to evaluate the derivative of the pressure with respect to y and rearranging, we obtain

$$\frac{\partial p^{(2)}}{\partial x} = \frac{\partial p^{(1)}}{\partial x} + (\rho_1 - \rho_2)\, g_y \,\frac{\partial y_I}{\partial x} + \gamma\,\frac{\partial^3 h}{\partial x^3}. \tag{9.4.13}$$

Substituting this expression in (9.4.9) to eliminate $p^{(2)}$, we derive an alternative expression for the interfacial velocity,

$$u_I = \frac{1}{\delta+\lambda}\Big[\frac{h_1 h_2}{2\mu_1}\Big(-(1+\delta)\,\frac{\partial p^{(1)}}{\partial x} - \delta\,(\rho_1-\rho_2)\, g_y\,\frac{\partial y_I}{\partial x} - \delta\gamma\,\frac{\partial^3 h}{\partial x^3}$$

$$+\rho_1\, g_x\,(1+\beta\,\delta)\Big) + \delta\, U_1 + \lambda\, U_2 + \frac{h_2}{\mu_1}\,\frac{\partial\gamma}{\partial x}\Big], \tag{9.4.14}$$

involving the pressure gradient in the lower layer and the instantaneous shape of the interface.

Flow rates and mass conservation

We proceed by integrating the velocity profiles (9.4.3) and (9.4.4) with respect to y over their domain of definition to derive expressions for the streamwise flow rates,

$$Q_1 \equiv \int_{-h}^{y_I} u_x^{(1)}\,\mathrm{d}y = -\frac{h_1^3}{6\mu_1}\Big(-\frac{\partial p^{(1)}}{\partial x} + \rho_1\, g_x\Big) - \frac{1}{2}\,\xi_1\, h_1^2 + u_I\, h_1,$$

$$Q_2 \equiv \int_{y_I}^{h} u_x^{(2)}\,\mathrm{d}y = -\frac{h_2^3}{6\mu_2}\Big(-\frac{\partial p^{(2)}}{\partial x} + \rho_2\, g_x\Big) + \frac{1}{2}\,\xi_2\, h_2^2 + u_I\, h_2. \tag{9.4.15}$$

Using expressions (9.4.6) to eliminate the shear rates ξ_1 and ξ_2 from the right-hand sides, we find

$$Q_1 = \frac{h_1^3}{12\mu_1}\Big(-\frac{\partial p^{(1)}}{\partial x} + \rho_1\, g_x\Big) + \frac{1}{2}\,(u_I + U_1)\, h_1,$$

$$Q_2 = \frac{h_2^3}{12\mu_2}\Big(-\frac{\partial p^{(2)}}{\partial x} + \rho_2\, g_x\Big) + \frac{1}{2}\,(u_I + U_2)\, h_2. \tag{9.4.16}$$

A mass balance over a control area confined by (a) two parallel planes that are normal to the channel walls and are separated by an infinitesimal distance, (b) the enclosed sections of the walls, and (c) the enclosed section of the evolving interface, requires that the rate of accumulation of mass of each layer inside this control area should be equal to the difference in the mass flow rates into and out from the control volume. In differential form,

$$\frac{\partial h_1}{\partial t} = \frac{\partial y_I}{\partial t} = -\frac{\partial Q_1}{\partial x}, \qquad \frac{\partial h_2}{\partial t} = -\frac{\partial y_I}{\partial t} = -\frac{\partial Q_2}{\partial x}. \tag{9.4.17}$$

Since $h_1 + h_2 = 2h$ is constant, $\partial h_1/\partial t + \partial h_2/\partial t = 0$, and

$$\frac{\partial Q_1}{\partial x} + \frac{\partial Q_2}{\partial x} = 0 \qquad \text{or} \qquad Q_1 + Q_2 = f(t), \tag{9.4.18}$$

where $f(t)$ is an unspecified function of time. To compute this function, we use the expressions for the flow rates and interfacial velocity given in equations (9.4.16) and (9.4.14). Eliminating the pressure gradient in the second layer using expression (9.4.13), we find

$$\frac{\partial p^{(1)}}{\partial x} = \frac{N + \mu_1 f(t)}{D}, \tag{9.4.19}$$

where

$$\begin{aligned} N = &-(\rho_1 - \rho_2)\, g_y\, (h_2 + \frac{6\,\lambda}{\lambda + \delta}\, h)\, h_2^2\, \frac{\partial h_1}{\partial x} - \gamma\, (h_2 + \frac{6\,\lambda}{\lambda + \delta}\, h)\, h_2^2\, \frac{\partial^3 h_1}{\partial x^3} \\ &+\rho_1\, g_x (\lambda\, h_1^3 + \beta\, h_2^3 + \frac{12\,\lambda}{\lambda + \delta}\, h\, \frac{1 + \lambda\delta}{1 + \delta}\, h^2\, h_2) \\ &+12\, \mu_2\, h\, (\frac{U_1 + \delta\, U_2}{1 + \delta} + \frac{\delta\, U_1 + \lambda\, U_2}{\lambda + \delta}) \\ &+h_2 \left(12 \frac{\lambda h}{\delta + \lambda} - h_2\, (\frac{6\, h\lambda}{\delta + \lambda} + h_2)\, \frac{\partial^2 y_I}{\partial x^2}\right) \frac{\partial \gamma}{\partial x}, \end{aligned} \tag{9.4.20}$$

and

$$D = \lambda\, h_1^3 + h_2^3 + \frac{12\,\lambda}{\lambda + \delta}\, h^2\, h_2. \tag{9.4.21}$$

Integrating (9.4.19) with respect to x over the length L and solving for $f(t)$, we find

$$f(t) = \frac{\Delta p - \int_0^L \frac{N}{D}\, \mathrm{d}x}{\mu_1 \int_0^L \frac{1}{D}\, \mathrm{d}x}, \tag{9.4.22}$$

where $\Delta p \equiv p(x = L) - p(x = 0)$ is the negative of the pressure drop over the length L. The integrals on the right-hand side of (9.4.22) can be computed from knowledge of the instantaneous shape of the interface using numerical methods.

Evolution equation

Having obtained the function $f(t)$, we evaluate the streamwise pressure gradient from (9.4.19), compute the interfacial velocity from (9.4.14), recover the flow rate Q_1 from the first equation in (9.4.16), and use the first equation in (9.4.17) to derive an expression for the rate of change of the lower film thickness or interface position. Symbolically, we write

$$\frac{\partial y_I}{\partial t} = F(y_I, \frac{\partial y_I}{\partial x}, \frac{\partial^2 y_I}{\partial x^2}, \frac{\partial^3 y_I}{\partial x^3}, \frac{\partial^4 y_I}{\partial x^4}), \tag{9.4.23}$$

where F is a nonlinear function of its arguments defined implicitly in terms of the aforementioned substitutions. The concentration of an insoluble surfactant is governed by the convection–diffusion equation (4.10.19),

$$\frac{\partial \Gamma}{\partial t} + \frac{\partial (u_I \Gamma)}{\partial x} = \frac{\partial}{\partial x}\Big(D_s \frac{\partial \Gamma}{\partial x}\Big). \tag{9.4.24}$$

The two evolution equations can be integrated in time using a standard finite difference method based on a one-dimensional grid whose nodes are deployed over one period along the x axis.

Explicit numerical integration

To implement an explicit method, we apply (9.4.23) at a grid point at time t and approximate the time derivatives on the left-hand side with a forward finite difference over a small time step, Δt, obtaining

$$\frac{y_I(x, t + \Delta t) - y_I(x, t)}{\Delta t} = F(t). \tag{9.4.25}$$

Evaluating the right-hand side by numerical differentiation and solving for $y_I(x, t + \Delta t)$, we obtain the position of the interface at the new time $t + \Delta t$.

The method is implemented in the following code *chan2l_exp* residing in directory *05_lub* of FDLIB for uniform surface tension:

```
h = 0.2;
th0 = 0.125*pi;    % inclination angle
L = 1.0;      % wave length
yIunp = 0.0;    % unperturbed position
a0 = 0.1;
rho1 = 1.0; rho2 = 1.0;
mu1 = 1.0; mu2 = 0.4;
gamma = 0.4;    % surface tension
U1 = -0.2; U2 = 1.0;
pd = 0.3;    % streamwise pressure gradient
gac = 1.0;    % acceleration of gravity
NSG = 32;
Dt = 0.001;    % time step nstep = 128*128;    % number of steps
method = 2;  % backward differences
method = 1;    % central differences

%----
% prepare
%----

sn0 = sin(th0);
cs0 = cos(th0);
ROT = [cs0,sn0;-sn0,cs0];    % rotation matrix for plotting
```

```
gx = gac*sn0;
gy = -gac*cs0;
hs = h*h;
vsr = mu2/mu1;   % viscosity ratio
Drho = rho1-rho2;   % density difference
delta = rho2/rho1;   % density ratio
wn = 2*pi/L;   % wave number
Dx = L/NSG;
Dx2 = 2.0*Dx;
Dx23 = 2.0*Dx*Dx*Dx;
Dxs = Dx*Dx;

%----------------
% initial profile
%----------------

for i=1:NSG+4
  x(i) = (i-1.0)*Dx;
  arg = wn*x(i);
  y(i) = yIunp + a0*cos(arg);
  srtn(i) = gamma;
end

%==============
% time stepping
%==============

for step=1:nstep

for i=3:NSG+2
  ib = i-2; ia = i-1; i1 = i+1; i2 = i+2;
  DyDx1(i) = (y(i1)-y(ia))/Dx2;
  DyDx2(i) = (y(i1)-2.0*y(i)+y(ia))/Dxs;
  DyDx3(i) = (y(i2)-2.0*y(i1)+2.0*y(ia)-y(ib))/Dx23;
  h1(i) = h+y(i); h2(i) = h-y(i); r(i) = h2(i)/h1(i);
end

DyDx1(1) = DyDx1(NSG+1);
DyDx2(1) = DyDx2(NSG+1);
DyDx3(1) = DyDx3(NSG+1);
h1(1) = h1(NSG+1); h2(1) = h2(NSG+1); r(1) = r(NSG+1);
DyDx1(2) = DyDx1(NSG+2);
DyDx2(2) = DyDx2(NSG+2);
DyDx3(2) = DyDx3(NSG+2);
h1(2) = h1(NSG+2); h2(2) = h2(NSG+2); r(2) = r(NSG+2);

%---------------
% compute dP1/dx
%---------------
```

```
for i=1:NSG
  tmp = 1.0/(vsr+r(i));
  Den = vsr*h1(i)^3 + h2(i)^3  ...
    + 12.0*vsr*tmp*hs*h2(i);
  t1 = -(Drho*gy/mu1)*(h2(i)+6.0*vsr*tmp*h) ...
    *h2(i)^2*DyDx1(i);
  t2 = - srtn(i)/mu1 *(h2(i)+6.0*vsr*h*tmp) ...
    *h2(i)*h2(i)*DyDx3(i);
  t3 =  (rho1*gx/mu1)* (vsr*h1(i)^3 + delta*h2(i)^3 ...
    +6.0*vsr*tmp*(1.0+delta*r(i))*h*h1(i)*h2(i));
  t4 = 6.0*vsr*(U1*h1(i)+U2*h2(i)) ...
    + 12.0D0*vsr*h*tmp*(vsr*U2+r(i)*U1);
  Dp1Dx(i) = mu1*(t1+t2+t3+t4)/Den;
  store(i) = mu1/Den;
end

%--------------------------
% compute the function f(t)
%--------------------------

sum1 = 0.0; sum2 = 0.0;
for i=1:NSG
  sum1 = sum1 + Dp1Dx(i);
  sum2 = sum2 + store(i);
end
sum1 = sum1*Dx; sum2 = sum2*Dx;
f = (pd-sum1)/sum2;

%-----------------------------------
% complete the computation of dp1/dx
%-----------------------------------

for i=1:NSG
  Dp1Dx(i) = Dp1Dx(i) + f*store(i);
end
Dp1Dx(NSG+1) = Dp1Dx(1);
Dp1Dx(NSG+2) = Dp1Dx(2);

%------------
% compute u_I
%------------

for i=1:NSG
  ri1 = 1.0+r(i);
  ui(i) = h*h2(i)/mu1 *(-Dp1Dx(i) ...
    +rho1*gx*(1.0+delta*r(i))/ri1 ) ...
        +0.5*h2(i)^2/mu1*(-Drho*gy*DyDx1(i) ...
    -srtn(i)*DyDx3(i))+ r(i)*U1+vsr*U2;
```

```
    ui(i) = ui(i)/(vsr+r(i));
end

ui(NSG+1) = ui(1);
ui(NSG+2) = ui(2);

%------------------------------------
% compute dQ1/dx
%
% then set dy/dt = - dQ1/dx
%------------------------------------

pr1 = 12.0*Dx2*mu1; pr2 = 2.0*Dx2;
pr3 = 12.0*Dx*mu1; pr4 = 2.0*Dx;

for i=2:NSG+1

  ia = i-1; i1 = i+1;

  %---
  if(method==1)   % central differences
  %---
    dq1dx = ( h1(i1)^3 *(-Dp1Dx(i1)+rho1*gx) ...
             -h1(ia)^3 *(-Dp1Dx(ia)+rho1*gx))/pr1 ...
      + ( (ui(i1)+U1)*h1(i1) -(ui(ia)+U1)*h1(ia))/pr2;
  %---
  else   % backward differences
  %---

     dq1dx = ( h1(i)^3 *(-Dp1Dx(i) +rho1*gx) ...
           -h1(ia)^3 *(-Dp1Dx(ia)+rho1*gx))/pr3 ...
      + ( (ui(i) +U1)*h1(i) -(ui(ia)+U1)*h1(ia))/pr4;
  %---
  end
  %---

DyDt(i) =-dq1dx;

end

for i=2:NSG+1
  y(i) = y(i)+Dt*DyDt(i);
end

y(1) = y(NSG+1);
y(NSG+2) = y(2);
y(NSG+3) = y(3);
y(NSG+4) = y(4);
```

```
%---
% plotting
%---

for k=1:NSG+1
  xx = ROT*[x(k) y(k)]';
  xplot1(k) = xx(1)/L;
  yplot1(k) = xx(2)/L;
end

if(step==1)
  Handle1 = plot(xplot1,yplot1,'r.-');
  xlabel('x/L','fontsize',15)
  ylabel('y/L','fontsize',15)
  set(gca,'fontsize',15)
  axis equal
end

set(Handle1,'XData',xplot1,'YData',yplot1)
drawnow

%=====
end   % of time stepping
%=====
```

A sequence of evolving profiles is shown in figure 9.4.2. Surface tension and gravity for stably stratified fluids dampen interfacial deviations from the flat shape.

Implicit numerical integration

The explicit method requires a small time step to prevent the onset of numerical instability manifested by growing oscillations in the nodal position of the interface. This restriction can be overcome by applying equations (9.4.23) at a grid point at a certain time $t + \Delta t$ and approximating the time derivatives on the left-hand side with a backward finite difference using a small time step Δt, to obtain

$$\frac{y_I(x, t + \Delta t) - y_I(x, t)}{\Delta t} = F(t + \Delta t). \tag{9.4.26}$$

Evaluating the right-hand side by numerical differentiation, we obtain a nonlinear system of algebraic equations for the interfacial elevation at the nodes at time $t + \Delta t$.

The method is implemented in code *chan2l_imp* residing in directory *05_lub* of FDLIB. A drawback of the implicit method is that computing the solution of the nonlinear algebraic system at each time step requires significant computational effort.

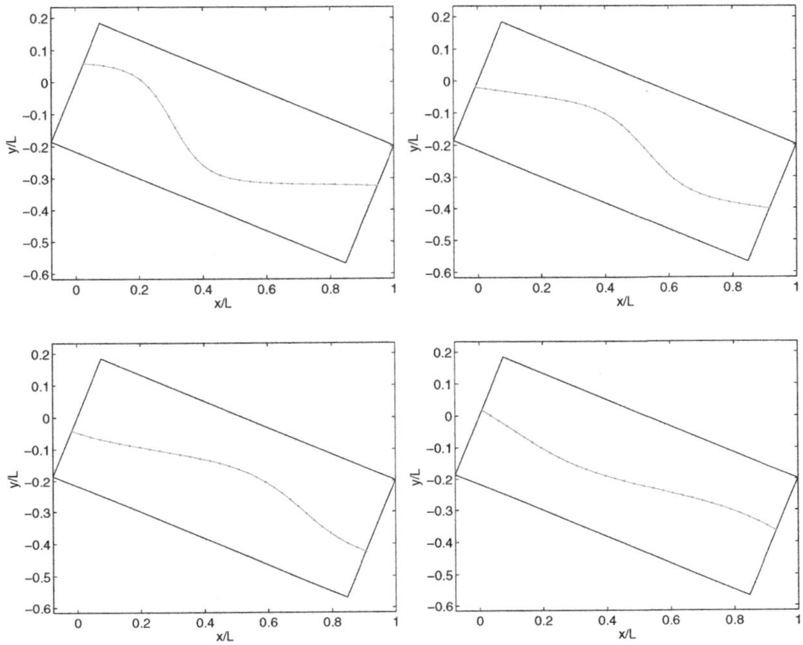

Figure 9.4.2 Snapshots of an evolving interface between two layers in a channel computed by code *chan2l_exp* of FDLIB.

Problem

9.4.1 *Single-layer flow in a channel*

Demonstrate that, when the densities and viscosities of the fluids are matched, the equations of two-layer flow reduce to those for single-layer flow.

Computer Problem

9.4.2 *Two-layer flow in a channel*

Run the code *chan2l_exp* for a set of conditions of your choice. Display sequences of evolving profiles, investigate the effect of the time step on the numerical stability, and discuss the nature of the motion.

(a)

(b)

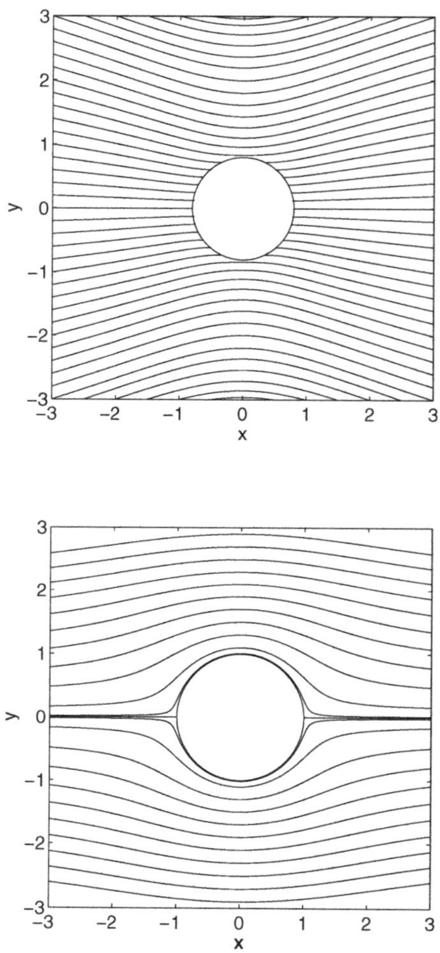

Figure 9.5.1 Streamline pattern in a meridional plane of axisymmetric Stokes flow due to the translation of a sphere in an infinite fluid. Streamlines are drawn (*a*) in a stationary frame of reference, and (*b*) in a frame of reference moving with the sphere.

9.5 Flow due to the motion of a sphere

A variety of natural and engineering applications involve particle motion in a viscous fluid. An elementary configuration involves a spherical particle settling with constant velocity due to its weight in a virtually infinite quiescent ambient fluid, thereby generating an axisymmetric flow, as illustrated in figure 9.5.1.

When the radius of the spherical particle, a, is small, or the fluid viscosity, μ, is high, or the fluid density, ρ, is low, or the particle velocity V is small, inertial forces near the particle are negligible and the left-hand side of the Navier-Stokes equation can be set to zero without introducing serious error. The result is the Stokes equation,

$$\nabla p = \mu \nabla^2 \mathbf{u} + \rho \mathbf{g}, \tag{9.5.1}$$

which, together with the continuity equation,

$$\frac{\partial u_x}{\partial x} + \frac{\partial u_y}{\partial y} + \frac{\partial u_z}{\partial z} = 0, \tag{9.5.2}$$

comprise the equations of Stokes or creeping flow. The three scalar components of the Stokes equation are

$$\begin{aligned}
\frac{\partial p}{\partial x} &= \mu \,(\frac{\partial^2 u_x}{\partial x^2} + \frac{\partial^2 u_x}{\partial y^2} + \frac{\partial^2 u_x}{\partial y^2}) + \rho\, g_x, \\
\frac{\partial p}{\partial y} &= \mu \,(\frac{\partial^2 u_y}{\partial x^2} + \frac{\partial^2 u_y}{\partial y^2} + \frac{\partial^2 u_y}{\partial y^2}) + \rho\, g_y, \\
\frac{\partial p}{\partial z} &= \mu \,(\frac{\partial^2 u_z}{\partial x^2} + \frac{\partial^2 u_z}{\partial y^2} + \frac{\partial^2 u_z}{\partial y^2}) + \rho\, g_z.
\end{aligned} \tag{9.5.3}$$

The formal requirement for fluid inertia to be negligible is that the Reynolds number defined with respect to the particle radius, $Re = a\rho V/\mu$, is small. However, we will see that this assumption does not guarantee that inertia will be negligible uniformly throughout the domain of the flow and, in particular, far from the sphere. Having noted this exception, we proceed to compute the solution by assuming that inertia is negligible throughout the domain of flow, and then return to assess the validity of our assumption.

9.5.1 Formulation in terms of the stream function

To simplify the analysis, we take advantage of the axial symmetry of the flow with respect to the direction of translation and refer to spherical polar coordinates with the x axis pointing in the direction of translation. To bypass the computation of the pressure, we work with the vorticity transport equation (6.6.9) for the component of the vorticity pointing in the meridional direction, ω_φ. In the absence of fluid inertia, we obtain the simplified form

$$E^2(\sigma\, \omega_\varphi) = 0, \tag{9.5.4}$$

where $\sigma = r \sin\varphi$ is the distance from the x axis. The second-order linear differential operator, E^2, is defined in (2.9.14) and (2.9.17).

Expressing the meridional component of the vorticity in terms of the stream function using (2.9.13), we find

$$E^4\psi \equiv E^2\, E^2\psi = 0. \tag{9.5.5}$$

Our task is to solve this fourth-order differential equation in the exterior of the sphere, subject to appropriate boundary and far-field conditions.

On the surface of the sphere, $r = a$, the no-slip and no-penetration boundary conditions require that the fluid velocity is equal to the velocity of the sphere,

$$\mathbf{u}(r = a) = V\,\mathbf{e}_x, \tag{9.5.6}$$

where $\mathbf{e}_x$ is the unit vector along the x axis. Thus,

$$u_r(r = a) = V\,\cos\theta, \qquad u_\theta(r = a) = -V\,\sin\theta, \tag{9.5.7}$$

The far-field condition requires that, as r tends to infinity, the velocity decays so that the fluid becomes quiescent.

Using the differential relations between the velocity and the stream function,

$$u_r = \frac{1}{r^2\sin\theta}\,\frac{\partial\psi}{\partial\theta}, \qquad u_\theta = -\frac{1}{r\sin\theta}\,\frac{\partial\psi}{\partial r}, \tag{9.5.8}$$

we find that, in terms of the stream function, the boundary and far-field conditions assume the forms

$$\psi(r = a) = \frac{1}{2}\,Va^2\,\sin^2\theta, \qquad \left(\frac{\partial\psi}{\partial r}\right)_{r=a} = Va\,\sin^2\theta, \tag{9.5.9}$$

and

$$\frac{\psi}{r^m} \to 0 \qquad \text{as} \qquad r \to \infty, \tag{9.5.10}$$

for $m \geq 2$. Condition (9.5.10) allows the stream function to diverge at infinity but requires that the rate of divergence is less than quadratic; otherwise, the velocity will not decay, as required.

Separation of variables

Motivated by the functional form of the boundary conditions (9.5.9), we search for a solution in separated variables, writing

$$\psi = q(r)\,\sin^2\theta. \tag{9.5.11}$$

The unknown function, $q(r)$, satisfies the boundary conditions

$$q(r = a) = \frac{1}{2}\,Va^2, \qquad \left(\frac{\mathrm{d}q}{\mathrm{d}r}\right)_{r=a} = Va, \tag{9.5.12}$$

and the far-field condition

$$\frac{q(r)}{r^m} \to 0 \qquad \text{as} \qquad r \to \infty, \tag{9.5.13}$$

for $m \geq 2$. Substituting (9.5.11) in (9.5.5) and carrying out the differentiations, we obtain a fourth-order linear ordinary differential equation,

$$\Big(\frac{\mathrm{d}^2}{\mathrm{d}r^2} - \frac{2}{r^2}\Big)\Big(\frac{\mathrm{d}^2}{\mathrm{d}r^2} - \frac{2}{r^2}\Big)\, q = 0, \tag{9.5.14}$$

which can be decomposed into two second-order component equations,

$$\frac{\mathrm{d}^2 w}{\mathrm{d}r^2} - \frac{2}{r^2}\, w = 0, \qquad \frac{\mathrm{d}^2 q}{\mathrm{d}r^2} - \frac{2}{r^2}\, q = w. \tag{9.5.15}$$

The second equation merely defines the intermediate function w.

The general solution of the first equation in (9.5.15) is readily found to be $w = Ar^2 + B/r$, where A and B are two constants. Substituting this expression in the second equation, we derive the inhomogeneous equation

$$\frac{\mathrm{d}^2 q}{\mathrm{d}r^2} - \frac{2q}{r^2} = A\, r^2 + \frac{B}{r}. \tag{9.5.16}$$

The solution of (9.5.16) is the sum of (a) the general solution of the homogeneous equation computed by setting the right-hand side to zero, given by $q = Cr^2 + D/r$, where C and D are two new constants, and (b) a particular solution found by inspection. The result is

$$q = \frac{A}{10}\, r^4 - \frac{B}{2} r + C\, r^2 + \frac{D}{r}. \tag{9.5.17}$$

To ensure the satisfaction of condition (9.5.13), we set A and C to zero. Substituting the remaining expression in the boundary conditions (9.5.12) and solving for the coefficients B and D, we find

$$B = -\frac{3}{2}\, V a, \qquad D = -\frac{1}{4}\, V a^3. \tag{9.5.18}$$

Substituting these values in (9.5.17) and the result in (9.5.11), we obtain the desired stream function,

$$\psi = V\, r^2\, \frac{1}{4}\, \frac{a}{r}\, \Big[\, 3 - \Big(\frac{a}{r}\Big)^2\, \Big]\, \sin^2\theta. \tag{9.5.19}$$

Velocity and pressure fields

The radial and azimuthal components of the velocity arise by substituting (9.5.19) in equations (9.5.8), yielding

$$u_r = \frac{V}{2}\,\frac{a}{r}\,[3 - \left(\frac{a}{r}\right)^2]\,\cos\theta, \qquad u_\theta = -\frac{V}{4}\,\frac{a}{r}\,[3 + \left(\frac{a}{r}\right)^2]\,\sin\theta. \tag{9.5.20}$$

Substituting these expressions in the r and θ spherical polar components of the Stokes equations, we derive expressions for the partial derivatives of the pressure $\partial p/\partial r$ and $\partial p/\partial \theta$. Integrating these expressions, we derive the pressure distribution

$$p = \frac{3}{2}\,\frac{\mu V}{a}\,\left(\frac{a}{r}\right)^2\,\cos\theta + \rho\,(g_x\,x + g_y\,y + g_z\,z) + c, \tag{9.5.21}$$

where $\mathbf{g} = (g_x, g_y, g_z)$ is the acceleration of gravity, and c is a constant determined by the pressure far from the sphere (problem 9.5.1).

Validity of the equations of creeping flow

We must now return to evaluate the assumptions under which the preceding analysis is valid. Equations (9.5.20) reveal that, in the absence of fluid inertia, the velocity decays like Va/r far from the sphere; the cubic term $V(a/r)^3$ decays much faster and can be neglected. Differentiating this asymptotic form with respect to r, we find that the derivative $\partial u_r/\partial r$ involved in the nonlinear convective term on the left-hand side of the Navier-Stokes equation decays like Va/r^2, and the derivative $\partial^2 u_r/\partial r^2$ involved in the viscous term on the right-hand side of the Navier-Stokes equation decays like Va/r^3.

Furthermore, we note that the time derivative $\partial u_r/\partial t$ scales with u_r/T, where T the characteristic time scale. In the absence of time dependence due to an external action, T is comparable to the ratio of the distance from the center of the sphere to the velocity of the sphere, $T = r/V$.

Using this scaling, we find that the ratio of the magnitude of the nonlinear inertial term to the magnitude of the viscous term in the equation of motion is on the order of

$$\frac{\rho\,V\,\frac{a}{r}\,V\,\frac{a}{r^2}}{\mu\,V\,\frac{a}{r^3}} = \frac{\rho a V}{\mu} \equiv Re, \tag{9.5.22}$$

and the ratio of the magnitude of the time-dependent inertial term to the magnitude of the viscous term is on the order of

$$\frac{\rho(V\,\frac{a}{r})/(\frac{r}{V})}{\mu V\,\frac{a}{r^3}} = \frac{\rho\,V r}{\mu} \equiv Re_r = Re\,\frac{r}{a}. \tag{9.5.23}$$

We have introduced the global Reynolds number, Re, defined with respect to the radius of the sphere, and the local Reynolds number, Re_r, defined with respect to distance from the center of the sphere.

Equations (9.5.22) and (9.5.23) reveal that, for inertial forces to be negligible, both Re and Re_r must be far less than unity. The former can be made arbitrarily small by adjusting one of the flow parameters involved in its definition, ρ, a, V, and μ. However, there is no way that the latter can be uniformly small throughout the domain of flow. Even if Re is exceedingly small, the ratio r/a becomes arbitrary large sufficiently far from the sphere and the local Reynolds number, Re_r, increases linearly with respect to distance, rendering the effect of fluid inertia significant.

We conclude that the approximation of creeping flow ceases to be accurate far from the sphere, with one disturbing consequence. If the governing equations are not valid all the way up to infinity, enforcing the far-field boundary condition expressed by (9.5.10) is not valid. Fortunately, a more detailed analysis using the method of matched asymptotic expansions shows that retaining the far-field condition introduces an error that is comparable to that introduced by dropping the inertial terms in the equation of motion, which is on the order of Re.

9.5.2 Traction, force, and the Archimedes-Stokes law

Having derived expressions for the velocity and pressure fields, given, respectively, in (9.5.20) and (9.5.21), we proceed to evaluate the traction on the surface of the sphere and then integrate it to obtain the force exerted on the sphere. For simplicity, we assume that gravity is directed along the x axis; that is, $g_x = g$, $g_y = 0$, and $g_z = 0$, where g is the magnitude of the acceleration of gravity.

Using the definition of the Newtonian stress tensor shown in equations (4.5.4), we derive an expression for the normal component of the traction

$$\sigma_{rr}(r = a) = (-p + 2\,\mu\,\frac{\partial u_r}{\partial r})_{r=a} = -\frac{3}{2}\,\frac{\mu V}{a}\,\cos\theta - \rho\,g\,x - c, \qquad (9.5.24)$$

and an expression for the tangential component of the traction,

$$\sigma_{r\theta}(r = a) = \sigma_{\theta r}(r = a) = [\,r\frac{\partial}{\partial r}(\frac{u_\theta}{r}) + \frac{1}{r}\frac{\partial u_r}{\partial \theta}\,]_{r=a} = \frac{3}{2}\,\frac{\mu V}{a}\,\sin\theta, \qquad (9.5.25)$$

where the constant c is associated with the unspecified pressure. Note that, because of the no-slip boundary condition, viscous stresses do not contribute to the normal component of the traction, in agreement with our discussion in section 4.7.

Combining (9.5.24) and (9.5.25), we find that the traction exerted on the sphere is

$$\mathbf{f} = \sigma_{rr}(r = a)\,\mathbf{e}_r + \sigma_{r\theta}(r = a)\,\mathbf{e}_\theta, \qquad (9.5.26)$$

where $\mathbf{e}_r$ and $\mathbf{e}_\theta$ are the unit vectors in the radial and azimuthal direction. Expressing $\mathbf{e}_r$ and $\mathbf{e}_\theta$ in terms of the Cartesian unit vectors using relations (1.3.35), substituting the results along with expressions (9.5.24) and (9.5.25) in (9.5.26) and simplifying the resulting expression, we derive the remarkably simple result

$$\mathbf{f} = -\frac{3}{2}\,\frac{\mu V}{a}\,\mathbf{e}_x - (\rho\, g\, x + c)\,\mathbf{e}_r, \tag{9.5.27}$$

which shows that, hydrostatic contributions aside, the traction over the surface of the sphere points opposite to the direction of translation and has a uniform magnitude.

To compute the force exerted on the sphere, $\mathbf{F}$, we integrate the traction over the surface of the sphere. Substituting $x = r\cos\theta$, we find

$$\mathbf{F} = \iint_{Sphere} \mathbf{f}\,\mathrm{d}S = \int_0^{2\pi}\int_0^{\pi} \mathbf{f}\, a^2 \sin\theta\,\mathrm{d}\theta\,\mathrm{d}\varphi = -6\pi\mu V a\,\mathbf{e}_x - \frac{4}{3}\,\pi a^3 \rho g\,\mathbf{e}_x. \tag{9.5.28}$$

The first term on the right-hand side expresses Stokes's law. The second term is Archimedes's buoyancy force familiar from our discussion in chapter 5 of hydrostatics, stating that the force exerted on an immersed body is equal in magnitude and opposite in direction to the weight of the fluid displaced by the body.

Terminal velocity of a settling sphere

As an application, we compute the terminal velocity of a solid sphere with density ρ_s settling in an infinite ambient fluid along the x axis. Balancing the weight of the sphere and the force given in (9.5.28), we find

$$\frac{4}{3}\,\pi a^3 \rho_s g\,\mathbf{e}_x - 6\pi\mu V a\,\mathbf{e}_x - \frac{4}{3}\,\pi a^3 \rho g\,\mathbf{e}_x = \mathbf{0}. \tag{9.5.29}$$

Solving for the velocity of the sphere, we find

$$V = \frac{2}{9}\,\frac{a^2(\rho_s - \rho)g}{\mu}. \tag{9.5.30}$$

In practice, this equation can be used to estimate the viscosity of a fluid from observation of the terminal velocity of a sphere in a device called the falling-ball viscometer.

Problems

9.5.1 *Pressure distribution around a moving sphere*

Derive the pressure distribution (9.5.21).

9.5.2 *Flow past a stationary sphere*

Consider steady uniform flow with velocity U along the x axis past a stationary sphere of radius a. Derive expressions for the stream function, the velocity, the pressure field, the traction and the force on the surface of the sphere. Show that, for the fluid inertia to be negligible throughout the domain of flow, both Re and Re_r defined in equations (9.5.22) and (9.5.23) must be considerably less than unity.

9.6 Point forces and point sources in Stokes flow

Consider the stream function of the flow due to a translating sphere given in equation (9.5.19). The right-hand side involves two terms that decay like the inverse or inverse cubic power of the scaled distance from the center of the sphere, r/a. To make this distinction clear, we recast equation (9.5.19) into the form

$$\psi = s_x \, \psi^{3D-ST-x} + d_x \, \psi^{3D-PSD-x}, \tag{9.6.1}$$

where

$$s_x = \frac{3}{4} V a, \qquad d_x = -\pi V a^3 \tag{9.6.2}$$

are two constant coefficients,

$$\psi^{3D-ST-x} \equiv r \sin^2\theta \tag{9.6.3}$$

is the stream function associated with a fundamental solution of the equations of Stokes flow called the three-dimensional Stokeslet, and

$$\psi^{3D-PSD-x} \equiv \frac{1}{4\pi r} \sin^2\theta \tag{9.6.4}$$

is a stream function representing irrotational flow due to a three-dimensional point-source dipole pointing along the x axis, as discussed in section 3.6.

The expressions for the velocity, pressure, and stress can be resolved into analogous components corresponding to the Stokeslet and point-source dipole. For example, the Cartesian components of the velocity are given by

$$\begin{bmatrix} u_x \\ u_y \\ u_z \end{bmatrix} = s_x \begin{bmatrix} \frac{1}{r} + \frac{x^2}{r^3} \\ \frac{xy}{r^3} \\ \frac{xz}{r^3} \end{bmatrix} + \frac{d_x}{4\pi} \begin{bmatrix} -\frac{1}{r^3} + \frac{3x^2}{r^5} \\ \frac{3xy}{r^5} \\ \frac{3xz}{r^5} \end{bmatrix}. \tag{9.6.5}$$

The first and second terms on the right-hand side of (9.6.5) represent, respectively, the velocity field due to a Stokeslet and a potential dipole both situated at the origin parallel to the x axis.

9.6.1 The Oseen tensor and the point force

Generalizing the expression for the Stokeslet given on the right-hand side of (9.6.5), we derive the velocity at the point $\mathbf{x} = (x, y, z)$ due to a three-dimensional Stokeslet with arbitrary vectorial strength, $\mathbf{s} = (s_x, s_y, s_z)$, situated at the point $\mathbf{x}_0 = (x_0, y_0, z_0)$, given by

$$\begin{bmatrix} u_x \\ u_y \\ u_z \end{bmatrix} (\mathbf{x}, \mathbf{x}_0) = \mathbf{S}(\mathbf{x}, \mathbf{x}_0) \cdot \begin{bmatrix} s_x \\ s_y \\ s_z \end{bmatrix}, \tag{9.6.6}$$

where $\mathbf{S}$ is the 3×3 Oseen tensor for three-dimensional flow defined as

$$\mathbf{S}(\mathbf{x}, \mathbf{x}_0) = \begin{bmatrix} \frac{1}{r} + \frac{(x-x_0)^2}{r^3} & \frac{(x-x_0)(y-y_0)}{r^3} & \frac{(x-x_0)(z-z_0)}{r^3} \\ \frac{(y-y_0)(x-x_0)}{r^3} & \frac{1}{r} + \frac{(y-y_0)^2}{r^3} & \frac{(y-y_0)(z-z_0)}{r^3} \\ \frac{(z-z_0)(x-x_0)}{r^3} & \frac{(z-z_0)(y-y_0)}{r^3} & \frac{1}{r} + \frac{(z-z_0)^2}{r^3} \end{bmatrix}, \tag{9.6.7}$$

and

$$r = [\,(x - x_0)^2 + (y - y_0)^2 + (z - z_0)^2\,]^{1/2} \tag{9.6.8}$$

is the distance of the field point, $\mathbf{x}$, from the location of the Stokeslet, $\mathbf{x}_0$. The three columns of the Oseen tensor represent, respectively, the x, y, and z components of the velocity associated with a Stokeslet of unit strength pointing in the direction of the x, y, or z axis. In index notation, the Oseen tensor takes the form

$$S_{ij}(\mathbf{x}, \mathbf{x}_0) = \frac{\delta_{ij}}{r} + \frac{\hat{x}_i\,\hat{x}_j}{r^3}, \tag{9.6.9}$$

where $\hat{\mathbf{x}} = \mathbf{x} - \mathbf{x}_0$ and δ_{ij} is Kronecker's delta representing the identity matrix.

The corresponding pressure field is

$$p(\mathbf{x}, \mathbf{x}_0) = \mu \frac{2}{r^3} \begin{bmatrix} x - x_0 \\ y - y_0 \\ z - z_0 \end{bmatrix} \cdot \begin{bmatrix} s_x \\ s_y \\ s_z \end{bmatrix}, \tag{9.6.10}$$

where the dot denotes the inner vector product.

It can be shown by direct substitution that the velocity (9.6.6) and accompanying pressure field (9.6.10) satisfy the equations of Stokes flow (9.5.3) and (9.5.2) with the gravity term absent, for any Stokeslet strength represented by the vector $\mathbf{s}$, everywhere except at the point $\mathbf{x}_0$ where the velocity and pressure become infinite. The streamline pattern in the xy plane induced by a Stokeslet pointing in the x direction is shown in figure 9.6.1.

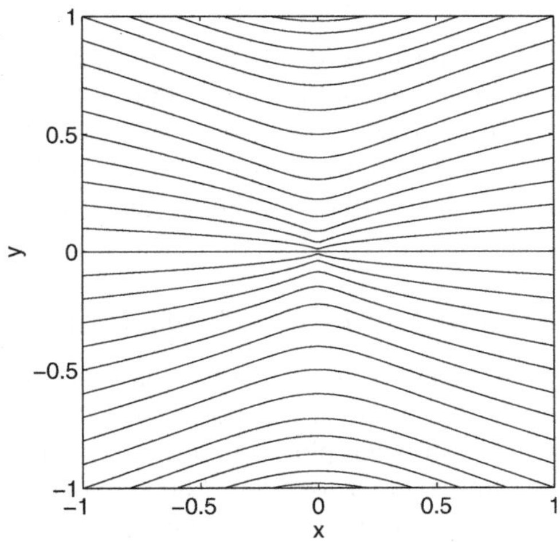

Figure 9.6.1 Streamline pattern in the xy plane of the axisymmetric flow induced by a three-dimensional point force pointing along the x axis.

Force on a surface enclosing the Stokeslet

Consider a surface, D, enclosing the singular point $\mathbf{x}_0$ of a Stokeslet. It can be shown that the force exerted on this closed surface is given by

$$\mathbf{F} \equiv \iint_D \boldsymbol{\sigma} \cdot \mathbf{n} \, \mathrm{d}S = -8\pi\mu \, \mathbf{s}, \tag{9.6.11}$$

independent of the geometry of the surface, where $\boldsymbol{\sigma}$ is the stress tensor and $\mathbf{n}$ is the unit vector normal to D pointing outward. The corresponding torque with respect to the singular point $\mathbf{x}_0$ is zero,

$$\mathbf{T} \equiv \iint_D (\mathbf{x} - \mathbf{x}_0) \times (\boldsymbol{\sigma} \cdot \mathbf{n}) \, \mathrm{d}S = \mathbf{0}. \tag{9.6.12}$$

The force and torque exerted on a closed surface that does not enclose the singular point $\mathbf{x}_0$, are both zero.

Equation (9.6.11) reveals that the Stokeslet expresses the flow due to a three-dimensional point force applied at a singular point, $\mathbf{x}_0$. In physical terms, this flow can be induced by a small moving particle located at the point $\mathbf{x}_0$. The strength of the point force counterbalances the force exerted on the particle due, for example, to gravity.

Point source and point-source dipoles

It can be shown by straightforward substitution that the irrotational velocity fields due to a point source or point-source dipole discussed in section 3.6 satisfy the equations of Stokes flow (9.5.3) and (9.5.2) with the gravity term absent and a corresponding constant pressure. The force and torque on a closed surface are zero, independent of whether the singular point $\mathbf{x}_0$ lies inside or outside the surface.

9.6.2 Flow representation in terms of singularities

We have discussed three singular fundamental solutions of the equations of Stokes flow, including the point force, the point source, and the point-source dipole. These singularities can be employed in a more general context to generate desired solutions by linear superposition. The key idea is to express a flow of interest in terms of a linear combination of singularities, and then compute the strengths of the singularities to satisfy the required boundary conditions. Superposition is permissible in light of the linearity of the equations of Stokes flow.

For example, we have already seen that the flow due to the translation of a sphere may be represented exactly by a point force and a point-source dipole placed at the center of the sphere, where the coefficients of the singularities are given in equations (9.6.2). Using equation (9.6.11), we find that the hydrodynamic force exerted on the sphere is

$$\mathbf{F} = -8\pi\mu\,\mathbf{s} = -6\pi\mu U a\,\mathbf{e}_x, \tag{9.6.13}$$

which is identical with expression (9.5.28) derived by detailed integration.

However, exact representation are rare and we must compromise by deriving approximate solutions. Illustrative examples are discussed in the remainder of this section.

9.6.3 A sphere moving inside a circular tube

Consider a spherical particle of radius a moving with velocity V along the axis of cylindrical tube of radius b under the action of pressure-driven flow, as illustrated in figure 9.6.2. In the absence of the sphere, the velocity profile is given by the Poiseuille profile

$$u_x^p = U_{max}^p\,(1 - \frac{\sigma^2}{b^2}), \tag{9.6.14}$$

where the superscript p denotes the unperturbed parabolic flow, U_{max}^p is the maximum velocity at the centerline, and

$$\sigma = \sqrt{y^2 + z^2} = r\,\sin\theta \tag{9.6.15}$$

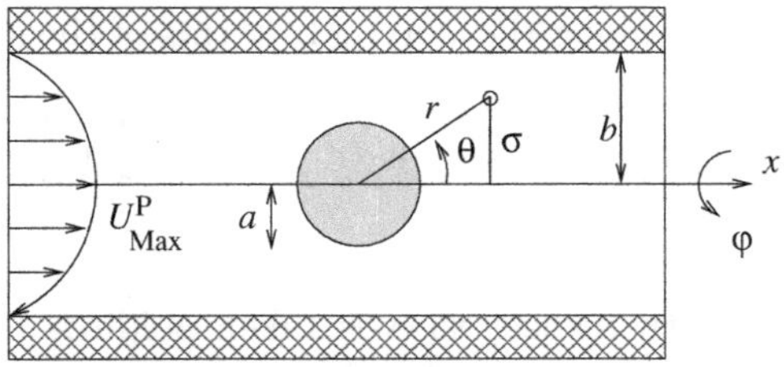

Figure 9.6.2 A spherical particle moving along the axis of a circular tube in Poiseuille flow.

is the distance from the x axis. The corresponding stream function is defined by the equation

$$\frac{1}{\sigma}\frac{\mathrm{d}\psi^P}{\mathrm{d}\sigma} = u_x^p. \tag{9.6.16}$$

Integrating with respect to σ, we derive the stream function

$$\psi^p = U^p_{max}\,\frac{1}{2}\sigma^2\Big(1 - \frac{1}{2}\frac{\sigma^2}{b^2}\Big) + c, \tag{9.6.17}$$

where c is an unspecified constant.

An approximate representation of the flow in the presence of the sphere emerges by superposing the unperturbed parabolic flow, the flow due to a point force, and the flow due to a point-source dipole, where the two singularities are located at the center of the sphere and point along the x axis. The composite stream function is given by

$$\psi = \psi^p + s_x\,\psi^{3D-ST-x} + d_x\,\psi^{3D-PSD-x}, \tag{9.6.18}$$

where s_x and d_x are the unknown strengths of the singularities. Expressing (9.6.3) and (9.6.4) in cylindrical polar coordinates, we find

$$\psi^{3D-ST-x} \equiv r\,\sin^2\theta = \frac{\sigma^2}{(x^2+\sigma^2)^{1/2}}, \tag{9.6.19}$$

and

$$\psi^{3D-PSD-x} \equiv \frac{1}{4\pi r}\sin^2\theta = \frac{1}{4\pi}\frac{\sigma^2}{(x^2+\sigma^2)^{3/2}}, \tag{9.6.20}$$

where $r = \sqrt{x^2+\sigma^2}$.

The boundary conditions require that the velocity at the surface of the sphere is equal to $\mathbf{u} = V\,\mathbf{e}_x$, where V is the velocity of translation of the sphere and $\mathbf{e}_x$ is the unit vector along the x axis. In terms of the stream function,

$$u_x(r=a) = \frac{1}{\sigma}\left(\frac{\partial\psi}{\partial\sigma}\right)_{r=a} = V, \qquad u_\sigma(r=a) = -\frac{1}{\sigma}\left(\frac{\partial\psi}{\partial x}\right)_{r=a} = 0, \tag{9.6.21}$$

where the derivative with respect to σ is taken holding x constant, and the derivative with respect to x is taken holding σ constant.

Substituting (9.6.17), (9.6.19), and (9.6.20) in the right-hand side of (9.6.18) and the resulting expression in the boundary conditions (9.6.21) and simplifying, we find

$$\left[U^p_{max}(1-\frac{\sigma^2}{b^2}) + \frac{s_x}{a}\,(2-\frac{\sigma^2}{a^2}) + \frac{d_x}{4\,\pi\,a^3}\,(2-3\frac{\sigma^2}{a^2})\right]_{r=a} = V, \tag{9.6.22}$$

and

$$s_x + \frac{3}{4\pi a^2}\,d_x = 0. \tag{9.6.23}$$

Using (9.6.23) to eliminate d_x in favor of s_x from (9.6.22), we find

$$\left[U^p_{max}\,(1-\frac{\sigma^2}{b^2}) + \frac{4}{3}\frac{s_x}{a}\right]_{r=a} = V. \tag{9.6.24}$$

Since $\sigma = a\sin\theta$ over the surface of the sphere, it is impossible to satisfy the boundary condition (9.6.24) over the whole surface of the sphere, except in the absence of the parabolic flow. This difficulty underlines the limitations of the approximate representation (9.6.18).

As a compromise, we require the satisfaction of (9.6.24) integrated over the surface of the sphere. Substituting $\sigma = a\sin\theta$, recalling that the integral of an axisymmetric function $f(\theta)$ over the surface of the sphere is given by

$$\iint f(\theta)\,\mathrm{d}S = 2\pi\,a^2 \int_0^\pi f(\theta)\sin\theta\,\mathrm{d}\theta, \tag{9.6.25}$$

and noting that the surface area of the sphere is $4\pi a^2$, we find

$$U^p_{max}(4\pi a^2 - \frac{2\pi a^4}{b^2}\int_0^\pi \sin^3\theta\,\mathrm{d}\theta\,) + \frac{4}{3}\,\frac{s_x}{a}\,4\pi a^2 = V\,4\pi a^2. \tag{9.6.26}$$

Computing $\int_0^\pi \sin^3\theta\,d\theta = \frac{4}{3}$ and simplifying, we obtain

$$U^p_{max}\left[1-\frac{2}{3}\left(\frac{a}{b}\right)^2\right] + \frac{4}{3}\frac{s_x}{a} = V. \tag{9.6.27}$$

According to (9.6.13), the x component of the hydrodynamic force exerted on the sphere is given by

$$F_x = -8\pi\mu s_x. \tag{9.6.28}$$

Using this expression to eliminate s_x in favor of F_x in (9.6.27), we obtain

$$V = U^p_{max}\left[1 - \frac{2}{3}\left(\frac{a}{b}\right)^2\right] - \frac{F_x}{6\pi\mu a}, \tag{9.6.29}$$

which provides us with an expression for the velocity of the sphere, V, in terms of the velocity profile of the Poiseuille flow determined by U^p_{max} and b, and the force exerted on the sphere F_x. The second term on the left-hand side of (9.6.29) is consistent with Stokes's law expressed by (9.5.28). If the tube is vertical and the x axis points downward in the direction of gravity, a force balance over the sphere requires that F_x is equal and opposite to the weight of the sphere reduced by the buoyancy force, $F_x = -\frac{4\pi}{3}a^3(\rho_s - \rho)g$, where ρ_s is the density of the sphere and g is the magnitude of the acceleration of gravity.

Freely suspended particle

In the case of a freely suspended particle, $F_x = 0$, equation (9.6.29) yields

$$V = U^p_{max}\left[1 - \frac{2}{3}\left(\frac{a}{b}\right)^2\right]. \tag{9.6.30}$$

This expression shows that the velocity of the sphere lags behind the local fluid velocity, U^p_{max}, by a factor that is determined by the square of the ratio of the sphere to tube radius, a^2/b^2.

Boundary effects

A tacit approximation underlies the preceding derivation: the no-slip and no-penetration conditions at the surface of the tube have not been enforced. As a consequence, expression (9.6.29) is strictly valid for a sphere immersed in infinite parabolic flow. The presence of the tube in the preceding discussion is relevant only insofar as to establish the curvature of the parabolic velocity profile.

In spite of the approximate satisfaction of the boundary condition on the surface of the sphere implemented by (9.6.26), expression (9.6.29) turns out to be exact for infinite parabolic flow, but not for wall-bounded parabolic flow.

9.6.4 Boundary integral representation

Let us consider an incident flow with velocity $\mathbf{u}^\infty$ past a stationary, translating, or rotating particle, and discretize the surface of the particle into a collection of N surface elements, such as curved quadrilaterals or triangles, as illustrated

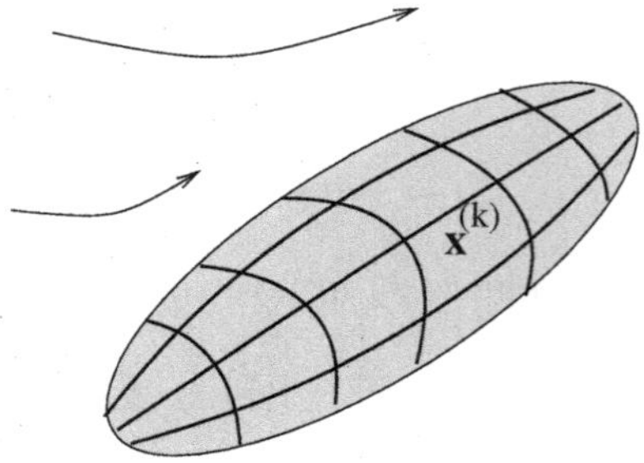

Figure 9.6.3 Stokes flow past a particle can be represented by a superposition of point forces located at designated centers of boundary elements.

in figure 9.6.3. An approximate representation of the flow can be obtained by adding to the incident flow the flow due to a collection of point forces located at designated centers of the surface elements.

Expressing the Stokeslet in terms of the Oseen tensor, as shown in equation (9.6.6), we find that the velocity at a point $\mathbf{x}$ that lies in the fluid is given by

$$\mathbf{u}(\mathbf{x}) = \mathbf{u}^\infty(\mathbf{x}) - \frac{1}{8\pi\mu} \sum_{k=1}^{N} \mathbf{S}(\mathbf{x}, \mathbf{x}^{(k)}) \cdot \mathbf{F}^{(k)}, \tag{9.6.31}$$

where $\mathbf{x}^{(k)}$ is the designated center of the kth element, and $\mathbf{F}^{(k)}$ is the corresponding vectorial strength of the point force. Rearranging the representation (9.6.31), we find

$$\mathbf{u}(\mathbf{x}) = \mathbf{u}^\infty(\mathbf{x}) - \frac{1}{8\pi\mu} \sum_{k=1}^{N} \mathbf{S}(\mathbf{x}, \mathbf{x}^{(k)}) \cdot \mathbf{f}^{(k)} \Delta S^{(k)}, \tag{9.6.32}$$

where $\Delta S^{(k)}$ is the surface area of the kth element, and we have defined the average traction

$$\mathbf{f}^{(k)} \equiv \frac{1}{\Delta S^{(k)}} \mathbf{F}^{(k)}. \tag{9.6.33}$$

As the number of elements, N, tends to infinity, and correspondingly the surface areas $\Delta S^{(k)}$ tend to zero, the sum on the right-hand side of (9.6.32) reduces to a surface integral, yielding the integral representation

$$\mathbf{u}(\mathbf{x}) = \mathbf{u}^\infty(\mathbf{x}) - \frac{1}{8\pi\mu} \iint_{Particle} \mathbf{S}(\mathbf{x}, \mathbf{x}') \cdot \mathbf{f}(\mathbf{x}') \, dS(\mathbf{x}'), \tag{9.6.34}$$

which expresses the flow in terms of a distribution of point forces over the particle surface. It can be shown that the density of the distribution, $\mathbf{f}$ is, in fact, the hydrodynamic traction exerted on the particle surface, $\mathbf{f} = \boldsymbol{\sigma} \cdot \mathbf{n}$, where $\mathbf{n}$ is the unit vector normal to the particle surface pointing into the fluid.

The integral representation (9.6.34) suggests a practical method of computing the flow: apply this equation at a point $\mathbf{x}$ on the particle surface, enforce the boundary conditions for the velocity to evaluate the left-hand side, and then solve the resulting integral equation for the traction, $\mathbf{f}$. In practice, the solution is found numerically by approximating (9.6.34) with a discrete form in terms of boundary elements, as shown in equation (9.6.32). Identifying the point $\mathbf{x}$ in (9.6.32) with the designated center of each element, we obtain a system of linear equations for the strengths of the point forces $\mathbf{F}^{(k)}$. The procedure is described as a boundary-element/collocation method. Other methods for solving the integral equation are available.

Problem

9.6.1 *Force on a surface enclosing a Stokeslet or point source*

(*a*) Verify equation (9.6.11) for a spherical surface centered at the singular point, $\mathbf{x}_0$.

(*b*) Verify that the force exerted on a spherical surface centered at a point source is zero. Recall that the pressure field associated with a point is uniform throughout the domain of a Stokes flow.

9.7 Two-dimensional Stokes flow

It might appear that our discussion of three-dimensional Stokes flow in sections 9.4 and 9.5 carries over to two-dimensional flow without any further difficulties or added considerations. While this is generally true, there is one important exception: the flow due to the motion of a two-dimensional body in an infinite and otherwise quiescent fluid is generally not defined, in that a solution that satisfies the condition of vanishing velocity at infinity cannot be obtained if the force exerted on the body is nonzero. We will consider this problematic case first and then discuss well-posed flows.

9.7.1 *Flow due to the motion of a cylinder*

Let us consider the flow due to a circular cylinder of radius a translating along the x axis with velocity U. If the Reynolds number is small, $Re = \rho a U/\mu < 1$, then sufficiently close to the cylinder inertial forces are negligible compared to pressure and viscous forces, and the motion of the fluid is governed by the linear equations of Stokes flow. In section 9.4, we emphasized that the smallness of the Reynolds number does not guarantee that inertial forces are insignificant sufficiently far from the cylinder. Thus, the consistency of a solution derived working exclusively in the context of Stokes flow should be verified.

To bypass the computation of the pressure, we seek a solution based on the vorticity transport equation for the non-vanishing vorticity component, ω_z,

$$\rho\,(\frac{\partial \omega_z}{\partial t} + u_x \frac{\partial \omega_z}{\partial x} + u_y \frac{\partial \omega_z}{\partial y}) = \mu\, \nabla^2 \omega_z, \tag{9.7.1}$$

where $\nabla^2 \equiv \partial^2/\partial x^2 + \partial^2/\partial y^2$ is the two-dimensional Laplacian operator. Setting the right-hand side of (9.7.1) to zero to eliminate inertial effects and substituting $\omega_z = -\nabla^2 \psi$, we derive a fourth-order linear differential equation for the stream function,

$$\nabla^2 \nabla^2 \psi \equiv \nabla^4 \psi = 0, \tag{9.7.2}$$

where

$$\begin{aligned} \nabla^4 \equiv \nabla^2 \nabla^2 &= (\frac{\partial^2}{\partial x^2} + \frac{\partial^2}{\partial y^2})(\frac{\partial^2}{\partial x^2} + \frac{\partial^2}{\partial y^2}) \\ &= \frac{\partial^4}{\partial x^4} + 2\frac{\partial^4}{\partial x^2 \partial y^2} + \frac{\partial^4}{\partial y^4} \end{aligned} \tag{9.7.3}$$

is the biharmonic operator in two dimensions. Our task is to solve the fourth-order differential equation (9.7.3) in the flow regime confined between $r = a$ and infinity, subject to appropriate boundary and far-field conditions.

Repeating the analysis of section 9.4 for the analogous problem of flow due to a translating sphere, we find that, in plane polar coordinates, (r, θ), the boundary and far-field conditions take the form

$$\psi(r = a) = U a \,\sin\theta, \qquad \Big(\frac{\partial \psi}{\partial r}\Big)_{r=a} = U\, \sin\theta, \tag{9.7.4}$$

and

$$\frac{\psi}{r^m} \to 0 \quad \text{as} \quad r \to \infty, \tag{9.7.5}$$

for $m \geq 1$. Condition (9.7.5) allows the stream function to diverge at infinity, but requires that the rate of growth is less than linear for the velocity to decay.

Motivated by the functional form of the boundary conditions (9.7.4), we seek a solution for the stream function by separation of variables in plane polar coordinates, setting

$$\psi = q(r)\, \sin\theta. \tag{9.7.6}$$

The unknown function $q(r)$ satisfies the boundary conditions

$$q(r = a) = U a, \qquad \Big(\frac{\partial q}{\partial r}\Big)_{r=a} = U, \tag{9.7.7}$$

and the far-field condition

$$\frac{q(r)}{r^m} \to 0 \qquad \text{as} \qquad r \to \infty, \tag{9.7.8}$$

for $m \geq 1$.

Equation (3.2.23) provides us with the Laplacian operator in planar polar coordinates. The corresponding biharmonic operator is

$$\nabla^4 = \Big(\frac{1}{r}\frac{\partial}{\partial r}(r\frac{\lambda}{\partial r}) + \frac{1}{r^2}\frac{\partial^2}{\partial \theta^2} \Big) \Big(\frac{1}{r}\frac{\partial}{\partial r}(r\frac{\partial}{\partial r}) + \frac{1}{r^2}\frac{\partial^2}{\partial \theta^2} \Big). \tag{9.7.9}$$

Substituting (9.7.6) in (9.7.2), expressing the biharmonic operator in the form provided by (9.7.9) and carrying out the differentiations, we derive a fourth-order linear ordinary differential equation,

$$(\frac{\mathrm{d}^2}{\mathrm{d}r^2} + \frac{1}{r}\frac{\mathrm{d}}{\mathrm{d}r} - \frac{1}{r^2})(\frac{\mathrm{d}^2}{\mathrm{d}r^2} + \frac{1}{r}\frac{\mathrm{d}}{\mathrm{d}r} - \frac{1}{r^2})\, q = 0. \tag{9.7.10}$$

Working as in section 9.4 for flow due to the motion of a sphere, we derive the general solution

$$q(r) = A\,r^3 + B\,r\,\ln r + C\,r + \frac{D}{r}, \tag{9.7.11}$$

where A, B, C and D, are four constants. For the far-field condition (9.7.8) to be satisfied, A, B, and C must all be zero, and this leaves us with only one coefficient to satisfy the two remaining boundary conditions (9.7.7), which is not possible. Thus, the problem of flow due to the motion of a cylinder in Stokes flow does not admit a solution.

Significance of inertia

To probe the origin of this catastrophe, we examine the magnitude of inertial and viscous terms in the equation of motion, as discussed in section 9.4. Cursory inspection reveals that the approximations underlying the notion of creeping flow cease to be accurate far from the cylinder. As a consequence, imposing the far field condition (9.7.8) leads to inconsistencies.

In the case of three-dimensional flow due to the motion of a sphere, these difficulties were shielded by the decay of the flow due to a three-dimensional Stokeslet expressing a point force. In contrast, in the case of two-dimensional flow, the velocity field due to a point force exhibits a logarithmic divergence contributed by the second term on the right-hand side of (9.7.11), and a decaying solution cannot be found.

Matched asymptotic expansions

To remedy the situation, we use the method of matched asymptotic expansions, dividing the flow into two regimes: (*a*) an inner regime where the motion of the fluid is governed by the equations of creeping flow subject to the no-slip and no-penetration boundary conditions on the surface of the cylinder; and (*b*) an outer regime where the motion of the fluid is governed by another simplified system of equations, called the equations of Oseen flow, subject to the far-field condition (9.7.8). Matching conditions arise by inspecting the functional form of the inner and outer solution, respectively, for large and small distances. The analysis involves sophisticated arguments that lie outside the scope of this text.

Not all problems of two-dimensional infinite flow are ill-posed, in that a solution that satisfies the boundary and far-field conditions cannot be found. If the force exerted on each internal boundary is zero, then a perfectly acceptable solution can be obtained.

9.7.2 Rotation of a circular cylinder

As an example, we consider the flow generated by a circular cylinder of radius a rotating with angular velocity Ω around its center. The induced flow is identical to that due to a point vortex with strength $\kappa = 2\pi\,\Omega\,a^2$ placed at the center of cylinder, as discussed in section 3.7. Using expressions (3.7.1), we derive the velocity field

$$u_r = 0, \qquad u_\theta = \frac{\Omega a^2}{r}. \tag{9.7.12}$$

The corresponding pressure field is uniform throughout the domain of flow.

An external torque must be applied to counterbalance the hydrodynamic torque due to the motion of the fluid and thus sustain the rotation of the cylinder. The torque with respect to the center of the cylinder is given by

$$T_z = \oint_{Cylinder} \sigma_{r\theta}\, a\, \mathrm{d}l. \tag{9.7.13}$$

Substituting (9.7.12) in the Newtonian constitutive equation (4.7.22), we find the stress component $\sigma_{r\theta} = -2\mu\Omega a^3/r^2$. Substituting this expression in (9.7.13), setting $\mathrm{d}l = a\,\mathrm{d}\theta$, and integrating with respect to θ from 0 to 2π, we find

$$T_z = -4\pi\mu\,\Omega\,a^2. \tag{9.7.14}$$

9.7.3 Simple shear flow past a circular cylinder

Consider simple shear flow along the x axis past a circular cylinder with radius a centered at the origin. The incident velocity field is given by $\mathbf{u}^\infty = (ky, 0)$, where k is the shear rate. The cylinder is allowed to rotate about its center with angular velocity $\Omega = -\frac{1}{2}\,k$. If k is positive, the cylinder rotates in the clockwise direction; if k is negative, the cylinder rotates in the counter-clockwise direction. Analysis shows that, if the torque exerted on the cylinder vanishes, the velocity field is given by

$$u_r = \frac{k}{2}\Big(r - \frac{2}{r} + \frac{1}{r^3} \Big) \sin 2\theta, \qquad u_\theta = \frac{k}{2}\Big(r + \big(r - \frac{1}{r^3} \cos 2\theta \big) \Big). \tag{9.7.15}$$

These results demonstrate that a freely-suspended cylinder rotates at an angular velocity that is equal to the negative of half the shear rate of the incident simple shear flow.

9.7.4 The Oseen tensor and the point force

The second term on the right-hand side of (9.7.11), involving the logarithm, represents the flow due to a two-dimensional point force oriented along the x axis. Differentiating with respect to x or y, we obtain the corresponding velocity field. Repeating the derivation for the y direction, we obtain the velocity field at a point, $\mathbf{x} = (x, y)$, due to a two-dimensional Stokeslet with vectorial strength $\mathbf{s} = (s_x, s_y)$ situated at the point $\mathbf{x}_0 = (x_0, y_0)$, expressed by

$$\begin{bmatrix} u_x \\ u_y \end{bmatrix} (\mathbf{x}, \mathbf{x}_0) = \mathbf{S}(\mathbf{x}, \mathbf{x}_0) \cdot \begin{bmatrix} s_x \\ s_y \end{bmatrix}, \tag{9.7.16}$$

where $\mathbf{S}$ is the 2×2 Oseen tensor for two-dimensional flow defined as

$$\mathbf{S}(\mathbf{x}, \mathbf{x}_0) = \begin{bmatrix} -\ln r + \frac{(x-x_0)^2}{r^2} & \frac{(x-x_0)(y-y_0)}{r^2} \\ \frac{(y-y_0)(x-x_0)}{r^2} & -\ln r + \frac{(y-y_0)^2}{r^2} \end{bmatrix}, \tag{9.7.17}$$

and $r = [(x - x_0)^2 + (y - y_0)^2]^{1/2}$ is the distance of the field point, $\mathbf{x}$, from the location of the Stokeslet, $\mathbf{x}_0$.

The two columns of the Oseen tensor represent, respectively, the x and y velocity components associated with a Stokeslet of unit strength pointing in the direction of the x or y axis. The corresponding pressure field is given by

$$p(\mathbf{x}, \mathbf{x}_0) = \mu\, \frac{2}{r^2} \begin{bmatrix} x - x_0 \\ y - y_0 \end{bmatrix} \cdot \begin{bmatrix} s_x \\ s_y \end{bmatrix}, \tag{9.7.18}$$

where a dot denotes the inner vector product. The streamline pattern of the flow induced by a two-dimensional Stokeslet oriented along the x axis is shown in figure 9.7.1.

Consider a closed contour, C, enclosing a singular point, $\mathbf{x}_0$. In can be shown that the force exerted on the contour is given by

$$\mathbf{F} = \oint_C \boldsymbol{\sigma} \cdot \mathbf{n}\, \mathrm{d}l = -4\pi\mu\, \mathbf{s}, \tag{9.7.19}$$

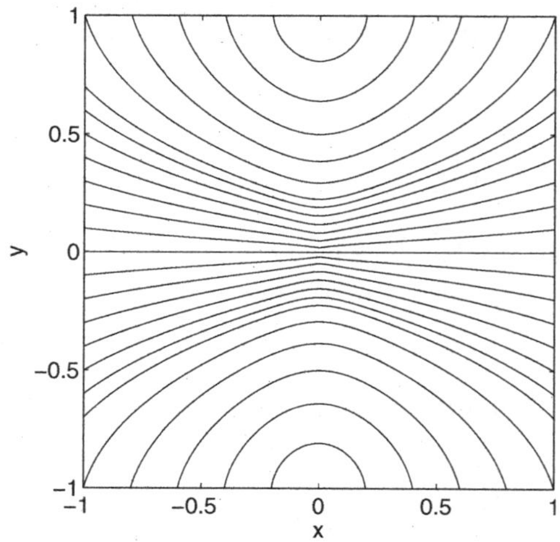

Figure 9.7.1 Streamline pattern of the flow induced by a two-dimensional point force pointing along the x axis.

independent of the contour shape, where $\boldsymbol{\sigma}$ is the stress tensor, $\mathbf{n}$ is the unit vector normal to C pointing outward, and l is the arc length. The torque with respect to the point $\mathbf{x}_0$ is zero. The force and torque exerted on a closed contour that does not enclose the singular point, $\mathbf{x}_0$, are also zero. These results are consistent with our earlier assertion that the two-dimensional Stokeslet expresses the flow due to a two-dimensional point force.

Problem

9.7.1 *Simple shear flow past a freely-suspended cylinder*

Compute the pressure field corresponding to the velocity field given in (9.7.15). Confirm that the torque exerted on the cylinder is zero.

9.8 Local solutions

The boundaries of a flow may involve walls, interfaces, and dividing streamlines meeting at sharp corners or cusps. Deep into the corners and cusps, the magnitude of the velocity decays rapidly, fluid inertia becomes negligible, and the Reynolds number of the local flow is small. The structure of the flow may then be studied in the context of Stokes flow regarding the far flow as a driving mechanism that determines the intensity and selects the geometrical mode and spatial structure of the local flow.

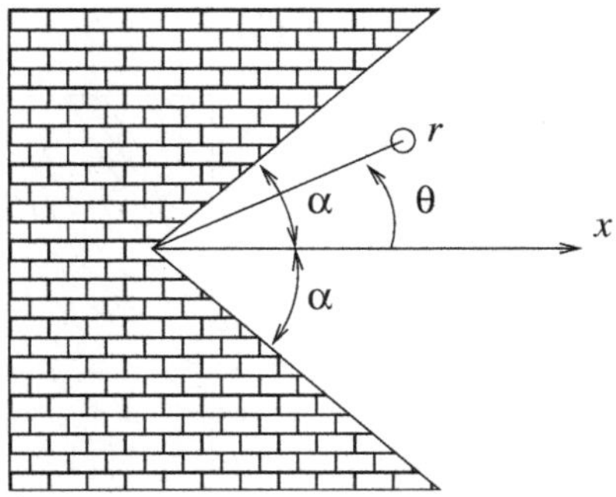

Figure 9.8.1 Because the local Reynolds number of the flow near a corner is small, the motion of the fluid is governed by the equations of Stokes flow. Similarity solutions can be derived working in local plane polar coordinates.

9.8.1 Separation of variables

Consider a two-dimensional flow between two planes intersecting at an angle 2α, as illustrated in figure 9.8.1. To facilitate the implementation of the boundary conditions, we introduced plane polar coordinates, (r, θ), with origin at the apex. To bypass the computation of the pressure, we introduce the stream function, ψ, and express it in the separated form

$$\psi = q(r)\, f(\theta). \tag{9.8.1}$$

Note that (9.8.1) is a generalization of (9.7.6) describing flow due to the motion of a cylinder.

We begin by stipulating that the component functions $q(r)$ and $f(\theta)$ exhibit, respectively, a power-law and an exponential dependence on their respective arguments,

$$q(r) = r^{\lambda}, \qquad f(\theta) = A \exp(\kappa\theta), \tag{9.8.2}$$

where λ and κ are two constants, and A is a constant coefficient. The exponent, κ, is allowed to be complex, $\kappa = \kappa_r + \mathrm{i}\kappa_i$, where i is the imaginary unit, $\mathrm{i}^2 = -1$, and κ_r, κ_i are two real constituents. Using the Euler decomposition of the complex exponential, we write

$$f(\theta) = A\, \exp(\kappa_r\theta)\, [\, \cos(\kappa_i\theta) + \mathrm{i} \sin(\kappa_i\theta)\,], \tag{9.8.3}$$

with the understanding that either the real or the imaginary part may be selected on the right-hand side of (9.8.1).

Substituting (9.8.2) in (9.8.1) and the result in the biharmonic equation $\nabla^4\psi = 0$ expressed in plane polar coordinates, as shown in (9.7.9), we derive a

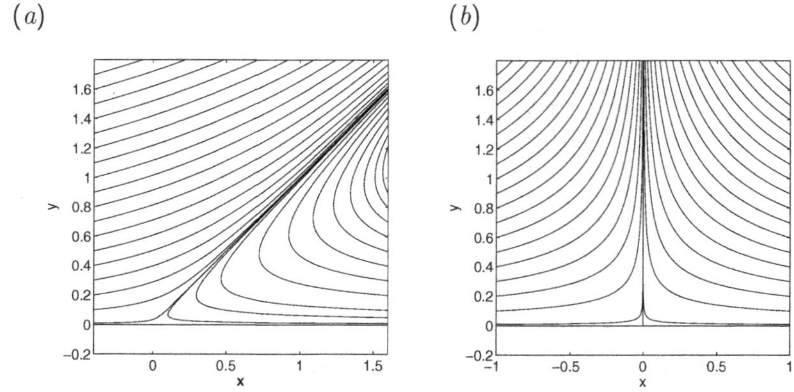

Figure 9.8.2 Streamline pattern of flow in the neighborhood of a stagnation point on a plane wall with a rectilinear dividing streamline. The angle subtended between the dividing streamline and the wall is (a) $\alpha = \pi/4$, and (b) $\pi/2$.

bi-quadratic algebraic equation for κ parametrized by the exponent λ,

$$\kappa^4 + 2\,(\lambda^2 - 2\,\lambda + 2)\,\kappa^2 + \lambda^2\,(\lambda - 2)^2 = 0. \tag{9.8.4}$$

The roots can be found readily using the quadratic formula. Solving for κ, substituting the result in (9.8.3), and rearranging the emerging expression, we obtain the general solution

$$f(\theta) = \begin{cases} B\,\sin(\lambda\theta - \beta) + C\,\sin[(\lambda - 2)\theta - \gamma] & \text{if } \lambda \neq 0, 1, 2, \\ B\,\sin(2\theta - \beta) + C\,\theta + D & \text{if } \lambda = 0, 2, \\ B\,\sin(\theta - \beta) + C\,\theta\sin(\theta - \gamma) & \text{if } \lambda = 1, \end{cases} \tag{9.8.5}$$

where B, C are complex constants, and β, γ are real constants.

A variety of flows can be generated by selecting, or solving for, the exponent λ. Illustrative examples are discussed in the remainder of this section.

Stagnation-point flow on a plane wall

First, we consider flow near a stagnation point on a plane wall, as illustrated in figure 9.8.2. The no-slip and no-penetration conditions require $f = 0$ and $\mathrm{d}f/\mathrm{d}\theta = 0$ at $\theta = 0$ and π. The structure of the flow is determined by specifying that $f = 0$ along the dividing streamline located at $\theta = \alpha$, where the angle α is a free parameter.

Making the judicious selection $\lambda = 3$ and using the general solution given by the first equation in (9.8.5), we find

$$f(\theta) = B\,\sin(3\theta - \beta) + C\,\sin(\theta - \gamma). \tag{9.8.6}$$

Enforcing the no-slip and no-penetration boundary conditions and specifying the orientation of the dividing streamline provides us with a system of three homogeneous equations for the constants B and C involving the unknown parameters β and γ,

$$f(0) = f(\pi) = -B\,\sin\beta - C\,\sin\gamma = 0,$$

$$\left(\frac{\mathrm{d}f}{\mathrm{d}\theta}\right)_{\theta=0} = -\left(\frac{\mathrm{d}f}{\mathrm{d}\theta}\right)_{\theta=\pi} = 3\,B\,\cos\beta + C\,\cos\gamma = 0, \tag{9.8.7}$$

$$f(\alpha) = B\sin(3\alpha - \beta) + C\sin(\alpha - \gamma) = 0.$$

The first two equations require

$$\tan\beta = -3\,\tan\gamma. \tag{9.8.8}$$

The second and third equations require

$$3\,\frac{\cos\beta}{\sin(3\alpha-\beta)} = -\frac{\cos\gamma}{\sin(\alpha-\gamma)}. \tag{9.8.9}$$

The system of equations (9.8.8) and (9.8.9) can be used to express β and γ in terms of α. Using any one of equations (9.8.7) to express B in terms of C, substituting the result in (9.8.6) and making use of the trigonometric identity $\sin 3a = 3\sin a - 4\,\sin^3 a$, where a is arbitrary, we derive the final form

$$f(\theta) = G\,\sin^2\theta\,\sin(\theta - \alpha), \tag{9.8.10}$$

where G is an arbitrary constant. Substituting (9.8.10) in (9.8.1), we obtain the desired solution for the stream function,

$$\psi = G\,r^3\sin^2\theta\,\sin(\theta-\alpha). \tag{9.8.11}$$

The value of the constant G is determined by the strength of the flow far from the stagnation point. Streamline patterns for $\alpha = \pi/4$ and $\pi/2$ are depicted in figure 9.8.2. A distinguishing feature of Stokes flow is that the dividing streamlines are straight lines emanating from the stagnation point on the wall.

9.8.2 Flow near a corner

In a second application, we consider flow inside a corner confined between two intersecting stationary planes located at $\theta = \pm\alpha$, as illustrated in figure 9.8.1. The no-slip and no-penetration boundary conditions require $f = 0$ and $\mathrm{d}f/\mathrm{d}\theta = 0$ at $\theta = \pm\alpha$. We confine our attention to flow that is antisymmetric will respect to the mid-plane located at $\theta = 0$, and require the condition $\mathrm{d}f/\mathrm{d}\theta = 0$ at $\theta = 0$, stating that the radial velocity vanishes at the mid-plane.

The value of the exponent λ cannot be specified *a priori* as in the case of stagnation-point flow discussed earlier in this section, but must be found instead as part of the solution. Assuming that $\lambda \neq 0, 1, 2$, we set $\beta = \pi/2$ and $\gamma = \pi/2$ in the first equation of (9.8.5) to satisfy the condition $\mathrm{d}f/\mathrm{d}\theta = 0$ at $\theta = 0$, and find

$$f(\theta) = Q\cos(\lambda\theta) + G\cos[(\lambda - 2)\theta], \tag{9.8.12}$$

where Q and G are two constants. Enforcing the boundary conditions $f = 0$ and $\mathrm{d}f/\mathrm{d}\theta = 0$ at $\theta = \pm\alpha$, we obtain a homogeneous system of two equations for the constants Q and G,

$$\begin{bmatrix} \cos(\lambda\alpha) & \cos[(\lambda-2)\alpha] \\ \lambda\sin(\lambda\alpha) & (\lambda-2)\,\sin[(\lambda-2)\alpha] \end{bmatrix} \cdot \begin{bmatrix} Q \\ G \end{bmatrix} = \begin{bmatrix} 0 \\ 0 \end{bmatrix}. \tag{9.8.13}$$

For a nontrivial solution to exist, the determinant of the coefficient matrix on the left-hand side must be zero, yielding

$$\cos(\lambda\alpha)(\lambda-2)\,\sin[(\lambda-2)\alpha] - \lambda\sin(\lambda\alpha)\cos[(\lambda-2)\alpha] = 0, \tag{9.8.14}$$

which can be restated as a nonlinear algebraic equation,

$$\sin[2\alpha(\lambda - 1)] = (1-\lambda)\sin(2\alpha). \tag{9.8.15}$$

An obvious solution is $\lambda = 1$. However, for this value, the third instead of the first equation in (9.8.5) should have been selected, and this disqualifies the obvious choice.

We anticipate that equation (9.8.15) will have a generally complex solution for λ, and write

$$\lambda = \lambda_r + \mathrm{i}\lambda_i, \tag{9.8.16}$$

where i is the imaginary unit. Substituting (9.8.16) in (9.8.15) and using standard formulas of complex calculus, we derive a system of two real equations for the real components λ_r and λ_i,

$$\begin{aligned} \sin[2\alpha(\lambda_r - 1)]\,\cosh(2\alpha\lambda_i) &= (1-\lambda_r)\sin(2\alpha), \\ \sin[2\alpha(\lambda_r - 1)]\,\sinh(2\alpha\lambda_i) &= -\lambda_i\sin(2\alpha). \end{aligned} \tag{9.8.17}$$

To simplify the notation, we introduce the auxiliary variables

$$\xi = 2\alpha(\lambda_r - 1), \qquad \eta = 2\alpha\lambda_i \qquad \kappa = \frac{\sin(2\alpha)}{2\alpha}, \tag{9.8.18}$$

and express the system (9.8.17) in the more convenient form

$$\begin{aligned} f_1(\xi,\eta) &\equiv \sin\xi\cosh\eta + \kappa\,\xi = 0, \\ f_2(\xi,\eta) &\equiv \cos\xi\sinh\eta + \kappa\,\eta = 0. \end{aligned} \tag{9.8.19}$$

Using the definition of ξ and η, we find

$$\lambda = \frac{1+\xi}{2\alpha} + \mathrm{i}\,\frac{\eta}{2\alpha}. \tag{9.8.20}$$

Our task is to solve the system of equations (9.8.19) for ξ and η, subject to a specified value for κ; that is, subject to a specified angle, α.

Newton's method

The solution of the nonlinear algebraic equations must be found by iteration. The numerical procedure involves guessing values for ξ and η that wishfully satisfy equations (9.8.19), and then improving the guess in a sensible fashion. In Newton's method, the solution is found by guessing the values ξ^G and η^G, replacing the functions $f_1(\xi,\eta)$ and $f_2(\xi,\eta)$ with their linearized Taylor series expansion, and then solving a linear system of equations for the unknowns using elementary analytical or numerical methods. The linearized Taylor series expansion provides us with the approximations

$$\begin{aligned} f_1(\xi,\eta) &\simeq f_1(\xi^G,\eta^G) + \left(\frac{\partial f_1}{\partial \xi}\right)_{\xi^G,\eta^G}(\xi-\xi^G) + \left(\frac{\partial f_1}{\partial \eta}\right)_{\xi^G,\eta^G}(\eta-\eta^G),\\ f_2(\xi,\eta) &\simeq f_2(\xi^G,\eta^G) + \left(\frac{\partial f_2}{\partial \xi}\right)_{\xi^G,\eta^G}(\xi-\xi^G) + \left(\frac{\partial f_2}{\partial \eta}\right)_{\xi^G,\eta^G}(\eta-\eta^G). \end{aligned} \tag{9.8.21}$$

Setting the right-hand sides to zero to satisfy (9.8.19), we find

$$\left(\begin{array}{cc} \frac{\partial f_1}{\partial \xi} & \frac{\partial f_1}{\partial \eta} \\ \frac{\partial f_2}{\partial \xi} & \frac{\partial f_2}{\partial \eta} \end{array}\right)_{\xi=\xi^G,\eta=\eta^G} \cdot \left(\begin{array}{c} \xi-\xi^G \\ \eta-\eta^G \end{array}\right) = \left(\begin{array}{c} -f_1(\xi^G,\eta^G) \\ -f_2(\xi^G,\eta^G) \end{array}\right). \tag{9.8.22}$$

The coefficient matrix on the left-hand side is the Jacobian matrix of the algebraic system (9.8.19). In summary, Newton's algorithm involves the following steps:

1. First, we guess the values ξ^G and η^G.

2. Second, we evaluate the right-hand side of (9.8.22) and the coefficient matrix on the left-hand side.

3. Third, we solve the system of two linear equations (9.8.22) for the differences $\Delta\xi \equiv \xi - \xi^G$ and $\Delta\eta \equiv \eta - \eta^G$; we recall that ξ and η are unknown.

4. Fourth, we improve the guesses by replacing the guessed values with the new values

$$\xi^N = \xi^G + \Delta\xi, \qquad \eta^N = \eta^G + \Delta\eta. \tag{9.8.23}$$

5. We may now return to step 2 and repeat the procedure with new values for $\xi^G = \xi^N$ and $\eta^G = \eta^N$.

The iterations converge as long as the initial guesses are sufficiently close to the exact solution.

System (9.8.19) has a family of solution branches obtained by making different selections for the initial guess. The most physically relevant branch is associated with the smallest value of λ_r, corresponding to a flow that decays at the lowest possible rate with distance from the apex, r, yielding the strongest possible flow.

Figure 9.8.3(a) shows graphs of the real and imaginary parts of λ plotted against the semi-angle α. When $0.41\pi < \alpha < \pi$, the exponent λ is real; the fluid moves uninterrupted along the walls and regions of recirculating flow do not develop, as depicted in figure 9.8.3(b). When $0 < \alpha < 0.41\pi$, the exponent λ is complex; the flow develops an infinite sequence of regions of recirculating fluid called eddies, as shown in figure 9.8.3(c).

Problem

9.8.1 *Flow near a scraper*

Derive a local solution for flow near two intersecting plates, where one of the plates is held stationary while the second plate moves parallel to itself with constant velocity. This idealized configuration is a model of the flow due to a plate scrapping a fluid off a flat surface, viewed in a frame of reference moving with the scraper.

Computer Problem

9.8.2 *Computation of the exponent*

Write a computer program that uses Newton's method to solve the nonlinear algebraic system (9.8.19) and reproduce the graph shown in figure 9.8.3(a).

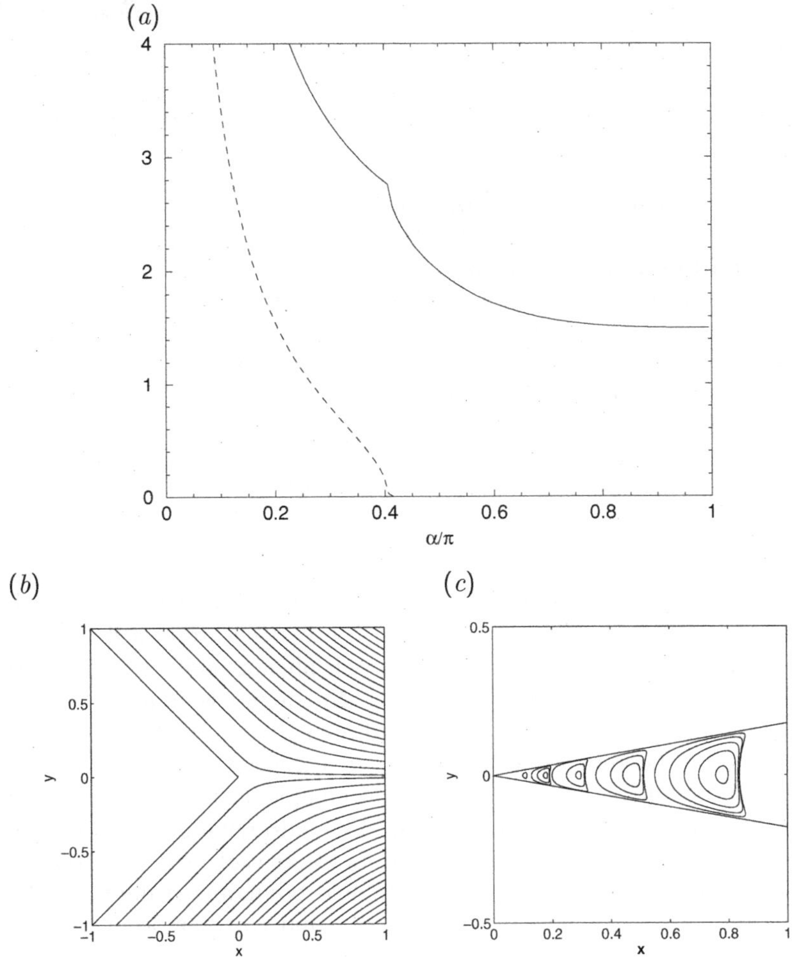

Figure 9.8.3 (a) Graphs of the exponent λ with the smallest real part for antisymmetric flow between two walls intersecting at the angle 2α. When $0.41\pi < \alpha < \pi$, λ is real; when $0 < \alpha < 0.41\pi$, λ is complex. (b, c) Streamline pattern of antisymmetric flow in a corner with half-angle (b) $\alpha = 135°$, and (c) $10°$.

High Reynolds Number Flow

10

Having discussed viscous flow at low Reynolds numbers, we now turn our attention to the diametrically opposite limit of inertia-dominated flow at high Reynolds numbers. Inspecting the changes in the structure of a flow with increasing Reynolds number, we encounter a rich phenomenology and a variety of transition protocols. In all cases, when the Reynolds number exceeds a critical threshold, flow instability occurs and a small-scale turbulent motion is established. The theoretical study of the structure and dynamics of flows at high Reynolds numbers encompasses several complementary topics including potential flow theory, boundary layer analysis, hydrodynamic stability, turbulent motion, and dynamics of vortex motion. Potential flow was considered earlier in this book; boundary layer theory, hydrodynamic stability, and turbulent motion will be discussed in this chapter; and vortex motion will be the exclusive topic of chapter 11.

10.1 Changes in the structure of a flow with increasing Reynolds number

As the Reynolds number of a flow increases by raising, for example, the magnitude of the velocity, the structure of the flow changes in a way that strongly depends on the particular flow configuration. Even though a general statement regarding the anticipated protocol of change cannot be made, several pervasive

C. Pozrikidis, *Fluid Dynamics: Theory, Computation, and Numerical Simulation*,
DOI: 10.1007/978-0-387-95871-2_10,

features can be identified:

- As the Reynolds number increases, diffusion of vorticity away from the boundaries and into the bulk of the flow is hindered by strong convection along and toward the boundaries. As a result, the vorticity tends to be confined inside boundary layers and free shear layers developing along fluid interfaces and free surfaces. Viscous forces are significant inside the boundary and free shear layers, but can be neglected outside their edges.

- Vorticity that has entered the flow by diffusion across the boundaries may accumulate to form compact regions of rotational motion identified as vortices. Vortices are typically generated behind bluff bodies and at the trailing edge of streamlined objects.

- Vortex interaction causes the flow to become unsteady in a deterministic or chaotic fashion. Spatial, temporal, and spatio-temporal chaos can be established even at moderate Reynolds numbers.

- When the Reynolds number exceeds a critical threshold, small disturbances amplify, altering the local and sometimes the global structure of a flow.

- Instability is followed by transition leading to turbulent flow, whereupon a small-scale unsteady motion is superimposed on a large-scale steady or unsteady flow. The small-scale motion affects and is affected significantly by the large-scale structure and global properties of the flow.

Flow past a cylinder

Figure 10.1.1(a) illustrates the main changes in the structure of streaming (uniform) flow past a circular cylinder with increasing Reynolds number, $Re \equiv \rho DU/\mu$, where D is the cylinder diameter, and U is the velocity of the incoming stream far from the cylinder. When the Reynolds number is less than unity, inertial forces are negligible near the cylinder, the motion of the fluid is governed by the equations of Stokes flow, and the streamline pattern is symmetric with respect to the vertical plane that is normal to the incident stream and passes through the cylinder mid-plane.

As the Reynolds number increases, inertial forces become significant and a boundary layer identified as a region of increased vorticity is established around the surface of the cylinder. The rotational fluid inside the boundary layer is convected along the cylinder and accumulates behind the cylinder to form two regions of recirculating flow leading to a slender wake.

A further increase in the Reynolds number causes the flow to become unstable. The pair of vortices developed behind the cylinder are shed downstream in an alternating fashion at a frequency that depends weakly on the Reynolds number, only to be replenished by newly formed vortices. The frequency of

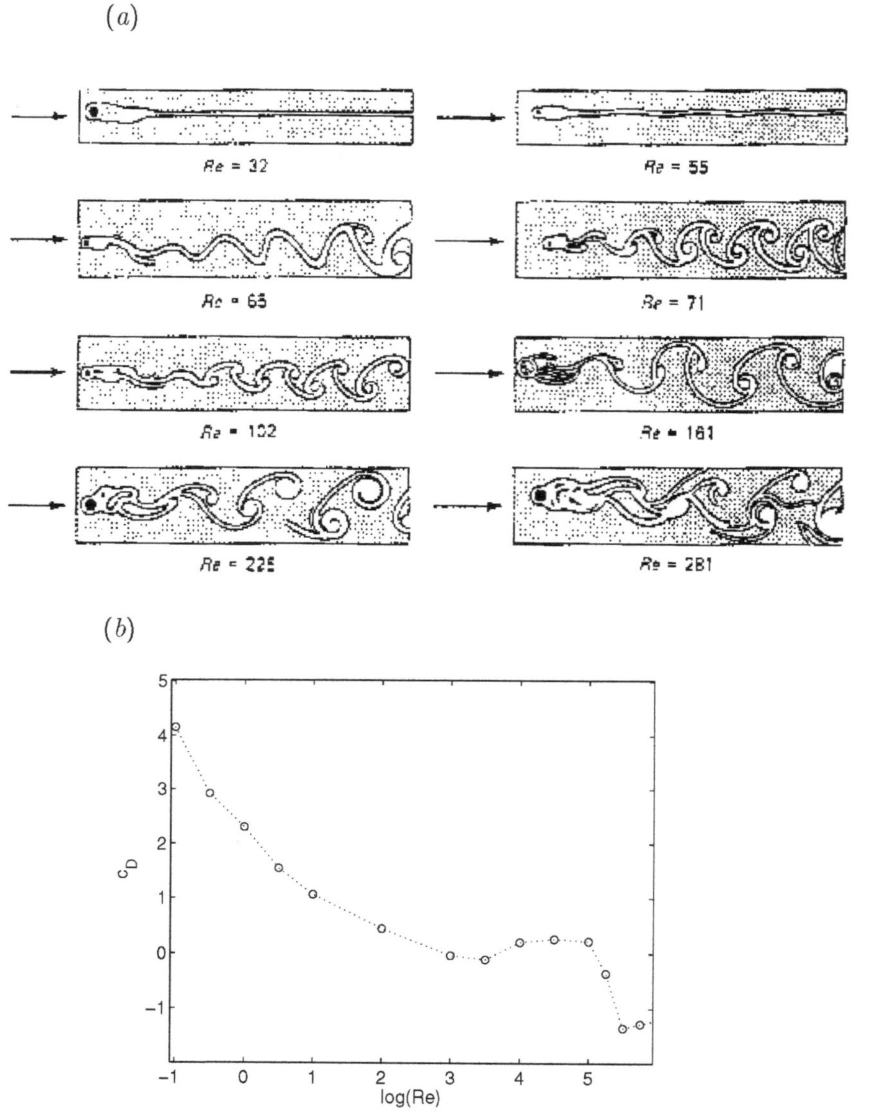

Figure 10.1.1 (*a*) Illustration of changes in the structure of streaming flow past a circular cylinder with increasing Reynolds numbers, showing boundary layer separation and the development of a wake, after Homann, F., Einfluss grosser Zähigkeit bei Strömung um Zylinder. *Forschg. Ing.-Wes.* **7**, 1–10 (1936). (*b*) Drag coefficient plotted against the Reynolds number defined with respect to the cylinder diameter.

shedding, f, is expressed by the dimensionless Strouhal number $St = fD/U$; in the case of a circular cylinder, laboratory observations show that $St \simeq 0.2$. Far from the cylinder, the wake consists of two rows of counter-rotating vortices arranged in a staggered formation known as the von Kàrmàn vortex street. At even higher Reynolds numbers, turbulent motion is established and the edges of the wake become blurred and poorly defined.

The changes in the structure of the flow described previously in this section have a strong influence on the drag force exerted on the cylinder. Figure 10.1.2 shows a graph of the dimensionless drag coefficient,

$$c_D \equiv \frac{2\,F}{\rho U^2 D}, \tag{10.1.1}$$

where F is the drag force per unit length exerted on the cylinder, plotted against the Reynolds number on a log-log scale. In the limit of vanishing Reynolds number, theoretical analysis shows that the drag force is given by the modified Stokes law

$$F \simeq \frac{4\pi\mu U}{\ln \frac{7.4}{Re}}. \tag{10.1.2}$$

Correspondingly, the drag coefficient is given by

$$c_D \simeq \frac{8\pi}{Re\,\log \frac{7.4}{Re}}. \tag{10.1.3}$$

The change in the functional form of the drag coefficient at a critical Reynolds number on the order of 10^3, shown in figure 10.1.1(b), is due to the detachment of the boundary layer from the surface of the cylinder at a certain point on the rear surface of the cylinder, as will be discussed in section 10.4, in a process described as flow separation. When the Reynolds number becomes on the order of 10^5, the flow becomes turbulent and the boundary layer reattaches, causing a sudden decline in the drag coefficient.

The non-monotonic dependence of the force drag coefficient on the Reynolds number illustrated in figure 10.1.2 is a manifestation of the complexity of fluid motion in high Reynolds number flow.

Problem

10.1.1 *Flow in a channel through an expansion*

Consider flow through a channel with a sudden expansion, as illustrated in figure 6.2.1. Discuss the expected changes in the structure of the flow with increasing a properly defined Reynolds number.

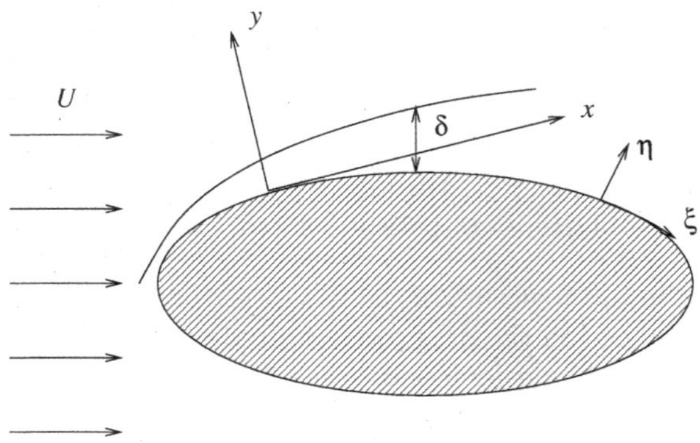

Figure 10.2.1 Schematic illustration of a Prandtl boundary layer developing around the surface of a two-dimensional curved body.

10.2 Prandtl boundary layer analysis

In section 10.1, we identified a boundary layer as a region of increased vorticity developing over a solid boundary in a high Reynolds number flow. To make the concept of a boundary layer more specific, we consider a model flow consisting of (a) an outer regime where the curl of the vorticity or the vorticity itself vanishes and the motion of the fluid is described by the equations of inviscid flow, including Euler's equation and the continuity equation, and (b) a boundary layer where the curl of the vorticity is substantial and viscous forces are significant, as illustrated in figure 10.2.1. Wakes and regions of recirculating flow are allowed, but are significant only insofar as to determine the structure of the outer flow.

The slenderness of the boundary layer compared to the typical size of the boundaries of the flow allows us to simplify the equation of motion for the flow inside the boundary layer, and thereby derive approximate solutions valid in the asymptotic limit of high Reynolds numbers. To illustrate the physical arguments involved in the formulation of boundary layer theory, and simultaneously demonstrate the salient mathematical simplifications, we consider the boundary layer developing along a mildly-curved two-dimensional rigid body held stationary in an incident irrotational flow, as depicted in figure 10.2.1. Extensions to axisymmetric and three-dimensional configurations are straightforward.

Continuity equation

We begin the analysis by introducing Cartesian coordinates where the x axis is tangential to the body at a point, and the y axis is perpendicular to the body at that point, as shown in figure 10.2.1, and apply the continuity equation at a point in the vicinity of the origin,

$$\frac{\partial u_x}{\partial x} + \frac{\partial u_y}{\partial y} = 0. \tag{10.2.1}$$

If L is the typical dimension of the body and U represents the magnitude of the velocity of the incident irrotational flow, we expect that the magnitude of the derivative $\partial u_x/\partial x$ inside the boundary layer will be comparable to the ratio U/L. Moreover, if δ is the designated thickness of the boundary layer, defined as the region around the body across which the vorticity undergoes a rapid transition and the magnitude of viscous forces is significant, and if V is typical of the magnitude of component of the velocity normal to the body at the edge of the boundary layer, we expect that the magnitude of the derivative $\partial u_y/\partial y$ will be comparable to the ratio V/δ. The continuity equation (10.2.1) requires

$$\frac{U}{L} \sim \frac{V}{\delta} \qquad \text{or} \qquad V \sim U\,\frac{\delta}{L}. \tag{10.2.2}$$

x component of the equation of motion

Next, we examine the two components of the equation of motion in the vicinity of the origin written for the dynamic pressure incorporating the hydrostatic variation, defined as $P \equiv p - \rho\,\mathbf{g}\cdot\mathbf{x}$.

Considering the x component of the Navier-Stokes equation,

$$\begin{array}{ccccccc} \dfrac{\partial u_x}{\partial t} + u_x\,\dfrac{\partial u_x}{\partial x} & + u_y\,\dfrac{\partial u_x}{\partial y} & = -\dfrac{1}{\rho}\,\dfrac{\partial P}{\partial x} & + \nu\,(\dfrac{\partial^2 u_x}{\partial x^2} & + \dfrac{\partial^2 u_x}{\partial y^2}), \\ \downarrow & \downarrow & & \downarrow & \downarrow \\ \dfrac{U^2}{L} & \dfrac{U^2}{L} & & \nu\,\dfrac{U}{L^2} & \nu\,\dfrac{U}{\delta^2} \end{array} \tag{10.2.3}$$

we scale u_x with U, $\partial u_x/\partial x$ with U/L, u_y with V, $\partial u_x/\partial y$ with U/δ, the second derivative $\partial^2 u_x/\partial x^2$ with U/L^2, and $\partial^2 u_x/\partial y^2$ with U/δ^2. Moreover, we use the scaling shown in (10.2.2) to eliminate V in favor of U, and find that the magnitude of the various terms is as shown by the arrows underneath equation (10.2.3). The scaling of the first term involving the time derivative on the left-hand side is determined by the temporal variation of the outer flow, which is left unspecified. At this point, there is no obvious way of scaling the x derivative of the dynamic pressure gradient on the right-hand side of (10.2.3) on the basis of kinematics.

The scalings shown underneath equation (10.2.3), combined with the assumption $\delta < L$, have two important consequences. First, the penultimate viscous term on the right-hand side is small compared to the last viscous term and may be neglected, yielding the boundary layer equation

$$\frac{\partial u_x}{\partial t} + u_x\,\frac{\partial u_x}{\partial x} + u_y\,\frac{\partial u_x}{\partial y} = -\frac{\partial P}{\partial x} + \nu\,\frac{\partial^2 u_x}{\partial y^2}. \tag{10.2.4}$$

Second, the magnitude of the last viscous term must be comparable to the magnitude of the inertial terms on the left-hand side, and this requires

$$\frac{U^2}{L} \sim \nu \, \frac{U}{\delta^2} \tag{10.2.5}$$

or

$$\delta \sim \left(\frac{\nu \, L}{U}\right)^{1/2} = \frac{L}{Re^{1/2}}, \tag{10.2.6}$$

where $Re = UL/\nu$ is the Reynolds number defined with respect to the size of the boundaries.

y component of the equation of motion

Next, we consider the individual terms in the y component of the equation of motion,

$$\begin{array}{cccccc} \dfrac{\partial u_y}{\partial t} + u_x \, \dfrac{\partial u_y}{\partial x} & + \, u_y \, \dfrac{\partial u_y}{\partial y} & = -\dfrac{1}{\rho} \, \dfrac{\partial P}{\partial y} & + \, \nu \, (\dfrac{\partial^2 u_y}{\partial x^2} & + \, \dfrac{\partial^2 u_y}{\partial y^2}), \\ \downarrow & \downarrow & & \downarrow & \downarrow \\ \dfrac{U^2 \, \delta}{L^2} & \dfrac{U^2 \, \delta}{L^2} & & \dfrac{U^2 \, \delta}{L^2} & \dfrac{U^2 \, \delta}{L^2} \end{array} \tag{10.2.7}$$

and scale u_y with V, $\partial u_y/\partial x$ with V/L, u_x with U, $\partial u_y/\partial y$ with V/δ, the second derivative $\partial^2 u_y/\partial x^2$ with V/L^2, and $\partial^2 u_y/\partial y^2$ with V/δ^2. Moreover, we express the kinematic viscosity ν in terms of δ using the first equation in (10.2.6), replacing it with $U\delta^2/L$, and find that the magnitude of the various terms is as shown by the arrows underneath equation (10.2.7).

The magnitude of all nonlinear convective and viscous terms is of order δ. Unless the magnitude of the temporal derivative on the left-hand side is of order unity, the dynamic pressure gradient across the boundary layer must also be of order δ, $\partial P/\partial y \simeq \delta$, and this suggests the leading-order approximation

$$\frac{\partial P}{\partial y} \simeq 0. \tag{10.2.8}$$

Thus, non-hydrostatic pressure variations across the boundary layer are negligible, and the dynamic pressure inside the boundary layer is primarily a function of arc length along the boundary.

10.2.1 Boundary-layer equations

To compute the streamwise pressure gradient, we evaluate the x component of the Euler equation (6.4.2) at the edge of the boundary layer, obtaining

$$-\frac{1}{\rho} \frac{\partial P}{\partial x} = \frac{\partial U_x}{\partial t} + U_x \, \frac{\partial U_x}{\partial x}, \tag{10.2.9}$$

where U_x is the tangential component of the velocity of the outer flow. The boundary layer equation (10.2.4) then becomes

$$\frac{\partial u_x}{\partial t} + u_x \frac{\partial u_x}{\partial x} + u_y \frac{\partial u_x}{\partial y} = \frac{\partial U_x}{\partial t} + U_x \frac{\mathrm{d}U_x}{\mathrm{d}x} + \nu \frac{\partial^2 u_x}{\partial y^2}. \tag{10.2.10}$$

Equations (10.2.1) and (10.2.10) provide us with a system of two second-order, nonlinear partial-differential equations for u_x and u_y, to be solved subject to two conditions: (a) the no-slip and no-penetration boundary conditions requiring that u_x and u_y vanish along the boundary, and (b) the far-field condition requiring that, as y/δ tends to infinity, u_x tends to the tangential component of the outer velocity, U_x. Because the boundary layer equations do not involve the second partial derivative of u_y with respect to y, a far-field condition for u_y is not required. The pressure follows from knowledge of the structure of the outer flow and plays the role of a forcing function computed by solving the equations governing the structure of the outer irrotational flow.

Favorable and adverse pressure gradient

Evaluating (10.2.10) at the origin and enforcing the no-slip and no-penetration conditions for a steady flow, we obtain

$$\left(\frac{\partial^2 u_x}{\partial y^2}\right)_{y=0} = -\frac{1}{\nu}\, U_x \frac{\mathrm{d}U_x}{\mathrm{d}x}, \tag{10.2.11}$$

which shows that the sign of the curvature of the velocity profile at the boundary is opposite to that of the streamwise acceleration of the outer flow, $\mathrm{d}U_x/\mathrm{d}x$. Thus, the flow inside the boundary layer in a decelerating outer flow, corresponding to $\mathrm{d}U/\mathrm{d}x < 0$, reverses direction, and this causes convection of vorticity away from the boundary and the consequent formation of vortices inside the bulk of the fluid.

When $\mathrm{d}U_x/\mathrm{d}x > 0$, the pressure gradient is negative, $\mathrm{d}P/\mathrm{d}x < 0$, and the boundary layer is subjected to a favorable pressure gradient. In the opposite case where $\mathrm{d}U_x/\mathrm{d}x < 0$, the pressure gradient is positive, $\mathrm{d}P/\mathrm{d}x > 0$, and the boundary layer is subjected to an adverse pressure gradient. Equation (10.2.11) shows that an adverse pressure gradient promotes flow separation.

10.2.2 Surface curvilinear coordinates

The Prandtl boundary layer equation (10.2.10) was developed with reference to the local Cartesian axes shown in figure 10.2.1, and is strictly valid near the origin. To avoid redefining the Cartesian axes at every point along the boundary, we introduce a curvilinear coordinate system where the ξ axis is tangential to the boundary and the η axis is perpendicular to the boundary, as shown in figure 10.2.1. The corresponding velocity components are denoted by u_ξ and u_η.

Repeating the boundary layer analysis, we find that the boundary layer equations (10.2.1), (10.2.8), and (10.2.10) remain valid to leading-order approximation, provided that the Cartesian x and y coordinates are replaced by corresponding arc lengths in the ξ and η direction denoted, respectively, by l_ξ and l_η. Equation (10.2.8) becomes

$$\frac{\partial P}{\partial l_\eta} = \frac{\rho U_\xi^2}{R}, \tag{10.2.12}$$

where R is the radius of curvature of the boundary. Thus, the dynamic pressure drop across the boundary layer is of order δ, provided that R is not too small; that is, provided that the boundary is not too sharply curved. For simplicity, in the remainder of this chapter we denote l_ξ and l_η, respectively, by x and y.

10.2.3 Parabolization

The absence of a second partial derivative with respect to x renders the boundary layer equation (10.2.10) a parabolic partial differential equation in streamwise position, x. This classification has important consequences on the nature of the solution and chosen numerical method for computing the solution. Most important, the system of equations (10.2.1) and (10.2.10) can be solved using a marching method with respect to x, beginning from a particular x station where the structure of the boundary layer is somehow known; examples will be discussed later in this chapter. In contrast, the Navier-Stokes equation is an elliptic partial differential equation with respect to x and y, and the solution must by found simultaneously at every point in the flow, even when the velocity and pressure at the inlet are specified.

The parabolic nature of (10.2.10) with respect to x implies that, if a perturbation is introduced at some point along the boundary layer, it will modify the structure of the flow downstream but will leave the upstream flow unaffected. The absence of the second partial derivative with respect to x due to the boundary layer approximation precludes a mechanism for upstream signal propagation.

10.2.4 Flow separation

Boundary-layer analysis for laminar flow is based on two key assumptions: the Reynolds number is sufficiently large, but not so large that the flow becomes turbulent; and the vorticity remains confined inside boundary layers that wrap around the boundaries.

The physical relevance of the second assumption depends on the structure of the incident flow and the geometry of the boundaries. Streamlined bodies allow laminar boundary layers to develop over a large distance, whereas bluff bodies cause the vorticity to concentrate inside compact regions forming steady or unsteady wakes. For example, the alternating ejection of vortices of opposite sign into a wake is responsible for the von Kàrmàn vortex street illustrated in figure 10.1.1. These limitations should be born in mind when carrying out a boundary layer analysis.

Problem

10.2.1 *Oscillatory flow over a plate*

Compute the boundary layer along an infinite plate driven by a streaming oscillatory outer flow with velocity $U_x = \bar{U}_x \cos(\omega t)$, where $\bar{U}_x$ is a constant amplitude and ω is the angular frequency.

10.3 Blasius boundary layer on a semi-infinite plate

Having established simplified equations of motion for the flow inside a boundary layer over a solid surface, we proceed to derive solutions by analytical and numerical methods.

In this section, we study the boundary layer established over a semi-infinite flat plate that is held stationary parallel to an incident streaming flow with velocity U, as illustrated in figure 10.3.1. Since the tangential velocity of the outer flow is constant, $\mathrm{d}U_x(x)/\mathrm{d}x = 0$, the boundary layer equation (10.2.10) for steady flow simplifies to the convection–diffusion equation,

$$u_x \frac{\partial u_x}{\partial x} + u_y \frac{\partial u_x}{\partial y} = \nu \frac{\partial^2 u_x}{\partial y^2}. \qquad (10.3.1)$$

The problem has been reduced to solving the simplified equation of motion (10.3.1) together with the continuity equation (10.2.1) for the velocity components u_x and u_y, subject to no-slip, no-penetration, and far-field conditions.

Because the length of the plate is infinite, our only choice of a characteristic length scale, L, introduced in section 10.2 is the streamwise distance, x. The second relation in (10.2.6) then provides us with an expression for the boundary layer thickness in terms of the local Reynolds number, $Re_x \equiv Ux/\nu$,

$$\delta(x) \sim \left(\frac{\nu x}{U}\right)^{1/2} = \frac{x}{Re_x^{1/2}}. \qquad (10.3.2)$$

Recall that this scaling has arisen by balancing the magnitude of inertial and viscous forces inside the boundary layer.

10.3.1 Self-similarity and the Blasius equation

Blasius discovered that computing the solution of the system of partial differential equations (10.2.1) and (10.3.1) requires solving a single ordinary differential

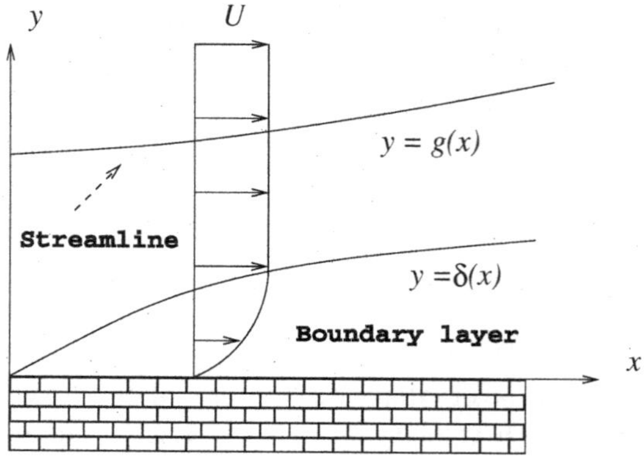

Figure 10.3.1 Schematic illustration of a boundary layer developing along a semi-infinite flat plate placed parallel to a uniform incident stream.

equation. To carry out this reduction, we assume that the flow develops in a self-similar manner, such that the streamwise velocity profile across the boundary layer is a function of a scaled dimensionless transverse position expressed by the similarity variable

$$\eta \equiv \frac{y}{\delta(x)} = y\left(\frac{U}{\nu x}\right)^{1/2}, \tag{10.3.3}$$

according to the functional form

$$u_x(x,y) = U\,F(\eta), \tag{10.3.4}$$

where $F(\eta)$ is an *a priori* unknown function. A key observation is that this self-similar streamwise profile derives from the stream function

$$\psi(x,y) = \sqrt{\nu\,U x}\,f(\eta), \tag{10.3.5}$$

where f is the indefinite integral or anti-derivative of F, satisfying $\mathrm{d}f/\mathrm{d}\eta = F$. The principal advantage of using the stream function is that the continuity equation is satisfied automatically and does not need to be further considered.

As a preliminary, we compute

$$\frac{\partial\eta}{\partial x} = -\frac{y}{2x}\left(\frac{U}{\nu x}\right)^{1/2}, \qquad \frac{\partial\eta}{\partial y} = \left(\frac{U}{\nu x}\right)^{1/2}. \tag{10.3.6}$$

Differentiating (10.3.5) with respect to y or x and setting

$$\frac{\partial f}{\partial x} = \frac{\mathrm{d}f}{\mathrm{d}\eta}\frac{\partial\eta}{\partial x}, \qquad \frac{\partial f}{\partial y} = \frac{\mathrm{d}f}{\mathrm{d}\eta}\frac{\partial\eta}{\partial y}, \tag{10.3.7}$$

we derive the velocity components

$$u_x(x,y) = \frac{\partial \psi}{\partial y} = U\sqrt{\nu U x}\,\frac{\mathrm{d}f}{\mathrm{d}\eta}\,\frac{\mathrm{d}\eta}{\mathrm{d}y} = U\sqrt{\nu U x}\,\frac{\mathrm{d}f}{\mathrm{d}\eta}\sqrt{\frac{U}{\nu x}} = U\,\frac{\mathrm{d}f}{\mathrm{d}\eta} \tag{10.3.8}$$

and

$$\begin{aligned} u_y(x,y) &= -\frac{\partial \psi}{\partial x} = -\sqrt{\nu U}\,\frac{\partial[\sqrt{x}\,f(\eta)]}{\partial x} \\ &= -\frac{1}{2}\sqrt{\frac{\nu U}{x}}\,f - \sqrt{\nu U x}\,\frac{\mathrm{d}f}{\mathrm{d}\eta}\,\frac{\partial \eta}{\partial x} = -\frac{1}{2}\sqrt{\frac{\nu U}{x}}\,(f - \eta\,\frac{\mathrm{d}f}{\mathrm{d}\eta}). \end{aligned} \tag{10.3.9}$$

Further differentiation yields

$$\frac{\partial u_x}{\partial x} = U\,\frac{\mathrm{d}^2 f}{\mathrm{d}\eta^2}\,\frac{\partial \eta}{\partial x}, \qquad \frac{\partial u_x}{\partial y} = U\,\frac{\mathrm{d}^2 f}{\mathrm{d}\eta^2}\,\frac{\partial \eta}{\partial y}, \qquad \frac{\partial^2 u_x}{\partial y^2} = U\,\frac{\mathrm{d}^3 f}{\mathrm{d}\eta^3}\left(\frac{\partial \eta}{\partial y}\right)^2. \tag{10.3.10}$$

Substituting these expressions in the boundary layer equation (10.3.1) and simplifying, we obtain a third-order nonlinear ordinary differential equation for the function $f(\eta)$,

$$\frac{\mathrm{d}^3 f}{\mathrm{d}\eta^3} + \frac{1}{2}\,f\,\frac{\mathrm{d}^2 f}{\mathrm{d}\eta^2} = 0, \tag{10.3.11}$$

derived by Blasius in 1908. Enforcing the no-slip and no-penetration conditions and requiring that the flow in the boundary layer reduces to the outer uniform flow far from the plate, we obtain

$$f = 0 \quad \text{and} \quad \frac{\mathrm{d}f}{\mathrm{d}\eta} = 0 \quad \text{at} \quad \eta = 0, \tag{10.3.12}$$

and

$$\frac{\mathrm{d}f}{\mathrm{d}\eta} \to 1 \quad \text{as} \quad \eta \to \infty. \tag{10.3.13}$$

These equations provide us with boundary and far-field conditions to be used in solving the Blasius equation (10.3.11).

Before proceeding to compute the solution, it is instructive to apply the Blasius equation (10.3.11) at the plate where $\eta = 0$. Using the first boundary condition in (10.3.12), we find $\mathrm{d}^3 f/\mathrm{d}\eta^3(\eta = 0) = 0$. This result shows that the curvature of the streamwise velocity profile vanishes at the wall, in agreement with equation (10.2.11).

10.3.2 Numerical solution

To solve the Blasius equation (10.3.11), we rename $x_1 = f$, denote the first and second derivative of the function f by $x_2 \equiv \mathrm{d}f/\mathrm{d}\eta$ and $x_3 \equiv \mathrm{d}x_2/\mathrm{d}\eta = \mathrm{d}^2 f/\mathrm{d}\eta^2$, and decompose the third-order equation into a system of three first-order nonlinear equations,

$$\frac{\mathrm{d}x_1}{\mathrm{d}\eta} = x_2, \qquad \frac{\mathrm{d}x_2}{\mathrm{d}\eta} = x_3, \qquad \frac{\mathrm{d}x_3}{\mathrm{d}\eta} = -\frac{1}{2}\, x_1\, x_3. \tag{10.3.14}$$

This system is accompanied by three boundary conditions, $x_1(\eta = 0) = 0$, $x_2(\eta = 0) = 0$, and $x_3(\eta = \infty) = 1$, originating from (10.3.12) and (10.3.13).

Since boundary conditions are provided at both ends of the solution domain with respect to η extending from 0 to ∞, we are presented with a two-point boundary value problem involving three first-order differential equations. The solution can be computed using a shooting method according to the following steps:

1. First, we guess the value of $x_3(0) \equiv f''(\eta = 0)$.

2. Second, we integrate equations (10.3.14) from $\eta = 0$ to $\eta = \infty$ subject to the initial conditions (10.3.12), using the guessed value of $x_3(\eta = 0)$. In practice, integrating up to $\eta = 10$ yields satisfactory accuracy.

3. Third, we check whether the far-field condition $x_3(\eta = \infty) = 1$ is satisfied; if not, we improve the guess for $q(\eta = 0)$, and return to step 2. The improvement can be made using several methods.[1]

The following MATLAB code *blasius* performs the integration in the second step using the modified Euler method:

```
%========
% Blasius boundary layer equation
%=======

Nstep = 64;   % integration steps
etamax = 10.0;

deta = etamax/Nstep;

eta = 0.0;
x(1) = 0.0;
x(2) = 0.0;
x(3)=0.332;

xplot(1) = eta;
```

[1] POZRIKIDIS, C., 2008, *Numerical Computation in Science and Engineering*, Second Edition, Oxford University Press.

```
yplot1(1) = x(1);
yplot2(1) = x(2);
yplot3(1) = x(3);

%---
for i=2:Nstep+1

  xp = blasius_ode(x);
  for j=1:3
    xsave(j) = x(j);
    xpsave(j) = xp(j);
    x(j) = x(j) + xp(j)*deta;
  end
  eta = eta + deta;
  xp = blasius_ode(x);
  for j=1:3
    x(j) = xsave(j) + 0.5*(xpsave(j)+xp(j))*deta;
  end
  xplot(i) = eta;
  yplot1(i) = x(1); yplot2(i) = x(2); yplot3(i) = x(3);

end
%---

%---
% plotting
%---

hold on
plot(yplot1,xplot,'--')
plot(yplot2,xplot)
plot(yplot3,xplot,':')
ylabel('\eta','fontsize',15)
```

The program calls the following MATLAB function to evaluate the right-hand side of the differential equations:

```
function xp = blasius_ode(x)

%---
% Blasius odes
%---

  xp(1) = x(2);
  xp(2) = x(3);
  xp(3) = -0.50*x(1)*x(3);

end
```

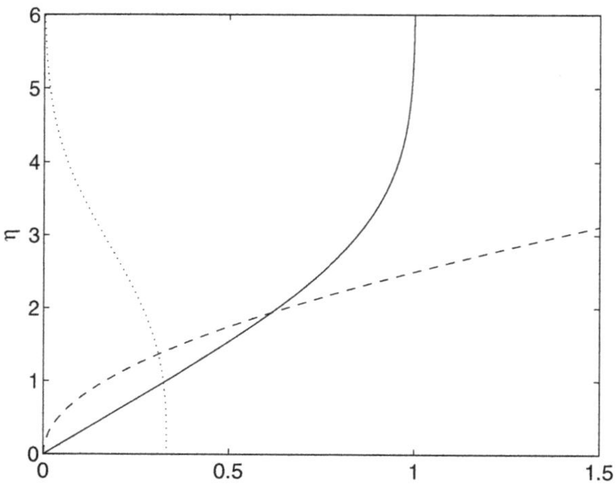

Figure 10.3.2 Graphs of the Blasius self-similar streamwise velocity profile $u/U = f'$ (solid line), and its integral and derivative functions f and f'' (dashed and dotted line).

Numerical computations show that the far-field boundary condition is satisfied when $f''(0) \simeq 0.332$. The corresponding profile of the streamwise velocity, $u/U = f' \equiv \mathrm{d}f/\mathrm{d}\eta$, is drawn with the solid line in figure 10.3.2, along with the profiles of f and $f'' \equiv \mathrm{d}^2 f/\mathrm{d}\eta^2$, drawn with the dashed and dotted lines.

The numerical solution shows that $u/U = 0.99$ when $\eta \simeq 4.9$. Based on this result, we may define the 99% boundary layer thickness

$$\delta_{99} = 4.9 \sqrt{\frac{\nu x}{U}} \qquad \text{or} \qquad \frac{\delta_{99}}{x} = \frac{4.9}{Re_x}, \tag{10.3.15}$$

where $Re_x \equiv Ux/\nu$. The 99.5% boundary layer thickness is defined in a similar fashion. The numerical solution shows that the corresponding coefficient on the right-hand side of equations (10.3.15) is approximately equal to 5.3.

10.3.3 Wall shear stress and drag force

The wall shear stress and drag force exerted on a boundary are of particular interest in the engineering design of equipment for high-speed flow. According to the Blasius similarity solution, the wall shear stress is given by

$$\tau_{xy}(x) = \mu \left(\frac{\partial u_x}{\partial y} \right)_{y=0} = \frac{f''(0)}{\sqrt{Re_x}} \rho U^2 = \frac{0.332}{\sqrt{Re_x}} \rho U^2. \tag{10.3.16}$$

We observe that the wall shear stress takes an infinite value at the leading edge and decreases like the inverse square root of the streamwise distance or local

Reynolds number, Re_x, along the plate. However, the physical significance of the singular behavior at the origin is undermined by the breakdown of the assumptions that led us to the boundary layer equations at the leading edge.

Even though the shear stress is infinite at the leading edge, the inverse-square-root singularity is integrable and the drag force exerted on any finite section of the plate extending from the leading edge up to an arbitrary point is finite. Using the similarity solution, we find that the drag force exerted on *both* sides of the plate over a length extending from the leading edge up to a certain distance x, is given by

$$D(x) \equiv 2 \int_0^x \tau_{xy}(\xi)\, \mathrm{d}\xi = 0.664\, \frac{\rho\, U^{3/2}}{\nu^{1/2}} \int_0^x \frac{\mathrm{d}\xi}{\sqrt{\xi}}. \tag{10.3.17}$$

Performing the integration, we find

$$D(x) = \frac{1.328}{\sqrt{Re_x}}\, \rho\, U^2 x. \tag{10.3.18}$$

Based on this expression, we define the dimensionless drag coefficient

$$c_D \equiv \frac{D}{\frac{1}{2}\rho U^2 x} = \frac{2.656}{\sqrt{Re_x}}. \tag{10.3.19}$$

The predictions of equations (10.3.18) and (10.3.19) agree well with laboratory measurements up to about $Re_x \simeq 1.2 \times 10^5$. At that point, the flow inside the boundary layer develops a wavy pattern and ultimately becomes turbulent. Above the critical value of Re_x, the function $c_D(Re_x)$ jumps to a different branch with significantly higher values.

10.3.4 Vorticity transport

Neglecting the velocity component along the y axis, we find that the z component of the vorticity inside the boundary layer is given by

$$\omega_z(x, y) \simeq -\frac{\partial u_x}{\partial y} = -\frac{f''(\eta)}{\sqrt{Re_x}}\, \frac{U^2}{\nu} = -f''(\eta)\, \frac{U}{\delta(x)}. \tag{10.3.20}$$

We observe that the strength of the vorticity at a particular location, η, decreases like the inverse of the local boundary layer thickness due to the broadening of the velocity profile.

The streamwise rate of convection of vorticity across a plane that is perpendicular to the plate is given by

$$\int_0^\infty u_x(x, y)\, \omega_z(x, y)\, \mathrm{d}y \simeq -\int_0^\infty u_x(x, y)\, \frac{\partial u_x}{\partial y}\, \mathrm{d}y = -\frac{1}{2}\, U^2, \tag{10.3.21}$$

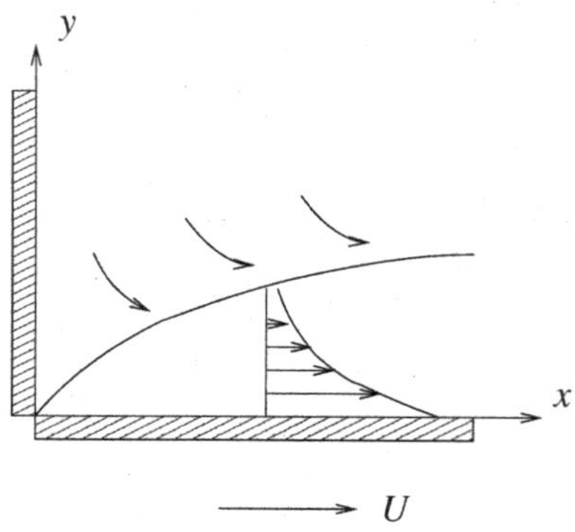

Figure 10.3.3 Illustration of the Sakiadis boundary layer developing over a semi-infinite translating belt.

independent of the downstream position, x. Thus, the flux of vorticity across the plate is zero and viscous diffusion of vorticity does not occur at the wall, in agreement with our earlier observation that the gradient of the vorticity vanishes at the wall, $U\, \partial f'''/\partial y = \partial^2 u_x/\partial y^2 = -\partial \omega/\partial y = 0$. Consequently, all convected vorticity is generated at the leading edge where the boundary layer approximation ceases to be valid. Viscous stresses at the leading edge somehow generate the proper amount of vorticity necessary to establish the Blasius self-similar flow.

Computer Problems

10.3.1 *Blasius solution*

Use the shooting method to compute $f''(0)$ for the Blasius boundary layer, accurate to the sixth decimal place.

10.3.2 *Sakiadis' boundary layer*

Consider flow due to the translation of a semi-infinite belt along the x axis with velocity U normal to a vertical stationary wall, as illustrated in figure 10.3.3. A boundary layer identified by Sakiadis is established along the belt.[2]

(a) Show that the flow in the boundary layer is governed by the Blasius equation (10.3.11), subject to the modified boundary conditions $f = 0$ and $f' = 0$ at $\eta = 0$, and $f' \to 0$ as η tends to infinity.

(b) Obtain the solution using the shooting method, plot the streamwise velocity

[2]SAKIADIS, B. C., 1961, Boundary-layer behavior on continuous solid surfaces: I. Boundary-layer equations for two-dimensional and axisymmetric flow, *A.I.Ch.E. J.* **7**, 26–28; II. The boundary layer on a continuous flat surface, *A.I.Ch.E. J.* **7**, 221–225.

profile and discuss the structure of the flow field far from the moving belt. *Hint*: the iterations converge when $f''(0) = 0.44375$.

10.4 Displacement and momentum thickness

Because of the broadening of the streamwise velocity profile in the streamwise direction, the streamlines inside the Blasius boundary layer are deflected upward and away from the plate, as shown in figure 10.3.1.

Displacement thickness

Let us consider a streamline outside the boundary layer, described by the equation $y = g(x)$, and write a mass balance over a control area that is enclosed by (a) the streamline, (b) a vertical plane at $x = 0$, (c) a vertical plane located at a certain distance x, and (d) the plate. Since the streamwise velocity profile at the leading edge located at $x = 0$ is flat, we obtain

$$\int_0^{g(0)} U \, \mathrm{d}y = \int_0^{g(x)} u_x(x, y) \, \mathrm{d}y. \tag{10.4.1}$$

Straightforward rearrangement yields

$$U \, [g(x) - g(0)] = \int_0^{g(x)} [U - u_x(x, y)] \, \mathrm{d}y. \tag{10.4.2}$$

Taking the limit as the streamline under consideration moves farther from the plate, we find

$$\lim_{x \to \infty} [\, g(x) - g(0) \,] = \delta^*(x), \tag{10.4.3}$$

where

$$\delta^*(x) \equiv \int_0^{\infty} (1 - \frac{u_x}{U}) \, \mathrm{d}y \tag{10.4.4}$$

is the the displacement thickness. Using the numerical solution of the Blasius equation to evaluate the integral on the right-hand side of (10.4.4), we derive the precise relation

$$\delta^*(x) = \sqrt{\frac{\nu \, x}{U}} \int_0^{\infty} \left(1 - \frac{\mathrm{d}f}{\mathrm{d}\eta} \right) \mathrm{d}\eta = 1.721 \, \sqrt{\frac{\nu \, x}{U}}, \tag{10.4.5}$$

which shows that the displacement thickness, like the 99% boundary layer thickness, increases like the square root of the streamwise position.

Physically, the displacement thickness represents the vertical displacement of the streamlines far from from the plate with respect to their elevation at the leading edge. Laboratory experiments have shown that the laminar boundary layer undergoes a transition from the laminar to the turbulent state when the displacement thickness reaches the value $\delta^* \sim 600\,\nu/U$. At that point, turbulent shear stresses become significant and the present analysis based on the assumption of laminar flow ceases to be valid.

The displacement thickness describes the surface of a fictitious impenetrable but slippery body that is held stationary in the incident irrotational flow. An improved boundary layer theory can be developed by replacing the tangential velocity of the outer flow along the plate, U, with the corresponding tangential component of the velocity of the irrotational flow past the fictitious body. The irrotational flow past the fictitious body must be computed after the displacement thickness has been established, as discussed in this section. This iterative improvement provides us with a basis for describing the flow in the context of asymptotic expansions.

Momentum thickness

It is illuminating to perform a momentum integral balance over the control area previously used to define the displacement thickness. Since the upper boundary of the control volume is a streamline, it does not contribute to the rate of momentum input. Assuming that the normal stresses on the vertical sides are equal in magnitude and opposite in sign, which is justified by the assumption that the pressure drop across the boundary layer is negligibly small, and neglecting the traction along the top streamline, we obtain

$$\int_0^{g(0)} U\,(\rho\,U)\,\mathrm{d}y - \int_0^{g(x)} u_x\,(\rho\,u_x)\,\mathrm{d}y - \frac{1}{2}\,D(x) = 0, \qquad (10.4.6)$$

where $D(x)$ is the drag force exerted on *both* sides of the plate, defined in equation (10.3.18). Now we make the upper limits of integration equal by recasting (10.4.6) into the form

$$-\rho\,U^2\,[g(x) - g(0)] - \rho\int_0^{g(x)} (U^2 - u_x^2)\,\mathrm{d}y - \frac{1}{2}\,D(x) = 0. \qquad (10.4.7)$$

Finally, we take the limit as the streamline defining the top of the control area moves far from the plate and use the definitions (10.4.3) and (10.4.4) to obtain the relation

$$D(x) = 2\,\rho\,U^2\Theta(x), \qquad (10.4.8)$$

where Θ is the momentum thickness defined as

$$\Theta(x) \equiv \int_0^{\infty} \frac{u_x}{U}\,(1 - \frac{u_x}{U})\,\mathrm{d}y. \qquad (10.4.9)$$

Using the numerical solution of the Blasius equation, we find

$$\Theta(x) = \sqrt{\frac{\nu x}{U}}\int_0^{\infty} f'(\eta)\,[1 - f'(\eta)]\,\mathrm{d}\eta = 0.664\sqrt{\frac{\nu x}{U}}, \qquad (10.4.10)$$

where $f'(\eta) = \mathrm{d}f/\mathrm{d}\eta$.

Shape factor

The ratio between the displacement and the momentum thickness is called the shape factor,

$$H \equiv \frac{\delta^*}{\Theta}. \tag{10.4.11}$$

Substituting the right-hand sides of expressions (10.4.5) and (10.4.9) in (10.4.11), we find that, for the boundary layer over a flat plate, $H = 2.591$.

Inspecting the definitions of δ^* and Θ given in equations (10.4.4) and (10.4.8), we find that the shape factor is greater than unity as long as the streamwise velocity u_x is less than U over a substantial portion of the boundary layer. The satisfaction of this restriction is consistent with physical intuition. The smaller the value of H, the more blunt the velocity profile across the boundary layer.

Relation between the wall shear stress and momentum thickness

The momentum thickness is related to the wall shear stress, and *vice versa*, by the integral momentum balance expressed by equation (10.4.8). Differentiating (10.3.18) with respect to x, we find

$$\frac{\mathrm{d}D(x)}{\mathrm{d}x} = 2\,\tau_{xy}(x). \tag{10.4.12}$$

Expressing the drag force in terms of the momentum thickness using (10.4.8), we obtain

$$\tau_{xy}(x) = \rho\, U^2 \frac{\mathrm{d}\Theta(x)}{\mathrm{d}x}. \tag{10.4.13}$$

Thus, if the shear stress is known, the momentum thickness can be computed by integration. Conversely, if the momentum thickness is known, the shear stress can be computed by differentiation.

10.4.1 Von Kàrmàn's approximate method

Given the velocity profile across the boundary layer, we have two ways of computing the wall shear stress: directly by differentiation, and indirectly by evaluating the momentum thickness and then differentiating it with respect to streamwise position x to obtain the shear stress according to equation (10.4.13). The indirect method is less sensitive to the structure of the velocity profile near the wall. For the velocity profile that arises by solving the Blasius equation, the two methods are equivalent (problem 10.4.1).

Next, we introduce a self-similar velocity profile with some reasonable form involving an unspecified function that is either stipulated by physical intuition or suggested by laboratory observation. Our goal is to adjust the unspecified function so that the two methods of computing the wall shear stress discussed in the preceding paragraph are equivalent.

A reasonable velocity profile is

$$\frac{u_x}{U} = \frac{\mathrm{d}f(\eta)}{\mathrm{d}\eta} = \begin{cases} \sin\frac{\pi y}{2\,\Delta(x)} & \text{for } 0 < y < \Delta(x), \\ 1 & \text{for } y > \Delta(x), \end{cases} \tag{10.4.14}$$

where

$$\eta \equiv \frac{y}{\Delta(x)}, \tag{10.4.15}$$

and $\Delta(x)$ is an unspecified function playing the role of a boundary layer thickness, similar to the δ_{99} thickness introduced in equation (10.3.15). Note that the velocity distribution (10.4.14) conforms with the required boundary conditions $f'(0) = 0$, $f'''(0) = 0$, and $f'(\infty) = 1$, but does not satisfy the Blasius equation; a prime denotes a derivative with respect to η.

Differentiating the profile (10.4.14) with respect to y, we obtain the wall shear stress

$$\tau_{xy}(x) = \frac{\pi\mu U}{2\,\Delta(x)}. \tag{10.4.16}$$

The displacement thickness, momentum thicknesses, and shape factor defined in (10.4.4), (10.4.8), and (10.4.11), are found to be

$$\delta^*(x) = (1 - \frac{2}{\pi})\,\Delta(x) = 0.363\,\Delta(x),$$

$$\Theta(x) = (\frac{2}{\pi} - \frac{1}{2})\,\Delta(x) = 0.137\Delta(x), \qquad H = 2.660. \tag{10.4.17}$$

It is reassuring to observe that the shape factor is remarkably close to that arising from the exact solution of the Blasius equation, $H = 2.591$.

Substituting the expressions for the momentum thickness and wall shear stress in the integral momentum balance (10.4.13), we derive an ordinary differential equation for $\Delta(x)$,

$$\frac{\pi\mu U}{2\,\Delta(x)} = 0.137\,\rho U^2\,\frac{\mathrm{d}\Delta(x)}{\mathrm{d}x}. \tag{10.4.18}$$

Rearranging and integrating with respect to x subject to the initial condition $\Delta = 0$ at $x = 0$, we find

$$\Delta(x) = 4.80\sqrt{\frac{\nu x}{U}}. \tag{10.4.19}$$

Substituting this expression back in (10.4.16) and in the first and second relations in (10.4.17), we find

$$\tau_{xy}(x) = \frac{0.327}{\sqrt{Re_x}}\,\rho U^2, \qquad \delta^*(x) = 1.743\,\sqrt{\frac{\nu x}{U}},$$

$$\Theta(x) = 0.665\,\sqrt{\frac{\nu x}{U}}. \tag{10.4.20}$$

These expressions are in excellent agreement with their exact counterparts shown in equations (10.3.16), (10.4.5), and (10.4.9). However, this agreement is fortuitous and atypical of the accuracy of the approximate method (problem 10.4.2).

Problems

10.4.1 *Wall shear stress*

Confirm that the wall shear stress computed directly by differentiating the velocity profile across the Blasius boundary layer is the same as that arising by differentiating the momentum thickness with respect to streamwise position, x, according to equation (10.4.13).

10.4.2 *Von Kàrmàn's method*

Assume that the velocity profile across the Blasius boundary layer is described by a hyperbolic tangent function, $u_x = U \tanh[y/\Delta(x)]$. Show that the effective boundary layer thickness, wall shear stress, displacement thickness, and momentum thicknesses are given by the right-hand sides of equations (10.4.19) and (10.4.20), except that the numerical coefficients are equal, respectively, to 2.553, 0.392, 1.770, and 0.783. Discuss the accuracy of these results with reference to the exact solution.

Computer Problem

10.4.3 *Sakiadis' boundary layer*

Compute the displacement thickness, momentum thickness, and shape factor of the Sakiadis boundary layer discussed in problem 10.3.2.

10.5 Boundary layers in accelerating and decelerating flow

In section 10.3, we discussed in detail the Blasius boundary layer developing along a flat plate that is aligned with a uniform incident stream, distinguished by the uniformity of the tangential velocity of the outer flow along the plate. In this

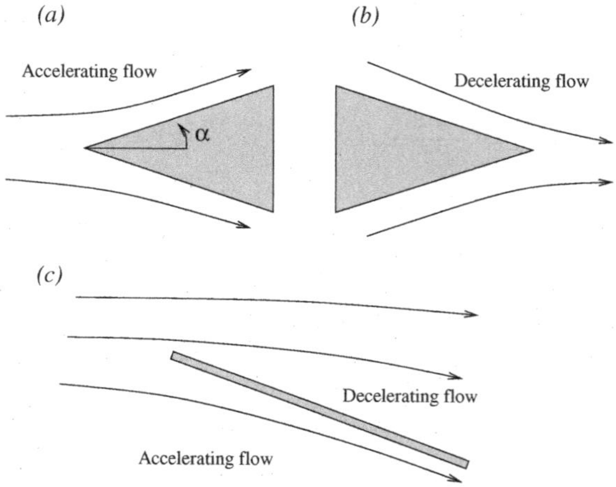

Figure 10.5.1 Boundary layers developing in accelerating or decelerating flow: (a, b) flow past a wedge, and (c) uniform flow past a flat plate at a non-zero angle of attack.

section, we proceed to consider the more general case of Falkner-Skan boundary layers developing in accelerating or decelerating incident flow. Examples of physical situations where these boundary layers occur are illustrated in figure 10.5.1.

Consider an outer flow whose tangential velocity, $U_x(x)$, exhibits a power-law dependence on the streamwise distance, x, along a flat plate,

$$U_x(x) = c\,x^m, \tag{10.5.1}$$

where c is a positive coefficient and m is a positive or negative exponent. When $m = 0$, we recover the Blasius boundary layer over a flat plate at zero angle of attack. When $m = 1$, we obtain a boundary layer in orthogonal stagnation-point flow. Intermediate values of m correspond to symmetric flow past a wedge with semi-angle $\alpha = \pi m/(m+1)$, as illustrated in figure 10.5.1(a).

Differentiating (10.5.1) with respect to x, we obtain the streamwise acceleration or deceleration of the outer flow,

$$\frac{\mathrm{d}U_x}{\mathrm{d}x} = c\,m\,x^{m-1}. \tag{10.5.2}$$

Thus, the outer flow accelerates when $m > 0$ and decelerates when $m < 0$. In the first case, conservation of mass expressed by the continuity equation requires that the y derivative of the velocity component normal to the wall, U_y, is negative, $\partial U_y/\partial y < 0$. Since U_y is zero on the wall, it must be negative at the edge of the boundary layer. The associated motion of the outer fluid toward the wall confines the vorticity and reduces the thickness of the boundary layer with respect to that in a non-accelerating flow.

Substituting (10.5.2) in the boundary layer equation (10.2.10) at steady state, we obtain the specific form

$$u_x \frac{\partial u_x}{\partial x} + u_y \frac{\partial u_x}{\partial y} = c^2 \, m \, x^{2m-1} + \nu \, \frac{\partial^2 u_x}{\partial y^2}. \tag{10.5.3}$$

Working as in section 10.3 for the Blasius boundary layer, we identify the characteristic length L with the current streamwise position, x, and use (10.3.2) to define the boundary layer thickness

$$\delta(x) \sim \sqrt{\frac{\nu \, x}{U_x(x)}} = \sqrt{\frac{\nu}{c \, x^{m-1}}}. \tag{10.5.4}$$

10.5.1 Self-similarity

To this end, we assume that the velocity profile across the boundary is self-similar. The means that the streamwise velocity profile, u_x/U_x, is a function of the dimensionless similarity variable

$$\eta \equiv \frac{y}{\delta(x)} = y \sqrt{\frac{U_x(x)}{\nu x}} = y \sqrt{\frac{c}{\nu x^{1-m}}}, \tag{10.5.5}$$

so that

$$u_x(x, y) = U_x(x) \, F(\eta), \tag{10.5.6}$$

where $F(\eta)$ is an *a priori* unknown function. A key observation is that this self-similar profile can be derived from the stream function

$$\psi(x, y) = \sqrt{\nu \, U_x(x) \, x} \, f(\eta) = \sqrt{\nu \, c \, x^{m+1}} \, f(\eta), \tag{10.5.7}$$

where $F = \mathrm{d}f/\mathrm{d}\eta$.

As a preliminary, we compute the derivatives

$$\frac{\partial \eta}{\partial x} = \frac{m-1}{2} \, \frac{y}{x} \sqrt{\frac{c \, x^{m-1}}{\nu}}, \qquad \frac{\partial \eta}{\partial y} = \sqrt{\frac{c}{\nu x^{1-m}}}. \tag{10.5.8}$$

Differentiating (10.5.4) with respect to x, we derive expressions for the two velocity components,

$$\begin{aligned} u_x(x, y) &= \frac{\partial \psi}{\partial y} = U_x(x) \, \frac{\mathrm{d}f}{\mathrm{d}\eta} = c \, x^m \, \frac{\mathrm{d}f}{\mathrm{d}\eta}, \\ u_y(x, y) &= -\frac{\partial \psi}{\partial x} = \frac{1}{2} \sqrt{\nu c x^{m-1}} \, [(1-m) \, \eta \, \frac{\mathrm{d}f}{\mathrm{d}\eta} - (1+m) \, f]. \end{aligned} \tag{10.5.9}$$

Further differentiation yields

$$\frac{\partial u_x}{\partial x} = c\,m\,x^{m-1}\frac{\mathrm{d}f}{\mathrm{d}\eta} + U\,\frac{\mathrm{d}^2 f}{\mathrm{d}\eta^2}\,\frac{\partial \eta}{\partial x},$$

$$\frac{\partial u_x}{\partial y} = U\,\frac{\mathrm{d}^2 f}{\mathrm{d}\eta^2}\,\frac{\partial \eta}{\partial y}, \qquad \frac{\partial^2 u_x}{\partial y^2} = U\,\frac{\mathrm{d}^3 f}{\mathrm{d}\eta^3}\left(\frac{\partial \eta}{\partial y}\right)^2. \tag{10.5.10}$$

Substituting these expressions in (10.5.3), we derive the Falkner-Skan ordinary differential equation for the function f,

$$\frac{\mathrm{d}^3 f}{\mathrm{d}\eta^3} + \frac{1}{2}\,(m+1)\,f\,\frac{\mathrm{d}^2 f}{\mathrm{d}\eta^2} - m\left(\frac{\mathrm{d}f}{\mathrm{d}\eta}\right)^2 + m = 0, \tag{10.5.11}$$

which is to be solved subject to the boundary conditions expressed by (10.3.12) and (10.3.13). When $m = 0$, we recover the Blasius equation (10.3.11). Since boundary conditions are specified at both ends of the solution domain $(0, \infty)$, we are presented with a two-point boundary-value problem involving three first-order differential equations.

10.5.2 Numerical solution

The solution of the boundary-value problem can be found by the shooting method described in section 10.3.2 for the Blasius equation. The counterpart of the Blasius system of differential equations (10.3.14) is

$$\frac{\mathrm{d}x_1}{\mathrm{d}\eta} = x_2, \qquad \frac{\mathrm{d}x_2}{\mathrm{d}\eta} = x_3, \tag{10.5.12}$$

$$\frac{\mathrm{d}x_3}{\mathrm{d}\eta} = -\frac{1}{2}\,(m+1)\,x_1\,x_3 + m\,x_2^2 - m.$$

The boundary conditions require $x_1(\eta = 0) = 0$, $x_2(\eta = 0) = 0$, and $x_3(\eta = \infty) = 1$. Convergence is achieved when $f''(0) = 1.491$ for $m = 1.5$, $f''(0) = 1.231$ for $m = 1.0$, $f''(0) = 0.675$ for $m = 0.25$, $f''(0) = 0.594$ for $m = 0.176$, $f''(0) = 0.332$ for $m = 0$, and $f''(0) = 0$ for $m = -0.0904$.

Figure 10.5.2 illustrates streamwise velocity profiles expressed by the derivative $f'(\eta)$ for several values of m. The profiles for $m < 0$, corresponding to decelerating flow, exhibit an inflection point near the wall. The wall shear stress vanishes when $m = -0.0904$.

Problems

10.5.1 *Orthogonal stagnation-point flow.*

The Falkner-Skan profile with $m = 1$ corresponds to irrotational orthogonal stagnation-point flow against a flat plate.

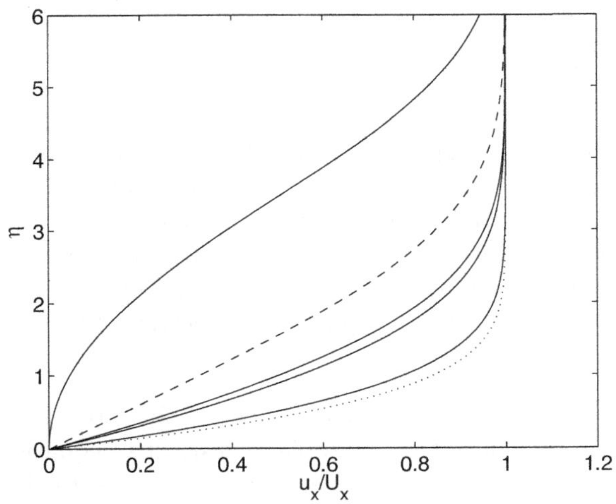

Figure 10.5.2 Velocity profiles across Falkner-Skan boundary layers for several values of the acceleration parameter; from bottom to top, $m = 1.5$ (dotted line), 1.0, 0.25, 0.176, 0.0 (dashed line), and -0.0904.

(*a*) Derive the velocity field of the outer irrotational flow.

(*b*) Show that the Falkner-Skan boundary layer solution satisfies the unsimplified Navier-Stokes equation, and thus it provides us with an exact solution.

Computer Problem

10.5.2 *Falkner-Skan equations*

Compute the velocity profile across a Falkner-Skan boundary layer with $m = -0.05$.

10.6 Momentum integral method

We have discussed the solution of the boundary layer equations over a flat surface subject to a constant, accelerating, or decelerating outer flow with a power-law dependence on streamwise position. To compute boundary layers developing over curved surfaces and for more general types of outer flow, we resort to approximate methods. Von Kàrmàn developed an elegant method for computing the flow inside a two-dimensional boundary layer developing over a surface with arbitrary shape, based on an integral momentum balance. The formulation culminates in an ordinary differential equation for a properly defined boundary layer thickness with respect to arc length measured along the surface.

Flow over a flat plate

To illustrate the method, we consider the boundary layer developing over a flat surface located at $y = 0$, and introduce a control area confined between two vertical planes located at x_1 and x_2, the flat surface, and a horizontal plane located at the elevation $y = h$. Consistent with our previous notation, we denote the tangential component of the outer flow along the surface by $U_x(x)$. For simplicity, we assume that the physical properties of the fluid are uniform throughout the domain of flow.

We begin by considering the x component of the integral momentum balance (6.2.16), and introduce the Newtonian constitutive equation for the stress tensor. Neglecting the normal viscous stresses over the vertical and top planes, and assuming that gravitational effects are negligibly small, we find

$$\begin{aligned}&\int_{x_1}^{x_2}\int_0^h \rho\,\frac{\partial u_x}{\partial t}(x,y)\,\mathrm{d}y\,\mathrm{d}x - \int_0^h [u_x\,(\rho\,u_x)](x=x_1,y)\,\mathrm{d}y \\ &+\int_0^h [u_x\,(\rho\,u_x)](x=x_2,y)\,\mathrm{d}y + \int_{x_1}^{x_2} [u_y\,(\rho\,u_x)](x,y=h)\,\mathrm{d}x \qquad (10.6.1)\\ &= \int_0^h (-p)(x=x_1,y)\,\mathrm{d}y - \int_0^h (-p)(x=x_2,y)\,\mathrm{d}y - \int_{x_1}^{x_2} \tau_{xy}(x)\,\mathrm{d}x.\end{aligned}$$

Next, we take the limit as x_1 tends to x_2, recall that the pressure remains constant across the boundary layer, set $u_x(x, y = h) = U_x(x)$, and rearrange to obtain an integro-differential relation,

$$\begin{aligned}\rho\int_o^h \frac{\partial u_x}{\partial t}\,\mathrm{d}y = h\left(\frac{\partial p}{\partial x}\right)_{y=h} - \rho\,\frac{\partial}{\partial x}\int_0^h u_x^2(x,y)\,\mathrm{d}y\\ -\rho\,U_x(x)\,u_y(x,y=h) - \tau_{xy}(x). \qquad (10.6.2)\end{aligned}$$

To reduce the number of unknowns, we eliminate $u_y(x, y = h)$ in favor of u_x using the continuity equation, setting

$$u_y(x,y=h) = -\int_0^h \frac{\partial u_x}{\partial x}(x,y')\,\mathrm{d}y'. \qquad (10.6.3)$$

Moreover, we use the x component of Euler's equation (6.4.3) to evaluate the streamwise pressure gradient, finding

$$\frac{\partial p}{\partial x} = -\rho\,(\,\frac{\partial U_x}{\partial t} + U_x\,\frac{\partial U_x}{\partial x}\,). \qquad (10.6.4)$$

Substituting expressions (10.6.3) and (10.6.4) in (10.6.2) and rearranging, we find

$$\rho \int_0^h \frac{\partial (U_x - u_x)}{\partial t}\, dy = -\rho \frac{\partial}{\partial x} \int_0^h u_x\,(U_x - u_x)\, dy \tag{10.6.5}$$

$$-\rho \frac{\partial}{\partial x} \int_0^h U_x\,(U_x - u_x)\, dy + \rho\, U_x \frac{\partial}{\partial x} \int_0^h (U_x - u_x)\, dy + \tau_{xy}(x),$$

which can be interpreted as an evolution law for the momentum deficit expressed by the term $\rho\,(U_x - u_x)$.

Now letting the reduced height h/δ tend to infinity and using the definitions of the displacement and momentum thickness stated in equations (10.4.4) and (10.4.9), we derive the von Kàrmàn integral momentum balance

$$\rho \frac{\partial (U_x\, \delta^*)}{\partial t} + \rho \frac{\partial (U_x^2\, \Theta)}{\partial x} + \rho \frac{\partial (U_x^2\, \delta^*)}{\partial x} - \rho\, U_x \frac{\partial (U_x\, \delta^*)}{\partial x} - \tau_{xy} = 0. \tag{10.6.6}$$

Rearranging, we derive an expression for the wall shear stress in terms of the displacement and momentum thickness,

$$\frac{\tau_{xy}}{\rho\, U_x^2} = \frac{\partial (U_x\, \delta^*)}{\partial t} + \frac{\partial \Theta}{\partial x} + (2\,\Theta + \delta^*)\, \frac{1}{U_x} \frac{\partial U_x}{\partial x}. \tag{10.6.7}$$

If the flow is steady, the first term on the right-hand side does not appear. It is reassuring to confirm that, when U_x is constant, equation (10.6.7) reduces to (10.4.13) describing the boundary layer developing over a flat plate that is held stationary in an incident stream at zero angle of attack.

If fluid is injected into the flow or withdrawn from the flow through a porous wall with normal velocity V, the right-hand side of (10.6.7) contains the additional term $-V/U_x$, where V is positive in the case of injection and negative in the case of suction.

10.6.1 The von Kàrmàn-Pohlhausen method

Von Kàrmàn and Pohlhausen developed an approximate method for computing the boundary layer thickness and associated structure of the flow based on the momentum integral balance (10.6.7). The main idea is to assume a sensible velocity profile across the boundary layer in the form $u_x = F(\eta)$, where $\eta \equiv y/\Delta(x)$ and $\Delta(x)$ is a boundary layer thickness similar to the δ_{99} boundary layer thickness. At the second stage, we compute $\Delta(x)$ to satisfy the integral momentum balance (10.6.7).

The implementation of the method for flow over a flat plate at zero angle of attack where $F(\eta)$ is a quarter of a period of a sinusoidal function, as shown in (10.4.15), was discussed in section 10.4. In the remainder of this section, we illustrate the implementation of the method for a general steady flow.

10.6.2 Pohlhausen polynomials

Pohlhausen described the velocity profile $F(\eta)$ across the boundary layer with a fourth-order polynomial,

$$\frac{u_x}{U_x} = F(\eta) = \begin{cases} a(x)\,\eta + b(x)\,\eta^2 + c(x)\,\eta^3 + d(x)\,\eta^4 & \text{for } 0 < \eta < 1, \\ 1 & \text{for } \eta > 1, \end{cases} \tag{10.6.8}$$

where $a(x), b(x), c(x)$, and $d(x)$ are four position-dependent coefficients to be computed as part of the solution. Note that the functional form (10.6.8) satisfies the no-slip boundary condition at the wall corresponding to $\eta = 0$. To compute the four coefficients a, b, c and d, we require four equations.

First, we demand that the overall velocity profile is continuous and has smooth first and second derivatives at the edge of the boundary layer corresponding to $\eta = 1$, and thus obtain three conditions,

$$F(\eta = 1) = 1, \qquad \frac{\mathrm{d}F}{\mathrm{d}\eta}(\eta = 1) = 0, \qquad \frac{\mathrm{d}^2 F}{\mathrm{d}\eta^2}(\eta = 1) = 0. \tag{10.6.9}$$

A fourth condition arises by applying the boundary layer equation (10.2.4) at the wall located at $y = 0$, and then using the no-slip and no-penetration boundary conditions to set the left-hand side to zero. Evaluating the streamwise pressure gradient from (10.6.4), where the time derivative on the right-hand side is set to zero, we find

$$\Big(\frac{\partial u_x^2}{\partial y^2}\Big)_{y=0} = -\frac{1}{\nu}\,U_x\,\frac{\mathrm{d}U_x}{\mathrm{d}x}. \tag{10.6.10}$$

Next, we express the velocity in terms of the function $F(\eta)$ introduced in (10.6.8), and obtain

$$\Big(\frac{\mathrm{d}^2 F}{\mathrm{d}\eta^2}\Big)_{\eta=0} = -\Lambda, \tag{10.6.11}$$

where

$$\Lambda(x) \equiv \frac{\Delta^2(x)}{\nu}\,\frac{\mathrm{d}U_x}{\mathrm{d}x}, \tag{10.6.12}$$

is a dimensionless function expressing the ratio of the magnitude of the inertial acceleration forces of the outer irrotational flow, to the magnitude of the viscous forces developing inside the boundary layer; if $\mathrm{d}U_x/\mathrm{d}x = 0$, then $\Lambda = 0$. By definition, the effective boundary layer thickness $\Delta(x)$ is related to $\Lambda(x)$ by

$$\Delta(x) \equiv \sqrt{\frac{\nu\,\Lambda}{U_x'}}, \tag{10.6.13}$$

where $U_x' \equiv \mathrm{d}U_x/\mathrm{d}x$.

Requiring that the Pohlhausen profile (10.6.8) satisfies equations (10.6.9) and (10.6.11), we obtain

$$a = 2 + \frac{\Lambda}{6}, \quad b = -\frac{\Lambda}{2}, \quad c = -2 + \frac{\Lambda}{2}, \quad d = 1 - \frac{\Lambda}{6}. \tag{10.6.14}$$

Substituting these expressions in (10.6.8) and rearranging, we obtain the velocity profile in terms of the parameter Λ,

$$\frac{u}{U_x} = F(\eta) = \begin{cases} \eta\,(2 - 2\,\eta^2 + \eta^3) + \frac{\Lambda}{6}\,\eta\,(1-\eta)^3 & \text{for } 0 < \eta < 1, \\ 1 & \text{for } \eta > 1. \end{cases} \tag{10.6.15}$$

Program *pohl_pol* inside directory *10_bl* of FDLIB computes polynomial profiles by evaluating the right-hand side of (10.6.15). A family of profiles for $\Lambda = 20$, 12 (heavy line), 6, 0, -6, -12, and -15, is shown in figure 10.6.1.

When $\Lambda > 12$, corresponding to a strongly accelerating external flow according to (10.6.12), the profile exhibits overshooting, and this places a limit on the physical relevance of the fourth-order polynomial expansion. When $\Lambda = -12$, the slope of the velocity profile is zero at the wall, and this suggests that the flow is on the verge of reversal. At that point, the approximations that led us to the boundary layer equations cease to be valid, and the boundary layer is expected to separate from the wall and develop regions of recirculating fluid attached to the wall.

The displacement thickness, momentum thickness, and wall shear stress can be computed in terms of $\Delta(x)$ and $\Lambda(x)$ using the profiles (10.6.15), and are found to be

$$\delta^* = \frac{\Delta}{10}\,(3 - \frac{\Lambda}{12}),$$

$$\Theta = \frac{\Delta}{315}\,(37 - \frac{\Lambda}{3} - \frac{5}{144}\,\Lambda^2), \tag{10.6.16}$$

$$\sigma_{xy}(x, y=0) = \frac{\mu\,U_x}{\Delta}\,(2 + \frac{\Lambda}{6}).$$

Expressing $\Delta(x)$ in terms of $\Lambda(x)$ using the definition (10.6.13), we obtain corresponding expressions in terms of Λ alone.

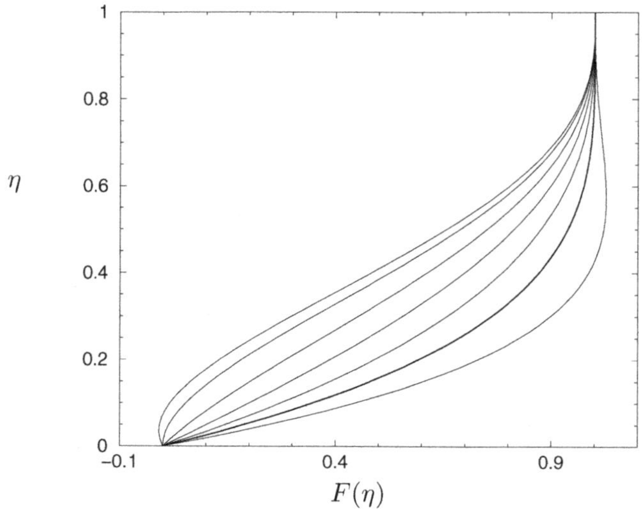

Figure 10.6.1 Profiles of Pohlhausen polynomials for $\Lambda = 20$, 12 (heavy line), 6, 0, -6, -12, and -15.

Next, we substitute expressions (10.6.16) in the momentum integral balance (10.6.7), and thus derive a first-order nonlinear ordinary differential equation for $\Lambda(x)$ with respect to x. Having solved this equation, we recover the boundary layer thickness, $\Delta(x)$, from the definition (10.6.13).

10.6.3 Numerical solution

It is convenient to introduce the Holstein-Bohlen parameter

$$\lambda(x) \equiv \frac{\Theta^2(x)}{\Delta^2(x)}\,\Lambda(x) = \frac{\Theta^2(x)}{\nu}\,\frac{\mathrm{d}U_x}{\mathrm{d}x}, \tag{10.6.17}$$

whose physical interpretation is similar to that of Λ discussed after the definition (10.6.12). Using the expression for the momentum thickness given in the second equation of (10.6.16), we obtain a relationship between λ and Λ,

$$\lambda = \frac{\Lambda}{315^2}\,(37 - \frac{\Lambda}{3} - \frac{5}{144}\,\Lambda^2)^2. \tag{10.6.18}$$

The value $\Lambda = -12$ corresponds to $\lambda = -0.15673$ where the boundary layer is expected to separate according to figure 10.6.1.

To expedite the solution, we multiply both sides of the momentum integral balance (10.6.7) at steady state by Θ, and rearrange to obtain

$$\frac{\mathrm{d}}{\mathrm{d}x}(\frac{\lambda}{\mathrm{d}U_x/\mathrm{d}x}) \equiv \frac{1}{\nu}\,\frac{\mathrm{d}^2\Theta}{\mathrm{d}x^2} = 2\,\frac{S(\lambda) - [2 + H(\lambda)]\,\lambda}{U_x}, \tag{10.6.19}$$

where H is the shape factor defined in (10.4.11), and S is the shear function defined as

$$S(\lambda) \equiv \frac{\Theta\, \sigma_{xy}(x, y=0)}{\mu\, U_x}. \tag{10.6.20}$$

Physically, the shear function expresses the ratio of the wall shear stress to the average value of the shear stress across the boundary layer, and is thus another measure of the sharpness of the velocity profile across the boundary layer. Using expressions (10.6.16), we find

$$H = \frac{315}{10} \frac{3 - \frac{\Lambda}{12}}{37 - \frac{\Lambda}{3} - \frac{5}{144}\Lambda^2},$$
$$S = \frac{1}{315}\,(2 + \frac{\Lambda}{6})\,(37 - \frac{\Lambda}{3} - \frac{5}{144}\,\Lambda^2), \tag{10.6.21}$$

where Λ can be expressed in terms of λ using equation (10.6.18).

The numerical procedure involves the following steps:

1. Given the value of λ at a particular position, x, we compute the corresponding value of Λ by solving the nonlinear algebraic equation (10.6.18).
2. Second, we evaluate the functions S and H using expressions (10.6.21).
3. Third, we compute the right-hand side of (10.6.19) to obtain the rate of change of the ratio on the left-hand side with respect to x.
4. Now we advance the value of λ over a small increment, Δx.
5. We return to step 1 and repeat for another cycle.

The following MATLAB function solves the nonlinear equation (10.6.18) required in step 1 using Newton's method:

```
function Lam = root(lam,Lam)

%---
% given lambda, solve for Lambda
%---

if(abs(lam)<0.000005)
  Lam = 0.00001;
  return
end

%---
% Newton's method
%---
```

```
for i=1:50
  fnc = ltoL(lam,Lam);
  fnc1 = ltoL(lam,Lam+0.0001);
  fncp = (fnc1-fnc)/0.0001;
  corr = -fnc/fncp;
  Lam = Lam+corr;
  if(abs(corr)<0.000001) return; end
end

  disp 'root:  could not find a root in 50 iterations'

return

%-------------

function f = ltoL(lam,Lam)

  f = Lam*((37.0-Lam/3.0-5.0*Lam*Lam/144.0)/315.0)^2-lam;

return
```

Evaluation at a stagnation point

The numerical integration typically begins at a stagnation point where the tangential velocity U_x vanishes and the right-hand side of (10.6.19) is undefined. To prevent a singularity, we require that the numerator is zero at that point, and thus obtain a nonlinear algebraic equation for Λ. A physically acceptable solution is $\Lambda = 7.0523231$ corresponding to $\lambda = 0.0770356$. These values are used to initialize the computation.

To evaluate the right-hand side of (10.6.19) at a stagnation point located at $x = 0$, we denote the expression enclosed by the angular brackets on the right-hand side by $Q(\lambda)$. Applying the l'Hôspital rule to evaluate the fraction, we find

$$\frac{\mathrm{d}}{\mathrm{d}x}\Big(\frac{\lambda}{\mathrm{d}U_x/\mathrm{d}x}\Big)_{x=0} = \Big(\frac{\mathrm{d}Q/\mathrm{d}x}{\mathrm{d}U_x/\mathrm{d}x}\Big)_{x=0} = \Big(\frac{\mathrm{d}Q/\mathrm{d}\lambda}{\mathrm{d}U_x/\mathrm{d}x}\,\frac{\mathrm{d}\lambda}{\mathrm{d}x}\Big)_{x=0}$$

$$= \Big(\frac{\mathrm{d}Q}{\mathrm{d}\lambda}\Big)_{x=0}\Big(\frac{\mathrm{d}}{\mathrm{d}x}\big(\frac{\lambda}{\mathrm{d}U_x/\mathrm{d}x}\big) + \lambda\,\frac{\mathrm{d}^2U_x/\mathrm{d}x^2}{(\mathrm{d}U_x/\mathrm{d}x)^2}\Big)_{x=0}. \tag{10.6.22}$$

Combining the left-hand side with the first term inside the square brackets on the right-hand side, we find

$$\frac{\mathrm{d}}{\mathrm{d}x}\Big(\frac{\lambda}{\mathrm{d}U_x/\mathrm{d}x}\Big)_{x=0} = \Big(\frac{\lambda}{1-\mathrm{d}Q/\mathrm{d}\lambda}\,\frac{\mathrm{d}Q}{\mathrm{d}\lambda}\,\frac{\mathrm{d}^2U_x/\mathrm{d}x^2}{(\mathrm{d}U_x/\mathrm{d}x)^2}\Big)_{x=0}. \tag{10.6.23}$$

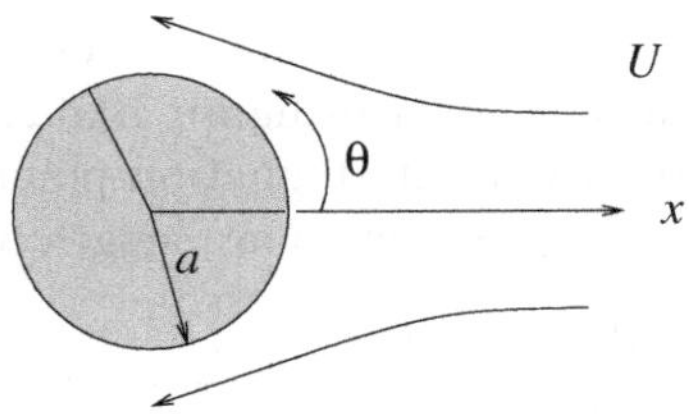

Figure 10.6.2 Illustration of the Prandtl boundary layer around a circular cylinder of radius a held stationary in an incident streaming flow with velocity U.

Evaluating the expression on the right-hand side using the definition of $Q(\lambda)$, we obtain the required initial value

$$\frac{\mathrm{d}}{\mathrm{d}x}\Big(\frac{\lambda}{\mathrm{d}U_x/\mathrm{d}x}\Big)_{x=0} = -0.0652\,\Big(\frac{\mathrm{d}^2U_x/\mathrm{d}x^2}{(\mathrm{d}U_x/\mathrm{d}x)^2}\Big)_{x=0}. \tag{10.6.24}$$

10.6.4 Boundary layer around a curved body

The Kàrmàn-Pohlhausen method was developed with reference to a planar boundary where the x coordinate increases along the boundary in the direction of the velocity of the outer flow. To tackle the more general case of a curved boundary, we simply replace x with the arc length l measured along the boundary in the direction of the tangential velocity of the incident flow, and begin the integration from a stagnation point.

A difficulty arises at the critical point where the acceleration $\mathrm{d}U_l/\mathrm{d}l$ becomes zero or infinite. However, the ambiguity can be removed by carrying out the integration at that point using the Falkner-Skan similarity solution with a proper value for the exponent m.

Boundary layer around a circular cylinder

As an application, we consider streaming (uniform) flow past a stationary circular cylinder of radius a, with vanishing circulation around the cylinder, as shown in figure 10.6.2. Far from the cylinder, the flow occurs toward the negative direction of the x axis and the velocity tends to the uniform value $-U\,\mathbf{e}_x$, where $U > 0$ is the magnitude of the streaming flow, and $\mathbf{e}_x$ is the unit vector along the x axis.

Using the velocity potential for irrotational flow past a circular cylinder given by equation (3.5.32) with $V_x = -U$, we find that the tangential component of velocity of the outer flow is

$$U_\theta(\theta) = \Big(\frac{\partial\phi}{\partial\theta}\Big)_{r=a} = 2\,U\,\sin\theta, \tag{10.6.25}$$

where θ is the polar angle measured around the center of the cylinder in the counterclockwise direction, as shown in figure 10.6.2. The arc length around the cylinder measured from the front stagnation point is $l = a\theta$. The required derivatives of the velocity with respect to arc length are

$$\frac{\mathrm{d}U_\theta}{\mathrm{d}l} = \frac{1}{a}\,\frac{\mathrm{d}U_\theta}{\mathrm{d}\theta} = 2\,\frac{U}{a}\,\cos\theta,$$
$$\frac{\mathrm{d}^2U_\theta}{\mathrm{d}l^2} = \frac{1}{a^2}\,\frac{\mathrm{d}^2U_\theta}{\mathrm{d}\theta^2} = -2\,\frac{U}{a^2}\,\sin\theta. \tag{10.6.26}$$

Equation (10.6.24) yields

$$\frac{\mathrm{d}}{\mathrm{d}l}\Big(\frac{\lambda}{\mathrm{d}U_\theta/\mathrm{d}l}\Big)_{l=0} = 0, \tag{10.6.27}$$

which is used to start up the computations.

The following MATLAB function evaluates the right-hand side of (10.6.19):

```
function f = phase_vel (U0,theta,Lam,lam)

  tmp = 37.0-Lam/3.0-5.0*Lam*Lam/144.0;
  H = 31.5*(3.0-Lam/12.0)/tmp;
  S = (2.0+Lam/6.0)*tmp/315.0;
  U = 2.0*U0*sin(theta);
  f = 2.0*(S-(2.0+H)*lam)/U;

return
```

The Kàrmàn-Pohlhausen method is implemented in the following MATLAB code *kp_cc* located in directory *10_bl* of FDLIB:

```
a = 1.0;    % cylinder radius
U0 = 1.0;    % incident velocity
nu = 1.0;    % kinematic viscosity
mu = 1.0;    % kinematic viscosity

Ndiv = 2*128;    % number of divisions
angle = 0.60*pi;    % maximum integration angle

%---
% prepare
%---

Dtht = angle/Ndiv;    % differential angle theta
Dx = a*Dtht;    % differential arc length
```

```
%---
% initial values
%---

tht = 0.0;
lam = 0.0770;
Lam = 7.052;
Up = 2.0*U0*cos(tht)/a;
z = lam/Up;

%---
% integrate by the modified Euler method
%---

for i=1:Ndiv

  U = 2.0*U0*sin(tht);    % velocity
  Up = 2.0*U0*cos(tht)/a;    % acceleration
  tmp = 37.0-Lam/3.0-5.0*Lam*Lam/144.0;
  H = 31.5*(3.0-Lam/12.0)/tmp;    % (10.6.21)
  S = (2.0+Lam/6.0)*tmp/315.0;    % (10.6.21)
  Del = sqrt(abs(nu*Lam/Up));    % (10.6.13)
  del = Del*(3.0-Lam/12.0)/10.0;    % (10.6.16)
  sw = mu*U/Del*(2.0+Lam/6.0);    % (10.6.16)
  THeta = 0.0;
  if(abs(sw)>0.0000001)
    THeta = mu*U*S/sw;    % (10.6.20)
  end
  if(i==1)
    fv = 0.0;
  else
    Lam = root(lam,Lam);
    fv = phase_vel (U0,tht,Lam,lam);
  end
  zsave = z; fvsave = fv;
  tht = tht+Dtht;
  z = z + fv*Dx;
  Up = 2.0*U0*cos(tht)/a;
  lam = z*Up;
  Lam = root(lam,Lam);    % solve for Lambda
  fv = phase_vel (U0,tht,Lam,lam);
  z = zsave + 0.5*(fv+fvsave)*Dx;
  lam = z*Up;

end
```

Graphs of the solution are shown in figure 10.6.3. The velocity profile across the boundary layer at different stations around the cylinder can be inferred from the scaled Pohlhausen profiles shown in figure 10.6.1 using the local value of Λ.

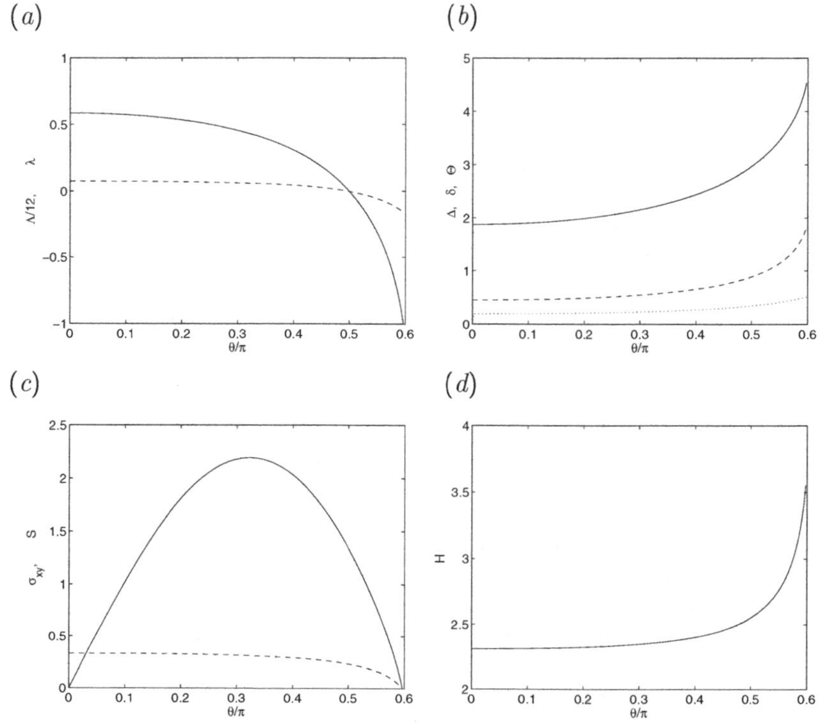

Figure 10.6.3 Features of the Prandtl boundary layer around a circular cylinder of radius a held stationary in an incident streaming flow with velocity U, computed by the von Kàrmàn-Pohlhausen method. (a) Distribution of the dimensionless parameters $\frac{1}{12}\,\Lambda$ (solid line) and λ (dashed line). (b) Boundary-layer thickness Δ (solid line), displacement thickness δ^* (dashed line), and momentum thickness Θ (dotted line); all are reduced by $\sqrt{\nu\, a/U}$. (c) Distribution of the wall shear stress reduced by $\mu\, U/a$ (solid line), and shape factor S (dashed line). (d) Distribution of the shear function, H.

The numerical solution reveals that $\Lambda = -12$ when $\theta = 109.5°$. At that point, the shear stress becomes zero and the boundary layer is expected to separate. Comparing this result with the experimentally observed value $\theta = 80.5°$, we find a serious disagreement attributed to the deviation of the actual outer flow from the idealized potential flow distribution described by (10.6.25) due to the presence of a wake.

To improve the solution, we may describe the tangential velocity distribution U_θ by interpolation based on data collected in the laboratory. When this is done, the predictions of the boundary layer analysis are in excellent agreement with laboratory observation.

Problem

10.6.1 *Von Kàrmàn method for the Blasius boundary layer*

Assume that the velocity profile across the Blasius boundary layer is given by the Pohlhausen polynomial (10.6.15). Show that the effective boundary layer thickness, wall shear stress, displacement thickness, and momentum thicknesses are given by the right-hand sides of equations (10.4.19) and (10.4.20), except that the numerical coefficients are equal, respectively, to 5.863, 0.343, 1.751, and 0.685. Discuss the accuracy of these results with reference to the exact solution obtained by numerical methods.

Computer Problem

10.6.2 *Boundary layer around a circular cylinder*

Plot velocity profiles u_θ/U_0 across the boundary layer around a circular cylinder at a sequence of angles, θ.

10.7 Instability of shear flows

In chapter 7, we derived exact solutions of the governing equations for channel and tube flow, assuming unidirectional motion with rectilinear or circular streamlines. The physical relevance of these assumptions is corroborated by laboratory observation at small and moderate Reynolds numbers. However, at high Reynolds numbers, small perturbations inherent in any real flow amplify to initiate an unsteady motion, possibly leading to a new steady state that is different than that computed under the assumption of unidirectional flow.

Linear stability analysis

Two questions naturally arise: what is the threshold value of the Reynolds number above which a flow becomes unstable? and what are the salient modes of amplification? To develop insight into the answers, we carry out a linear stability analysis. Our strategy is to consider a flow of interest at steady state, introduce small perturbations, and describe the time evolution of the perturbations by solving simplified versions of the governing equations that arise by linearization. If all perturbations decay, the flow is stable; if some perturbations amplify, the flow is unstable; if some perturbations stay constant in time and all other perturbations decay, the flow is neutrally stable. Perturbations that grow or decay exponentially in time represent normal modes.

10.7.1 Stability analysis of shear flow

To illustrate the procedures, we consider a steady unidirectional flow along the x axis with velocity, pressure, and vorticity given by

$$u_x^B = U(y), \qquad u_y^B = 0, \qquad p^B = -G\,x, \qquad \omega_z^B = -\frac{\mathrm{d}U(y)}{\mathrm{d}y}. \tag{10.7.1}$$

The superscript B designates the base flow whose stability is to be examined, $U(y)$ is the unperturbed velocity profile, and G is the negative of the streamwise pressure gradient. It should be noted that, unless the velocity profile is parabolic, the base flow will not satisfy the steady version of the equation of motion. However, we assume that the base flow evolves at a rate that is much slower than that of the perturbations, and may thus be considered to be in a quasi-steady state.

Next, we introduce a two-dimensional perturbation with velocity, pressure, and vorticity described by

$$\begin{aligned} &u_x^P(x,y,t) = \epsilon\, u_x^D(x,y,t), \qquad u_y^P(x,y,t) = \epsilon\, u_y^D(x,y,t), \\ &p^P(x,y,t) = \epsilon\, p^D(x,y,t), \\ &\omega_z^P(x,y,t) = \epsilon\, \omega^D(x,y,t) = \epsilon\,(\frac{\partial u_y^D}{\partial x} - \frac{\partial u_x^D}{\partial y})(x,y,t), \end{aligned} \tag{10.7.2}$$

where the superscript P designates the perturbation, the superscript D designates the disturbance, and ϵ is a dimensionless coefficient whose magnitude is much smaller than unity.

The complete flow arises by adding corresponding variables of the base and perturbation flows shown in (10.7.1) and (10.7.2). For example, the x velocity component of the perturbed flow is

$$u_x(x,y,t) = u_x^B(y) + \epsilon\, u_x^D(x,y,t). \tag{10.7.3}$$

Substituting this sum and its counterparts for all other variables in the vorticity transport equation for two-dimensional flow stated in equation (6.6.8), we obtain

$$\begin{aligned} &\epsilon\,\frac{\partial \omega_z^D}{\partial t} + \epsilon\, u_x^B\,\frac{\partial \omega_z^D}{\partial x} + \epsilon^2\, u_x^D\,\frac{\partial \omega_z^D}{\partial x} + \epsilon\, u_y^D\,\frac{\partial \omega_z^B}{\partial y} + \epsilon^2\, u_y^D\,\frac{\partial \omega_z^D}{\partial y} \\ &\qquad = \nu\,(\frac{\partial^2 \omega^B}{\partial x^2} + \frac{\partial^2 \omega^B}{\partial y^2} + \epsilon\left(\frac{\partial^2 \omega^D}{\partial x^2} + \frac{\partial^2 \omega^D}{\partial y^2}\right)), \end{aligned} \tag{10.7.4}$$

where ν is the kinematic viscosity of the fluid. Since under the quasi-steady approximation the vorticity of the base flow satisfies the steady version of the vorticity transport equation for unidirectional flow, the sum of the first two terms on the right-hand side is zero.

Because ϵ has been assumed small, quadratic terms that are proportional to ϵ^2 are small compared to linear terms that are proportional to ϵ, and may be discarded from both sides of (10.7.4). Collecting the linear terms and setting their sum to zero, we obtain the linearized vorticity transport equation

$$\frac{\partial \omega_z^D}{\partial t} + u_x^B \frac{\partial \omega_z^D}{\partial x} + u_y^D \frac{\partial \omega_z^B}{\partial y} = \nu \, (\frac{\partial^2 \omega_z^D}{\partial x^2} + \frac{\partial^2 \omega_z^D}{\partial y^2}). \tag{10.7.5}$$

Substituting the expressions for the base flow stated (10.7.1) in (10.7.5), we derive the more specific form

$$\frac{\partial \omega_z^D}{\partial t} + U(y) \frac{\partial \omega_z^D}{\partial x} - u_y^D \frac{d^2 U}{dy^2} = \nu \, (\frac{\partial^2 \omega_z^D}{\partial x^2} + \frac{\partial^2 \omega_z^D}{\partial y^2}). \tag{10.7.6}$$

The problem has been reduced to solving the linear equation (10.7.6) for the disturbance flow, subject to a specified initial condition and appropriate boundary conditions.

Disturbance stream function

It is convenient to express the disturbance flow in terms of a disturbance stream function, $\psi^D(x, y, t)$, defined by the equations

$$u_x^D = \frac{\partial \psi^D}{\partial y}, \qquad u_y^D = -\frac{\partial \psi^D}{\partial x}. \tag{10.7.7}$$

The disturbance vorticity is given by

$$\omega_z^D = -\Big(\frac{\partial^2 \psi^D}{\partial x^2} + \frac{\partial^2 \psi^D}{\partial y^2} \Big) = -\nabla^2 \psi^D, \tag{10.7.8}$$

where $\nabla^2 \equiv \partial^2/\partial x^2 + \partial^2/\partial y^2$ is the Laplacian operator in two dimensions. Substituting these expressions in (10.7.6) and rearranging, we find

$$\frac{\partial \, \nabla^2 \psi^D}{\partial t} + U(y) \frac{\partial \, \nabla^2 \psi^D}{\partial x} - \nu \, \nabla^4 \, \psi^D = \frac{\partial \psi^D}{\partial x} \frac{\mathrm{d}^2 U}{\mathrm{d}y^2}, \tag{10.7.9}$$

where $\nabla^4 = \nabla^2 \nabla^2$ is the biharmonic operator in two dimensions.

10.7.2 Normal-mode analysis

To study the evolution of each and every possible disturbance is practically impossible. As an alternative, we exploit the linearity of equation (10.7.9) and deduce the nature of the general solution corresponding to an arbitrary initial condition from the behavior of an infinite family of solutions corresponding to disturbances that are sinusoidal functions of the streamwise position, x, and exponential functions of time, t, called normal modes. The general solution can be constructed by linear superposition.

Let us consider a normal mode with wave length L and corresponding wave number $k = 2\pi/L$. The disturbance stream function is expressed in the form

$$\psi^D(x, y, t) = \chi_R(y, t)\, \cos(kx) + \chi_I(y, t)\, \sin(kx), \tag{10.7.10}$$

where $\chi_R(y, t)$ and $\chi_I(y, t)$ are two real functions. To simplify the notation, we introduce the complex function

$$\chi(y, t) \equiv \chi_R(y, t) - \mathrm{i}\, \chi_I(y, t), \tag{10.7.11}$$

where i is the imaginary unit, $\mathrm{i}^2 = -1$. Using the Euler decomposition of the imaginary exponential, $\exp(ikx) = \cos(kx) + \mathrm{i}\, \sin(kx)$, we recast (10.7.10) into the form

$$\psi^D(x, y, t) = \mathrm{Real}\{\Psi^D(x, y, t)\}, \tag{10.7.12}$$

where Real designates the real part of the complex quantity enclosed by the angular brackets, and

$$\Psi^D(x, y, t) \equiv \chi(y, t)\, \exp(\mathrm{i}kx) \tag{10.7.13}$$

is a complex stream function. To simplify the analysis, we require that the imaginary part of Ψ^D also satisfies equation (10.7.9). Substituting (10.7.13) in (10.7.9), carrying out the differentiation with respect to x, and noting that

$$\nabla^2 \Psi^D = (-k^2 \chi + \frac{\partial^2 \chi}{\partial y^2})\, \exp(\mathrm{i}kx), \tag{10.7.14}$$

we derive the equation

$$-k^2 \frac{\partial \chi}{\partial t} + \frac{\partial^3 \chi}{\partial y^2\, \partial t} + \mathrm{i}\, k\, U(y)\, (-k^2 \chi + \frac{\partial^2 \chi}{\partial y^2}) \\ -\nu\, (k^4 \chi - 2k^2\, \frac{\partial^2 \chi}{\partial y^2} + \frac{\partial^4 \chi}{\partial y^4}) = \mathrm{i}\, k\, \chi\, \frac{\mathrm{d}^2 U}{\mathrm{d}y^2}. \tag{10.7.15}$$

Growth rate

A solution of (10.7.15) can be found by expressing $\chi(x, t)$ in the separated form

$$\chi(y, t) = f(y)\, \exp(-\mathrm{i}\, \sigma\, t) = f(y)\, \exp(-\mathrm{i}kct), \tag{10.7.16}$$

where $f(y)$ is a complex function, σ is a complex constant called the complex growth rate, and $c \equiv \sigma/k$ is another complex constant called the complex phase velocity. Substituting (10.7.16) in (10.7.13), we derive the corresponding complex disturbance stream function

$$\Psi(x, y, t) = f(y)\, \exp[\mathrm{i}\, (kx - \sigma t)] = f(y)\, \exp[\mathrm{i}k(x - ct)]. \tag{10.7.17}$$

Next, we decompose σ and c into their real and imaginary parts, writing $\sigma = \sigma_R + \mathrm{i}\,\sigma_I$ and $c = c_R + \mathrm{i}\,c_I$, where the subscripts R and I stand for real and imaginary, and obtain

$$\Psi(x, y, t) = f(y) \exp[\mathrm{i}\, k\, (x - c_R t)] \exp(\sigma_I t). \qquad (10.7.18)$$

Expression (10.7.18) illustrates two important features:

- c_R is the real phase velocity of the disturbance. The crests and troughs of the sinusoidal perturbation, but not the fluid itself, travel along the x axis with velocity c_R.

- σ_I is the growth rate of the disturbance; if σ_I is positive, the disturbance grows at an exponential rate in time; if σ_I is negative, the disturbance decays at an exponential rate in time; if $\sigma_I = 0$, the amplitude of the disturbance remains constant in time. In the first case, the flow is unstable; in the second case, the flow is stable; and in the third case, the flow is neutrally stable.

Orr-Sommerfeld equation

Substituting (10.7.16) in (10.7.15) and rearranging the resulting expression, we derive the Orr-Sommerfeld equation

$$k^4 f - 2\, k^2 \frac{\mathrm{d}^2 f}{\mathrm{d}y^2} + \frac{\mathrm{d}^4 f}{\mathrm{d}y^4} = \frac{\mathrm{i}\, k}{\nu} \Big([U(y) - c]\, (\frac{\mathrm{d}^2 f}{\mathrm{d}y^2} - k^2 f) - \frac{\mathrm{d}^2 U}{\mathrm{d}y^2}\, f \Big). \qquad (10.7.19)$$

A trivial solution is $f = 0$. Nontrivial solutions expressing normal modes exist for certain values of c that are the eigenvalues of the Orr-Sommerfeld equation. The main objective of linear stability analysis is to compute these eigenvalues and associated eigenfunctions, and thereby assess whether the amplitude of a normal mode will grow, decay, or remain constant in time.

Rayleigh equation

When viscous forces are negligible, the left-hand side of the Orr-Sommerfeld equation (10.7.19) may be set equal to zero yielding the Rayleigh equation

$$(U(y) - c)\, (\frac{\mathrm{d}^2 f}{\mathrm{d}y^2} - k^2 f) - \frac{\mathrm{d}^2 U}{\mathrm{d}y^2}\, f = 0, \qquad (10.7.20)$$

also available in the alternative form

$$\frac{\mathrm{d}^2 f}{\mathrm{d}y^2} - (k^2 + \frac{1}{U(y) - c}\, \frac{\mathrm{d}^2 U}{\mathrm{d}y^2})\, f = 0. \qquad (10.7.21)$$

Because of the absence of viscous forces, the Rayleigh equation is a second-order differential equation, whereas the inclusive Orr-Sommerfeld equation is a fourth-order differential equation. Both equations are linear, but the coefficients multiplying the derivatives of the unknown complex function f are not necessarily constant.

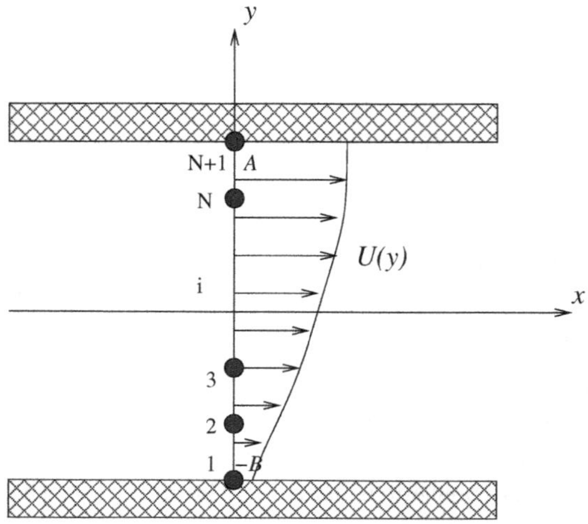

Figure 10.7.1 Illustration of a finite-difference grid used to solve the Rayleigh equation determining the growth rate of two-dimensional perturbations in inviscid unidirectional shear flow.

10.7.3 Finite-difference solution

Analytical solutions to the Orr-Sommerfeld and Rayleigh equations are possible only for a limited class of purely viscous or idealized inviscid flows. To study the stability of more general flows, we resort to numerical methods.

Rayleigh equation

To illustrate the implementation of a finite-difference method, we study the stability of a shear flow in a channel confined between two parallel walls located at $y = -A$ and B, as illustrated in figure 10.7.1. Assuming that the effect of viscosity is negligible, we express Rayleigh's equation (10.7.20) in the form

$$U(y)\,\frac{\mathrm{d}^2 f}{\mathrm{d}y^2} - (\,U(y)\,k^2 + \frac{\mathrm{d}^2 U}{\mathrm{d}y^2}\,)\,f = c\,(\frac{\mathrm{d}^2 f}{\mathrm{d}y^2} - k^2\,f). \qquad (10.7.22)$$

Note that the unknown eigenvalue, c, has been moved to the right-hand side.

Next, we introduce a one-dimensional uniform grid of nodes separated by the grid spacing Δy, located at y_i, $i = 0, 1, \ldots, N+1$, where $y_0 = -A$ and $y_{N+1} = B$, as illustrated in figure 10.7.1. For simplicity, we denote the value of f at the ith node by f_i. To satisfy the no-penetration condition, we require that the stream function is constant at the lower and upper wall,

$$f_0 = 0 \qquad \text{and} \qquad f_{N+1} = 0. \qquad (10.7.23)$$

Because the values of the stream function over the two walls are equal, the disturbance flow will not generate a net flow rate in the streamwise direction.

Applying equation (10.7.22) at the ith node, where $i = 1, \ldots, N$, and approximating the second derivative $\mathrm{d}^2 f/\mathrm{d}y^2$ with a centered difference, we obtain the difference equation

$$U(y_i)\,\frac{f_{i+1} - 2\,f_i + f_{i-1}}{\Delta y^2} - (\,U(y_i)\,k^2 + \frac{\mathrm{d}^2 U}{\mathrm{d}y^2}(y_i)\,)\,f_i = c\,(\frac{f_{i+1} - 2\,f_i + f_{i-1}}{\Delta y^2} - k^2\,f_i). \qquad (10.7.24)$$

Denoting $U_i \equiv U(y_i)$ and $U_i^{''} \equiv (\mathrm{d}^2 U/\mathrm{d}y^2)(y_i)$, and rearranging, we obtain

$$U_i\,f_{i-1} - [2\,U_i + \Delta y^2\,(U_i\,k^2 + U_i^{''})]\,f_i + U_i\,f_{i+1} = c\,[f_{i-1} - (2 + k^2\,\Delta y^2)\,f_i + f_{i+1}]. \qquad (10.7.25)$$

Applying equation (10.7.25) for $i = 1, \ldots, N$, we derive a system of linear equations compiled into the matrix form

$$\mathbf{A} \cdot \mathbf{f} = c\,\mathbf{B} \cdot \mathbf{f}, \qquad (10.7.26)$$

where

$$\mathbf{f} \equiv \begin{bmatrix} f_1 \\ f_2 \\ \vdots \\ f_{N-1} \\ f_N \end{bmatrix} \qquad (10.7.27)$$

is the N-dimensional solution vector, $\mathbf{A}$ is an $N \times N$ tridiagonal matrix,

$$\mathbf{A} \equiv \begin{bmatrix} -2\,U_1 - \Delta y^2\,(U_1\,k^2 + U_1^{''}) & U_1 & 0 & 0 & \cdots & 0 \\ U_2 & -2\,U_2 - \Delta y^2\,(U_2\,k^2 + U_2^{''}) & U_2 & 0 & \cdots & 0 \\ \cdots & \cdots & \cdots & \cdots & \cdots & \cdots \\ 0 & 0 & 0 & 0 & U_N & -2\,U_N - \Delta y^2\,(U_N\,k^2 + U_N^{''}) \end{bmatrix}, \qquad (10.7.28)$$

and $\mathbf{B}$ is another $N \times N$ tridiagonal matrix defined as

$$\mathbf{B} \equiv \begin{bmatrix} -2 - k^2\,\Delta y^2 & 1 & 0 & 0 & \cdots & 0 \\ 1 & -2 - k^2\,\Delta y^2 & 1 & 0 & \cdots & 0 \\ \cdots & \cdots & \cdots & \cdots & \cdots & \cdots \\ 0 & 0 & 0 & \cdots & 1 & -2 - k^2\,\Delta y^2 \end{bmatrix}. \qquad (10.7.29)$$

Note that the elements of the matrix $\mathbf{B}$ are independent of the velocity profile, $U(y)$.

Generalized eigenvalue problem

Equation (10.7.26) expresses a generalized algebraic eigenvalue problem that can be stated as follows: compute a value of c so that (10.7.26) has a nontrivial solution for the vector $\mathbf{f}$; that is, a solution other than the null vector. To compute the eigenvalues, c, we restate equation (10.7.26) as a homogeneous equation,

$$\mathbf{E} \cdot \mathbf{f} = \mathbf{0}, \tag{10.7.30}$$

where $\mathbf{E}$ is a tridiagonal matrix given by

$$\mathbf{E} \equiv \begin{bmatrix} -2 - \Delta y^2 (k^2 + \frac{U_1''}{U_1 - c}) & 1 & 0 & 0 \\ 1 & -2 - \Delta y^2 (k^2 + \frac{U_2''}{U_2 - c}) & 1 & 0 \\ \cdots & \cdots & \cdots & \cdots \\ 0 & 0 & 0 & 0 \end{bmatrix}$$

$$\begin{matrix} 0 & \cdots & 0 \\ 0 & \cdots & 0 \\ \cdots & \cdots & \cdots \\ 0 & 1 & -2 - \Delta y^2 (k^2 + \frac{U_N''}{U_N - c}) \end{bmatrix}. \tag{10.7.31}$$

For system (10.7.30) to have a nontrivial solution, the coefficient matrix $\mathbf{E}$ must be singular. Phrased differently, c must be such that the determinant of the complex matrix $\mathbf{E}$ is zero. This observation provides us with a basis for a numerical method involving the following steps:

1. We begin by guessing a complex value for c.

2. We compute the determinant of $\mathbf{E}$ using the algorithm discussed in the following section.

3. We improve c to reduce the magnitude of the determinant. The improvement can be made using Newton's method, setting

$$c^{new} = c^{old} - \frac{\det[\mathbf{E}(c^{old})]}{(\frac{\mathrm{d} \det[\mathbf{E}(c)]}{\mathrm{d}c})_{c=c^{old}}}. \tag{10.7.32}$$

4. We now return to step 2 and repeat if necessary.

The derivative in the denominator on the right-hand side of (10.7.32) can be approximated with a finite difference,

$$\frac{\mathrm{d}\det[\mathbf{E}(c^{old})]}{\mathrm{d}c}\Big)_{c=c^{old}} \simeq \frac{\det[\mathbf{E}(c^{old}+\epsilon)]-\det[\mathbf{E}(c^{old})]}{\epsilon}, \tag{10.7.33}$$

where ϵ is a real or complex increment with small magnitude.

Determinant of a tridiagonal matrix

To compute the determinant of the tridiagonal matrix $\mathbf{E}$, we use an efficient algorithm applicable to general tridiagonal matrices of the form

$$\mathbf{T} \equiv \begin{bmatrix} a_1 & b_1 & 0 & 0 & \cdots & 0 & 0 & 0 \\ c_2 & a_2 & b_2 & 0 & \cdots & 0 & 0 & 0 \\ \cdots & \cdots & \cdots & \cdots & \cdots & \cdots & \cdots & \cdots \\ 0 & 0 & 0 & 0 & \cdots & c_{N-1} & a_{N-1} & b_{N-1} \\ 0 & 0 & 0 & 0 & \cdots & 0 & c_N & a_N \end{bmatrix}, \tag{10.7.34}$$

where a_i, b_i, and c_i are real or complex constants. The algorithm involves computing a sequence of numbers, P_i, based on the recursion relation

$$\begin{aligned} &P_0 = 0, \\ &P_1 = a_1, \\ &\cdots \\ &P_i = a_i\,P_{i-1} - b_{i-1}\,c_i\;P_{i-2}, \end{aligned} \tag{10.7.35}$$

for $i = 2, \ldots, N$; then $\det(\mathbf{T}) = P_N$.

Instability of a shear flow with hyperbolic tangent profile

Computer code *sf1* inside directory *08_stab* of FDLIB computes the complex phase velocity of perturbations using the finite-difference method discussed in this section. As an example, we consider an inviscid shear flow with velocity profile

$$U(y) = U_0\,\tanh\frac{y}{b}, \tag{10.7.36}$$

where U_0 is the uniform velocity of the flow far above and below the shear layer, and b is half the shear layer thickness.

The dashed lines in figure 10.7.2 represent the dimensionless imaginary part of the phase velocity of unstable perturbations, $\hat{c}_I \equiv c_I/U_0$, for a sequence of channel widths with $A = B$. The solid lines represent the reduced growth rate $\hat{\sigma}_I \equiv 4\,b\,\sigma_I/U_0$. When the reduced wave number, kb, is larger than a critical threshold, $(kb)_{cr}$, determined by the ratio A/b, the growth rate is zero and the perturbations are neutrally stable. As A/b tends to infinity, we obtain an

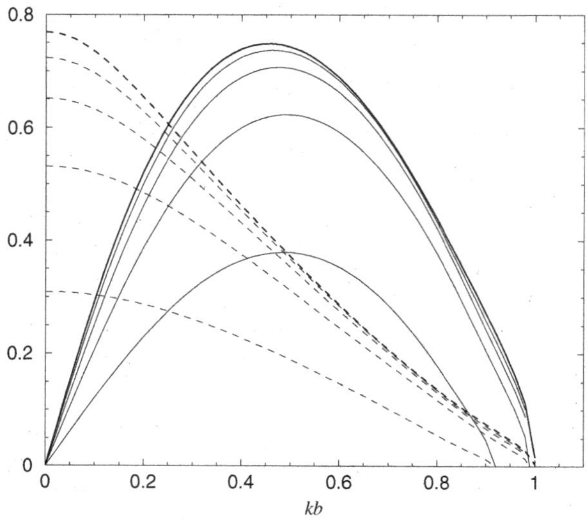

Figure 10.7.2 Instability of an inviscid shear flow whose velocity profile is described by equation (10.7.36). Graphs of the reduced imaginary part of the phase velocity $\hat{c}_I \equiv c_I/U_0$ (dashed lines) and reduced growth rate $\hat{\sigma}_I \equiv 4b\sigma_I/U_0$ (solid lines) for $A/b = B/b = 2.0$, 2.5, 3.0, 4.0, and 6.0 (heavy lines), in the regime of unstable wave numbers.

infinite shear flow in the absence of side walls. In this limit, the critical wave number for neutral stability is known to be $(kb)_{cr} = 1.0$. The results in figure 10.7.2 demonstrate that the walls reduce the growth rate of perturbations by restricting the lateral extent over which fluid motion is allowed to develop. Maximum growth rate occurs at a certain wave number $(kb)_{max} \simeq 0.50$. The corresponding perturbation is expected to dominate the instability and therefore spontaneously arise in a randomly perturbed flow.

Orr-Sommerfeld equation

Finite-difference methods for the Orr-Sommerfeld equation can be developed working in a similar fashion. The no-penetration and no-slip boundary conditions over a stationary solid surface require that the boundary values of f and its first derivative with respect to transverse distance y are zero. Discretizing the Orr-Sommerfeld equation yields a system of linear equations similar to that shown in (10.7.26). However, in the case of viscous flow, because of the presence of the fourth derivative $f^{(iv)}$, the matrix $\mathbf{A}$ is complex pentadiagonal; *penta* derives from the Greek word $\pi\epsilon\nu\tau\epsilon$ which means five. The algebraic system descending from the finite-difference discretization can be recast into a form that is analogous to that shown in (10.7.30), where the matrix $\mathbf{E}$ is now pentadiagonal. Unfortunately, the determinant of this matrix may no longer be

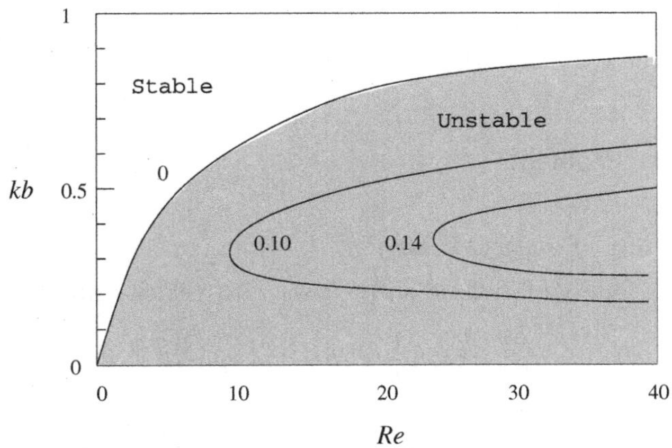

Figure 10.7.3 Contour plot of the reduced growth rate $\hat{\sigma}_I \equiv b\sigma_I/U_0$ for viscous shear flow whose velocity profile is described by equation (10.7.36). The Reynolds number is defined as $Re \equiv U_0\, b/\nu$.

computed using an efficient numerical method. Better methods for computing the eigenvalues are available.

Instability of a shear flow with hyperbolic tangent profile

To illustrate the effect of viscosity, we consider an infinite shear flow whose velocity profile is described by equation (10.7.36). Figure 10.7.3 shows a contour plot of the reduced growth rate $\hat{\sigma}_I \equiv b\sigma_I/U_0$ in the $kb-Re$ (wave number versus Reynolds number) plane, where $Re \equiv U_0\, b/\nu$. The contour $\hat{\sigma}_I = 0$ corresponds to neutrally stable perturbations. A perturbation whose wave number lies in the shaded area below this contour is unstable, whereas a perturbation whose wave number lies in the unshaded area above this contour are stable.

As the Reynolds number tends to infinity, we recover the results presented in figure 10.7.2 for inviscid flow. Figure 10.7.3 reveals that the flow is unstable even at arbitrarily small Reynolds numbers. The destabilizing effect of inertia becomes evident by observing that, as the Reynolds number increases, the range of unstable wave numbers $[0, (kb)_{cr}]$ widens and a broader range of perturbations grow at higher growth rates.

Problem

10.7.1 *Instability of an inviscid shear flow*

Consider an infinite shear flow whose velocity profile is described by (10.7.36). Show that an eigenvalue and the corresponding eigenfunction of Rayleigh's equation describing a neutrally stable perturbation are $c = 0$ and $f(y) = d\,\mathrm{sech}(y/b)$, where d is an arbitrary constant.

Computer Problem

10.7.1 *Instability of inviscid shear flow*

(a) Consider a family of inviscid shear flows with velocity profile

$$U(y) = U_0 \left(\delta \tanh \frac{y}{b} + (\delta - 1) \exp[-(\tfrac{y}{b})^2] \right), \qquad (10.7.37)$$

where the dimensionless parameter δ takes values in the range $[0, 1]$. The limiting value $\delta = 1$ corresponds to a shear layer with a hyperbolic tangent velocity profile, whereas the limiting value $\delta = 0$ corresponds to a symmetric wake with a Gaussian velocity profile. Assume that the flow is confined between two parallel walls located at $y = \pm A$ where the no-penetration condition is required.

Use the program *sf1* to generate a graph of a properly defined dimensionless growth rate plotted against the reduced wave number kb, for $\delta = 0$, 0.50, and 1.0, in each case for $A/b = 2.0$, 3.0, and 4.0. Discuss the results of your computations.

(b) Repeat (a) for a shear flow with velocity profile

$$U(y) = U_0 \left(\delta \tanh \frac{y}{b} + (1 - \delta)\,\mathrm{sech}^2 \frac{y}{b} \right), \qquad (10.7.38)$$

where the parameter δ takes values in the range $[0, 1]$. The limiting values $\delta = 1$ and 0 correspond, respectively, to a shear layer with a hyperbolic tangent velocity profile and to the Bickley jet.

10.8 Turbulent flow

Turbulent flow is established when the Reynolds number, defined in an appropriate fashion for the particular flow under consideration, $Re = \rho V L/\mu$, exceeds a certain threshold, usually on the order of 10^3. In the definition of the Reynolds number, V is a characteristic macroscopic velocity, and L is a characteristic macroscopic length scale typically associated with the size of the boundaries. Both V and L are classified as *external scales*. For example, in the case of pipe flow, V can be identified either with the mean velocity or with the maximum velocity occurring at the centerline, and L can be identified with the tube radius or diameter.

Turbulence is characterized by random motion in *both* time and space. Thus, a graph of a velocity component plotted against time at a particular location in a turbulent flow reveals random fluctuations, as illustrated in figure 10.8.1(a). An analogous graph of the velocity against the spatial coordinate x

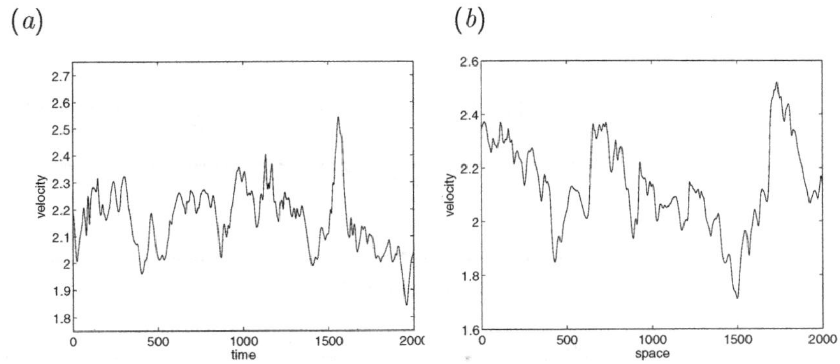

Figure 10.8.1 (a) Temporal, and (b) spatial variation of a velocity component in a turbulent flow.

at a particular instant in time reveals similar random fluctuations, as illustrated in figure 10.8.1(b).

10.8.1 Transition to turbulence

The transition from laminar to turbulent flow with increasing Reynolds number is not sudden, but occurs through a sequence of events eventually leading to randomly fluctuating motion. For example, pressure-driven flow in a circular tube is laminar when the Reynolds number defined with respect to the tube radius and the maximum velocity at the centerline is less than 1,100; transition occurs when the Reynolds number lies in the range between 1,100 and 1,500; and fully developed turbulent motion is established at higher Reynolds numbers. Wall roughness and entrance conditions affect the precise thresholds for transition.

Figure 10.8.2 shows a recording of the streamwise velocity component in uniform (streaming) flow past a flat plate of length L, at a sequence of increasing Reynolds numbers defined with respect to the length of the plate.[3] The velocity probe was placed 0.02 in above the plate and 56 in behind the leading edge. The graphs illustrate the onset of oscillations, the development of turbulent spots, and the ultimate establishment of fully turbulent motion.

Logistic mapping

A simple model illustrating the process of transition from simple to complex behavior is provided by the logistic mapping. Given a number, $x^{(0)}$, logistic

[3]Cebeci T., & Smith, A. M. O., 1974, *Analysis of Turbulent Boundary Layers*, Academic Press

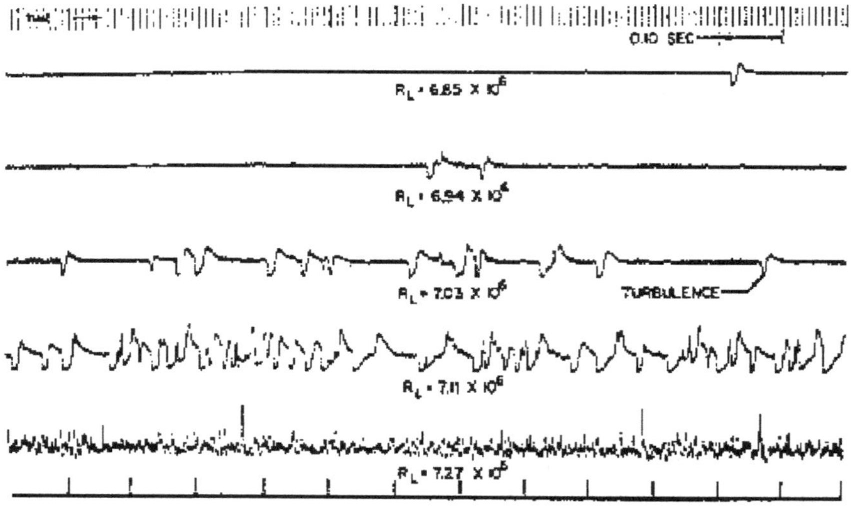

Figure 10.8.2 Recording of the streamwise component of the velocity in flow past a flat plate at a sequence of increasing Reynolds numbers, R_L, defined with respect to the length of the plate, L. The signal shows the onset of oscillations, the development of turbulent spots, and the ultimate establishment of fully turbulent flow.

mapping generates a sequence of numbers, $x^{(1)}, x^{(2)}, \ldots,$ computed by the recursion formula

$$x^{(k+1)} = \lambda\, x^{(k)}\, (1 - x^{(k)}), \tag{10.8.1}$$

for $k = 0, 1, \ldots,$ where λ is a specified positive constant. The special choices $x^{(0)} = 0$ and $(\lambda - 1)/\lambda$ are the fixed points of the mapping; for these choices, $x^{(k)} = x^{(0)}$ for all k, and the logistic sequence is stationary.

To illustrate the transition, we introduce the λx plane and perform a series of computations according to the following steps:

1. Choose a value for λ.
2. Select a value for $x^{(0)}$ that lies between 0 and 1, but is not exactly equal to 0 or 1.
3. Compute a few hundred terms based on the logistic mapping (10.8.1).

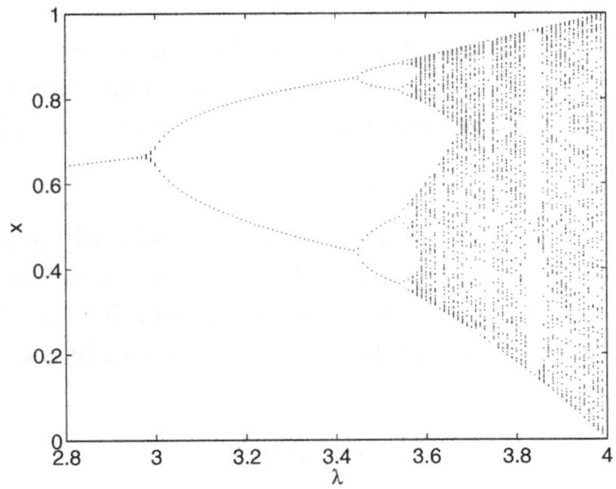

Figure 10.8.3 Behavior of sequences generated by the logistic mapping equation (10.8.1), illustrating the process of transition by way of a cascade of bifurcations in the λx plane.

4. Skip the first one hundred terms, and graph the rest of the terms in the λx plane with dots.
5. Return to step 1 and repeat.

The result of this computation is shown in figure 10.8.3. As λ increases, a cascade of bifurcations and a random behavior reminiscent of turbulent motion emerges.

10.8.2 Lagrangian turbulence

Point particles in a certain class of unsteady two-dimensional laminar flows and steady or unsteady three-dimensional laminar flows have been observed to move in a random fashion exhibiting a Lagrangian turbulent motion. However, fluid motion in a turbulent flow should be distinguished from the seemingly random motion of point particles in these laminar yet chaotic flows. A distinguishing feature of a turbulent flow is that strong three-dimensional vorticity is present and significant kinematic and molecular diffusion occurs accompanied by high levels of viscous dissipation.

10.8.3 Features of turbulent motion

Turbulence has defied a simple physical interpretation in terms of elementary fluid motions. In the traditional approach, a turbulent flow is regarded as a stochastic random process amenable to statistical analysis. However, several important features distinguish the turbulent motion from a generic random process, including the following.

Intermittency

Turbulence is intermittent. A record of the velocity at a certain point in a turbulent flow may appear regular for a period of time, only to be interrupted by periods of violent turbulent motion in an intermittent fashion.

Coherent structures

A turbulent flow contains small-scale short-lived and large-scale long-lived coherent structures associated with eddies and vortices with a well-defined structure. Examples include vortex billows developing in shear layers, and horseshoe vortices developing near boundary layers and in regions of high shear rates.

Vortex motion

An intimate connection exists between the dynamics of a turbulent flow and vortex dynamics discussed in chapter 11. However, it is not clear how studies of vortex dynamics can be used to precisely delineate the physics or mathematical properties of a turbulent flow.

Eddy motion

Eddies in a turbulent flow carry turbulent kinetic energy distributed over a broad range of scales; from the external scale, L, to the energy dissipating Kolmogorov scale, η, defined later in this section. Energy is transferred across the scales, forward and backward, from large to small and small to large, and a balance is achieved at dynamic equilibrium. A net transfer of energy occurs toward the small scales.

Dependence on the type of flow

The dynamics of turbulent flow associated with eddy motion and the distribution of energy among the different length scales is not universal but depends on the particular flow under consideration. Thus, the properties of wall-bounded turbulent shear flow are different from those of unbounded shear flow, and different from those of grid-turbulent flow generated behind a grid intercepting a high-speed flow. This diversity reflects differences in the physical mechanism by which energy is supplied into a turbulent flow, ultimately to be dissipated by small-scale motion.

Sedation near a wall

The no-slip boundary condition sedates the turbulent motion near a wall where a viscous sublayer of unsteady laminar flow is established. A buffer zone separates the viscous sublayer from the regime of fully-developed turbulent flow. Figure 10.8.4 shows the temporal velocity signal of a turbulent flow at different radial positions inside a circular pipe of radius 15 cm, illustrating the cessation of the turbulent motion near the wall. [4]

[4] CORRSIN, S., 1943, Investigation of flow in an axially asymmetric heated jet-air. *NACA Rep. 3L23.*

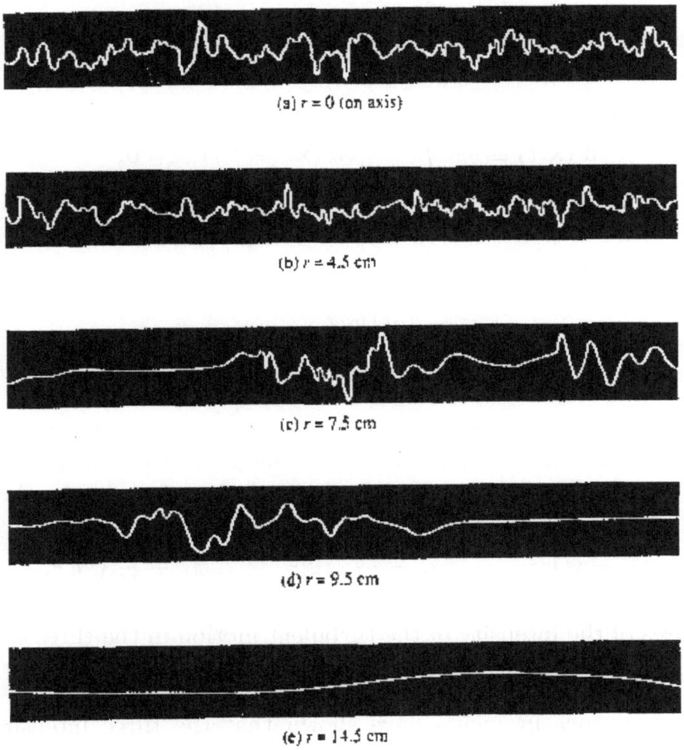

Figure 10.8.4 Temporal velocity fluctuations in turbulent pipe flow at different distances from the wall, showing the presence of the laminar sublayer and buffer zone near the wall.

10.8.4 Decomposition into mean and fluctuating components

It is useful to decompose a turbulent flow variable into a smoothly varying or mean component, and a rapidly fluctuating component. The smoothly varying component can be identified with the time averaged value over a period of time, t_0, that is large compared to the time scale of the fluctuations, but small compared to the external time scale L/V. The mean velocity at the position $\mathbf{x}$, designated by an overbar, is defined as

$$\bar{\mathbf{u}}(\mathbf{x}, t) \equiv \frac{1}{t_0} \int_{t-t_0/2}^{t+t_0/2} \mathbf{u}(\mathbf{x}, t+t') \, dt'. \tag{10.8.2}$$

The fluctuating velocity, designated by a prime, is then defined by the decomposition

$$\mathbf{u}(\mathbf{x}, t) = \bar{\mathbf{u}}(\mathbf{x}, t) + \mathbf{u}'(\mathbf{x}, t). \tag{10.8.3}$$

Definition (10.8.2) implies that the time-averaged value of the fluctuating velocity is zero by construction,

$$\overline{\mathbf{u}'}(\mathbf{x}, t) \equiv \frac{1}{t_0} \int_{t-t_0/2}^{t+t_0/2} \mathbf{u}'(\mathbf{x}, t+t') \, \mathrm{d}t' = \mathbf{0}. \tag{10.8.4}$$

In contrast, the time averaged value of the square of the x component of the fluctuating velocity,

$$\overline{u_x'^2}(\mathbf{x}, t) \equiv \frac{1}{t_0} \int_{t-t_0/2}^{t+t_0/2} u_x'^2(\mathbf{x}, t+t') \, \mathrm{d}t', \tag{10.8.5}$$

is not zero; the y and z components of the fluctuating velocity are also nonzero. The square root of these time averages, called the root-mean-square (RMS) values, normalized by an external velocity scale V, expressed by the ratios

$$i_x \equiv \frac{1}{V} \sqrt{\overline{u_x'^2}}, \qquad i_y \equiv \frac{1}{V} \sqrt{\overline{u_y'^2}}, \qquad i_z \equiv \frac{1}{V} \sqrt{\overline{u_z'^2}}, \tag{10.8.6}$$

are measures of the intensity of the turbulent motion in the three spatial directions.

Laboratory measurements have shown that the three intensities defined in (10.8.6) have different magnitudes, except in the idealized case of isotropic turbulence occurring in the absence of boundaries. Nearly isotropic turbulence can be realized in the laboratory by placing eight fans at the vertices of a cube, and turning the fans toward the cube center. In the case of channel or tube flow, the turbulence intensity in the direction of the flow is significantly greater than that in directions perpendicular to the flow, especially near the walls.

A single measure of the magnitude of the turbulent velocity fluctuations is provided by the velocity scale

$$u \equiv \left(\frac{\overline{u_x'^2} + \overline{u_y'^2} + \overline{u_x'^2}}{3} \right)^{1/2}. \tag{10.8.7}$$

In the case of isotropic turbulence, the three terms in the numerator on the right-hand side are equal.

Flow in a pipe

As an example, we consider pressure-driven turbulent flow through a circular tube of radius a. Figure 10.8.5 shows a schematic illustration of the mean velocity profile (solid line), distribution of the streamwise turbulence intensity (dashed line), and distribution of the lateral turbulence intensity (dotted line).

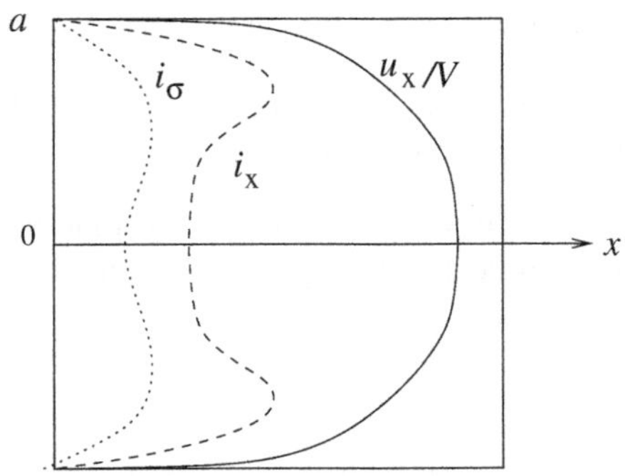

Figure 10.8.5 Schematic illustration of the mean velocity profile (solid line), and distribution of the streamwise (dashed line) and lateral (dotted line) turbulence intensity reduced by the maximum mean velocity at the centerline, for turbulent flow in a circular tube.

The intensities have been normalized by the maximum mean velocity occurring at the centerline. The mean velocity profile can be approximated with the algebraic form

$$\bar{u}_x(\sigma) = V \left(1 - \frac{\sigma}{a}\right)^{1/7}, \tag{10.8.8}$$

where $V \equiv (\bar{u}_x)_{max}$ is the maximum mean velocity occurring at the centerline. This profile should be contrasted with its parabolic counterpart shown in equation (7.3.4) for laminar flow.

10.8.5 Inviscid scales

A turbulent flow contains an infinite collection of interacting eddies defined and regarded as elementary fluid motions. Inspecting the turbulence signal shown in figure 10.8.1(*b*), we identify spatial scales with a broad range of magnitudes. One important scale, classified as inviscid, is the scale of the energy containing turbulent motion, denoted by l. Using l and the magnitude of the velocity fluctuations u defined in (10.8.7), we deduce that the time scale of the energy containing eddies is comparable to l/u.

The actual size of l varies according to the particular flow under consideration, as follows:

- In the case of boundary layer flow, l is comparable to the local boundary layer thickness.
- In the case of turbulent jet flow, l is comparable to the local jet diameter.
- In the case of a wake behind a body, l is comparable to the local width of the wake.

- In the case of pipe flow, l is comparable to the tube diameter.
- In the case of infinite shear flow over a wall, l in the buffer zone is proportional to the distance from the wall.
- In the case of free turbulence generated by placing a grid in a uniform stream, l is comparable to the grid size behind the grid, and increases with downstream position.

10.8.6 Viscous scales

Energy is dissipated in a turbulent flow even if the mean flow has a uniform velocity profile, requiring a continuous external power supply to sustain the motion. The rate of viscous dissipation, with units of kinetic energy per mass per time, is denoted by

$$\epsilon[=]\frac{L^2}{T^3}, \tag{10.8.9}$$

where the symbol [=] denotes dimensional equivalence, L is length, and T is time. One distinguishing property of turbulent flow is that dissipation mainly occurs due to small-scale motion. Combining ϵ with the kinematic viscosity, ν, we form the Kolmogorov length scale

$$\eta \equiv \left(\frac{\nu^3}{\epsilon}\right)^{1/4}, \tag{10.8.10}$$

and accompanying Kolmogorov velocity scale

$$u_K \equiv (\nu\,\epsilon)^{1/4}. \tag{10.8.11}$$

The two scales are related by

$$\frac{u_K\,\eta}{\nu} = 1. \tag{10.8.12}$$

The left-hand side of (10.8.12) defines the Reynolds number of the energy dissipating motion. By design, this is equal to unity, underlying the dominance of viscous forces responsible for converting kinetic to thermal energy in a viscous flow.

10.8.7 Relation between inviscid and viscous scales

Energy conservation requires that the rate of viscous dissipation, ϵ, scales as

$$\epsilon \simeq \frac{u^3}{l}. \tag{10.8.13}$$

Substituting this estimate in (10.8.10) and rearranging, we find

$$\frac{\eta}{l} \simeq \left(\frac{\nu}{u\,l}\right)^{3/4}. \tag{10.8.14}$$

The magnitude of the velocity fluctuations, u, is typically comparable to the external velocity scale, V, and the inviscid length scale l is comparable to the external length scale, L. As a result, the inverse of the ratio on the right-hand side of (10.8.14) is comparable to the Reynolds number $Re = VL/\nu$, yielding the scaling

$$\frac{\eta}{l} \simeq Re^{-3/4}. \tag{10.8.15}$$

Working in a similar fashion with (10.8.11), we derive the scaling law

$$\frac{u_K}{u} \simeq Re^{-1/4}. \tag{10.8.16}$$

These equations allow us to estimate the scales of the energy-dissipating motion from measurable or observable outer scales and the Reynolds number of the flow.

10.8.8 Fourier analysis

To analyze the distribution of energy across the scales of a turbulent flow, we may decompose a recording of the velocity at a particular location into a Fourier series with respect to time, and then examine the magnitude of the Fourier coefficients. In the laboratory, the velocity is typically measured by two methods: hot-wire anemometry based on a calibration that associates velocity to heat loss from a small heated wire probe placed in a flow; and laser-Doppler velocimetry based on light scattering from a patterned laser beam caused by small particles seeded in the flow.

To develop the Fourier decomposition, we consider a times series of a function, $f(t)$, comprised of a sequence of values of the function recorded at evenly spaced time intervals separated by a sampling time, Δt. Suppose that the time series contains N records corresponding to times 0, Δt, $2\,\Delta t$, $\ldots, (N-1)\Delta t$, and define $T = N\,\Delta t$. Using Fourier representation theory, we assume that $f(N\,\Delta t) = f(0)$ and express the function $f(t)$ over the time interval $(0, T)$ in the form of a complete Fourier series

$$f(t) \simeq \frac{1}{2}\,a_0 + \sum_{p=1}^{M} a_p \cos\left(\frac{2\pi p}{N}\,\frac{t}{\Delta t}\right) + \sum_{p=1}^{M} b_p \sin\left(\frac{2\pi p}{N}\,\frac{t}{\Delta t}\right), \tag{10.8.17}$$

where M is a specified truncation level, p is an integer, a_p are cosine Fourier coefficients, and b_p are sine Fourier coefficients. The complex Fourier coefficients are defined by

$$c_p \equiv \frac{1}{2}\,(a_p + \mathrm{i}\,b_p), \tag{10.8.18}$$

for $p = 0, 1, \ldots,$ where i is the imaginary unit, $\mathrm{i}^2 = -1$. In terms of the complex Fourier coefficients, the Fourier series (10.8.17) can be recast into the compact form

$$f(t) \simeq \sum_{p=-M}^{M} c_p \exp(-\mathrm{i}\,\frac{2\pi p}{N}\,\frac{t}{\Delta t}), \tag{10.8.19}$$

where the negative-indexed complex Fourier coefficients are given by

$$c_{-p} = c_p^* = \frac{1}{2}\,(a_p - \mathrm{i}\,b_p), \tag{10.8.20}$$

and an asterisk denotes the complex conjugate. Fourier theory provides us with a remarkably simple method for evaluating the Fourier coefficients.[5] Denoting the data by $f_l \equiv f(t_l)$, where $t_1 = 0, t_2 = \Delta t, \ldots, t_N = (N-1)\,\Delta t$, we find

$$a_p = \frac{2}{N}\,[f_1 + \cos(2\pi\,\frac{p}{N})\,f_2 + \cos(2\pi\,\frac{2p}{N})\,f_3 + \cdots + \cos(2\pi\,\frac{(N-1)p}{N})\,f_N],$$

$$b_p = \frac{2}{N}\,[f_1 + \sin(2\pi\,\frac{p}{N})\,f_2 + \sin(2\pi\,\frac{2p}{N})\,f_3 + \cdots + \sin(2\pi\,\frac{(N-1)p}{N})\,f_N]\,. \tag{10.8.21}$$

In practice, the number of data points, N, can be on the order of several thousand or even higher, and the direct evaluation of the sums on the right-hand sides of expressions (10.8.21) requires a prohibitive amount of computational time. Fortunately, the computations can be expedited considerably by use of an ingenious algorithm for computing the Fourier coefficients, known as the fast Fourier transform (FFT).

Function *fft* in directory *stats* inside directory *13_turbo* of FDLIB performs the FFT of a time series with size $N = 2^q$, where q is an integer.

Power spectrum

Now we take the square of both sides of (10.8.19), expand the square of the product on the right-hand side, integrate the resulting expression with respect to time from $t = 0$ to $T \equiv N\,\Delta t$, and use trigonometric identities to set the integral of a large number of terms equal to zero, finding

$$\int_0^T f^2(t)\,\mathrm{d}t \simeq T \sum_{p=-M}^{M} c_p c_p^*, \tag{10.8.22}$$

which can be rearranged into

[5] POZRIKIDIS, C., 2008, *Numerical Computation in Science and Engineering*, Second Edition, Oxford University Press.

$$\overline{f^2} \equiv \frac{1}{T} \int_0^T f^2(t)\, \mathrm{d}t \simeq \sum_{p=-M}^{M} |c_p|^2 = c_0^2 + 2 \sum_{p=1}^{M} |c_p|^2, \qquad (10.8.23)$$

where

$$|c_p|^2 = c_p c_p^* = \frac{1}{4} (a_p^2 + b_p^2) \qquad (10.8.24)$$

is the square of the magnitude of the pth complex Fourier coefficient.

A graph of the coefficients $2\,|c_p|^2$ against the angular frequency $\omega_p \equiv 2\pi p/\Delta t$ for $p = 1, \ldots,$ is the discrete temporal power spectrum of the function $f(t)$. Of particular interest if the behavior at high values of p corresponding to high angular frequencies. A graph of the coefficients $2\,|c_p|^2$ against the spatial wave number $k_p \equiv \omega_p V = 2\pi p\, V/\Delta t$ is the discrete power spectrum of the function $f(t)$, where V is a specified velocity. Of particular interest is the behavior at high values of p corresponding to high wave numbers.

Identifying the generic function $f(t)$ with the x, y, or z component of the velocity, we obtain the discrete energy spectrum of a turbulent flow, providing us with information on how kinetic energy is distributed among the different scales. In practice, the discrete power spectrum is computed by taking the Fourier transform of a time series comprised of sets of data points on the order of $2^{12} = 4096$. The power spectrum computed using one data set shows large fluctuations. To obtain a smooth spectrum, we average the Fourier coefficients over sets corresponding to different realizations or different time periods for the same flow conditions.

Energy density function

As the sample size, N, and total sampling time, $T = N\Delta t$, tend to infinity, the sum on the right-hand side of (10.8.19) reduces to a Fourier integral. Correspondingly, the right-hand side of (10.8.23) takes the form

$$\overline{f^2} = \int_0^\infty E_t(\omega)\, \mathrm{d}\omega = \int_0^\infty E(k)\, \mathrm{d}k, \qquad (10.8.25)$$

where $E_t(\omega)$ and $E(k)$ are temporal and spatial energy density functions. Now making a correspondence between (10.8.25) and (10.8.23), we obtain the relations

$$E_t(\omega_p) = \frac{\Delta t}{2\pi}\, 2\,|c_p|^2, \qquad E(k_p) = \frac{V\Delta t}{2\pi}\, 2\,|c_p|^2, \qquad (10.8.26)$$

which allow us to prepare graphs and study the shape of the energy density function. Taylor's frozen-field hypothesis amounts to setting V equal to the local mean value of the streamwise velocity.

Problems

10.8.1 *Turbulent mean velocity profile in pipe flow*

Consider turbulent flow in a pipe whose mean velocity profile is described by (10.8.8). Derive a relationship between the mean flow rate and the mean velocity at the centerline. Compare this relationship with its counterpart for laminar flow.

10.8.2 *Kolmogorov length scale for pipe flow*

Laboratory data for turbulent flow in a circular tube with diameter D has shown that: (a) the length scale of the energy containing turbulent motion, l, is comparable to D, and (b) the magnitude of the turbulent velocity fluctuations in comparable to the maximum mean velocity occurring at the centerline, V. Based on this information, compute the Kolmogorov length and velocity scale for a tube with diameter $D = 10$ cm, at Reynolds number $Re \equiv U\,D/\nu = 10^6$.

Computer Problem

10.8.3 *Stratified shear flow*

Figure 10.8.1 illustrates a stratified turbulent shear flow behind a vertical grid with mean velocity profile $\bar{u}_x(y)$ and mean temperature field $\bar{T}(y)$. Kurt Keller recorded a time series of the x and y velocity components and temperature at a point located 457.2 cm behind the grid, at the sampling frequency 5 Khz corresponding to sampling time 0.0002 s^{-1}, for the following conditions: grid spacing 2.54 cm; mean shear rate $\text{d}\bar{u}_x/\text{d}y = -7.63\ \text{s}^{-1}$; mean temperature gradient $\text{d}\bar{T}/\text{d}y = 35.8\ \text{K m}^{-1}$; local microscale Reynolds number $Re \equiv \lambda u/\nu = 91.2$.

In the definition of the microscale Reynolds number, $u \equiv \sqrt{\overline{u_x'^2}}$ is the RMS value of the fluctuations of the streamwise component of the velocity, and λ is the Taylor microscale defined by the relation

$$\frac{u^2}{\lambda^2} = \overline{\left(\frac{\text{d}u'}{\text{d}y}\right)^2}. \tag{10.8.27}$$

The data are arranged in the three columns of file *keller.dat* in directory *stats* inside directory *13_turbo* of FDLIB.

(a) Compute and plot the mean and RMS values of the velocity and temperature, and investigate their dependence on the sample size.

(b) Compute and plot the discrete power spectrum of the x and y velocity components and temperature. Discuss the results of your computation.

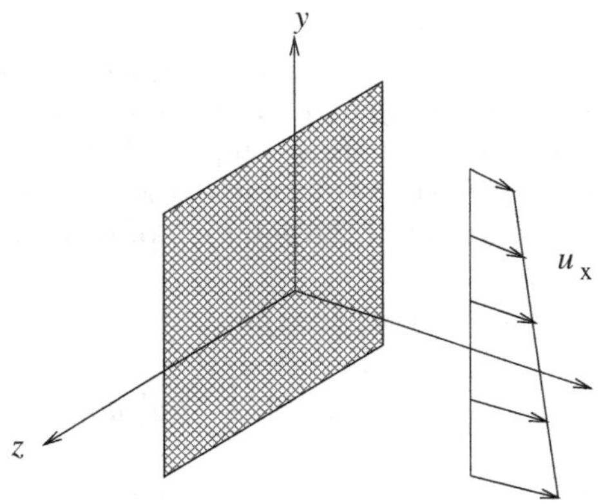

Figure 10.8.1 Stratified turbulent shear flow behind a vertical grid. A time series of the velocity and temperature recorded by Kurt Keller can be found in file *keller.dat* in directory *stats* inside directory *13_turbo* of FDLIB.

10.9 Analysis and modeling of turbulent flow

Deriving exact solutions of the Navier-Stokes equation for turbulent flow is out of the question. Direct numerical simulation (DNS) is prohibited by pragmatic constraints associated with the finite grid size, requiring sub-grid modeling: to capture the dynamics of a turbulent flow, we must resolve a prohibitively broad range of length scales. Progress can be made by developing approximate models and phenomenological theories based on empirical correlations inspired by laboratory observation.

10.9.1 Reynolds stresses

A point of departure for developing phenomenological theories is the decomposition of the velocity into a mean and a fluctuating component, as shown in equation (10.8.3). A similar decomposition of the pressure yields

$$p(\mathbf{x}, t) = \bar{p}(\mathbf{x}, t) + p'(\mathbf{x}, t), \tag{10.9.1}$$

where

$$\bar{p}(\mathbf{x}, t) \equiv \frac{1}{t_0} \int_{t-t_0/2}^{t+t_0/2} p(\mathbf{x}, t + t')\,\mathrm{d}t' \tag{10.9.2}$$

is the mean pressure. The continuity equation for an incompressible fluid becomes

$$\nabla \cdot \bar{\mathbf{u}} = 0, \tag{10.9.3}$$

where a bar denotes the time-averaged value.

Next, we substitute the decompositions (10.8.3) and (10.9.2) in Cauchy's equation of motion (6.3.15), expand the derivatives of the products, take the time average of both sides, and simplify to derive a modified equation of motion for the mean component. The fluctuating component appears as an effective inertial hydrodynamic volume force. For example, the term $\partial(\rho\, u_x\, u_y)/\partial y$ on the left-hand side of the x component of the equation of motion (6.3.15), becomes

$$\frac{\partial}{\partial y}(\rho\, u_x u_y) = \frac{\partial}{\partial y}[\rho\,(\bar{u}_x + u'_x)\,(\bar{u}_y + u'_y)] \tag{10.9.4}$$
$$= \frac{\partial}{\partial y}(\rho\,\bar{u}_x\,\bar{u}_y) + \frac{\partial}{\partial y}(\rho\,\bar{u}_x\,u'_y) + \frac{\partial}{\partial y}(\rho\,u'_x\,\bar{u}_y) + \frac{\partial}{\partial y}(\rho\,u'_x\,u'_y).$$

Taking the time average of both sides, defined in equations (10.8.2) and (10.9.2), and interchanging the order of time-averaging and space differentiation, we find

$$\frac{\partial}{\partial y}(\overline{\rho\, u_x u_y}) = \frac{\partial}{\partial y}(\overline{\rho\,\bar{u}_x \bar{u}_y}) + \frac{\partial}{\partial y}(\overline{\rho\,\bar{u}_x u'_y}) + \frac{\partial}{\partial y}(\overline{\rho\,u'_x \bar{u}_y}) + \frac{\partial}{\partial y}(\overline{\rho\,u'_x u'_y})$$
$$= \frac{\partial}{\partial y}(\rho\,\bar{u}_x \bar{u}_y) + \frac{\partial}{\partial y}(\rho\,\bar{u}_x \bar{u'_y}) + \frac{\partial}{\partial y}(\rho\,\bar{u'_x} \bar{u}_y) + \frac{\partial}{\partial y}(\overline{\rho\,u'_x u'_y}). \tag{10.9.5}$$

Because of (10.8.4), the second and third terms in the second line of (10.9.5) are zero, leaving the simplified expression

$$\frac{\partial}{\partial y}(\overline{\rho\, u_x u_y}) = \frac{\partial}{\partial y}(\rho\,\bar{u}_x\,\bar{u}_y) + \frac{\partial}{\partial y}(\overline{\rho\,u'_x u'_y}). \tag{10.9.6}$$

Working in a similar fashion with the other terms on the left-hand side of the Navier-Stokes equation, we derive a the Reynolds-averaged Navier-Stokes (RANS) equation

$$\rho\,(\frac{\partial\,\bar{\mathbf{u}}}{\partial t} + \bar{\mathbf{u}}\cdot\nabla\bar{\mathbf{u}}) = -\nabla\bar{p} + \mu\,\nabla^2\bar{\mathbf{u}} + \nabla\cdot\boldsymbol{\sigma}^R + \rho\,\mathbf{g}, \tag{10.9.7}$$

where $\boldsymbol{\sigma}^R$ is the Reynolds stress tensor with components

$$\sigma^R_{ij} = -\overline{\rho\,u'_i u'_j}, \tag{10.9.8}$$

expressing the transfer of momentum from the ith to the $j$the direction, and *vice versa*, by turbulence fluctuations. An alternative form of (10.9.7) is

$$\frac{\partial\,(\rho\,\bar{\mathbf{u}})}{\partial t} + \nabla\cdot(\rho\,\bar{\mathbf{u}}\,\bar{\mathbf{u}}) = -\nabla\bar{p} + \mu\,\nabla^2\bar{\mathbf{u}} + \nabla\cdot\boldsymbol{\sigma}^R + \rho\,\mathbf{g}. \tag{10.9.9}$$

Phenomenological theories seek to establish a relationship between the Reynolds stresses and the structure of the time-averaged flow. Once this has been done, the averaged equation of motion (10.9.7) or (10.9.9) can be solved along with the time-averaged continuity equation (10.9.3) to produce the velocity distribution in a turbulent flow.

Boussinèsq's law

Boussinèsq proposed the constitutive law

$$\boldsymbol{\sigma}^R = -(\frac{3}{2}\,\rho k)\,\mathbf{I} + \mu_T\,2\bar{\mathbf{E}}, \tag{10.9.10}$$

where

$$k = \frac{1}{2}\,(u_x'^2 + u_y'^2 + u_z'^2) \tag{10.9.11}$$

is the turbulent kinetic energy, μ_T is an eddy viscosity, $\mathbf{I}$ is the identity matrix, and $\bar{\mathbf{E}}$ is the time-averaged rate of deformation tensor. The term inside the parentheses on the right-hand side of (10.9.10) plays the role of a positive turbulence pressure.

In the $k - \epsilon$ model, the turbulent kinetic energy is related to the rate of viscous dissipation, ϵ, by

$$\nu_T\,\epsilon = C_\mu\,\rho\,k^2, \tag{10.9.12}$$

where

$$\epsilon = \nu_T\,\overline{\Big(\frac{\partial u_i}{\partial x_j}\Big)\Big(\frac{\partial u_i}{\partial x_j}\Big)}, \tag{10.9.13}$$

summation is implied over the repeated indices i and j, $\nu_T = \mu_T/\rho$ is the turbulent kinematic viscosity, and C_μ is a dimensionless coefficient.

10.9.2 Prandtl's mixing length model

Prandtl proposed an intuitive model that relates the Reynolds stresses to the velocity profile of the mean flow. Motivation is provided by a tentative analogy between eddy motion in a turbulent flow and molecular motion in a gas. The derivation is similar to that discussed in section 4.4.5 concerning the fluid viscosity.

Consider a unidirectional turbulent shear flow along the x axis with mean velocity profile $\bar{u}_x = U(y)$, as illustrated in figure 10.9.1. Suppose that, because of the turbulent motion, a small fluid parcel with volume δV_p initially located at $y = y_1$ is displaced to the position $y = y_2$ where it travels in the streamwise direction with the new local velocity. The change in the x component of the momentum is

$$\begin{aligned}\delta M_x &= \rho\,\delta V_p\,[u_x(y=y_2,t) - u_x(y=y_1,t)]\\ &\simeq \rho\,\delta V_p\,[U(y=y_2) - U(y=y_1)].\end{aligned} \tag{10.9.14}$$

Because the turbulent fluctuations have been assumed small, the total velocity has been approximated with the mean velocity to yield the expression shown in the second line of (10.9.14).

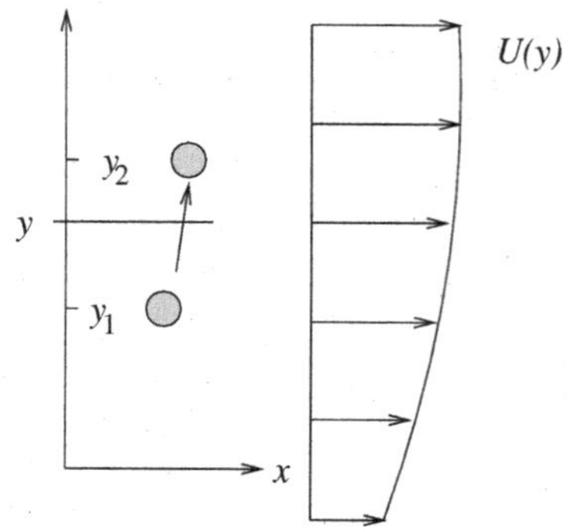

Figure 10.9.1 Profile of turbulent shear flow along the x axis illustrating the random displacement of a fluid parcel by a distance that is comparable to the Prandtl mixing length.

Next, we consider the transport of momentum across a horizontal line at the y elevation shown in figure 10.9.1. During a small period of time, Δt, all parcels residing inside a layer of thickness $u_y(y,t)\,\Delta t$ adjacent to this line cross the line to find themselves on the other side. The transport of x momentum across a horizontal length Δx due to this motion is

$$\sum \delta M_x \simeq \rho\,[U(y=y_2) - U(y=y_1)] \sum \delta V_p \simeq \rho\,(y_2 - y_1)\,(\frac{\partial U}{\partial y})_y \sum \delta V_p. \tag{10.9.15}$$

Setting

$$\sum \delta V_p = u_y(y,t)\,\Delta t\,\Delta x, \tag{10.9.16}$$

we obtain

$$\sum \delta M_x \simeq \rho\, u_y(y,t)\,(y_2 - y_1)\left(\frac{\partial U}{\partial y}\right)_y \Delta t\,\Delta x. \tag{10.9.17}$$

Averaging this expression over all time intervals Δt, we find

$$\overline{\sum \delta M_x} \simeq \rho\,\overline{u_y(y,t)\,(y_2 - y_1)}\left(\frac{\partial U}{\partial y}\right)_y \Delta t\,\Delta x. \tag{10.9.18}$$

According to Newton's second law of motion, this averaged transfer of momentum is tantamount to a force pointing in the x direction, expressed by the Reynolds shear stress, σ^R_{xy}. Setting

$$\frac{\overline{\sum \delta M_x}}{\Delta t} = \sigma^R_{xy}\,\Delta x, \tag{10.9.19}$$

and rearranging, we obtain

$$\sigma_{xy}^R = \rho \, \overline{u_y(y,t) \, \Delta y} \left(\frac{\partial U}{\partial y}\right)_y, \tag{10.9.20}$$

where $\Delta y \equiv y_2 - y_1$.

In the next key step, we introduce the scaling

$$\overline{u_y(y,t) \, \Delta y} \simeq \sqrt{\overline{u_y'^2}} \, l_P \simeq \left|\frac{\partial U}{\partial y}\right| l_P^2, \tag{10.9.21}$$

where l_P is Prandtl's mixing length. Substituting (10.9.21) in (10.9.20), we obtain the targeted constitutive equation

$$\sigma_{xy}^R(y) = \rho \, l_P^2 \left|\frac{\partial U}{\partial y}\right| \left(\frac{\partial U}{\partial y}\right). \tag{10.9.22}$$

The product of the first three terms on the right-hand side of (10.9.22) plays the role of the eddy viscosity introduced by Boussinèsq.

Readers who are overwhelmed by uneasiness regarding the physical relevance of the various steps involved in the preceding derivation are not alone. Numerous constitutive relations similar to that shown in (10.9.22) have been proposed based on tentative analogies and laboratory observation.

10.9.3 Logarithmic law for wall-bounded shear flow

As an application, we use Prandtl's mixing length model expressed by equation (10.9.22) to deduce the functional form of the velocity profile in wall-bounded shear flow, away from the viscous sublayer and the buffer zone. Assuming that the sign of dU/dy is positive, neglecting the viscous shear stress $\mu \, dU/dy$ in comparison to the Reynolds shear stress, and ignoring the variation of the shear stress in the y direction due to a streamwise pressure drop, we find

$$\sigma_{xy}^R(y) = \rho \, l_P^2 \left(\frac{dU}{dy}\right)^2 = \tau_W, \tag{10.9.23}$$

where τ_W is the wall shear stress. Rearranging, we obtain

$$l_P^2 \left(\frac{dU}{dy}\right)^2 = \frac{\tau_W}{\rho} \equiv u_*^2, \tag{10.9.24}$$

where u_* is the friction velocity.

Next, we set the Prandtl mixing length proportional to the distance from the wall, $l_P = \kappa y$, where κ is a dimensionless constant. For flow in a circular tube, measurements suggest $\kappa \simeq 0.36$. Substituting in (10.9.24), taking the square root of the emerging equation and rearranging, we find

$$\frac{dU}{dy} = \frac{u_*}{\kappa y}. \tag{10.9.25}$$

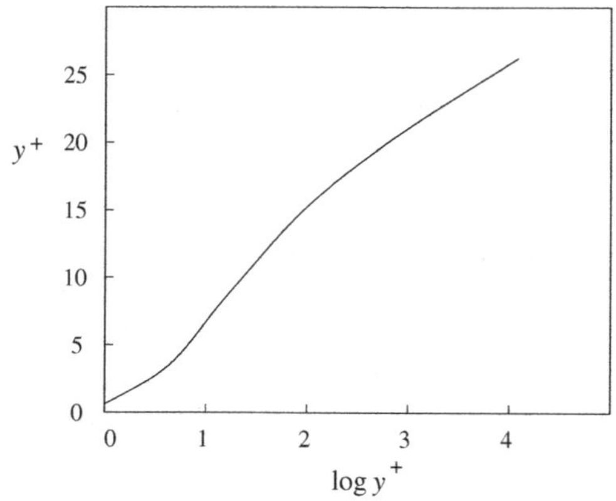

Figure 10.9.2 Schematic illustration of the distribution of the mean velocity in the dimensionless variables defined in equations (10.9.28), for wall-bounded turbulent shear flow.

Integrating with respect to y, we derive the logarithmic relationship

$$\frac{U(y)}{u_*} = \frac{1}{\kappa} \ln y + A, \tag{10.9.26}$$

where A is a dimensionless constant. In dimensionless variables, expression (10.9.26) takes the form

$$u^+ = \frac{1}{\kappa} \ln y^+ + B, \tag{10.9.27}$$

where B is a dimensionless constant, and we have defined

$$u^+ \equiv \frac{\bar{u}_x(y)}{u_*} = \frac{U(y)}{u_*}, \qquad y^+ \equiv \frac{y\, u_*}{\nu}. \tag{10.9.28}$$

Laboratory measurements have shown that equation (10.9.27) with $B = 3.6$ accurately describes the velocity profile for $y^+ > 26$, as illustrated in figure 10.9.2. In the viscous sublayer attached to the wall, laboratory data suggest the linear relation $u^+ = y^+$ for $0 < y^+ < 5$. In the buffer zone, $5 < y^+ < 26$, a more involved relation is required.

10.9.4 Correlations

Space-time correlations can be used to extract information on the small-scale structure of a turbulent flow. Consider the ith component of the fluctuating velocity at a point, $\mathbf{x}_1$, at time t, and the jth component of the fluctuating velocity at another point, $\mathbf{x}_2$, at time $t+\tau$, where τ is the time delay. The corresponding second-order space-time correlation is defined as

$$R_{ij}(\mathbf{x}_1, \mathbf{x}_2, t, \tau) \equiv \overline{u_i'(\mathbf{x}_1, t)\, u_j'(\mathbf{x}_2, t+\tau)}$$

$$= \frac{1}{t_0} \int_{t-t_0/2}^{t+t_0/2} u_i'(\mathbf{x}_1, t+t')\, u_j'(\mathbf{x}_2, t+\tau+t')\, \mathrm{d}t'. \qquad (10.9.29)$$

Two special correlations are of particular interest: the spatial correlation corresponding to $\tau = 0$, and the time-delayed correlation arising when the points $\mathbf{x}_1$ and $\mathbf{x}_2$ coincide.

Taylor's frozen-field hypothesis provides us with a relationship between the time-delayed and spatial correlation. The underlying premise is that, in a low-intensity turbulent flow, the mean velocity sweeps the turbulence so fast that the eddies do not evolve significantly during the time it takes them to cross a fixed point in space. Physically, the velocity vector field appears to be frozen in time. If the mean velocity is in the direction of the x axis, we may write

$$R_{ij}(\mathbf{x}_1, \mathbf{x}_2, t, \tau) \simeq R_{ij}(\mathbf{x}_1, \mathbf{x}_1 + \Delta x\, \mathbf{e}_x, t, \tau) = R_{ij}(\mathbf{x}_1, \mathbf{x}_1, t, \tau = -\frac{\Delta x}{\bar{u}_x}), \qquad (10.9.30)$$

where $\mathbf{e}_x$ is the unit vector along the x axis. This expression provides us with a convenient method for obtaining a spatial correlation in terms of a more accessible time-delayed correlation.

The usefulness of second-order correlations lies in their ability to produce information on the geometrical structure and dynamics of eddy motion in a turbulent flow. As the point $\mathbf{x}_2$ tends to the point $\mathbf{x}_1$, local fluid motions are coordinated and the correlations are significant. As the points $\mathbf{x}_1$ and $\mathbf{x}_2$ are moved far apart, fluid motions become independent or decorrelated and the correlations decay to zero.

In the case of homogeneous turbulent flow, the correlations depend on the vectorial distance between the points $\mathbf{x}_1$ and $\mathbf{x}_2$, but not on the absolute position of these two points. To signify this dependence, we write $R_{ij}(\mathbf{x}_2 - \mathbf{x}_1, t, \tau)$. In the case of isotropic turbulent flow, the correlations depend on the scalar distance between the two points $\mathbf{x}_1$ and $\mathbf{x}_2$. To signify this dependence, we write $R_{ij}(|\mathbf{x}_2 - \mathbf{x}_1|, t, \tau)$.

Evolution equations for second-order correlations can be derived departing from the Navier-Stokes equation. Just as the averaged Navier-Stokes equation (10.9.7) involves the Reynolds stresses, evolution equations for second-order correlations involve third-order correlations defined as the time averages of products of three scalar fluctuating components. An important field of study in turbulent flow seeks to relate high-order correlations to the structure of the mean flow, thereby achieving closure.

Problem

10.9.1 *Deissler correlation*

Deissler replaced Prandtl's constitutive equation (10.9.22) with a more involved equation inspired by laboratory measurements,

$$\sigma^R_{xy}(y) = n^2 \, \rho \, y \, U(y) \left(\frac{\partial U}{\partial y}\right)_y [1 - \exp(-\frac{n^2 U \, y}{\nu})], \tag{10.9.31}$$

where $n = 0.124$ is an experimentally determined dimensionless constant. Substitute this relation in (10.9.23), integrate to compute the velocity profile, and then compare the profile with that shown in (10.9.27).

Computer Problem

10.9.2 *Stratified shear flow*

Compute, plot, and discuss the form of the time-delayed correlation of the velocity components and temperature recorded in file *keller.dat* residing in directory *stats* inside directory *13_turbo* of FDLIB.

11 Vortex Motion

Flows at high Reynolds numbers develop islands of concentrated vorticity, concisely called vortices, embedded in a low-vorticity or virtually irrotational ambient fluid. The velocity field can be decomposed into two constituents: an irrotational component prevailing in the absence of the vortices, and a rotational component associated with the localized vorticity distribution. The latter can be conveniently expressed as an integral over the volume of fluid occupied by the vortices. At high Reynolds numbers, viscous forces are insignificant away from flow boundaries and the vortices evolve according to simplified rules dictated by the vorticity transport equation.

In this chapter, we derive an integral representation for the velocity in terms of the vorticity, discuss the simplified laws governing vortex motion in a flow with negligible viscous forces, and develop numerical methods for describing the dynamics of a prototypical class of vortex flows with specifically chosen vorticity distributions. The study of these flows allow us to develop insights into the dynamics of more general flows at high Reynolds number dominated by vortex interactions.

11.1 Vorticity and circulation in two-dimensional flow

The circulation around a closed loop in a two-dimensional flow is defined as the line integral of the tangential component of the velocity with respect to arc length around the loop, as shown in equation (3.7.10), repeated here for convenience,

$$C \equiv \oint_L u_t \, \mathrm{d}l = \oint_L \mathbf{u} \cdot \mathbf{t} \, \mathrm{d}l, \qquad (11.1.1)$$

C. Pozrikidis, *Fluid Dynamics: Theory, Computation, and Numerical Simulation*,
DOI: 10.1007/978-0-387-95871-2_11,

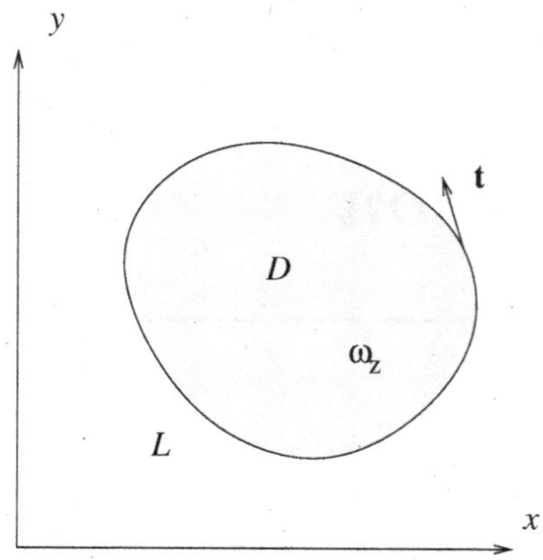

Figure 11.1.1 A closed loop in the xy plane, denoted by L, enclosing the area D. The circulation around the loop is equal to the areal integral of the strength of the vorticity, ω_z, over D.

where L denotes the loop, $\mathrm{d}l = (\mathrm{d}x^2 + \mathrm{d}y^2)^{1/2}$ is the differential arc length around the loop, and the unit tangent vector $\mathbf{t}$ points in the counterclockwise direction along L, as illustrated in figure 11.1.1.

If the loop is reducible, meaning that it can be shrunk to a point without crossing flow boundaries or singular points, we may use the Stokes circulation theorem to express the circulation around the loop as the areal integral of the z component of the vorticity over the area D enclosed by the loop, as shown in equation (3.7.11), repeated here for convenience,

$$C = \iint_D \omega_z \, \mathrm{d}x \, \mathrm{d}y, \tag{11.1.2}$$

(problem 11.1.1). With this expression as a point of departure and using the vorticity transport equation discussed in section 6.6, we proceed to derive an important theorem that considerably facilitates the study and computation of two-dimensional vortex flow.

We begin by considering a reducible material loop consisting of a fixed collection of point particles with a permanent identity, as illustrated in figure 11.1.1. The fluid enclosed by the loop also has a permanent identity; that is, it is composed of the same collection of point particles at all times. The vorticity transport equation for a flow with uniform density and negligible viscous forces requires that the point particles maintain their vorticity as they move about the domain of flow; that is, $\mathrm{D}\omega_z/\mathrm{D}t = 0$, as shown in equation (6.6.7), where $\mathrm{D}/\mathrm{D}t$ is the material derivative. Moreover, because the fluid has been assumed incompressible, the area $\mathrm{d}x\mathrm{d}y$ occupied by an infinitesimal patch of fluid located inside the loop remains constant in time, $\mathrm{D}(\mathrm{d}x \, \mathrm{d}y)/\mathrm{D}t = 0$.

Combining these observations, we find that the integral on the right-hand side of (11.1.2) remains constant in time. Formally, we write

$$\frac{dC}{dt} = \frac{d}{dt}\iint_D \omega_z \, dx\, dy = \iint_D \frac{D(\omega_z \, dx\, dy)}{Dt}$$
$$= \iint_D \frac{D\omega_z}{Dt} \, dx\, dy + \iint_D \omega_z \frac{D(dx\, dy)}{Dt} = 0. \qquad (11.1.3)$$

We have shown that, when viscous forces are negligible, the circulation around a reducible material loop remains constant in time. In section 11.4, we will see that the circulation around an irreducible loop also remains constant in time. Consequently, the vorticity of point particles residing inside any loop in a two-dimensional flow is preserved during the motion.

Problem

11.1.1 *Stokes circulation theorem*

Prove that the circulation around a reducible loop can be expressed in terms of the vorticity, as shown in equation (11.1.2). *Hint:* Apply Gauss's divergence theorem stated in equation (2.6.24) for the vector functions $\mathbf{h} = (u_y, 0)$ and $\mathbf{h} = (0, u_x)$.

11.2 Point vortices

We begin the study of vortex dynamics by considering the motion of point vortices in a flow with negligible viscous forces. Expressions (3.7.1) provide us with the plane polar component of the velocity induced at a point $\mathbf{x} = (x, y)$ by a point vortex with strength κ located at the point $\mathbf{x}_0 = (x_0, y_0)$. The corresponding Cartesian components of the velocity are given by

$$u_x(x,y) = -\frac{\kappa}{2\pi}\frac{y-y_0}{(x-x_0)^2+(y-y_0)^2},$$
$$u_y(x,y) = \frac{\kappa}{2\pi}\frac{x-x_0}{(x-x_0)^2+(y-y_0)^2}. \qquad (11.2.1)$$

The denominator on the right-hand sides of equations (11.2.1) is the square of the distance of the field point, $\mathbf{x}$, from the location of the point vortex, $\mathbf{x}_0$. Since the numerator is a linear function of the difference in the x or y coordinates, the velocity due to a point vortex decays as the inverse of the distance from the point vortex, in agreement with (3.7.1).

We can readily verify by straightforward differentiation that the strength of the vorticity, $\omega_z \equiv \partial u_y/\partial x - \partial u_x/\partial y$, vanishes everywhere in the flow, except at the location of the point vortex where the right-hand sides of equations (11.2.1)

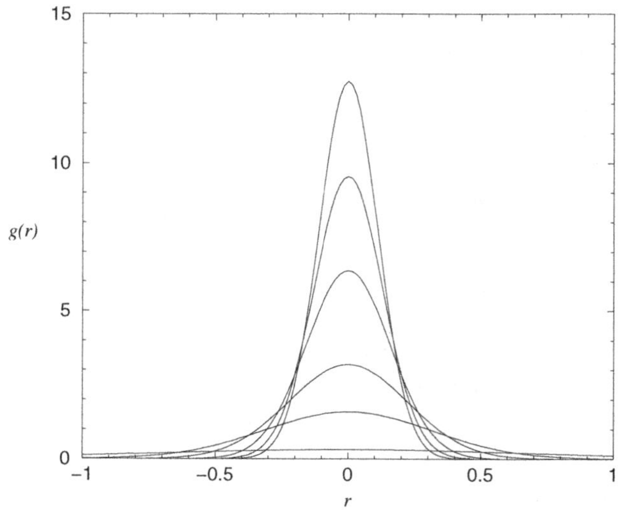

Figure 11.2.1 A family of test functions described by equation (11.2.2), for $\lambda =$ 1, 5, 10, 20, 30, and 40. In the limit as λ tends to infinity, we obtain Dirac's delta function in two dimensions.

and their derivatives are not defined. Dirac's delta function in two dimensions provides us with a convenient device for expressing this singular vorticity distribution in compact form using the concept of generalized functions.

11.2.1 Dirac's delta function in a plane

To construct the Dirac delta function in two dimensions, we introduce a family of test functions, $g_\lambda(x, y)$, parametrized by the variable λ. The test functions are radially symmetric with respect to a specified point (x_0, y_0); that is, they depend only on the distance of the field point, $\mathbf{x}$, from the chosen point, $\mathbf{x}_0$; they peak at the point (x_0, y_0); they rapidly decay to zero with distance from this point; and their areal integral over the entire xy plane is equal to unity.

One such family of test functions is given by

$$g_\lambda(r) = \frac{\lambda}{\pi} \exp(-\lambda r^2), \tag{11.2.2}$$

where $r = [(x - x_0)^2 + (y - y_0)^2]^{1/2}$. Graphs of several such functions corresponding to different values of λ are displayed in figure 11.2.1. As λ tends to infinity, the support of these functions shrinks to zero, yielding Dirac's delta function in the xy plane, denoted by $\delta_2(x - x_0, y - y_0)$. By construction then, $\delta_2(x - x_0, y - y_0)$ is endowed with the following properties:

1. $\delta_2(x - x_0, y - y_0)$ vanishes everywhere, except at the point $x = x_0$ and $y = y_0$ where it takes an infinite value.

2. The integral of the delta function over an area D that contains the point (x_0, y_0) is equal to unity

$$\iint_D \delta_2(x - x_0, y - y_0)\, dx\, dy = 1. \tag{11.2.3}$$

   This property reveals that the delta function in two dimensions has units of inverse squared length.

3. The integral of the product of an arbitrary function, $f(x, y)$, and the delta function over an area D that contains the point (x_0, y_0) is equal to value of the function at the singular point,

$$\iint_D \delta_2(x - x_0, y - y_0)\, f(x, y)\, dx\, dy = f(x_0, y_0). \tag{11.2.4}$$

   Note that identity (11.2.3) arises from (11.2.4) by setting $f(x, y) = 1$. The integral of the product of an arbitrary function $f(x, y)$ and the delta function over an area D that does *not* contain the point (x_0, y_0) is zero.

Vorticity associated with a point vortex

The vorticity distribution associated with the velocity field (11.2.1) can be expressed in terms of the two-dimensional delta function in the compact form

$$\omega_z(x, y) = \kappa\, \delta_2(x - x_0, y - y_0) = \kappa\, \delta_2(\mathbf{x} - \mathbf{x}_0). \tag{11.2.5}$$

The strength of the point vortex, κ, has units of circulation, velocity multiplied by length, and the two-dimensional delta function has units of inverse length squared. The product has units of velocity over length, which is consistent with the definition of the vorticity in terms of spatial derivatives of the velocity.

Reviewing the process by which the delta function arose from a family of smooth functions with increasingly narrow support and high peak, we interpret a point vortex as an idealized vortex structure emerging in the limit as the size of a compact vortex region in the xy plane tends to zero, while the circulation around the vortex is held constant.

It is instructive to confirm equation (3.7.12) stating that the circulation around a closed loop enclosing a point vortex is equal to the strength of the point vortex. Substituting (11.2.5) in the integrand on the right-hand side of (11.1.2) and using property (11.2.3), we find $C = \kappa$.

11.2.2 Evolution of the point vortex strength

When viscous forces are negligible, the circulation around any material loop that encloses a point vortex, and therefore the strength of the point vortex, remain constant in time,

$$\frac{\mathrm{d}\kappa}{\mathrm{d}t} = 0. \tag{11.2.6}$$

Thus, a point vortex maintains its strength as it moves with the fluid velocity in the domain of flow.

11.2.3 Velocity of a point vortex

The computation of the fluid velocity at the position of a point vortex is frustrated by the singular nature of the right-hand side of (11.2.1). To circumvent this difficulty, we observe that, although the fluid in the vicinity of a point vortex spins about the point vortex with a velocity that increases as the inverse of the distance from the point vortex, radial symmetry prevents it from exhibiting a net translational motion. This observation suggests that the self-induced velocity of a point vortex in an infinite domain vanishes, and the point vortex is convected with a velocity other than that associated with its own vorticity distribution. For example, a point vortex embedded in uniform flow simply translates with the velocity of the uniform flow.

11.2.4 Motion of a collection of point vortices

A collection of N point vortices move under the influence of their mutually-induced velocities. The rate of change of position of the ith point vortex, $\mathbf{X}_i = (X_i, Y_i)$, is governed by the differential equations

$$\frac{\mathrm{d}X_i}{\mathrm{d}t} = -\sum_{j=1}^{N}{}' \frac{\kappa_j}{2\pi} \frac{Y_i - Y_j}{(X_i - X_j)^2 + (Y_i - Y_j)^2},$$

$$\frac{\mathrm{d}Y_i}{\mathrm{d}t} = \sum_{j=1}^{N}{}' \frac{\kappa_j}{2\pi} \frac{X_i - X_j}{(X_i - X_j)^2 + (Y_i - Y_j)^2}, \tag{11.2.7}$$

for $i = 1, \ldots, N$; the prime after the sum indicates that the term $j = i$ is excluded from the summation to bypass the singular self-induced contribution. Equation (11.2.6) requires that the strength of each point vortex remains constant during the motion.

Equations (11.2.7) comprise a system of $2N$ first-order differential equations for the point vortex coordinates, (X_i, Y_i). Having specified the initial positions, we may compute the subsequent motion using a standard method for solving ordinary equations, such as the explicit modified Euler method discussed in earlier sections. Program *pvm* inside directory *09_vortex* of FDLIB simulates the motion of a collection of point vortices in an unbounded domain of flow.

11.2.5 Effect of boundaries

When the domain of flow is bounded by impermeable surfaces, the right-hand sides of equations (11.2.1) and (11.2.7) must be modified with the addition of a complementary flow whose purpose is to ensure the satisfaction of the no-penetration boundary condition. For simple boundary geometries, the complementary flow can be identified with the flow induced by point vortices located at image positions.

Directory *pv* inside directory *09_vortex* of FDLIB contains functions that return the velocity field induced by a point vortex for several boundary geometries.

Point vortex near a plane wall

The complementary flow for a point vortex near an infinite plane located at $y = w$ is generated by reflecting the point vortex with respect to the wall. If a primary point vortex with strength κ is located at the point (x_0, y_0), then an image point vortex with strength $-\kappa$ is located at $(x_0, 2w - y_0)$. The velocity field induced by the primary point vortex and its image is

$$u_x(x,y) = -\frac{\kappa}{2\pi}\Big[\frac{y-y_0}{(x-x_0)^2+(y-y_0)^2} - \frac{y-2w+y_0}{(x-x_0)^2+(y-2w+y_0)^2}\Big],$$
$$u_y(x,y) = \frac{\kappa}{2\pi}\Big[\frac{x-x_0}{(x-x_0)^2+(y-y_0)^2} - \frac{x-x_0}{(x-x_0)^2+(y-2w+y_0)^2}\Big]. \tag{11.2.8}$$

The streamline pattern induced by the vortex pair is illustrated in figure 11.2.2(a).

The x velocity component induced by the image vortex at the location of the primary vortex is $v_x \equiv u(x_0, y_0) = \kappa/[4\pi(y_0 - w)]$, while the y component vanishes, $v_y \equiv u_y(x_0, y_0) = 0$. Thus, the primary point vortex translates parallel to the wall with constant velocity.

Point vortex inside or outside a circular cylinder

The complementary flow of a point vortex located inside or outside a circular cylinder of radius a centered at the point (x_c, y_c) is generated by an image point vortex located at the inverse point of the primary point vortex with respect to the cylinder. If a primary point vortex with strength κ is located at (x_0, y_0), then an image point vortex with strength $-\kappa$ is located at

$$x_0^I = x_c + (x_0 - x_c)\,\frac{a^2}{d^2}, \qquad y_0^I = y_c + (y_0 - y_c)\,\frac{a^2}{d^2}, \tag{11.2.9}$$

where $d = [(x_0 - x_c)^2 + (y_0 - y_c)^2]^{1/2}$ is the distance of the primary point vortex from the center of the cylinder. The velocity field induced by the primary point

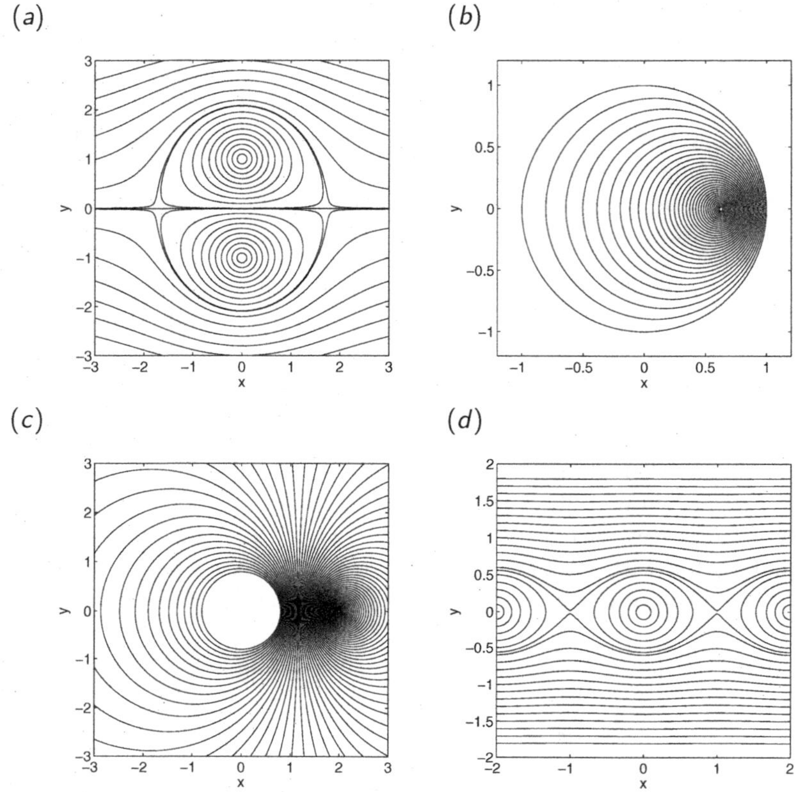

Figure 11.2.2 Streamline pattern of the flow due to (a) a pair of point vortices with opposite strength, (b) a point vortex inside a circular cylinder, (c) a point vortex outside a circular cylinder, and (d) a periodic array of point vortices. The streamline patterns were generated using the program *strml* in directory *04_various* of FDLIB.

vortex and its image is

$$u_x(x,y) = -\frac{\kappa}{2\pi}\Big[\frac{y-y_0}{(x-x_0)^2+(y-y_0)^2} - \frac{y-y_0^I}{(x-x_0^I)^2+(y-y_0^I)^2}\Big],$$

$$u_y(x,y) = \frac{\kappa}{2\pi}\Big[\frac{x-x_0}{(x-x_0)^2+(y-y_0)^2} - \frac{x-x_0^I}{(x-x_0^I)^2+(y-y_0^I)^2}\Big]. \qquad (11.2.10)$$

The streamline pattern induced by the vortex pair inside and outside the cylinder is illustrated in figure 11.2.2$(b,\ c)$. Examining the velocity induced by the image point vortex at the location of the primary vortex, we find that the primary vortex rotates around the center of the cylinder in the direction of the polar angle θ measured around the center of the cylinder in the counterclockwise direction, with velocity

$$v_\theta = \frac{\kappa}{2\pi} \frac{d}{a^2 - d^2}. \tag{11.2.11}$$

11.2.6 A periodic array of point vortices

Next, we consider a periodic array of point vortices with identical strengths separated by the distance a along the x axis, as illustrated in figure 11.2.2(d). The mth point vortex is located at the position $x_m = x_0 + ma$, $y_m = y_0$, where (x_0, y_0) is the position of an arbitrary point vortex labeled 0, and m is an integer label. If we attempt to compute the velocity induced by the array simply by summing the individual contributions, we will encounter unphysical divergent sums.

To overcome this difficulty, we consider the stream function corresponding to the velocity field induced by the individual point vortices, defined by the equations $u_x = \partial\psi/\partial y$ and $u_y = -\partial\psi/\partial x$, and express it in the form

$$\psi_0(x, y) = -\frac{\kappa}{2\pi} \ln \frac{r_0}{a}, \qquad \psi_m(x, y) = -\frac{\kappa}{2\pi} \ln \frac{r_m}{|m|\, a}, \tag{11.2.12}$$

for $m = \pm 1, \pm 2, \ldots$, where $r_m \equiv [\,(x - x_m)^2 + (y - y_m)^2\,]^{1/2}$ is the distance of the field point (x, y) from the location of the mth point vortex. The denominators of the fractions in the arguments of the logarithms on the right-hand sides of (11.2.12) have been chosen judiciously to facilitate the forthcoming algebraic manipulations.

It is important to observe that, as m tends to $\pm\infty$, the fraction on the right-hand side of the second equation in (11.2.12) tends to unity and its logarithm tends to vanish, thereby ensuring that remote point vortices make small contributions. In contrast, if the denominators were not included, remote point vortices would make contributions that are proportional to the logarithm of the distance of a point vortex from the point (x, y) where the stream function is evaluated.

Next, we express the stream function due to the infinite array as the sum of (a) a constant expressed by the term after the first equal sign in equation (11.2.14), and (b) the individual stream functions stated in expressions (11.2.12), obtaining

$$\begin{aligned} \psi(x, y) &= -\frac{\kappa}{2\pi} \ln(\sqrt{2}\pi) + \sum_{m=-\infty}^{\infty} \psi_m(x, y) \\ &= -\frac{\kappa}{2\pi} \ln \frac{\sqrt{2}\,\pi\, r_0}{a} - \frac{\kappa}{2\pi} \sum_{m=\pm 1, \pm 2, \ldots} \ln \frac{r_m}{|m|\, a}, \end{aligned} \tag{11.2.13}$$

which can be restated as

$$\psi(x,y) = -\frac{\kappa}{2\pi} \ln[\, \frac{\sqrt{2}\,\pi\, r_0}{a} \prod_{m=\pm1,\pm2,\ldots} \frac{r_m}{|m|\, a}\,], \tag{11.2.14}$$

where Π denotes the product. An identity allows us to compute the infinite product on the right-hand side of (11.2.14) in closed form, obtaining

$$\frac{\sqrt{2}\,\pi\, r_0}{a} \prod_{m=\pm1,\pm2,\ldots} \frac{r_m}{|m|\, a} = \{\cosh[k(y-y_0)] - \cos[k(x-x_0)]\}^{1/2}. \tag{11.2.15}$$

Substituting the right-hand side of (11.2.15) in (11.2.14), we derive the desired stream function

$$\psi(x,y) = -\frac{\kappa}{4\pi} \ln\Big(\cosh[k(y-y_0)] - \cos[k(x-x_0)]\Big), \tag{11.2.16}$$

where $k = 2\pi/a$ is the wave number. Differentiating the right-hand side of (11.2.16) with respect to x or y, we obtain the corresponding velocity components

$$\begin{aligned} u_x(x,y) &= -\frac{\kappa}{2a}\, \frac{\sinh[k(y-y_0)]}{\cosh[k(y-y_0)] - \cos[k(x-x_0)]}, \\ u_y(x,y) &= \frac{\kappa}{2a}\, \frac{\sin[k(x-x_0)]}{\cosh[k(y-y_0)] - \cos[k(x-x_0)]}. \end{aligned} \tag{11.2.17}$$

The streamline pattern due to the periodic array is illustrated in figure 11.2.2(d).

Because of symmetry, the velocity at the location of one point vortex induced by all other point vortices is zero, and the array is stationary. Far above or below the array, the x component of the velocity tends to the value $-\kappa/a$ or κ/a, while the y component decays at an exponential rate. This behavior renders the infinite array a reasonable model of the flow generated by the instability of a shear layer separating two streams that merge at different velocities. The Kelvin-Helmholtz instability causes the shear layer to roll up into compact vortices represented by the point vortices of the periodic array.

Motion of a collection of point vortices in a periodic arrangement

The motion of a periodic collection of N point vortices, each repeated in the x direction with period a, is governed by the counterpart of equations (11.2.7) for periodic flow. Using the velocity field (11.2.17), we find

$$\begin{aligned} \frac{dX_i}{dt} &= -\sum_{j=1}^{N}{}' \frac{\kappa_j}{2a}\, \frac{\sinh[k(Y_i-Y_j)]}{\cosh[k(Y_i-Y_j)] - \cos[k(X_i-X_j)]}, \\ \frac{dY_i}{dt} &= \sum_{j=1}^{N}{}' \frac{\kappa_j}{2a}\, \frac{\sin[k(X_i-X_j)]}{\cosh[k(Y_i-Y_j)] - \cos[k(X_i-X_j)]}, \end{aligned} \tag{11.2.18}$$

for $i = 1, \ldots, N$; the prime after the sum indicates that the term $j = i$, corresponding to the vanishing velocity induced by the host array, is excluded from the summation. The strength of each point vortex remains constant during the motion.

Program *pvm_pr* inside directory *09_vortex* of FDLIB simulates the motion of a periodic row of point vortices perturbed from the planar configuration. The motion is known to suffer from severe numerical instabilities that cause the appearance of small-scale irregularities at an early stage of the motion. One way to suppress these instabilities is by smoothing the coordinates of the point vortices, replacing them with weighted averages involving the neighbors. The five-point formula of Longuett-Higgins and Cokelet replaces the old positions with the new positions according to the formula

$$f_i^{new} = \frac{1}{16}\,(-f_{i-2} + 4\,f_{i-1} + 10 f_i + 4\,f_{i+1} - f_{i+2}), \tag{11.2.19}$$

where f stands for x or y. Results of a simulation with smoothing applied after each time step are shown in figure 11.2.3.

11.2.7 A point vortex between two parallel walls

Consider a point vortex between two parallel walls separated by the distance h, as illustrated in figure 11.2.4(a). The image system consists of the reflections of the point vortex, and the reflections of the reflections of the point vortex with respect to both walls. The result is an image system consisting of two infinite periodic arrays of point vortices separated by the distance $2h$. One array contains the primary point vortex, and the second array contains the reflection of the primary array with respect to one of the walls. The strength of the point vortices in the second array is equal in magnitude and opposite in sign to that of the point vortices in the first array. The stream function and velocity field may be deduced from expressions (11.2.16) and (11.2.17).

Consideration of the velocity field reveals that the point vortex travels parallel to the walls with velocity

$$v = \frac{\kappa}{4h}\,\frac{\sin(2\pi b/h)}{1 - \cos(2\pi b/h)}, \tag{11.2.20}$$

where b is the distance of the point vortex from the upper or lower wall. As b tends to zero, v tends to the limiting value $\kappa/(4\pi b)$ corresponding to a point vortex above a plane wall.

11.2.8 A point vortex in a semi-infinite strip

Consider a point vortex between two parallel walls separated by the distance h and intersecting at a right angle a third plane wall, thereby forming a semi-infinite rectangular strip, as illustrated in figure 11.2.4(b). The image system

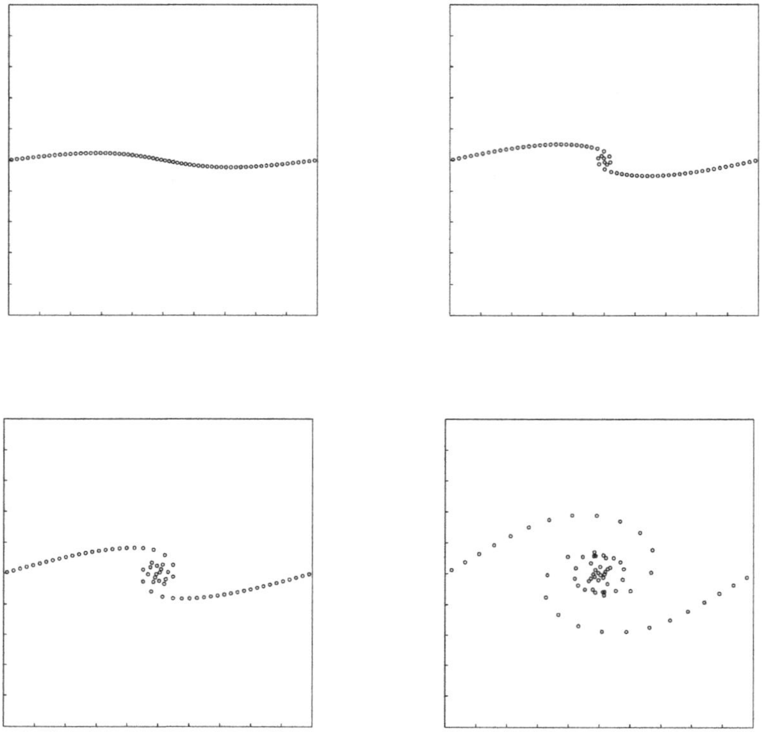

Figure 11.2.3 Stages in the evolution of one period of a perturbed periodic array of point vortices, showing periodic roll up and the eventual occurrence of disorganized motion.

consists of the image system associated with the two parallel walls discussed in section 11.2.8, and the reflection of the image system with respect to the third intersecting wall. The strength of the reflected point vortices is the negative of that of their images. The motion of the point vortex must be computed by numerical integration.

Problems

11.2.1 *Dirac delta function in one dimension*

The Dirac delta function in one dimension, denoted by $\delta_1(x - x_0)$, is distinguished by the following properties:

1. $\delta_1(x - x_0)$ vanishes everywhere, except at the point $x = x_0$ where it becomes infinite.

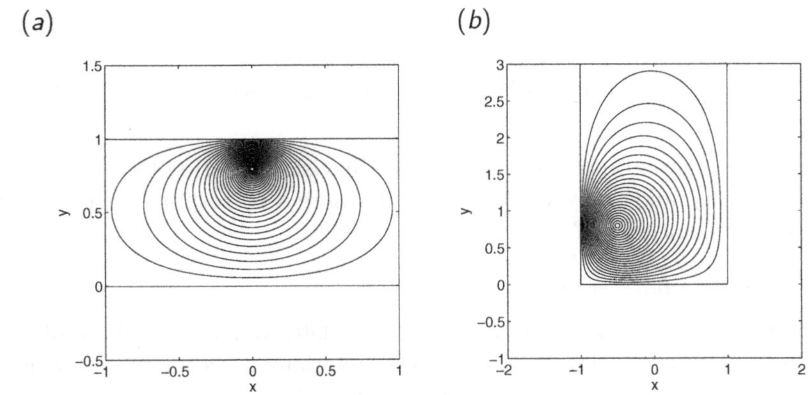

Figure 11.2.4 Streamline pattern of the flow due to (a) a point vortex between two parallel plane walls, and (b) a point vortex in a semi-infinite rectangular strip. The streamline patterns were generated using the program *strml* inside directory *04_various* of FDLIB.

2. The integral of the delta function over an interval I that contains the point x_0 is equal to unity,

$$\int_I \delta_1(x - x_0)\,\mathrm{d}x = 1. \tag{11.2.21}$$

This property reveals that the delta function in one dimension has units of inverse length.

3. The integral of the product of an arbitrary function $f(x)$ and the delta function over an interval I that contains the point x_0 is equal to value of the function at the singular point,

$$\int_D \delta_1(x - x_0)\, f(x)\,\mathrm{d}x = f(x_0). \tag{11.2.22}$$

The integral of the product of an arbitrary function $f(x)$ and the delta function over an interval I that does *not* contain the point x_0 is zero. Note that identity (11.2.21) arises from (11.2.22) by setting $f(x) = 1$.

(a) Show that δ_1 arises from the family of test functions

$$q_\lambda(|x - x_0|) = \left(\frac{\lambda}{\pi}\right)^{1/2} \exp[-\lambda\,(x - x_0)^2], \tag{11.2.23}$$

in the limit as the parameter λ tends to infinity.

(b) Show that the test functions $g_\lambda(r)$ defined in (11.2.2) derive from the test functions q_λ defined in (11.2.23) as

$$g_\lambda(r) = q_\lambda(|x - x_0|)\, q_\lambda(|y - y_0|). \tag{11.2.24}$$

Explain why we may write

$$\delta_2(x - x_0, y - y_0) = \delta_1(x - x_0)\,\delta_1(y - y_0). \tag{11.2.25}$$

(*c*) Show that the integral of $\delta_1(x - x_0)$ is the Heaviside function that is equal to zero when $x < x_0$, and unity when $x > x_0$. Is there a corresponding Heaviside function in two dimensions?

11.2.2 *Dirac delta function in three dimension*

State the distinguishing properties of the Dirac delta function in three dimensions. Devise an appropriate family of test functions corresponding to those discussed in the text for the delta function in two dimensions.

11.2.3 *A point vortex near a corner*

Show that the image flow associated with a point vortex located between two semi-infinite walls intersecting at a right-angle is represented by three point vortices located at the reflections, and the reflection of the reflections of the primary vortex with respect to the two walls. Introduce plane polar coordinates (r, θ) with origin at the apex and the walls located at $\theta = 0$ and $\pi/2$, and show that the primary point vortex moves along a path described by $r = c/\sin(2\theta)$, where the constant c is determined by the initial position.

11.2.4 *Point vortex in a rectangular box*

Discuss how the image system for a point vortex located in a semi-infinite rectangular strip discussed in section 11.2.9 can be extended to describe flow inside a rectangular box.

Computer Problems

11.2.5 *Motion of point vortices*

(*a*) Run program *pvm* to simulate the motion of a collection of N point vortices with identical strength placed at the vertices of an N-sided regular polygon of radius a. Carry out simulations for $N = 8$, 16, and 32, and discuss the nature of the motion at long times.

(*b*) Modify the program *pvm* to include the presence of a boundary of your choice. Compute the motion of a collection of point vortices of your choice and discuss the nature of the motion.

11.2.6 *Motion of a periodic collection of point vortices*

Run the program *pv_pr* to compute the motion of a periodic arrangement of point vortices using an initial condition of your choice, with and without smoothing. Discuss the nature of the motion in each case.

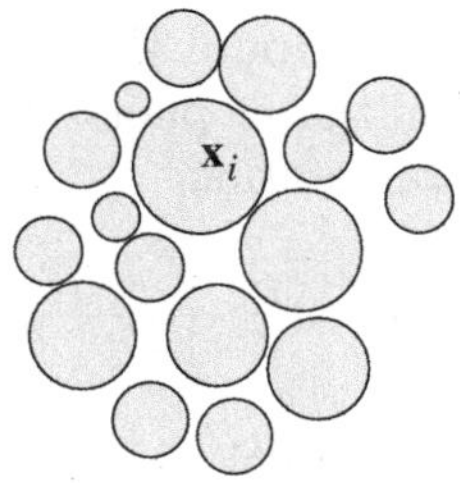

Figure 11.3.1 A vortex in a two-dimensional flow is discretized into small parcels. Each parcel is then condensed into a point vortex located at a designated parcel center. As the number of parcels tends to infinity, the velocity induced by the point vortices is expressed by the integral representation (11.3.2).

11.3 Two-dimensional flow with distributed vorticity

We now broaden the scope of our discussion by considering the more general case of two-dimensional flow in the xy plane containing a region of concentrated vorticity embedded in an otherwise perfectly or nearly irrotational fluid, as illustrated in figure 11.3.1. To deduce the velocity induced by the vorticity, we subdivide the vortex region into a collection of N parcels with small areas, δA_i, where $i = 1, \ldots, N$.

Let $\omega_i \equiv \omega_z(\mathbf{x}_i)$ be the strength of the vorticity at a designated center of the ith parcel, $\mathbf{x}_i$. The strength of the parcel, defined as the integral of the vorticity over the area of the parcel, is approximately equal to $\kappa_i \equiv \omega_i \, \delta A_i$. For the purpose of evaluating the velocity at a point in the flow, we now replace the parcels with point vortices located at the designated parcel centers. Using equations (11.2.1), we find that the velocity induced by the collection of the point vortices is given by

$$u_x(x,y) = -\frac{1}{2\pi} \sum_{i=1}^{N} \frac{y - y_i}{(x-x_i)^2 + (y-y_i)^2} \, \omega_i \, \delta A_i,$$

$$u_y(x,y) = \frac{1}{2\pi} \sum_{i=1}^{N} \frac{x - x_i}{(x-x_i)^2 + (y-y_i)^2} \, \omega_i \, \delta A_i. \tag{11.3.1}$$

As the number of parcels, N, tends to infinity, the sums reduce to areal integrals yielding an integral representation for the velocity in terms of the vorticity,

$$u_x(x,y) = -\frac{1}{2\pi} \iint_{Vortex} \frac{y - y'}{(x-x')^2 + (y-y')^2} \, \omega_z(x',y') \, dx' \, dy',$$

$$u_y(x,y) = \frac{1}{2\pi} \iint_{Vortex} \frac{x - x'}{(x-x')^2 + (y-y')^2} \, \omega_z(x',y') \, dx' \, dy'. \tag{11.3.2}$$

Equations (11.3.2) allow us to compute the velocity field associated with a specified vorticity distribution in the xy plane. The stream function is given by the corresponding integral representation

$$\psi(x,y) = -\frac{1}{4\pi}\iint_{Vortex} \ln[(x-x')^2 + (y-y')^2]\,\omega_z(x',y')\,dx'\,dy', \quad (11.3.3)$$

which is a solution of the Poisson equation,

$$\nabla^2\psi = -\omega_z, \quad (11.3.4)$$

arising from the definitions of the vorticity and stream function.

Recovering point vortices

It is instructive to observe that the velocity field due to a point vortex arises by substituting the singular vorticity distribution (11.2.5) in expressions (11.3.2), and then using property (11.2.4) to evaluate the integrals. The delta function simply switches x' to x_0 and y' to y_0, and thereby produces the flow due to a point vortex expressed by (11.2.1).

11.3.1 Vortex patches with uniform vorticity

Next, we consider a compact vortex with uniform vorticity, Ω, enclosed by the closed contour C, as illustrated in figure 11.3.2(a). Extracting the vorticity from the integral on the right-hand side of (11.3.3), we obtain the stream function

$$\psi(x,y) = -\frac{\Omega}{4\pi}\iint_{Vortex} \ln[(x-x')^2 + (y-y')^2]\,dx'\,dy'. \quad (11.3.5)$$

The x and y velocity components derive from the stream function as $u_x = \partial\psi/\partial y$ and $u_y = -\partial\psi/\partial x$. Differentiating (11.3.5) with respect to x or y, transferring the derivatives on the right-hand sides into the integrals, and then writing

$$\frac{\partial}{\partial x}\ln[(x-x')^2 + (y-y')^2] = -\frac{\partial}{\partial x'}\ln[(x-x')^2 + (y-y')^2],$$
$$\frac{\partial}{\partial y}\ln[(x-x')^2 + (y-y')^2] = -\frac{\partial}{\partial y'}\ln[(x-x')^2 + (y-y')^2], \quad (11.3.6)$$

we obtain

$$u_x(x,y) = \frac{\Omega}{4\pi}\iint_{Vortex}\frac{\partial}{\partial y'}\ln[(x-x')^2 + (y-y')^2]\,dx'\,dy',$$
$$u_y(x,y) = -\frac{\Omega}{4\pi}\iint_{Vortex}\frac{\partial}{\partial x'}\ln[(x-x')^2 + (y-y')^2]\,dx'\,dy'. \quad (11.3.7)$$

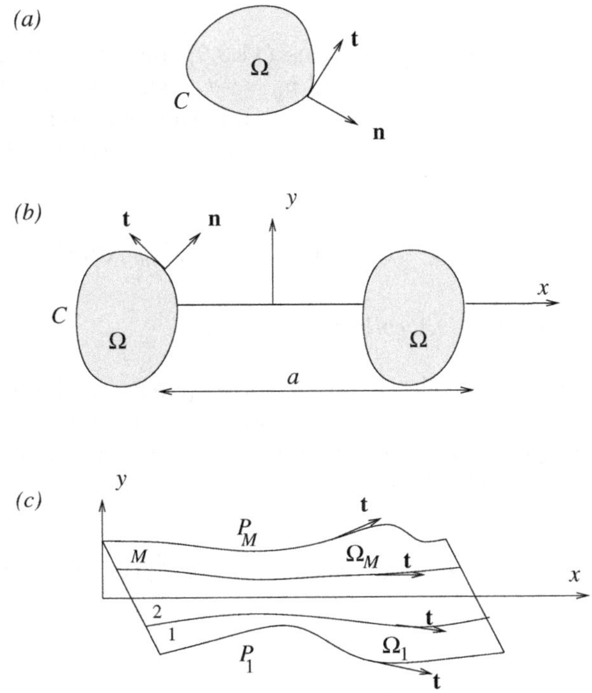

Figure 11.3.2 Patches of constant vorticity in the xy plane of a two-dimensional flow: (a) solitary, (b) periodic, and (c) in the form of periodic vortex layers.

These manipulations were motivated by our ability to convert the areal integral of the x or y derivative of a function over the region occupied by the vortex into a line integral along the boundary of the vortex. The conversion is done using the Gauss divergence theorem stated in equation (2.6.24) for an arbitrary vector function, $\mathbf{h} = (h_x, h_y)$. Setting

$$h_x = 0, \qquad h_y = \ln[(x - x')^2 + (y - y')^2], \tag{11.3.8}$$

we find

$$u_x(x, y) = \frac{\Omega}{4\pi} \oint_C \ln[(x - x')^2 + (y - y')^2] \, n_y(x', y') \, \mathrm{d}l(x', y'), \tag{11.3.9}$$

where n_y is the y component of the unit vector normal to the vortex contour C pointing outward from the vortex, and $\mathrm{d}l$ is the differential arc length along C. Working in a similar fashion with the y component of the velocity, we find

$$u_y(x, y) = -\frac{\Omega}{4\pi} \oint_C \ln[(x - x')^2 + (y - y')^2] \, n_x(x', y') \, \mathrm{d}l(x', y'). \tag{11.3.10}$$

To recast the component equations (11.3.9) and (11.3.10) into a unified vector form, we note that $n_x = t_y$ and $n_y = -t_x$, where $\mathbf{t} = (t_x, t_y)$ is the unit vector tangent to the vortex contour pointing in the counterclockwise direction, as shown in figure 11.3.2(a). The velocity induced by the patch may now be expressed in the vector form

$$\mathbf{u}(x,y) = -\frac{\Omega}{4\pi} \oint_C \ln[(x-x')^2 + (y-y')^2]\, \mathbf{t}(x',y')\, \mathrm{d}l(x',y'). \tag{11.3.11}$$

A collection of patches

If the flow contains a collection of M vortex patches with uniform vorticity Ω_l, where $l = 1, \ldots M$, the velocity field arises by integrating around the contour of each individual patch, and then summing the contributions according to the generalized version of (11.3.11)

$$\mathbf{u}(\mathbf{x},\mathbf{y}) = -\frac{1}{4\pi} \sum_{l=1}^{M} \Omega_l \oint_{C_l} \ln[(x-x')^2 + (y-y')^2]\, \mathbf{t}(x',y')\, \mathrm{d}l(x',y'), \tag{11.3.12}$$

where C_l is the contour of the lth patch.

Periodic patches

To develop a contour integral representation of the flow induced by a vortex patch that is repeated periodically in the x direction with period a, as illustrated in figure 11.3.2(b), we repeat the preceding analysis using the stream function of the flow induced by a periodic array of point vortices given in (11.2.16). Straightforward generalization provides us with the counterpart of (11.3.12) for a flow containing a collection of M periodic vortex patches,

$$\mathbf{u}(x,y) = -\frac{1}{4\pi} \sum_{l=1}^{M} \Omega_l \oint_{C_l} \ln\Big(\cosh[k(y-y')] - \cos[k(x-x')]\Big) \times \mathbf{t}(\mathbf{x}',\mathbf{y}')\, \mathrm{d}l(x',y'), \tag{11.3.13}$$

where $k = 2\pi/a$ is the common patch wave number. For the configuration depicted in figure 11.3.2(b), $M = 1$.

Periodic layers

A judicious rearrangement of (11.3.13) allows us to obtain a contour integral representation of the flow induced by a number of adjacent periodic vortex layers with constant vorticity, as illustrated in figure 11.3.2(c). To develop this representation, we identify one period of a vortex layer with a periodic patch, and note that the contour integrals over periodic segments cancel. The result is the integral representation

$$\mathbf{u}(x, y) = -\frac{1}{4\pi} \sum_{l=1}^{M} (\Omega_l - \Omega_{l+1}) \int_{P_l} \ln\Big(\cosh[k(y - y')] - \cos[k(x - x')]\Big) \times \mathbf{t}(x', y')\, \mathrm{d}l(x', y'), \qquad (11.3.14)$$

with the understanding that $\Omega_{M+1} = 0$, where P_l is one period of the lth contour, as illustrated in figure 11.3.2(c). A group of adjacent layers with gradually varying vorticity is a model of a shear layer with smoothly varying vorticity separating two streams.

11.3.2 Contour dynamics

The vorticity transport equation (6.6.7) requires that, if viscous forces are insignificant, the vorticity inside a patch with uniform vorticity remains constant in time. To compute the evolution of the flow, it suffices then to pursue the motion of the vortex contour. In numerical practice, this is done by tracing the contour with a collection of point particles, evaluating the velocity at the position of the point particles using the contour integral representation, and then advancing the position of the point particles using a numerical method for integrating ordinary differential equations.

Consider a solitary vortex patch immersed in an infinite fluid, as depicted in figure 11.3.2(a). The numerical method involves the following steps:

1. In the first step, we trace the vortex contour with $N + 1$ point particles located at $\mathbf{X}_i = (X_i, Y_i)$, where $i = 1, \ldots, N + 1$; points numbered 1 and $N + 1$ coincide.

2. In the second step, we describe the shape of the contour by interpolation. In the simplest method, the contour is approximated with the polygonal line connecting successive marker points, concisely called a polyline.

3. In the third step, we compute the velocity of each marker point by evaluating the integral on the right-hand side of (11.3.11) for $x = X_i$ and $y = Y_i$, where $i = 1, \ldots, N$.

4. In the fourth step, we compute the motion of the marker points by integrating in time the ordinary differential equations

$$\frac{\mathrm{d}X_i}{\mathrm{d}t} = u_x(X_i, Y_i), \qquad \frac{\mathrm{d}Y_i}{\mathrm{d}t} = u_y(X_i, Y_i). \qquad (11.3.15)$$

To evaluate the velocity in the third step, we replace the contour integral with the sum of integrals over the individual straight elements connecting adjacent point particles. Observing that the unit tangent vector over the jth element is constant and equal to

$$\mathbf{t}^{(j)} = \frac{1}{\Delta l_j}(\mathbf{X}_{j+1} - \mathbf{X}_j), \tag{11.3.16}$$

we write

$$\mathbf{u}(X_i, Y_i) \simeq -\frac{\Omega}{4\pi}\sum_{j=1}^{N} \mathbf{t}^{(j)} \int_{E_j} \ln[(X_i - x')^2 + (Y_i - y')^2]\, \mathrm{d}l(x', y'), \tag{11.3.17}$$

where E_j denotes the jth element, and Δl_j is the element length. To facilitate the logistics, we recast equation (11.3.17) into the compact form

$$\mathbf{u}(X_i, Y_i) = -\frac{\Omega}{2\pi}\sum_{j=1}^{N} A_j(X_i, Y_i)\,(\mathbf{X}_{j+1} - \mathbf{X}_j), \tag{11.3.18}$$

where

$$A_j(X_i, Y_i) \equiv \frac{1}{\Delta l_j}\int_{E_j} \ln[(X_i - x')^2 + (Y_i - y')^2]^{1/2}\, \mathrm{d}l(x', y'). \tag{11.3.19}$$

are dimensionless influence coefficients. If the flow contains a collection of vortex patches, the right-hand side of (11.3.18) is summed over all contours to account for all induced contributions.

Singular elements

As the integration point, (x', y'), approaches the evaluation point, (X_i, Y_i), the integrand of the contour integral diverges at a logarithmic rate, and the integrals over the elements numbered $i-1$ and i are singular. Fortunately, these integrals can be computed analytically, yielding the influence coefficients

$$A_j(X_i, Y_i) = \ln r_{i,j} - 1, \tag{11.3.20}$$

for $j = i-1$ and i, where

$$r_{i,j} \equiv [\,(X_i - X_j)^2 + (Y_i - Y_j)^2\,]^{1/2} \tag{11.3.21}$$

is the distance between the marker points (problem 11.3.1).

Regular elements

The rest of the integrals defining the influence coefficients are non-singular and can be computed by standard numerical methods. For example, choosing the trapezoidal rule, we replace the integrand in (11.3.19) with the average of the values corresponding to the end points of the integration domain, finding

$$A_j(X_i, Y_i) = \frac{1}{2}\,(\ln r_{i,j} + \ln r_{i,j+1}), \tag{11.3.22}$$

for $j = 1, \ldots, i-2, i+1, \ldots, N$. Note that we have skipped the singular elements numbered $i-1$ and i.

Periodic flow

To compute the evolution of a periodic flow, we replace equation (11.3.11) with its counterpart originating from equation (11.3.13). In the case of a vortex patch that is repeated in the x direction with period a, corresponding to $M = 1$, we find

$$\mathbf{u}(X_i, Y_i) = -\frac{\Omega}{4\pi} \sum_{j=1}^{N} \mathbf{t}^{(j)} \int_{E_j} \ln\Big(\cosh[k(Y_i - y')] - \cos[k(X_i - x')] \Big)\, dl(\mathbf{x}), \tag{11.3.23}$$

where $k = 2\pi/a$ is the wave number.

The periodic integrand in (11.3.23) exhibits a logarithmic singularity over the two elements hosting a marker point. To remove this singularity, we add and subtract the non-periodic kernel corresponding to a solitary point vortex, writing

$$\begin{aligned} &\ln\Big(\cosh[k(Y_i - y')] - \cos[k(X_i - x')] \Big) \\ &= \ln \frac{\cosh[k(Y_i - y')] - \cos[k(X_i - x')]}{(X_i - x')^2 + (Y_i - y')^2} + \ln[(X_i - x')^2 + (Y_i - y')^2]. \end{aligned} \tag{11.3.24}$$

As x' tends to X_i and y' tends to Y_i, the fraction on the right-hand side of (11.3.24) tends to a finite value. The integral of the corresponding logarithmic term may then be computed using a standard numerical method. The improper integral of the last term on the right-hand side of (11.3.24) can be computed analytically, as discussed previously in this section for non-periodic flow.

11.3.3 Gauss integration quadrature

Various modifications of the basic contour dynamic algorithm outlined in this section can be made to improve the accuracy of the numerical method.

Consider the evaluation of the influence coefficients A_j defined in equation (11.3.18) over non-singular elements. The trapezoidal rule expressed by (11.3.22) replaces the integral with a weighted average of the values of integrand at the two end points; both weights are equal to 1/2. Generalizing this approximation, we evaluate the integral by a quadrature expressing a weighted average of the integrand at craftily selected quadrature points,

$$\int_{E_j} \ln[(X_i - x')^2 + (Y_i - y')^2]^{1/2} \, dl(x', y')$$
$$\simeq \frac{\Delta l_j}{2} \sum_{k=1}^{N_Q} w_k \ln[(X_i - x_k)^2 + (Y_i - y_k)^2]^{1/2}, \tag{11.3.25}$$

where N_Q is a chosen number of quadrature points, and w_k are integration weights. The base points (x_k, y_k) lie on element E_j, and their position is given by

$$x_k = \frac{1}{2}(X_{j+1} + X_j) + \frac{1}{2}(X_{j+1} - X_j)\, t_k,$$
$$y_k = \frac{1}{2}(Y_{j+1} + Y_j) + \frac{1}{2}(Y_{j+1} - Y_j)\, t_k, \tag{11.3.26}$$

where the scaled base-point positions t_k take values in the interval $[-1, 1]$. The left extreme value $t_k = -1$ corresponds to $x_k = X_j$, $y_k = Y_j$, and the right extreme value $t_k = 1$ corresponds to $x_k = X_{j+1}$, $y_k = Y_{j+1}$. The trapezoidal rule expressed by (11.3.22) corresponds to $N_Q = 2$ with $t_1 = -1$, $t_2 = 1$, and equal weights $w_1 = w_2 = 1.0$.

Handbooks of mathematical functions provide us with tables of the optimal positioning of the base points, t_k, and corresponding weights, w_k, for a specified number of quadrature base points, N_Q, where $k = 1, \ldots, N_Q$. The base points are the zeros of a selected class of orthogonal polynomials, and the weights arise by integrating Lagrange interpolating polynomials defined in terms of the base points. For smooth integrands that do not exhibit singularities, the tables come under the header of the Gauss-Legendre quadrature. For integrands with integrable singularities, or integrals over infinite domains, the tables come under different headers.

Subdirectory *107_integration* inside directory *01_num_meth* of FDLIB includes subroutines that tabulate base points and weights. Other programs in that directory perform numerical integration.

11.3.4 Representation with circular arcs

To account for the curvature of a vortex contour, we replace the straight segments connecting adjacent marker points with circular arcs, as illustrated in figure 11.3.5. The backward arc passes through a trio of points numbered $j-1, j$, and $j+1$, and the forward arc passes through a trio of points numbered $j, j+1$, and $j+2$. Each arc is specified by the coordinates of its center and radius, computed by solving a system of three linear equations using interpolation

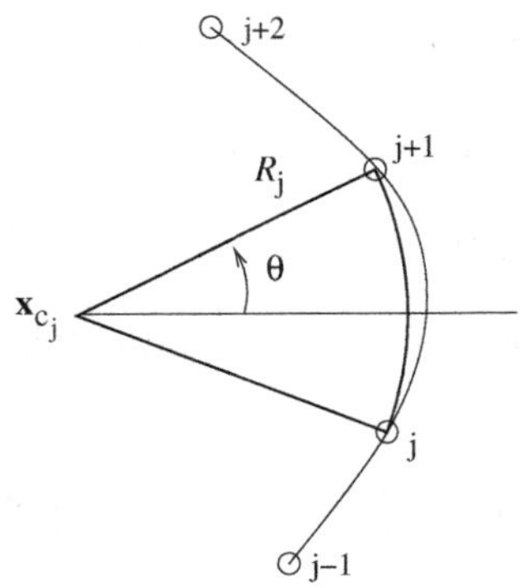

Figure 11.3.5 Local approximation of a vortex contour with a circular arc passing through three marker points.

constraints. The blended arc arises by averaging the geometrical properties of the forward and backward arc on either side of an interval.

The position of a point on the jth arc is described in parametric form by the equations

$$x = x_{c_j} + R_j \cos\theta, \qquad y = y_{c_j} + R_j \sin\theta, \tag{11.3.27}$$

where (x_{c_j}, y_{c_j}) are the coordinates of the arc center, R_j is the arc radius, and the polar angle θ varies between two limits corresponding to the arc end points. The components of the unit tangent vector are given by

$$t_x(\theta) = - \pm \sin\theta, \qquad t_y(\theta) = \pm \cos\theta, \tag{11.3.28}$$

and the differential arc length along the arc is given by $\mathrm{d}l = \pm R_j \,\mathrm{d}\theta$. The plus sign is chosen when the arc is traced in the counterclockwise direction from point j to point $j+1$, and the minus sign otherwise.

The x and y components of the integral over the j arc on the right-hand side of (11.3.11) evaluated at the point (X_i, Y_i) are given by

$$-R_j \int_{\theta_j}^{\theta_{j+1}} \ln[(X_i - x_{c_j} - R_j\cos\theta)^2 + (Y_i - y_{c_j} - R_j \sin\theta)^2] \sin\theta \,\mathrm{d}\theta,$$

$$R_j \int_{\theta_j}^{\theta_{j+1}} \ln[(X_i - x_{c_j} - R_j\cos\theta)^2 + (Y_i - y_{c_j} - R_j \sin\theta)^2] \cos\theta \,\mathrm{d}\theta. \tag{11.3.29}$$

The integrals with respect to θ on the right-hand sides can be computed using the Gauss-Legendre quadrature.

When the evaluation point, (X_i, Y_i), lies on the jth arc, the integrands in (11.3.29) exhibit a logarithmic singularity. To compute the integrals, we express the coordinates, (X_i, Y_i), in local plane polar coordinates, writing

$$X_i = x_{c_j} + R_j \cos\theta_i, \qquad Y_i = y_{c_j} + R_j \sin\theta_i. \tag{11.3.30}$$

Substituting these expressions in (11.3.29), simplifying and rearranging, we find

$$\begin{aligned} &-R_j \int_{\theta_j}^{\theta_{j+1}} \ln\{2\,R_j^2\,[1-\cos(\theta-\theta_i)]\}\, \sin\theta\,\mathrm{d}\theta, \\ &R_j \int_{\theta_j}^{\theta_{j+1}} \ln\{2\,R_j^2\,[1-\cos(\theta-\theta_i)]\}\, \cos\theta\,\mathrm{d}\theta. \end{aligned} \tag{11.3.31}$$

As the integration angle, θ, tends to the evaluation angle, θ_i, the integrands in (11.3.31) exhibit a logarithmic singularity. To remove this singularity, we manipulate the x component by writing

$$\begin{aligned} I_x &\equiv \int_{\theta_j}^{\theta_{j+1}} \ln\{2\,R_j^2\,[1-\cos(\theta-\theta_i)]\}\, \sin\theta\,\mathrm{d}\theta \\ &= \ln(2R_j^2) \int_{\theta_j}^{\theta_{j+1}} \sin\theta\, d\theta + \int_{\theta_j}^{\theta_{j+1}} \ln[1-\cos(\theta-\theta_i)]\, \sin\theta\,\mathrm{d}\theta, \end{aligned} \tag{11.3.32}$$

and then

$$\begin{aligned} I_x = &- \ln(2R_j^2)\,(\cos\theta_j - \cos\theta_{j+1}) + \int_{\theta_j}^{\theta_{j+1}} \ln\frac{1-\cos(\theta-\theta_i)}{(\theta-\theta_i)^2}\, \sin\theta\,\mathrm{d}\theta \\ &+ \int_{\theta_j}^{\theta_{j+1}} \ln(\theta-\theta_i)^2\,(\sin\theta - \sin\theta_i)\,\mathrm{d}\theta + \sin\theta_i \int_{\theta_j}^{\theta_{j+1}} \ln(\theta-\theta_i)^2\,\mathrm{d}\theta. \end{aligned} \tag{11.3.33}$$

Using the Taylor series expansion of the cosine with respect to θ about the point θ_i, we find that, as θ tends to θ_i, the fraction in the first integral on the right-hand side of (11.3.33) tends to 1/2 (problem 11.3.2). Thus, the integral is nonsingular and can be computed by a standard numerical method. The second integral on the right-hand side is also nonsingular. The singularity has been moved to the third integral, which can be evaluated by analytical methods. The y component expressed by the second of equations (11.3.31) is manipulated in a similar fashion.

An alternative method of computing the singular integrals in (11.3.31) employs a quadrature that is specifically designed for integrals with a logarithmic singularity, as discussed in texts on numerical methods cited in the bibliography.

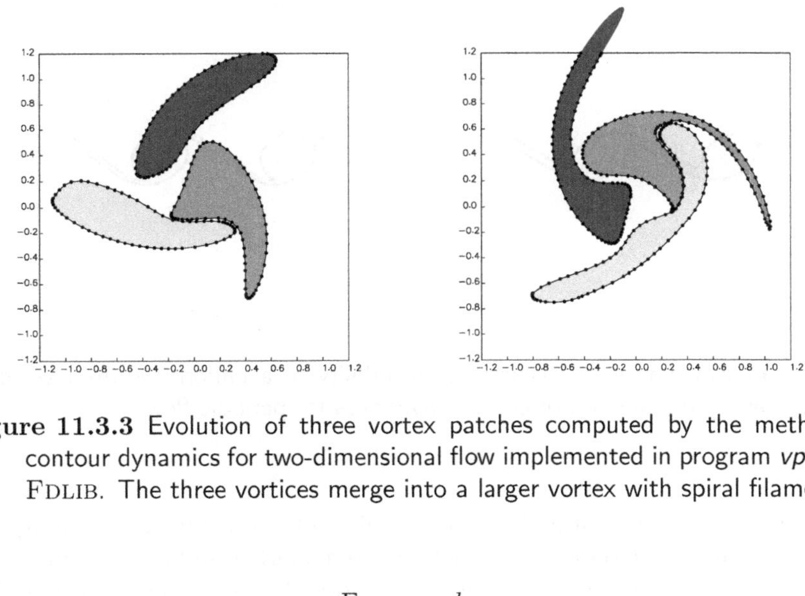

Figure 11.3.3 Evolution of three vortex patches computed by the method of contour dynamics for two-dimensional flow implemented in program *vp_2d* of FDLIB. The three vortices merge into a larger vortex with spiral filaments.

FDLIB *codes*

Program *vp_2d* inside directory *09_vortex* of FDLIBcomputes the evolution of a collection of vortex patches, where the contour of each patch is approximated with a collection of blended circular arcs. The marker points are redistributed adaptively during the motion to capture the development of regions of large curvature and prevent point clustering and dilution. Figure 11.3.3 illustrates stages in the evolution of three vortex patches in close proximity. The dots around the contours mark adaptively redistributed marker points. The results reveal spontaneous vortex merger under the influence of the mutually induced velocity.

Program *vp_2d_pr* inside directory *09_vortex* FDLIB computes the evolution of a collection of adjacent vortex layers. The numerical method is similar to that described previously for vortex patches. Figure 11.3.4 illustrates stages in

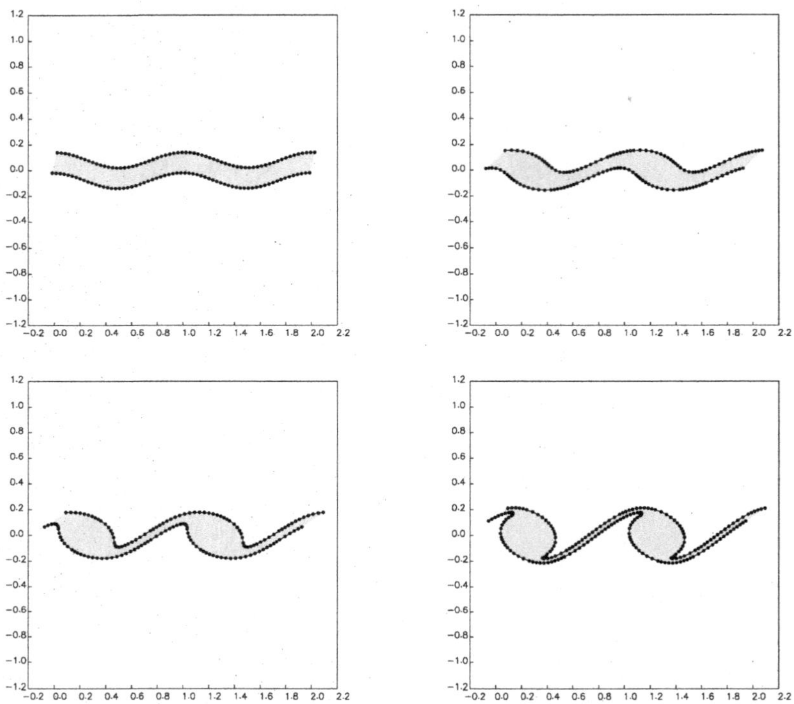

Figure 11.3.4 The Kelvin-Helmholtz instability of a uniform vortex layer computed by the method of contour dynamics for periodic flow.

the Kelvin-Helmholtz instability of a periodic vortex layer separating a uniform stream moving to the right above the layer from another uniform stream moving to the left below the layer. The initially sinusoidal vortex contours roll up into a periodic sequence of compact vortices connected by thinning braids. In the final stages of the motion, the vortex layer transforms into a periodic array of billows.

Problems

11.3.1 *Influence coefficient for a singular element*

(*a*) Derive the influence coefficients for a singular element shown in (11.3.20). *Hint:* Work in local Cartesian coordinates where the x axis is tangential to the singular element.

(*b*) Compute in analytical form the last integral on the right-hand side of (11.3.33).

(*c*) Derive the counterpart of (11.3.33) for the y component of the induced velocity.

11.3.2 *Periodic kernel*

Evaluate the limit of the fraction after the logarithm on the right-hand side of (11.3.33) as the integration point tends to the evaluation point.

Computer Problem

11.3.3 *Motion of vortex patches and vortex layers*

(*a*) Run the code *vp_2d* to simulate the evolution of a vortex arrangement of your choice. Discuss the nature of the motion.

(*b*) Run the code *vp_2d_pr* to simulate the motion of a deformed vortex layer. Discuss the nature of the motion.

11.4 Vorticity and circulation in three-dimensional flow

The circulation around a loop in a three-dimensional flow is defined as the line integral of the tangential component of the velocity around the loop, L,

$$C \equiv \oint_L u_t \, \mathrm{d}l = \oint_L (u_x \, t_x + u_y \, t_y + u_z \, t_z) \, \mathrm{d}l = \oint_L \mathbf{u} \cdot \mathbf{t} \, \mathrm{d}l, \qquad (11.4.1)$$

where $\mathbf{t} = (t_x, t_y, t_z)$ is the unit vector tangent to the loop, and

$$\mathrm{d}l = (\mathrm{d}x^2 + \mathrm{d}y^2 + \mathrm{d}z^2)^{1/2} \qquad (11.4.2)$$

is the differential arc length around the loop measured from an arbitrary point, as depicted in figure 11.4.1. Th definition (11.4.1) is a generalization of that stated in equation (11.4.1) for two-dimensional flow.

If the loop is reducible, meaning that it can be shrunk to a point without crossing flow boundaries or singular lines, we may use the Stokes circulation theorem to express the circulation around the loop as an integral of the component of the vorticity vector normal to any three-dimensional surface patch D bounded by the loop,

$$C = \iint_D \boldsymbol{\omega} \cdot \mathbf{n} \, \mathrm{d}S, \qquad (11.4.3)$$

where $\mathrm{d}S$ is the differential surface area of D. The orientation of the unit normal vector, $\mathbf{n}$, is such that, as we view the patch from the positive direction of the normal vector, the unit tangent vector $\mathbf{t}$ points in a direction corresponding to counterclockwise rotation.

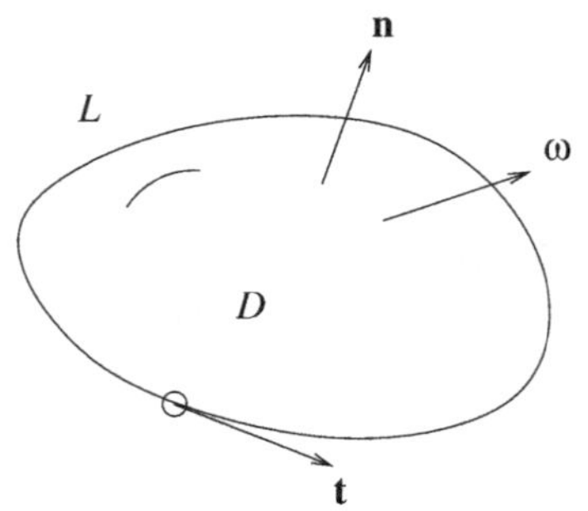

Figure 11.4.1 Illustration of a loop in a three-dimensional flow, L, enclosing a patch, D. The circulation around the loop is equal to the surface integral of the normal component of the vorticity, $\boldsymbol{\omega} \cdot \mathbf{n}$, over D.

In the case of two-dimensional flow in the xy plane, the loop L and surface D enclosed by L may be chosen to lie in the xy plane. The unit normal vector $\mathbf{n}$ is then parallel to the z axis, and expression (11.4.3) reduces to (11.1.2) where $\boldsymbol{\omega} = \omega_z \, \mathbf{n}$.

11.4.1 Preservation of circulation

To compute the rate of change of circulation around a material loop consisting of a fixed collection of point particles with permanent identity, we take the time derivative of the definition (11.4.1) and find

$$\frac{\mathrm{d}C}{\mathrm{d}t} = \frac{\mathrm{d}}{\mathrm{d}t} \oint_L \mathbf{u} \cdot \mathbf{t} \, \mathrm{d}l = \oint_L \frac{\mathrm{D}\mathbf{u}}{\mathrm{D}t} \cdot \mathbf{t} \, \mathrm{d}l + \oint_L \mathbf{u} \cdot \frac{\mathrm{D}(\mathbf{t} \, \mathrm{d}l)}{\mathrm{D}t}, \tag{11.4.4}$$

where $\mathrm{D}/\mathrm{D}t$ is the material derivative.

Focusing on the second integral on the right-hand side of (11.4.4), we express it in the form

$$\oint_L \mathbf{u} \cdot \frac{\mathrm{D}(\mathbf{t} \, \mathrm{d}l)}{\mathrm{D}t} = \oint_L \mathbf{u} \cdot \frac{\mathrm{D}\mathrm{d}\mathbf{X}}{\mathrm{D}t}. \tag{11.4.5}$$

The material derivative $\mathrm{D}(\mathrm{d}\mathbf{X})/\mathrm{D}t$ expresses the rate of change of the components of an infinitesimal material vector beginning at a certain point particle and ending at another point particle.

If the two point particles move with the same velocity, the material vector will simply translate, and $\mathrm{D}(\mathrm{d}\mathbf{X})/\mathrm{D}t = \mathbf{0}$. This observation suggests that $\mathrm{D}(\mathrm{d}\mathbf{X})/\mathrm{D}t$ is proportional to the local rate of change of the velocity with respect to arc length along the material vector. Using Taylor series expansions, we find

$$\frac{\mathrm{D}\,\mathrm{d}X}{\mathrm{D}t} = \mathrm{d}X\,\frac{\partial u_x}{\partial x} + \mathrm{d}Y\,\frac{\partial u_x}{\partial y} + \mathrm{d}Z\,\frac{\partial u_x}{\partial z},$$

$$\frac{\mathrm{D}\,\mathrm{d}Y}{Dt} = \mathrm{d}X\,\frac{\partial u_y}{\partial x} + \mathrm{d}Y\,\frac{\partial u_y}{\partial y} + \mathrm{d}Z\,\frac{\partial u_y}{\partial z}, \tag{11.4.6}$$

$$\frac{\mathrm{D}\,\mathrm{d}Z}{\mathrm{D}t} = \mathrm{d}X\,\frac{\partial u_z}{\partial x} + \mathrm{d}Y\,\frac{\partial u_z}{\partial y} + \mathrm{d}Z\,\frac{\partial u_z}{\partial z}.$$

Considering the fist term in the inner product on the right-hand side of (11.4.5), we write

$$u_x\,\frac{\mathrm{D}\mathrm{d}X}{\mathrm{D}t} = \mathrm{d}X\,u_x\,\frac{\partial u_x}{\partial x} + \mathrm{d}Y\,u_x\,\frac{\partial u_x}{\partial y} + \mathrm{d}Z\,u_x\,\frac{\partial u_x}{\partial z}$$

$$= \mathrm{d}X\,\frac{1}{2}\,\frac{\partial u_x^2}{\partial x} + \mathrm{d}Y\,\frac{1}{2}\,\frac{\partial u_x^2}{\partial y} + \mathrm{d}Z\,\frac{1}{2}\,\frac{\partial u_x^2}{\partial z} = \frac{1}{2}\,\frac{\mathrm{d}u_x^2}{\mathrm{d}l}. \tag{11.4.7}$$

Because the right-hand side of (11.4.7) is an exact differential, its line integral over the closed loop is zero. Working in a similar fashion with the projections of the y and z components, we find that the second integral on the right-hand side of (11.4.4) is also zero, yielding

$$\frac{\mathrm{d}C}{\mathrm{d}t} = \oint_L \frac{\mathrm{D}\mathbf{u}}{\mathrm{D}t}\cdot\mathbf{t}\,\mathrm{d}l. \tag{11.4.8}$$

We have found that the rate of change of the circulation around a material loop is equal to the instantaneous circulation of the acceleration field around the loop.

Kelvin circulation theorem

If viscous forces are negligible, we may use Euler's equation (6.4.2) to recast the integral on the right-hand side of (11.4.8) into the form

$$\oint_L \frac{\mathrm{D}\mathbf{u}}{\mathrm{D}t}\cdot\mathbf{t}\,\mathrm{d}l = \oint_L \Big(-\frac{1}{\rho}\,\nabla p + \mathbf{g}\Big)\cdot\mathbf{t}\,\mathrm{d}l. \tag{11.4.9}$$

The unit tangent vector is

$$\mathbf{t} = \frac{1}{\mathrm{d}l}\mathrm{d}\mathbf{X}, \tag{11.4.10}$$

where $\mathbf{X}$ is the position of a point particle around the loop. Using this expression and assuming that the fluid density is uniform throughout the domain of flow, we find

$$\oint_L \frac{\mathrm{D}\mathbf{u}}{\mathrm{D}t}\cdot\mathbf{t}\,\mathrm{d}l = -\frac{1}{\rho}\oint_L \mathrm{d}\mathbf{X}\cdot\nabla p + \mathbf{g}\cdot\oint_L \mathrm{d}\mathbf{X} = -\frac{1}{\rho}\oint_L \mathrm{d}p + \mathbf{g}\cdot\oint_L \mathrm{d}\mathbf{X}. \tag{11.4.11}$$

Because the loop is closed, the domain of integration with respect to $\mathbf{X}$ is periodic. Since the two integrands on the right-hand side of (11.4.11) are exact differentials, their integrals vanish yielding Kelvin's circulation theorem expressed by the conservation law

$$\frac{\mathrm{d}C}{\mathrm{d}t} = 0. \tag{11.4.12}$$

When viscous forces are negligible, the circulation around any closed material loop remains constant in time.

11.4.2 Flow induced by vorticity

Given the velocity field, we may compute the associated vorticity field using the definition of the vorticity by analytical or numerical differentiation,

$$\boldsymbol{\omega} \equiv \nabla \times \mathbf{u}. \tag{11.4.13}$$

Is there a way of doing the inverse, that is, expressing the velocity in terms of the vorticity?

In the case of two-dimensional flow, the velocity field associated with a specified vorticity distribution in the xy plane arises from the integral representation (11.3.2). The corresponding formula for three-dimensional flow is expressed by the Biot-Savart integral

$$\mathbf{u}(x,y,z) = -\frac{1}{4\pi}\iiint_{Flow} \frac{(\mathbf{x}-\mathbf{x}')\times\boldsymbol{\omega}(x',y',z')}{[(x-x')^2+(y-y')^2+(z-z')^2]^{3/2}}\,\mathrm{d}x'\mathrm{d}y'\mathrm{d}z'. \tag{11.4.14}$$

In index notation,

$$u_i(x,y,z) = -\frac{1}{4\pi}\iiint_{Flow} \frac{\epsilon_{ijk}\,(x_j-x_j')\,\omega_k(x',y',z')}{[(x-x')^2+(y-y')^2+(z-z')^2]^{3/2}}\,\mathrm{d}x'\,\mathrm{d}y'\,\mathrm{d}z'. \tag{11.4.15}$$

The second set of parentheses in the numerators on the right-hand side of equations (11.4.14) and (11.4.15) enclose the arguments of the vorticity.

The numerator of the integrand in (11.4.14) consists of the cross product of (a) the vectorial distance of the evaluation point, $\mathbf{x} = (x,y,z)$, from the integration point, $\mathbf{x}' = (x',y',z')$, and (b) the vorticity. The denominator is the cubic power of the scalar distance between the evaluation point and the integration point, $|\mathbf{x}-\mathbf{x}'|^3$. Thus, far from the integration point, $\mathbf{x}'$, the integrand decays like $1/r^2$, where

$$r = [(x-x')^2+(y-y')^2+(z-z')^2]^{1/2}. \tag{11.4.16}$$

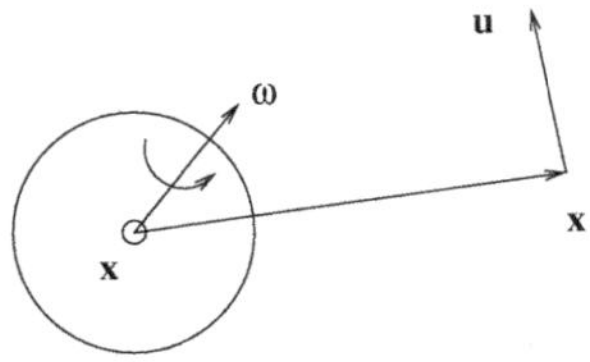

Figure 11.4.2 The flow induced by the rotation of a small fluid parcel is expressed by the Biot-Savart integral shown in (11.4.15).

The physical interpretation of (11.4.14) becomes evident by replacing the integral with a sum of integrals over the volumes of elementary fluid parcels. The velocity induced by each individual parcel due to its rotation is illustrated in figure 11.4.2. A direct analogy with the magnetic field induced by an electrical current explains why the integral on the right-hand side of (11.4.14) is known as the Biot-Savart integral of vortex dynamics.

The integral representation (11.4.14) is applicable only for unbounded flow extending to infinity in all directions and decaying at infinity. In the presence of boundaries, an additional complementary flow must be included to ensure the solenoidality of the velocity field and the satisfaction of the no-penetration and no-slip boundary conditions.

Problem

11.4.1 *Change of circulation*

Assume that the viscous force in a fluid is $\mu \nabla^2 \mathbf{u} = \kappa \mathbf{u}$, where κ is a resistance coefficient. What are the dimensions of κ? Derive an equation for the rate of change of circulation around a material loop.

11.5 Axisymmetric flow induced by vorticity

The vorticity of an axisymmetric vortex flow without swirling motion points in the direction of the meridional angle, φ, as illustrated in figure 11.5.1. The known orientation of the vorticity field and accompanying axial symmetry of the velocity field allow us to simplify the volume integral on the right-hand side of (11.4.14) and thereby derive representations that are amenable to analytical and numerical computation.

In the cylindrical polar coordinates depicted in figure 11.5.1, the vorticity vector takes the form $\boldsymbol{\omega} = \omega_\varphi(x, \sigma)\, \mathbf{e}_\varphi$, where ω_φ is the scalar strength of the vorticity, and $\mathbf{e}_\varphi$ is the unit vector pointing in the direction of the meridional

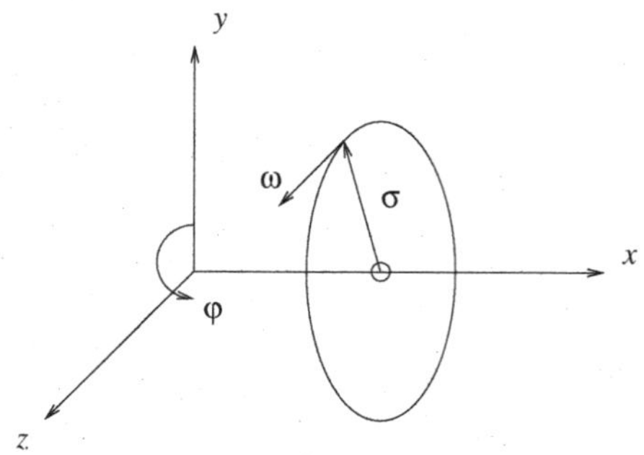

Figure 11.5.1 Illustration of an axisymmetric flow without swirling motion. The vorticity vector points in the direction of the meridional angle, φ.

angle, φ. The Cartesian components of the vorticity are

$$\omega_x = 0, \qquad \omega_y = -\omega_\varphi \sin\varphi, \qquad \omega_z = \omega_\varphi \cos\varphi. \tag{11.5.1}$$

Hill's spherical vortex

Hill's spherical vortex is a paradigm of an axisymmetric vortex patch with distributed vorticity. Inside Hill's vortex, the strength of the vorticity is proportional to the distance from the x axis,

$$\omega_\varphi = \Omega\,\sigma, \tag{11.5.2}$$

where Ω is a constant. Outside Hill's vortex, the flow is irrotational. The streamline pattern in a stationary frame of reference and in a frame of reference translating with the vortex are shown in figure 11.5.2.

In the moving frame of reference corresponding to figure 11.5.2(b), the axisymmetric stream function of the flow inside Hill's vortex is given by

$$\psi^{in} = \frac{1}{10}\,\Omega\,\sigma^2\,(a^2 - x^2 - \sigma^2). \tag{11.5.3}$$

Outside the vortex, the stream function describes irrotational flow past a sphere,

$$\psi^{out} = -\frac{1}{15}\,\Omega\,a^2\,\sigma^2\,(1 - \frac{a^3}{(x^2+\sigma^2)^{3/2}}). \tag{11.5.4}$$

Accordingly, Hill's vortex shown in figure 11.5.2(a), translates along the x axis with velocity

$$U = \frac{2}{15}\,\Omega\,a^2. \tag{11.5.5}$$

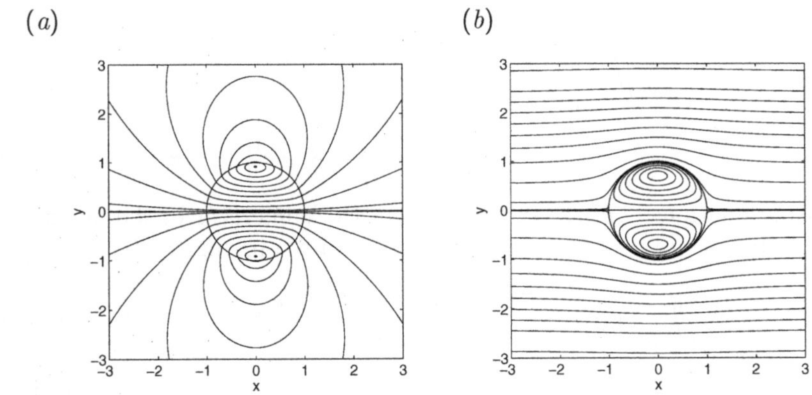

Figure 11.5.2 Streamline pattern associated with Hill's spherical vortex, (*a*) in a stationary frame of reference, and (*b*) in a frame of reference translating with the vortex.

Hill's vortex is the extreme member of a family of vortex rings arising in the limit as the core of the rings spreads out and the ring contour touches and then extends over the x axis. The opposite extreme member in this family is a line vortex ring with a core of infinitesimal radius discussed later in this section.

11.5.1 Biot-Savart integral for axisymmetric flow

Referring to the integral representation (11.4.14), we express the y and z coordinates of the evaluation point, $\mathbf{x}$, and integration point, $\mathbf{x}'$, in terms of their cylindrical polar coordinates,

$$y = \sigma\,\cos\varphi, \qquad z = \sigma\,\sin\varphi, \qquad y' = \sigma'\,\cos\varphi', \qquad z' = \sigma'\,\sin\varphi'. \tag{11.5.6}$$

The square of the distance between these two points is

$$\begin{aligned} r^2 &\equiv (x - x')^2 + (y - y')^2 + (z - z')^2 \\ &= (x - x')^2 + (\sigma\,\cos\varphi - \sigma'\,\cos\varphi')^2 + (\sigma\,\sin\varphi - \sigma'\,\sin\varphi')^2. \end{aligned} \tag{11.5.7}$$

Expanding the squares and using elementary trigonometric identities, we find

$$\begin{aligned} r^2 &= (x - x')^2 + (\sigma + \sigma')^2 - 2\sigma\sigma'[1 + \cos(\varphi - \varphi')] \\ &= (x - x')^2 + (\sigma + \sigma')^2 - 4\,\sigma\sigma'\,\cos^2\frac{\hat{\varphi}}{2}, \end{aligned} \tag{11.5.8}$$

where $\hat{\varphi} \equiv \varphi - \varphi'$.

Next, we consider the x component of (11.4.15), and express the differential volume $\mathrm{d}x'\,\mathrm{d}y'\,\mathrm{d}z'$ in the alternative form $\mathrm{d}x'\,\mathrm{d}\sigma'\,\sigma'\,\mathrm{d}\varphi'$ to obtain

$$u_x(x,\sigma) = -\frac{1}{4\pi}\iint_{Flow}\Big[\int_0^{2\pi}\frac{1}{r^3}\,[(y-y')\,\omega_z(x',y',z')$$

$$-(z-z')\,\omega_y(x',y',z')]\,\sigma'\,\mathrm{d}\varphi'\Big]\mathrm{d}x'\,\mathrm{d}\sigma'. \qquad (11.5.9)$$

Substituting relations (11.5.1) and (11.5.6) in the numerator of the integrand, simplifying by using trigonometric identities and rearranging, we find

$$u_x(x,\sigma) = \frac{1}{4\pi}\iint_{Flow}[\int_0^{2\pi}\frac{-\sigma\cos\hat{\varphi}+\sigma'}{r^3}\,\mathrm{d}\hat{\varphi}\,]\,\omega_\varphi(x',\sigma')\,\sigma'\,\mathrm{d}x'\,\mathrm{d}\sigma'. \quad (11.5.10)$$

To compute the σ component of the velocity, we work in a similar fashion departing from equation $u_\sigma = u_y\,\cos\varphi + u_z\,\sin\varphi$. The final result is

$$u_\sigma(x,\sigma) = \frac{1}{4\pi}\iint_{Flow}[\int_0^{2\pi}\frac{(x-x')\cos\hat{\varphi}}{r^3}\,\mathrm{d}\hat{\varphi}]\,\omega_\varphi(x',\sigma')\,\sigma'\,\mathrm{d}x'\,\mathrm{d}\sigma'. \quad (11.5.11)$$

To simplify the notation, we recast equations (11.5.10) and (11.5.11) into the forms

$$u_x(x,\sigma) = \frac{1}{4\pi}\iint_{Flow}[-\sigma\,I_{31}(x-x',\sigma,\sigma')$$

$$+\sigma'\,I_{30}(x-x',\sigma,\sigma')]\,\omega_\varphi(x',\sigma')\,\sigma'\,\mathrm{d}x'\,\mathrm{d}\sigma',$$

$$u_\sigma(x,\sigma) = \frac{1}{4\pi}\iint_{Flow}(x-x')\,I_{31}(x-x',\sigma,\sigma')\,\omega_\varphi(x',\sigma')\,\sigma'\,\mathrm{d}x'\,\mathrm{d}\sigma'.$$

$$(11.5.12)$$

We have introduced the integrals

$$I_{30}(x-x',\sigma,\sigma') \equiv \int_0^{2\pi}\frac{\mathrm{d}\hat{\varphi}}{r^3} = \int_0^{2\pi}\frac{\mathrm{d}\hat{\varphi}}{[(x-x')^2+(\sigma+\sigma')^2-4\,\sigma\,\sigma'\,\cos^2\frac{\hat{\varphi}}{2}]^{3/2}}$$

$$= \frac{4}{[(x-x')^2+(\sigma+\sigma')^2]^{3/2}}\int_0^{\pi/2}\frac{\mathrm{d}\eta}{(1-k^2\,\cos^2\eta)^{3/2}}, \qquad (11.5.13)$$

and

$$I_{31}(x-x',\sigma,\sigma') \equiv \int_0^{2\pi}\frac{\cos\hat{\varphi}}{r^3}\,\mathrm{d}\hat{\varphi}$$

$$= \int_0^{2\pi}\frac{\cos\hat{\varphi}}{[(x-x')^2+(\sigma+\sigma')^2-4\,\sigma\,\sigma'\,\cos^2\frac{\hat{\varphi}}{2}]^{3/2}}\,\mathrm{d}\hat{\varphi}$$

$$= \frac{4}{[(x-x')^2+(\sigma+\sigma')^2]^{3/2}}\int_0^{\pi/2}\frac{\cos(2\eta)}{(1-k^2\,\cos^2\eta)^{3/2}}\,\mathrm{d}\eta, \qquad (11.5.14)$$

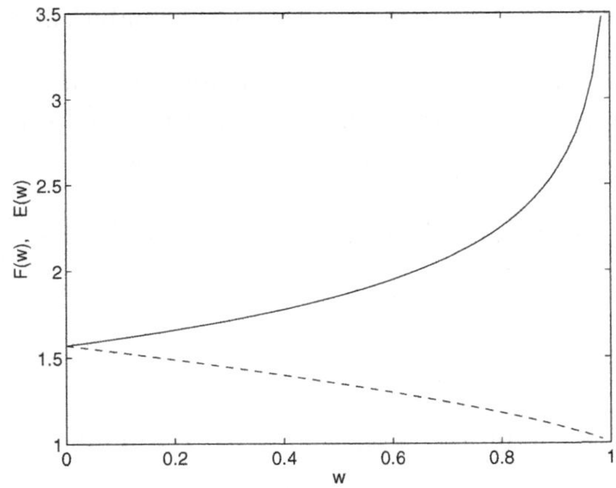

Figure 11.5.3 Graphs of the complete elliptic integral of the first kind, $F(k)$ (solid line), and second kind, $E(k)$ (dashed line).

where $\eta \equiv \hat{\varphi}/2$, and

$$k^2 \equiv \frac{4\,\sigma\,\sigma'}{(x-x')^2+(\sigma+\sigma')^2}. \tag{11.5.15}$$

Complete elliptic integrals

The integrals on the right-hand side of (11.5.13) and (11.5.14) can be expressed in terms of complete elliptic integrals of the first and second kind denoted, F and E, defined as

$$F(k) \equiv \int_0^{\pi/2} \frac{\mathrm{d}\eta}{(1-k^2\sin^2\eta)^{1/2}}, \qquad E(k) \equiv \int_0^{\pi/2} (1-k^2\sin^2\eta)^{1/2}\mathrm{d}\eta. \tag{11.5.16}$$

Graphs of the complete elliptic integrals over their domain of definition are shown in figure 11.5.3. Note that $F(k)$ diverges as k tends to unity.

Referring to standard tables of integrals, we find

$$\int_0^{\pi/2} \frac{\mathrm{d}\eta}{(1-k^2\,\cos^2\eta)^{3/2}} = \frac{E(k)}{1-k^2}, \tag{11.5.17}$$

and

$$\int_0^{\pi/2} \frac{\cos(2\eta)}{(1-k^2\,\cos^2\eta)^{3/2}}\,\mathrm{d}\eta = -\frac{2}{k^2}\,F(k) + \frac{2-k^2}{k^2\,(1-k^2)}\,E(k). \tag{11.5.18}$$

The derivation of these expressions concludes the analytical part of our work.

The complete elliptic integrals can be computed efficiently by numerical approximation or iteration. An efficient method is based on the following expressions involving rapidly converging infinite products,

$$F(k) = \frac{\pi}{2}(1+K_1)(1+K_2)\ldots, \qquad E(k) = F(k)\,(1 - \frac{k^2}{2}\,P), \qquad (11.5.19)$$

where

$$P = 1 + \frac{K_1}{2}[1 + \frac{K_2}{2}(1 + \frac{K_3}{3}\ldots)]. \qquad (11.5.20)$$

The sequence K_p is computed using the recursive formula

$$K_p = \frac{1-(1-K_{p-1}^2)^{1/2}}{1+(1-K_{p-1}^2)^{1/2}}, \qquad (11.5.21)$$

beginning with $K_0 = k$. MATLAB computes complete elliptic integrals using the function *ellipke*.

11.5.2 Line vortex ring

Now we are in a position to evaluate the velocity field associated with a specified axisymmetric vorticity distribution based on the integral representation (11.5.12).

The simplest vorticity distribution is associated with a line vortex ring, which is the axisymmetric counterpart of the point vortex in two-dimensional flow. The strength of the vorticity in a meridional plane is expressed by the generalized distribution

$$\omega_\varphi(x,\sigma) = \kappa\,\delta_2(x - x_r, \sigma - \sigma_r), \qquad (11.5.22)$$

where δ_2 is the two-dimensional delta function, κ is the strength of the ring, σ_r is the ring radius, and x_r is the axial position of the ring. The strength of the ring is equal to the circulation around any irreducible loop that encloses the trace of the ring in a meridional plane.

Substituting (11.5.22) in expressions (11.5.12) and using the distinctive properties of the delta function discussed in section 11.2 to evaluate the integrals, we find

$$u_x(x,\sigma) = \frac{\kappa\,\sigma_r}{4\pi}\,[-\sigma_r\,I_{31}(x - x_r,\sigma,\sigma_r) + \sigma'\,I_{30}(x - x_r,\sigma,\sigma_r)],$$

$$u_\sigma(x,\sigma) = \frac{\kappa\,\sigma_r}{4\pi}(x - x_r)\,I_{31}(x - x_r,\sigma,\sigma_r). \qquad (11.5.23)$$

The streamline pattern of the flow induced by the ring is shown in figure 11.5.4(*a*).

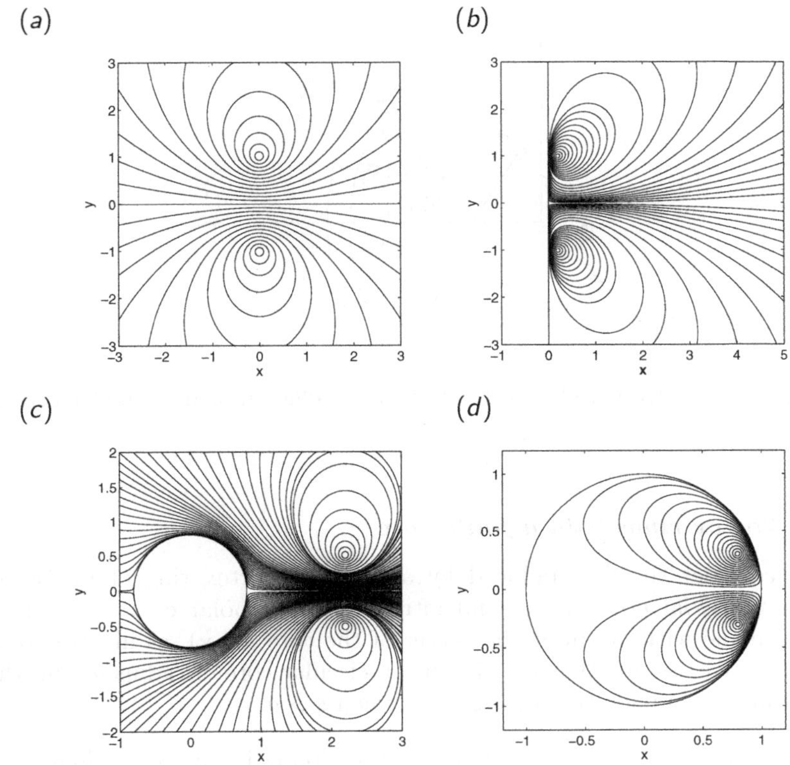

Figure 11.5.4 Streamline pattern of the flow induced by a line vortex ring in a meridional plane for (a) an unbounded domain of flow, (a) a semi-infinite domain of flow bounded by a plane wall, (c) outside a sphere, and (d) inside a sphere. These streamline patterns were produced by program *strml* residing in directory *04_various* of FDLIB.

Effect of boundaries

In the presence of axisymmetric boundaries, the velocity field induced by a vortex ring expressed by (11.5.23) must be enhanced with a complementary velocity field that ensures the satisfaction of the no-penetration boundary condition. For simple boundary geometries involving one or two parallel walls, the complementary flow can be expressed in terms of image vortex rings. Programs that evaluate the velocity field are can be found in directory *lvr* inside directory *09_vortex* of FDLIB.

Figure 11.5.4$(b$–$d)$ illustrates the streamline pattern in a meridional plane of the axisymmetric flow induced by a vortex ring in front of a plane wall and in the exterior or interior of a sphere.

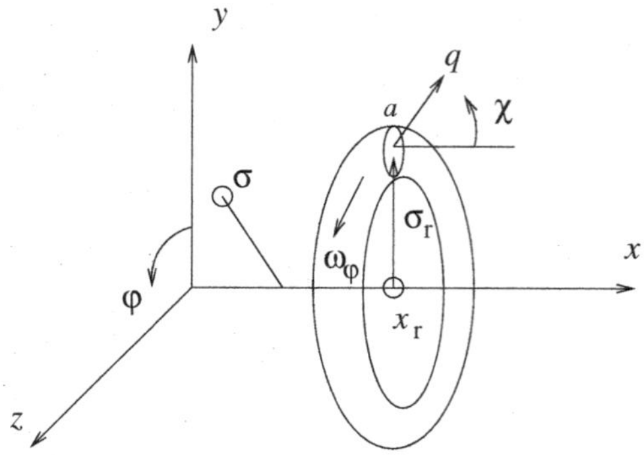

Figure 11.5.5 Illustration of a vortex ring with radius σ_r and a circular core of radius a.

11.5.3 Vortex rings with a finite core

Next, we consider the flow induced by a compact vortex ring of radius σ_r located at the axial position x_r, and introduce plane polar coordinates in a meridional plane with origin at the center of the core, (q, χ), as illustrated in figure 11.5.5. We assume that the ring has a circular core of radius a, and the vorticity vanishes outside the core, $\omega_\varphi = 0$ when $q > a$.

Using expression (11.4.3), we find that the circulation around a loop that encloses the core is

$$C = \iint_{core} \omega_\varphi(x', \sigma')\, \mathrm{d}x'\, \mathrm{d}\sigma' = \int_0^{2\pi} \int_0^a \omega_\varphi(x', \sigma')\, q\, \mathrm{d}q\, \mathrm{d}\chi. \qquad (11.5.24)$$

Our earlier discussion allows us to identify the circulation with the strength of the vortex ring, κ.

Resorting to the integral representation for the velocity given in (11.5.12), we perform the integration in the centerline plane polar coordinates, (q, χ), and find

$$u_x(x, \sigma) = \frac{1}{4\pi} \int_0^{2\pi} \int_0^a (-\sigma\, I_{31} + \sigma'\, I_{30})\, \omega_\varphi(x', \sigma')\, \sigma'\, q\, \mathrm{d}q\, \mathrm{d}\chi,$$

$$u_\sigma(x, \sigma) = \frac{1}{4\pi} \int_0^{2\pi} \int_0^a (x - x')\, I_{31}\, \omega_\varphi(x', \sigma')\, \sigma'\, q\, \mathrm{d}q\, \mathrm{d}\chi. \qquad (11.5.25)$$

Note that, because of the singular behavior of the complete elliptic integral $F(k)$, the integrands in (11.5.25) exhibit a logarithmic singularity when the evaluation point (x, σ) lies inside the core.

The simplest way of evaluating the double integrals on the right-hand sides of (11.5.25) is by the double-trapezoidal rule. The numerical method involves

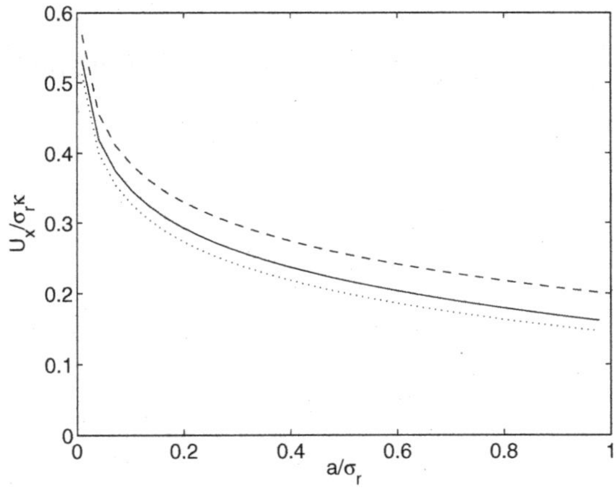

Figure 11.5.6 Self-induced velocity of a vortex ring with radius σ_r and circulation κ, plotted against the ratio of the core to the ring radius, a/σ_r. The solid and dashed lines correspond, respectively, to a ring with uniform or diffuse vorticity distribution described by equations (11.5.26) and (11.5.27). The dotted line represents Helmholtz's asymptotic prediction for small core radius expressed by equation (11.5.28).

dividing the integration domain with respect to q and χ into evenly spaced intervals defining circular sectors, approximating the integrand over each sector with the value at the center-point, and adding all contributions.

Self-induced velocity

The self-induced velocity of a vortex ring propagating along the x axis with velocity U_x, can be identified with the x component of the fluid velocity at the center of the vortex core. Thus, the ring velocity can be computed by evaluating the first expression in (11.5.25) for $x = x_r$ and $\sigma = \sigma_r$.

Figure 11.5.6 shows a graph of the dimensionless self-induced velocity, $U_x \sigma_r/\kappa$, plotted against the reduced core radius a/σ_r, evaluated using the double-trapezoidal rule. The computation neglects the weak logarithmic singularity of the integrand. The solid line corresponds to a core with uniform vorticity

$$\omega_\varphi = \frac{\kappa}{\pi a^2}, \tag{11.5.26}$$

and the dashed line corresponds to a core with distributed vorticity described by

$$\omega_\varphi = \frac{\pi\kappa}{(\pi^2 - 4)a^2}\,(1 + \cos\frac{\pi q}{a}), \tag{11.5.27}$$

for $0 < q < a$, and $\omega_\varphi = 0$ for $q > a$.

The numerical method is implemented in the following program *ring* residing inside directory *09_vortex* of FDLIB:

```
rr = 1.0;    % ring radius
circ = 1.0;    % circulation
nrad = 64;   %radial divisions
nchi = 64;   % azimuthal divisions
nrepeat = 32;

%---
% prepare
%---

dchi = 2*pi/nchi;    % angular divisions

%===========
for repeat=1:nrepeat
%===========

cr = 0.01+(repeat-1.0)/nrepeat;
dcr = cr/nrad;    % radial divisions

%-----------------------------------------
% discretize the vortex core into elements
% and compute the area of each element:  ar
%-----------------------------------------

for i=1:nrad
  for j=1:nchi
    ar(i,j) = 0.5*dchi*dcr*dcr*(2.0*i-1);
  end
end

%-----------
% initialize
%-----------

arintv1 = 0.0;    % areal integral of the vorticity
arintv2 = 0.0;

u1 = 0.0; v1 = 0.0;
u2 = 0.0; v2 = 0.0;

%----------------------
% compute the self-induced velocity
% integrate by the trapezoidal rule
%----------------------
```

```
for i=1:nrad   % loop over radial elements

  rad = (i-0.5)*dcr;   % center of the ring elements
  % uniform core:
  w1 = circ/(pi*cr*cr);
  % normalized bell shape:
  w2 = circ*pi/(cr*cr)/(pi*pi-4.0)*(1.0+cos(pi*rad/cr));

  for j=1:nchi   % loop over azimuthal elements
    chi = (j-0.5)*dchi;
    sig = rr+rad*sin(chi);
    Dx = rad*cos(chi);
    Dxs = Dx*Dx;
    ks = 4.0*rr*sig/(Dxs+(rr+sig)^2);
    [F, E]= ellipke(ks);
    RJ30 = E/(1.0-ks);
    RJ31 = (-2.0*F + E*(2.0-ks)/(1.0-ks))/ks;
    cf = 4.0/sqrt((Dxs+(rr+sig)^2)^3);
    RI30 = cf*RJ30;
    RI31 = cf*RJ31;
    cf1 = ar(i,j)*sig*w1/(4*pi);
    u1 = u1 + cf1*(-rr*RI31+sig*RI30);
    v1 = v1 + cf1* Dx*RI31;
    cf2 = ar(i,j)*sig*w2/(4*pi);
    u2 = u2 + cf2*(-rr*RI31+sig*RI30);
    v2 = v2 + cf2*Dx*RI31;
    arintv1 = arintv1 + w1*ar(i,j);
    arintv2 = arintv2 + w2*ar(i,j);
  end   % of azimuthal loop

end   % of radial loop

%-----------------------
% Helmholtz approximation
% for comparison purposes
%-----------------------

uh = circ/(4*pi*rr)*(log(8.0*rr/cr)-0.25);

xplot(repeat) = cr/rr;
velr1(repeat) = u1; velr2(repeat) = u2; velr3(repeat) = uh;

%==
end
%==

plot(xplot,velr1);
plot(xplot,velr2,'--');
```

```
plot(xplot,velr3,':');
xlabel('a/\sigma_r','fontsize',15)
ylabel('U_x/\sigma_r\kappa','fontsize',15)
```

The precise vorticity distribution over the core has little effect on the predicted ring velocity. As the size of the core tends to zero while the strength of the ring is held constant, the self-induced velocity takes increasingly large values irrespective of the form of the vorticity distribution over the core. Thus, the self-induced velocity of a line vortex ring with infinitesimal core, which is the axisymmetric counterpart of a point vortex, is not defined. We shall see later in this chapter that this is a more general result applicable to vortex filaments with non-rectilinear shapes.

An asymptotic analysis of the self-induced velocity in the limit of small core size shows that, to leading order, the self-induced velocity of a slender vortex ring is given by Helmholtz's formula

$$U_x = \frac{C}{4\pi\sigma_r}(-\ln\frac{8\,a}{\sigma_r} - \frac{1}{4}). \tag{11.5.28}$$

This formula explicitly demonstrates that the self-induced velocity diverges at a logarithmic rate with respect to the reduced core radius a/σ_r. The predictions of (11.5.28) are represented by the dotted line in figure 11.5.6. The results are in good agreement with the exact values represented by the solid and dashed lines even when the core radius, a, is not small compared to the ring radius, σ_r.

11.5.4 Motion of a collection of vortex rings

A vortex ring belonging to a collection of coaxial vortex rings translates under the influence of its self-induced velocity as well as of the velocity induced by its peers. When the ring core size is small, the self-induced velocity is accurately predicted by Helmholtz's formula (11.5.28).

The vorticity transport equation (6.6.9) requires that, as the radius of a ring, σ_r changes during the motion, the strength of the vorticity, ω_φ, increases or decreases in the same proportion. In response to this change, the radius of the core, a, is adjusted to preserve the ring strength and core volume,

$$\frac{\mathrm{d}}{\mathrm{d}t}[(2\pi\sigma_r)\,\pi a^2] = 0. \tag{11.5.29}$$

Expanding out the derivative on the left-hand side and rearranging we obtain an evolution equation for the core radius,

$$\frac{\mathrm{d}a}{\mathrm{d}t} = -\frac{1}{2}\frac{a}{\sigma_r}\frac{\mathrm{d}\sigma_r}{\mathrm{d}t} = -\frac{1}{2}\frac{a}{\sigma_r}U_\sigma, \tag{11.5.30}$$

where $U_\sigma \equiv \mathrm{d}\sigma_r/\mathrm{d}t$ is the rate of expansion of the circular centerline.

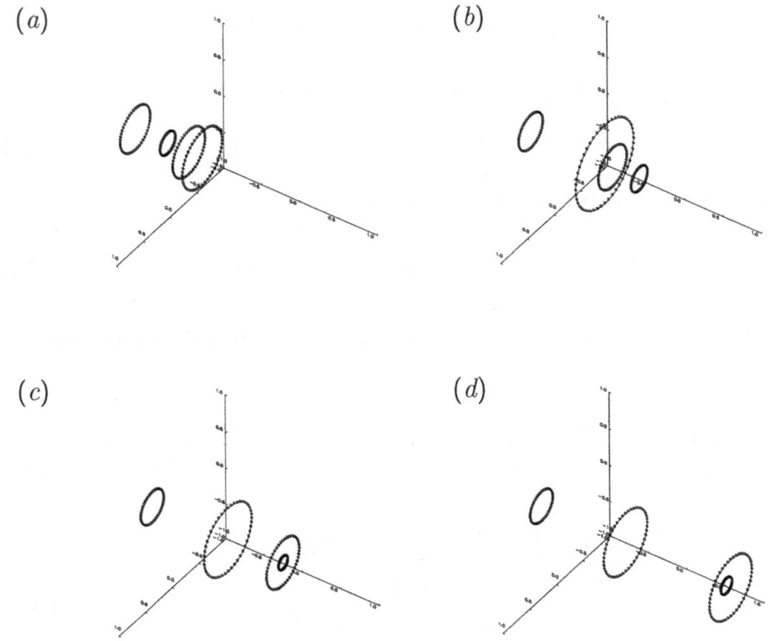

Figure 11.5.6 Motion of a collection of coaxial vortex rings demonstrating leap-frogging.

Appending equation (11.5.30) to the equation of motion for the ring centerline, we obtain a system of differential equations governing the ring axial position, centerline, and core radius. A numerical method for integrating this system from a given initial condition is implemented in program *lvrm* residing inside directory *09_vortex* of FDLIB. Figure 11.5.6 shows stages in the motion of four vortex rings computed using this program. The simulation reveals that neighboring vortex rings pass through one another in an alternating fashion, exhibiting a type of motion that can be described as leap-frogging.

11.5.5 Vortex patch in axisymmetric flow

Next, we consider an axisymmetric flow containing an annular vortex whose vorticity is proportional to the distance from the x axis, $\omega_\varphi = \Omega\,\sigma$, where Ω is a constant. The flow outside the vortex is irrotational. Substituting the linear vorticity distribution in the integral representation (11.5.12) and using the Gauss divergence theorem, we derive a contour integral representation that is analogous to that for a two-dimensional vortex patch with uniform vorticity discussed in section 11.2. The axial velocity component is given by

$$u_x(x,\sigma) = -\frac{\Omega}{4\pi}\int_C [(x-x')I_{10}(x-x',\sigma,\sigma')\, n_x(x',\sigma') \\ +\sigma\, I_{11}(x-x',\sigma,\sigma')\, n_\sigma(x',\sigma')]\,\sigma'\,\mathrm{d}l(x',\sigma'), \qquad (11.5.31)$$

and the radial velocity component is given by

$$u_\sigma(x,\sigma) = \frac{\Omega}{4\pi}\int_C I_{11}(x-x',\sigma,\sigma')\, n_x(x',\sigma')\,\sigma'^2\,\mathrm{d}l(x',\sigma'), \qquad (11.5.32)$$

where C is the trace of the vortex contour in a meridional plane. We have introduced the integrals

$$I_{10}(x-x',\sigma,\sigma') \equiv \int_0^{2\pi}\frac{\mathrm{d}\hat{\varphi}}{r} = \int_0^{2\pi}\frac{\mathrm{d}\hat{\varphi}}{[(x-x')^2+(\sigma+\sigma')^2-4\,\sigma\,\sigma'\,\cos^2\frac{\hat{\varphi}}{2}]^{1/2}}$$

$$= \frac{4}{[(x-x')^2+(\sigma+\sigma')^2]^{1/2}}\int_0^{\pi/2}\frac{\mathrm{d}\eta}{(1-k^2\,\cos^2\eta)^{1/2}} \\ = \frac{4\,F(k)}{[(x-x')^2+(\sigma+\sigma')^2]^{1/2}}, \qquad (11.5.33)$$

and

$$I_{11}(x-x',\sigma,\sigma') \equiv \int_0^{2\pi}\frac{\cos\hat{\varphi}}{r}\,\mathrm{d}\hat{\varphi} \\ = \int_0^{2\pi}\frac{\cos\hat{\varphi}}{[(x-x')^2+(\sigma+\sigma')^2-4\,\sigma\,\sigma'\,\cos^2\frac{\hat{\varphi}}{2}]^{1/2}}\,\mathrm{d}\hat{\varphi} \\ = \frac{4}{[(x-x')^2+(\sigma+\sigma')^2]^{1/2}}\int_0^{\pi/2}\frac{\cos(2\eta)}{(1-k^2\,\cos^2\eta)^{1/2}}\,\mathrm{d}\eta, \qquad (11.5.34)$$

where $\eta \equiv \hat{\varphi}/2$, k^2 is defined in (11.5.15), and $F(k)$ is the complete elliptic integral of the first kind defined in the first equation of (11.5.16). The integral on the right-hand side of (11.5.34) can be expressed in terms of complete elliptic integral of the first and second kind,

$$\int_0^{\pi/2}\frac{\cos(2\eta)}{(1-k^2\,\cos^2\eta)^{1/2}}\,\mathrm{d}\eta = \frac{2-k^2}{k^2}\,F(k) - \frac{2}{k^2}\,E(k). \qquad (11.5.35)$$

The vorticity transport equation for axisymmetric flow with negligible viscous forces stated in equation (6.6.10) requires that the vorticity inside an axisymmetric patch whose vorticity distribution is linear with respect to distance from the x axis, σ, remains linear at all times. To compute the evolution of the flow, it suffices to simulate the motion of the vortex contour using the counterpart of the contour dynamics formulation for two-dimensional flow discussed in section 11.3.

Program *vp_ax* inside directory *09_vortex* of FDLIB simulates the motion of axisymmetric vortex patches using the method of contour dynamics for axisymmetric flow. In the numerical implementation, the contour of each patch is approximated with blended circular arcs, as discussed in section 11.3. The marker points defining each contour are redistributed adaptively during the motion to capture the development of regions of high curvature and prevent point clustering or dilution.

Problem

11.5.1 *Velocity induced by vorticity*

Derive the representation (11.5.11).

11.5.2 *Hill's spherical vortex*

Show that the velocity corresponding to the stream functions (11.5.3) and (11.5.4) is continuous across the spherical interface of Hill's vortex.

11.5.3 *Contour dynamics for three-dimensional flow*

Is it possible to derive a contour dynamics method for a three-dimensional (non-axisymmetric) vortex flow?

Computer Problems

11.5.4 *Motion of line vortex rings*

Run the program *lvrm* to simulate the motion of a collection of rings of your choice. Discuss the nature of the motion.

11.5.5 *Motion of vortex patches*

Run the program *vp_ax* to simulate the motion of a vortex patch of your choice. Discuss the nature of the motion.

11.6 Three-dimensional vortex motion

Previously in this chapter, we have studied the dynamics of two-dimensional and axisymmetric vortex flow based on an integral representation for the velocity in terms of the vorticity expressed by the Biot-Savart integral, combined with simplified versions of the vorticity transport equation for inviscid flow. Extensions to three dimensions are straightforward in principle but subtle in numerical implementation.

11.6.1 Vortex particles

The vortex-particle method arises by replacing the Biot-Savart integral (11.4.14) with a sum of integrals over parcels of rotational fluid, and then condensing the vorticity of each parcel into a designated center represented by point particles. This approximation replaces the rotational fluid with a collection of three-dimensional singularities called rotlets or vortons. The strength of the vortons evolves in time according to rules stemming from the vorticity transport equation for three-dimensional flow, expressing stretching and reorientation.

The vorton discretization is analogous to the point-vortex discretization of two-dimensional flow and to the circular-line-vortex discretization of axisymmetric flow. However, because the three-dimensional discretization breaks the continuity of the vortex lines, some fundamental difficulties arise in three-dimensional flow. These difficulties, combined with high computational cost, explain why the vorton method is less attractive compared to its counterparts for two-dimensional and axisymmetric flow.

11.6.2 Line vortices and the local induction approximation (LIA)

A simplified model of the self-induced motion of a three-dimensional vortex filament with small core size illustrated in figure 11.6.1 can be devised using our earlier results for vortex rings with small but non-infinitesimal core radius.

To compute the motion of the filament centerline, we evaluate the velocity at the position of point particles distributed over the centerline using the Biot-Savart integral (11.4.14), and then advance the position of the point particles using a standard numerical method. Our earlier discussion of vortex rings suggests that the self-induced velocity of the filament is determined primarily by the ratio of the local filament core radius and the local radius of curvature of the filament centerline, as shown in (11.5.28). This observation provides us with a basis for computing the motion of the centerline according to the local induction approximation (LIA).

In this approximation, the velocity at a point along the centerline is assumed to be parallel to the local unit binormal vector, $\mathbf{b}$, defined as the cross product of the unit tangent vector, $\mathbf{t}$, and the unit normal vector, $\mathbf{n}$,

$$\mathbf{b} = \mathbf{t} \times \mathbf{n}. \tag{11.6.1}$$

The unit tangent vector arises by differentiating the position vector, $\mathbf{x}$, with respect to arc length, l, along the centerline,

$$\mathbf{t} = \frac{\mathrm{d}\mathbf{x}}{\mathrm{d}l}. \tag{11.6.2}$$

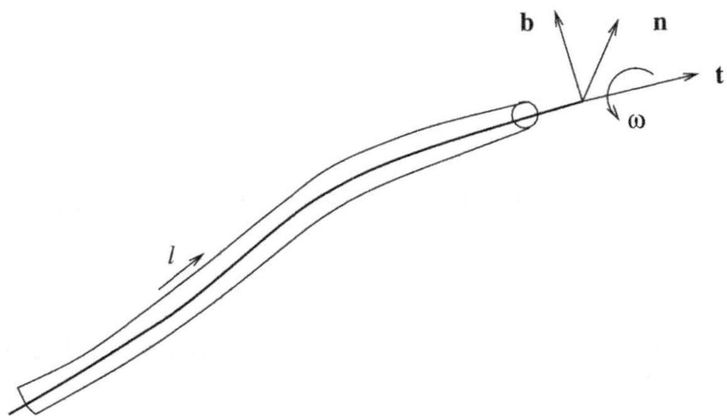

Figure 11.6.1 The motion of a thin vortex filament can be computed using the local induction approximation expressed by equations (11.6.1) and (11.6.2).

The unit normal vector arises by further differentiating the unit tangent vector with respect to arc length,

$$\mathbf{n} = -\frac{1}{R}\frac{\mathrm{d}\mathbf{t}}{\mathrm{d}l}. \tag{11.6.3}$$

The positive or negative scalar coefficient R is the radius of curvature of the filament centerline. In the local induction approximation, the self-induced velocity of the filament is

$$\mathbf{u} = u_b\,\mathbf{b}, \tag{11.6.4}$$

where κ is the strength of the filament. The magnitude of the velocity, u_b, derives from Helmholtz's formula (11.5.28) as

$$u_b = \frac{\kappa}{4\pi R}(-\ln\frac{8a}{R} - \frac{1}{4}), \tag{11.6.5}$$

The local filament core radius, a, is allowed to vary with position along the centerline.

The vorticity transport equation requires that the vorticity in the filament core increases or decreases depending on whether the filament centerline undergoes stretching or compression. An evolution equation for the filament core radius may then be derived by requiring conservation of volume of rotational fluid residing inside the core. Let δl be an infinitesimal arc length along the centerline corresponding to a material segment that begins and ends at two material point particles. Conservation of volume requires

$$\frac{\mathrm{d}}{\mathrm{d}t}(\delta l\,\pi\,a^2) = 0, \tag{11.6.6}$$

expressing the counterpart of equation (11.5.29) for axisymmetric flow. Expanding the derivative of the product on the right-hand side and rearranging, we obtain the evolution equation

$$\frac{\mathrm{d}a}{\mathrm{d}t} = -\frac{1}{2}\frac{a}{\delta l}\frac{\mathrm{d}\delta l}{\mathrm{d}t}, \tag{11.6.7}$$

expressing the counterpart of equation (11.5.30) for axisymmetric flow.

Numerical method

A numerical method for computing the evolution of the centerline of a closed filament involves the following steps:

1. In the first step, we trace the centerline with $N+1$ marker points and assign initial values to the core radius. Point 1 coincides with point $N+1$.

2. In the second step, we approximate the centerline over the interval subtended between points numbered $i-1$ and $i+1$ with a circular arc passing through the three points labeled $i-1$, i, and $i+1$, and compute the arc center and radius for $i = 1, \ldots N$.

3. In the third step, we compute the unit tangent, normal, and binormal vector at the ith marker point using equations (11.6.1)–(11.6.3), for $i = 1, \ldots, N$.

4. In the fourth step, we set the radius of curvature of the line vortex at each point equal to the radius of the arc, R.

5. In the fifth step, we advance the position of the marker points with the velocity computed from (11.6.4).

6. In the sixth step, we update the core radius by requiring conservation of volume expressed by (11.6.6).

The motion of marker points computed by this method suffers from numerical instabilities that cause the onset of strong oscillations due to violent amplification of numerical or round-off error. Smoothing the position of the marker points using the five-point formula (11.2.19) applied for the Cartesian coordinates of the marker points filters out the oscillations and allows the motion to be pursued for an extended period of time.

Program *lv_lia* inside directory *09_vortex* of FDLIB simulates the motion of a closed line vortex using the local induction approximation. Stages in the evolution of a closed line vortex computed by this program are shown in figure 11.6.2. The code features adaptive point redistribution to capture the development of regions of high curvature. The simulation shows that the ring travels while exhibiting wobbly oscillations familiar to the cigar smoker.

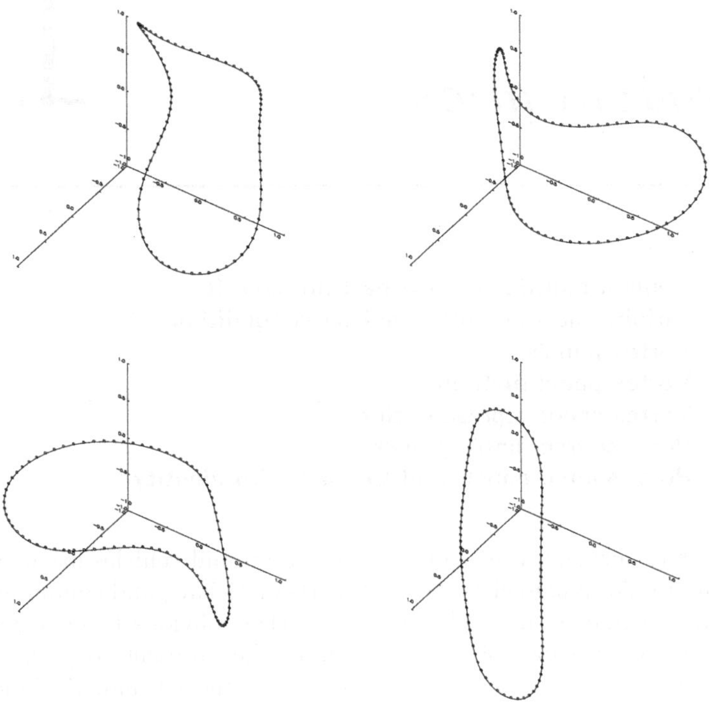

Figure 11.6.2 Stages in the evolution of a closed line vortex resembling a smoke ring, computed by program *lv_lia* of FDLIB.

Problem

11.6.1 *LIA for a circular line vortex ring*

Confirm that the local induction approximation consistently describes the self-induced motion of a circular line vortex ring discussed in section 11.5.

Computer Problem

11.6.2 *Motion of line vortex*

Run program *lv_lia* to simulate the evolution of a line vortex with initial shape and core radius of your choice. Discuss the nature of the motion.

12 Aerodynamics

Flow past airplane wings and high-speed ground vehicles has captured the attention of fluid dynamicists, applied mathematicians, and computational scientists and engineers, not only because of their obvious technological significance, but also because of the opportunity they present to perform elegant mathematical analysis and develop realistic and efficient numerical models. Although these flows occur at high Reynolds numbers and often at transonic or supersonic speeds that are comparable to or even exceed the speed of sound, the effect of viscosity is important in two ways. First, viscous stresses determine the drag force exerted on moving surfaces and thus the energy required to sustain the motion. Second, viscous stresses are responsible for the production of vorticity which generates circulation and thereby induces a lift force.

A comprehensive analysis of high-speed flow in aerodynamics incorporates the effects of fluid compressibility and turbulent motion, and accounts for the presence of boundary layers and regions of recirculating flow. In this chapter, we discuss the basic configuration that arises by neglecting the presence of boundary layers and wakes, and thus assuming inviscid and incompressible flow. The simplified model involving irrotational flow in the presence of global circulatory motion is amenable to efficient numerical methods that illustrate the importance of computational fluid dynamics in the practical field of aerodynamics.

12.1 General features of flow past an aircraft

Figure 12.1.1 shows a schematic illustration of flow past an aircraft that has taken off and traveled by a certain distance, viewed in a frame of reference

C. Pozrikidis, *Fluid Dynamics: Theory, Computation, and Numerical Simulation*,
DOI: 10.1007/978-0-387-95871-2_12,

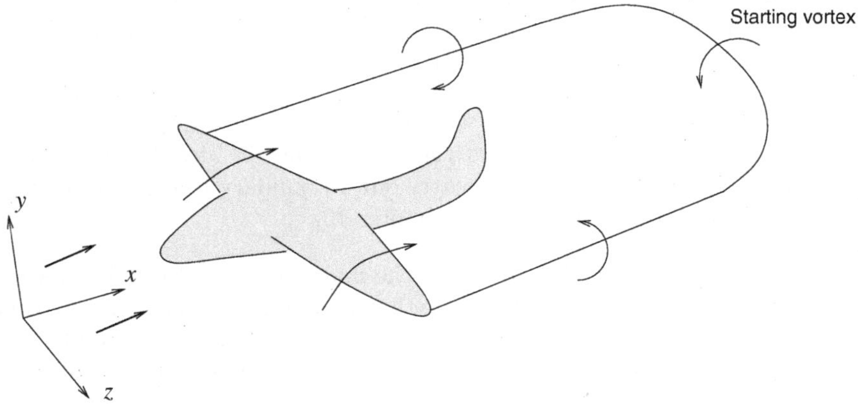

Figure 12.1.1 Schematic illustration of flow past an aircraft in a frame of reference moving with the aircraft.

moving with the aircraft. If the flow is subsonic, the following features are most significant:

- A thin vortex loop resembling a line vortex is established behind the aircraft. The loop extends from the left wing tip, back at a length that is comparable to the distance that the aircraft has traveled, and then forward up to the right wing tip. The trailing vortex at the back of the loop was generated when the airplane started moving, and is accordingly called the starting vortex. The vortex loop can be extended artificially into the wings to form a closed line vortex.

  The circulation around any closed loop that encloses the line vortex is constant, independent of the shape and location of the loop. Thus, the circulation around a simple loop that encloses a wing is equal to the circulation around a simple loop that encloses the trailing vortex.

- Viscous stresses cause the vortex loop to diffuse and its vortex core to be smeared out. However, the circulation around any loop that encloses the smeared vortex loop is equal to the circulation around a loop that encloses a wing, no matter how far the vorticity has spread out.

- The circulation around a loop that encloses a wing is determined by the speed of the aircraft and the geometry and orientation of the wings with respect to the incoming wind, as discussed in section 12.2.

- If the aircraft suddenly changes its speed or direction of flight, a new vortex loop will be ejected contributing an additional amount of circulation around the wings.

- Each wing experiences a lift force normal to the direction of flight, and a drag force parallel to the direction of flight. The later must be compensated by the thrust produced by the engine.

The lift force can be computed with surprising accuracy by neglecting the effects of viscosity and assuming that the flow around the airfoil is irrotational. To compute the drag force, we must carry out a boundary-layer analysis of the basic irrotational flow, as discussed in chapter 10.

It is important to bear in mind that the main features of the flow past an aircraft discussed in this chapter assume that the wings are only slightly tilted with respect to the direction of the incoming wind. When this condition is not met, large regions of recirculating flow develop over the upper surface of the airfoil, seriously affecting the structure of the flow and performance of the aircraft.

To study the flow past the wings and compute the lift force per unit span exerted on them, we may assume that the flow is locally two-dimensional occurring in the xy plane that is normal to the line connecting the wing tips. It turns out that neglecting the third dimension provides us with a theoretical model whose predictions are in good and sometimes excellent agreement with laboratory measurements taken in wind tunnels. In the remainder of this chapter, we concentrate on the two-dimensional flow. The three-dimensional flow is an advanced topic suitable for a second course in aerodynamics.

12.2 Airfoils and the Kutta-Joukowski condition

An airfoil is a section of a wing, as depicted in figure 12.2.1(a). The shape of an airfoil is determined by the following geometrical parameters:

- The chord line, defined as the straight line connecting the leading edge to the trailing edge.

- The chord, defined as the distance from the leading edge to the trailing edge.

- The camber line, defined as the locus of points located halfway between the upper and lower surface of the airfoil.

- The camber, defined as the maximum distance of the camber line from the chord line. If the camber is zero, the airfoil is symmetric.

- The airfoil thickness along the camber line.

The angle subtended between the incoming wind and the chord line, α, is defined as the angle of attack.

(a)

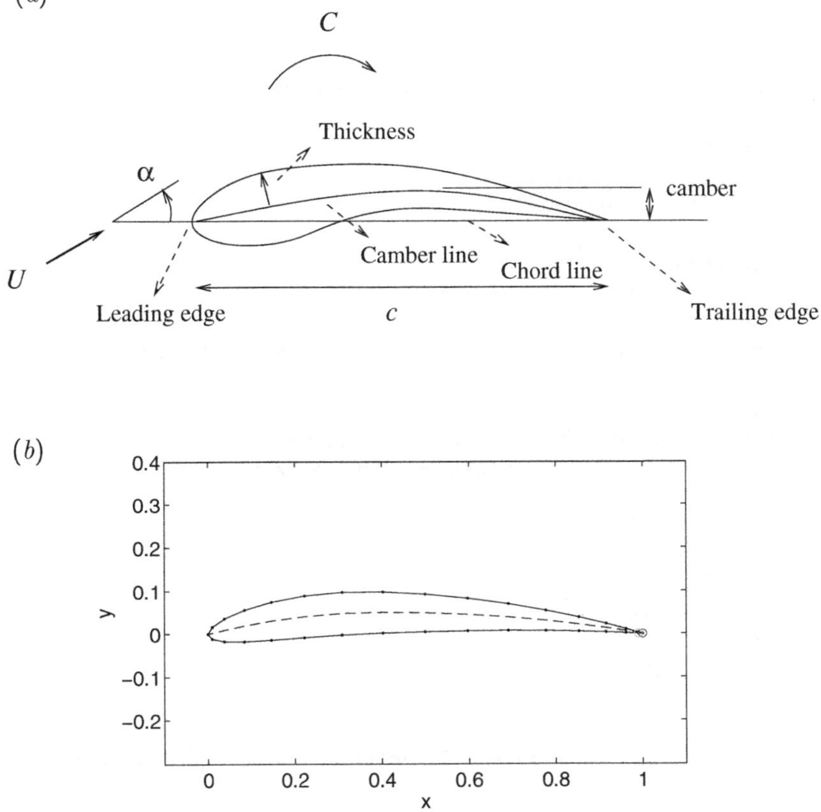

Figure 12.2.1 (a) An airfoil and its standard geometrical properties. (b) Contour of a four-digit NACA airfoil.

NACA airfoils

The National Advisory Committee for Aeronautics of the United States (NACA), the ancestor of today's NASA, has standardized airfoil shapes to facilitate engineering design. NACA airfoils are produced by specifying the geometry of the camber line, and then wrapping around the camber line the airfoil contour to obtain a desired distribution of half-thickness.

The dated four-digit NACA *mnlk* airfoils, where m, n, k, l are four integers, have a camber of $0.0f \times c$, occurring at a distance $0.0f \times c$ from the leading edge, where c is the chord. The maximum airfoil thickness is $0.gh \times c$. The following MATLAB program *NACA4* residing in directory *airf_2d* inside directory *07_ptf* of FDLIB produces a four-digit NACA airfoil:

```
chord = 1.0; xcam = 0.4;
cam = 0.05; thick = 0.1;

n = 32;    % number of nodes around the airfoil

for i=1:n
  psi = (i-1.0)*2*pi/n;
  x = 0.5*(1.0+cos(psi));
% camber line:
  if(x<xcam)
    yc(i) = cam*(2.0*xcam*x-x*x)/xcam^2;
  else
    yc(i) = cam/(1.0-xcam)^2*((1.0-2.0*xcam)+2.0*xcam*x-x*x);
  end
% thickness:
  yt = 5.0*thick *(0.2969*sqrt(x) -0.1260*x...
    -0.3516*x*x +0.2843*x*x*x -0.1036*x*x*x*x );
% contour:
  if(i<(n/2+1))
    ya(i) = (yc(i)-yt)*chord;
  else
    ya(i) = (yc(i)+yt)*chord;
  end
  xa(i) = x*chord;
end

xa(n+1) = chord; ya(n+1) = 0.0; yc(n+1) = 0.0;    % trailing edge

%----
% plot
%----

plot(xa,ya,'k.-'); plot(xa,yc,'k--');
axis equal
xlabel('x','fontsize',15)
ylabel('y','fontsize',15)
```

The graphics produced by the code is shown in figure 12.2.1(b).

Modern five- and six-digit airfoils are encoded with additional geometrical and flow properties.

Airfoil shapes by mapping

Airfoil shapes can be produced by mapping a closed contour in an auxiliary parametric (ξ, η) plane to the airfoil contour in the physical xy plane, using an appropriate mapping function. In theoretical aerodynamics, the mapping function arises from a function of a complex variable $f(\zeta)$, by setting $z = f(\zeta)$,

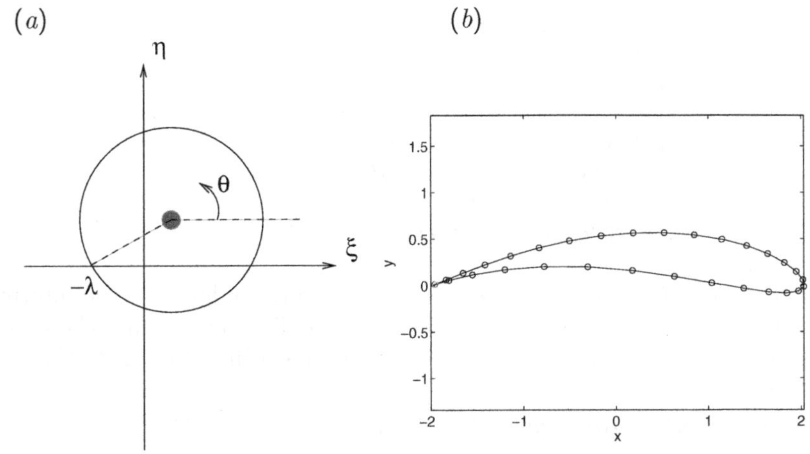

Figure 12.2.2 A Joukowski airfoil produced by conformal mapping.

where $z \equiv x + \mathrm{i}\,y$, $\zeta \equiv \xi + \mathrm{i}\,\eta$ are two complex variables, and i is the imaginary unit satisfying $\mathrm{i}^2 = -1$.

Joukowski's transformation employs the mapping function

$$f(\zeta) = \zeta + \frac{\lambda^2}{\zeta}, \tag{12.2.1}$$

where λ is a real constant. A circle in the ζ plane passing through the point $(-\lambda, 0)$ and enclosing the reflected point $(\lambda, 0)$, shown in figure 12.2.2(a), is mapped to a cusped airfoil whose camberline and camber are determined by the location of the center of the circle in the ζ plane, as shown in figure 12.2.2(b). Different shapes can be produced by fixing the value of λ and varying the position of the center of the circle in the $\xi\eta$ plane.

The following MATLAB code *joukowski* residing in directory *airf_2d* inside directory *07_ptf* of FDLIB produces produces airfoil shapes for $\lambda = \pi$:

```
xc = 0.1; yc = 0.2;    % circle center
n = 32;      % number of contour nodes

rad = sqrt((xc+1.0)^2+yc*yc);
th1 = -pi+acos((xc+1)/rad);

for i=1:n+1
  th = th1+(i-1)*2*pi/n;
  xi = xc+rad*cos(th);
  et = yc+rad*sin(th);
  r2 = xi*xi+et*et;
```

```
  x(i) = xi*(1+1/r2);
  y(i) = et*(1-1/r2);
end

plot(x,y,'o-') axis equal
xlabel('x','fontsize',15)
ylabel('y','fontsize',15)
set(gca,'fontsize',15)
```

The first and last points are mapped to the trailing edge, and the intermediate points are distributed in the counterclockwise direction around the airfoil. The graphics produced by the code is shown as shown in figure 12.2.2(*b*).

12.2.1 The Kutta-Joukowski theorem

It can be shown that the lift force per unit span exerted on an airfoil, L, is determined by the incoming wind speed, U, and by the fluid circulation around the airfoil, C, by means of the Kutta-Joukowski theorem expressed by the equation

$$L = -\rho U C, \tag{12.2.2}$$

where ρ is the fluid density. Note that positive lift requires negative circulation associated with clockwise rotation around the airfoil, as illustrated in figure 12.2.1. If the circulation vanishes, the lift force is zero.

The Kutta-Joukowski theorem can be proved most readily working under the auspices of analytic function theory in complex variables, as discussed in the texts cited in the bibliography.

Flow past a cylinder

To confirm the Kutta-Joukowski theorem, we consider uniform flow past a circular cylinder of radius a discussed in section 3.7. Applying Bernoulli's equation (6.4.14) at infinity and at a point on the surface of the cylinder, and evaluating the tangential velocity on the surface of the cylinder using formula (3.7.9), we derive an expression for the surface pressure,

$$p(r = a) = p_\infty - 2\rho\, V_x^2\, (\sin\theta + \beta)^2, \tag{12.2.3}$$

where p_∞ is the pressure at infinity and β is the dimensionless circulation parameter defined in equation (3.7.8). In the absence of viscous stresses, the force per unit span exerted on the cylinder is given by

$$\mathbf{F} = \oint [-p(r = a)]\, \mathbf{n}\, \mathrm{d}l = -\int_0^{2\pi} p(r = a)\, \mathbf{n}\, a\, \mathrm{d}\theta, \tag{12.2.4}$$

where $\mathrm{d}l = a\,\mathrm{d}\theta$ is the differential arc length around the cylinder, and $\mathbf{n} = (\cos\theta, \sin\theta)$ is the unit vector normal to the cylinder pointing into the fluid.

Substituting (12.2.3) in (12.2.4) and carrying out the integration with respect to θ, we find that the x component of the force vanishes, while the y component of the force is given by $L = 4\pi a \rho V_x^2 \beta$. Recalling the definition of β given in (3.7.8), we find

$$L = -\rho \, V_x \, \kappa, \tag{12.2.5}$$

which agrees with the Kutta-Joukowski theorem expressed by (12.2.2).

12.2.2 The Kutta-Joukowski condition

In the context of irrotational flow theory, the circulation around an airfoil, or any two-dimensional body, is arbitrary. Kutta and Joukowski independently observed that, in practice, when the angle of attack α is small, the flow on the upper side of an airfoil joins smoothly with the flow on the lower side of the airfoil at the trailing edge. This observation provides us with a physical basis for the Kutta-Joukowski condition stipulating that the circulation established around an airfoil is such that a singular flow does not arise at the trailing edge. Thus, in the context of irrotational flow theory, the Kutta-Joukowski condition requires that the fluid does not turn around a cornered or cusped trailing edge. Referring to the Kutta-Joukowski theorem expressed by (12.2.2), we see that a well-designed airfoil should be able to produce a high degree of circulation while minimizing the drag force exerted on the airfoil.

Problem

12.2.1 *Flow past a cylinder*

Carry out the integration in (12.2.4) with the pressure given in (12.2.3) to derive expression (12.2.5).

Computer Problem

12.2.2 *Joukowski airfoils*

Run the program *joukowski* to generate and graph several airfoil shapes of your choice. Investigate the effect of the mapped circle center.

12.3 Vortex panels

We begin the study of two-dimensional flow past an airfoil by introducing a class of elementary flows associated with vortex panels. Our ultimate objective is to use these elementary flows as fundamental building blocks for describing the flow past an airfoil with arbitrary shape, where the circulation around the airfoil is determined by the Kutta-Joukowski condition.

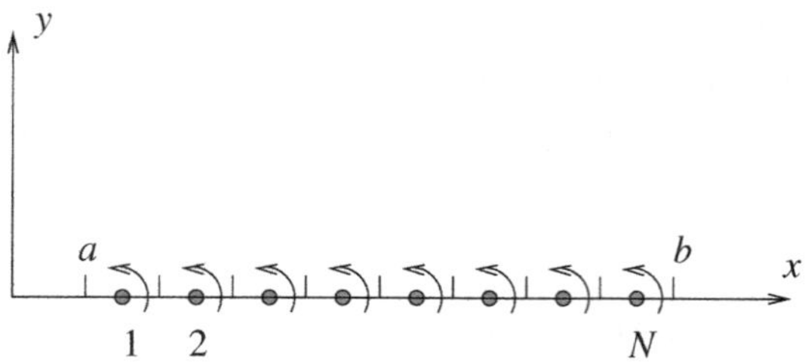

Figure 12.3.1 A collection of N point vortices are deployed along the x axis inside the interval (a, b). As the number of point vortices becomes increases, we obtain a continuous distribution yielding a vortex panel.

In section 3.7, we introduced the two-dimensional irrotational flow with circulatory motion induced by a point vortex. The x and y components of the velocity at a point, $\mathbf{x} = (x, y)$, due to a point vortex with strength κ located at the point $\mathbf{x}_0 = (x_0, y_0)$ were given in equations (11.2.1), repeated here for convenience,

$$u_x(x, y) = -\frac{\kappa}{2\pi} \frac{y - y_0}{(x - x_0)^2 + (x - y_0)^2},$$
$$u_y(x, y) = \frac{\kappa}{2\pi} \frac{x - x_0}{(x - x_0)^2 + (x - y_0)^2}. \qquad (12.3.1)$$

The corresponding stream function and velocity potential are given by

$$\psi(x, y) = -\frac{\kappa}{4\pi} \log[(x - x_0)^2 + (y - y_0)^2], \qquad (12.3.2)$$

and

$$\phi(x, y) = \frac{\kappa}{2\pi} \arctan \frac{y - y_0}{x - x_0}. \qquad (12.3.3)$$

12.3.1 From point vortices to vortex panels

Consider a collection of N point vortices distributed evenly along the x axis inside the interval (a, b), separated by the distance $\Delta x = (b - a)/N$, as illustrated in figure 12.3.1. The ith point vortex is situated at the position $x_i = a + \frac{i}{2}\,\Delta x$, $y_i = 0$, where $i = 1, \ldots, N$, and its strength is equal to κ_i; the first point vortex is located at $x_i = a + \frac{1}{2}\,\Delta x$, and the last point vortex is located at $x_N = b - \frac{1}{2}\,\Delta x$.

Superposing the stream functions associated with the individual point vortices, we find that the stream function of the flow induced by the collection is given by

$$\psi(x, y) = -\sum_{i=1}^{N} \frac{\kappa_i}{4\pi} \log[(x - x_i)^2 + (y - y_i)^2], \tag{12.3.4}$$

which can be recast into the form

$$\psi(x, y) = -\sum_{i=1}^{N} \frac{1}{4\pi} \log[(x - x_i)^2 + (y - y_i)^2] \frac{\kappa_i}{\Delta x} \Delta x, \tag{12.3.5}$$

with the understanding that $y_i = 0$.

In the limit as N tends to infinity, and correspondingly Δx tends to zero, while the strength of the point vortices is reduced so that the ratio $\gamma_i \equiv \kappa_i/\Delta x$ remains constant, the sum on the right-hand side of (12.3.5) reduces to a line integral over the domain of distribution of the point vortices, yielding the integral representation

$$\psi^{VP}(x, y) = -\frac{1}{4\pi} \int_a^b \log[(x - x')^2 + (y - y')^2] \, \gamma(x') \, dx', \tag{12.3.6}$$

with the understanding that $y' = 0$. The right-hand side of (12.31236) expresses the flow due to a two-dimensional finite vortex sheet or vortex panel (VP) with strength density $\gamma(x)$, subtended between the points $x = a$ and b, as indicated by the superscript VP. The circulation around the panel, defined as the strength of the panel, is equal to the integral of the strength density,

$$\Gamma^{VP} \equiv \int_a^b \gamma(x) \, dx. \tag{12.3.7}$$

Following the discussion in section 3.7, we find that the circulation around a closed loop that does not enclose the panel is zero, whereas the circulation around a simple loop that wraps around the panel is equal to the strength of the panel, Γ_p. If the strength of the panel is zero, the circulation vanishes.

12.3.2 Vortex panels with uniform strength

Next, we consider a vortex panel with uniform strength density equal to $\gamma^{(0)}$. According to (12.3.7), the circulation around the panel is equal to $\Gamma_p = (b - a)\,\gamma^{(0)}$. Using (12.3.6) with $y' = 0$, we find that the stream function of the induced flow is given by

$$\psi^{(0)}(x, y) = -\frac{\gamma^{(0)}}{4\pi} \int_a^b \log[(x - x')^2 + y^2] \, dx'. \tag{12.3.8}$$

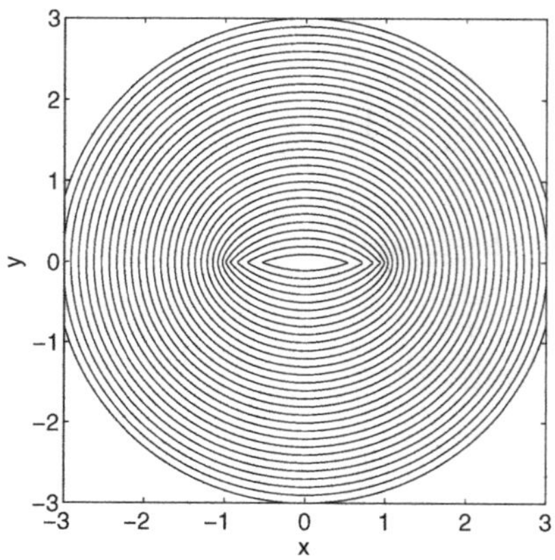

Figure 12.3.2 Streamline pattern of the flow induced by a vortex panel with uniform strength situated on the x axis in the interval $-1 < x < 1$.

The integral on the right-hand side of (12.3.8) can be computed in closed form with the aid of standard mathematical tables, and is found to be

$$\psi^{(0)}(x,y) = -\frac{\gamma^{(0)}}{4\pi}\Big(-(x-b)\log[(x-b)^2+y^2] + (x-a)\log[(x-a)^2+y^2] + 2y\,(\arctan\frac{y}{x-b} - \arctan\frac{y}{x-a}) - 2(b-a)\Big). \quad (12.3.9)$$

The velocity components arise by straightforward differentiation, and are found to be

$$u_x^{(0)}(x,y) = \frac{\partial\psi^{(0)}}{\partial y} = -\frac{\gamma^{(0)}}{2\pi}\,(\arctan\frac{y}{x-b} - \arctan\frac{y}{x-a}), \quad (12.3.10)$$

and

$$u_y^{(0)}(x,y) = -\frac{\partial\psi^{(0)}}{\partial x} = -\frac{\gamma^{(0)}}{4\pi}\ln\frac{(x-b)^2+y^2}{(x-a)^2+y^2}. \quad (12.3.11)$$

The streamline pattern induced by a vortex panel subtended between the points $x = \pm b$ is shown in figure 12.3.2. Far from the panel, the flow resembles that due to a point vortex with strength

$$\kappa = \Gamma_{VP} = \gamma^{(0)}\,(b-a) = 2\,b\,\gamma^{(0)}, \quad (12.3.12)$$

situated at the origin.

Jump in velocity across the panel

Expression (12.3.11) shows that the y component of the velocity is continuous throughout the domain of flow as well as across the vortex panel. In contrast, because of the inverse tangent function on the right-hand side of (12.3.10), the x velocity component undergoes a discontinuity across the vortex panel. To demonstrate this jump, we evaluate the velocity at a point on the upper surface of the panel, for $a < x < b$ and $y = +\epsilon$ where ϵ is a small positive number, and find

$$u_x^{(0)}(x, y \to 0+) = -\frac{\gamma^{(0)}}{2\pi}\,[-\arctan(-\infty) + \arctan(+\infty)\,] = -\frac{\gamma^{(0)}}{2\pi}\,(\frac{\pi}{2} + \frac{\pi}{2}) = -\frac{\gamma^{(0)}}{2}, \qquad (12.3.13)$$

independent of x, as long as $a < x < b$. The x velocity component at a point on the lower surface of the panel, for $a < x < b$ and $y = -\epsilon$, is given by

$$u_x^{(0)}(x, \to 0-) = -\frac{\gamma^{(0)}}{2\pi}\,[-\arctan(+\infty) + \arctan(-\infty)] = -\frac{\gamma^{(0)}}{2\pi}\,(-\frac{\pi}{2} - \frac{\pi}{2}) = \frac{\gamma^{(0)}}{2}, \qquad (12.3.14)$$

independent of x, as long as $a < x < b$. Thus, as we cross the vortex panel from the upper to the lower side, the velocity undergoes a discontinuity whose magnitude is equal to the strength of the vortex sheet,

$$u_x(x, -\epsilon) - u_x(x, +\epsilon) = \gamma^{(0)}. \qquad (12.3.15)$$

In contrast, a discontinuity does not occur beyond the edges of the vortex panel.

These observations suggest that a vortex panel or more generally a vortex sheet can be identified with a surface across which the tangential component of the velocity undergoes a discontinuity. The magnitude of the discontinuity is the strength of the vortex panel or vortex sheet.

12.3.3 Vortex panel with linear strength density

Next, we consider a vortex panel with linear strength density distribution given by

$$\gamma(x) = \gamma^{(0)} + \gamma^{(1)}\,(x - a). \qquad (12.3.16)$$

Using (12.3.7), we find that the circulation around the panel is

$$\Gamma_p = \int_a^b \gamma\,\mathrm{d}x = (b - a)\,[\gamma^{(0)} + \frac{1}{2}\gamma^{(1)}(b - a)] = (b - a)\,\gamma(\frac{a + b}{2}). \qquad (12.3.17)$$

Applying (12.3.6) with $y' = 0$ and rearranging, we find that the stream function of the induced flow is given by

$$\psi^{(01)}(x,y) = -\frac{\gamma^{(0)} + \gamma^{(1)}(x-a)}{4\pi} \int_a^b \ln[(x-x')^2 + y^2]\, dx'$$
$$-\frac{\gamma^{(1)}}{4\pi} \int_a^b (x'-x)\, \ln[(x-x')^2 + y^2]\, dx'. \qquad (12.3.18)$$

The first integral on the right-hand side of (12.3.18) is equal to the expression enclosed by the tall parentheses following the fraction on the right-hand side of (12.3.9). The second integral can be found with the help of standard tables of integrals, yielding

$$\int_a^b (x'-x)\, \ln[(x-x')^2 + y^2]\, dx' = \frac{1}{2}\,[(x-b)^2 + y^2] \ln[(x-b)^2 + y^2]$$
$$-\frac{1}{2}[(x-a)^2 + y^2] \ln[(x-a)^2 + y^2] - \frac{1}{2}(x-b)^2 + \frac{1}{2}(x-a)^2. \qquad (12.3.19)$$

Combining (12.3.8), (12.3.19), and (12.3.6), and consolidating various terms, we find

$$\psi^{(01)}(x,y) = \psi^{(0)}(x,y) + \psi^{(1)}(x,y), \qquad (12.3.20)$$

where $\psi^{(0)}$ is given by (12.3.9), and

$$\psi^{(1)}(x,y) = -\frac{\gamma^{(1)}}{4\pi}\Big(\frac{1}{2}\,[y^2 - (x-b)(x+b-2a)] \ln[(x-b)^2 + y^2]$$
$$-\frac{1}{2}\,[y^2 - (x-a)^2] \ln[(x-a)^2 + y^2]$$
$$+2y\,(x-a)\,(\arctan\frac{y}{x-b} - \arctan\frac{y}{x-a})$$
$$-\frac{1}{2}(x-b)^2 + \frac{1}{2}(x-a)^2 - 2(x-a)(b-a)\Big). \qquad (12.3.21)$$

The velocity components can be resolved into corresponding forms,

$$u_x^{(01)}(x,y) = u_x^{(0)}(x,y) + u_x^{(1)}(x,y), \qquad (12.3.22)$$

and

$$u_y^{(01)}(x,y) = u_y^{(0)}(x,y) + u_y^{(1)}(x,y), \qquad (12.3.23)$$

where $u_x^{(0)}$ and $u_y^{(0)}$ are given in (12.3.10) and (12.3.11). Tedious differentiation yields the final forms

$$u_x^{(1)}(x,y) = \frac{\partial \psi^{(1)}}{\partial y} = -\frac{\gamma^{(1)}}{4\pi}\Big(y\,\ln\frac{(x-b)^2 + y^2}{(x-a)^2 + y^2}$$
$$+2\,(x-a)\,(\arctan\frac{y}{x-b} - \arctan\frac{y}{x-a})\Big), \qquad (12.3.24)$$

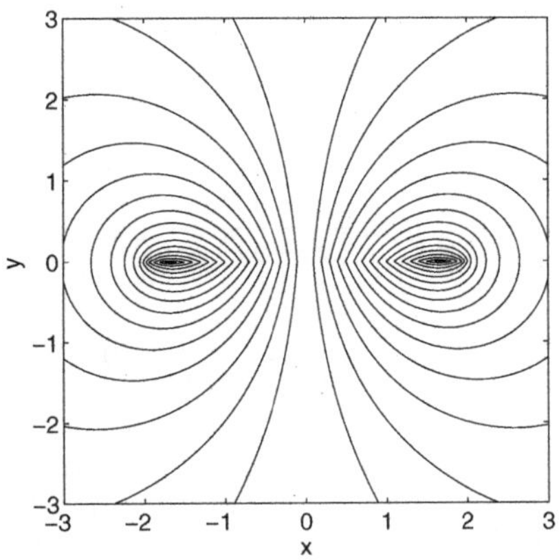

Figure 12.3.3 Streamline pattern of the flow due to a vortex panel with linear strength extending between $x = \pm b$, for vanishing circulation around the panel.

and

$$u_y^{(1)}(x,y) = -\frac{\partial \psi^{(1)}}{\partial x} = -\frac{\gamma^{(1)}}{4\pi}\Big((x-a)\,\ln\frac{(x-b)^2+y^2}{(x-a)^2+y^2} -2\,y\,(\arctan\frac{y}{x-b} - \arctan\frac{y}{x-a}) + 2\,(b-a)\Big). \tag{12.3.25}$$

Using these expressions, we find that, as we cross the vortex panel normal to the x axis, the x component of the velocity undergoes a discontinuity whose magnitude is equal to the local strength of the vortex sheet.

The streamline pattern of the flow induced by a vortex panel subtended between the points $x = \pm b$ with linear strength density distribution given by $\gamma^{(1)} = -\gamma^{(0)}/b$, is shown in figure 12.3.3. Because the circulation around the panel vanishes, as required by (12.3.17), far from the panel the flow reduces to that due to a point-vortex dipole situated at the center of the panel and pointing along the x axis. This is identical to the flow due to a point-source dipole placed at the same location pointing along the y axis.

Problems

12.3.1 *Velocity potential due to vortex panels.*

Confirm that the velocity potentials corresponding to the stream functions

(12.3.9) and (12.3.21) are given, respectively, by

$$\phi^{(0)}(x,y) = -\frac{\gamma^{(0)}}{2\pi}[(x-b)\ \arctan\frac{y}{x-b} - (x-a)\ \arctan\frac{y}{x-a} + \frac{1}{2}\,y\,\ln\frac{(x-b)^2+y^2}{(x-a)^2+y^2}\ ], \tag{12.3.26}$$

and

$$\phi^{(1)}(x,y) = -\frac{\gamma^{(1)}}{4\pi}\{\ (x-a)\,y\,\ln\frac{(x-b)^2+y^2}{(x-a)^2+y^2} + [(x-a)^2-(b-a)^2-y^2]\ \arctan\frac{y}{x-b} - [(x-a)^2-y^2]\ \arctan\frac{y}{x-a} + y(b-a)\ \}. \tag{12.3.27}$$

12.4 Vortex panel method

In section 12.3, we discussed the flow induced by a vortex panel with constant or linear density distribution. Having derived these elementary panel flows, we proceed to develop a vortex-panel method that allows us to compute an irrotational flow past a two-dimensional airfoil, where the Kutta-Joukowski condition satisfied at the trailing edge. The key idea is to express the flow as a superposition of an incident streaming flow and a collection of flows induced by vortex panels with *a priori* unknown strength densities deployed around the contour of an airfoil. The panel strength densities are then computed to satisfy the no-penetration boundary condition around the airfoil.

Contour discretization

Figure 12.4.1 illustrates uniform flow with velocity $\mathbf{U} = (U_x, U_y)$ past an airfoil at angle of attack α. As a preliminary, we trace the contour of the airfoil with $N+1$ marker points distributed around the airfoil in the clockwise sense, where points numbered 1 and $N+1$ coincide with the trailing edge. A pair of successive points $\mathbf{x}^{(i)}$ and $\mathbf{x}^{(i+1)}$ define a flat vortex panel labeled i. The union of the N vortex panels defines a polygonal contour which is an approximation to the generally curved contour of the airfoil.

Flow representation

In the vortex-panel method, the velocity at a field point, $\mathbf{x} = (x,y)$, is expressed by the superposition

$$\mathbf{u}(x,y) = \mathbf{U} + \sum_{i=1}^{N} \mathbf{u}^{(i)}(x,y), \tag{12.4.1}$$

where $\mathbf{u}^{(i)}(x,y)$ is the velocity induced by the ith vortex panel.

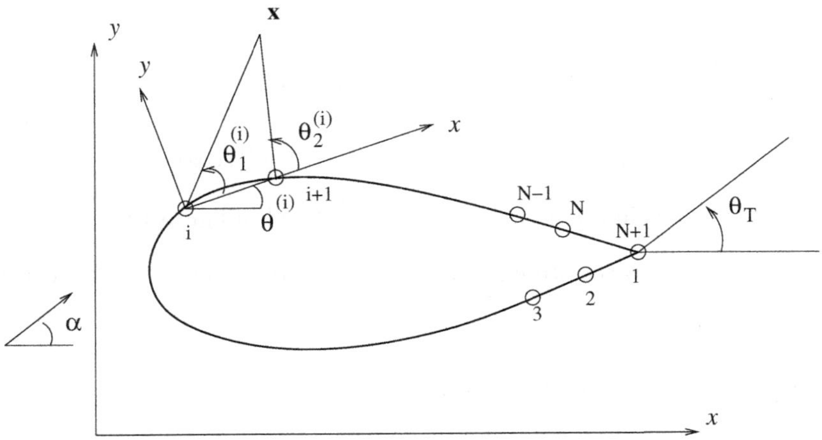

Figure 12.4.1 Discretization of the contour of an airfoil into flat panels defined by a sequence of marker points.

In the implementation discussed in this section, we use panels with linear strength density. The strength density of the ith panel varies linearly from the value γ_i assigned to the ith marker point, which is the first point of the ith panel, to the value γ_{i+1} assigned to the $i+1$ marker point, which is the second point of the ith panel. The $N+1$ unknown values γ_i, where $i = 1, \ldots, N+1$ must be computed to satisfy the no-penetration condition around the airfoil in some approximate sense.

We begin by observing that the velocity induced by the ith panel, $\mathbf{u}^{(i)}(x, y)$, is determined by the position of, and strength density of the panel at the two end points. To signify this dependence, we write

$$\mathbf{u}^{(i)}(x, y) = \mathbf{u}^{LVP}(x, y; \mathbf{x}^{(i)}, \mathbf{x}^{(i+1)}, \gamma^{(i)}, \gamma^{(i+1)}), \tag{12.4.2}$$

where the superscript LVP stands for "linear vortex panel."

Panel-induced velocity in global coordinates

Our first task is to develop a numerical method for computing $\mathbf{u}^{LVP}$ as a function of its arguments listed in (12.4.2), using the formulas developed in section 12.3. For this purpose, we introduce a local coordinate system (x', y') with the x' axis passing through the two end points of the ith panel, $\mathbf{x}^{(i)}$ and $\mathbf{x}^{(i+1)}$, and set the origin at the first end point $\mathbf{x}^{(i)}$, as depicted in figure 12.4.1.

The coordinates of a point in the local system, (x', y'), are related to those in the global system, (x, y), by the equations

$$x' = (x - x^{(i)}) \cos\theta^{(i)} + (y - y^{(i)}) \sin\theta^{(i)},$$
$$y' = -(x - x^{(i)}) \sin\theta^{(i)} + (y - y^{(i)}) \cos\theta^{(i)}, \qquad (12.4.3)$$

where $\theta^{(i)}$ is the inclination of the ith panel defined in figure 12.4.1.

Next, we express the strength density of the ith panel in the linear form

$$\gamma(x') = \gamma^{(0)} + \gamma^{(1)}(x' - x'^{(i)}), \qquad (12.4.4)$$

with the understanding that $x'^{(i)} = 0$, and require

$$\gamma(x' = x'^{(i)}) = \gamma^{(0)} = \gamma^{(i)},$$
$$\gamma(x' = x'^{(i+1)}) = \gamma^{(0)} + \gamma^{(1)}(x'^{(i+1)} - x'^{(i)}) = \gamma^{(i+1)}, \qquad (12.4.5)$$

to find

$$\gamma^{(0)} = \gamma^{(i)}, \qquad \gamma^{(1)} = \frac{\gamma^{(i+1)} - \gamma^{(i)}}{x'^{(i+1)} - x'^{(i)}}. \qquad (12.4.6)$$

The length of the panel is

$$x'^{(i+1)} = [\,(x_{i+1} - x_i)^2 + (y_{i+1} - y_i)^2\,]^{1/2}. \qquad (12.4.7)$$

Reviewing the results of section 12.3, we find that the x' and y' components of the velocity induced by the ith panel are given by equations (12.3.22) and (12.3.23), subject to the following substitutions:

$$\begin{array}{ccc} a & \rightarrow & 0 \\ b & \rightarrow & x'^{(i+1)} \\ \gamma^{(0)} & \rightarrow & \gamma^{(i)} \\ \gamma^{(1)} & \rightarrow & (\gamma^{(i+1)} - \gamma^{(i)})/(x'^{(i+1)} - x'^{(i)}) \\ x & \rightarrow & x' \\ y & \rightarrow & y' \end{array} \qquad (12.4.8)$$

After carrying out some algebra, we find that the velocity components of the flow induced by the ith panel in the local frame are given by

$$u_{x'}^{(i)}(x', y') = a_{x'}^{(i,1)} \gamma^{(i)} + a_{x'}^{(i,2)} \gamma^{(i+1)},$$
$$u_{y'}^{(i)}(x', y') = a_{y'}^{(i,1)} \gamma^{(i)} + a_{y'}^{(i,2)} \gamma^{(i+1)}, \qquad (12.4.9)$$

where $a_{x'}^{(i,1)}, a_{x'}^{(i,2)}, a_{y'}^{(i,1)}$, and $a_{y'}^{(i,2)}$ are local influence coefficients given by

$$a_{x'}^{(i,1)} = \frac{1}{2\pi x_2'} \, [\, y' \, c + (x' - x_2') \, \Delta\theta' \,],$$

$$a_{y'}^{(i,1)} = \frac{1}{2\pi x_2'} \, [\, (x' - x_2') \, c - y' \, \Delta\theta' + x_2' \,],$$

$$a_{x'}^{(i,2)} = -\frac{1}{2\pi x_2'} \, [\, y' \, c + x' \, \Delta\theta' \,], \tag{12.4.10}$$

$$a_{y'}^{(i,2)} = -\frac{1}{2\pi x_2'} \, [\, x' \, c - y' \, \Delta\theta' + x_2' \,].$$

We have defined

$$x_2' \equiv x'^{(i+1)}, \qquad \Delta\theta' \equiv \theta_2' - \theta_1', \qquad \theta_2' \equiv \arctan \frac{y'}{x' - x_2'},$$

$$\theta_1' \equiv \arctan \frac{y'}{x'}, \qquad c \equiv \log \Big(\frac{(x' - x_2')^2 + y'^2}{x'^2 + y'^2} \Big)^{1/2}. \tag{12.4.11}$$

To obtain the velocity components in the global frame, we use the inverse of the coordinate transformation shown in (12.4.3), finding

$$u_x^{(i)}(x, y) = u_{x'}^{(i)}(x', y') \, \cos\theta^{(i)} - u_{y'}^{(i)}(x', y') \, \sin\theta^{(i)},$$

$$\tag{12.4.12}$$

$$u_y^{(i)}(x, y) = u_{x'}^{(i)}(x', y') \, \sin\theta^{(i)} + u_{y'}^{(i)}(x', y') \, \cos\theta^{(i)}.$$

Substituting expressions (12.4.9) in the right-hand sides of equations (12.4.12) and rearranging, we derive explicit relations in terms of the strength of the vortex panel at the end points,

$$u_x^{(i)}(x, y) = a_x^{(i,1)} \gamma^{(i)} + a_x^{(i,2)} \gamma^{(i+1)},$$

$$\tag{12.4.13}$$

$$u_y^{(i)}(x, y) = a_y^{(i,1)} \gamma^{(i)} + a_{y'}^{(i,2)} \gamma^{(i+1)},$$

where $a_x^{(i,1)}, a_x^{(i,2)}, a_y^{(i,1)}$, and $a_y^{(i,2)}$ are global influence coefficients given by

$$a_x^{(i,1)}(x, y) = a_{x'}^{(i,1)}(x', y') \cos\theta^{(i)} - a_{y'}^{(i,1)}(x', y') \sin\theta^{(i)},$$

$$a_x^{(i,2)}(x, y) = a_{x'}^{(i,2)}(x', y') \cos\theta^{(i)} - a_{y'}^{(i,2)}(x', y') \sin\theta^{(i)},$$

$$a_y^{(i,1)}(x, y) = a_{x'}^{(i,1)}(x', y') \sin\theta^{(i)} + a_{y'}^{(i,1)}(x', y') \cos\theta^{(i)}, \tag{12.4.14}$$

$$a_y^{(i,2)}(x, y) = a_{x'}^{(i,2)}(x', y') \sin\theta^{(i)} + a_{y'}^{(i,2)}(x', y') \cos\theta^{(i)}.$$

Given the coordinates of the evaluation point (x, y), we evaluate these coefficients by carrying out the following steps:

1. Compute the panel inclination angle, $\theta^{(i)}$, and the panel length,
$$x'_2 \equiv x'^{(i+1)} = [\,(x_{i+1} - x_i)^2 + (y_{i+1} - y_i)^2\,]^{1/2}. \qquad (12.4.15)$$
2. Compute the local coordinates (x', y') using (12.4.3).
3. Compute the local influence coefficients using (12.4.10).
4. Compute the global influence coefficients using (12.4.14).

12.4.1 Velocity in terms of the panel strength

Substituting expressions (12.4.13) in the right-hand side of (12.4.1), we obtain explicit expressions for the global components of the velocity in terms of the strength of the vortex panels at the nodes,

$$\begin{aligned} u_x(x, y) &= U_x + \sum_{i=1}^{N} [\, a_x^{(i,1)}(x, y)\, \gamma^{(i)} + a_x^{(i,2)}(x, y)\, \gamma^{(i+1)}\,] \\ &\equiv U_x + \sum_{i=1}^{N+1} b_x^{(i)}(x, y)\, \gamma^{(i)} \end{aligned} \qquad (12.4.16)$$

and

$$\begin{aligned} u_y(x, y) &= U_y + \sum_{i=1}^{N} [\, a_y^{(i,1)}(x, y)\, \gamma^{(i)} + a_y^{(i,2)}(x, y)\gamma^{(i+1)}\,] \\ &\equiv U_y + \sum_{i=1}^{N+1} b_y^{(i)}(x, y)\, \gamma^{(i)}. \end{aligned} \qquad (12.4.17)$$

We have introduced the new influence coefficients

$$\begin{aligned} b_x^{(1)}(x, y) &= a_x^{(1,1)}(x, y), \\ b_x^{(2)}(x, y) &= a_x^{(1,2)}(x, y) + a_x^{(2,1)}(x, y), \\ b_x^{(3)}(x, y) &= a_x^{(2,2)}(x, y) + a_x^{(3,1)}(x, y), \\ &\dots \\ b_x^{(N)}(x, y) &= a_x^{(N-1,2)}(x, y) + a_x^{(N,1)}(x, y), \\ b_x^{(N+1)}(x, y) &= a_x^{(N,2)}(x, y). \end{aligned} \qquad (12.4.18)$$

The coefficients $b_y^{(i)}$ are defined by corresponding expressions.

12.4.2 Point collocation

If we knew the strength of the vortex panels at the nodes, we would be able to use equations (12.4.17) to evaluate the velocity at any point in the flow. The fundamental idea underlying the vortex-panel method is that the $N+1$ unknown values, $\gamma^{(i)}$, where $i = 1, \ldots, N$, should be computed to satisfy the no-penetration condition,

$$\mathbf{u}(\mathbf{x}) \cdot \mathbf{n}(\mathbf{x}) = 0, \tag{12.4.19}$$

where the point $\mathbf{x}$ is located at the contour of the airfoil, and $\mathbf{n}$ is the unit vector normal to the contour of the airfoil.

In the panel-collocation method, N equations emerge by requiring that (12.4.19) is satisfied at the mid-point of each panel located at

$$\mathbf{x}_M^{(j)} = \frac{1}{2}\,(\mathbf{x}^{(j)} + \mathbf{x}^{(j+1)}), \tag{12.4.20}$$

where $j = 1, \ldots, N$. Using (12.4.17) to express the velocity in terms of the strength of the vortex sheet and rearranging, we find

$$\sum_{i=1}^{N+1} A_{i,j}\,\gamma^{(i)} = -U_x\, n_x(x_M^{(j)}, y_M^{(j)}) - U_y\, n_y(x_M^{(j)}, y_M^{(j)}), \tag{12.4.21}$$

where we have defined

$$A_{i,j} \equiv b_x^{(i)}(x_M^{(j)}, y_M^{(j)})\, n_x(x_M^{(j)}, y_M^{(j)}) + b_y^{(i)}(x_M^{(j)}, y_M^{(j)})\, n_y(x_M^{(j)}, y_M^{(j)}). \tag{12.4.22}$$

It is important to bear in mind that, when $i = j$, the self-induced velocity is evaluated at the mid-point of the panel on the side of the flow. With reference to equations (12.4.10), this means that $y' = 0$, $c = 0$, and $\Delta\theta' = \pi$. Applying equation (12.4.21) for $j = 1, \ldots, N$, we obtain a system of N linear algebraic equations for the $N+1$ unknowns $\gamma^{(i)}$.

Kutta-Joukowski condition

One degree of freedom is available and can be used to arbitrarily specify the circulation around the airfoil. In section 12.4.2, we saw that, in practice, the circulation established around the airfoil is such that the Kutta-Joukowski condition is fulfilled at the trailing edge. In our formulation, this condition is implemented by requiring that the strength of the vortex sheet on the upper side of the airfoil at the trailing edge is equal in magnitude and opposite in sign to the strength of the vortex sheet on the lower side of the airfoil at the trailing edge. The mathematical statement of the Kutta-Joukowski condition is then

$$\gamma^{(1)} = -\gamma^{(N+1)}. \tag{12.4.23}$$

Appending this equation to equation (12.4.21) written for $j = 1, \ldots, N$, we obtain the desired system of $N+1$ equations for the $N+1$ unknowns $\gamma^{(i)}$. The solution can be computed using, for example, the method of Gauss elimination discussed in section 3.4.

12.4.3 Circulation and pressure coefficient

Once the strength of the vortex sheet is available, the tangential velocity, $u_t = \mathbf{u} \cdot \mathbf{t}$, can be evaluated from the discrete representation (12.4.17), where $\mathbf{t}$ is the unit vector tangent to the airfoil pointing along the local x' axis.

The circulation around the airfoil can be approximated with either one of the expressions

$$C = -\sum_{i=1}^{N} u_t(\mathbf{x}_M^{(i)})\, \Delta l^{(i)} = \frac{1}{2} \sum_{i=1}^{N} (\gamma^{(i)} + \gamma^{(i+1)})\, \Delta l_i, \tag{12.4.24}$$

where Δl_i is the length of the ith panel. The second expression implements the trapezoidal rule for integrating the strength of the vortex sheet with respect to arc length around the airfoil.

The dimensionless pressure coefficient at the panel mid-points is defined by the expression

$$c_p^{(i)}(\mathbf{x}_M^{(i)}) \equiv \frac{p(\mathbf{x}_M^{(i)}) - p_\infty}{\frac{1}{2}\,\rho U^2} = 1 - \frac{u_t^2(\mathbf{x}_M^{(i)})}{U^2}, \tag{12.4.25}$$

where p_∞ is the pressure at infinity, and $U^2 \equiv U_x^2 + U_y^2$.

Figure 12.4.2(a) shows the streamline pattern and distribution of the pressure coefficient around a NACA airfoil. High pressure coefficient corresponds to the lower surface of the airfoil, and low pressure coefficient corresponds to the upper surface of the airfoil; the difference produces a lift force.

12.4.4 Lift

In the absence of viscous stresses, the force exerted on the airfoil is given by the pressure integral

$$\mathbf{F} = \oint (-p)\, \mathbf{n}\, \mathrm{d}l = \frac{1}{2}\,\rho\, U^2 - \oint (c_p + \frac{p^\infty}{\frac{1}{2}\rho U^2})\, \mathbf{n}\, \mathrm{d}l, \tag{12.4.26}$$

where $\mathbf{n}$ is the unit vector normal to the airfoil pointing into the fluid. Since the integral of the normal vector, $\mathbf{n}$, around a closed contour is zero,

$$\mathbf{F} = -\frac{1}{2}\,\rho\, U^2 \oint c_p\, \mathbf{n}\, \mathrm{d}l, \tag{12.4.27}$$

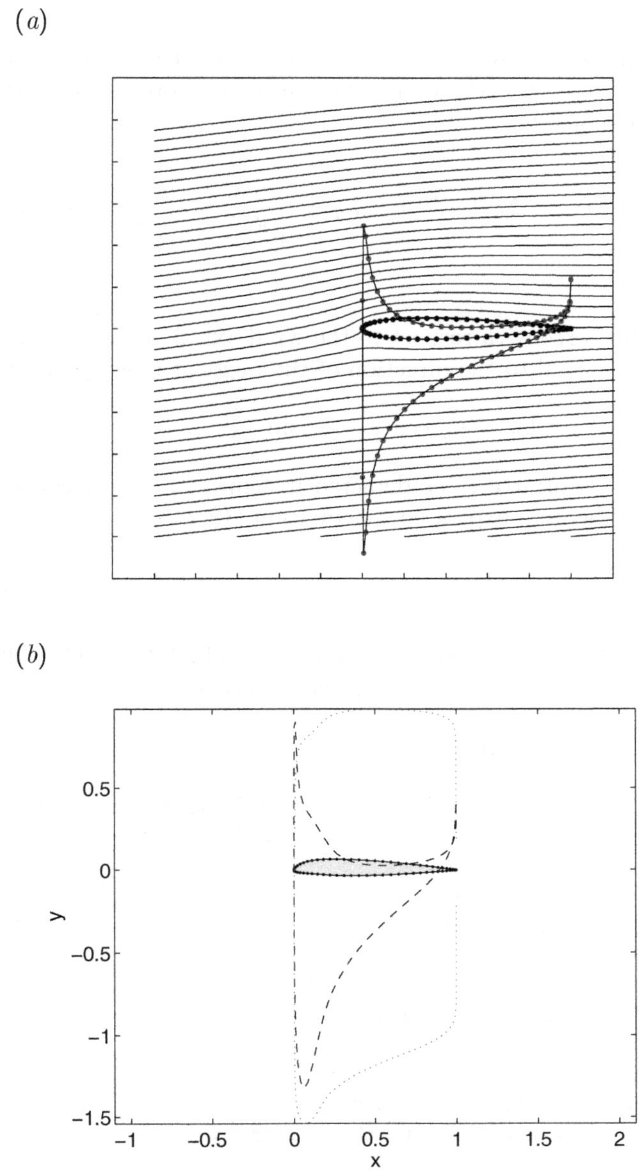

Figure 12.4.2 (*a*) Streamline pattern and distribution of the pressure coefficient around the upper and lower surface of an airfoil for angle of attack 5°, computed using the linear vortex panel method. (*b*) Distribution of the pressure coefficient (dashed line) and panel strength (dotted line) around another airfoil. Both solutions were generated by the FDLIB program *airf_2d_lvp*.

Using the trapezoidal rule to approximate the two scalar components of the vector integral in (12.4.27), we obtain the following expressions for the scaled x and y force components,

$$\hat{F}_x \equiv \frac{F_x}{\frac{1}{2}\rho U^2} = -\sum_{i=1}^{N} c_p^{(i)} \Delta l^{(i)} \, n_x^{(i)}, \qquad \hat{F}_y \equiv \frac{F_y}{\frac{1}{2}\rho U^2} = -\sum_{i=1}^{N} c_p^{(i)} \Delta l^{(i)} \, n_y^{(i)}. \tag{12.4.28}$$

The scaled lift force with respect to the wind axis is defined as the component of the force normal to the direction of the incident flow, given by

$$\widehat{L}_w = \widehat{F}_y \, \cos\alpha - \widehat{F}_x \, \sin\alpha, \tag{12.4.29}$$

where α is the angle of attack, as depicted in figure 12.4.1. According to the Kutta-Joukowski theorem expressed by equation (12.2.2),

$$\widehat{L}_w = -2\,\frac{C}{U}. \tag{12.4.30}$$

The difference in the values of the lift produced by (12.4.29) or (12.4.30) serves as a measure of the accuracy of the numerical method.

12.4.5 Vortex panel code

The following MATLAB program *airf_2d_lvp* residing in directory *07_ptf* of FDLIB implements the linear vortex panel method:

```
%============================
% linear vortex panel method
%============================

Npl = 64;    % number of panels
chord = 1.0;    % chord
thick = 0.1;    % airfoil thickness
Umag = 1.0;   % incident velocity
alpha = 4.0;    % angle of attack in degrees

%----
% generate the NACA 23012 airfoil contour
% ep are the panel end points
%----

ep = naca(chord,thick,Npl);

%---
% coordinates of the panel end points
%---
```

```
for i=1:Npl
  pt1(i,1) = ep(i,1);
  pt1(i,2) = ep(i,2);
  pt2(i,1) = ep(i+1,1);
  pt2(i,2) = ep(i+1,2);
end

%--------------------------------------
% compute:
%
% 1.  panel inclination angle th(j)
% (between -pi and pi)
% 2.  panel lengths
% 3.  tangential and normal unit vectors
%--------------------------------------

for i=1:Npl
  dx = pt2(i,1)-pt1(i,1);
  dy = pt2(i,2)-pt1(i,2);
  dl(i) = sqrt( (pt2(i,1)-pt1(i,1))^2 ...    % panel length
      + (pt2(i,2)-pt1(i,2))^2);
  th(i) = atan2(dy,dx);
  tnx(i) = cos(th(i));    % tangential vector
  tny(i) = sin(th(i));
  vnx(i) = -tny(i);    % normal vector points into flow
  vny(i) = tnx(i);
end

%---
% collocation points are panel mid-points
%---

for i=1:Npl
  co(i,1) = 0.5*(pt1(i,1)+pt2(i,1));
  co(i,2) = 0.5*(pt1(i,2)+pt2(i,2));
end

%------------------------------------
% compute the influence coefficients
% and right-hand side of linear system
%------------------------------------

for i=1:Npl   % OVER COLLOCATION POINTS; ENDS AT THE RABBIT

%---
% compute influence of jth panel
% on ith collocation point
%---
```

```
for j=1:Npl   % OVER PANELS; ENDS AT THE DOG

%---
% Compute the local panel coordinates (x, y).
% The first panel point is located at the shifted origin;
% the panel becomes horizontal
%---

xt = co(i,1)-pt1(j,1);   % shift collocation point
yt = co(i,2)-pt1(j,2);
x = xt*tnx(j)+yt*tny(j);   % and rotate
y = -xt*tny(j)+yt*tnx(j);
x1 = 0.0; y1 = 0.0;
x2t = pt2(j,1)-pt1(j,1);   % shift second-end point
y2t = pt2(j,2)-pt1(j,2);
x2 = x2t*tnx(j) + y2t*tny(j);   % and rotate y2 = 0.0;

%---
% Compute radial distances:  r1, r2
% subtended angles:  th1, th2
%---

r1 = sqrt(x*x+y*y);
r2 = sqrt((x-x2)*(x-x2)+y*y);
th1 = atan2(y,x);
th2 = atan2(y,x-x2);

%--------------------------------------------
% gamma1, gamma2 are the vortex sheet strengths
% at the first and second panel points.
%
% Compute influence coefficients associated with
% the velocity components (ux, uy)
% corresponding to gamma1 and gamma, so that:
%
% ux = ax1*gamma1 + ax2*gamma2
% uy = ay1*gamma1 + ay2*gamma2
%
% These velocities are in
% the jth-panel reference frame.
%--------------------------------------------

if(i==j)   % self-induced velocity
  ax1 = 0.5*(x/x2-1.0);
  ay1 = 1.0/(2*pi);
  ax2 =-0.5*x/x2;
  ay2 =-1.0/(2*pi);
else
```

```
  dth = th2-th1;
  rrt = r2/r1;
  rrtl = log(rrt);
  fcc = 1.0/(2*pi*x2);
  ax1 = fcc*( y*rrtl + (x-x2)*dth );
  ay1 = fcc*((x-x2)*rrtl - y*dth + x2);
  ax2 = -fcc*(y*rrtl + x*dth );
  ay2 = -fcc*(x*rrtl - y*dth + x2);
end

%---
% transform the influence coefficient into the
% global reference frame by rotation
%---

ux1 = ax1*tnx(j) - ay1*tny(j);
uy1 = ax1*tny(j) + ay1*tnx(j);
ux2 = ax2*tnx(j) - ay2*tny(j);
uy2 = ax2*tny(j) + ay2*tnx(j);

%-----------------------------------------------
% Compute the coefficients of gamma in the
% master influence matrix.
%
% These are the velocity influence coefficients
% projected onto the normal vector of ith panel
%-----------------------------------------------

if(j==1)
  a(i,1)= ux1*vnx(i) + uy1*vny(i);
  holda = ux2*vnx(i) + uy2*vny(i);   % hold for the next panel
elseif(j==Npl)
  a(i,Npl) = ux1*vnx(i) + uy1*vny(i) + holda;
  a(i,Npl+1) = ux2*vnx(i) + uy2*vny(i);
else
  a(i,j)= ux1*vnx(i) + uy1*vny(i) + holda;
  holda = ux2*vnx(i) + uy2*vny(i);   % hold for the next panel
end

%-----------------------------------------------------
% b are tangential velocity influence coefficients
%
% These are the velocity influence coefficients
% projected onto the tangential vector of the ith panel
%-----------------------------------------------------

if(j==1)
  b(i,1)= ux1*tnx(i) + uy1*tny(i);
  holdb = ux2*tnx(i) + uy2*tny(i);   % hold for the next panel
```

```
elseif(j==Npl)
  b(i,Npl) = ux1*tnx(i) + uy1*tny(i) + holdb;
  b(i,Npl+1) = ux2*tnx(i) + uy2*tny(i);
else
  b(i,j)= ux1*tnx(i) + uy1*tny(i) + holdb;
  holdb = ux2*tnx(i) + uy2*tny(i);   % hold for the next panel
end

end % OVER PANELS - DOG

end % OVER COLLOCATION POINTS - RABBIT

%--------------------------------
% Add the Kutta condition
% expressed by the Npl+1 equation:
%
% The strength of the sheet at the first point
% of the first panel is equal and opposite
% to the strength of the sheet at the last point
% of the last panel,
% so that the mean value vanishes
%--------------------------------

a(Npl+1,1) = 1.0;
a(Npl+1,Npl+1) = 1.0;

%--------------------
% set right-hand side
%--------------------

al = alpha*pi/180.0;
cal = cos(al); sal = sin(al);
Ux = Umag*cal; Uy = Umag*sal;

for i=1:Npl
  rhs(i) = -Ux*vnx(i)-Uy*vny(i);
end
rhs(Npl+1)=0.0;

%----------------------
% solve the linear system
%----------------------

gamma = rhs/a';

%--------------------------------
% compute c_p and the circulation
%--------------------------------
```

```
circul = 0.0;    % circulation
circulg = 0.0;    % circulation in terms of gamma

for i=1:Npl
  velt = Ux*tnx(i)+Uy*tny(i); % tangential velocity
  for j=1:Npl+1
    velt = velt + b(i,j)*gamma(j);
  end
  circul = circul - velt*dl(i);    % circulation
  circulg = circulg ...
    +0.5*(gamma(i)+gamma(i+1))*dl(i);    % circulation
  cp(i) = 1.0-velt*velt/(Umag*Umag);    % press coeff
end

cp(Npl+1)=cp(1);

%---
% plotting
%---

patch(ep(:,1),ep(:,2),'y')
plot(ep(:,1),ep(:,2),'k.-')
plot(ep(:,1),cp,'--')
plot(ep(:,1),gamma,'r:')
xlabel('x','fontsize',15)
ylabel('y','fontsize',15)
```

The airfoil profile is generated by the following MATLAB function:

```
function [ep]=naca(chord,thick,Npl)

%----
% generate the NACA 23012 airfoil contour
% ep are the panel end points
%----

dpsi = 2*pi/Npl;

ep(1,1) = chord;
ep(1,2) = 0.0;

for i=2:Npl

  psi = (i-1.0)*dpsi;
  x = 0.5*(1.0+cos(psi));

%---
% camber line:
%---
```

```
  if(x<0.2025)
    yc = 2.6595*x*(0.1147+x*(-0.6075+x));
  else
    yc = 0.02208*(1.0-x);
  end

%---
% thickness:
%---

  yt = 5.0*thick*(0.2969*sqrt(x) ...
    -0.1260*x -0.3516*x*x +0.2843*x*x*x ...
    -0.1036*x*x*x*x );
  if(i<=(Npl/2+1))
    y = yc-yt;
  else
    y = yc+yt;
  end

  ep(i,1) = x*chord;
  ep(i,2) = y*chord;
end

ep(Npl+1,1) = chord;   % trailing edge
ep(Npl+1,2) = 0.0;    % trailing edge

%---
% done
%---

return
```

The graphics generated by the code is shown in figure 12.4.2(*b*). The dotted line shows the distribution of the panel strength, and the dashed line shows the distribution of the pressure coefficient.

Computer Problem

12.4.1 *Linear vortex panel method*

(*a*) Run the code *airf_2d_lvp* for an airfoil of your choice. Prepare graphs and discuss the distribution of the panel strength and pressure coefficient.

(*b*) Evaluate the velocity at several points inside the airfoil and discuss your results.

12.5 Vortex sheet representation

Consider the vortex panel method for flow past a two-dimensional airfoil discussed in section 12.4. In the limit as the number of panels, N, tends to infinity, the piecewise linear strength distributions over the individual panels join to yield a smooth distribution defined over the airfoil contour. Correspondingly, the sum on the right-hand side of equation (12.4.1) reduces to an integral with respect to arc length around the airfoil contour, representing the velocity induced by a vortex sheet with a generally curved shape.

Generalizing expression (12.3.6), we find that the stream function associated with this vortex sheet is given by

$$\psi^{VS}(x, y) = -\frac{1}{4\pi} \oint \ln[(x - x')^2 + (y - y')^2]\, \gamma(x')\, \mathrm{d}l', \tag{12.5.1}$$

where $\mathrm{d}l' = (\,\mathrm{d}x'^2 + \mathrm{d}y'^2\,)^{1/2}$ is the differential arc length around the airfoil measured in the clockwise direction starting at the trailing edge, and the integration is performed around the airfoil contour.

Conversely, the vortex panel representation can be viewed as the result of discretizing the integral on the right-hand side of (12.5.1) into geometrical elements represented by the vortex panels. In section 12.4, we discussed straight elements with linear strength distribution. In more advanced implementations, curved elements including sections of parabolas and circular arcs, and quadratic or higher-order strength density distributions are employed.

Internal flow

Although the vortex sheet representation is physically meaningful only when it is used to evaluate the stream function and velocity at a point in the flow, nothing prevents us from performing corresponding evaluations at a point inside the airfoil. When this is done, we find that the stream function is constant and the velocity vanishes inside the airfoil (problem 12.5.2).

To explain this curiosity, we observe that the strength of the vortex sheet is computed to satisfy the no-penetration boundary condition. Because the normal component of the velocity is continuous across the vortex sheet, the interior flow occurs under vanishing normal boundary velocity. Since tangential velocity on the interior side of the airfoil is prohibited by the condition of irrotational motion in the absence of singularities, the fictitious internal flow must vanish.

12.5.1 Thin airfoil theory

Next, we consider flow past a thin cambered airfoil, as illustrated in figure 12.5.1, and introduce a system of coordinates such that the leading edge lies at

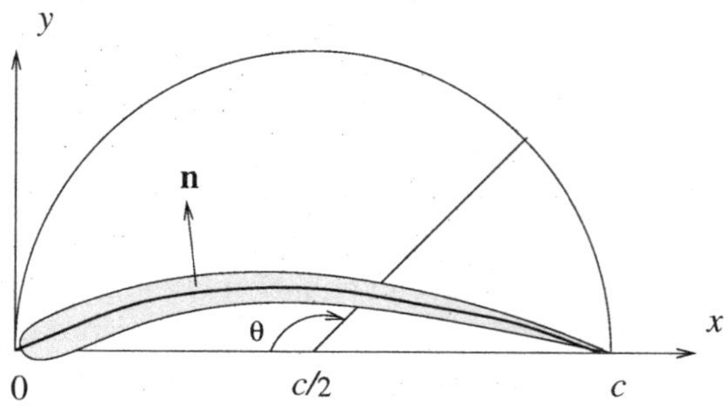

Figure 12.5.1 Illustration of the camberline of a thin airfoil, and definition of variables used to develop the slender-airfoil theory.

the origin of the x axis, and the trailing edge lies at the point $x = c$ along the x axis. The camberline is described by the equation

$$y = \epsilon\, \eta_c(x), \tag{12.5.2}$$

where ϵ is a small dimensionless number, and $\eta_c(x)$ is the camberline shape function required to satisfy the geometrical constraints $\eta_c(0) = 0$ and $\eta_c(c) = 0$. If $\eta_c = 0$ for all x, the camberline is flat.

Because both sides of the airfoil are close to the x axis, the corresponding line integrals in (12.5.1) can be approximated with integrals with respect to x from $x = 0$ to c. Setting $y' = 0$, tracing the airfoil in the clockwise direction beginning at the trailing edge, and noting that on the upper side of the airfoil $\mathrm{d}l' = \mathrm{d}x'$ while on the lower side $\mathrm{d}l' = -\mathrm{d}x'$, we find

$$\psi^{VS}(x, y) \simeq -\frac{1}{4\pi} \int_0^c \ln[(x - x')^2 + y^2]\, (\gamma^+ - \gamma^-)(x')\, \mathrm{d}x', \tag{12.5.3}$$

where the superscripts + and - denote, respectively, the upper and lower side. Defining $\chi \equiv \gamma^+ - \gamma^-$, we obtain a representation in terms of an effective flat vortex sheet subtended between the leading and trailing edge,

$$\psi^{VS}(x, y) \simeq -\frac{1}{4\pi} \int_0^c \log[(x - x')^2 + y^2]\, \chi(x')\, \mathrm{d}x'. \tag{12.5.4}$$

The corresponding velocity potential is given by

$$\phi^{VS}(x, y) \simeq \frac{1}{2\pi} \int_0^c \arctan(\frac{y}{x - x'})\, \chi(x')\, \mathrm{d}x'. \tag{12.5.5}$$

Our goal is to compute the strength of the effective vortex sheet, χ, so as to satisfy the no-penetration condition around the airfoil.

Velocity on either side of the vortex sheet

As a preliminary, we consider the velocity induced by the effective vortex sheet on the upper and lower side of the airfoil. To begin, we consider the limit of the velocity potential as the evaluation point $\mathbf{x} = (x, y)$ approaches the vortex sheet from the upper side; that is, $y \to 0+$ with $0 < x < c$. In this limit, the inverse tangent function on the right-hand side of (12.5.5) is zero when $x' < x$, or π when $x' > x$. Consequently, the potential takes the value

$$\phi^{VS}(x, y \to 0+) = \frac{1}{2} \int_x^c \chi(x')\, dx'. \tag{12.5.6}$$

Differentiating both sides of (12.5.6) with respect to x, we find that the x velocity component is given by

$$u_x^{VS}(x, y \to 0+) = -\frac{1}{2}\, \chi(x). \tag{12.5.7}$$

Working in a similar fashion for the lower side, we find

$$u_x^{VS}(x, y \to 0-) = \frac{1}{2}\, \chi(x). \tag{12.5.8}$$

The last two equations illustrate once again that the velocity undergoes a discontinuity whose magnitude is equal to the strength of the vortex sheet.

The y component of the velocity over the airfoil can be found by differentiating either (12.5.4) with respect to x or (12.5.5) with respect to y. Either way, we find

$$u_y(x, y = 0) = \frac{1}{2\pi} \int_0^c \frac{\chi(x')}{x - x'} dx'. \tag{12.5.9}$$

Unlike the x component of the velocity, the y component remains continuous across the vortex sheet.

Asymptotics

To compute the strength of the effective vortex sheet, χ, we implement the no-penetration condition through a series of approximations that may appear drastic but nevertheless have a solid theoretical foundation.

First, we replace the disturbance flow due to the airfoil with the flow due to the effective vortex sheet expressed by (12.5.4) or (12.5.5). Enforcing the no-penetration boundary condition at the camberline, we obtain

$$\mathbf{n} \cdot [\, \mathbf{U} + \mathbf{u}^{VS}(x = \epsilon\, \eta_c)\,] = 0, \tag{12.5.10}$$

where $\mathbf{n}$ is the unit vector normal to the camberline on the upper side, pointing into the fluid. Using (12.5.2), we find

$$n_x = -\frac{\epsilon\, \eta_c'}{\sqrt{1 + \epsilon^2\, \eta_c'^2}}, \qquad n_y = \frac{1}{\sqrt{1 + \epsilon^2\, \eta_c'^2}}. \tag{12.5.11}$$

where $\eta'_c = \mathrm{d}\eta_c/\mathrm{d}x$. Because ϵ has been assumed small, the denominators can be replaced by unity, yielding the simplified expressions

$$n_x \simeq -\epsilon \frac{\mathrm{d}\eta_c}{\mathrm{d}x}, \qquad n_y \simeq 1. \tag{12.5.12}$$

A second approximation arises by replacing the velocity on the upper side of the camberline with the velocity on the upper side of the effective vortex sheet given in (12.5.7) and (12.5.9).

Substituting (12.5.12), (12.5.7), and (12.5.9) in (12.5.10), we find

$$-\epsilon \frac{\mathrm{d}\eta_c}{\mathrm{d}x}(x)\,[\,U_x - \frac{1}{2}\chi(x)\,] + U_y + \frac{1}{2\pi}\int_0^c \frac{\chi(x')}{x - x'}\,\mathrm{d}x' = 0. \tag{12.5.13}$$

Next, we confine our attention to flow that is nearly parallel to the x axis, meaning that U_y is small compared to U_x. Writing

$$U_y = \tan\alpha\, U_x \simeq \alpha\, U_x, \tag{12.5.14}$$

we obtain

$$-\epsilon \frac{\mathrm{d}\eta_c}{\mathrm{d}x}(x)\,[\,U_x - \frac{1}{2}\chi(x)\,] + \alpha\, U_x + \frac{1}{2\pi}\int_0^c \frac{\chi(x')}{x - x'}\,\mathrm{d}x' = 0, \tag{12.5.15}$$

where α is the angle of attack.

Inspecting the various terms on the left-hand side of equation (12.5.15), we find that the magnitude of χ is on the order of ϵ or α; both have been assumed small. Since χ is negligible compared to U_x, it can be discarded from the expression enclosed by the square brackets, leaving the equation

$$\frac{1}{2\pi U_x}\int_0^c \frac{\chi(x')}{x - x'}\,\mathrm{d}x' = \epsilon \frac{\mathrm{d}\eta_c}{\mathrm{d}x}(x) - \alpha, \tag{12.5.16}$$

where $0 < x < c$. The problem has been reduced to computing a function, $\chi(x)$, that satisfies the integral equation of the first kind (12.5.16).

Solution by Fourier expansion

One way to solve equation (12.5.16) is by expanding the unknown function χ in a Fourier series with respect to the angle θ varying between 0 and π, defined such that

$$x = \frac{c}{2}(1 - \cos\theta), \tag{12.5.17}$$

as illustrated in figure 12.5.1. The Kutta-Joukowski condition requires that the strength of the vortex sheet vanishes at the trailing edge, and wind tunnel evidence reveals a large peak at the leading edge. Accordingly, we express χ as the sum of the singular function, $\cot\frac{\theta}{2}$, and a sine Fourier series in the form

$$\chi = 2\,U_x\,[\,a_0\,\cot\frac{\theta}{2} + \sum_{i=1}^{\infty} a_i \sin(i\theta)\,], \tag{12.5.18}$$

where a_i are constant coefficients. Substituting (12.5.17) and (12.5.18) in (12.5.16), we find

$$\frac{1}{\pi}\int_0^{\pi} \frac{\sin\theta'}{\cos\theta' - \cos\theta}\,[\,a_0\,\cot\frac{\theta'}{2} + \sum_{i=1}^{\infty} a_i \sin(i\theta')\,]\,\mathrm{d}\theta' = \epsilon\,\frac{\mathrm{d}\eta_c}{\mathrm{d}x}(x) - \alpha. \tag{12.5.19}$$

The integrals on the left-hand side of (12.5.19) can be evaluated with the help of standard tables. First, we write

$$\cot\frac{\theta'}{2} = \frac{1+\cos\theta'}{\sin\theta'}, \tag{12.5.20}$$

and find

$$\int_0^{\pi} \frac{\sin\theta'}{\cos\theta' - \cos\theta}\,\cot\frac{\theta'}{2}\,\mathrm{d}\theta' = \int_0^{\pi} \frac{1+\cos\theta'}{\cos\theta' - \cos\theta}\,\mathrm{d}\theta' = \pi. \tag{12.5.21}$$

Second, we note that

$$\int_0^{\pi} \frac{\sin\theta'}{\cos\theta' - \cos\theta}\,\sin(i\theta')\,\mathrm{d}\theta' = -\pi\cos(i\theta). \tag{12.5.22}$$

Substituting these results in (12.5.19), we derive the remarkably simple expression

$$a_0 - \sum_{i=1}^{\infty} a_i\,\cos(i\theta) = \epsilon\,\frac{\mathrm{d}\eta_c}{\mathrm{d}x}(x) - \alpha. \tag{12.5.23}$$

The left-hand side of (12.5.23) is the cosine Fourier expansion of the right-hand side with respect to θ. Multiplying both sides by $\cos(j\theta)$, where j is an integer, integrating with respect to θ from 0 to π, and using the identity

$$\int_0^{\pi} \cos(i\theta)\,\cos(j\theta)\,\mathrm{d}\theta = \begin{cases} 0 & \text{if } i \neq j, \\ \pi & \text{if } i = j = 0, \\ \pi/2 & \text{if } i = j \neq 0, \end{cases} \tag{12.5.24}$$

we find

$$a_0 = -\alpha + \frac{\epsilon}{\pi}\int_0^{\pi} \frac{\mathrm{d}\eta_c}{\mathrm{d}x}\,\mathrm{d}\theta, \qquad a_i = -\frac{2\epsilon}{\pi}\int_0^{\pi} \frac{\mathrm{d}\eta_c}{\mathrm{d}x}\,\cos(i\theta)\,\mathrm{d}\theta, \tag{12.5.25}$$

for $i = 1, 2, \ldots$.

The camberline slope, $\mathrm{d}\eta_c/\mathrm{d}x$, can be expanded in a cosine Fourier series with respect to θ,

$$\frac{\mathrm{d}\eta_c}{\mathrm{d}x}(x) = \frac{1}{2}\,b_0 + \sum_{i=1}^{\infty} b_i \cos(i\theta), \tag{12.5.26}$$

where b_i are dimensionless coefficients. Substituting the right-hand side of (12.5.26) in (12.5.23) and setting the sum of like Fourier coefficients to zero, we obtain

$$a_0 = -\alpha + \frac{\epsilon}{2}\,b_0, \qquad a_i = -\epsilon\, b_i, \tag{12.5.27}$$

for $i = 1, 2, \ldots$. These relations illustrate that only the leading coefficient, a_0, depends on the angle of attack; the rest of the coefficients are determined by the geometry of the camberline.

Lift, lift coefficient, and lift slope

To compute the lift force per unit span exerted on the airfoil, we use the Kutta-Joukowski theorem expressed by equation (12.2.2), obtaining

$$L = -\rho\, U_x \int_0^c \chi(x)\,\mathrm{d}x. \tag{12.5.28}$$

Substituting expansion (12.5.18) in the right-hand side and evaluating the integrals, we find that only two terms make a non-zero contribution, yielding the lift force

$$L = -\rho\, U_x^2\, c\,\pi\,(a_0 + \frac{1}{2}\,a_1). \tag{12.5.29}$$

Using (12.5.27), we find

$$L = \rho\, U_x^2\, c\,\pi\,(\alpha - \frac{\epsilon}{2}\sum_{i=1}^{\infty} a_i). \tag{12.5.30}$$

The lift coefficient is

$$c_L \equiv \frac{L}{\frac{1}{2}\,\rho\, U_x^2\, c} = -2\pi\,(a_0 + \frac{1}{2}a_1) = 2\pi\,(\alpha - \frac{\epsilon}{2}\sum_{i=1}^{\infty} a_i). \tag{12.5.31}$$

In practical aerodynamics, the performance of an airfoil is characterized by the lift slope, defined as the slope $\mathrm{d}c_L/\mathrm{d}\alpha$. Our analysis has shown that the lift slope of a thin airfoil is constant and equal to 2π, independent of the camber.

Pressure difference

The difference in the pressure on either side of the vortex sheet representing the airfoil,

$$\Delta p \equiv p(x, y \to 0-) - p(x, y \to 0+), \tag{12.5.32}$$

can by computed using Bernoulli's equation (6.4.14). Using expressions (12.5.7) and (12.5.8), we find that the velocity on the upper or lower side of the vortex sheet is, respectively, equal to $U_x - \frac{1}{2}\chi$ and $U_x + \frac{1}{2}\chi$. Substituting these expressions in Bernoulli's equation, we obtain

$$\Delta p = [p_\infty - \frac{1}{2}\rho\,(U_x + \frac{1}{2}\chi)^2] - [p_\infty - \frac{1}{2}\rho\,(U_x - \frac{1}{2}\chi)^2] = -\rho\,U_x\,\chi. \tag{12.5.33}$$

Pressure moment

The moment of the pressure force with respect to the leading edge is expressed by the integral

$$M = -\int_o^c x\,\Delta p\,\mathrm{d}x. \tag{12.5.34}$$

Substituting (12.5.33) and expansion (12.5.18) in the right-hand side and evaluating the emerging definite integrals with the help of standard mathematical tables, we find

$$M = \frac{1}{4}\,\rho\,U_x^2\,c^2\,\pi\,(a_0 + a_1 - \frac{1}{2}\,a_2), \tag{12.5.35}$$

which shows that only three coefficients contribute to the moment. The moment coefficient is defined by

$$c_M \equiv \frac{M}{\frac{1}{2}\,\rho\,U_x^2\,c^2} = \frac{\pi}{2}\,(a_0 + a_1 - \frac{1}{2}a_2). \tag{12.5.36}$$

The moment of the pressure forces with respect to an arbitrary point, $x = d$, is given by

$$M_d = -\int_o^c (x - d)\,\Delta p\,\mathrm{d}x = M + d\,L. \tag{12.5.37}$$

Substituting (12.5.29) and (12.5.35) in the right-hand side of (12.5.37), we find that, when $d = c/4$, the coefficient a_0 disappears and M_d becomes independent of the angle of attack, α.

Accordingly, the quarter chord moment, $M_{c/4}$, and associated moment coefficient ,

$$c_{M_{c/4}} \equiv \frac{M_{c/4}}{\frac{1}{2}\,\rho\,U_x^2\,c^2}, \tag{12.5.38}$$

are used to characterize the performance of the airfoil.

Symmetric airfoils

Since the camber of a symmetric airfoil vanishes, we may set in the preceding equations either $\epsilon = 0$ or $\eta_c(x) = 0$. Equations (12.5.27) yield $a_0 = -\alpha$ and $a_i = 0$ for $i = 1, 2, \ldots$, and expansion (12.5.18) reduces to

$$\chi = -2\,\alpha\,U_x \cot\frac{\theta}{2}. \tag{12.5.39}$$

The lift force and lift coefficient computed from equations (12.5.29) and (12.5.31) are given by

$$L = \pi\,\alpha\,c\,\rho\,U_x^2, \qquad c_L = 2\,\pi\,\alpha. \tag{12.5.40}$$

The moment and moment coefficient computed from equations (12.5.35) and (12.5.36) are given by

$$M = -\pi\alpha c^2\,\rho\,U_x^2, \qquad c_M = -\frac{\pi}{2}\,\alpha. \tag{12.5.41}$$

Problems

12.5.1 *Thin airfoil with parabolic camber*

Consider a thin airfoil with parabolic camberline described by the shape function $\eta_c(x) = 4\,x\,(1 - x/c)$. The camber is equal to $\epsilon\,\eta_c(c/2) = \epsilon c$. Show that the lift and moment coefficients are given by

$$c_L = 2\pi\,(\alpha + 2\,\epsilon), \qquad c_M = -\frac{\pi}{2}\,(\alpha + 4\epsilon). \tag{12.5.42}$$

Note that the lift vanishes when $\alpha = -2\epsilon$.

12.5.2 *NACA 23012 airfoil*

The camberline of the NACA 23012 airfoil is described by

$$\hat{y} = \begin{cases} 2.6595\,(\hat{x}^3 - 0.6075\hat{x}^2 + 0.1147\,\hat{x}) & \text{for } 0 \le \hat{x} \le 0.2025, \\ 0.02208\,(1 - \hat{x}) & \text{for } 0.2025 \le \hat{x} \le 1\ , \end{cases} \tag{12.5.43}$$

where $\hat{x} \equiv x/c$ and $\hat{y} \equiv y/c$. Derive expressions for the lift and moment coefficients in terms of the angle of attack. Compare your results with experimental measurements for $\alpha = 4°$, showing that $c_L = 0.55$ and $c_{M_{c/4}} = -0.01$.

Computer Problem

12.5.3 *Comparison of asymptotics with the vortex panel method*

Consider the NACA 23012 airfoil discussed in problem 12.5.2. Run the code *airf_2d_lvp* in directory *07_ptf* of FDLIB to compute the lift coefficient in wind axes. Compare the numerical results with the asymptotic predictions for small airfoil thickness.

12.6 Point-source-dipole panels

Consider the stream function of the flow induced by a vortex panel situated along the x axis between the point $x = a$ and b, as shown in equation (12.3.6). We define the distribution of circulation along the panel, $\mu(x)$, as the integral of the strength density of the vortex sheet with respect to x from an arbitrary point $x = d$ up to a point x, where $a < x < b$,

$$\mu(x) \equiv \int_d^x \gamma(x')\,\mathrm{d}x'. \tag{12.6.1}$$

Using the rules of integral differentiation, we find

$$\frac{\mathrm{d}\mu}{\mathrm{d}x} = \gamma. \tag{12.6.2}$$

Substituting the left-hand side of (12.6.2) in the integral of (12.3.6) and integrating by parts, we find

$$\begin{aligned} \psi^{VP}(x,y) = {} & \frac{\mu(x=a)}{2\pi}\,\ln[(x-a)^2+(y-y')^2] \\ & -\frac{\mu(x=b)}{2\pi}\,\ln[(x-b)^2+(y-y')^2] \\ & +\frac{1}{4\pi}\int_a^b \frac{\mathrm{d}\ln[(x-x')^2+(y-y')^2]}{\mathrm{d}x'}\,\mu(x')\,\mathrm{d}x', \end{aligned} \tag{12.6.3}$$

with the understanding that $y' = 0$. Carrying out the differentiation under the integral sign, we derive the final form

$$\begin{aligned} \psi^{VP}(x,y) = {} & \frac{\mu(x=a)}{2\pi}\,\ln[(x-a)^2+(y-y')^2] \\ & -\frac{\mu(x=b)}{2\pi}\,\ln[(x-b)^2+(y-y')^2] \\ & -\frac{1}{2\pi}\int_a^b \frac{x-x'}{(x-x')^2+(y-y')^2}\,\mu(x')\,\mathrm{d}x'. \end{aligned} \tag{12.6.4}$$

The first term represents the flow due to a point vortex with strength $-\mu(x = a)$ placed at the first panel end point. The second term represents the flow due to a point vortex with strength $\mu(x = b)$ placed at the second panel end point.

Recalling that, by definition, $u_x = \partial\psi/\partial y$ and $u_y = -\partial\psi/\partial x$, and comparing the third term on the right-hand side of (12.6.4) with expressions (3.5.24), we find that this term represents the flow due to a distribution of point-source dipoles with strength density $\mu(x)$ oriented normal to the panel. If $\mu(x')$ is positive, the dipole at x' points toward the positive direction of the y axis. If $\mu(x')$ is negative, the dipole at x' points toward the negative direction of the y axis.

Now we denote the stream function due to the point-source dipole distribution by ψ^{SDP}, where SDP stands for source-dipole panel, and rearrange expression (12.6.4) to obtain

$$\begin{aligned}\psi^{SDP}(x,y) &\equiv -\frac{1}{2\pi}\int_a^b \frac{x-x'}{(x-x')^2+(y-y')^2}\,\mu(x')\,\mathrm{d}x' \\ &= -\frac{\mu(x=a)}{2\pi}\ln[(x-a)^2+(y-y')^2] \\ &+\frac{\mu(x=b)}{2\pi}\ln[(x-b)^2+(y-y')^2]+\psi^{VP}(x,y),\end{aligned} \tag{12.6.5}$$

which establishes a correspondence between the flow due to a source-dipole panel and the flow due to a vortex panel, subject to the differential relation (12.6.2). The first term on the right-hand side of (12.6.5) represents the flow due to a point vortex with strength $\mu(x = a)$ placed at the first panel end point. The second term represents the flow due to a point vortex with strength $-\mu(x = b)$ placed at the second panel end point.

Panels with constant source-dipole strength density

When the strength density of the source-dipole distribution is constant, the strength of the vortex sheet vanishes and the last term on the right-hand side of (12.6.5) disappears. In that case, the flow due to the panel is identical to the flow induced by two point vortices with strengths $\mu(0)$ and $-\mu(0)$ situated, respectively, at the first and second panel end point.

12.6.1 Source-dipole panel method

The flow due to a point-source dipole panel can be used as a fundamental building block for representing and subsequently computing a flow past an airfoil. To develop the source-dipole panel method, we work by analogy with the vortex panel method discussed in section 12.4, with some differences. In the first step, we trace the contour of the airfoil with $N+1$ marker points distributed in the clockwise sense, as illustrated in figure 12.4.1. A pair of successive marker points, $\mathbf{x}^{(i)}$ and $\mathbf{x}^{(i+1)}$, defines the ith flat source-dipole panel.

In the source-dipole panel method, the velocity at a point in the flow is described by superposing the incident flow and the flows induced by the N source-dipole panels. An additional degree of freedom is required to specify the circulation around the airfoil, and is provided by enhancing the source-dipole representation with an additional contribution mediated by a point vortex with strength κ placed at the trailing edge. The composite representation is

$$\mathbf{u}(x,y) = \mathbf{U} + \sum_{i=1}^{N} \mathbf{u}^{(i)}(x,y) + \mathbf{u}^{PV}(x,y), \tag{12.6.6}$$

where $\mathbf{u}^{(i)}(x,y)$ is the velocity induced by the ith panel, and $\mathbf{u}^{PV}(x,y)$ is the velocity induced by the point vortex.

Uniform panels

We have shown that, if the source-dipole strength is uniform over each panel, the flow induced by the ith panel is identical to the flow induced by two point vortices with strength $\mu^{(i)}$ and $-\mu^{(i)}$ located at the first and second panel end points, where $\mu^{(i)}$ is the constant value of the source dipole strength density over the ith panel. Consequently, the ith marker point hosts two point-vortices, where $i = 2, \ldots, N$. One point vortex with strength $-\mu^{(i-1)}$ is contributed by the $i-1$ panel, and a second point vortex with strength $\mu^{(i)}$ is contributed by the i panel; the combined strength is $\mu^{(i)} - \mu^{(i-1)}$.

The first node located at the trailing edge hosts three point vortices: one due to the first panel, a second due to the last panel, and a third circulation-producing point vortex placed at the trailing edge. The Kutta-Joukowski condition requires that the net strength of the trailing edge point vortex is zero,

$$\mu^{(1)} - \mu^{(N)} + \kappa = 0. \tag{12.6.7}$$

The $N+1$ unknowns, including $\mu^{(i)}$ for $i = 1, \ldots, N$, and κ, can be computed by the collocation method discussed in section 12.3, incorporating the Kutta-Joukowski condition (12.6.7).

Linear panels

When the strength density of a source dipole panel varies linearly with respect to arc length, all three terms on the right-hand side of (12.6.5) make a non-zero contribution. Expression (12.6.2) shows that the strength density of the equivalent vortex sheet is constant and equal to the slope of the dipole density over the panel.

In the linear source-dipole panel method, the flow is represented by a superposition of the three components shown in equation (12.6.6), where the dipole density over the ith panel varies linearly with respect to arc length from the initial value $\mu^{(i)}$ to the final value $\mu^{(i+1)}$. Summing the flows due to the individual panels expressed by the right-hand side of (12.6.5) and consolidating the left- and right-panel point vortices at the panel end points, we derive an equivalent representation in terms of point vortices and vortex panels with constant strength. The uniform strength density of the ith vortex panel is

$$\gamma^{(i)} \equiv \frac{\mu^{(i+1)} - \mu^{(i)}}{\Delta l_i}, \tag{12.6.8}$$

where Δl_i is the panel length.

Next, we observe that, because the dipole strength is continuous around the approximate polygonal contour of the airfoil described by the panels, the strength of the point vortices vanishes at all but the first point where it takes the value $\mu^{(1)} - \mu^{(N+1)} + \kappa$. We conclude that the linear source-dipole panel representation is equivalent to the constant vortex panel representation supplemented by a point vortex at the trailing edge. The strength of the point vortex must vanish to satisfy the Kutta-Joukowski condition at the trailing edge, as required by equation (12.6.7).

12.6.2 Source-dipole representation

As the number of panels, N, tends to infinity, the individual panel strength density distributions join to yield a smooth distribution defined around the airfoil. Correspondingly, the sum on the right-hand side of (12.6.6) reduces to an integral with respect to arc length around the airfoil, yielding an integral representation in terms of a source-dipole sheet.

Equation (12.6.5) provides us with an expression for the stream function associated with a source-dipole panel situated over the x axis, where the source-dipoles point along the y axis. Generalizing this expression, we find that the stream function associated with source-dipole distribution around the airfoil is given by

$$\psi^{SD}(x, y) = -\frac{1}{2\pi} \oint \frac{(x - x')\, n_y - (y - y')\, n_x}{(x - x')^2 + (y - y')^2}\, \mu(x')\, \mathrm{d}l', \tag{12.6.9}$$

where $\mathrm{d}l' = (\mathrm{d}x'^2 + \mathrm{d}y'^2)^{1/2}$ is the differential arc length around the airfoil measured in the clockwise direction from a designated origin. The associated velocity potential

$$\phi^{SD}(x, y) = -\frac{1}{2\pi} \oint \frac{(x - x')\, n_x(\mathbf{x}') + (y - y')\, n_y(\mathbf{x}')}{(x - x')^2 + (y - y')^2}\, \mu(x')\, \mathrm{d}l'. \tag{12.6.10}$$

The counterpart of the panel representation (12.6.6) is

$$\mathbf{u}(x, y) = \mathbf{U} + \mathbf{u}^{SD}(x, y) + \mathbf{u}^{PV}(x, y), \tag{12.6.11}$$

where $\mathbf{u}^{SD}(x, y)$ is the velocity corresponding to (12.6.9) and (12.6.10). The stream function and velocity potential are given by the corresponding expressions

$$\psi(x, y) = U_x\, y - U_y\, x + \psi^{SD}(x, y) + \psi^{PV}(x, y), \tag{12.6.12}$$

and

$$\phi(x, y) = U_x\, x + U_y\, y + \phi^{SD}(x, y) + \phi^{PV}(x, y). \tag{12.6.13}$$

Conversely, the source-dipole panel representation can be regarded as the result discretizing the integral on the right-hand side of (12.6.9) or (12.6.10) into geometrical elements representing the source-dipole panels. In this section, we have discussed straight elements with constant and linear strength density distributions. In more advanced implementations, curved elements including sections of a parabola and circular arcs, and quadratic or higher-order strength density distributions are employed.

12.6.3 Solution of the interior problem

Assume that the strength of the source-dipole sheet satisfying the no-penetration condition around the airfoil is available. For reasons discussed in section 12.3, if we evaluate the right-hand sides of equations (12.6.9) – (12.6.11) at a point located inside the airfoil, we will find that the velocity vanishes and the stream function and potential take constant values.

This observation suggests an alternative method for computing the strength density of the source dipoles. Instead of using the Neumann no-penetration boundary condition, we use the Dirichlet condition requiring that the potential and stream function are constant on the *interior* side of the airfoil.

Panels with constant strength density

To illustrate the implementation of the method, we discretize the airfoil contour into N flat panels with constant strength density, as illustrated in figure 12.4.1. Identifying the flow induced by each panel with the flow induced by two point vortices located at the end points, as discussed earlier in this section, we obtain the potential

$$\phi(x, y) = U_x\, x + U_y\, y + \frac{1}{2\pi} \sum_{i=1}^{N} [\mu^{(i)} (\theta_1^{(i)} - \theta_2^{(i)})] + \frac{\kappa}{2\pi}\, \theta_T, \tag{12.6.14}$$

where the angles $\theta_1^{(i)}$, $\theta_2^{(i)}$, and θ_T are defined in figure 12.6.1.

To implement the collocation method, we evaluate (12.6.14) at the mid-point of the j panel, $(x_M^{(j)}, y_M^{(j)})$, on the interior side of the airfoil. Now we note that $\theta_1^{(j)} = 0$ and $\theta_2^{(j)} = -\pi$, and assign to the potential the reference value of zero to derive the algebraic equation

$$0 = U_x \, x_M^{(j)} + U_y \, y_M^{(j)} + \frac{1}{2\pi} \sum_{i=1}^{N}{}' \, [\mu^{(i)} (\theta_1^{(i)} - \theta_2^{(i)})] + \frac{1}{2} \, \mu^{(j+1)} + \frac{\kappa}{2\pi} \, \theta_T, \tag{12.6.15}$$

where the primed sum signifies that the term $i = j$ is excluded from the summation. Rearranging (12.6.15), we derive a linear equation relating the panel source-dipole densities to the strength of the trailing-edge point vortex,

$$\frac{1}{2} \, \mu^{(j+1)} + \frac{1}{2\pi} \sum_{i=1}^{N}{}' [\mu^{(i)} (\theta_1^{(i)} - \theta_2^{(i)})] + \frac{\kappa}{2\pi} \, \theta_T = -U_x \, x_M^{(j)} - U_y \, y_M^{(j)}. \tag{12.6.16}$$

Applying equation (12.6.16) for $j = 1, \ldots N$, and appending to the resulting system of equations the Kutta-Joukowski condition expressed by (12.6.7), we obtain a linear system for the $N + 1$ unknowns $\mu^{(i)}, i = 1, \ldots N$, and κ.

Distribution of the potential over the airfoil

Inspecting the third term on the right-hand side of (12.6.14), we find that the potential undergoes a discontinuity of magnitude -$\mu^{(i)}$ across the ith panel. Since the potential inside the airfoil is constant and equal to zero, the potential on the exterior side of the panel must be equal to $-\mu^{(i)}$.

This result is valid in a more general context: the potential at the outer side of the airfoil is equal to the negative of the strength density of the source dipoles. The tangential velocity may then be computed by numerically differentiating μ with respect to arc length along the airfoil contour.

Computer Problem

12.6.1 *Constant strength dipole panels*

Program *airf_2d_cdp* inside directory *07_ptf* of FDLIB computes flow past an airfoil using the constant-strength source-dipole-panel method.

(*a*) Run the code for an airfoil of your choice. Prepare graphs, and discuss the distribution of the pressure coefficient.

(*b*) Evaluate the velocity at several points inside the airfoil and discuss your findings.

12.7 Point-source panels and Green's third identity

In previous sections, we have discussed flow representations in terms of vortex and point-source dipole panels expressed, respectively, by equations (12.5.1) and (12.6.10) or (12.6.11). In this section, we introduce a new representation in terms of a point source distribution. Working by analogy with (12.6.11), we find that the harmonic potential of the induced flow is given by

$$\phi^S(x,y) = \frac{1}{4\pi} \oint \ln[(x-x')^2 + (y-y')^2]\, \sigma(x')\, \mathrm{d}l(\mathbf{x}'), \qquad (12.7.1)$$

where σ is the strength density of the distribution.

The point-source representation carries an important restriction: conservation of mass requires that the total strength of the point sources, defined as the integral of the strength density σ with respect to arc length around the airfoil, vanishes; otherwise, a net radial flow due to an effective point source will arise. This restriction is satisfied automatically only in the case of symmetric flow past a symmetric non-lifting airfoil at zero angle of attack, and in the absence circulatory motion.

In spite of this limitation, the point-source representation is not without its merits. Its usefulness stems predominantly from Green's third identity discussed in section 12.7.2, stating that a judicious combination of the point source and source-dipole representation ensures the satisfaction of the condition of zero flow rate, and also endows the strength densities of the distributions with simple physical interpretations.

12.7.1 Source panels with constant density

Consider a flat source panel with uniform strength density equal to $\sigma^{(0)}$ situated on the x axis, and subtended between the points $x = a$ and b. Applying (12.7.1) with $y' = 0$, we find that the corresponding velocity potential is given by

$$\phi^{(0)}(x,y) = \frac{\sigma^{(0)}}{4\pi} \int_a^b \ln[(x-x')^2 + y^2]\, \mathrm{d}x'. \qquad (12.7.2)$$

Note that this expression is identical to that for the stream function due to a vortex panel with constant strength density given in equation (12.3.8), subject to the substitution $\gamma^{(0)} = -\sigma^{(0)}$. Using (12.3.9), we find

$$\begin{aligned}\phi^{(0)}(x,y) = \frac{\sigma^{(0)}}{4\pi}\Big(&-(x-b)\, \ln[(x-b)^2 + y^2] \\ &+(x-a)\, \ln[(x-a)^2 + y^2] \\ &+2\, y\, (\arctan\frac{y}{x-b} - \arctan\frac{y}{x-a}) - 2(b-a)\Big).\end{aligned} \qquad (12.7.3)$$

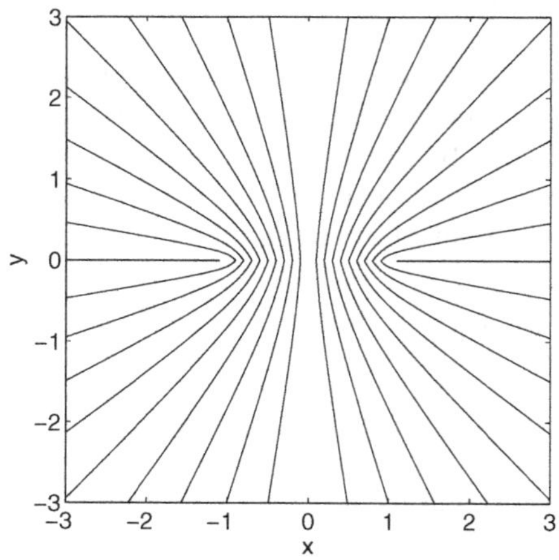

Figure 12.7.1 Streamline pattern of the flow due to a point-source panel with uniform strength subtended between the points $x = \pm 1$.

The components of the velocity are found by straightforward differentiation,

$$u_x^{(0)}(x, y) = \frac{\partial \phi^{(0)}}{\partial x} = -\frac{\sigma^{(0)}}{4\pi} \ln \frac{(x-b)^2 + y^2}{(x-a)^2 + y^2}, \tag{12.7.4}$$

and

$$u_y^{(0)}(x, y) = \frac{\partial \phi^{(0)}}{\partial y} = \frac{\sigma^{(0)}}{2\pi} \left(\arctan \frac{y}{x-b} - \arctan \frac{y}{x-a} \right). \tag{12.7.5}$$

The streamline pattern of the flow induced by a uniform panel is shown in figure 12.7.1. Far from the panel, the flow resembles that due to a point source with strength $\sigma^{(0)}\,(b-a)$ situated at the origin.

Jump in velocity across the panel

Expression (12.7.4) shows that the x velocity component is continuous throughout the domain of flow and across the source panel. In contrast, because of the inverse tangent functions on the right-hand side of (12.7.5), the y velocity component undergoes a discontinuity of magnitude $\sigma^{(0)}$ across the panel. Specifically, the y velocity component on the upper or lower side of the panel, $a < x < b$, is given by

$$u_y^{(0)}(x, y \to \pm 0) = \pm \frac{\sigma^{(0)}}{2}. \tag{12.7.6}$$

Inverse tangent functions also appear on the right-hand side of (12.7.4). However, because these functions are multiplied by y which is zero over the panel, a discontinuity in the potential does not arise.

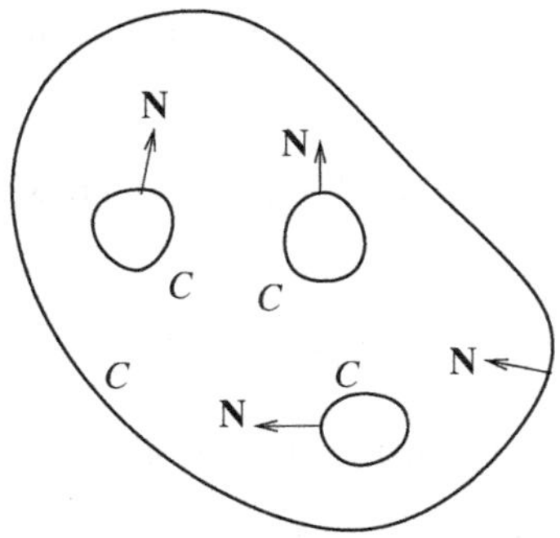

Figure 12.7.2 A control area in the xy plane confined by a collection of curves, C, is used to establish Green's third identity; $\mathbf{N}$ is the unit vector normal to the boundaries pointing into the fluid.

12.7.2 Green's third identity

The source panels can be used in the familiar way to develop a representation of the flow past a symmetric airfoil at zero angle of attack. The numerical implementation is analogous to that of the vortex panel method discussed in section 12.3. However, a more interesting and more general representation is possible thanks to Green's third identity.

Consider a control area in the xy plane bounded by a collection of boundaries C, as illustrated in figure 12.7.2. Green's third identity states that the harmonic potential at a point $\mathbf{x}$ that lies inside the control area can be represented in terms of a combined point-source / dipole-source distribution, in the form

$$\phi(x,y) = \frac{1}{4\pi}\oint \log[(x-x')^2+(y-y')^2]\,\mathbf{N}(\mathbf{x}')\cdot\nabla\phi(\mathbf{x}')\,\mathrm{d}l'$$

$$+\frac{1}{2\pi}\oint \frac{(x-x')\,N_x(\mathbf{x}')+(y-y')\,N_y(\mathbf{x}')}{(x-x')^2+(y-y')^2}\,\phi(\mathbf{x}')\,\mathrm{d}l', \qquad (12.7.7)$$

where $\mathbf{N}$ is the unit vector normal to the boundaries pointing into the control area, as shown in figure 12.7.2. Comparing the two terms on the right-hand side of (12.7.7) with (12.7.1) and (12.6.11), we identify the strength density of the point-source distribution with the boundary values of the potential, and the strength density of the source-dipole distribution with the boundary values of the normal derivative of the potential.

Let us now identify the control area with the area occupied by an airfoil, and apply Green's identity for the potential of an incident streaming flow. Writing $\mathbf{N} = -\mathbf{n}$, where $\mathbf{n}$ is the unit vector normal to the airfoil pointing into the exterior and $\mathbf{N}$ is the unit vector normal to the airfoil pointing into the interior, we obtain

$$\phi_\infty(x,y) = -\frac{1}{4\pi}\oint \ln[(x-x')^2+(y-y')^2]\,\mathbf{n}(\mathbf{x}')\cdot\nabla\phi_\infty(\mathbf{x}')\,\mathrm{d}l' \\ -\frac{1}{2\pi}\oint \frac{(x-x')\,n_x(\mathbf{x}')+(y-y')\,n_y(\mathbf{x}')}{(x-x')^2+(y-y')^2}\,\phi_\infty(\mathbf{x}')\,\mathrm{d}l', \qquad (12.7.8)$$

where the point $\mathbf{x}=(x,y)$ lies inside the airfoil.

Equation (12.6.14) embodies a double-layer representation in the form

$$\phi(x,y) = \phi_\infty(x,y) + \frac{1}{2\pi}\oint \frac{(x-x')\,n_x(\mathbf{x}')+(y-y')\,n_y(\mathbf{x}')}{(x-x')^2+(y-y')^2}\,\phi^{(+)}(\mathbf{x}')\,\mathrm{d}l' \\ +\phi^{PV}(x,y), \qquad (12.7.9)$$

where $\phi^{(+)}$ is the potential on the exterior side of the airfoil. Since the point (x,y) is located in the interior of the airfoil, the left-hand side is a constant that may be set equal to zero.

Combining equations (12.7.8) and (12.7.9), we obtain

$$\frac{1}{4\pi}\oint \ln[(x-x')^2+(y-y')^2]\,\mathbf{n}(\mathbf{x}')\cdot\nabla\phi_\infty(\mathbf{x}')\,\mathrm{d}l' \qquad (12.7.10) \\ = -\frac{1}{2\pi}\oint \frac{(x-x')\,n_x(\mathbf{x}')+(y-y')\,n_y(\mathbf{x}')}{(x-x')^2+(y-y')^2}\,\mu(\mathbf{x}')\,\mathrm{d}l' + \phi^{PV}(x,y),$$

where

$$\mu \equiv -\phi^{(+)} + \phi_\infty. \qquad (12.7.11)$$

Equation (12.7.10) is an integral equation for the dipole density μ. Once this equation has been solved using, for example, a panel method, the potential on the exterior side of the airfoil can be computed from equation (12.7.11) as

$$\phi^{(+)} = \phi_\infty - \mu. \qquad (12.7.12)$$

The advantages of this representation over the vortex panel representation are now apparent.

Computer Problems

12.7.1 *Source-dipole panel method*

Program *airf_2d_csdp* inside directory *07_ptf* of FDLIB computes flow past an airfoil using a panel method based on equation (12.7.10). Run the code for an airfoil of your choice, prepare graphs and discuss the distribution of the pressure coefficient around the airfoil.

12.7.2 *Constant source panel method*

Write a code that computes flow past a symmetric airfoil at zero angle of attack using the point-source panel method. Run the code for an airfoil of your choice, prepare graphs, and discuss the distribution of the pressure coefficient around the airfoil.

FDLIB Software Library

A

The software library FDLIB contains a collection of FORTRAN 77, MATLAB, and other programs that solve a broad range of problems in fluid dynamics and related disciplines using a variety of numerical methods. FDLIB consist of thirteen main directories listed in table A.1.

Each main directory of FDLIB contains a multitude of nested subdirectories that include main programs, assisting subroutines, and utility subroutines. Linked with drivers, the utility subroutines become stand-alone modules; all drivers are provided. A list of subdirectories and a brief statement of their contents are given in this appendix. Further information can be found at the FDLIB Internet site.

Download

The source code of FDLIB containing program and data files is available from the Internet site: `http://dehesa.freeshell.org/FDLIB`. The directories have been archived using the `tar` Unix facility into the compressed *FDLIB.tgz* file. To unravel the directories on a Unix system, issue the Unix command: `tar xzf FDLIB.tgz`. This will generate the directory *FDLIB* containing nested subdirectories. To unravel the directories on a Windows or Macintosh system, double-click on the archived tar file and follow the on-screen instructions of the invoked application.

Installation and compilation

MATLAB programs are executed as interpreted scripts. The downloaded FDLIB package does not contain FORTRAN 77 object files or executables. To compile and link FORTRAN 77 programs, follow the instructions of your compiler.

An application can be built using the makefile provided in each subdirectory. A makefile is a script interpreted by the *make* utility that instructs the operating system how to compile the main program and subroutines, and then link the object files into an executable file using a compiler.

	Subject	Directory
1	Numerical methods	*01_num_meth*
2	Grids	*02_grids*
3	Hydrostatics	*03_hydrostat*
4	Various	*04_various*
5	Lubrication	*05_lub*
6	Stokes flow	*06_stokes*
7	Potential flow	*07_ptf*
8	Hydrodynamic stability	*08_stab*
9	Vortex motion	*09_vortex*
10	Boundary layers	*10_bl*
11	Finite difference methods	*11_fdm*
12	Boundary element methods	*12_bem*
13	Turbulence	*13_turbo*

Table A.1 FDLIB is arranged in thirteen directories according to physical or numerical classification.

To compile an application named *nea_krini*, navigate to the subdirectory where the application resides and type: `make nea_krini`.

To remove the object files, output files, and executable of an application named *polihni*, navigate to the subdirectory where the application resides and issue the command: `make clean`.

To compile the FORTRAN 77 programs using a FORTRAN 90 compiler, simply make appropriate compiler call substitutions in the makefiles.

CFDLAB

A subset of FDLIB has been combined with the X11 graphics library *vogle* into the integrated application CFDLAB that visualizes the results of simulations and performs interactive animation. The source code of CFDLAB can be downloaded from the Internet site: `http://dehesa.freeshell.org/CFDLAB`.

BEMLIB

A subset of FDLIB containing boundary-element codes have been arranged in the library BEMLIB.[1] The source code of BEMLIB can be downloaded from the Internet site: `http://dehesa.freeshell.org/BEMLIB`.

[1]This library accompanies the book: POZRIKIDIS, C., 2002, *A Practical Guide to Boundary-Element Methods with the Software Library BEMLIB,* Chapman & Hall/CRC.

FDLIB directory contents

The public FDLIB directories are listed in the following tables along with a brief description. Further information can be found at the FDLIB Internet site.

01_num_meth

This directory contains a suite of general-purpose programs on general numerical methods and differential equations.[2]

Subdirectory	*Topic*
01_num_comp	General aspects of numerical computation
02_lin_calc	Linear algebra and linear calculus
03_lin_eq	Systems of linear algebraic equations
04_nl_eq	Nonlinear algebraic equations
05_eigen	Eigenvalues and eigenvectors of matrices
06_interp_diff	Function interpolation and differentiation
07_integration	Function integration
08_approximation	Function approximation
09_ode	Ordinary differential equations
10_ode_ddm	Ordinary differential equations; domain discretization methods
11_pde_diffusion	Partial differential equations; unsteady diffusion
12_pde_poisson	Partial differential equations; poisson equation
13_pde_cd	Partial differential equations; convection--diffusion equation
14_bem	Boundary-element methods
15_fem	Finite-element methods
99_spec_fnc	Special functions

02_grids

This directory contains programs that perform grid generation, adaptive discretization, parametrization, representation, and meshing of planar lines, three-dimensional lines, and three-dimensional surfaces.

Subdirectory	*Topic*
grid_2d	Discretization of a planar line into a mesh of straight or circular elements
prd_2d	Adaptive parametrization of a closed line
prd_3d	Adaptive parametrization of a closed three-dimensional line

[2]This directory accompanies the book: POZRIKIDIS, C., 2008, *Numerical Computation in Science and Engineering*, Second edition, Oxford University Press.

prd_ax	Adaptive parametrization of a planar line representing the trace of an axisymmetric surface in a meridional plane
rec_2d	Interpolation through a Cartesian grid
trgl	Triangulation of a closed surface
trgl_sqr	Triangulation of a square patch

03_hydrostat

This directory contains codes that compute interfacial shapes in hydrostatics.

Subdirectory	*Topic*
drop_2d	Shape of a two-dimensional pendant or sessile drop on a horizontal or inclined plane
drop_ax	Shape of an axisymmetric pendant or sessile drop on a horizontal plane
flsphere	Position of a sphere floating on an interface with a curved meniscus
men_2d	Shape of a two-dimensional meniscus between two parallel plates
men_2d_plate	Shape of a two-dimensional meniscus attached to an inclined plate
men_3d	Shape of a three-dimensional meniscus in the exterior of an ellipse
men_ax	Shape of an axisymmetric meniscus inside a vertical circular tube

04_various

This directory contains miscellaneous codes that compute the structure and kinematics of various flows.

Subdirectory	*Topic*
chan_2d	Steady flow in a channel
chan_2d_2l	Steady two-layer flow in a channel
chan_2d_imp	Impulsive flow in a channel
chan_2d_ml	Multi-layer flow in a channel
chan_2d_osc	Oscillatory flow in a channel
chan_2d_trans	Transient flow in a channel
chan_2d_wom	Pulsating flow in a channel
chan_brush	Steady flow in a brush channel
film	Film flow down an inclined plane
films	Multi-film flow down an inclined plane
flow_1d	Steady unidirectional flow in a tube with arbitrary cross-section

flow_1d_1p	Steady unidirectional flow over a periodic array of cylinders with arbitrary cross-section
flow_1d_osc	Oscillatory unidirectional flow in a tube with arbitrary cross-section
path_lines	Computation of path lines
plate_imp	Flow due to the impulsive motion of a plate
plate_osc	Flow due to the oscillations of a plate
spf	Similarity solutions for stagnation-point flow
strml	Streamline patterns of a broad range of flows offered in a menu
tube_ann	Steady annular flow
tube_ann_ml	Steady multi-layer annular flow
tube_ann_sw	Steady swirling annular flow
tube_ann_sw_ml	Steady multi-layer swirling annular flow
tube_crc	Steady flow through a circular tube
tube_crc_ml	Steady multi-layer flow through a circular tube
tube_sec	Steady flow through a circular tube due to the translation of a sector
tube_sw	Transient swirling flow in a circular tube
tube_sw_trans	Transient flow through a circular tube
tube_sw_wom	Pulsating flow through a circular tube
tube_ell	Steady flow through a tube with elliptical cross-section
tube_rec	Steady flow through a tube with rectangular cross-section
tube_trgl_eql	Steady flow through a tube with triangular cross-section

05_lub

This directory contains codes that solve problems involving lubrication flows.

Subdirectory	*Topic*
bear_2d	Dynamical simulation of the motion of a slider bearing pressing against a wall
chan_2l_exp	Dynamical simulation of the evolution of two superposed viscous layers in a horizontal or inclined channel computed by an explicit finite-difference method
chan_2l_imp	Same as chan_2l_exp but with an implicit finite-difference method
films	Evolution of an arbitrary number of superposed films on a horizontal or inclined wall

06_stokes

This directory contains codes that compute viscous flows at vanishing Reynolds numbers.

Subdirectory	*Topic*
bump_3d	Shear flow over a spherical bump on a plane wall
flow_2d	Two-dimensional flow in a domain with arbitrary geometry
prtcl_2d	Flow past a fixed bed of two-dimensional particles with arbitrary shapes for a variety of flow configurations computed by a boundary-element method
prtcl_3d	Flow past or due to the motion of a three-dimensional particle for a variety of configurations computed by a boundary-element method
prtcl_3d_mob	Same as prtcl_3d but for the mobility problem where the force and torque on the particle are specified and the particle velocity is computed in the solution
prtcl_ax	Flow past or due to the motion of a collection of axisymmetric particles computed by a boundary-element method
prtcl_sw	Swirling flow produced by the rotation of an axisymmetric particle computed by a boundary-element method
sgf_2d	Green's functions of two-dimensional flow
sgf_3d	Green's functions of three-dimensional flow
sgf_ax	Green's functions of axisymmetric flow

07_ptf

This directory contains codes that solve problems involving potential flows.

Subdirectory	*Topic*
airf_2d	Airfoil shapes
airf_2d_cdp	Flow past an airfoil computed by the constant-dipole-panel method
airf_2d_csdp	Flow past an airfoil computed by the constant-source-dipole-panel method
airf_2d_lvp	Flow past an airfoil computed by the linear-vortex-panel method
body_2d	Flow past or due to the motion of a two-dimensional body computed by a boundary element method

body_ax	Flow past or due to the motion of an axisymmetric body computed by a boundary element method
cvt_2d	Flow in a rectangular cavity computed by a finite difference method
flow_2d	Two-dimensional flow in an arbitrary domain computed by a boundary element method
lgf_2d	Green and Neumann functions of Laplace's equation in two dimensions
lgf_3d	Green and Neumann functions of Laplace's equation in three dimensions
lgf_ax	Green and Neumann functions of Laplace's equation in axisymmetric domains
tank_2d	Dynamical simulation of liquid sloshing in a rectangular tank computed by a boundary-element method

08_stab

This directory contains codes that perform stability analysis of miscellaneous flows.

Subdirectory	*Topic*
ann_21	Capillary instability of two annular layers placed between two concentric cylinders in the presence of an insoluble surfactant
ann_210	Same as ann_21 for Stokes flow
ann_21el	Same as ann_21 for an elastic interface
ann_21el0	Same as ann_210 for an elastic interface
chan_210	Instability of two-layer flow in a channel under conditions of Stokes flow
chan_210_s	Instability of two-layer flow in a channel in the presence of an insoluble surfactant under conditions of Stokes flow
coat0_s	Instability of a liquid film resting on a plane wall in the presence of an insoluble surfactant under conditions of Stokes flow
film0	Instability of a liquid film down an inclined plane for Stokes flow
film0_s	Instability of a liquid film down an inclined plane in the presence of an insoluble surfactant for Stokes flow
if0	Instability of a horizontal interface between two semi-infinite fluids for Stokes flow
layer0	Instability of a horizontal liquid layer resting on a horizontal wall under a semi-infinite fluid for Stokes flow

layer0sf_s	Instability of a sheared liquid layer coated on a horizontal plane in the presence of an insoluble surfactant for Stokes flow
prony	Prony fitting of a times series with a sum of complex exponentials
sf1	Instability of an inviscid shear flow with an arbitrary velocity profile
thread0	Instability of an infinite viscous thread suspended in an ambient viscous fluid for Stokes flow
thread1	Instability of an inviscid thread suspended in an inert ambient fluid
vl	Instability of a vortex layer
vs	Instability of a vortex sheet

09_vortex

This directory contains codes that compute vortex motion.

Subdirectory	*Topic*
lv_lia	Dynamical simulation of the motion of a three-dimensional line vortex computed by the local-induction approximation (LIA)
lvr	Velocity induced by line vortex rings
lvrm	Dynamical simulation of the motion of a collection of coaxial line vortex rings
pv	Velocity induced by point vortices
pvm	Dynamic simulation of the motion of a collection of point vortices
pvm_pr	Dynamical simulation of the motion of a periodic collection of point vortices
ring	Self-induced velocity of a vortex ring with core of finite size
vl_2d	Dynamical simulation of the evolution of compound periodic vortex layers
vp_2d	Dynamical simulation of the evolution of a collection of two-dimensional vortex patches
vp_ax	Dynamical simulation of the evolution of a collection of axisymmetric vortex rings and vortex patches

10_bl

This directory contains codes that solve boundary-layers flows.

Subdirectory	*Topic*
blasius	Computation of the Blasius boundary layer
falskan	Computation of the Falkner-Skan boundary layer
kp_cc	Boundary layer around a circular cylinder computed by the Kàrmàn-Pohlhausen method
pohl_pol	Profiles of the Pohlhausen polynomials

11_fdm

This directory contains codes that solve problems using finite-difference methods.

Subdirectory	*Topic*
channel	Unidirectional flow in a channel
cvt_pm	Transient flow in a rectangular cavity computed by a projection method
cvt_stag	Steady Stokes flow in a rectangular cavity computed on a staggered grid
cvt_sv	Steady flow in a rectangular cavity computed by the stream function/vorticity formulation

12_bem

This directory contains codes that produce solutions to Laplace's equation by boundary-element methods.

Subdirectory	*Topic*
ldr_3d	Solution of Laplace's equation with Dirichlet boundary conditions in the interior or exterior of a three-dimensional region computed using the boundary-integral formulation
ldr_3d_2p	Solution of Laplace's equation with Dirichlet boundary conditions in a semi-infinite region bounded by a doubly-periodic surface computed using the double-layer formulation
ldr_3d_ext	Solution of Laplace's equation with Dirichlet boundary conditions in the exterior of a three-dimensional region computed using the double-layer formulation

ldr_3d_int	Solution of Laplace's equation with Dirichlet boundary conditions in the interior of a three-dimensional region computed using the double-layer formulation
lnm_3d	Solution of Laplace's equation with Neumann boundary conditions in the interior or exterior of a three-dimensional region computed using the boundary-integral formulation

13_turbo

This directory contains data and codes pertinent to turbulent flows.

Subdirectory	*Topic*
stats	Statistical analysis of a turbulent flow time series

B References

Further discussion of fluid mechanics, applied mathematics, numerical methods, and scientific computing can be found in the highly recommended texts listed in this appendix.

Introductory classical mechanics

MARION, J. B., 1970, *Classical Dynamics of Particles and Systems*, Harcourt Brace.

Introductory fluid dynamics

BIRD, R. B., STEWART, W. E. & LIGHTFOOT, E. N., 2006, *Transport Phenomena*, Second Edition, Wiley.

PAPANASTASIOU, T. C., GEORGIOU, G. & ALEXANDROU, A. N., 1999, *Viscous Fluid Flow*, CRC Press.

MORRISON, F. A., 2001, *Understanding Rheology*, Oxford University Press.

Advanced fluid dynamics

BATCHELOR, G. K., 1967. *An Introduction to Fluid Dynamics*, Cambridge University Press.

BRODKEY, R. S., 2004, *The Phenomena of Fluid Motions*, Dover.

POZRIKIDIS, C., 1997, *Introduction to Theoretical and Computational Fluid Dynamics*, Oxford University Press.

WARSI, Z. U. A., 2005, *Fluid Dynamics; Theoretical and Computational Approaches*, Third Edition, CRC Press.

WHITE, F. M., 2005, *Viscous Fluid Flow*, Third Edition, McGraw-Hill.

Computational fluid dynamics

FERZIGER, J. H. & PERIĆ, M., 1996, *Computational Methods for Fluid Dynamics*, Springer-Verlag.

HIRSCH, C., 2007, *Numerical Computation of Internal and External Flows*, Second Edition, Butterworth-Heinemann.

POZRIKIDIS, C., 1997, *Introduction to Theoretical and Computational Fluid Dynamics*, Oxford University Press.

Low Reynolds number flow

HAPPEL, J. & BRENNER, H., 1983, *Low Reynolds Number Hydrodynamics*, Springer.

KIM, S. & KARRILA, S. J., 1991, *Microhydrodynamics: Principles and Selected Applications,* Butterworth-Heinemann.

POZRIKIDIS, C., 1992, *Boundary Integral and Singularity Methods for Linearized Viscous Flow*, Cambridge University Press.

Aerodynamics

ANDERSON, J. D., 2002, *Modern Compressible flow with Historical Perspective*, Third Edition, McGraw-Hill.

ANDERSON, J. D., 2007, *Fundamentals of Aerodynamics*, Fourth Edition, McGraw-Hill.

KATZ, J. & PLOTKIN, A., 2001, *Low-Speed Aerodynamics; from Wing Theory to Panel Methods*, Second Edition, Cambridge University Press.

Numerical methods

POZRIKIDIS, C., 2008, *Numerical Computation in Science and Engineering*, Second Edition, Oxford University Press.

Calculus

BOAS, M. L., 2005, *Mathematical Methods in the Physical Sciences*, Third Edition, Wiley.

HILDEBRAND, F. B., 1976, *Advanced Calculus for Applications*, Second Edition, Prentice-Hall.

Mathematical resources

ABRAMOWITZ, M. & STEGUN, I. A., 1972, *Handbook of Mathematical Functions*, Dover.

GRADSHTEYN, I. S. & RYZHIK, I. M., 1980, *Table of Integrals, Series, and Products*, Academic Press.

KORN, G. A. & KORN, T. M., 1968, *Mathematical Handbook for Scientists and Engineers*, McGraw-Hill.

Computer Programming

POZRIKIDIS, C., 2007, *Introduction to C++ Programming and Graphics*, Springer.

C

Matlab Primer

MATLAB is a software package for interactive numerical computation and graphics visualization produced by The Mathworks corporation. The application was developed in the 1970s as a virtual laboratory for matrix calculus and linear algebra. Today, MATLAB can be described both as a programming language and as a computing environment.

As a programming language, MATLAB is roughly equivalent, in some ways superior and in some ways inferior to traditional upper-level languages such as FORTRAN 77, C, or C++. As a computing environment, MATLAB is able to run indefinitely in its own workspace. Thus, a session defined by the values of all initialized variables and graphical objects can be saved and reinstated at a later time. In this sense, MATLAB is an operating system running inside the operating system empowering the computer. Symbolic algebraic manipulation is available through an add-on library (toolbox) that uses a kernel borrowed from the mathematical software engine Maple.

An attractive feature of MATLAB is the availability of a broad range of utility commands, intrinsic functions, and computational toolboxes, especially graphics. A simplifying feature of MATLAB is that the size of vectors and matrices used in the calculations are automatically assigned and can be changed in the course of a session, thereby circumventing the need for variable declaration and memory allocation.

Matlab must be purchased and installed with a proper license. Inexpensive licenses are available for students and educators through educational editions. On Linux, the educational version of Matlab is installed as a stand-alone application. On Windows, the installation CD-ROM must be present in the drive.

C.1 Invoking Matlab

To invoke MATLAB in Windows, double-click on the MATLAB icon. This runs a starter program, currently a disk operating system (DOS) batch script, that launches the main MATLAB executable. Alternatively, we can start MATLAB from a DOS command line by a procedure similar to that discussed next for Unix.

To invoke MATLAB in Unix, we run the Unix script *matlab* by issuing the command:

```
matlab
```

Assuming that the script is in the executable path, this will launch the MATLAB executable in some graphical user interface (GUI) or command line mode. To suppress the memory consuming GUI, we issue either the command:

```
matlab -nodesktop
```

or the command:

```
matlab -nojvm
```

`jvm` is an acronym for java virtual machine. Starting MATLAB by issuing the command:

```
matlab -nodesktop -nosplash
```

suppresses both the GUI and the splash screen at start-up. In Unix, MATLAB can be launched with a number of options. To obtain a list, we request help by issuing the command:

```
matlab -help
```

MATLAB uses a number of shared libraries, parameters, and environmental variables. To obtain a complete list, we issue the command:

```
matlab -n
```

C.2 Matlab programming

Only elementary computer programming skills are necessary to read and write MATLAB code. The code is written in one file or a collection of files, called the source or program files, using a standard file editor, such as the *vi* editor. The source code includes the main program, also called a script, and the necessary user-defined functions. The names of these files must be suffixed with `.m` . Execution begins by typing the name of the file containing the main program in the MATLAB environment.

Alternatively, the code can be typed one line at a time followed by the RETURN keystroke in the MATLAB environment. MATLAB is an interpreted language, which means that the instructions are translated into machine language and executed in real time, one at a time. In contrast, a source code written in FORTRAN, C, or C++ must first be compiled to produce the object files, which are then linked together with the necessary system and user-defined libraries to produce the executable binary file.

C.3 Matlab Grammar and syntax

Following is a list of general rules regarding the grammar and syntax of MATLAB. When confronted with an error after issuing a command or during execution, this list should serve as a first check point:

- MATLAB *variables are (lower and upper) case-sensitive:*
  The variable `echidna` is different from the variable `echiDna`. Similarly, the MATLAB command `return` is not equivalent to the erroneous command `Return`; the latter will not be recognized by the interpreter.

- MATLAB *variables must start with a letter:*
  A variable name is described by a string of up to thirty-one characters, including letters, digits, and the underscore; punctuation marks are not allowed.

- MATLAB *string variables are enclosed by a single quote:*
  For example, we may define the string variable:

```
melody_764 = 'Agapanthos'
```

- *Beginning and end of a command line:*
  A MATLAB command can begin at any position in a line, and may continue practically indefinitely in the same line.

- *Line continuation:*
  To continue a command onto the next line, we put three dots at the end of the line.

- *Multiple commands in a line:*
  Two or more commands can be placed in the same line provided they are separated with a semicolon (;).

- *Display:*
  When a command is executed directly or by running a MATLAB code, MATLAB displays the numerical value assignment or the result of a calculation. To suppress the output, we put a semicolon (;) at the end of the command.

- *White space:*
  More than one empty space between words are ignored by the compiler. However, numbers cannot be split in sections separated by blank spaces.

- *Range of indices:*
  Vectors and arrays must have positive and nonzero indices; the vector entry `v(-3)` is unacceptable in MATLAB. This annoying restriction can be circumvented in clever ways by redefining the indices.

- *Comments:*
  A line beginning with the % character, or the tail-end of a line after the % character, is a comment, and is ignored by the MATLAB interpreter.

- *Mathematical symbols and special characters:*
  Table C.3.1 lists mathematical symbols and special characters used in MATLAB interactive dialog and programming.

- *Logical control flow commands:*
  Table C.3.2 lists the basic logical control flow commands.

- *Input/output commands:*
  Tables C.3.3–5 list basic input/output (I/O) commands, functions, and formatting statements. Once the output format is set, it remains in effect until changed.

C.4 Precision

MATLAB stores all numbers in the long format of the floating point representation. This means that real numbers have a finite precision of roughly sixteen significant digits, and a range of definition roughly varying between 10^{-308} and 10^{+308} in absolute value. Numbers smaller than 10^{-308} or larger than 10^{+308} in absolute value cannot be accommodated. MATLAB performs all computations in double precision. However, this should not be confused with the ability to view and print numbers with a specified number of significant figures.

C.5 Matlab commands

Once invoked, MATLAB responds interactively to various commands, statements, and declarations issued by the user in the MATLAB window. These are implemented by typing the corresponding name, single- or multi-line syntax, and then pressing the ENTER key. Table C.5.1 lists general utility and interactive-input MATLAB commands. Issuing the command *demos* initiates various demonstrations and illustrative examples. A session can be saved using the command *save* and reinstated at a later time.

To obtain a full explanation of a MATLAB command, statement, or function, we may use the MATLAB `help` facility, which is the counterpart of the Unix manual `man` facility. For example, issuing the command `help break` in the MATLAB environment produces the description:

```
BREAK Terminate execution of WHILE or FOR loop.
   BREAK terminates the execution of FOR and WHILE loops.
   In nested loops, BREAK exits from the innermost loop only.

   BREAK is not defined outside of a FOR or WHILE loop.
```

+	Plus
-	Minus
*	Matrix multiplication
.*	Array multiplication
^	Matrix power
.^	Array power
kron	Kronecker tensor product
\	Backslash or left division
/	Slash or right division
./	Array division
:	Colon
()	Parentheses
[]	Brackets
.	Decimal point
..	Parent directory
...	Line continuation
,	Comma
;	Semicolon, used to suppress the screen display
%	Indicates that the rest of the line is a comment
!	Exclamation point
'	Matrix transpose
"	Quote
.'	Nonconjugated transpose
=	Set equal to
==	Equal
∼=1	Not equal
<	Less than
<=	Less than or equal to
>	Greater than
>=	Greater than or equal to
&	Logical *and*
|	Logical *or*
∼	Logical *not*
xor	Logical *exclusive or*
i, j	Imaginary unit
pi	number $\pi = 3.14159265358\ldots$

Table C.3.1 MATLAB operators, symbols, special characters, and constants.

```
Use RETURN in this context instead.

See also FOR, WHILE, RETURN, CONTINUE.
```

The command `clear` is especially important, as it resets all variables to the

break	Terminate the execution
else	Use with the `if` statement
elseif	Use with the `if` statement
end	Terminate a `for` loop, a `while` loop, or an `if` block
error	Display a message and abort
for	Loop over commands a specific number of times
if	Conditionally execute commands
pause	Wait for user's response
return	Return to the MATLAB environment, invoking program or function
while	Repeat statements an indefinite number of times until a specified condition is met

Table C.3.2 MATLAB logical control flow commands and construct components.

disp	Display numerical values or text Use as: `disp` `disp()` `disp('text')`
fclose	Close a file
fopen	Open a file
fread	Read binary data from a file
fwrite	Write binary data to a file
fgetl	Read a line from a file, discard newline character
fgets	Read a line from a file, keep newline character
fprintf	Write formatted data to a file using C language conventions
fscanf	Read formatted data from a file
feof	Test for end-of-file (EOF)
ferror	Inquire the I/O error status of a file
frewind	Rewind a file
fseek	Set file position indicator
ftell	Get file position indicator
sprintf	Write formatted data to string
sscanf	Read formatted string from file
csvread	Read from a file values separated by commas
csvwrite	Write into file values separated by commas
uigetfile	Retrieve the name of a file to open through dialog box
uiputfile	Retrieve the name of a file to write through dialog box

Table C.3.3 MATLAB input/output (I/O) commands.

"uninitialized" status, and thereby prevents the use of improper values defined or produced in a previous calculation. A detailed explanation of this command can be obtained by typing `help clear` .

input	Prompt for user input
keyboard	Invoke keyboard as though it were a script file
menu	Generate menu of choices for user input

Table C.3.4 MATLAB interactive input commands.

format short	Fixed point with 4 decimal places (default)
format long	Fixed point with 14 decimal places
format short e	Scientific notation with 4 decimal places
format long e	Scientific notation with 15 decimal places
format hex	Hexadecimal format
format +	+, -, and space are printed for positive, negative, and zero elements
format compact	Suppress extra line feeds
format loose	Reinstates extra line feeds

Table C.3.5 MATLAB formatting commands.

clear	Clear variables and functions from memory
demo	Run demos
exit	Terminate a MATLAB session
help	Online documentation
load	Retrieve variables from a specified directory
save	Save workspace variables to a specified directory
saveas	Save figure or model using a specified format
size	Reveal the size of matrix
who	List current variables
quit	Terminate a MATLAB session

Table C.5.1 General utility MATLAB commands.

C.6 Matlab examples

In the following examples, we demonstrate the interactive usage of MATLAB with simple sessions. A line beginning with two "greater than" signs (>>) denotes the MATLAB command line where we enter a definition or issue a statement. Unless stated otherwise, a line that does not begin with >> is MATLAB output. Recall that the command `clear` clears the memory from previous data to prevent misappropriation.

- Numerical value assignment and addition:

```
>> a=1
a =
```

```
       1
>> b=2
b =
       2
>> c=a+b
c =
       3
```

- Numerical value assignment and subtraction:

```
>> clear
>> a=1; b=-3; c=a-b
c =
     4
```

- Number multiplication:

```
>> clear
>> a=2.0; b=-3.5; c=a*b;
>> c
c =
    -7
```

Typing the variable c displays its current value, in this case -7.

- Vector definition:

```
>> clear
>> v = [2 1]
v =
     2 1
>> v(1)
ans =
     2
>> v' % transpose
ans =
     2
     1
```

Typing v(1) produces the first component of the vector v as an answer. The comment "transpose" is ignored since it is preceded by the comment delimiter "%." The answer ans is, in fact, a variable evaluated by MATLAB.

- Vector addition:

```
>> v = [1 2]; u = [-1, -2]; u+v
ans =
     0 0
```

- Matrix definition, addition, and multiplication:

```
>> a = [1 2; 3 4]
a =
     1 2
     3 4
>> b = [ [1 2]' [2 4]' ]
b =
     1 2
     2 4
>> a+b
ans =
     2 4
     5 8
>> c=a*b
c =
      5 10
     11 22
```

- Multiply a complex matrix by a complex vector:

```
>> a = [1+2i 2+3i; -1-i 1+i]
a =
      1.0000 + 2.0000i 2.0000 + 3.0000i
     -1.0000 - 1.0000i 1.0000 + 1.0000i
>> v = [1+i 1-i]
v =
     1.0000 + 1.0000i 1.0000 - 1.0000i
>> c=a*v'
c =
      2.0000 + 6.0000i
     -2.0000 + 2.0000i
```

By taking the transpose indicated by a prime, the row vector, v, becomes a column vector that is conformable with the square matrix, a.

- For loop:

```
>> for j=-1:0
   j
   end
j =
    -1
j =
    0
```

In this example, the first three lines are entered by the user.

- If statement:

```
>> j=0;
>> i=1;
>> if i==j+1, disp 'case 1', end
case 1
```

- For loop:

```
>> n=3;
>> for i=n:-1:2
disp 'i='; disp(i), end
i=
     3
i=
     2
```

The loop is executed backward, starting at n, with step of -1.

- If loop:

```
>> i=1; j=2;
>> if i==j+1; disp 'case 1'
   elseif i==j; disp 'case2'
   else; disp 'case3'
   end
case3
```

In this example, all but the last line are entered by the user.

- While loop:

```
>> i=0;
>> while i<2, i=i+1; disp(i), end
     1
     2
```

The four statements in the while loop could have been typed in separate lines; that is, the commas can be replaced by the ENTER keystroke.

C.7 Matlab functions

MATLAB comes with an extensive library of internal functions for numerical computation and data visualization. Table C.7.1 lists general and specialized mathematical functions. The Matlab *help* facility provides detailed information on the proper function usage, arguments, and parameters. If the proper syntax or applicability of a function is unclear, it is best to code the numerical method in a user-defined function, working from first principles. It is both rewarding and instructive to create a personal library of user-defined functions based on control-flow commands.

Numerical methods

MATLAB includes a comprehensive library of numerical methods whose functions perform numerical linear algebra, solve algebraic equations, perform minimization, carry out function integration, solve differential equations, and execute a variety of other tasks. Special-purpose libraries of interest to a particular discipline are accommodated in toolboxes. Table C.7.2 shows selected MATLAB numerical methods functions.

The following Matlab session illustrates the solution of a linear system, $\mathbf{A}\mathbf{x} = \mathbf{b}$,

```
>> A=[1 1; 3 2];
>> b(1)=0; b(2)=1;
>> x=b/A'
x =
1.0000 -1.0000
```

C.8 User-defined functions

In MATLAB, a user-defined function is written in a file whose name defines the calling name of the function. The file name must be suffixed with the MATLAB identifier: `.m` . Thus, a function named *koulourtzis* must reside in a file named *koulourtzis.m*, whose general structure is:

```
function [output1, output2, ...] = fanouropita(input1, input2,...)
......
return
```

The three dots indicate additional input and output variables separated by commas, and the six dots indicate additional lines of code. The output list, `output1, output2, ...`, consists of numbers, vectors, matrices, and string variables evaluated by the function by performing operations involving the input string, `input, input2, ...`. A variable may appear both in the input and output list.

To execute this function in the MATLAB environment or invoke it from a program file, we issue the command:

```
[evaluate1, evaluate2, ...] = koulourtzis(parameter1, parameter2, ...)
```

Function	*Purpose*
abs	Absolute value
acos	Inverse cosine
acosh	Inverse hyperbolic cosine
acot	Inverse cotangent
acoth	Inverse hyperbolic cotangent
acsc	Inverse cosecant
acsch	Inverse hyperbolic cosecant
angle	Phase angle
asec	Inverse secant
asech	Inverse hyperbolic secant
asin	Inverse sine
asinh	Inverse hyperbolic sine
atan	Inverse tangent
atan2	Four quadrant inverse tangent
atanh	Inverse hyperbolic tangent
ceil	Round toward plus infinity.
cart2pol	Cartesian-to-polar coordinate conversion
cart2sph	Cartesian-to-spherical coordinate conversion
conj	Complex conjugate
cos	Cosine
cosh	Hyperbolic cosine
cot	Cotangent
coth	Hyperbolic cotangent
csc	Cosecant
csch	Hyperbolic cosecant
exp	Exponential
expm	Matrix exponential
fix	Round toward zero
floor	Round toward minus infinity
gcd	Greatest common divisor
imag	Complex imaginary part
lcm	Least common multiple
log	Natural logarithm
log10	Common logarithm
pol2cart	Polar-to-Cartesian coordinate conversion
real	Real part
sec	Secant

Table C.7.1 Common and specialized MATLAB mathematical functions (*Continuing.*)

Function	*Purpose*
sech	Hyperbolic secant
sign	Signum function
sin	Sine
sinh	Hyperbolic sine
sqrt	Square root
tan	Tangent
tanh	Hyperbolic tangent
Specialized	
bessel	Bessel functions
besseli	Modified Bessel functions of the first kind
besselj	Bessel functions of the first kind
besselk	Modified Bessel functions of the second kind
bessely	Bessel functions of the second kind
beta	Beta function
betainc	Incomplete beta function
betaln	Logarithm of the beta function
ellipj	Jacobi elliptic functions
ellipke	Complete elliptic integral
erf	Error function
erfc	Complementary error function
erfinv	Inverse error function
expint	Exponential integral
gamma	Gamma function
gammainc	Incomplete gamma function
gammaln	Logarithm of gamma function
legendre	Associated Legendre functions
log2	Dissect floating point numbers
pow2	Scale floating point numbers
Initialization	
eye	Identity matrix
ones	Matrix of ones
rand	Uniformly distributed random numbers and arrays
randn	Normally distributed random numbers and arrays
zeros	Matrix of zeros

Table C.7.1 (*Continued.*)

Function	*Purpose*
cat	Concatenate arrays
cond	Condition number of a matrix
det	Matrix determinant
eig	Matrix eigenvalues and eigenvectors
inv	Matrix inverse
lu	LU decomposition of a matrix
ode23	Solution of ordinary differential equations by the second/third-order Runge-Kutta method
ode45	Solution of ordinary differential equations by the fourth/fifth-order Runge-Kutta-Fehlberg method
qr	QR decomposition of a matrix
poly	Characteristic polynomial of a matrix
quad	Function integration by Simpson's rule
root	Polynomial root finder
svd	Singular-value decomposition
trapz	Function integration by the trapezoidal rule
x = A\b	Solves the linear system $\mathbf{A}\mathbf{x} = \mathbf{b}$, where $\mathbf{A}$ is an $N \times N$ matrix, and $\mathbf{b}, \mathbf{x}$ are N-dimensional column vectors Also solves an overdetermined system of equations
x = b/A	Solves the linear system $\mathbf{x}\mathbf{A} = \mathbf{b}$, where $\mathbf{A}$ is an $N \times N$ matrix, and $\mathbf{b}, \mathbf{x}$ are N-dimensional row vectors Also solves an overdetermined system of equations
x = b/A'	Solves the linear system $\mathbf{A}\mathbf{x} = \mathbf{b}$, where $\mathbf{A}$ is an $N \times N$ matrix, and $\mathbf{b}, \mathbf{x}$ are N-dimensional row vectors Also solves an overdetermined system of equations
fsolve	Solves a system of nonlinear equations
fminunc	Performs unconstrained minimization
fmincon	Performs constrained minimization

Table C.7.2 A partial list of numerical methods functions encapsulated in MATLAB.

After the function has been successfully executed, `evaluate1` takes the value of `output1`, `evaluate2` takes the value of `output2`, and the rest of the output variables take corresponding values.

If a function evaluates only one number, vector, matrix, character string, entity, or object, then the function statement and corresponding function declaration can be simplified to:

```
function evaluate = kastanas(input1, input2, ...)
.....
return
```

An example of a simple function residing in the file `bajanakis.m` is:

```
function bname = bajanakis(isel)

if(isel==1)
  bname = 'sehoon';
elseif(isel==2)
  bname = 'phaethon';
else
  bname = 'alkiviadis';
end
%---
% done
%---

return
```

C.9 Matlab graphics

A powerful feature of MATLAB is the ability to produce professional graphics, including animation. Graphics are displayed in dedicated windows appearing in response to graphics commands. Graphics functions are listed in tables C.9.1 in several categories. The Matlab *help* facility provides a detailed description of the various graphics function and their arguments and parameters. Some useful tips are:

- To generate a new graphics window, use the command: *figure*
- To produce a graphics file, use the *export* or *save* option under the *file* pull-down menu.
- To manipulate axis properties, use the function *axis* with appropriate arguments.
- To superimpose graphs, use the command: *hold.*
- To close a graphics window, use the command: *close.*

Two-dimensional graphs	
bar	Bar graph
comet	Animated comet plot
compass	Compass plot
errorbar	Error bar plot
fplot	Plot a function
fill	Draw filled two-dimensional polygons
hist	Histogram plot
loglog	Log-log scale plot
plot	Linear plot
polar	Polar coordinate plot
semilogx	Semi-log scale plot, x-axis logarithmic
semilogy	Semi-log scale plot, y-axis logarithmic
stairs	Stair-step plot
stem	Stem plot for discrete sequence data
Graph annotation and operations	
grid	Grid lines
gtext	Mouse placement of text
legend	Add legend to plot
text	Text annotation
title	Graph title
xlabel	x-axis label
ylabel	y-axis label
zoom	Zoom in and out of a two-dimensional plot
Line and fill commands	
fill3	Draw filled three-dimensional polygons
plot3	Plot lines and points
Two-dimensional graphs of three-dimensional data	
clabel	Contour plot elevation labels
comet3	Animated comet plot
contour	Contour plot
contour3	Three-dimensional contour plot

Table C.9.1 Elementary and specialized MATLAB graphics functions and procedures. (*Continuing.*)

contourc	Contour plot computation (used by contour)
image	Display image
imagesc	Scale data and display as image
pcolor	Pseudocolor (checkerboard) plot
quiver	Quiver plot
slice	Volumetric slice plot
Surface and mesh plots	
mesh	Three-dimensional mesh surface
meshc	Combination mesh/contour plot
meshgrid	Generate x and y arrays
meshz	Three-dimensional mesh with zero plane
slice	Volumetric visualization plot
surf	Three-dimensional shaded surface
surfc	Combined surf/contour plot
surfl	Shaded surface with lighting
trimesh	Triangular mess plot
trisurf	Triangular surface plot
waterfall	Waterfall plot
Three-dimensional objects	
cylinder	Generate a cylinder
sphere	Generate a sphere
Graph appearance	
axis	Axis scaling and appearance
caxis	Pseudocolor axis scaling
colormap	Color lookup table
hidden	Mesh hidden line removal
shading	Color shading
view	Graph viewpoint specification
viewmtx	View transformation matrices
Graph annotation	
grid	Grid lines
legend	Add legend to plot

Table C.9.1 (*Continuing.*)

text	Text annotation
title	Graph title
xlabel	x-axis label
ylabel	y-axis label
zlabel	z-axis label for three-dimensional plots
Graphics control	
capture	Screen capture of current figure in Unix
clf	Clear current figure
close	Abandon figure
figure	Create a figure in a new graph window
gcf	Get handle to current figure
graymon	Set default figure properties for grayscale monitors
newplot	Determine correct axes and figure for new graph
refresh	Redraw current figure window
whitebg	Toggle figure background color
Axis control	
axes	Create axes at arbitrary position
axis	Control axis scaling and appearance
caxis	Control pseudo-color axis scaling
cla	Clear current axes
gca	Get handle to current axes
hold	Hold current graph
ishold	True if hold is on
subplot	Create axes in tiled positions
Graphics objects	
figure	Create a figure window
image	Create an image
line	Generate a line
patch	Generate a surface patch
surface	Generate a surface
text	Create text
uicontrol	Create user interface control
uimenu	Create user interface menu

Table C.9.1 (*Continuing.*)

Graphics operations	
delete	Delete object
drawnow	Flush pending graphics events
findobj	Find object with specified properties
gco	Get handle of current object
get	Get object properties
reset	Reset object properties
rotate	Rotate an object
set	Set object properties
Hard copy and storage	
orient	Set paper orientation
print	Print graph or save graph to file
printopt	Configure local printer defaults
Movies and animation	
getframe	Get movie frame
movie	Play recorded movie frames
moviein	Initialize movie frame memory
Miscellaneous	
ginput	Graphical input from mouse
ishold	Return hold state
rbbox	Rubber-band box for region selection
waitforbuttonpress	Wait for key/button press over figure
Color controls	
caxis	Pseudocolor axis scaling
colormap	Color lookup table
shading	Color shading mode

Table C.9.1 (*Continuing.*)

Color maps	
bone	Grayscale with a tinge of blue color map
contrast	Contrast enhancing grayscale color map
cool	Shades of cyan and magenta color map
copper	Linear copper-tone color map
flag	Alternating RGB and black color map
gray	Linear grayscale color map
hsv	Hue-saturation-value color map
hot	Black-red-yellow-white color map
jet	Variation of HSV color map (no wrap)
pink	Pastel shades of pink color map
prism	Prism-color color map
white	All white monochrome color map
Color map functions	
brighten	Brighten or darken color map
colorbar	Display color map as color scale
hsv2rgb	Hue-saturation-value to RGB equivalent
rgb2hsv	RGB to hue-saturation-value conversion
rgbplot	Plot color map
spinmap	Spin color map
Lighting models	
diffuse	Diffuse reflectance
specular	Specular reflectance
surfl	Three-dimensional shaded surface with lighting
surfnorm	Surface normals

Table C.9.1 (*Continued.*)

In the remainder of this section, we present several graphics sessions followed by the graphics output.

- Graph of the function: $f(x) = \sin^3(\pi x)$

```
>> x=-1.0:0.01:1.0; % define an array of abscissae
>> y = sin(pi*x).^3; % note the array power operator .^ (table F.3.1)
>> plot(x,y)
>> set(gca,'fontsize',15)
>> xlabel('x','fontsize',15)
>> ylabel('y','fontsize',15)
```

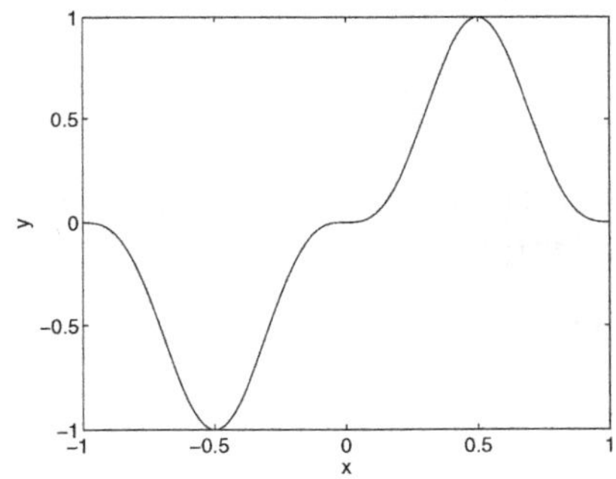

- Graph of the Gaussian function: $f(x) = e^{-x^2}$

```
>> fplot('exp(-x^2)',[-5, 5])
>> set(gca,'fontsize',15)
>> xlabel('x','fontsize',15)
>> ylabel('y','fontsize',15)
```

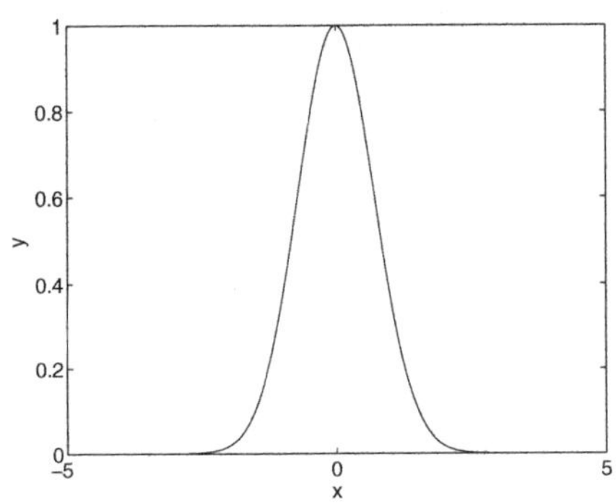

- Paint a polygon in yellow:

```
>> x =[0.0 1.0 1.0]; y=[0.0 0.0 1.0]; c='y';
>> fill(x,y,c)
>> set(gca,'fontsize',15)
```

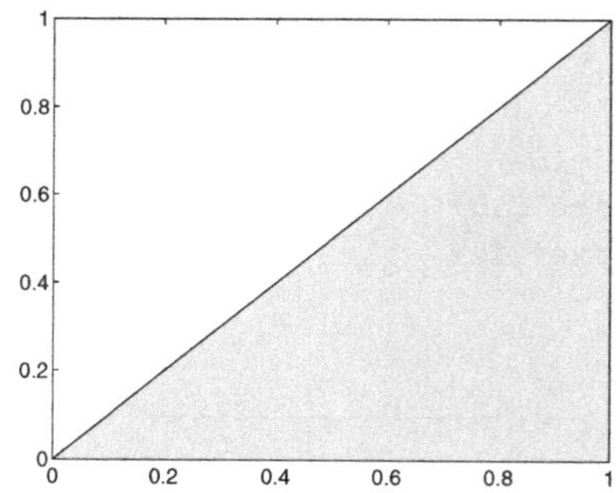

- Mesh plot:

```
>> [x, y] = meshgrid(-1.0:0.10:1.0, -2.0:0.10:2.0);
>> z = sin(pi*x+pi*y);
>> mesh(z)
>> set(gca,'fontsize',15)
```

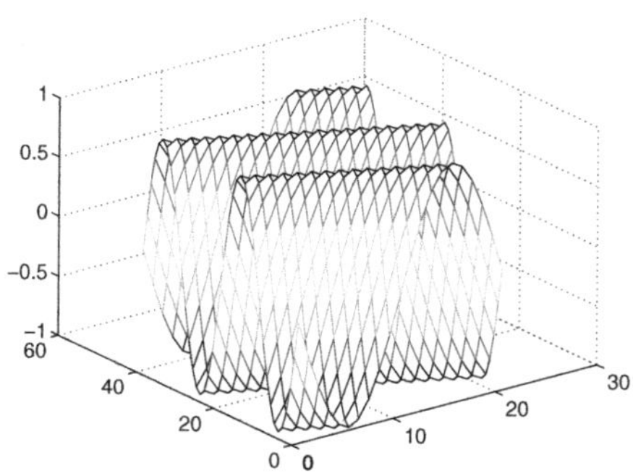

Index

CPSIA information can be obtained at www.ICGtesting.com
Printed in the USA
LVOW01*2103150915

454279LV00010B/82/P

9 780387 958699